Birkhäuser

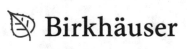

Frontiers in Mathematics

Frontiers in Elliptic and Parabolic Problems

The goal of this series is to reflect the impressive and ongoing evolution of dealing with initial and boundary value problems in elliptic and parabolic PDEs. Recent developments include fully nonlinear elliptic equations, viscosity solutions, maximal regularity, and applications in finance, fluid mechanics, and biology, to name a few. Many very classical notions have been revisited, such as degree theory or Sobolev spaces. Books in this series present the state of the art keeping applications in mind wherever possible.

The series is curated by the Series Editor.

Arnaud Rougirel

Unified Theory for Fractional and Entire Differential Operators

An Approach via Differential Quadruplets and Boundary Restriction Operators

Birkhäuser

Arnaud Rougirel
Laboratoire de Mathématiques
University of Poitiers
Poitiers, France

ISSN 1660-8046 ISSN 1660-8054 (electronic)
Frontiers in Mathematics
ISSN 2730-549X ISSN 2730-5503 (electronic)
Frontiers in Elliptic and Parabolic Problems
ISBN 978-3-031-58355-1 ISBN 978-3-031-58356-8 (eBook)
https://doi.org/10.1007/978-3-031-58356-8

This book is published under the imprint Birkhäuser, www.birkhauser-science.com by the registered company Springer Nature Switzerland AG
The registered company address is: Gewerbestrasse 11, 6330 Cham, Switzerland

Contents

Introduction

1

1.1 Motivations

It appears that the methods for solving entire order and fractional ordinary differential equations are similar [Die10, Chap. 5 & 6]. Regarding partial differential equations, there are also analog results for usual evolutionary partial differential equations and for the corresponding time-fractional equations. In addition, the solvability methods may be similar for equations with a first-order time derivative and for the corresponding fractional equations involving a Riemann–Liouville derivative [ORB21]. When we consider fractional equations involving a Caputo derivative, the solvability methods are more involved than in standard cases, but remain quite similar: see [Zac09, Zac12, FRW22, FKR+21]. Hence, we may expect to set up *a unified theory* for the analysis of entire and fractional derivatives. This is our first motivation.

Let us be a little more explicit about such a theory. Being given a positive real number b and a function $h : (0, b) \to \mathbb{R}$, consider the following three basic differential equations.

$$\mathrm{D}^1 u = u + h \tag{1.1.1}$$

$$^{\mathrm{RL}}\mathrm{D}^\alpha v = v + h \tag{1.1.2}$$

$$^{\mathrm{C}}\mathrm{D}^\alpha w = w + h \tag{1.1.3}$$

where the unknowns u, v and w are scalar functions defined on $(0, b)$. Here $\mathrm{D}^1 u$ denotes the first-order derivative of u, and $^{\mathrm{RL}}\mathrm{D}^\alpha v$, $^{\mathrm{C}}\mathrm{D}^\alpha_{L^p} w$ are respectively, the Riemann–Liouville derivative of v and the Caputo derivative of w. We seek an abstract framework in which the latter equations can be rewritten under the form

A. Rougirel, *Unified Theory for Fractional and Entire Differential Operators*, Frontiers in Elliptic and Parabolic Problems,
https://doi.org/10.1007/978-3-031-58356-8_1

$$\mathrm{D}u = u + h, \tag{1.1.4}$$

and (1.1.4) is solvable. Otherwise said, we are interested in a framework allowing to solve the three Eqs. (1.1.1)–(1.1.3) in one shot. Of course, an issue is to define the symbol D appearing in (1.1.4). In general, the cost to pay for an unified theory is an increase of the level of abstraction.

Let us give a second motivation. Starting from the heat equation

$$\frac{\mathrm{d}}{\mathrm{d}t}u = \Delta u$$

where the unknown $u : [0, \infty) \times \mathbb{R}^3 \to \mathbb{R}$ is a function of the time variable t and the space variable x, we consider the following more general evolution equation.

$$\frac{\mathrm{d}}{\mathrm{d}t}u = Lu \tag{1.1.5}$$

where L is an operator on an abstract space $\mathcal{X}$, and $u : (0, \infty) \to \mathcal{X}$. In order to solve (1.1.5), it is assumed that L keeps some properties of the Laplace operator Δ. For example, it is assumed that L is (at least) sectorial [Hen81] or dissipative [Paz83] or self-adjoint [Bre11].

Now our aim is to go one step further by introducing the equation

$$Au = Lu \tag{1.1.6}$$

where A and L are operators acting on another abstract space $\mathcal{Z}$ and u lies in the domains of A and L. Proceeding for A as we have done for L, we would like that A keeps some properties of the operator $\frac{\mathrm{d}}{\mathrm{d}t}$. There are many works in that direction (see for instance [dPG75, Sho97]). However, we would like that A keeps also some properties of fractional operators. Moreover, we look for an framework allowing to solve boundary value problems at an abstract level.

1.2 Some Issues in Fractional Calculus

How many initial conditions? If we consider a first-order initial value problem, it is clear that only one initial condition is needed for well posedness. Also, a second-order initial value problem requires two initial conditions. Now consider a fractional initial value problem of order α where α lies in $(0, 1)$. Then, the issue is to find the suitable number of initial conditions to get a well-posed problem. As we will see in Chap. 7, this number is equal to the dimension of the kernel of the corresponding fractional operator D^α.

What are the boundary values of functions in the domain of D^α? Since we consider only operators acting on functions of the one variable x, if $\alpha = 1$ and x lies in the interval

$(0, b)$, then a function u in the domain of D^1 has two boundary values, namely $u(0)$ and $u(b)$. The issue is to define and compute the boundary values in the case where α is not a integer.

This issue is central in this book and is addressed mainly in Chap. 6. Altough the precise representation of these boundary values depends of course on the specific fractional differential operator under consideration, there are some invariants. More precisely, there are always $2n$ boundary values where n is the finite dimension of the kernel of D^α. Furthermore, these boundary values may be splitted into two parts of equal size n. A first part is obtained by evaluations of some functions at $x = 0$, whereas the second part is obtained by evaluations at $x = b$.

Another issue is to classify the dozens of fractional time derivatives available in the literature [DGGS20, Sil20]. We have started this task by classifying derivatives built upon the usual first-order derivative $\frac{\mathrm{d}}{\mathrm{d}x}$ and the Riemann–Liouville derivative (see Sect. 6.4).

1.3 Unified Theories

1.3.1 How to Construct a Unified Theory?

The starting point is to gather well-chosen significant examples. Secondly, identify *the invariants* among these examples, and define classes of objects from these invariants. At this stage, we proceed by induction in the sense that we go from *particular* situations to a *general* case. Notice that there is nothing to prove here.

Then, deduce the properties of these objects, to setup a mathematical theory. Finally, apply the abstract results of the theory to any suitable situations and in particular to the previous examples.

Unified theories are very common in mathematics. As a first example, let us quote *linear algebra*. Abstract vector spaces constitute a class of objects defined from the invariants. Linear maps between vector spaces form another class. Basically, computations in $\mathbb{R}^n$, $\mathbb{C}^n$, or in the space of polynomials are replaced by computations in an abstract vector space. *The theory of symmetric operators* and *the groups theory* are also unified theories.

1.3.2 Outline of Our Unified Theory

This theory offers to unify the calculus of standard and fractional derivatives by means of an abstract operator-theoretic approach. Significant examples involve usual first-order derivatives, Riemann–Liouville and Caputo derivatives. We consider the *operators* induced by these derivatives, denoted respectively by D^1, ${}^{\mathrm{RL}}\mathrm{D}^\alpha$ and ${}^{\mathrm{C}}\mathrm{D}^\alpha$. This basic change of point of view (i.e., operators versus derivatives) is of fundamental importance in our theory.

Then, the analysis of these operators (in particular the computation of their adjoint) allows to identify some invariants. The first invariant is that D^1, ${}^{\mathrm{RL}}\mathrm{D}^\alpha$, and ${}^{\mathrm{C}}\mathrm{D}^\alpha$ possess a right inverse. For D^1, this property is well known and is called *the fundamental theorem of calculus*. Then, we consider these operators as instances of an unbounded operator acting on some abstract space. Let us call $\mathbb{A}$ this operator, and suppose for simplicity that $\mathbb{A}$ acts on a (abstract) Hilbert space $\mathcal{H}$. We assume that $\mathbb{A}$ possesses a right inverse, which means that $\mathbb{A}\mathbb{B} = \mathrm{id}_{\mathcal{H}}$ for some operator $\mathbb{B} : \mathcal{H} \to \mathcal{H}$. This statement, which was seen as property in concrete examples, now becomes an axiom of the theory.

The second invariant turns out to concern $\mathbb{B}$ and is the following. The adjoint $\mathbb{B}^*$ of $\mathbb{B}$ is conjugate to $\mathbb{B}$ in the sense that

$$\mathbb{B}^* = \mathbb{S}\mathbb{B}\mathbb{S},$$

where $\mathbb{S}$ is a basic operator on $\mathcal{H}$, namely *a symmetry* (see Sect. 4.2.1). These two invariants led us to introduce the so-called class of *differential triplets* $(\mathbb{A}, \mathbb{B}, \mathbb{S})$ on the Hilbert space $\mathcal{H}$.

When the underlying abstract space is a reflexive Banach space, we have to consider *a differential quadruplet* (see Chap. 5). Then, things are more involved, but the fundamental ideas remain the same.

Let us give two topics developed in this theory. Let us go back for simplicity to the Hilbertian setting where $(\mathbb{A}, \mathbb{B}, \mathbb{S})$ is a differential triplet acting on a Hilbert space $\mathcal{H}$. The first aim is to define abstract boundary values, which we call *(abstract) endogenous boundary values*. Let us first explain the meaning of the adjective *endogenous* in the present context. If we consider a partial differential equations set on an open ball Ω of $\mathbb{R}^3$, then its solutions live typically in a subspace of $L^2(\Omega)$. However, the boundary values of these solutions lie in general in (a subspace of) $L^2(\partial\Omega)$, which is disjoint from $L^2(\Omega)$. So, in order to define boundary values at an abstract level, it is natural to introduce extra spaces [Cal39]. In this sense, usual boundary values may be described as *heterogeneous*. On the contrary, we introduce a kind of boundary values directly linked to the ambient space $\mathcal{H}$. That is why they are called *endogenous*.

In the case where the kernel of $\mathbb{A}$ has positive (finite) dimension (let us say n), it appears that any element in the domain of $\mathbb{A}$ possesses $2n$ *(abstract) endogenous boundary values*, which are coordinates on a subspace of $D(A)$ isomorphic to $\ker \mathbb{A} \times \ker \mathbb{A}$. See Sect. 4.4 and in particular Definition 4.4.2. Moreover, these $2n$ endogenous boundary values may be splitted in two parts. A first part is obtained by projection on the kernel of $\mathbb{A}$ (Sect. 7.2.3), and in applications correspond to initial values. See (4.1.3) and Sect. 7.3. The other n endogenous boundary values of the second part are a little bit more difficult to detect even in a Hilbertian setting. They correspond in applications to "final" boundary values. We refer the reader to Sects. 4.4.2, 5.7.1.3, and 6.4 for more details.

This splitting reveals a main difference between *linear initial value problems* and *boundary values problems*. The former are *unconditionally solvable* (Definition 7.2.9) and use only endogenous boundary values of the first part. On the other hand, the latter are in general *not unconditionally solvable* and use endogenous boundary values of the second part. There results that roughly speaking only *sublinear initial value problems* are solvable by iterative methods (Sect. 7.6).

A second topic is to establish a quantitative theory of linear endogenous boundary value problems. That is, we will give existence and uniqueness results for linear systems of abstract differential equations supplemented with inhomogeneous endogenous boundary conditions. The equations have the form

$$\mathbb{A}u = Lu + h$$

where L is a bounded operator on $\mathcal{H}$ commuting with $\mathbb{B}$, h lies in $\mathcal{H}$, and the unknown u belongs to the domain of $\mathbb{A}$. It should be emphasized that there is no time variable here since $\mathcal{H}$ is a generic Hilbert space. This quantitative theory is featured in Chap. 7.

The abstract results of these two topics apply to the first-order derivative and to the Riemann–Liouville and Caputo derivatives of order α. For example, Eqs. (1.1.1)–(1.1.3) are particular cases of the latter equation. Moreover, these results apply also to more general derivatives built upon *Sonine kernels*. See for instance, Chap. 6 and Sects. 7.3–7.5.

1.4 About This Book

1.4.1 Preliminary Remarks

The author has tried to ensure that the skills allowing a self-contained reading of this book are that of a postgraduate diploma in mathematical analysis. That said, any additional information should be freely accessible online.

Since this book establishes the axiomatic foundations of a certain branch of the fractional calculus, there results that in principle no a priori knowledge on fractional calculus is needed. However, acculturation to fractional calculus will make reading easier.

One of the main ideas of this book is to state and prove ready-to-use results at an abstract level and to apply these results to entire and fractional derivatives. That shows the deep unity between entire and fractional derivatives.

To make the new objects introduced in this book more easily accessible, we do not always go from general to particular cases. For instance, regarding the unified theory, the Hilbertian setting is covered first, although in many aspects, it is a particular case of the Banach space setting. This approach obviously impacts the way to read this book (see Sect. 1.4.3).

This book is organized into chapters, sections, subsections, paragraphs, and sub-paragraphs. For instance Sect. 5.7.2.1 refers to the first paragraph in the second subsection of the 7th section in Chap. 5. Moreover, Equation (1.2.17) is the 17th equation in the second section of Chap. 1. Proposition 2.3.5 takes place in the third section of Chap. 2.

1.4.2 Outline of the Book

The unified theory is featured in Chaps. 4 and 5. The former deals with differential triplets and boundary restriction operators in a Hilbertian setting. The latter is concerned with differential quadruplets and boundary restriction operators on Banach spaces. The next chapter deals with applications. It is a catalog of differential quadruplets. The boundary values of functions in the domain of various fractional differential operators are computed, and some (fractional) integration by parts formulae are given. Also, adjoints of fractional differential operators are computed.

Chapter 7 is concerned with fractional boundary values problems. That is, in applications, we consider systems of fractional differential equations supplemented with (endogenous) boundary conditions. In accordance with the precited main idea, almost all the existence and uniqueness results are proved at an abstract level. The exception concerns *non linear* problems (Sect. 7.6). The last chapter has a flavor of partial differential equations and deals with abstract and fractional Laplace operators.

Prerequisites are given in Chaps. 2 and 3. These chapters should contain no new material; however, some results are not usual.

1.4.3 How to Read This Book?

The answer depends of the aims of the reader. In the sequel, we list some aims and give a corresponding way to proceed.

(i) To get a quick overview of the unified theory and its applications, it is simpler to consider only the Hilbertian setting. So we suggest to read Chap. 4, which deals with the theory of abstract differential triplets and boundary restriction operators. The ideas underlying the formal definitions of these objects are explained. Also, we give some simple examples built upon the usual first-order derivative. There is no fractional calculus in this chapter. Basic applications are given in Sect. 6.2. Regarding differential equations, only the introductory example of Chap. 7.1 takes place on a Hilbert space. However, almost all the theory leading to *initial* values problems may be read without reference to differential quadruplets. It comprises Sects. 7.2.1–7.2.5 and 7.3, 7.4.

(ii) To get a comprehensive overview of the unified theory and its applications, we suggest to read Chaps. 5–8. Notice, however, that the underlying ideas of Chap. 4 are not repeated in Chap. 5. Also the latter contains less examples.

1.5 Notation and Conventions

1.5.1 Sets of Numbers

The set

$$\{0, 1, 2, 3, \ldots\}$$

of non negative integers is denoted by $\mathbb{N}$. *An index* is a non-negative integer. For all integers i and j, *the Kronecker symbol* is

$$\delta_{i,j} := \begin{cases} 1 & \text{if } i = j \\ 0 & \text{otherwise.} \end{cases} \tag{1.5.1}$$

The convention of repeated indexes means that $\sum_{i=1}^{d} x_i y_i$ is abbreviate by $x_i y_i$. That is to say, we set

$$x_i y_i := \sum_{i=1}^{d} x_i y_i .$$

The fields of real and complex numbers are denoted by $\mathbb{R}$ and $\mathbb{C}$, respectively. Real constants whose value is irrelevant for our purpose are generically denoted by C. Any complex number z may be written under the form $z = a + ib$ where a, b lie in $\mathbb{R}$, and $\mathrm{i}^2 = -1$.

The symbol $\mathbb{K}$ denotes any element of the set $\{\mathbb{R}, \mathbb{C}\}$, that-is-to-say $\mathbb{K}$ is equal to $\mathbb{R}$ or $\mathbb{C}$.

Let m and n be integers. If $n < m$ then any sum of the form

$$\sum_{i=m}^{n} \cdots$$

is set to zero.

1.5.2 Miscellaneous Notation

Let b be a positive real number and X be a set. For any function $f : (0, b) \to X$, we denote by $f(b - \cdot)$ the function

$$(0, b) \to X, \qquad x \mapsto f(b - x).$$

The symbols := are used to introduce an affectation or a notation. For instance, if x is any number in $\mathbb{K}$, then $x := 1$ means that the value of x is set to 1. Also

$$E := \{x \in \mathbb{R} \mid x + 3 \geq 0\},$$

means that the latter set is denoted by E.

1.5.3 Matrices

Let m and n be a positive integers. For each index i in $[1, m]$ and j in $[1, n]$, let $a_{i,j}$ belong to $\mathbb{K}$. By

$$(a_{i,j})_{1 \leq i \leq m,\, 1 \leq j \leq n},$$

we mean the matrix with m lines and n columns, whose entry located at the intersection of the ith line and j^{th} column is $a_{i,j}$. The space of such matrices is denoted by $\mathcal{M}(m \times n, \mathbb{K})$. We abbreviate $\mathcal{M}(n \times n, \mathbb{K})$ by $\mathcal{M}(n, \mathbb{K})$, and $(a_{i,j})_{1 \leq i \leq n,\, 1 \leq j \leq n}$ by $(a_{i,j})_{i,j=1,\ldots n}$.

If

$$M := (a_{i,j})_{1 \leq i \leq m,\, 1 \leq j \leq n}$$

is any element of $\mathcal{M}(m \times n, \mathbb{K})$ then *its transpose* denoted by M^t, is the matrix

$$(a_{j,i})_{1 \leq i \leq n,\, 1 \leq j \leq m}.$$

Notice that M^t lies in $\mathcal{M}(n \times m, \mathbb{K})$. Moreover, $\mathbb{K}^n$ is identified with $\mathcal{M}(n \times 1, \mathbb{K})$.

1.5.4 Spaces

If $\mathcal{X}$ is a vector space and $x_1, \ldots, x_n$ lie in $\mathcal{X}$ then

$$\langle x_1, \ldots, x_n \rangle$$

denotes the subspace of $\mathcal{X}$ generated by $x_1, \ldots, x_n$.

Let $(\mathcal{X}, \|\cdot\|)$ be a normed space over $\mathbb{K}$. If $(x_n)_{n \in \mathbb{N}}$ is a sequence in $\mathcal{X}$ and x lies in $\mathcal{X}$ then the notation

$$x_n \xrightarrow{\mathcal{X}} x$$

means that $(x_n)_{n\in\mathbb{N}}$ converges toward x in $\mathcal{X}$, that-is-to-say

$$\|x_n - x\| \xrightarrow[n\to\infty]{} 0.$$

If no confusion may occur, we will write simply $x_n \to x$.

Background on Functional Analysis

2

2.1 Vector Spaces

Let $\mathcal{X}$, $\mathcal{Y}$, and $\mathcal{Z}$ be three sets. In general, a map defined on $\mathcal{X} \times \mathcal{Y}$ with values in $\mathcal{Z}$ is denominated by a letter, let us say, f. The image of any ordered pair (x, y) of $\mathcal{X} \times \mathcal{Y}$ is denoted by $f(x, y)$. However, in some situations, this image is designated by $\langle x, y\rangle$ or by (x, y). For the basic maps on *vector spaces*, the image of (x, y) is $x + y$ or xy. These maps, which we will call *operations*, are denoted by $+$ and . (i.e. the dot symbol). Let $\mathbb{K}$ be equal to $\mathbb{R}$ or $\mathbb{C}$.

2.1.1 Basic Definitions

Definition 2.1.1 Being given a set $\mathcal{X}$, let $+ : \mathcal{X} \times \mathcal{X} \to \mathcal{X}$ and $. : \mathbb{K} \times \mathcal{X} \to \mathcal{X}$ be two maps. The triplet $(X, +, .)$ is called a *vector space over* $\mathbb{K}$ provided the following properties hold true.

(i) $(\mathcal{X}, +)$ is an Abelian group. That is to say, the map $+$ is commutative, associative and possesses an *identity element* denoted by $0_{\mathcal{X}}$ or simply be 0 if no confusion may occur. Moreover, for each x in $\mathcal{X}$, there exists an element x' in $\mathcal{X}$ such that $x + x' = 0_{\mathcal{X}}$.

(ii) For each (λ, μ) in $\mathbb{K} \times \mathbb{K}$ and each x in $\mathcal{X}$, one has $(\lambda\mu)x = \lambda(\mu x)$. Moreover, $1x = x$.

(iii) For each (λ, μ) in $\mathbb{K} \times \mathbb{K}$ and each (x, y) in $\mathcal{X} \times \mathcal{X}$, the following distributivity properties hold:

$$(\lambda + \mu)x = \lambda x + \mu x, \qquad \lambda(x + y) = \lambda x + \lambda y.$$

A. Rougirel, *Unified Theory for Fractional and Entire Differential Operators*,
Frontiers in Elliptic and Parabolic Problems,
https://doi.org/10.1007/978-3-031-58356-8_2

A vector space over $\mathbb{K}$ is also called a $\mathbb{K}$-*vector space*. If $(X, +, .)$ is $\mathbb{K}$-*vector space* then the operations $+$ and . are called *(vector) addition* and *scalar multiplication*. □

It is customary to skip the dot symbol for scalar multiplication, that is, we have written and we will write λx instead of $\lambda .x$. Also, the identity element, which we will call *the zero vector of* $\mathcal{X}$, is unique. Most of the time, a $\mathbb{K}$-*vector space* $(\mathcal{X}, +, .)$ will be designated simply by $\mathcal{X}$.

Definition 2.1.2 Let $\mathcal{X}$ be a $\mathbb{K}$-vector space. A subset M of $\mathcal{X}$ is called a *subspace of* $\mathcal{X}$ if M is non-empty and M is closed under addition and scalar multiplication. That is,

$$x + y \in M, \qquad \lambda x \in M,$$

for each (λ, x, y) in $\mathbb{K} \times M \times M$. □

A subspace of a vector space is a vector space. More precisely, if $(X, +, .)$ is a $\mathbb{K}$-vector space and M is a subspace of $\mathcal{X}$, then the restriction of the addition $+$ to the set $M \times M$ induces a map, still labeled $+$, from $M \times M$ into M. Similarly, the restriction of the scalar multiplication to $\mathbb{K} \times M$ induces an operation, still labeled . with values into M. With these notations, $(M, +, .)$ is a $\mathbb{K}$-vector space.

By restricting the scalar multiplication of a complex vector space, we may pass easily from a complex space to a real space.

Definition 2.1.3 Let $(\mathcal{X}, +, .)$ be a complex vector space. Denoting by $\circ$ the restriction of the scalar multiplication to $\mathbb{R} \times \mathcal{X}$, the triplet $(\mathcal{X}, +, \circ)$ is a real vector space, which will be designated by $\mathcal{X}_{\mathbb{R}}$. We will call it the *real space associated to* $\mathcal{X}$. □

Let us notice that $\mathcal{X}_{\mathbb{R}}$ contains the same vectors than $\mathcal{X}$. In particular, $\mathrm{i}x$ lies in $\mathcal{X}_{\mathbb{R}}$ for each x in $\mathcal{X}$. However, if $x \neq 0$, then $\mathrm{i}x$ and x are linearly independent in $\mathcal{X}_{\mathbb{R}}$.

The *power set of a set E* is the set of all subsets of E. The power set of E is denoted by $\mathcal{P}(E)$. The empty set is an element of $\mathcal{P}(E)$. If we want to remove this set from the power set, we will consider the set $\mathcal{P}(E) \setminus \{\emptyset\}$ instead of $\mathcal{P}(E)$.

Starting from a $\mathbb{K}$-vector space $\mathcal{X}$, operations on $\mathcal{P}(\mathcal{X}) \setminus \{\emptyset\}$ may be inferred from vector addition and scalar multiplication.

Definition 2.1.4 Let $\mathcal{X}$ be a $\mathbb{K}$-vector space. The addition of sets denoted again by $+$ and defined on $\mathcal{P}(\mathcal{X}) \setminus \{\emptyset\} \times \mathcal{P}(\mathcal{X}) \setminus \{\emptyset\}$ with values in $\mathcal{P}(\mathcal{X}) \setminus \{\emptyset\}$ maps any ordered pair (U, V) of non-empty subsets of $\mathcal{X}$ into the set

$$U + V := \{u + v \mid u \in U,\ v \in V\}.$$

The image $U + V$ is called the *sum of U and V*. If U is a singleton, namely if $U = \{x\}$ for some x in $\mathcal{X}$, then we will write $x + U$ instead of $\{x\} + U$.

In a same way, the operation . defined on $\mathbb{K} \times \mathcal{P}(\mathcal{X}) \setminus \{\emptyset\}$ with values in $\mathcal{P}(\mathcal{X}) \setminus \{\emptyset\}$ maps any ordered pair (λ, U) into the set

$$\lambda U := \{\lambda u \mid u \in U\}. \qquad \square$$

By using commutativity and associativity of the vector addition, and the stability of subspaces with respect to addition, this proposition is easily proved.

Proposition 2.1.1 *Let $\mathcal{X}$ be a $\mathbb{K}$-vector space, and U, V, W be non-empty subsets of $\mathcal{X}$. Then,*

(i) $U + V = V + U$;
(ii) $(U + V) + W = U + (V + W)$;
(iii) *If* $U \subseteq V$ *then* $U + W \subseteq V + W$;
(iv) *If* M *is a subspace of* $\mathcal{X}$ *then* $M + M = M$.

In the same way, by using properties of scalar multiplication, the following results are easily proved.

Proposition 2.1.2 *Being given a $\mathbb{K}$-vector space $\mathcal{X}$, let U, V be non-empty subsets of $\mathcal{X}$, and λ, μ belong to $\mathbb{K}$. Then,*

(i) $\lambda(U + V) = \lambda U + \lambda V$;
(ii) $(\lambda + \mu)U = \lambda U + \mu U$;
(iii) *If* M *is a subspace of* $\mathcal{X}$, *then* $\lambda M = M$.

2.1.2 Quotient Spaces

Quotient space is a important tool is this chapter since it will be useful for studying the product of operators (see Sect. 2.4.3). Let $\mathcal{X}$ be a $\mathbb{K}$-vector space.

Definition 2.1.5 Let M be a subspace of $\mathcal{X}$. *The quotient space* of $\mathcal{X}$ by M is the set of all elements $\mathcal{A}$ of $\mathcal{P}(\mathcal{X})$ satisfying $\mathcal{A} = x + M$ for some x in $\mathcal{X}$. Denoting by $\mathcal{X}/M$ this quotient space, we have in symbol

$$\mathcal{X}/M = \{\mathcal{A} \in \mathcal{P}(\mathcal{X}) \mid \exists x \in \mathcal{X},\ \mathcal{A} = x + M\}. \qquad \square$$

Clearly, $\mathcal{X}/M$ is a subset of $\mathcal{P}(\mathcal{X})$. If $M := \mathcal{X}$, then $\mathcal{X}/M = \mathcal{X}/\mathcal{X} = \{\mathcal{X}\}$, hence $\mathcal{X}/\mathcal{X}$ contains only one element. On the other hand, if $M = \{0_{\mathcal{X}}\}$, then

$$\mathcal{X}/\{0_{\mathcal{X}}\} = \big\{\{x\} \in \mathcal{P}(\mathcal{X}) \mid x \in \mathcal{X}\big\}.$$

Let us observe that there are one-to-one correspondences between $\mathcal{X}/\mathcal{X}$ and $\{0_{\mathcal{X}}\}$, and between $\mathcal{X}/\{0_{\mathcal{X}}\}$ and $\mathcal{X}$. Since $\{0_{\mathcal{X}}\}$ and $\mathcal{X}$ are vector spaces, we may equip $\mathcal{X}/\mathcal{X}$ and $\mathcal{X}/\{0_{\mathcal{X}}\}$ with a vector space structure. More generally, we will see in Theorem 2.1.5 that every quotient space may be turned into a vector space.

Proposition 2.1.3 *Let M be a subspace of a vector space $\mathcal{X}$, and let x, x' be any vectors of $\mathcal{X}$. Then, the following assertions hold.*

(i) $x' \in x + M \Longleftrightarrow x' + M = x + M$.
(ii) *For each x in $\mathcal{X}$, there exists a unique $\mathcal{A}$ in $\mathcal{X}/M$ such that x lies in $\mathcal{A}$. This unique element $\mathcal{A}$ in $\mathcal{X}/M$, denoted by $[x]_{\mathcal{X}/M}$ or simply by $[x]$, is called the* class of x in *$\mathcal{X}/M$.*
(iii) *For each x in $\mathcal{X}$, $[x] = x + M$.*
(iv) *Let $\mathcal{A}$ belong to $\mathcal{X}/M$. Then for each x in $\mathcal{A}$, one has $[x] = \mathcal{A}$.*

Proof In order to prove (i), let us assume that $x' \in x + M$. Thus, Proposition 2.1.1 (iii) yields that $x' + M \subseteq (x + M) + M$. By Proposition 2.1.1 (ii) and (iv), we have $(x + M) + M = x + M$. Then, $x' + M \subseteq x + M$. By the definition of $x + M$, we infer that $x \in x' + M$, thus we have also $x + M \subseteq x' + M$, so that $x' + M = x + M$. Conversely, since $x' = x' + 0_{\mathcal{X}}$, we derive that $x' \in x' + M$. Assuming $x' + M = x + M$, we deduce $x' \in x + M$, which proves (i).

Since $x \in x + M$, we get the existence part for Item (ii) by choosing $\mathcal{A} := x + M$. If $\mathcal{B} \in \mathcal{X}/M$ contains also x, then writing $\mathcal{B}$ under the form $y + M$ for some $y \in \mathcal{X}$, we get that $x \in y + M$. Thus $\mathcal{A} = \mathcal{B}$ by (i). That shows the uniqueness of $\mathcal{A}$.

Item (iii) follows from the existence part in the proof of (ii). Finally, we write any $\mathcal{A}$ in $\mathcal{X}/M$ under the form $x' + M$ for some x' in $\mathcal{X}$, and consider any element x of $\mathcal{A}$. Then $x \in x' + M$. Thus $x + M = x' + M$ by (i), and $[x] = \mathcal{A}$ due to Item (iii). □

Notice that Proposition 2.1.3 may be proved by using *the equivalence relation*: $x \mathscr{R} x' \Longleftrightarrow x' \in x + M$.

Let us now equip $\mathcal{X}/M$ with a vector space structure. For each $\mathcal{A}$, $\mathcal{B}$ in $\mathcal{X}/M$ and λ in $\mathbb{K}$, $\mathcal{A} + \mathcal{B}$ and $\lambda\mathcal{A}$ are defined through Definition 2.1.4. A priori, these sets belong to $\mathcal{P}(\mathcal{X}) \setminus \{\emptyset\}$, hence we have to check more specifically that they live in $\mathcal{X}/M$.

Proposition 2.1.4 *Let M be a subspace of a vector space $\mathcal{X}$. Then, for each $\mathcal{A}$, $\mathcal{B}$ in $\mathcal{X}/M$ and $\lambda \in \mathbb{K}$, the sets $\mathcal{A} + \mathcal{B}$ and $\lambda\mathcal{A}$ belong to $\mathcal{X}/M$. Moreover,*

$$\mathcal{A} + \mathcal{B} = [x + x'], \qquad \lambda\mathcal{A} = [\lambda x],$$

for each (x, x') belonging to $\mathcal{A} \times \mathcal{B}$.

Proof It is enough to prove the two equalities stated in the proposition. In order to establish the first one, let us consider any x in $\mathcal{A}$, and any x' in $\mathcal{B}$. By Proposition 2.1.3 (iv) and (iii), one has $x + M = \mathcal{A}$, $x' + M = \mathcal{B}$. Whence

$$\begin{aligned} \mathcal{A} + \mathcal{B} &= x + x' + M && \text{(by Proposition 2.1.1)} \\ &= [x + x'] && \text{(by Proposition 2.1.3 (iii)).} \end{aligned}$$

In the same way, Propositions 2.1.2 and 2.1.3 entail

$$\lambda \mathcal{A} = \lambda x + \lambda M = \lambda x + M = [\lambda x].$$

□

Proposition 2.1.4 yields that the restriction of the addition to the set $\mathcal{X}/M \times \mathcal{X}/M$ induces a map, still labeled $+$, from $\mathcal{X}/M \times \mathcal{X}/M$ into $\mathcal{X}/M$. Similarly, the restriction of the scalar multiplication to $\mathbb{K} \times \mathcal{X}/M$ induces an operation, still labeled . with values in $\mathcal{X}/M$. This construction leads to the following result.

Theorem 2.1.5 *Let M be a subspace of a $\mathbb{K}$-vector space $\mathcal{X}$. Then with the above mentioned operations on $\mathcal{X}/M$, $(\mathcal{X}/M, +, .)$ is a $\mathbb{K}$-vector space. Besides, $0_{\mathcal{X}/M} = M$.*

Proof By Proposition 2.1.1, the addition is commutative and associative on $\mathcal{X}/M$. Moreover, for each $\mathcal{A}$ in $\mathcal{X}/M$, one has $\mathcal{A} = x + M$ for some x in $\mathcal{A}$. Thus, Proposition 2.1.1 yields

$$\mathcal{A} + M = (x + M) + M = x + M = \mathcal{A}.$$

Whence M is the *identity element*, that is $0_{\mathcal{X}/M} = M$. Besides, $\mathcal{A} = [x]$ by Proposition 2.1.3 (iii), and $-\mathcal{A} = [-x]$ by Proposition 2.1.4. Thus

$$\begin{aligned} \mathcal{A} + (-\mathcal{A}) &= [0_{\mathcal{X}}] && \text{(by Proposition 2.1.4)} \\ &= M && \text{(by Proposition 2.1.3 (iii)).} \end{aligned}$$

There results that $(\mathcal{X}/M, +)$ is an Abelian group. Finally, the Items (ii) and (iii) in Definition 2.1.1 are easily obtained from Proposition 2.1.2. □

Example 2.1.6 Let Ω be a non-empty set, $\mathscr{F}$ be a σ-algebra on Ω, and $\mu : \mathscr{F} \to [0, \infty]$ be a measure on $\mathscr{F}$. We denote by $\mathcal{X} := \mathscr{L}^0_{\mathscr{F}}(\Omega, \mathbb{K})$ the $\mathbb{K}$-vector space of all *measurable* functions $f : \Omega \to \mathbb{K}$. By *measurable*, we mean that $f^{-1}(U)$ lies in $\mathscr{F}$, for each open subset U of $\mathbb{K}$.

For any f, g in $\mathscr{L}^0_{\mathscr{F}}(\Omega, \mathbb{K})$, we recall that $f = g$ *almost everywhere on* Ω (and abbreviate $f = g$ *a.e. on* Ω) provided there exists some N in $\mathscr{F}$ such that $\mu(N) = 0$ and

$$f = g \qquad \text{on } \Omega \setminus N.$$

Let us now consider the following subset of $\mathscr{L}^0_{\mathscr{F}}(\Omega, \mathbb{K})$

$$M := \{f \in \mathscr{L}^0_{\mathscr{F}}(\Omega, \mathbb{K}) \mid f = 0 \text{ a.e. on } \Omega\}.$$

We claim that M is a subspace of $\mathscr{L}^0_{\mathscr{F}}(\Omega, \mathbb{K})$. Indeed, it is clear that M is non-empty and stable under scalar multiplication. In order to check the stability under addition, we notice that for each f, g in $\mathscr{L}^0_{\mathscr{F}}(\Omega, \mathbb{K})$, there exist N_f and N_g in $\mathscr{F}$ of zero measure and such that

$$f = 0 \qquad \text{on } \Omega \setminus N_f, \qquad g = 0 \qquad \text{on } \Omega \setminus N_g.$$

Thus,

$$f + g = 0 \qquad \text{on } \Omega \setminus (N_f \cup N_g).$$

Since

$$\mu(N_f \cup N_g) \leq \mu(N_f) + \mu(N_g),$$

we deduce that $f + g$ lies in M, which completes the proof of the claim.

We put

$$L^0_\mu(\Omega, \mathbb{K}) := \mathscr{L}^0_{\mathscr{F}}(\Omega, \mathbb{K})/M.$$

Proposition 2.1.3 entails that $[f] = [g]$ if and only if $f = g$ a.eon Ω. It is customary to write f instead of $[f]$. Beside, by Theorem 2.1.5, $L^0_\mu(\Omega, \mathbb{K})$ is a $\mathbb{K}$-vector space.

In the sequel, we will only consider the case where Ω is an open subset of the Euclidean space $\mathbb{R}^d$ (where d is a positive integer), $\mathscr{F} := \mathscr{F}_{\mathrm{L},\Omega}$ is the σ-algebra of *Lebesgue measurable subsets of* Ω, and $\mu := \mu_{\mathrm{L},\Omega}$ is the *Lebesgue measure on* Ω (see, for instance, [Rud87] for more details). Then, the space $L^0_{\mu_{\mathrm{L},\Omega}}(\Omega, \mathbb{K})$ will be denoted by $L^0(\Omega)$. In the particular case where $d = 1$ and Ω is the interval (a, b) of $\mathbb{R}$, we will write $L^0(\Omega)$ under the simplified form $L^0(a, b)$. □

Theorem 2.1.5 allows linear algebra between quotient spaces. We will give now a basic example of a linear map between some quotient spaces.

Proposition 2.1.6 *Let M and V be subspaces of a $\mathbb{K}$-vector space $\mathcal{X}$. For each $\mathscr{A}$ in $V/(V \cap M)$, there exists a unique $\mathcal{A}$ in $\mathcal{X}/M$ such that $\mathscr{A} \subseteq \mathcal{A}$. Also, $\mathcal{A} = [x]_{\mathcal{X}/M}$ for each x in $\mathscr{A}$.*

Proof Let $\mathscr{A}$ be in $V/(V \cap M)$ and x_0 be in $\mathscr{A}$. In order to show uniqueness, let $\mathcal{A}$, $\mathcal{B}$ be elements of $\mathcal{X}/M$ containing $\mathscr{A}$. Then, x_0 lies in $\mathcal{A} \cap \mathcal{B}$, so that $\mathcal{A} = \mathcal{B}$ by Proposition 2.1.3 (ii). The existence goes as follows: using Proposition 2.1.3 once again, we get

$$\mathscr{A} = [x_0]_{V/(V\cap M)} = x_0 + V \cap M \subseteq x_0 + M = [x_0]_{\mathcal{X}/M}. \qquad \square$$

Under the assumptions and notation of Proposition 2.1.6, we define the map i_V by

$$i_V : V/(V \cap M) \to \mathcal{X}/M, \qquad \mathscr{A} \mapsto \mathcal{A}, \tag{2.1.1}$$

where $\mathcal{A}$ is the unique element of $\mathcal{X}/M$ containing $\mathscr{A}$. Then, one has

$$i_V\big([x]_{V/(V\cap M)}\big) = [x]_{\mathcal{X}/M}, \qquad \forall x \in V. \tag{2.1.2}$$

Let us notice that (2.1.2) could be used for the definition of i_V. However, this mapping is a priory multivalued. Thus, we would have to prove a posteriori that i_V is univoque, i.e., that $[x]_{\mathcal{X}/M}$ is independent of the choice of x in $[x]_{V/(V\cap M)}$. In general, we prefer to avoid, if possible, such reasoning and thus, define maps in a direct way.

Let us now give some properties of i_V.

Proposition 2.1.7 *Let M and V be subspaces of a vector space $\mathcal{X}$. Then, the map i_V defined by* (2.1.1) *is linear and injective.*

Proof Let us start to show that i_V is additive. For each x and x' in $\mathcal{V}$,

$$\begin{aligned} i_V\big([x]_{V/(V\cap M)} + [x']_{V/(V\cap M)}\big) \\ &= i_V\big([x + x']_{V/(V\cap M)}\big) \\ &= [x + x']_{\mathcal{X}/M} && \text{(by (2.1.2))} \\ &= [x]_{\mathcal{X}/M} + [x']_{\mathcal{X}/M} \\ &= i_V\big([x]_{V/(V\cap M)}\big) + i_V\big([x']_{V/(V\cap M)}\big) && \text{(by (2.1.2))}. \end{aligned}$$

The identity for the scalar multiplication goes in a same way, so its proof is skipped. Finally, let x belong to V and satisfy

$$i_V\big([x]_{V/(V\cap M)}\big) = 0_{\mathcal{X}/M}.$$

Since $0_{\mathcal{X}/M} = M$ by Theorem 2.1.5, we deduce with (2.1.2) that x lies in M. Since x belongs to V as well, we derive

$$[x]_{V/(V\cap M)} = 0_{V/(V\cap M)}.$$

Hence i_V is injective. □

Theorem 2.1.5 allows us to introduce the notion of *co-dimension of a subspace*, which turns out to be a purely algebraic object.

Definition 2.1.7 Let $\mathcal{X}$ be a $\mathbb{K}$-vector space. We say that a subspace M of $\mathcal{X}$ has *finite co-dimension in* $\mathcal{X}$ if the dimension of $\mathcal{X}/M$ is finite. In that case, the dimension of $\mathcal{X}/M$ is called the *co-dimension of* M *in* $\mathcal{X}$ and is denoted by $\operatorname{codim}_{\mathcal{X}} M$. □

Proposition 2.1.8 *Let $\mathcal{X}$ be a $\mathbb{K}$-vector space, and M, V be subspaces of $\mathcal{X}$. We assume that M is contained in V, and that M has finite co-dimension in $\mathcal{X}$. Then, V has also finite co-dimension in $\mathcal{X}$.*

Proof Any $\mathcal{A}$ in $\mathcal{X}/V$ reads $\mathcal{A} = x + V$ for some x in $\mathcal{X}$. By considering a basis $([e_1]_{\mathcal{X}/M}, \ldots, [e_n]_{\mathcal{X}/M})$ of $\mathcal{X}/M$, there exist $\lambda_1, \ldots, \lambda_n$ in $\mathbb{K}$ such that

$$[x]_{\mathcal{X}/M} = \sum_{i=1}^{n} \lambda_i [e_i]_{\mathcal{X}/M}.$$

Since $M \subseteq V$,

$$x \in \sum_{i=1}^{n} \lambda_i e_i + V,$$

so that Proposition 2.1.3 (iv) entails that

$$[x]_{\mathcal{X}/V} = \sum_{i=1}^{n} \lambda_i [e_i]_{\mathcal{X}/V}.$$

Recalling that $\mathcal{A} = [x]_{\mathcal{X}/V}$, we deduce that the vectors $[e_1]_{\mathcal{X}/V}, \ldots, [e_n]_{\mathcal{X}/V}$ generate $\mathcal{X}/V$. □

2.1.3 Direct Sums

Direct sum is a fundamental tool in the analysis of *differential triplets and quadruplets. A direct sum decomposition* generalizes the notion of basis and works in infinite dimensional spaces.

Definition 2.1.8 Let $\mathcal{X}$ be a $\mathbb{K}$-vector space, n be a positive integer, and $M_1, \ldots, M_n$ be subspaces of $\mathcal{X}$. If

$$M_i \bigcap \Big(\sum_{j\neq i} M_j\Big) = \{0\}, \quad \forall i = 1, \ldots, n,$$

then, $M_1, \ldots, M_n$ are said to be *in direct sum*, and the set

$$M_1 \oplus \cdots \oplus M_n := \Big\{ \sum_{i=1}^{n} m_i \mid (m_1, \ldots, m_n) \in M_1 \times \cdots \times M_n \Big\}$$

is called the *direct sum of the* M_i*'s*. In the particular case where $n = 2$, M_1 and M_2 are in direct sum if $M_1 \cap M_2 = \{0\}$. □

Proposition 2.1.9 *Let $\mathcal{X}$ be a $\mathbb{K}$-vector space, n be a positive integer, and $M_1, \ldots, M_n$ be subspaces of $\mathcal{X}$. Then, the following assertions are equivalent.*

(i) *$M_1, \ldots, M_n$ are in direct sum.*
(ii) *For each x in $\sum_{i=1}^{n} M_i$, there exists a unique n-tuple $(m_1, \ldots, m_n)$ in $M_1 \times \cdots \times M_n$ such that*

$$x = \sum_{i=1}^{n} m_i.$$

(iii) *If $(m_1, \ldots, m_n)$ lies in $M_1 \times \cdots \times M_n$ and*

$$\sum_{i=1}^{n} m_i = 0,$$

then $m_1 = \cdots = m_n = 0$.

Proof By linearity, (ii) and (iii) are equivalent. Let us prove that (i) implies (iii). For, let $(m_1, \ldots, m_n)$ belong to $M_1 \times \cdots \times M_n$ and satisfy

$$\sum_{i=1}^{n} m_i = 0.$$

Then for each $i = 1, \ldots, n$, one has

$$m_i = \sum_{j \neq i} (-m_j).$$

Thus $m_i = 0$ since the M_i's are in direct sum. Hence (iii) holds true. Conversely let i be any index in $\{1, \ldots, n\}$. Any vector m_i in

$$M_i \bigcap (\sum_{j \neq i} M_j)$$

reads

$$m_i = \sum_{j \neq i} (-m_j),$$

where m_j lies in M_j for each $j \neq i$. By (iii), there results that $m_i = 0$. □

Proposition 2.1.10 *Let $\mathcal{X}$ be a $\mathbb{K}$-vector space, and V, M_1, and M_2 be subspaces of $\mathcal{X}$. We assume that*

(i) *M_1, M_2 are in direct sum;*
(ii) *V and $M_1 \oplus M_2$ are in direct sum.*

Then V, M_1, M_2 are in direct sum.

Proof Let (v, m_1, m_2) in $V \times M_1 \times M_2$ be such that

$$v + m_1 + m_2 = 0.$$

By Assumption (ii), $v = m_1 + m_2 = 0$. Next, (i) implies that $m_1 = m_2 = 0$. We conclude thanks Proposition 2.1.9. □

Of course, the direct sum of subspaces is commutative, that is, $M \oplus V = V \oplus M$. An useful situation is when M is given and we can find a subspace V for which $V \oplus M$ is equal to the whole space $\mathcal{X}$. In that case, V is called a *complementary subspace of* M. More precisely, we give the following definition.

Definition 2.1.9 Let M be a subspace of a vector space $\mathcal{X}$. A subspace V of $\mathcal{X}$ is *a complementary subspace of M in $\mathcal{X}$* if M and V are in direct sum and $\mathcal{X} = M \oplus V$. Besides, M and V are said to be *complementary subspaces in $\mathcal{X}$* if V is a complementary subspace of M in $\mathcal{X}$. □

The next result gives a sufficient condition for the existence of a complementary subspace.

Proposition 2.1.11 *Let M be a subspace of a vector space $\mathcal{X}$. If M has finite co-dimension, then M admits a complementary subspace V in $\mathcal{X}$. Moreover, the dimension of V is finite and equal to the co-dimension of M.*

Proof If $M = \mathcal{X}$ then, $V := \{0\}$ is a complementary subspace of M, and the dimension of V (i.e., 0) is equal to the co-dimension of M. Otherwise, $\mathcal{X}/M$ has positive (finite) dimension, let us say n. Then, there exist $v_1, \dots, v_n$ in $\mathcal{X}$ such that $([v_1]_{\mathcal{X}/M}, \dots, [v_n]_{\mathcal{X}/M})$ is a basis of $\mathcal{X}/M$. We claim that the subspace V of $\mathcal{X}$ generated by $v_1, \dots, v_n$ has dimension n. For, it is enough to show that the latter vectors are linearly independent in $\mathcal{X}$, which is easy to prove since $([v_1]_{\mathcal{X}/M}, \dots, [v_n]_{\mathcal{X}/M})$ is a basis of $\mathcal{X}/M$. That proves the claim.

In order to prove that M and V are in direct sum, let x be in $M \cap V$. Then,

$$x = \sum_{i=1}^{n} \lambda_i v_i,$$

for some scalars $\lambda_1, \dots, \lambda_n$. Thus,

$$\begin{aligned} \sum_{i=1}^{n} \lambda_i [v_i]_{\mathcal{X}/M} &= [x]_{\mathcal{X}/M} && \text{(by Proposition 2.1.4)} \\ &= 0_{\mathcal{X}/M} && \text{(since } x \text{ lies in } M). \end{aligned}$$

Since $([v_1]_{\mathcal{X}/M}, \dots, [v_n]_{\mathcal{X}/M})$ is a basis of $\mathcal{X}/M$, we derive that all the λ_i's vanish. Thus $x = 0$, which proves that $M \cap V$ is trivial; hence, M and V are in direct sum. There remains to show that $\mathcal{X} = M + V$. For, let x be any vector in $\mathcal{X}$. Using again the latter basis of $\mathcal{X}/M$, we get for some (new) scalars still labeled λ_i,

$$[x]_{\mathcal{X}/M} = \sum_{i=1}^{n} \lambda_i [v_i]_{\mathcal{X}/M}.$$

Then, Proposition 2.1.3 (i) yields that

$$x \in \sum_{i=1}^{n} \lambda_i v_i + M.$$

Hence, x lies in $V + M$. □

Conversely, the following result tells us, on the one hand, that if M admits a finite dimensional complementary subspace, then M has finite co-dimension. On the other hand, all finite dimensional complementary subspaces of M have the same dimension.

Proposition 2.1.12 *Let M and V be complementary subspaces in a real or complex vector space $\mathcal{X}$. If V has finite dimensional, then*

(i) *the co-dimension of M in $\mathcal{X}$ is finite;*
(ii) *the dimension of V is equal to the co-dimension of M. In symbol,*

$$\dim V = \operatorname{codim} M.$$

Proof By using $M \cap V = \{0\}$ and recalling (2.1.1), the map

$$i_V : V \simeq V/\{0\} \to \mathcal{X}/M$$

is linear and injective according to Proposition 2.1.4. Moreover, since $\mathcal{X} = V + M$, it is easily seen that i_V is onto. Thus V and $\mathcal{X}/M$ have the same (finite) dimension. We conclude with Definition 2.1.7 of the co-dimension of M. □

2.2 Normed Vector Spaces

The additional structure induced by *norms* allows us to perform *analysis* on vector spaces. Our fundamental tools are two *Hahn–Banach theorems* stated in Sect. 2.2.2. Let us start by introducing basic definitions and results about normed vector spaces.

2.2.1 Generalities on Normed Spaces

Definition 2.2.1 Let $\mathcal{X}$ be a $\mathbb{K}$-vector space. A map $\|\cdot\| : X \to [0, \infty)$ is a *norm on* $\mathcal{X}$ if for each x, y in $\mathcal{X}$ and λ in $\mathbb{K}$, one has

(i) $\|x + y\| \leq \|x\| + \|y\|$;
(ii) $\|\lambda x\| = |\lambda| \, \|x\|$;
(iii) if $\|x\| = 0$, then $x = 0$.

The ordered pair $(\mathcal{X}, \|\cdot\|)$ or simply $\mathcal{X}$, is called a *normed (vector) space.* □

Let $(\mathcal{X}, \|\cdot\|_{\mathcal{X}})$ be a normed space. If M is a subspace of the vector space $\mathcal{X}$ (in the sense of Definition 2.1.2), then the restriction of the map $\|\cdot\|_{\mathcal{X}}$ to M is clearly a norm on M. By a slight abuse of notation, this restriction will also be denoted by $\|\cdot\|_{\mathcal{X}}$. Hence $(M, \|\cdot\|_{\mathcal{X}})$ is a normed space. If no confusion may occur, $(M, \|\cdot\|_{\mathcal{X}})$ will be simply denoted by M.

The next definition depicts a more general situation.

Definition 2.2.2 Let $(\mathcal{X}, \|\cdot\|_{\mathcal{X}})$ and $(\mathcal{Y}, \|\cdot\|_{\mathcal{Y}})$ be $\mathbb{K}$-normed spaces. We say that $\mathcal{X}$ is *continuously embedded in* $\mathcal{Y}$ if $\mathcal{X}$ is included in $\mathcal{Y}$ and this inclusion induces a continuous map from $\mathcal{X}$ into $\mathcal{Y}$. □

The following basic result will be useful to prove density result.

Proposition 2.2.1 *Let $\mathcal{X}$, $\mathcal{Y}$ be two $\mathbb{K}$-normed spaces, and U be a subset of $\mathcal{X}$. If $f : \mathcal{X} \to \mathcal{Y}$ is an isomorphism between the normed spaces $\mathcal{X}$ and $\mathcal{Y}$ then*

$$f(\overline{U}) = \overline{f(U)}.$$

Anticipating Definition 2.2.13, by an *isomorphism between normed spaces,* we mean a bijective, continuous, and linear map between these spaces whose inverse is continuous as well. Here, $\overline{U}$ denotes *the closure of U in $\mathcal{X}$*, whereas $\overline{f(U)}$ is the closure of $f(U)$ in $\mathcal{Y}$.

Proof of Proposition 2.2.1 Since f is continuous, one has

$$f(\overline{U}) \subseteq \overline{f(U)}.$$

Then, by continuity of f^{-1},

$$f^{-1}\big(\overline{f(U)}\big) \subseteq \overline{f^{-1}\big(f(U)\big)} = \overline{U}.$$

Hence

$$\overline{f(U)} \subseteq f(\overline{U}).$$

The equality follows. □

We will now defined new normed spaces built upon a priori given normed spaces. This is a standard process in mathematics. Let us start with quotient spaces.

Let M be a closed subspace of a normed space $(\mathcal{X}, \|\cdot\|)$. We claim that the function

$$\|\cdot\|_{\mathcal{X}/M} : \mathcal{X}/M \to [0, \infty), \qquad \mathcal{A} \mapsto \inf_{x \in \mathcal{A}} \|x\|, \tag{2.2.1}$$

is a norm on $\mathcal{X}/M$. Indeed, only the third property of Definition 2.2.1 is not obvious. In order to prove it, we observe that $\|\mathcal{A}\|_{\mathcal{X}/M} = 0$ implies the existence of some sequence $(x_n)_{n\in\mathbb{N}}$ in $\mathcal{A}$ which converges toward 0 in $\mathcal{X}$. Since $x_n - x_0$ lies in M, and M is closed, we deduce that x_0 lies in $M \cap \mathcal{A}$. Since $M = 0_{\mathcal{X}/M}$, Proposition 2.1.3 (iv) yields that $\mathcal{A} = 0_{\mathcal{X}/M}$. That proves the claim.

Quotient spaces (by closed subspace) will always be equipped with the above norm.

Item (i) in the forthcoming proposition gives a geometric insight of the norm on a quotient space and is illustrated by Fig. 2.1. The other items are technical results, which will be useful in Sects. 2.3.1 and 2.4.3.

Proposition 2.2.2 *Let M be a closed subspace of a normed space $(\mathcal{X}, \|\cdot\|)$ and $\|\cdot\|_{\mathcal{X}/M}$ be defined by* (2.2.1)*. Then, the following assertions hold true.*

(i) *For each x in $\mathcal{X}$,*

$$\|[x]\|_{\mathcal{X}/M} = d(x, M),$$

where $d(x, M) := \inf_{m \in M} \|x - m\|$ denotes the distance between x and M. See Fig. 2.1.

(ii) *Let ε be a positive number and $\mathcal{A}$, $\mathcal{B}$ in $\mathcal{X}/M$ be such that*

$$\|\mathcal{B} - \mathcal{A}\|_{\mathcal{X}/M} < \varepsilon.$$

Then, for each x in $\mathcal{A}$, there exists x' in $\mathcal{B}$ such that $\|x' - x\| < \varepsilon$.

(iii) *If V is a dense subspace of $\mathcal{X}$ then V/M is dense in $\mathcal{X}/M$.*

Proof

(i) For each y in $[x]$, the vector $m := x - y$ lies in M. Then

$$\|y\| = \|x - m\| \geq d(x, M).$$

Hence, by (2.2.1), $\|[x]\|_{\mathcal{X}/M} \geq d(x, M)$. The reserve inequality is obtained along the same lines. Indeed, for each m in M, the vector $y := x - m$ lies in $[x]$, and

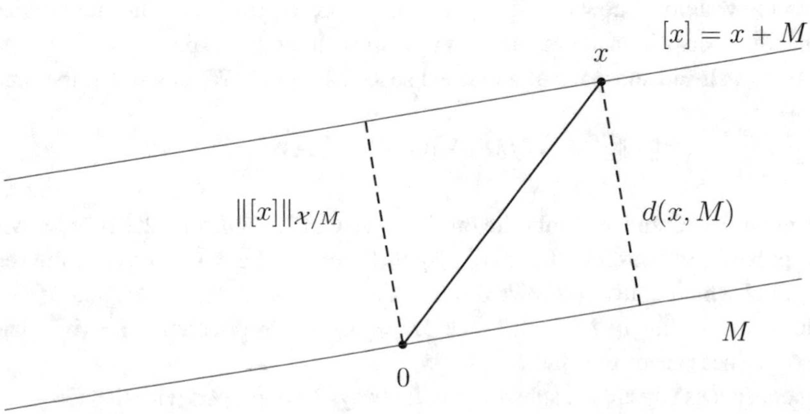

Fig. 2.1 An illustration of the equality $\|[x]\|_{\mathcal{X}/M} = d(x, M)$ stated in Proposition 2.2.2

$$\|x - m\| = \|y\| \geq \|[x]\|_{\mathcal{X}/M}.$$

Hence, $d(x, M) \geq \|[x]\|_{\mathcal{X}/M}$. The equality is proved.

(ii) Let x lie in $\mathcal{A}$ and z belong to $\mathcal{B}$. Observing that

$$\|[z - x]\|_{\mathcal{X}/M} < \varepsilon,$$

we infer from (2.2.1), the existence of some y in $[z - x]$ satisfying $\|y\| < \varepsilon$. Now, $y = z - x + m$ for some m in M. Setting $x' := z + m$, we get that x' belongs to $\mathcal{B}$ and $\|x' - x\| < \varepsilon$.

(iii) Let x be any vector of $\mathcal{X}$. By the density of $\mathcal{V}$, there exists a sequence (v_n) in V converging toward x in $\mathcal{X}$. Then,

$$\|[v_n] - [x]\|_{\mathcal{X}/M} = \|[v_n - x]\|_{\mathcal{X}/M} \leq \|v_n - x\|,$$

by (2.2.1). Whence V/M is dense in $\mathcal{X}/M$. □

Definition 2.2.3 Let $(\mathcal{X}, \|\cdot\|)$ be a normed vector space over $\mathbb{K}$. A function $f : \mathcal{X} \to \mathbb{K}$ is *a linear form on* $\mathcal{X}$ if for each x, y in $\mathcal{X}$ and λ in $\mathbb{K}$, one has

(i) $f(x + y) = f(x) + f(y)$;
(ii) $f(\lambda x) = \lambda f(x)$.

The *dual space of* $\mathcal{X}$ is the $\mathbb{K}$-vector space, denoted by $\mathcal{X}'$, whose elements are the continuous linear forms on $\mathcal{X}$. Equipped with the *dual norm*

$$\|\cdot\|_{\mathcal{X}'} : \mathcal{X} \to [0, \infty), \qquad f \mapsto \sup_{\|x\| \leq 1} |f(x)|,$$

$\mathcal{X}'$ becomes a normed space. □

The next standard result of linear algebra allows to compute the dimension of certain spaces defined by linear equations. It is used in the proof of Proposition 4.4.6.

Proposition 2.2.3 *Let $\mathcal{X}_{\mathrm{f}}$ be a finite dimensional space over $\mathbb{K}$, and let $F_1, \ldots, F_d$ be linear forms on $\mathcal{X}_{\mathrm{f}}$. If the F_j's are linearly independent in the dual space $\mathcal{X}_{\mathrm{f}}'$ of $\mathcal{X}_{\mathrm{f}}$ then,*

$$\dim \bigcap_{j=1}^{d} \ker F_j = \dim \mathcal{X}_{\mathrm{f}} - d.$$

In the particular case where $d = 1$, we claim that the kernel of a non trivial linear form F_1 has co-dimension one. Indeed Proposition 2.2.3 tells us that the dimension of this kernel is equal to $\dim \mathcal{X}_{\mathrm{f}} - 1$. Thus, $\ker F_1$ admits a one-dimensional complementary subspace. By applying Proposition 2.1.12 with $M := \ker F_1$, we derive the claim.

Proof of Proposition 2.2.3 For the sake of concision, we denote by n the dimension of $\mathcal{X}_{\mathrm{f}}$. Let $F : \mathcal{X}_{\mathrm{f}} \to \mathbb{K}^d$ be the linear map whose components are $F_1, \ldots, F_d$. Then,

$$\ker F = \bigcap_{j=1}^{d} \ker F_j.$$

By *the rank-nullity theorem*,

$$\dim \ker F = \dim \mathcal{X}_{\mathrm{f}} - \dim R(F),$$

Thus, there remains to show that the range $R(F)$ of F has dimension d. For, we consider a basis $(e_1, \ldots, e_n)$ of $\mathcal{X}_{\mathrm{f}}$ and we put $r := \dim R(F)$. Then, r is equal to the rank of $F(e_1), \ldots, F(e_n)$, which is equal to the rank of the matrix with entries

$$\big(F_i(e_j)\big)_{1 \le i \le d,\, 1 \le j \le n}.$$

By tranposition, r is also equal to the rank of the matrix

$$\big(F_j(e_i)\big)_{1 \le i \le n,\, 1 \le j \le d}.$$

For each $j = 1, \ldots, d$, denote by X_j the jth column of the latter matrix; that is, X_j is the vector of $\mathbb{K}^n$ defined by

$$X_j := \big(F_j(e_1), \ldots, F_j(e_n)\big)^t.$$

Then r is equal to the rank of the X_j's. We will be done if these d vectors are linearly independent in $\mathbb{K}^n$. Let $\lambda_1, \ldots, \lambda_d$ be scalars satisfying

$$\sum_{j=1}^{d} \lambda_j X_j = 0.$$

Then

$$\sum_{j=1}^{d} \lambda_j F_j(e_i) = 0, \qquad \forall i = 1, \ldots, n.$$

Since $(e_1, \ldots, e_n)$ is a basis of $\mathcal{X}_{\mathrm{f}}$, the latter equalities imply that

$$\sum_{j=1}^{d} \lambda_j F_j = 0 \qquad \text{in } \mathcal{X}_{\mathrm{f}}^{'} .$$

Since by assumption, the F_j's are linearly independent, there results that the X_j's are linearly independent as well. □

Definition 2.2.4 Let $(\mathcal{X}, \|\cdot\|)$ be a normed vector space over $\mathbb{K}$. A function $f : \mathcal{X} \to \mathbb{K}$ is *an anti-linear form on* $\mathcal{X}$ if for each x, y in $\mathcal{X}$ and λ in $\mathbb{K}$, one has

(i) $f(x + y) = f(x) + f(y)$;
(ii) $f(\lambda x) = \overline{\lambda} f(x)$.

The anti-dual space of $\mathcal{X}$, denoted by $\mathcal{X}^*$, is the $\mathbb{K}$-vector space whose elements are the continuous anti-linear forms on $\mathcal{X}$. Equipped with the *anti-dual norm*

$$\|\cdot\|_{\mathcal{X}^*} : \mathcal{X} \to [0, \infty), \qquad f \mapsto \sup_{\|x\| \le 1} |f(x)|,$$

$\mathcal{X}^*$ becomes a normed space. □

We may work with the *dual space of* $\mathcal{X}$ as well. However, we choose anti-dual spaces since the inner product of a complex Hilbert space $\mathcal{H}$ induces a natural isometry (see Definition 2.2.13) between $\mathcal{H}$ and its anti-dual $\mathcal{H}^*$. Of course, if $\mathbb{K} = \mathbb{R}$ then only dual spaces are involved, and $\mathcal{X}^*$ is the space of continuous *linear* forms on $\mathcal{X}$.

We may pass easily from $\mathcal{X}'$ to $\mathcal{X}^*$: for each F in $\mathcal{X}'$, we set

$$\overline{F} : \mathcal{X} \to \mathbb{K}, \qquad x \mapsto \overline{F(x)}. \tag{2.2.2}$$

Then, $\overline{F}$ belongs to $\mathcal{X}^*$, so that the map

$$C_{\mathcal{X}'} : \mathcal{X}' \to \mathcal{X}^*, \qquad F \mapsto \overline{F} \tag{2.2.3}$$

is well defined. Although easily proved, the following property of $C_{\mathcal{X}'}$ will be stated as a proposition for further references.

Proposition 2.2.4 *Let* $\mathcal{X}$ *be a* $\mathbb{K}$*-normed space. The map* $C_{\mathcal{X}'}$ *defined by* (2.2.3) *is an isometric anti-isomorphism between* $\mathcal{X}'$ *and* $\mathcal{X}^*$ *(in the sense of Definition 2.2.13).*

Let $\mathcal{X}$ be a $\mathbb{K}$-normed space. For each f in $\mathcal{X}^*$, we define the *real part of* f as the map

$$\operatorname{Re} f : \mathcal{X} \to \mathbb{R}, \qquad x \mapsto \frac{1}{2}\big(f(x) + \overline{f(x)}\big).$$

Observe that each f in $\mathcal{X}^*$ satisfies

$$f(x) = \mathrm{Re}\, f(x) + \mathrm{i}\mathrm{Re}\, f(\mathrm{i}x), \qquad \forall x \in \mathcal{X}. \tag{2.2.4}$$

It is easily seen that $\mathrm{Re}\, f$ lies is $(\mathcal{X}_{\mathbb{R}})'$, where according to Definition 2.1.3, $\mathcal{X}_{\mathbb{R}}$ stands for *the real space associated to* $\mathcal{X}$. Hence, $(\mathcal{X}_{\mathbb{R}})'$ is the real vector space of continuous linear forms on $\mathcal{X}_{\mathbb{R}}$. Thus, we have a well-defined map

$$\psi : (\mathcal{X}^*)_{\mathbb{R}} \to (\mathcal{X}_{\mathbb{R}})', \qquad f \mapsto \mathrm{Re}\, f, \tag{2.2.5}$$

whose properties are stated now.

Proposition 2.2.5 *Let* $(\mathcal{X}, \|\cdot\|)$ *be a normed space and* ψ *be defined by* (2.2.5). *Then,* ψ *is an isometric isomorphism between the real spaces* $(\mathcal{X}^*)_{\mathbb{R}}$ *and* $(\mathcal{X}_{\mathbb{R}})'$.

Proof Is is clear that ψ is a linear map between real spaces. For each f in $\mathcal{X}^*$, (2.2.4) entails that ψ is one-to-one and

$$\|\psi(f)\|_{(\mathcal{X}_{\mathbb{R}})'} \leq \|f\|_{(\mathcal{X}^*)_{\mathbb{R}}}.$$

In order to show the reverse inequality, for each x in $\mathcal{X}$, consider a complex number λ_0 satisfying $|\lambda_0| = 1$ and $\lambda_0 f(x) = |f(x)|$. Then, we compute

$$|f(x)| = f(\overline{\lambda_0} x) = \mathrm{Re}\, f(\overline{\lambda_0} x) \leq \|\psi(f)\|_{(\mathcal{X}_{\mathbb{R}})'} \|x\|.$$

Hence, $\|f\|_{(\mathcal{X}^*)_{\mathbb{R}}} \leq \|\psi(f)\|_{(\mathcal{X}_{\mathbb{R}})'}$, and ψ is an isometry.

There remains to check that ψ is onto. For each u in $(\mathcal{X}_{\mathbb{R}})'$, define $f : \mathcal{X} \to \mathbb{C}$ by

$$f(x) = u(x) + \mathrm{i}u(\mathrm{i}x), \qquad \forall x \in \mathcal{X}.$$

Then we claim that f is a continuous anti-linear form on $\mathcal{X}$, so that f lies in $\mathcal{X}^*$ and in $(\mathcal{X}^*)_{\mathbb{R}}$ since theses two spaces possess the same vectors. In order to prove the claim, let us consider any x in $\mathcal{X}$ and λ in $\mathbb{C}$. Writing λ under the form $a + ib$ where a and b are real numbers, we compute

$$\begin{aligned}
f(\lambda x) &= u(\lambda x) + \mathrm{i}u(\mathrm{i}\lambda x) && \text{(by definition of } f\text{)} \\
&= a\big(u(x) + \mathrm{i}u(\mathrm{i}x)\big) - ib\big(u(x) + \mathrm{i}u(\mathrm{i}x)\big) && \text{(since } (\mathcal{X}_{\mathbb{R}})' \text{ is a real vector space)} \\
&= \overline{\lambda} f(x).
\end{aligned}$$

Then, the claim follows easily. Since u is real-valued, $\psi(f) = u$, so that ψ is onto. The proof of the proposition is now completed. □

2.2.2 Two Hahn–Banach Theorems and Consequences

Hahn–Banach theorems reveal deep properties of linear and anti-linear forms. Starting from two classical Hahn–Banach theorems stated for *real* vector spaces and *linear* forms, we will prove their extension to the field of *complex* numbers and *anti-linear* forms. We will also derive some useful consequences of these theorems.

Our first Hahn–Banach theorem allows to extend "nicely" continuous forms defined on subspaces. By a nice extension, we mean a norm-preserving extension. There are more general statements than those stated here, and Theorem 2.2.6 is sometimes referred to in the literature as a corollary of a more general Hahn–Banach theorem. However, the framework of normed vector spaces is sufficient for our purposes.

In this subsection, we will give three consequences of these Hahn–Banach theorems, namely Corollaries 2.2.7, 2.2.9, and 2.2.10.

Theorem 2.2.6 (First Hahn–Banach Theorem) *Let $(\mathcal{X}, \|\cdot\|)$ be a $\mathbb{K}$-normed space, M be a subspace of $\mathcal{X}$ and g be a continuous anti-linear form on M. Then there exists f in $\mathcal{X}^*$ whose restriction on M coincide with g and such that*

$$\|f\|_{X^*} = \|g\|_{M^*}.$$

Proof In the case where $\mathbb{K} = \mathbb{R}$, we refer the reader to [Rud91, Th. 3.3] or [Bre11, Coro. I.2]. Let us now assume that $\mathbb{K} = \mathbb{C}$. For each g in M^*, the real part $\operatorname{Re} g$ of g lies in $(M_{\mathbb{R}})'$ and

$$\|\operatorname{Re} g\|_{(M_{\mathbb{R}})'} = \|g\|_{M^*},$$

according to Proposition 2.2.5. By the previous real case, there exists u in $(\mathcal{X}_{\mathbb{R}})'$ whose restriction to M is equal to $\operatorname{Re} g$ and such that

$$\|u\|_{(\mathcal{X}_{\mathbb{R}})'} = \|\operatorname{Re} g\|_{(M_{\mathbb{R}})'}.$$

With Proposition 2.2.5 again, we deduce that $f := \psi^{-1}(u)$ lies in $\mathcal{X}^*$ and $\|f\|_{\mathcal{X}^*} = \|g\|_{M^*}$. Moreover, by (2.2.4), $f = g$ on M (since $\operatorname{Re} f$ and $\operatorname{Re} g$ coincide on M). Hence, f matches the conditions of the theorem. □

Corollary 2.2.7 *Let $(\mathcal{X}, \|\cdot\|)$ be a $\mathbb{K}$-normed space, and let x_0 be a non-zero vector of $\mathcal{X}$. Then, there exists some f in $\mathcal{X}^*$ such that $\|f\|_{\mathcal{X}^*} = 1$ and $f(x_0) = \|x_0\|$.*

Proof We apply Theorem 2.2.6 with $M := \mathbb{K}x_0$, $g(\lambda x_0) := \overline{\lambda}\|x_0\|$ for each $\lambda \in \mathbb{K}$, so that $\|g\|_{M^*} = 1$ since x_0 is non-trivial. □

Let us notice that the assumption $x_0 \neq 0$ cannot be removed because if $\mathcal{X} = \{0\}$ is the trivial vector space, then its anti-dual space contains only the trivial form.

Our second Hahn–Banach theorem expresses a strict separation between two disjoint convex sets. This separation occurs through a well chosen linear form f. All the values taken by f on the first set are far away from all its values coming from the second set. Geometrically, this theorem states that there exists a linear manifold of co-dimension one, which separates strictly these two sets.

Theorem 2.2.8 (Second Hahn–Banach Theorem) *Let $(\mathcal{X}, \|\cdot\|)$ be a $\mathbb{K}$-normed space. Suppose that*

(i) *A and B are disjoint, non-empty convex subsets of $\mathcal{X}$;*
(ii) *A is compact and B is closed.*

Then, there exists a linear form f in $\mathcal{X}^*$ and two real numbers s and t such that

$$\operatorname{Re} f(x) < s < t < \operatorname{Re} f(y), \qquad \forall (x, y) \in A \times B.$$

Proof In the case where $\mathbb{K} = \mathbb{R}$, we refer the reader to [Rud91, Th. 3.2] or [Bre11, Th. I.7]. Let us now assume that $\mathbb{K} = \mathbb{C}$. By the previous real case, there exists u in $(\mathcal{X}_{\mathbb{R}})'$ and real numbers s and t such that

$$u(x) < s < t < u(y), \qquad \forall (x, y) \in A \times B.$$

Now Proposition 2.2.5 entails that $u = \operatorname{Re} f$ for a (unique) f in $\mathcal{X}^*$. The proof of the complex case is now completed. □

Corollary 2.2.9 *Let $\mathcal{X}$ be a $\mathbb{K}$-normed space and Y be a closed subspace of $\mathcal{X}$. If $Y \neq \mathcal{X}$, then for each x_0 in $\mathcal{X} \setminus Y$, there exists f in $\mathcal{X}^*$ such that*

$$f_{|Y} = 0 \quad \text{and} \quad f(x_0) \neq 0.$$

In the above statement, $f_{|Y}$ stands for the restriction of f to Y.

Proof of Corollary 2.2.9 We apply Theorem 2.2.8 with $A := \{x_0\}$ and $B := Y$. Thus, there exists some f in $\mathcal{X}^*$ such that

$$\operatorname{Re} f(x_0) < \operatorname{Re} f(y), \qquad \forall y \in Y.$$

Since Y is a vector space, we deduce that for each y in Y

$$\operatorname{Re} f(x_0) < t \operatorname{Re} f(y), \qquad \forall t \in \mathbb{R}.$$

Whence the restriction of Re f to Y must be zero. Then, $f_{|Y} = 0$ by Proposition 2.2.5, and $f(x_0) \neq 0$. □

The next result gives a convenient way for proving the density of subspaces.

Corollary 2.2.10 *Let $\mathcal{X}$ be a $\mathbb{K}$-normed space and M be a subspace of $\mathcal{X}$. Let us assume that M possesses the following property: for every f in $\mathcal{X}^*$,*

$$f_{|M} = 0 \Longrightarrow f = 0.$$

Then, M is dense in $\mathcal{X}$.

Proof We will prove the contrapositive. That is to say, assuming that the closure $\overline{M}$ of M is different from $\mathcal{X}$, we have to show that there exists a nontrivial form f in $\mathcal{X}^*$ vanishing on M. This assertion follows easily from Corollary 2.2.9. □

2.2.3 The Second Anti-Dual Space

Let $\mathcal{X}$ be a $\mathbb{K}$-normed space and $\mathcal{X}^*$ be its anti-dual space.

Definition 2.2.5 The anti-dual space of $\mathcal{X}^*$ is called the *second anti-dual space of* $\mathcal{X}$. It is denoted by $\mathcal{X}^{**}$, so that $\mathcal{X}^{**} = (\mathcal{X}^*)^*$. Equipped with the anti-dual norm, $\mathcal{X}^{**}$ becomes a normed space. □

Let x_0 lie in $\mathcal{X}$ and f_0 belong to $\mathcal{X}^*$. By the very definition of $\mathcal{X}^*$, we have two anti-linear forms

$$\begin{aligned} f_0 : \mathcal{X} &\to \mathbb{K} \\ x &\mapsto f_0(x) \end{aligned} \quad \text{and} \quad \begin{aligned} \mathcal{X}^* &\to \mathbb{K} \\ f &\mapsto \overline{f(x_0)} \end{aligned} \tag{2.2.6}$$

In the first map, x is the variable and f_0 is the function. In the second map, x_0 appears as a function, and f is the variable. If one wants to highlight this symmetry induced by the duality, we will use the notation

$$\langle f, x \rangle_{\mathcal{X}^*, \mathcal{X}}$$

or simply $\langle f, x \rangle$ instead of $f(x)$, for every (f, x) in $\mathcal{X}^* \times \mathcal{X}$.

The second form in (2.2.6) is continuous on $\mathcal{X}^*$. Thus, it belongs to the second anti-dual space $\mathcal{X}^{**}$. Denoting this form by $J_{\mathcal{X}}(x_0)$, one has

$$\langle J_{\mathcal{X}}(x_0), f\rangle_{\mathcal{X}^{**},\mathcal{X}^*} = \overline{\langle f, x_0\rangle}_{\mathcal{X}^*,\mathcal{X}}.$$

Then, we may introduce the map

$$J_{\mathcal{X}} : \mathcal{X} \to \mathcal{X}^{**}$$

which assigns to each x in $\mathcal{X}$ the element $J_{\mathcal{X}}(x)$ of $\mathcal{X}^{**}$ defined by

$$\langle J_{\mathcal{X}}(x), f\rangle_{\mathcal{X}^{**},\mathcal{X}^*} = \overline{\langle f, x\rangle}_{\mathcal{X}^*,\mathcal{X}}, \qquad \forall f \in \mathcal{X}^*. \tag{2.2.7}$$

Proposition 2.2.11 *Let $\mathcal{X}$ be a normed space and $J_{\mathcal{X}} : \mathcal{X} \to \mathcal{X}^{**}$ be defined by* (2.2.7). *Then, $J_{\mathcal{X}}$ is an isometry from $\mathcal{X}$ into $\mathcal{X}^{**}$(in the sense of Definition 2.2.13).*

Proof It is clear from (2.2.7) that $J_{\mathcal{X}}$ is linear and that

$$\|J_{\mathcal{X}}(x)\|_{\mathcal{X}^{**}} \le \|x\|_{\mathcal{X}}.$$

If $x = 0$, then the equality holds in the previous relation. If $x \neq 0$, then Corollary 2.2.7 yields the existence of some anti-linear form f_0 in $\mathcal{X}^*$ satisfying $\|f_0\|_{\mathcal{X}^*} = 1$ and $f_0(x) = \|x\|_{\mathcal{X}}$. Thus,

$$\|J_{\mathcal{X}}(x)\|_{\mathcal{X}^{**}} \ge \|x\|_{\mathcal{X}}.$$

Hence, $J_{\mathcal{X}}$ is an isometric map. □

In view of Proposition 2.2.11, we put the following definition.

Definition 2.2.6 Let $\mathcal{X}$ be a normed space. The map $J_{\mathcal{X}} : \mathcal{X} \to \mathcal{X}^{**}$ defined by (2.2.7) is called the *canonical isometry from $\mathcal{X}$ into $\mathcal{X}^{**}$*. □

In infinite dimensional spaces, an *isometry* is not always surjective. The subjectivity of the canonical isometry $J_{\mathcal{X}}$ forces the normed space $\mathcal{X}$ to be a Banach space (see Theorem 2.3.9). Therefore, surjective canonical isometries will be considered in the next section dedicated to Banach spaces.

2.2.4 Annihilators

Let $\mathcal{X}$ be a $\mathbb{K}$-normed space and $\mathcal{X}^*$ be its anti-dual space.

Definition 2.2.7 Let E is a non empty subset of $\mathcal{X}$. Then, *the annihilator of E in $\mathcal{X}^*$* is the set of all anti-linear forms f in $\mathcal{X}^*$ whose kernel contains E. The annihilator of E is denoted by $E^{\perp \mathcal{X}^*}$ or $E^{\perp}$ if no confusion may occur. In symbol,

$$E^{\perp \mathcal{X}^*} := \left\{ f \in \mathcal{X}^* \mid \langle f, x \rangle_{\mathcal{X}^*, \mathcal{X}} = 0 \ \forall x \in E \right\}.$$

In the case where E reduces to one vector, that is $E := \{x\}$ for some x in $\mathcal{X}$, we will abbreviate $\{x\}^{\perp}$ by the symbol $x^{\perp}$.

If $F \subseteq \mathcal{X}^*$ is non empty, then *the annihilator of F in $\mathcal{X}$* is the set of all x in $\mathcal{X}$ such that the kernel of $J_{\mathcal{X}}(x)$ contains F. The annihilator of F is denoted by $F^{\top, \mathcal{X}}$ or simply by $F^{\top}$. In symbol,

$$F^{\top \mathcal{X}} := \left\{ x \in \mathcal{X} \mid \langle f, x \rangle_{\mathcal{X}^*, \mathcal{X}} = 0 \ \forall f \in F \right\}. \qquad \square$$

If F is a non-empty subset of $\mathcal{X}^*$, then there are two annihilators: one living in $\mathcal{X}^{**}$ obtained by considering $\mathcal{X}^*$ and its anti-dual space $\mathcal{X}^{**}$, and another one in $\mathcal{X}$ obtained by considering $\mathcal{X}^*$ as the anti-dual space of $\mathcal{X}$. According to Definition 2.2.7, the former annihilator is denoted by $F^{\perp \mathcal{X}^{**}}$, and the latter annihilator is denoted by $F^{\top \mathcal{X}}$, so that no confusion may occur.

Proposition 2.2.12 *If M is a subspace of a normed space $\mathcal{X}$, then*

$$\left(M^{\perp}\right)^{\top} = \overline{M}.$$

Proof Let x lie in M. Each f in $M^{\perp}$ satisfies $\langle f, x \rangle_{\mathcal{X}^*, \mathcal{X}} = 0$. Thus, x belongs to the annihilator of $M^{\perp}$. That is $M \subseteq \left(M^{\perp}\right)^{\top}$. Since the annihilators are closed, we obtain

$$\overline{M} \subseteq \left(M^{\perp}\right)^{\top}$$

Conversely, it is enough to show that the complement of $\overline{M}$ is included in the complement of $\left(M^{\perp}\right)^{\top}$. For, let x_0 be in $\mathcal{X}$ with $x_0 \notin \overline{M}$. The Hahn–Banach Theorem 2.2.8 insures the existence of some f in $\mathcal{X}^*$ such that

$$\operatorname{Re} f(x_0) < \operatorname{Re} f(y), \qquad \forall y \in \overline{M}. \tag{2.2.8}$$

Arguing as in the proof of Corollary 2.2.10, we obtain that the restriction of f to M must be zero. Thus, f lies in $M^{\perp}$. Now, going back to (2.2.8), we get that $f(x_0) \neq 0$. That is, x_0 does not belong to $\left(M^{\perp}\right)^{\top}$. $\square$

Another consequence of the Hahn–Banach theorem 2.2.6 concerns the existence of *closed* complementary subspaces of a finite dimensional space. Recalling Definition 2.1.9 of a complementary subspace, the result stated in Proposition 2.2.14 relies on the following lemma.

Lemma 2.2.13 *Let $\mathcal{X}$ be a $\mathbb{K}$-normed space and n be a positive integer. Let also $e_1, \ldots, e_n$ be in $\mathcal{X}$, and $f_1, \ldots, f_n$ be in $\mathcal{X}^*$. Assume that the square matrix $M = (M_{i,j})_{i,j=1,\ldots,n}$ whose entries are defined by*

$$M_{i,j} := \langle f_i, e_j \rangle_{\mathcal{X}^*,\mathcal{X}},$$

is invertible. Then, $e_1, \ldots, e_n$ and $f_1, \ldots, f_n$ are linearly independent in $\mathcal{X}$ and $\mathcal{X}^*$ respectively, and

$$\mathcal{X} = \langle e_1, \ldots, e_n \rangle \oplus \langle f_1, \ldots, f_n \rangle^{\top \mathcal{X}} \tag{2.2.9}$$

$$\mathcal{X}^* = \langle f_1, \ldots, f_n \rangle \oplus \langle e_1, \ldots, e_n \rangle^{\perp \mathcal{X}^*}. \tag{2.2.10}$$

Proof If $\lambda_1 e_1 + \cdots + \lambda_n e_n = 0$ for scalars $\lambda_1, \ldots, \lambda_n$ then for each $i = 1, \ldots, n$,

$$\sum_{j=1}^{n} \overline{\lambda_j} \langle f_i, e_j \rangle = 0.$$

By invertibitily of the matrix M, we infer that the λ_i's vanish, so that the e_i's are linearly independent. In a same way, the f_i's are linearly independent. In order to prove (2.2.9), we start to show that the sum is direct. For, if $x = \lambda_1 e_1 + \cdots + \lambda_n e_n$ lies in $\langle f_1, \ldots, f_n \rangle^{\top}$, then

$$0 = \langle f_i, x \rangle = \sum_{j=1}^{n} \overline{\lambda_j} \langle f_i, e_j \rangle, \qquad \forall i = 1, \ldots, n.$$

By invertibitily of M, there results that $x = 0$. Thus the intersection of $\langle e_1, \ldots, e_n \rangle$ and $\langle f_1, \ldots, f_n \rangle^{\top}$ is trivial. Regarding the sum of these spaces, we observe that by invertibitily of M, for each x in $\mathcal{X}$, there exist n scalars $\lambda_1, \ldots, \lambda_n$ such that

$$\sum_{j=1}^{n} \overline{\lambda_j} \langle f_i, e_j \rangle = \langle f_i, x \rangle, \qquad \forall i = 1, \ldots, n.$$

Now the vector $y := \lambda_1 e_1 + \cdots + \lambda_n e_n$ lies in $\langle e_1, \ldots, e_n \rangle$, and by the previous relations

$$\langle f_i, x - y \rangle = \langle f_i, x \rangle - \sum_{j=1}^{n} \overline{\lambda_j} \langle f_i, e_j \rangle = 0, \qquad \forall i = 1, \ldots, n.$$

Hence $x - y$ lies in $\langle f_1, \dots, f_n \rangle^\top$, which proves (2.2.9). Finally, applying (2.2.9) with $\mathcal{X}^*$ in place of $\mathcal{X}$, we get from (2.2.9)

$$\mathcal{X}^* = \langle f_1, \dots, f_n \rangle \oplus \langle J_{\mathcal{X}}(e_1), \dots, J_X(e_n) \rangle^{\top \mathcal{X}^*},$$

where $J_{\mathcal{X}}$ is *the canonical isometry* given by Definition 2.2.6. Since

$$J_{\mathcal{X}}(E)^{\top \mathcal{X}^*} = E^{\perp \mathcal{X}^*},$$

for any non-empty subset E of $\mathcal{X}$, we end up with (2.2.10). □

Proposition 2.2.14 *Let N be a finite dimensional subspace of a normed space $\mathcal{X}$. Then, the following assertions hold true.*

(i) *N admits a closed complementary subspace V in $\mathcal{X}$.*
(ii) *$N^\perp$ and $V^\perp$ are closed complementary subspaces in $\mathcal{X}^*$.*
(iii) *$V^\perp$ has finite dimension, moreover $V^\perp$ and N have the same dimension.*
(iv) *If N has positive dimension n and $(e_1, \dots, e_n)$ is one of its basis then there exists $f_1, \dots, f_n$ in $\mathcal{X}^*$ such that*

$$\langle f_i, e_j \rangle = \delta_{i,j}, \qquad \forall i, j = 1, \dots n,$$

where $\delta_{i,j}$ is the Kronecker symbol.

Proof Let us start to prove Items (i) and (iv). Let n be the dimension of N. If $n = 0$, then (i) holds by choosing $V := \mathcal{X}$. Otherwise, we consider a basis $(e_1, \dots, e_n)$ of N. For each $i = 1, \dots, n$, let us introduce the following anti-linear form

$$g_i : N \to \mathbb{K}, \qquad x = \sum_{i=1}^{n} x_i e_i \mapsto \overline{x_i}. \tag{2.2.11}$$

By the Hahn–Banach Theorem 2.2.6, g_i may be extended into a continuous anti-linear form f_i on $\mathcal{X}$. We put

$$V := \langle f_1, \dots, f_n \rangle^{\top \mathcal{X}}.$$

Then clearly, V is closed in $\mathcal{X}$. Moreover, for all $i, j = 1, \dots, n$,

$$\langle f_i, e_j \rangle = \langle g_i, e_j \rangle = \delta_{i,j},$$

which proves (iv). With the notation of Lemma 2.2.13, M is the identity matrix of $\mathbb{K}^n$. Hence, Lemma 2.2.13 yields that N and V are closed complementary subspaces.

Regarding the proof of (ii) and (iii), (2.2.10) yields that

$$\mathcal{X}^* = \langle f_1, \dots, f_n \rangle \oplus N^\perp.$$

By lemma 2.2.13, the f_i's are linearly independent so that there remains to prove that

$$\langle f_1, \dots, f_n \rangle = V^\perp. \tag{2.2.12}$$

For, it is easily seen that the former space is contained in the latter. Conversely, for any f in $V^\perp$, we set

$$g := f - \sum_{i=1}^{n} \langle f, e_i \rangle f_i.$$

Then, g vanishes on N since $\langle f_i, e_j \rangle = \delta_{i,j}$. Moreover, g vanishes on V, indeed for each v in V, one has

$$\langle g, v \rangle = \langle f, v \rangle - \sum_{i=1}^{n} \langle f, e_i \rangle \langle f_i, v \rangle = 0,$$

since f and f_i belong to $V^\perp$. Due to $\mathcal{X} = N + V$, g is trivial, so that f lies in $\langle f_1, \dots, f_n \rangle$. Hence (2.2.12) holds true. □

Corollary 2.2.15 *Let N be a finite dimensional subspace of a real or complex normed space $\mathcal{X}$. Then, the co-dimension of $N^\perp$ in $\mathcal{X}^*$ is finite and*

$$\operatorname{codim} N^\perp = \dim N.$$

Proof By Proposition 2.2.14, N admits a complementary subspace V in $\mathcal{X}$,

$$\mathcal{X}^* = N^\perp \oplus V^\perp \tag{2.2.13}$$

and

$$\dim V^\perp = \dim N.$$

Besides, from Proposition 2.1.12, (2.2.13) entails that

$$\dim V^{\perp} = \operatorname{codim} N^{\perp}.$$

Thus $\operatorname{codim} N^{\perp} = \dim N$. □

Corollary 2.2.16 *Being given a real or complex normed space $\mathcal{X}$, let N and N^{s} be two non trivial finite dimensional subspaces of $\mathcal{X}$ and $\mathcal{X}^*$ respectively. We assume that*

$$\mathcal{X}^* = N^{\mathrm{s}} \oplus N^{\perp}. \tag{2.2.14}$$

Then,

$$\dim N = \dim N^{\mathrm{s}}.$$

In addition, if $(e_1, \ldots, e_{\dim N})$ is a basis of N and $(f_1, \ldots, f_{\dim N})$ is a basis of N^{s}, then the square matrix with entries

$$\langle f_i, e_j \rangle_{\mathcal{X}^*, \mathcal{X}}, \qquad \forall i, j = 1, \ldots, \dim N$$

is invertible.

Proof By Corollary 2.2.15,

$$\operatorname{codim} N^{\perp} = \dim N.$$

Applying Proposition 2.1.12 in $\mathcal{X}^*$, we get thanks (2.2.14)

$$\operatorname{codim} N^{\perp} = \dim N^{\mathrm{s}}.$$

Hence, $\dim N = \dim N^{\mathrm{s}}$. In order to prove the inversibility of the matrix, it is enough to prove that its transpose has a trivial kernel. So, let us assume that scalars $\lambda_1, \ldots, \lambda_{\dim N}$ satisfy (with the convention of repeated indexes)

$$\langle f_j, e_i \rangle \lambda_j = 0, \qquad \forall i = 1, \ldots, \dim N$$

Since $\langle f_j, e_i \rangle \lambda_j = \langle \lambda_j f_j, e_i \rangle$, we derive that $\lambda_j f_j$ lies in $N^{\mathrm{s}} \cap N^{\perp}$. Then (2.2.14) yields that $\lambda_j f_j = 0$, and all the λ_j's vanish since $(f_1, \ldots, f_{\dim N})$ is a basis. □

2.2.5 Operators on Normed Spaces

Boundary restriction operators establish a special class of operators. Thus, operators are of fundamental importance in this book. Being given $\mathcal{X}$ and $\mathcal{Y}$ two normed spaces over

$\mathbb{K} = \mathbb{R}$ or $\mathbb{C}$, with respective norms $\|\cdot\|_{\mathcal{X}}$ and $\|\cdot\|_{\mathcal{Y}}$, we will recall some usual definitions and notation. An *operator on $\mathcal{X}$ with values in $\mathcal{Y}$* is a linear map T defined on a subspace $D(T)$ of $\mathcal{X}$, with values in $\mathcal{Y}$. $D(T)$ is called the *domain of* T. The expression "*Let $T : D(T) \subseteq \mathcal{X} \to \mathcal{Y}$ be an operator*" means that T is an operator on $\mathcal{X}$ with values in $\mathcal{Y}$, and whose domain is $D(T)$.

Let $T : D(T) \subseteq \mathcal{X} \to \mathcal{Y}$ be an operator. Its *kernel* and *range* are denoted by $\ker T$ and $R(T)$ respectively. We recall that

$$\ker T := \{x \in D(T) \mid Tx = 0\}, \qquad R(T) := \{Tx \mid x \in D(T)\}.$$

Definition 2.2.8 Let $\mathcal{X}$ and $\mathcal{Y}$ be two normed spaces over $\mathbb{K}$. Let also $T : D(T) \subseteq \mathcal{X} \to \mathcal{Y}$ and $L : D(L) \subseteq \mathcal{X} \to \mathcal{Y}$ be operators. Then the *sum of the operators T and L* is the operator denoted by $T + L$, whose domain is

$$D(T + L) := D(T) \cap D(L),$$

and which maps any x of $D(T + L)$ into

$$(T + L)x := Tx + Lx. \qquad \square$$

Definition 2.2.9 Let $\mathcal{X}$, $\mathcal{Y}$, and $\mathcal{Z}$ be $\mathbb{K}$-normed spaces. Let also $T : D(T) \subseteq \mathcal{X} \to \mathcal{Y}$ and $L : D(L) \subseteq \mathcal{Y} \to \mathcal{Z}$ be operators. Then, the *product* of T and L is the operator denoted by LT or $L \circ T$, with domain

$$D(LT) := \{x \in D(T) \mid Tx \in D(L)\} = T^{-1}D(L),$$

and defined for each x in $D(LT)$ by

$$LTx := L(Tx). \qquad \square$$

Definition 2.2.10 Let $T : D(T) \subseteq \mathcal{X} \to \mathcal{Y}$ be an operator. T is *invertible* if there exists an operator $L : D(L) \subseteq \mathcal{Y} \to \mathcal{X}$ such that

$$TL = \mathrm{id}_{D(L)}, \qquad LT = \mathrm{id}_{D(T)}. \qquad \square$$

If T is invertible, then $D(L) = R(T)$, L is unique, and we put $T^{-1} := L$.

Remark 2.2.11 It is easily seen that an injective operator is invertible. $\square$

Let us introduce an useful notation, which avoids to introduce multiplicative constants with irrelevant value.

Notation 2.2.12 Let E and F be non-empty sets such that $E \subseteq F$. Let N and M be maps defined on parts of F containing E, and with values into $[0, \infty)$. The notation

$$N(v) \lesssim M(v), \qquad \forall v \in E,$$

means that there exists a positive constant C such that $N(v) \leq CM(v)$ for all v in E. □

An operator $T : D(T) \subseteq \mathcal{X} \to \mathcal{Y}$ is *bounded* if

$$\|Tx\|_Y \lesssim \|x\|_X, \qquad \forall x \in D(T).$$

This is equivalent to say that T is a *continuous* operator between the normed spaces $(D(T), \|\cdot\|_X)$ and Y.

We denote by $\mathcal{L}(\mathcal{X}, \mathcal{Y})$, the vector space of all continuous linear maps defined on $\mathcal{X}$ with values in $\mathcal{Y}$. Equipped with the so-called *operator-norm* defined for each T in $\mathcal{L}(\mathcal{X}, \mathcal{Y})$ by

$$\|T\|_{\mathcal{L}(\mathcal{X},\mathcal{Y})} := \sup_{\|x\|_{\mathcal{X}} \leq 1} \|Tx\|_{\mathcal{Y}},$$

$\mathcal{L}(\mathcal{X}, \mathcal{Y})$ becomes a normed space. If $\mathcal{X}$ is a nontrivial space, then we may take the supremum of the unit sphere of $\mathcal{X}$ only, that is,

$$\|T\|_{\mathcal{L}(\mathcal{X},\mathcal{Y})} = \sup_{\|x\|_{\mathcal{X}} = 1} \|Tx\|_{\mathcal{Y}}.$$

If $\mathcal{X} = \mathcal{Y}$ then $\mathcal{L}(\mathcal{X}, \mathcal{Y})$ will be simply designated by $\mathcal{L}(\mathcal{X})$. In a same way, we define the normed space of continuous *anti-linear* maps from $\mathcal{X}$ into $\mathcal{Y}$. This space will be denoted by ${}^{\mathrm{A}}\mathcal{L}(\mathcal{X}, \mathcal{Y})$. In particular, if $\mathcal{Y} = \mathbb{K}$, one has ${}^{\mathrm{A}}\mathcal{L}(\mathcal{X}, \mathbb{K}) = \mathcal{X}^*$.

Definition 2.2.13 Let $\mathcal{X}, \mathcal{Y}$ be $\mathbb{K}$-normed spaces and $T : \mathcal{X} \to \mathcal{Y}$ be a map defined on $\mathcal{X}$ with values in $\mathcal{Y}$. T is an *isomorphism from $\mathcal{X}$ onto $\mathcal{Y}$* if it lies in $\mathcal{L}(\mathcal{X}, \mathcal{Y})$, is invertible, and its inverse belongs to $\mathcal{L}(\mathcal{Y}, \mathcal{X})$. T is an *anti-isomorphism from $\mathcal{X}$ onto $\mathcal{Y}$* if T lies in ${}^{\mathrm{A}}\mathcal{L}(\mathcal{X}, \mathcal{Y})$, and admits an inverse belonging to ${}^{\mathrm{A}}\mathcal{L}(\mathcal{Y}, \mathcal{X})$.

The space of all isomorphisms from $\mathcal{X}$ onto $\mathcal{Y}$ is denoted by $\mathcal{L}\mathrm{is}(\mathcal{X}, \mathcal{Y})$, and $\mathcal{L}\mathrm{is}(\mathcal{X}, \mathcal{X})$ is abbreviated by $\mathcal{L}\mathrm{is}(\mathcal{X})$. A map $T : \mathcal{X} \to \mathcal{Y}$ is said to be *isometric* or *norm-preserving* if $\|Tx\| = \|x\|$ for each x in $\mathcal{X}$. An *isometry from $\mathcal{X}$ into $\mathcal{Y}$* is a norm-preserving linear map from $\mathcal{X}$ into $\mathcal{Y}$. □

Notice that a *surjective isometry* is an *isometric isomorphism*.

Let us go back to unbounded operators. Except otherwise stated, from now on the domain $D(T)$ of an operator $T : D(T) \subseteq \mathcal{X} \to \mathcal{Y}$ will always be equipped with the so-called *graph-norm* defined by

$$\|x\|_{D(T)} := \|u\|_{\mathcal{X}} + \|Tx\|_{\mathcal{Y}}, \qquad \forall x \in D(T).$$

Definition 2.2.14 Let $\mathcal{X}$ and $\mathcal{Y}$ be two normed spaces. An operator $T : D(T) \subseteq \mathcal{X} \to \mathcal{Y}$ is *closed* if for each sequence $(x_n)_{n\in\mathbb{N}}$ in $D(T)$, and every (x, y) in $\mathcal{X} \times \mathcal{Y}$ satisfying

$$x_n \to x, \qquad Tx_n \to y,$$

the following properties hold: x lies in $D(T)$ and $Tx = y$. □

An operator $T : D(T) \subseteq \mathcal{X} \to \mathcal{Y}$ is closed if and only if its *graph* is closed in $\mathcal{X} \times \mathcal{Y}$. The *graph* of T is, by definition the subspace of $\mathcal{X} \times \mathcal{Y}$ whose elements are the ordered pairs (x, Tx) where x runs over $D(T)$.

Example 2.2.15 Let $\mathcal{X}$ and $\mathcal{Y}$ be two normed spaces, and $T : D(T) \subseteq \mathcal{X} \to \mathcal{Y}$ be an operator. Our aim is to defined an induced operator on $\mathcal{X}/\ker T$ (equipped with the norm given by (2.2.1)) with values in $\mathcal{Y}$. Let $\mathcal{A}$ be in $D(T)/\ker T$. For each x, y in $\mathcal{A}$, one has $Tx = Ty$. Thus by setting $D(\hat{T}) := D(T)/\ker T$, we may introduce the operator

$$\hat{T} : D(\hat{T}) \subseteq \mathcal{X}/\ker T \to \mathcal{Y},$$

defined for every $\mathcal{A}$ in $D(T)/\ker T$, by

$$\hat{T}\mathcal{A} = Tx,$$

where x is any vector in $\mathcal{A}$.

Obviously, $\hat{T}$ is linear and one-to-one. Besides, we claim that *if T is closed then $\hat{T}$ is a closed operator as well.* Indeed, let $(\mathcal{A}_n)_{n\in\mathbb{N}}$ be a sequence in $D(\hat{T})$, and let $\mathcal{A}$ lie in $\mathcal{X}/\ker T$, y belong to $\mathcal{Y}$ such that

$$\mathcal{A}_n \to \mathcal{A}, \qquad \hat{T}\mathcal{A}_n \to y.$$

Let $\mathcal{A}_n = [x_n]$ and $\mathcal{A} = [x]$ where (x_n, x) belong to $D(T) \times \mathcal{X}$. Since $\mathcal{A}_n \to \mathcal{A}$, there exits a sequence (x_n^0) in $\ker T$ such that

$$x_n + x_n^0 \to x.$$

Moreover, $T(x_n + x_n^0) = Tx_n \to y$. Since T is closed, we derive that x lies in $D(T)$ and $Tx = y$. Thus,

$$[x] = \mathcal{A} \in D(\hat{T}), \qquad \hat{T}\mathcal{A} = y,$$

by definition of $\hat{T}\mathcal{A}$. That proves the claim. □

Definition 2.2.16 Given normed spaces $\mathcal{X}$ and $\mathcal{Y}$, let T and $\mathbb{T}$ be operators on $\mathcal{X}$ with values in $\mathcal{Y}$. We say that T is *a restriction of* $\mathbb{T}$, or that $\mathbb{T}$ is *an extension of* T if

$$D(T) \subseteq D(\mathbb{T}) \qquad \text{and} \qquad Tx = \mathbb{T}x, \qquad \forall x \in D(T).$$

In symbol, we write $T \subseteq \mathbb{T}$. □

Definition 2.2.17 Let M be a subspace of a normed space $\mathcal{X}$, and A be any operator on $\mathcal{X}$. The *realization of A on the subspace M*, denoted by A_M, is the operator on M with domain

$$D(A_M) := \big\{u \in M \cap D(A) \mid Au \in M\big\}, \tag{2.2.15}$$

and defined for each u in $D(A_M)$ by $A_M u = Au$. □

Proposition 2.2.17 *Let $\mathcal{X}$, $\mathcal{Y}$, $\mathcal{Z}$ be normed spaces, and m be a positive integer. For each $i = 1, \ldots, m$, let $T_i : D(T_i) \subseteq \mathcal{X} \to \mathcal{Y}$ and $L_i : D(L_i) \subseteq \mathcal{Y} \to \mathcal{Z}$ be operators. If*

$$T := \sum_{i=1}^{m} T_i, \qquad L := \sum_{i=1}^{m} L_i$$

then

$$\sum_{i,j=1}^{m} L_i T_j \subseteq LT.$$

Proof By Definitions 2.2.8 and 2.2.9,

$$D\Big(\sum_{i,j=1}^{m} L_i T_j\Big) = \bigcap_{i,j=1}^{m} T_j^{-1} D(L_i), \qquad D(LT) = T^{-1} \bigcap_{i=1}^{m} D(L_i).$$

If x lies in the domain of $\sum_{i,j=1}^{m} L_i T_j$ then for any i, j in $\{1, \ldots, m\}$, $T_j x$ belongs to $D(L_i)$. Thus, by linearity,

$$Tx \in \bigcap_{i=1}^{m} D(L_i).$$

Hence x lies in $D(LT)$. Moreover, it is clear that for each x in the domain of $\sum_{i,j=1}^{m} L_i T_j$, one has

$$\sum_{i,j=1}^{m} L_i T_j x = LTx.$$

□

2.2.6 Adjoint of Operators

2.2.6.1 Introduction and Definition

The concept of *adjoint* is fundamental in this book since it will allow to define boundary values at an abstract level (see Sect. 4.4). Roughly speaking, the ideas leading to the construction of the adjoint of an operator go as follows. If we forget boundary conditions, the *integration by parts formula* states that

$$\int v(x) \frac{\mathrm{d}}{\mathrm{d}x} u(x)\, \mathrm{d}x = \int -\frac{\mathrm{d}}{\mathrm{d}x} v(x)\, u(x)\, \mathrm{d}x.$$

If we consider the derivative $\frac{\mathrm{d}}{\mathrm{d}x}$ as an operator then the latter formula may be written under the form

$$\big(v, \frac{\mathrm{d}}{\mathrm{d}x} u\big) = \big(-\frac{\mathrm{d}}{\mathrm{d}x} v, u\big).$$

At an abstract level, the correspondence $\frac{\mathrm{d}}{\mathrm{d}x} \mapsto -\frac{\mathrm{d}}{\mathrm{d}x}$ is replaced by $T \mapsto T^*$, T^* is called the adjoint of T, and the identity

$$\big(u, Tv\big) = \big(T^* u, v\big)$$

serves somehow as a definition of T^*.

Let us now give a precise insight of the adjoint of an operator. For, we consider two normed spaces $\mathcal{X}, \mathcal{Y}$, and a *densely defined* operator $T : D(T) \subseteq \mathcal{X} \to \mathcal{Y}$. First we define $D(T^*)$ as the space of all y^* in $\mathcal{Y}^*$ for which there exists some x^* in $\mathcal{X}^*$ such that

$$\langle y^*, Tx \rangle_{\mathcal{Y}^*,\mathcal{Y}} = \langle x^*, x \rangle_{\mathcal{X}^*,\mathcal{X}}, \qquad \forall x \in D(T).$$

Since T is densely defined, x^* is unique. Then, it is easily seen that the map

$$T^* : D(T^*) \subseteq \mathcal{Y}^* \to \mathcal{X}^*, \qquad y^* \mapsto x^*$$

is linear, so that T^* is an operator on $\mathcal{Y}^*$ with values in $\mathcal{X}^*$. T^* is called *the adjoint operator of* T. The fundamental identity connecting T and T^* is

$$\langle y^*, Tx\rangle_{\mathcal{Y}^*,\mathcal{Y}} = \langle T^*y^*, x\rangle_{\mathcal{X}^*,\mathcal{X}}, \qquad \forall\,(x, y^*) \in D(T) \times D(T^*). \tag{2.2.16}$$

2.2.6.2 Properties

Proposition 2.2.18 *Let $\mathcal{X}$, $\mathcal{Y}$ be $\mathbb{K}$-normed spaces and $T : D(T) \subseteq \mathcal{X} \to \mathcal{Y}$ be a densely defined operator. Let also y^* belong to $\mathcal{Y}^*$. Then y^* lies in $D(T^*)$ if and only if*

$$|\langle y^*, Tx\rangle_{\mathcal{Y}^*,\mathcal{Y}}| \lesssim \|x\|_{\mathcal{X}}, \qquad \forall x \in D(T). \tag{2.2.17}$$

Proof If y^* lies in $D(T^*)$, then (2.2.16) implies (2.2.17). Conversely, (2.2.17) entails that the anti-linear form

$$D(T) \to \mathbb{K}, \qquad x \mapsto \langle y^*, Tx\rangle_{\mathcal{Y}^*,\mathcal{Y}}$$

is continuous on $(D(T), \|\cdot\|_{\mathcal{X}})$. By the Hahn–Banach Theorem 2.2.6, there exists some x^* in $\mathcal{X}^*$ extending the latter form, that is,

$$\langle y^*, Tx\rangle_{\mathcal{Y}^*,\mathcal{Y}} = \langle x^*, x\rangle_{\mathcal{X}^*,\mathcal{X}}, \qquad \forall x \in D(T).$$

Hence, by definition of $D(T^*)$, y^* belongs to $D(T^*)$. □

Here are other consequences of the fundamental identity (2.2.16).

Proposition 2.2.19 *Let $\mathcal{X}$, $\mathcal{Y}$ be $\mathbb{K}$-normed spaces and $T : D(T) \subseteq \mathcal{X} \to \mathcal{Y}$ be a densely defined operator. Then, T^* is a closed operator on $\mathcal{Y}^*$. Moreover,*

$$\ker T^* = R(T)^{\perp}.$$

Proof In order to show that T^* is a closed, we consider a sequence $(y_n^*)_{n\in\mathbb{N}}$ in $D(T^*)$, and x^* in $\mathcal{X}^*$, y^* in $\mathcal{Y}^*$ such that

$$y_n^* \to y^*, \qquad T^*y_n^* \to x^*.$$

Then, for each x in $D(T)$, (2.2.16) reads

$$\langle y_n^*, Tx\rangle_{\mathcal{Y}^*,\mathcal{Y}} = \langle T^*y_n^*, x\rangle_{\mathcal{X}^*,\mathcal{X}}.$$

Letting $n \to \infty$, we end up with

$$\langle y^*, Tx\rangle_{\mathcal{Y}^*,\mathcal{Y}} = \langle x^*, x\rangle_{\mathcal{X}^*,\mathcal{X}}, \qquad \forall x \in D(T).$$

Thus, by definition of T^*, y^* lies in the domain of T^* and $T^*y^* = x^*$. Hence, T^* is a closed.

Regarding the kernel of T^*, using (2.2.16) again, we readily obtain that $\ker T^*$ is a part of $R(T)^\perp$. Conversely, let y^* be in $R(T)^\perp$. Then,

$$\langle y^*, Tx\rangle_{\mathcal{Y}^*,\mathcal{Y}} = 0, \qquad \forall x \in D(T).$$

By definition of T^*, we deduce that y^* belongs to the domain of T^* and $T^*y^* = 0$. Hence $R(T)^\perp$ is included in $\ker T^*$. □

Let us now consider the case where T is a bounded operator defined on the whole the space $\mathcal{X}$.

Proposition 2.2.20 *Let $(\mathcal{X}, \|\cdot\|_{\mathcal{X}})$, $(\mathcal{Y}, \|\cdot\|_{\mathcal{Y}})$ be $\mathbb{K}$-normed spaces and T belong to $\mathcal{L}(\mathcal{X}, \mathcal{Y})$. Then, T^* lies in $\mathcal{L}(\mathcal{Y}^*, \mathcal{X}^*)$,*

$$\|T^*\|_{\mathcal{L}(\mathcal{Y}^*,\mathcal{X}^*)} \leq \|T\|_{\mathcal{L}(\mathcal{X},\mathcal{Y})}$$

and

$$\langle y^*, Tx\rangle_{\mathcal{Y}^*,\mathcal{Y}} = \langle T^*y^*, x\rangle_{\mathcal{X}^*,\mathcal{X}}, \qquad \forall (x, y^*) \in \mathcal{X} \times \mathcal{Y}^*.$$

Proof For any y^* in $\mathcal{Y}^*$, one has for every x in $\mathcal{X}$,

$$|\langle y^*, Tx\rangle_{\mathcal{X}^*,\mathcal{X}}| \leq \|y^*\|_{\mathcal{Y}^*}\|T\|_{\mathcal{L}(\mathcal{X},\mathcal{Y})}\|x\|_{\mathcal{X}} \tag{2.2.18}$$

Thus, (2.2.17) yields that $D(T^*) = \mathcal{Y}^*$. Besides, if in addition $\|x\|_{\mathcal{X}} \leq 1$, one has

$$\begin{aligned} |\langle T^*y^*, x\rangle_{\mathcal{X}^*,\mathcal{X}}| &= |\langle y^*, Tx\rangle_{\mathcal{Y}^*,\mathcal{Y}}| && \text{(by (2.2.16))} \\ &\leq \|T\|_{\mathcal{L}(\mathcal{X},\mathcal{Y})}\|y^*\|_{\mathcal{Y}^*} && \text{(by (2.2.18)).} \end{aligned}$$

Thus,

$$\|T^*y^*\|_{\mathcal{X}^*} \leq \|T\|_{\mathcal{L}(\mathcal{X},\mathcal{Y})}\|y^*\|_{\mathcal{Y}^*}.$$

There results that T^* lies in $\mathcal{L}(\mathcal{Y}^*, \mathcal{X}^*)$ and satisfies the estimate stated in the current proposition. Now the identity follows from (2.2.16). □

Proposition 2.2.21 *Let $\mathcal{X}$, $\mathcal{Y}$ be $\mathbb{K}$-normed spaces and $T : \mathcal{X} \to \mathcal{Y}$ be an isomorphism. Then,*

(i) *T^* is a isomorphism from $\mathcal{Y}^*$ onto $\mathcal{X}^*$, and $(T^*)^{-1} = (T^{-1})^*$;*
(ii) *if, in addition, T is isometric, then T^* is a surjective isometry.*

Proof Let us prove the first assertion. According to Proposition 2.2.20, we already know that T^* lies in $\mathcal{L}(\mathcal{Y}^*, \mathcal{X}^*)$ and that $(T^{-1})^*$ belongs to $\mathcal{L}(\mathcal{X}^*, \mathcal{Y}^*)$. Now, for any y^* in $\mathcal{Y}^*$, one has for each y in $\mathcal{Y}$

$$\begin{aligned}
\langle (T^{-1})^* T^* y^*, y\rangle_{\mathcal{Y}^*,\mathcal{Y}} &= \langle T^* y^*, T^{-1} y\rangle_{\mathcal{X}^*,\mathcal{X}} && \text{(by (2.2.16))}\\
&= \langle y^*, TT^{-1} y\rangle_{\mathcal{Y}^*,\mathcal{Y}} && \text{(by (2.2.16))}\\
&= \langle y^*, y\rangle_{\mathcal{Y}^*,\mathcal{Y}}.
\end{aligned}$$

Thus, $(T^{-1})^* T^* = \mathrm{id}_{\mathcal{Y}^*}$. Besides, for each x^* in $\mathcal{X}^*$, one has in a same way, for any x in $\mathcal{X}$

$$\langle T^*(T^{-1})^* x^*, x\rangle_{\mathcal{X}^*,\mathcal{X}} = \langle x^*, x\rangle_{\mathcal{X}^*,\mathcal{X}}.$$

Thus, $T^*(T^{-1})^* = \mathrm{id}_{\mathcal{X}^*}$. That proves the first assertion. In order to prove Item (ii), we consider any y^* in $\mathcal{Y}^*$. Then for each x in $\mathcal{X}$ satisfying $\|x\|_{\mathcal{X}} \leq 1$, one has

$$|\langle T^* y^*, x\rangle_{\mathcal{X}^*,\mathcal{X}}| = |\langle y^*, Tx\rangle_{\mathcal{Y}^*,\mathcal{Y}}| \leq \|y^*\|_{\mathcal{Y}^*} \|Tx\|_{\mathcal{Y}} \leq \|y^*\|_{\mathcal{Y}^*},$$

since by assumption $\|Tx\|_{\mathcal{Y}} = \|x\|_{\mathcal{X}} \leq 1$. Thus,

$$\|T^* y^*\|_{\mathcal{X}^*} \leq \|y^*\|_{\mathcal{Y}^*}. \tag{2.2.19}$$

Inversely, for each y^* in $\mathcal{Y}^*$, one has for any y in $\mathcal{Y}$ satisfying $\|y\|_{\mathcal{Y}} \leq 1$

$$\langle y^*, y\rangle_{\mathcal{Y}^*,\mathcal{Y}} = \langle y^*, TT^{-1} y\rangle_{\mathcal{Y}^*,\mathcal{Y}} = \langle T^* y^*, T^{-1} y\rangle_{\mathcal{X}^*,\mathcal{X}}.$$

Thus,

$$|\langle y^*, y\rangle_{\mathcal{Y}^*,\mathcal{Y}}| \leq \|T^* y^*\|_{\mathcal{X}^*} \|T^{-1} y\|_{\mathcal{X}} \leq \|T^* y^*\|_{\mathcal{X}^*},$$

since $\|T^{-1} y\|_{\mathcal{X}} = \|y\|_{\mathcal{Y}} \leq 1$. Thus,

$$\|y^*\|_{\mathcal{Y}^*} \leq \|T^* y^*\|_{\mathcal{X}^*}.$$

Recalling (2.2.19), we get that T^* is an isometry, which is surjective according to (i). □

2.2.7 Identifications

2.2.7.1 Introduction

According to [Ben],

> In each field of (pure) mathematics, the question “when are two objects to be treated as essentially the same” is one of the key questions and is answered in relation to the particular properties of interest in that field.

In functional analysis, identification consists, broadly speaking, of considering as *equal* a priori distinct objects. If we succeed to identify each element of a set X with one and only one element of a set Y, then X may be identified as a subset of Y. Let us notice that such identification may occur even if X and Y are a priori disjoint.

Let us consider a basic example of identification in topology.

Example 2.2.18 By considering the real interval $[0, 2\pi]$, we may identify its end-points 0 and 2π. Then, after identification, we do not write $0 = 2\pi$ (as an equality between real numbers), but rather consider the elements of $[0, 2\pi]$ as measures of angles or as points on a circle. This identification induces a deep transformation of the initial set $[0, 2\pi]$, which becomes a circle. □

Here, we do not want such drastic changes to occur. For instance, if X is a set of functions identified with another set Y, then we wish to consider X again has a set of functions after identification. For instance in [Bre11, Rem. 3, Chap. 5] (see Sect. 2.5.5), the author describes a situation where two identifications are not simultaneously suitable. We will give now an example where even one identification causes troubles.

Example 2.2.19 Being given a positive number b, consider the functions

$$g_1 : (0, b) \to \mathbb{R}, \quad x \mapsto 1, \qquad g_2 : (0, b) \to \mathbb{R}, \quad x \mapsto x.$$

Referring to Sect. 3.2.1.1 for the definition of the standard norm $\| \cdot \|$ of $L^2(0, b)$ and for the definition of $H^1(0, b)$, set

$${}_0H^1(0, b) := \{u \in H^1(0, b) \mid u(0) = 0\},$$

equip ${}_0H^1(0, b)$ with the norm $\| \frac{\mathrm{d}}{\mathrm{d}x} \cdot \|$, and define the operator

$$I : {}_0H^1(0, b) \to L^2(0, b), \qquad u \mapsto \frac{\mathrm{d}}{\mathrm{d}x}u.$$

Then, we may check easily that I is a surjective isometry whose inverse is

$$I^{-1} : L^2(0,b) \to {}_0H^1(0,b), \qquad f \mapsto g_1 * f,$$

where $g_1 * f$ maps each x of $(0, b)$ into $\int_0^x f(y)\,\mathrm{d}y$ (see Sect. 3.3). If we identify ${}_0H^1(0,b)$ and $L^2(0, b)$ by means of I, then $Ig_2 = g_1$, i.e.,

$$x = 1, \qquad \forall x \in (0,b).$$

Hence, we cannot identify ${}_0H^1(0,b)$ and $L^2(0,b)$ by means of I. □

Now let us be a little more specific about the kind of identifications we have in mind. By considering two normed spaces $(\mathcal{X}, \|\cdot\|_{\mathcal{X}})$ and $(\mathcal{Y}, \|\cdot\|_{\mathcal{Y}})$, we require that identification preserves the structure of these normed spaces by specifying that there exists a surjective isometry $I : \mathcal{X} \to \mathcal{Y}$ allowing to identify each element x of $\mathcal{X}$, which its image Ix in $\mathcal{Y}$. Thus, after identification, we take for granted the equalities

$$Ix = x, \qquad \forall x \in \mathcal{X}.$$

In Example 2.2.19, troubles come from the fact that the evaluation map at $x = 0$, namely the map

$${}_0H^1(0,b) \to \mathbb{K}, \qquad u \mapsto u(0)$$

takes different values for g_2 and g_1. More generally, troubles originate from the identification of two vectors belonging to the same set. To circumvent this difficulty, we will assume that the spaces $\mathcal{X}$ and $\mathcal{Y}$ to be identified are disjoint.

From the naive point of view adopted here, identification (in functional analysis) results in the addition of new axioms. Such a process must be done with great care in order to avoid any contradiction.

By ending this introduction, let us emphasize that in terms of identification, we know pretty much what not to do, but we cannot prove that what we are going to do is consistent. In any case, we do not pretend to exhaust the subject.

2.2.7.2 Identification by a Surjective Isometry

Let $\mathbb{K} := \mathbb{R}$ or $\mathbb{C}$. Let also $\mathcal{X}$, $\mathcal{Y}$ be two disjoint $\mathbb{K}$-normed spaces and $I : \mathcal{X} \to \mathcal{Y}$ be a surjective isometry between $\mathcal{X}$ and $\mathcal{Y}$. Then, *the identification of $\mathcal{X}$ and $\mathcal{Y}$ by means of I* consists in setting

$$\mathcal{X} = \mathcal{Y}$$

and

$$Ix = y, \qquad \forall x \in \mathcal{X}.$$

Thus, identification consists of considering as equal the isometric spaces $\mathcal{X}$ and $\mathcal{Y}$, and also to replace the isometry I by the identity of $\mathcal{X}$.

2.2.7.3 Identification Through a Linear Embedding

Let $(\mathcal{X}, \|\cdot\|_{\mathcal{X}})$ and $(\mathcal{Y}, \|\cdot\|_{\mathcal{Y}})$ be disjoint $\mathbb{K}$-normed spaces.

Definition 2.2.20 A map $i : \mathcal{X} \to \mathcal{Y}$ is *a linear embedding* if i is linear, continuous, and injective. □

Let $i : \mathcal{X} \to \mathcal{Y}$ be a linear embedding. We claim that $\mathcal{X}$ *can be identified with a continuously embedded subspace of* $\mathcal{Y}$ (in the sense of Definition 2.2.2). Accordingly, $\mathcal{X}$ becomes a subspace of $\mathcal{Y}$, and i becomes the (continuous) inclusion map of $\mathcal{X}$ into $\mathcal{Y}$ i.e.

$$i : \mathcal{X} \to \mathcal{Y}, \qquad x \mapsto x.$$

Let us prove the claim. Since i is a bijection from $\mathcal{X}$ onto its image $i(\mathcal{X})$, we may set

$$\|y\|_i := \|i^{-1}(y)\|_{\mathcal{X}}, \qquad \forall\, y \in i(\mathcal{X}).$$

Then, we readily check that

$$\mathcal{Y}_i := \big(i(\mathcal{X}), \|\cdot\|_i\big)$$

is a normed vector space and that the map

$$I : \mathcal{X} \to \mathcal{Y}_i, \qquad x \mapsto i(x)$$

is a surjective isometry. Next since i is a linear embedding, $\mathcal{Y}_i$ is continuously embedded in $\mathcal{Y}$. Hence, the claim follows from the identification of $\mathcal{X}$ and $\mathcal{Y}_i$ by the surjective isometry I.

2.2.7.4 Basic Examples of Identification

Let $\mathcal{X}$ and $\mathcal{Y}$ be $\mathbb{K}$-normed spaces such that $\mathcal{X}$ and $\mathcal{Y}^*$ are disjoint, and $I : \mathcal{X} \to \mathcal{Y}^*$ be a surjective isometry. Identifying $\mathcal{X}$ and $\mathcal{Y}^*$, there results that $\mathcal{X}$ is the anti-dual space of $\mathcal{Y}$. This anti-duality takes the form

$$\langle x, y\rangle_{\mathcal{X},\mathcal{Y}} := \langle Ix, y\rangle_{\mathcal{Y}^*,\mathcal{Y}}, \qquad \forall\,(x, y) \in \mathcal{X} \times \mathcal{Y}. \tag{2.2.20}$$

In the sequel, we will consider three particular cases.

2.2.7.4.1 Anti-Dual Space of Lebesgue Spaces

Let $\mathbb{K} = \mathbb{R}$ or $\mathbb{C}$. Basic instances of *Banach spaces* are *Lebesgue spaces on a real interval with exponent in* $[1, \infty]$. More precisely, let b be a positive number, p belongs to $[1, \infty]$, and $L^0(0, b)$ be the space of Lebesgue-measurable class of functions introduced in Example 2.1.6. The *Lebesgue space on* $(0, b)$ *with exponent* p is the space of all (classes of) functions v in $L^0(0, b)$ such that

$$\|v\|_{L^p(0,b)} := \begin{cases} \left(\int_0^b |v(x)|^p \,\mathrm{d}x\right)^{\frac{1}{p}} & \text{if } p \in [1, \infty) \\ \inf\left\{C \in [0, \infty] \mid |v| \leq C \text{ a.e. on } (0, b)\right\} & \text{if } p = \infty, \end{cases}$$

is finite. This *Lebesgue space* equipped with the norm $\|\cdot\|_{L^p(0,b)}$ is a Banach space which will denote by $L^p(0, b)$. We refer to [CR16], [Rud87, Th. 6.16] for more information on Lebesgue spaces. If no confusion occurs, we will write $\|\cdot\|_{L^p}$ instead of $\|\cdot\|_{L^p(0,b)}$

Let q be the *conjugate exponent* of p, that is, if $p > 1$ then q is the number of $(1, \infty)$ satisfying $\frac{1}{p} + \frac{1}{q} = 1$. If $p = 1$ then $q := \infty$. The *dual* space of $L^p(0, b)$ will be denoted by $L^p(0, b)'$. Classically, in the case where $\mathbb{K} = \mathbb{R}$, the *Riesz Representation Theorem* (see [Rud87, Th. 6.16]) states that the map

$$I^{\mathrm{r}}_{q,p} : L^q(0, b) \to L^p(0, b)'$$

defined for every f in $L^q(0, b)$ by

$$\langle I^{\mathrm{r}}_{q,p} f, h\rangle_{(L^p(0,b))', L^p(0,b)} = \int_0^b f(x)h(x)\,\mathrm{d}x, \qquad \forall\, h \in L^p(0, b),$$

is a surjective isometry.

This result may be extended to the anti-dual space of $L^p(0, b)$ as follows. Let us set for reading comfort, $L^p := L^p(0, b)$. Then, the map

$$I_{q,p} : L^q(0, b) \to L^p(0, b)^*$$

defined for every f in $L^q(0, b)$ by

$$\langle I_{q,p} f, h\rangle_{(L^p)^*, L^p} = \int_0^b f(x)\overline{h(x)}\,\mathrm{d}x, \qquad \forall\, h \in L^p(0, b).$$

satisfies

$$I_{q,p} f = C_{(L^p)'} I^{\mathrm{r}}_{q,p}(\overline{f}), \qquad \forall\, f \in L^q(0, b),$$

where $C_{(L^p)'}$ defined by (2.2.3) is an isometric anti-isomorphism between $(L^p)'$ and $(L^p)^*$ according to Proposition 2.2.4. Hence $I_{q,p}$ *is a surjective isometry between* $L^q(0, b)$ *and* $L^p(0, b)^*$. Therefore, in view of Sect. 2.2.7.2, we will *identify* $L^q(0, b)$ *and* $L^p(0, b)^*$. By (2.2.20), the anti-duality between $L^p(0, b)$ and its anti-dual space $L^q(0, b)$ takes the form:

$$\langle f, h\rangle_{L^q,L^p} := \int_0^b f(x)\overline{h(x)}\,\mathrm{d}x, \qquad \forall h \in L^p(0, b),\ \forall f \in L^q(0, b). \tag{2.2.21}$$

2.2.7.4.2 Identification of $\mathcal{X}$ and $\mathcal{X}^{**}$

We take $\mathcal{Y} := \mathcal{X}^*$. Then, $\mathcal{Y}^* := \mathcal{X}^{**}$, and by Proposition 2.2.11, the map $J_{\mathcal{X}}$ is an isometry between $\mathcal{X}$ and $\mathcal{X}^{**}$. Assuming that $J_{\mathcal{X}}$ is onto, $\mathcal{X}$ is identified with $\mathcal{X}^{**}$ by means of $J_{\mathcal{X}}$, and (2.2.20) reads

$$\langle x, x^*\rangle_{\mathcal{X},\mathcal{X}^*} := \langle J_{\mathcal{X}}x, x^*\rangle_{\mathcal{X}^{**},\mathcal{X}^*} \qquad \forall (x, x^*) \in \mathcal{X} \times \mathcal{X}^*.$$

With (2.2.7),

$$\langle x, x^*\rangle_{\mathcal{X},\mathcal{X}^*} = \overline{\langle x^*, x\rangle}_{\mathcal{X}^*,\mathcal{X}}, \qquad \forall (x, x^*) \in \mathcal{X} \times \mathcal{X}^*.$$

2.2.7.4.3 Anti-Dual Space of Cartesian Products

As a third particular case, let us assume that

$$\mathcal{X} := \mathcal{X}_1{}^* \times \mathcal{X}_2{}^*, \qquad \mathcal{Y} := \mathcal{X}_1 \times \mathcal{X}_2,$$

for some normed spaces $\mathcal{X}_1$ and $\mathcal{X}_2$. We equip $\mathcal{Y}$ with the norm

$$\mathcal{Y} \to [0, \infty), \qquad (x_1, x_2) \mapsto \|x_1\|_{\mathcal{X}_1} + \|x_2\|_{\mathcal{X}_2},$$

and $\mathcal{X}$ with the norm

$$\mathcal{X}_1{}^* \times \mathcal{X}_2{}^* \to [0, \infty), \qquad (x_1^*, x_2^*) \mapsto \max\left(\|x_1^*\|_{\mathcal{X}_1^*}, \|x_2^*\|_{\mathcal{X}_2^*}\right).$$

Then we may check that the map

$$I : \mathcal{X}_1{}^* \times \mathcal{X}_2{}^* \to \mathcal{Y}^*$$

define for each (x_1^*, x_2^*) in $\mathcal{X}$ by

$$\langle I(x_1^*, x_2^*), (x_1, x_2)\rangle_{\mathcal{Y}^*,\mathcal{Y}} := \langle x_1^*, x_1\rangle_{\mathcal{X}_1{}^*,\mathcal{X}_1} + \langle x_2^*, x_2\rangle_{\mathcal{X}_2{}^*,\mathcal{X}_2}$$

is well-defined (i.e., $I(x_1^*, x_2^*)$ lies indeed in $\mathcal{Y}^*$), and especially is an isometry. In order to show that I is onto, for each φ in $\mathcal{Y}^*$, we introduce its restrictions x_1^* and x_2^* to the spaces $\mathcal{X}_1 \times \{0\}$ and $\{0\} \times \mathcal{X}_2$, respectively. Then,

$$\varphi = I(x_1^*, x_2^*).$$

Accordingly (see (2.2.20)), we may *identify* $\mathcal{X}_1{}^* \times \mathcal{X}_2{}^*$ *with the anti-dual space of* $\mathcal{X}_1 \times \mathcal{X}_2$.

2.2.8 Operators on Cartesian Powers of a Space

Let $\mathbb{K}$ be equal to $\mathbb{R}$ or $\mathbb{C}$, $(\mathcal{X}, \|\cdot\|_{\mathcal{X}})$ be a $\mathbb{K}$-normed space, and n be a positive integer. The *Cartesian* n^{th}*-power of the space* $\mathcal{X}$ is the Cartesian product of n copies of $\mathcal{X}$. It is denoted by $\mathcal{X}^n$. Generic vectors of $\mathcal{X}^n$ will be designated by $\vec{x}$ or $(x_1, \ldots, x_n)$ or $(x_i)_{1 \le i \le n}$ or simply by (x_i) if no confusion may occur. Equipped with the norm

$$\|(x_1, \ldots, x_n)\|_{\mathcal{X}^n} := \|x_1\|_{\mathcal{X}} + \cdots + \|x_n\|_{\mathcal{X}},$$

$\mathcal{X}^n$ becomes a $\mathbb{K}$-normed space.

2.2.8.1 Induced Operators on $\mathcal{X}^n$

Definition 2.2.21 Let $T : D(T) \subseteq \mathcal{X} \to \mathcal{X}$ be an operator on $\mathcal{X}$. Then, the *operator induced by* T *on* $\mathcal{X}^n$ is the operator denoted by $\vec{T}$, whose domain is

$$D(\vec{T}) := D(T)^n,$$

and which maps any $(u_1, \ldots, u_n)$ in $D(\vec{T})$ into

$$\vec{T}(u_1, \ldots, u_n) := (Tu_1, \ldots, Tu_n).$$

□

Let us give some basic properties of induced operators. If T lies in $\mathcal{L}(\mathcal{X})$ then clearly, $\vec{T}$ belongs to $\mathcal{L}(\mathcal{X}^n)$ and

$$\|\vec{T}\|_{\mathcal{L}(\mathcal{X}^n)} \le \|T\|_{\mathcal{L}(\mathcal{X})}. \tag{2.2.22}$$

Moreover, *the adjoint of* $\vec{T}$ *satisfies* (2.2.23). Indeed, according to Sect. 2.2.7.4.3, the anti-dual space of $\mathcal{X}^n$ is $(\mathcal{X}^*)^n$. Thus, for each (x_i) in $\mathcal{X}^n$ and each (f_i) in $(\mathcal{X}^*)^n$, one has

$$\langle (f_i), (x_i) \rangle_{(\mathcal{X}^*)^n, \mathcal{X}^n} = \sum_{i=1}^{n} \langle f_i, x_i \rangle_{\mathcal{X}^*, \mathcal{X}}.$$

Then we readily check that

$$(\vec{T})^* = \overrightarrow{T^*}, \tag{2.2.23}$$

that-is-to-say the domain of $(\vec{T})^*$ is $D(T^*)^n$ and for each $(f_1, \ldots, f_n)$ in $D(T^*)^n$,

$$(\vec{T})^*(f_1, \ldots, f_n) = (T^* f_1, \ldots, T^* f_n).$$

Regarding the *product of induced operators*, let $T : D(T) \subseteq \mathcal{X} \to \mathcal{X}$ and $L : D(L) \subseteq \mathcal{X} \to \mathcal{X}$ be operators on $\mathcal{X}$. Then,

$$\vec{T}\vec{L} = \overrightarrow{TL}, \tag{2.2.24}$$

that-is-to-say the domain of $\vec{T}\vec{L}$ is equal to $D(TL)^n$ and for each $(u_1, \ldots, u_n)$ in $D(TL)^n$,

$$\vec{T}\vec{L}(u_1, \ldots, u_n) = (TLu_1, \ldots, TLu_n).$$

2.2.8.2 Tensor Products in $\mathcal{X}^n$

Tensor products will just be seen here as a convenient notation for particular vectors of $\mathcal{X}^n$. However, they are related to the general theory of tensor products since $\mathcal{X}^n$ can be identified with $\mathbb{K}^n \otimes \mathcal{X}$.

Definition 2.2.22 Let $\vec{\lambda} := (\lambda_1, \ldots, \lambda_n)$ belong to $\mathbb{K}^n$ and x lie in $\mathcal{X}$. *The tensor product of* $\vec{\lambda}$ *and* x is the vector of $\mathcal{X}^n$ denoted by $\vec{\lambda} \otimes x$ and defined by

$$\vec{\lambda} \otimes x := (\lambda_1 x, \ldots, \lambda_n x).$$

□

The following proposition shows that the tensor product and the usual product on $\mathbb{K}^2$ share some properties. This explains the name *product* in the expression "tensor product." The elementary proof of the following assertions will be omitted.

Proposition 2.2.22

(i) *For each* $\vec{\lambda}$ *in* $\mathbb{K}^n$, *the map*

$$\mathcal{X} \to \mathcal{X}^n, \qquad x \mapsto \vec{\lambda} \otimes x$$

is linear and continuous. Besides, $\|\vec{\lambda} \otimes x\|_{\mathcal{X}^n} \leq \|\vec{\lambda}\|_{\mathbb{K}^n} \|x\|_{\mathcal{X}}$.

(ii) *For each x in $\mathcal{X}$, the map*

$$\mathbb{K}^n \to \mathcal{X}^n, \quad \vec{\lambda} \mapsto \vec{\lambda} \otimes x$$

is linear and continuous.
(iii) *For each $\vec{\lambda}$ in $\mathbb{K}^n$ and each x in $\mathcal{X}$, $\vec{\lambda} \otimes x = 0$ if and only if $\vec{\lambda} = 0$ or $x = 0$.*
(iv) *For each $(\vec{\lambda}, x)$ in $\mathbb{K}^n \times \mathcal{X}$, and every scalar μ,*

$$(\mu\vec{\lambda}) \otimes x = \vec{\lambda} \otimes (\mu x).$$

Tensor products turn out to be a convenient tool to define linear map induced by scalar matrices.

Definition 2.2.23 Let M be a square scalar matrix of dimension n. Denoting by $\vec{c}_1, \ldots, \vec{c}_n$ its columns, the map

$$\vec{M} : \mathcal{X}^n \to \mathcal{X}^n, \qquad (x_1, \ldots, x_n) \mapsto \sum_{i=1}^{n} \vec{c}_i \otimes x_i$$

called the *map induced by the matrix M*. □

From Proposition 2.2.22 (i), $\vec{M}$ is linear and continuous on $\mathcal{X}^n$. Besides,

$$\|\vec{M}\|_{\mathcal{L}(\mathcal{X}^n)} \le \max_{1 \le i \le n} \|\vec{c}_i\|_{\mathbb{K}^n}. \tag{2.2.25}$$

Proposition 2.2.23 *Let M, L be two square scalar matrices of dimension n, and $T : D(T) \subseteq \mathcal{X} \to \mathcal{X}$ be an operator on $\mathcal{X}$. Let also $\vec{\lambda}$ belong to $\mathbb{K}^n$ and u be in $D(T)$. Then, the following assertions hold true.*

(i) $\vec{M}(\vec{\lambda} \otimes u) = (M\vec{\lambda}) \otimes u$.
(ii) *$\vec{\lambda} \otimes u$ lies in $D(\vec{T})$, and $\vec{T}(\vec{\lambda} \otimes u) = \vec{\lambda} \otimes Tu$.*
(iii) *$\vec{M}\vec{T} \subseteq \vec{T}\vec{M}$. Also $\vec{\lambda} \otimes u$ lies in $D(\vec{M}\vec{T})$, and*

$$\vec{M}\vec{T}(\vec{\lambda} \otimes u) = \vec{T}\vec{M}(\vec{\lambda} \otimes u) = (M\vec{\lambda}) \otimes Tx.$$

(iv) $\vec{M}\vec{L} = \overrightarrow{ML}$.
(v) *If M is an invertible matrix, then $\vec{M}$ is an isomorphism on $\mathcal{X}^n$ whose inverse is $\overrightarrow{M^{-1}}$, that is, $\vec{M}^{-1} = \overrightarrow{M^{-1}}$.*
(vi) *If M is an invertible matrix then $\vec{M}$ commute with $\vec{T}$, i.e., $\vec{M}\vec{T} = \vec{T}\vec{M}$.*
(vii) *If $\vec{M}$ is a bijection and $\mathcal{X}$ is a nontrivial vector space, then the matrix M is invertible.*

Proof

(i) Denoting by $\vec{c}_1, \ldots, \vec{c}_n$ the columns of M, we compute

$$\begin{aligned}
\vec{M}(\vec{\lambda} \otimes u) &= \sum_{i=1}^{n} \vec{c}_i \otimes (\lambda_i u) && \text{(by Definition 2.2.23)} \\
&= \sum_{i=1}^{n} (\lambda_i \vec{c}_i) \otimes u && \text{(by Proposition 2.2.22 (iv))} \\
&= \Big(\sum_{i=1}^{n} \lambda_i \vec{c}_i\Big) \otimes u && \text{(by Proposition 2.2.22 (ii))} \\
&= (M\vec{\lambda}) \otimes u.
\end{aligned}$$

(ii) There results from Definition 2.2.22 that $\vec{\lambda} \otimes u$ lies in $D(\vec{T})$, and the identity is obvious. (iii) Since $\vec{M}$ is continuous on $\mathcal{X}^n$, the domain of $\vec{M}\vec{T}$ is equal to $D(\vec{T})$. Moreover, for each $\vec{u} = (u_1, \ldots, u_n)$ in $D(\vec{T})$, the tensor $\vec{c}_i \otimes u_i$ lies in $D(\vec{T})$ by Item (ii). Thus, $\vec{M}\vec{u}$ lies in $D(\vec{T})$. Hence, $D(\vec{M}\vec{T}) \subseteq D(\vec{T}\vec{M})$. Next,

$$\vec{M}\vec{T}\vec{u} = \sum_{i=1}^{n} \vec{c}_i \otimes T u_i.$$

On the other hand,

$$\vec{T}\vec{M}\vec{u} = \vec{T}\Big(\sum_{i=1}^{n} \vec{c}_i \otimes u_i\Big) = \sum_{i=1}^{n} \vec{c}_i \otimes T u_i,$$

by Item (ii). Hence, $\vec{M}\vec{T} \subseteq \vec{T}\vec{M}$. Now, the identities follow from Items (i) and (ii). That proves Item (iii). Item (iv) is proved by using (i), and noticing that the ith column of ML is the product of M by the ith column of L. (v) By Item (iv),

$$\vec{M}\overrightarrow{M^{-1}} = \overrightarrow{MM^{-1}} = \overrightarrow{\mathrm{id}_{\mathbb{K}^n}} = \mathrm{id}_{\mathcal{X}^n}.$$

In the same way, $\overrightarrow{M^{-1}}\vec{M} = \mathrm{id}_{\mathcal{X}^n}$. Thus, the inverse of $\vec{M}$ is $\overrightarrow{M^{-1}}$. Since $\overrightarrow{M^{-1}}$ is continuous (see (2.2.25)), $\vec{M}$ is an isomorphism. (vi) By (iii), it is enough to show that $D(\vec{T}\vec{M}) \subseteq D(\vec{M}\vec{T})$. For, since M is assumed to be invertible, Item (v) yields that $\vec{M}$ is an isomorphism on $\mathcal{X}^n$, thus $D(\vec{T}\vec{M}) = D(\vec{T})$. Since $D(\vec{T}) = D(\vec{M}\vec{T})$, we are done. Eventually, it is enough to show that M is one-to-one. For, let $\vec{\lambda}$ belong to $\mathbb{K}^n$ and satisfy $M\vec{\lambda} = 0$. Then, choose a nontrivial vector x_1 of $\mathcal{X}$, and compute with Item (i),

$$\vec{M}(\vec{\lambda} \otimes x_1) = (M\vec{\lambda}) \otimes x_1 = 0.$$

Since $\vec{M}$ is a bijection, $\vec{\lambda} \otimes x_1 = 0$. Since $x_1 \neq 0$, Proposition 2.2.22 (iii) yields that $\vec{\lambda} = 0$. □

2.3 Banach Spaces

Banach spaces are convenient frameworks for functional analysis and its applications, in particular for the analysis of ordinary and partial differential equations. Let $\mathbb{K}$ be equal to $\mathbb{R}$ or $\mathbb{C}$.

2.3.1 Generalities on Banach Spaces

Definition 2.3.1 A $\mathbb{K}$-normed space $\mathcal{X}$ is *complete* if every *Cauchy sequence of* $\mathcal{X}$ has a limit in $\mathcal{X}$. A complete normed space is called a *Banach space.* □

As usual, we may construct new Banach spaces starting from well-identified Banach spaces. In order to illustrate this principle, let us consider two normed spaces $(\mathcal{X}, \|\cdot\|_{\mathcal{X}})$ and $(\mathcal{Y}, \|\cdot\|_{\mathcal{Y}})$. In view of Sect. 2.2.5, we recall that $\mathcal{L}(\mathcal{X}, \mathcal{Y})$ and ${}^{\mathrm{A}}\mathcal{L}(\mathcal{X}, \mathcal{Y})$ denote respectively the vector space of all continuous linear and anti-linear maps defined on $\mathcal{X}$ with values in $\mathcal{Y}$.

Proposition 2.3.1 *If $\mathcal{X}$ is a normed space and $\mathcal{Y}$ is a Banach space, then $\mathcal{L}(\mathcal{X}, \mathcal{Y})$ and ${}^{\mathrm{A}}\mathcal{L}(\mathcal{X}, \mathcal{Y})$ are Banach spaces.*

That result may appear surprising since the completeness of *both* spaces $\mathcal{X}$ and $\mathcal{Y}$ seems a more natural assumption. A striking consequence of Proposition 2.3.1 (which will be used in the proof of Theorem 2.3.9) is that the anti-dual space of a *normed* space is a Banach space.

Proof Since anti-linearity plays no role in this proof, we will only consider the space $\mathcal{L}(\mathcal{X}, \mathcal{Y})$. For any Cauchy sequence $(T_n)_{n\in\mathbb{N}}$ in $\mathcal{L}(\mathcal{X}, \mathcal{Y})$, set

$$\varepsilon_n := \sup_{p\geq 0} \|T_{n+p} - T_n\|_{\mathcal{L}(\mathcal{X},\mathcal{Y})}.$$

By definition of Cauchy sequences,

$$\varepsilon_n \xrightarrow[n\to\infty]{} 0. \tag{2.3.1}$$

Thus, for each x in $\mathcal{X}$, $(T_n x)_{n\in\mathbb{N}}$ is clearly a Cauchy sequence in $\mathcal{Y}$. Since $\mathcal{Y}$ is complete, $(T_n x)$ converges in $\mathcal{Y}$. Clearly, the map

$$T : X \to \mathcal{Y}, \qquad x \mapsto \lim_{n\to\infty} T_n x$$

is linear. There remains to show that T is continuous and $T_n \to T$ in $\mathcal{L}(\mathcal{X}, \mathcal{Y})$. For positive integers n and p, one has

$$\sup_{\|x\|_{\mathcal{X}}\leq 1} \|T_{n+p}x - T_n x\|_{\mathcal{Y}} \leq \varepsilon_n.$$

By a continuity argument, we obtain that $\|Tx - T_n x\|_{\mathcal{Y}} \leq \varepsilon_n$ for any x in $\mathcal{X}$ with norm less than or equal to 1. Thus,

$$\sup_{\|x\|_{\mathcal{X}}\leq 1} \|(T - T_n)x\|_{\mathcal{Y}} \leq \varepsilon_n.$$

From this estimate, we deduce on the one hand that $T - T_n$ belongs to $\mathcal{L}(\mathcal{X}, \mathcal{Y})$, so that, by difference, T lies in $\mathcal{L}(\mathcal{X}, \mathcal{Y})$. On the other hand, we have

$$\|T - T_n\|_{\mathcal{L}(\mathcal{X},\mathcal{Y})} \leq \varepsilon_n.$$

Hence (T_n) converges to T in $\mathcal{L}(\mathcal{X}, \mathcal{Y})$ by (2.3.1). □

In view of (2.2.1), normed quotient spaces may also be turned into Banach spaces.

Proposition 2.3.2 *Let M be a closed subspace of a Banach space $(\mathcal{X}, \|\cdot\|_{\mathcal{X}})$. Then $\mathcal{X}/M$ is a Banach space.*

Proof It is enough to show that every Cauchy sequence in $\mathcal{X}/M$ admits a converging subsequence. Let $(\mathcal{A}_n)_{n\in\mathbb{N}}$ be a Cauchy sequence in $\mathcal{X}/M$. Up to a subsequence, we may assume that

$$\|\mathcal{A}_{n+1} - \mathcal{A}_n\|_{\mathcal{X}/M} < 2^{-n}, \qquad \forall n \in \mathbb{N}.$$

Let x_0 belong to $\mathcal{A}_0$ be fixed. Thanks to Proposition 2.2.2 (ii), we may construct by induction a sequence $(x_n)_{n\geq 1}$ in $\mathcal{X}$ such that x_n lies in $\mathcal{A}_n$ and

$$\|x_n - x_{n-1}\|_{\mathcal{X}} < 2^{-n}, \qquad \forall n \geq 1.$$

It is easy to see that (x_n) is a Cauchy sequence in the Banach space $\mathcal{X}$. Thus, (x_n) has a limit, let us say x, in $\mathcal{X}$. There remains to show that $(\mathcal{A}_n)$ converges in $\mathcal{X}/M$. For, one has

$$\|\mathcal{A}_n - [x]\|_{\mathcal{X}/M} = \|[x_n - x]\|_{\mathcal{X}/M} \leq \|x_n - x\|_{\mathcal{X}},$$

by (2.2.1). Hence, $\mathcal{X}/M$ is complete. □

2.3.2 Projections

Let us now introduce the projections as a particular class of bounded operators on $\mathcal{X}$. Projections will be useful in the sequel since they allow to define *endogenous boundary conditions* for abstract differential equations (see Chap. 6). Let us notice that the two following propositions hold in normed spaces.

Definition 2.3.2 Let $\mathcal{X}$ be a $\mathbb{K}$-normed space. An element P of $\mathcal{L}(\mathcal{X})$ satisfying $P^2 = P$ is called a projection of $\mathcal{X}$. □

Proposition 2.3.3 *Let $\mathcal{X}$ be a $\mathbb{K}$-normed space, and P be a projection of $\mathcal{X}$. Then* $\ker P$ *and $R(P)$ are closed complementary subspaces of $\mathcal{X}$. Moreover,*

$$Pz = z, \qquad \forall z \in R(P). \tag{2.3.2}$$

Proof Each z in $R(P)$ reads $z = Px$ for some x in $\mathcal{X}$. Since $P^2 = P$, we get (2.3.2). Let us prove now that the intersection of $\ker P$ and $R(P)$ is trivial. For, let x belong to this intersection. Then, $0 = Px = x$ by (2.3.2). In order to compute $\ker P + R(P)$, we observe that $x - Px$ lies in $\ker P$ for each x in $\mathcal{X}$. Then, the decomposition $x = (x - Px) + Px$ entails that $\ker P + R(P) = \mathcal{X}$. Thus $\ker P$ and $R(P)$ are complementary subspaces of $\mathcal{X}$.

Now, $\ker P$ is closed since P is continuous on $\mathcal{X}$. Finally, let $(z_n)_{n\in\mathbb{N}}$ be a sequence in $R(P)$ converging toward some vector x of $\mathcal{X}$. Then, $Pz_n \to Px$ by continuity of P. Besides, by (2.3.2),

$$Pz_n = z_n \to x.$$

By uniqueness of the limit, $x = Px$, so that x lies in $R(P)$. Hence, $R(P)$ is closed in $\mathcal{X}$. □

In order to prove a kind of converse of the latter proposition, we need the following famous result that will be stated without proof.

Theorem 2.3.4 (Closed Graph Theorem) *Let $\mathcal{X}$ and $\mathcal{Y}$ be two Banach spaces and T be an operator on $\mathcal{X}$ with values in $\mathcal{Y}$. If the domain of T is the whole space $\mathcal{X}$, and T is closed then T is continuous on $\mathcal{X}$.*

We refer the reader, for instance, to [Bre11, Th 2.9] for a proof.

Proposition 2.3.5 *Let N and R be closed complementary subspaces of a Banach space $\mathcal{X}$. Then, there exists a unique projection of $\mathcal{X}$ whose kernel is N and range is R.*

Proof The existence part goes as follows. For each x in $\mathcal{X}$, there exists a unique ordered pair (y, z) in $N \times R$ such that $x = y + z$ (since N and R are complementary subspaces). Then, we define the map

$$P : \mathcal{X} \to \mathcal{X}, \qquad x \mapsto z.$$

Clearly, P is linear and satisfies

$$P^2 = P, \qquad \ker P = N, \qquad R(P) = R.$$

Let us show that P is a closed operator. Let $(x_n)_{n\in\mathbb{N}}$ be a sequence in $\mathcal{X}$, and x, z be in $\mathcal{X}$ such that

$$x_n \to x, \qquad Px_n \to z.$$

Then, z belongs to R since R is assumed to be closed. Moreover, the definition of P entails the existence of a sequence (y_n) in N such that

$$x_n = y_n + Px_n, \qquad \forall n \in \mathbb{N}.$$

Thus, (y_n) converges toward $x - z$. Since N is closed, $x - z$ lies in N. Thus $Px = z$ by definition of P, which proves that P is a closed operator on $\mathcal{X}$. Therefore, the *Closed Graph Theorem* yields that P is continuous on $\mathcal{X}$. Hence, P is a projection whose kernel and range are N and R.

In order to prove the uniqueness, let Q be another projection such that $\ker Q = N$ and $R(Q) = R$. Since $P^2 = P$, any x in $\mathcal{X}$ reads $x = (x - Px) + Px$ where $x - Px$ and Px belong to N and R respectively. We have

$$\begin{aligned} Qx &= QPx && \text{(since } \ker Q = N) \\ &= Px && \text{(by (2.3.2) and } R(Q) = R(P)). \end{aligned}$$

Then, $Q = P$. □

Definition 2.3.3 Let N and R be closed complementary subspaces of a Banach space $\mathcal{X}$. Then, the unique projection given by Proposition 2.3.5 is called *the projection on R along N*. □

Proposition 2.3.6 *Let $\mathcal{X}$, $\mathcal{Y}$ be Banach spaces, $T : \mathcal{X} \to \mathcal{Y}$ be an isomorphism, and P be a projection of $\mathcal{X}$. Then TPT^{-1} is the projection of $\mathcal{Y}$ on $TR(P)$ along $T(\ker P)$.*

Proof Clearly TPT^{-1} lies in $\mathcal{L}(\mathcal{Y})$, and its square is equal to the identity of $\mathcal{Y}$. Hence, TPT^{-1} is a projection. Moreover, $\ker TPT^{-1} = T \ker P$ and $R(TPT^{-1}) = TR(P)$. Hence we infer from Definition 2.3.3 that P is the projection on $TR(P)$ along $T(\ker P)$. □

Proposition 2.3.7 *Let M and N be closed complementary subspaces of a Banach space $\mathcal{X}$. Then, the map*

$$\eta : M \to \mathcal{X}/N, \qquad m \mapsto [m] := m + N$$

is a Banach space isomorphism between M and $\mathcal{X}/N$.

Proof η is obviously linear and continuous (see (2.2.1)). Let us show that it is one-to-one. For, let m be in M such that $\eta(m) = 0_{\mathcal{X}/N}$. Thus, $m + N = N$. Since M and N are in *direct sum*, we deduce that $m = 0$. Now, we will show that η is onto. Each $\mathcal{A}$ in $\mathcal{X}/N$ reads $\mathcal{A} = x + N$ for some x in $\mathcal{X}$. Since $\mathcal{X} = M \oplus N$, one has $x = m + n$ for some ordered pair (m, n) in $M \times N$. Thus $\mathcal{A} = m + N$, so that $\mathcal{A} = [m]$. Hence, η is onto.

There remains to show that η^{-1} is continuous. In view of Definition 2.3.3, let P be the projection on M along N. For each $\mathcal{A}$ in $\mathcal{X}/N$, let $m := \eta^{-1}(\mathcal{A})$. For any x in $\mathcal{A}$, $x - m$ lies in N, thus $Px = m$. By continuity of P,

$$\|m\| \leq \|P\|_{\mathcal{L}(\mathcal{X})}\|x\|, \qquad \forall x \in \mathcal{A}.$$

Thus, $\|\eta^{-1}(\mathcal{A})\| \leq \|P\|_{\mathcal{L}(\mathcal{X})}\|\mathcal{A}\|_{\mathcal{X}/N}$, so that η^{-1} is continuous. □

Roughly speaking, the following results state that if closed subspaces M, V satisfy $\mathcal{X} = M \oplus V$, then $\mathcal{X}^* = M^\perp \oplus V^\perp$.

Proposition 2.3.8 *Let M and V be closed complementary subspaces of a Banach space $\mathcal{X}$. Then, $M^\perp$ and $V^\perp$ are closed complementary subspaces of $\mathcal{X}^*$.*

Proof It is clear that $M^\perp$ and $V^\perp$ are closed. Since $(M + V)^\perp$ is equal to $M^\perp \cap V^\perp$, there results that $M^\perp$ and $V^\perp$ are in direct sum. Let us show that $\mathcal{X}^* = M^\perp + V^\perp$. For, by Definition 2.3.3, we may legitimately consider the projection P on M along V. Now for each x^* in $\mathcal{X}^*$, put $f_0 := x^* \circ P$. Since $Pv = 0$ for each v in V, f_0 lies in $V^\perp$. Moreover, since $Pm = m$ for each m in M, we derive that $x^* - f_0$ belongs to $M^\perp$. Hence x^* lies in $M^\perp + V^\perp$. □

2.3.3 Reflexive Banach Spaces

These spaces are of main importance in this book since they are the frame of *the theory of differential quadruplets and boundary restriction operators*.

Anticipating Definition 2.3.4, a reflexive Banach space $\mathcal{X}$ is a space for which *the canonical isometry* (see Definition 2.2.6) is onto. So its seems natural to consider the a priori more general category of reflexive *normed* spaces. However, these two categories coincide as shown by the forthcoming result.

Theorem 2.3.9 *Let $\mathcal{X}$ be a $\mathbb{K}$-normed space. If the canonical isometry*

$$J_{\mathcal{X}} : \mathcal{X} \to \mathcal{X}^{**}$$

is surjective then $\mathcal{X}$ is complete, that-is-to-say $\mathcal{X}$ is a Banach space.

Proof First, Proposition 2.3.1 yields that $\mathcal{X}^{**} = {}^{\mathrm{A}}\mathcal{L}(\mathcal{X}^*, \mathbb{R})$ is a Banach space. Secondly, the canonical isometry $J_{\mathcal{X}}$ allows to transport the Banach space structure of $\mathcal{X}^{**}$ on $\mathcal{X}$. More precisely, let $(x_n)_{n\in\mathbb{N}}$ be a Cauchy sequence in $\mathcal{X}$. Since $J_{\mathcal{X}}$ is norm-preserving and linear, we infer that $(J_{\mathcal{X}}x_n)_{n\in\mathbb{N}}$ is a Cauchy sequence in $\mathcal{X}^{**}$. Thus by completeness of $\mathcal{X}^{**}$,

$$J_{\mathcal{X}}x_n \xrightarrow[n\to\infty]{} \xi \qquad \text{in } \mathcal{X}^{**},$$

for some ξ in $\mathcal{X}^{**}$. Since $J_{\mathcal{X}}$ is a surjective isometry, its inverse belongs to $\mathcal{L}(\mathcal{X}^{**}, \mathcal{X})$. There results that (x_n) converges in $\mathcal{X}$, which shows that $\mathcal{X}$ is a Banach space. □

Definition 2.3.4 A Banach space $\mathcal{V}$ is *reflexive* if the canonical isometry $J_{\mathcal{V}} : \mathcal{V} \to \mathcal{V}^{**}$ given by Definition 2.2.6 is onto. □

The following result gives a sufficient condition for a space to be reflexive. It will be used in Example 2.3.6 for establishing the reflexivity of Lebesgue spaces and in the next section about Hilbert spaces.

Proposition 2.3.10 *Let $\mathcal{X}_1$ and $\mathcal{X}_2$ be $\mathbb{K}$-normed spaces and*

$$I_1 : \mathcal{X}_1 \to \mathcal{X}_2{}^*, \qquad I_2 : \mathcal{X}_2 \to \mathcal{X}_1{}^*$$

be isomorphisms between $\mathcal{X}_1$ and $\mathcal{X}_2{}^*$, and between $\mathcal{X}_2$ and $\mathcal{X}_1{}^*$ respectively. In addition, we suppose that for each $(x_1, x_2) \in \mathcal{X}_1 \times \mathcal{X}_2$,

$$\langle I_2x_2, x_1\rangle_{\mathcal{X}_1{}^*,\mathcal{X}_1} = \overline{\langle I_1x_1, x_2\rangle}_{\mathcal{X}_2{}^*,\mathcal{X}_2}. \tag{2.3.3}$$

Then,

$$J_{\mathcal{X}_1} = (I_2^{-1})^* \circ I_1, \tag{2.3.4}$$

and $\mathcal{X}_1$ is a reflexive Banach space.

Proof According to Proposition 2.2.21, $I_2{}^*$ is an isomorphism and

$$(I_2^{-1})^* = (I_2^*)^{-1}$$

Thus, in view of Theorem 2.3.9, it is enough to prove (2.3.4). For all purposes, let us notice that

$$(I_2^{-1})^* \circ I_1 : \mathcal{X}_1 \xrightarrow{I_1} \mathcal{X}_2{}^* \xrightarrow{(I_2^{-1})^*} \mathcal{X}_1{}''.$$

For each $(x_1, x_1^*) \in \mathcal{X}_1 \times \mathcal{X}_1{}^*$, we compute

$$\begin{aligned} \langle J_{\mathcal{X}_1}(x_1), x_1^* \rangle_{\mathcal{X}_1'', \mathcal{X}_1{}^*} &= \overline{\langle x_1^*, x_1 \rangle_{\mathcal{X}_1{}^*, \mathcal{X}_1}} && \text{(by (2.2.7))} \\ &= \overline{\langle I_2(I_2^{-1} x_1^*), x_1 \rangle_{\mathcal{X}_1{}^*, \mathcal{X}_1}} \\ &= \langle I_1 x_1, I_2^{-1} x_1^* \rangle_{\mathcal{X}_2{}^*, \mathcal{X}_2} && \text{(by (2.3.3))} \\ &= \langle (I_2^{-1})^* \circ I_1 x_1, x_1^* \rangle_{\mathcal{X}_1'', \mathcal{X}_1{}^*}. \end{aligned}$$

Whence (2.3.4) follows. □

Remark 2.3.5 Let $\mathcal{V}$ be a reflexive Banach space. In view of Sect. 2.2.7.2, $\mathcal{V}$ is identified with $\mathcal{V}^{**}$ by means of $J_{\mathcal{V}}$. Then, the duality between $\mathcal{V} = \mathcal{V}^{**}$ and $\mathcal{V}^*$ takes the form

$$\langle v, v^* \rangle_{\mathcal{V}, \mathcal{V}^*} := \langle J_{\mathcal{V}}(v), v^* \rangle_{\mathcal{V}^{**}, \mathcal{V}^*}, \qquad \forall\, (v, v^*) \in \mathcal{V} \times \mathcal{V}^*. \tag{2.3.5}$$

For all purposes, let us notice that the duality bracket $\langle \cdot, \cdot \rangle_{\mathcal{V}, \mathcal{V}^*}$ is linear with respect to its first argument and anti-linear with respect to its second argument. This observation follows from (2.3.5).

In the sequel, we will identify a reflexive Banach space with its second anti-dual space.

□

Example 2.3.6 (Reflexive Lebesgue Spaces) Let $\mathbb{K} = \mathbb{R}$ or $\mathbb{C}$. We will prove that *the Lebesgue space $L^p(0, b)$ is reflexive for each p in $(1, \infty)$*. For, with the notation of Sect. 2.2.7.4.1, we apply Proposition 2.3.10 with

$$\mathcal{X}_1 := L^p, \qquad \mathcal{X}_2 := L^q, \qquad I_1 := I_{p,q}, \qquad I_2 := I_{q,p}.$$

We check (2.3.3): for each h in $L^p(0, b)$ and f in $L^q(0, b)$, one has

$$\begin{aligned}
\langle I_{q,p} f, h \rangle_{(L^p)^*, L^p} &= \int_0^b f(x)\overline{h(x)}\, \mathrm{d}x && \text{(by definition of } I_{q,p} f\text{)} \\
&= \overline{\int_0^b h(x)\overline{f(x)}\, \mathrm{d}x} \\
&= \overline{\langle I_{p,q} h, f \rangle_{(L^q)^*, L^q}} && \text{(by definition of } I_{p,q} h\text{)}.
\end{aligned}$$

Whence (2.3.3) holds true, which proves that $L^p(0, b)$ is a reflexive Banach space.

According to Remark 2.3.5, we will identify $L^p(0, b)$ with its second anti-dual space by means of J_{L^p}. □

Proposition 2.3.11 *If $\mathcal{V}$ is a reflexive Banach space, then $\mathcal{V}^*$ is a reflexive Banach space as well. Moreover, for all non-empty subsets E of $\mathcal{V}$ and F of $\mathcal{V}^*$, their annihilators satisfy*

$$E^{\perp \mathcal{V}^*} = E^{\top \mathcal{V}^*}, \qquad F^{\top \mathcal{V}} = F^{\perp \mathcal{V}}.$$

Therefore, we will write $E^\perp$ instead of $E^{\top \mathcal{V}^*}$, and $F^\perp$ instead of $F^{\top \mathcal{V}}$.

Proof Since $\mathcal{V}$ is reflexive, the canonical isometry $J_\mathcal{V}$ is an isomorphism from $\mathcal{V}$ onto $\mathcal{V}^{**}$. We will apply Proposition 2.3.10 with

$$\mathcal{X}_1 = \mathcal{V}^*, \qquad \mathcal{X}_2 = \mathcal{V}, \qquad I_1 = \mathrm{id}_{\mathcal{V}^*}, \qquad I_2 = J_\mathcal{V}.$$

With these notations, (2.2.7) becomes (2.3.3). Hence, Proposition 2.3.10 entails that $\mathcal{V}^*$ is a reflexive Banach space.

Regarding annihilators, one has

$$\begin{aligned}
E^{\perp \mathcal{V}^*} &= \{v^* \in \mathcal{V}^* \mid \langle v^*, e \rangle_{\mathcal{V}^*, \mathcal{V}} = 0, \forall\, e \in E\} && \text{(by Definition 2.2.7)} \\
&= \{v^* \in \mathcal{V}^* \mid \langle e, v^* \rangle_{\mathcal{V}, \mathcal{V}^*} = 0, \forall\, e \in E\} && \text{(by (2.3.5))} \\
&= E^{\top \mathcal{V}^*} && \text{(since } E \subseteq \mathcal{V}^{**}\text{)}.
\end{aligned}$$

The remaining identity is be proved in the same way:

$$\begin{aligned}
F^{\perp \mathcal{V}} &= \{v \in \mathcal{V} \mid \langle v, f^* \rangle_{\mathcal{V}, \mathcal{V}^*} = 0, \forall\, f^* \in F\} && \text{(since } \mathcal{V} = \mathcal{V}^{**}\text{)} \\
&= \{v \in \mathcal{V} \mid \langle f^*, v \rangle_{\mathcal{V}^*, \mathcal{V}} = 0, \forall\, f^* \in F\} && \text{(by (2.3.5))} \\
&= F^{\top \mathcal{V}} && \text{(by Definition 2.2.7)}.
\end{aligned}$$

□

Proposition 2.3.12 *Let $\mathcal{V}$ be a reflexive Banach space and Y be a closed subspace of $\mathcal{V}$. Then, Y (endowed with the norm of $\mathcal{V}$) is a reflexive Banach space.*

Proof Let y'' be any fixed element of Y''. With Theorem 2.3.9, it is enough to show that there exists v_0 in Y such that $J_Y(v_0) = y''$. For each v^* in $\mathcal{V}^*$, the restriction of v^* to Y, namely $v^*{}_{|Y}$ lies in Y^* and $\|v^*{}_{|Y}\|_{Y^*} \leq \|v^*\|_{\mathcal{V}^*}$. Thus,

$$|\langle\, y'',\, v^*{}_{|Y}\,\rangle_{Y'',Y^*}| \leq \|y''\|_{Y''}\|v^*\|_{\mathcal{V}^*}.$$

Accordingly, the map

$$F(y''): \mathcal{V}^* \to \mathbb{K}, \qquad v^* \mapsto \langle\, y'',\, v^*{}_{|Y}\,\rangle_{Y'',Y^*}$$

lies in $\mathcal{V}^{**}$. Since $\mathcal{V}$ is reflexive, one has

$$F(y'') = J_{\mathcal{V}}(v_0) \tag{2.3.6}$$

for some v_0 in $\mathcal{V}$. We claim that v_0 lies in Y. Indeed otherwise, since Y is assumed to be closed in $\mathcal{V}$, Corollary 2.2.9 entails the existence of some v_0^* in $\mathcal{V}^*$ such that

$$v^*_{0\,|Y} = 0 \quad \text{and} \quad \langle\, v_0^*,\, v_0\,\rangle_{\mathcal{V}^*,\mathcal{V}} \neq 0.$$

Thus,

$$\begin{aligned}
0 \neq \overline{\langle\, v_0^*,\, v_0\,\rangle}_{\mathcal{V}^*,\mathcal{V}} &= \langle\, J_{\mathcal{V}}(v_0),\, v_0^*\,\rangle_{\mathcal{V}^{**},\mathcal{V}^*} && \text{(by (2.2.7))}\\
&= \langle\, F(y''),\, v_0^*\,\rangle_{\mathcal{V}^{**},\mathcal{V}^*} && \text{(by (2.3.6))}\\
&= \langle\, y'',\, v^*_{0\,|Y}\,\rangle_{Y'',Y^*} = 0.
\end{aligned}$$

This contradiction proves the claim. Finally, let us show that $J_Y(v_0) = y''$. For each y^* in Y^*, by the Hahn-Banach Theorem 2.2.6, there exists v^* in $\mathcal{V}^*$ such that $v^*{}_{|Y} = y^*$. Thus,

$$\begin{aligned}
\langle\, y'',\, y^*\,\rangle_{Y'',Y^*} &= \langle\, F(y''),\, v^*\,\rangle_{\mathcal{V}^{**},\mathcal{V}^*}\\
&= \langle\, J_{\mathcal{V}}(v_0),\, v^*\,\rangle_{\mathcal{V}^{**},\mathcal{V}^*} && \text{(by (2.3.6))}\\
&= \overline{\langle\, v^*,\, v_0\,\rangle}_{\mathcal{V}^*,\mathcal{V}}\\
&= \overline{\langle\, y^*,\, v_0\,\rangle}_{Y^*,Y} && \text{(since } v_0 \text{ lies in Y)}\\
&= \langle\, J_{\mathcal{V}}(v_0),\, y^*\,\rangle_{Y'',Y^*}.
\end{aligned}$$

□

2.4 Operators on Banach Spaces

2.4.1 Operators on Reflexive Spaces

Let us recall that, in view of Remark 2.3.5, a reflexive Banach space is identified with it second anti-dual space.

Theorem 2.4.1 *Let $\mathcal{V}$, $\mathcal{W}$ be reflexive Banach spaces and $T : D(T) \subseteq \mathcal{V} \to \mathcal{W}$ be a closed densely defined operator. Then $D(T^*)$ is dense in $\mathcal{W}^*$, so that the adjoint T^{**} of T^* is well defined. Moreover T^{**} is an operator on $\mathcal{V}$ and $T^{**} = T$.*

Proof Let us start to show the density of $D(T^*)$. For, by Corollary 2.2.10, it is enough to show that a non-zero vector w of $\mathcal{W}$ cannot belong to $D(T^*)^\perp$.

Since $w \neq 0$, the ordered pair $(0, w)$ does not lie in the graph $G(T)$ of T. Moreover, since T is closed, its graph is a closed subset of $\mathcal{V} \times \mathcal{W}$. Thus the Hahn-Banach Theorem 2.2.8 entails the existence of some ordered pair (v_0^*, w_0^*) in $\mathcal{V}^* \times \mathcal{W}^*$ such that

$$\operatorname{Re}\big\langle (v_0^*, w_0^*), (0, w)\big\rangle < \operatorname{Re}\big\langle (v_0^*, w_0^*), (v, Tv)\big\rangle, \qquad \forall\, v \in D(T).$$

Since $D(T)$ is a linear space, we derive classically that on the one hand,

$$\langle w_0^*, Tv\rangle_{\mathcal{W}^*,\mathcal{W}} = \langle -v_0^*, v\rangle_{\mathcal{V}^*,\mathcal{V}},$$

which gives that w_0^* lies in $D(T^*)$, and on the other hand that

$$\langle w_0^*, w\rangle_{\mathcal{W}^*,\mathcal{W}} \neq 0. \tag{2.4.1}$$

Since w_0^* lies in $D(T^*)$, (2.4.1) implies that w does not belong to the annihilator of $D(T^*)$. Whence $T^* : D(T^*) \subseteq \mathcal{W}^* \to \mathcal{V}^*$ is densely defined. Due to the identification of reflexive spaces with their second anti-dual space, T^{**} is an operator on $\mathcal{V}$ with values in $\mathcal{W}$.

Regarding the proof of $T^{**} = T$, we introduce the function

$$\rho : \mathcal{V} \times \mathcal{W} \to \mathcal{W} \times \mathcal{V}, \qquad (v, w) \mapsto (-w, v).$$

Then,

$$\rho^* : \mathcal{W}^* \times \mathcal{V}^* \to \mathcal{V}^* \times \mathcal{W}^*, \qquad (w^*, v^*) \mapsto (v^*, -w^*).$$

Let us start to show that

$$\rho^*\big[G(T^*)\big] = G(T)^\perp. \tag{2.4.2}$$

For all purposes, we recall that $G(T^*)$ is a subset of $\mathcal{W}^* \times \mathcal{V}^*$. Then, for any ordered pair (w^*, v^*) in $\mathcal{W}^* \times \mathcal{V}^*$, one has by definition of T^*,

$$\begin{aligned}
&(w^*, v^*) \in G(T^*) \\
&\iff \langle w^*, Tv\rangle_{\mathcal{W}^*,\mathcal{W}} = \langle v^*, v\rangle_{\mathcal{V}^*,\mathcal{V}}, \qquad \forall\, v \in D(T) \\
&\iff \big\langle (v^*, -w^*), (v, Tv)\big\rangle_{\mathcal{V}^*\times\mathcal{W}^*,\mathcal{V}\times\mathcal{W}} = 0, \qquad \forall\, v \in D(T) \\
&\iff \rho^*(w^*, v^*) \in G(T)^{\perp}.
\end{aligned}$$

Hence, (2.4.2) follows.

Arguing along the same lines, we may prove that

$$\rho\big[G(T^{**})\big] = G(T^*)^{\top}. \tag{2.4.3}$$

Our penultimate step consists of establishing that each non-empty subset F of $\mathcal{V}^* \times \mathcal{W}^*$ satisfies

$$\rho\big[F^{\top}\big] = \big[\rho^{*-1}(F)\big]^{\top}. \tag{2.4.4}$$

Just before to prove this identity, notice that $\rho^{*-1}(F)$ lies in the anti-dual space of $\mathcal{W} \times \mathcal{V}$. Thus by the definition of the annihilator, the right hand side of (2.4.4) lives in $\mathcal{W} \times \mathcal{V}$.

Let (w, v) be in $\mathcal{W} \times \mathcal{V}$. Then since $\rho^{*-1}(v^*, w^*) = (-w^*, v^*)$,

$$\begin{aligned}
&(w, v) \in \big[\rho^{*-1}(F)\big]^{\top} \\
&\iff \big\langle (-w^*, v^*), (w, v)\big\rangle_{\mathcal{W}^*\times\mathcal{V}^*,\mathcal{W}\times\mathcal{V}} = 0, \qquad \forall\, (v^*, w^*) \in F \\
&\iff \langle v^*, v\rangle = \langle w^*, w\rangle, \qquad \forall\, (v^*, w^*) \in F.
\end{aligned}$$

On the other hand, since $\rho^{-1}(w, v) = (v, -w)$,

$$\begin{aligned}
&(w, v) \in \rho\big(F^{\top}\big) \\
&\iff (v, -w) \in F^{\top} \\
&\iff \big\langle (v^*, w^*), (v, -w)\big\rangle_{\mathcal{V}^*\times\mathcal{W}^*,\mathcal{V}\times\mathcal{W}} = 0, \qquad \forall\, (v^*, w^*) \in F \\
&\iff \langle v^*, v\rangle = \langle w^*, w\rangle, \qquad \forall\, (v^*, w^*) \in F.
\end{aligned}$$

Hence, (2.4.4) follows.

Now we may conclude.

$$G(T^*)^{\top} = \Big(\rho^{*-1}\big[G(T)^{\perp}\big]\Big)^{\top} \qquad \text{(by (2.4.2))}$$

$$
\begin{aligned}
&= \rho\Big(\big[G(T)^{\perp}\big]^{\top}\Big) && \text{(by (2.4.4) with } F := G(T)^{\perp}\text{)}\\
&= \rho\big(\overline{G(T)}\big) && \text{(by Proposition 2.2.12)}\\
&= \rho\big(G(T)\big) && \text{(since } T \text{ is closed).}
\end{aligned}
$$

Then with (2.4.3), we end up with $T^{**} = T$. □

Let us remark in the proof of Theorem 2.4.1, the amazing use of the assumption $w \neq 0$. This trick turns out be to the cornerstone of the proof of the density of $D(T^*)$.

Proposition 2.4.2 *Let $\mathcal{V}$ be a reflexive Banach space and $T : D(T) \subseteq \mathcal{V} \to \mathcal{V}$ be a closed operator on $\mathcal{V}$. Then, $D(T)$ (endowed with its graph-norm) is a reflexive Banach space.*

Proof Since T is closed, its graph $G(T)$ is a closed subspace of $\mathcal{V} \times \mathcal{V}$. Since the anti-dual space of the product $\mathcal{X}_1 \times \mathcal{X}_2$ of two normed spaces is equal to $\mathcal{X}_1{}^* \times \mathcal{X}_2{}^*$, we derive easily that $\mathcal{V} \times \mathcal{V}$ is reflexive. Thus Proposition 2.3.12 yields that $G(T)$ is reflexive.

Moreover, the map

$$D(T) \to G(T), \qquad u \mapsto (u, Tu) \in \mathcal{V} \times \mathcal{V}$$

is obviously an isomorphism. Thus, Lemma 2.4.3 below ensures that $D(T)$ is reflexive. □

Lemma 2.4.3 *Let $T : \mathcal{V} \to \mathcal{Z}$ be an isomorphism between the normed spaces $\mathcal{V}$ and $\mathcal{Z}$. If $\mathcal{V}$ is reflexive then $\mathcal{Z}$ is a reflexive Banach space.*

Proof By Theorem 2.3.9, any reflexive normed space is a Banach space. Thus, in view of the following diagram

$$
\begin{array}{ccc}
\mathcal{V} & \xrightarrow{\;T\;} & \mathcal{Z}\\
{\scriptstyle J_{\mathcal{V}}}\big\downarrow & & \big\downarrow{\scriptstyle J_{\mathcal{Z}}}\\
\mathcal{V}^{**} & \xrightarrow[T^{**}]{} & \mathcal{Z}^{**}
\end{array}
$$

it is enough to show that $J_{\mathcal{Z}} = T^{**} \circ J_{\mathcal{V}} \circ T^{-1}$. For every (z, z^*) in $\mathcal{Z} \times \mathcal{Z}^*$, we compute

$$
\begin{aligned}
\langle T^{**} \circ J_{\mathcal{V}} \circ T^{-1}z, z^*\rangle_{\mathcal{Z}'',\mathcal{Z}^*} &= \langle J_{\mathcal{V}} \circ T^{-1}z, T^*z^*\rangle_{\mathcal{V}^{**},\mathcal{V}^*} && \text{(by (2.2.16))}\\
&= \overline{\langle T^*z^*, T^{-1}z\rangle_{\mathcal{V}^*,\mathcal{V}}} && \text{(by (2.2.7))}
\end{aligned}
$$

$$= \overline{\langle z^*, z\rangle_{\mathcal{Z}^*,\mathcal{Z}}} \qquad \text{(by (2.2.16))}$$

$$= \langle J_{\mathcal{Z}} z, z^*\rangle_{\mathcal{Z}'',\mathcal{Z}^*} \qquad \text{(by (2.2.7))}.$$

□

2.4.2 Operators with Closed Range

Let $\mathbb{K}$ be equal to $\mathbb{R}$ or $\mathbb{C}$.

Proposition 2.4.4 *Let $\mathcal{X}$ be a $\mathbb{K}$-normed space and $\mathcal{Y}$ be a Banach space over $\mathbb{K}$. Let $T : D(T) \subseteq \mathcal{X} \to \mathcal{Y}$ be a closed and injective operator.*

(i) *If $R(T)$ is closed then*

$$\|x\|_{\mathcal{X}} \lesssim \|Tx\|_{\mathcal{Y}}, \qquad \forall x \in D(T). \tag{2.4.5}$$

(ii) *If $\mathcal{X}$ is a Banach space, then the converse is true, that is* (2.4.5) *implies that the range of T is closed.*

Proof

(i) Since T is injective, its inverse T^{-1} is an operator on $\mathcal{Y}$ with values in $\mathcal{X}$, whose domain is $R(T)$. Since T is closed, it is easily proved that T^{-1} is closed as well. Moreover, by assumption $R(T)$ is closed, hence $(R(T), \|\cdot\|_{\mathcal{Y}})$ is a Banach space. The Closed Graph Theorem 2.3.4 yields that T^{-1} is continuous on $R(T)$. Then, the estimate (2.4.5) follows since the range of T^{-1} is equal to $D(T)$.
(ii) Let $(y_n)_{n\in\mathbb{N}}$ be any sequence in $R(T)$ converging toward some vector y of $\mathcal{Y}$. Then, for each index n, one has $y_n = Tx_n$ for some x_n in $D(T)$. By (2.4.5), $(x_n)_{n\in\mathbb{N}}$ is a Cauchy sequence in the Banach space $\mathcal{X}$. Thus, this sequence has a limit, let us say x, in $\mathcal{X}$. Since $Tx_n \to y$ and T is closed, we infer that x lies in $D(T)$ and $Tx = y$. Accordingly, y belongs to $R(T)$, which shows that the range of T is closed.

□

Proposition 2.4.5 *Let $\mathcal{X}$ and $\mathcal{Y}$ be Banach spaces, and $T : D(T) \subseteq \mathcal{X} \to \mathcal{Y}$ be a closed operator. If the co-dimension of $R(T)$ is finite, then the range of T is closed in $\mathcal{Y}$.*

The ideas of the proof go as follows. We prove the closedness of $R(T)$ by means of Proposition 2.4.4. In order to work with an injective operator, we replace T by $\hat{T}$, where $\hat{T}$ is defined in Example 2.2.15. Next $\hat{T}$ is extended into a surjective operator $\hat{T}_0$, so that the range of this operator becomes immediately closed. By a double use of Proposition 2.4.4, we finally get the closedness of $R(T)$.

Proof By Proposition 2.1.11, $R(T)$ admits a finite dimensional complementary subspace N, in $\mathcal{Y}$. Starting from $\hat{T}$, the induced operator introduced in Example 2.2.15, we put

$$D(\hat{T}_0) := D(\hat{T}) \times N$$

and define the operator

$$\begin{aligned} \hat{T}_0 : D(\hat{T}_0) \subseteq \mathcal{X}/\ker T \times \mathcal{Y} &\to \mathcal{Y} \\ ([x], y) &\mapsto \hat{T}[x] + y. \end{aligned}$$

Of course, $\hat{T}_0([x], y) = Tx + y$.

Let us show that $\hat{T}_0$ fulfills the assumption of Proposition 2.4.4 (i). In view of Example 2.2.15, $\hat{T}$ is one-to-one, and besides $R(T) \cap N$ is trivial. Thus $\hat{T}_0$ is injective. Let us prove that it is closed. For, let $(x_n)_{n\in\mathbb{N}}$ and $(y_n)_{n\in\mathbb{N}}$ be sequences in $\mathcal{X}$ and N respectively, and (x, y, z) be in $\mathcal{X} \times \mathcal{Y} \times \mathcal{Y}$ such that

$$[x_n] \overset{\mathcal{X}/\ker T}{\longrightarrow} [x], \qquad y_n \overset{\mathcal{Y}}{\longrightarrow} y, \qquad \hat{T}[x_n] + y_n \overset{\mathcal{Y}}{\longrightarrow} z.$$

Since N has finite dimension, it is a closed subspace of $\mathcal{Y}$. Hence y lies in N. Next,

$$\hat{T}[x_n] \to z - y,$$

and $\hat{T}$ is closed (see Example 2.2.15). Therefore, $[x]$ lies in $D(\hat{T})$ and $\hat{T}[x] = z - y$. Hence $([x], y)$ belongs to $D(\hat{T}_0)$ and

$$\hat{T}_0([x], y) = z.$$

That proves that $\hat{T}_0$ is closed. Moreover,

$$R(\hat{T}_0) = R(\hat{T}) + N = R(T) + N = \mathcal{Y}$$

since by construction, N is a complementary subspace of $R(T)$ in $\mathcal{Y}$. Hence $R(\hat{T}_0)$ is trivially closed in $\mathcal{Y}$.

Then, we are in position to apply Proposition 2.4.4 (i) to the operator $\hat{T}_0$, which gives

$$\|([x], y)\|_{\mathcal{X}/\ker T \times \mathcal{Y}} \lesssim \|\hat{T}_0([x], y)\|_{\mathcal{Y}}, \qquad \forall ([x], y) \in D(\hat{T}_0).$$

In particular, for $y = 0$, we get

$$\|[x]\|_{\mathcal{X}/\ker T} \lesssim \|\hat{T}[x]\|_{\mathcal{Y}}, \qquad \forall x \in D(T).$$

Since $\mathcal{X}/\ker T$ is a Banach space (by Proposition 2.3.2), Proposition 2.4.4 (ii) yields that $R(\hat{T})$ is closed. Since the latter is equal to $R(T)$, we get that $R(T)$ is closed. □

Corollary 2.4.6 *Let $\mathcal{X}$ and $\mathcal{Y}$ be Banach spaces, and $T : D(T) \subseteq \mathcal{X} \to \mathcal{Y}$ be a closed operator. We assume that T satisfies at least one of the following hypothesis.*

(i) *The range of T has finite co-dimension;*
(ii) *the range of T is closed in $\mathcal{Y}$.*

Then

$$\|[x]\|_{\mathcal{X}/\ker T} \lesssim \|Tx\|_{\mathcal{Y}}, \qquad \forall x \in D(T).$$

Proof Let us assume that $R(T)$ is closed. According to Example 2.2.15, the induced operator $\hat{T}$ is closed and one-to-one. Thus Proposition 2.4.4 (i) (applied to $\hat{T}$) gives the wished estimate since, by definition of $\hat{T}$, $\hat{T}[x] = Tx$ for each x in $D(T)$.

Now, if $R(T)$ has finite co-dimension, then $R(T)$ is closed in $\mathcal{Y}$ according to Proposition 2.4.5. Thus, the estimate holds in this case as well. □

2.4.3 Product of Operators

We recall Definition 2.2.10 of the product of two operators. The following deep results, and their proof come from combinations of [Gol66, Chap IV] and [Sch70]. See also [Sch01, Chap. 7].

Our aim is to study some properties of the product of two operators. Theorem 2.4.7 states sufficient conditions for this product to be a closed operator. Theorem 2.4.10 is concerned with the adjoint of the product of operators.

Theorem 2.4.7 *Let $\mathcal{X}$, $\mathcal{Y}$ and $\mathcal{Z}$ be Banach spaces over $\mathbb{K}$. Let also*

(i) *$L : D(L) \subseteq \mathcal{Z} \to \mathcal{X}$ be a closed operator;*
(ii) *$T : D(T) \subseteq \mathcal{X} \to \mathcal{Y}$ be a closed operator with finite dimensional kernel and closed range.*

Then TL is a closed operator on $\mathcal{Z}$ with values in $\mathcal{Y}$.

Proof Let $(z_n)_{n\in\mathbb{N}}$ be a sequence in $D(TL)$, and (z, y) lie in $\mathcal{Z} \times \mathcal{Y}$ such that

$$z_n \xrightarrow{\mathcal{Z}} z, \qquad TLz_n \xrightarrow{\mathcal{Y}} y.$$

By Corollary 2.4.6,

$$\|[x]\|_{\mathcal{X}/\ker T} \lesssim \|Tx\|_{\mathcal{Y}}, \qquad \forall x \in D(T).$$

Since $T(Lz_n) \to y$, we infer by completeness of $\mathcal{X}/\ker T$ that $([Lz_n])$ converges toward some $[x]$ in $\mathcal{X}/\ker T$. Then by Proposition 2.2.2 (ii), there exists a sequence (x_n^0) in $\ker T$ such that

$$Lz_n - x_n^0 \xrightarrow{\mathcal{X}} x. \tag{2.4.6}$$

We claim that (x_n^0) is bounded in $(\ker T, \|\cdot\|_{\mathcal{X}})$. Otherwise, up to a subsequence, we may assume that $x_n^0 \neq 0$ for each n and $\|x_n^0\|_{\mathcal{X}} \to \infty$. Hence, with (2.4.6),

$$\frac{Lz_n - x_n^0}{\|x_n^0\|_{\mathcal{X}}} \to 0.$$

Since $\ker T$ is finite dimensional, a compactness argument gives the existence of some vector x^0 in $\ker T$ such that

$$\|x^0\|_{\mathcal{X}} = 1 \tag{2.4.7}$$

and, up to a subsequence still labeled (x_n^0),

$$\frac{x_n^0}{\|x_n^0\|_{\mathcal{X}}} \to x^0.$$

Thus,

$$L\frac{z_n}{\|x_n^0\|_{\mathcal{X}}} \to x^0, \qquad \frac{z_n}{\|x_n^0\|_{\mathcal{X}}} \to 0,$$

since (z_n) converges and $\|x_n^0\|_{\mathcal{X}} \to \infty$. By closedness of L, we get $x^0 = 0$, in contradiction with (2.4.7). Hence, the claim follows.

Since (x_n^0) is bounded in a finite dimensional space, a compactness argument yields that, for a subsequence still labeled (x_n^0), one has $x_n^0 \to x^0$ for some x^0 in $\ker T$. Going back to (2.4.6), we obtain

$$Lz_n \to x + x^0.$$

Using also $z_n \to z$, the closedness of L entails that z lies in $D(L)$ and $Lz = x + x^0$.

Besides, T is closed and

$$Lz_n \to x + x^0, \qquad TLz_n \to y.$$

Thus, $x + x^0$ belongs to $D(T)$ and $T(x + x^0) = y$. Finally, with $Lz = x + x^0$, we get that z lies in $D(TL)$ and $TLz = y$. Hence, TL is a closed operator. □

Now we will study the adjoint of the product TL. Our aim is to give quite general sufficient conditions to recover the formula $(TL)^* = L^*T^*$, which is well-known in finite dimensional frameworks. The following lemma will be useful to prove that the domain of the product TL in dense, so that its adjoint is well defined.

Lemma 2.4.8 *Let $\mathcal{X}$ be a Banach space and M be a closed subspace of $\mathcal{X}$ with finite co-dimension. Let also V be a subspace of $\mathcal{X}$.*

(i) *Then there exists a finite dimensional subspace N of V such that $\overline{V} \cap M$ and N are in direct sum and*

$$\overline{V} = (\overline{V} \cap M) \oplus N.$$

(ii) *If V is dense in $\mathcal{X}$, then $V \cap M$ is dense in M.*

Proof

(i) In view of Proposition 2.1.7, the map

$$i_{\overline{V}} : \overline{V}/(\overline{V} \cap M) \to \mathcal{X}/M$$

(defined by (2.1.1)) is linear and injective. Since $\mathcal{X}/M$ has finite dimension, there results that the co-dimension of $\overline{V} \cap M$ in $\overline{V}$ is finite. Then Proposition 2.1.11 entails that $\overline{V} \cap M$ admits a finite dimensional complementary subspace W, in $\overline{V}$, that is

$$\overline{V} = (\overline{V} \cap M) \oplus W. \tag{2.4.8}$$

Since M is assumed to be closed, $\overline{V} \cap M$ is a closed subspace of the Banach space $(\overline{V}, \|\cdot\|_{\mathcal{X}})$. Thus, according to Definition 2.3.3, we may introduce the projection $P : \overline{V} \to \overline{V}$ on W along $\overline{V} \cap M$. One has

$$\begin{aligned} W = P\overline{V} &\subseteq \overline{PV} && \text{(since } P \text{ is continuous)} \\ &= PV && \text{(since } \dim PV < \infty). \end{aligned} \tag{2.4.9}$$

Let $(w_1, \dots, w_n)$ be a basis of W. By (2.4.9), for each $i = 1, \dots, n$, there exists a vector v_i in V such that $Pv_i = w_i$. Let N be the space generated by the v_i's. It is clear that N is a finite dimensional subspace of V. Thus there remains to show that N is a complementary subspace of $\overline{V} \cap M$ in $\overline{V}$. We will first prove that the intersection is trivial. For each x in $(\overline{V} \cap M) \cap N$, there exists scalars $\lambda_1, \ldots, \lambda_n$ satisfying

$$x = \sum_{i=1}^{n} \lambda_i v_i,$$

Since x lies in $\overline{V} \cap M$, one has

$$0 = Px = \sum_{i=1}^{n} \lambda_i w_i.$$

Since $(w_1, \ldots, w_n)$ is a basis, we deduce that $x = 0$. Next, we claim that

$$(\overline{V} \cap M) + N = \overline{V}.$$

For each x in $\overline{V}$, Px lies in W, thus there exists $\lambda_1, \ldots, \lambda_n$ in $\mathbb{K}$ such that

$$Px = \sum_{i=1}^{n} \lambda_i w_i.$$

We set $v := \sum_{i=1}^{n} \lambda_i v_i$. Then v belongs to N. Moreover, $P(x - v) = 0$ since $Pv_i = w_i$. Thus $x - v$ lies in $\overline{V} \cap M$. That proves the claim, and completes the proof of Item (i).

(ii) If V is dense in $\mathcal{X}$, then Item (i) yields that $\mathcal{X} = M \oplus N$. Since $N \subseteq V$, we deduce that $V = (V \cap M) \oplus N$. Next, since $M \cap N$ is trivial, Proposition 2.1.7 yields that the map

$$i_M : M \to \mathcal{X}/N, \qquad m \mapsto m + N$$

is linear and injective. Moreover, i_M is onto since $\mathcal{X} = M + N$, and continuous from the very definition of the norm of $\mathcal{X}/N$. Then, Corollary 2.4.9 of the open mapping theorem entails that i_M is an isomorphism between M and $\mathcal{X}/N$. Thus, by Proposition 2.2.1,

$$\overline{V \cap M} = {i_M}^{-1}\big(\overline{i_M(V \cap M)}\big). \tag{2.4.10}$$

We claim that $i_M(V \cap M) = V/N$. Indeed, it is clear that the former space is contained in the latter one. Conversely, since $V = (V \cap M) \oplus N$, any v in V reads $v = x + n$ for some x in $V \cap M$ and n in N. Thus $v + N = i_M(x)$, so that $V/N \subseteq i_M(V \cap M)$, and the claim follows.

Finally going back to (2.4.10) and using the claim, we get

$$\overline{V \cap M} = {i_M}^{-1}\big(\overline{V/N}\big).$$

Since V is dense in $\mathcal{X}$, V/N is dense in $\mathcal{X}/N$ according to Proposition 2.2.2 (iii). Thus, $\overline{V \cap M} = M$. □

The next result is a celebrated corollary of the open mapping theorem. It will be called in this book *the corollary of the open mapping theorem.* We refer to [Bre11, Coro. 2.7] for a proof. It has been used in the proof of Lemma 2.4.8 and will be used again many times in this book.

Corollary 2.4.9 *Let $\mathcal{X}$ and $\mathcal{Y}$ be two Banach spaces and let $T : \mathcal{X} \to \mathcal{Y}$ be a continuous linear operator from $\mathcal{X}$ into $\mathcal{Y}$ that is bijective. Then, its inverse T^{-1} is continuous from $\mathcal{Y}$ into $\mathcal{X}$.*

Theorem 2.4.10 *Let $\mathcal{X}$, $\mathcal{Y}$ and $\mathcal{Z}$ be Banach spaces over $\mathbb{K}$. Let also*

(i) *$L : D(L) \subseteq \mathcal{Z} \to \mathcal{X}$ be a densely defined closed operator whose range has finite co-dimension;*
(ii) *$T : D(T) \subseteq \mathcal{X} \to \mathcal{Y}$ be a densely defined operator.*

Then $D(TL)$ is dense in $\mathcal{Z}$, and $(TL)^* = L^*T^*$.

For all purposes, let us notice that T^* is an operator on $\mathcal{Y}^*$ with values in $\mathcal{X}^*$, L^* is an operator on $\mathcal{X}^*$ with values in $\mathcal{Z}^*$, and L^*T^* is an operator on $\mathcal{Y}^*$ with values in $\mathcal{Z}^*$. In symbol, we write

$$L^*T^* : \mathcal{Y}^* \xrightarrow{T^*} \mathcal{X}^* \xrightarrow{L^*} \mathcal{Z}^*.$$

In the same way,

$$(TL)^* : \mathcal{Y}^* \longrightarrow \mathcal{Z}^*.$$

Proof Let us first show that $D(TL)$ is dense. Since $D(L)$ is dense in $\mathcal{Z}$, it is enough to show that $D(TL)$ is dense in $(D(L), \|\cdot\|_{\mathcal{Z}})$. Being given any z in $D(L)$, we set $x := Lz$. Since $R(L)$ has finite co-dimension, $R(L)$ is closed in $\mathcal{X}$ by Proposition 2.4.5. Moreover T is assumed to be densely defined. Thus Lemma 2.4.8 (ii) entails that $D(T) \cap R(L)$ is dense in $(R(L), \|\cdot\|_{\mathcal{X}})$.

Hence, there exists a sequence $(x_n)_{n\in\mathbb{N}}$ in $D(T) \cap R(L)$ converging toward x with respect to the norm of $\mathcal{X}$. Since x_n lies in $R(L)$, we may write $x_n = Lz_n$ for some z_n in $D(L)$. Obviously,

$$Lz_n \xrightarrow{\mathcal{X}} Lz \tag{2.4.11}$$

Since $R(L)$ has finite co-dimension, Corollary 2.4.6 entails that

$$\|[z']\|_{\mathcal{Z}/\ker L} \lesssim \|Lz'\|_{\mathcal{X}}, \qquad \forall z' \in D(L).$$

With (2.4.11), we deduce that $([z_n])$ converges toward $[z]$. Thus there exists a sequence $(z_n^0)_{n\in\mathbb{N}}$ in $\ker L$ such that

$$z_n + z_n^0 \xrightarrow{\mathcal{Z}} z. \tag{2.4.12}$$

Moreover, z_n belongs to $D(TL)$ from the very definition of $D(TL)$. Thus, $z_n + z_n^0$ lies also in $D(TL)$ since $Lz_n^0 = 0$. With (2.4.12), we get that $D(TL)$ is dense in $D(L)$. That proves the density of $D(TL)$.

Since $D(TL)$ is dense in $\mathcal{Z}$, $(TL)^*$ is well defined. Now, we will first show that $L^*T^* \subseteq (TL)^*$. For, let y^* belong to $D(L^*T^*)$ and z be any element of $D(TL)$. Then

$$\begin{aligned}\langle L^*T^*y^*, z\rangle_{\mathcal{Z}^*,\mathcal{Z}} &= \langle T^*y^*, Lz\rangle_{\mathcal{X}^*,\mathcal{X}} && (\text{since } T^*y^* \in D(L^*),\ z \in D(L))\\ &= \langle y^*, TLz\rangle_{\mathcal{Y}^*,\mathcal{Y}} && (\text{since } y^* \in D(T^*),\ Lz \in D(T)).\end{aligned}$$

By definition of the adjoint of TL, y^* lies in $D((TL)^*)$ and $L^*T^*y^*$ is equal to $(TL)^*y^*$. Thus, $L^*T^* \subseteq (TL)^*$.

Conversely, let y^* be in $D((TL)^*)$. We claim that y^* lies in $D(T^*)$. Since $R(L)$ has finite co-dimension, Corollary 2.4.6 entails that

$$\|[z]\|_{\mathcal{Z}/\ker L} \lesssim \|Lz\|_{\mathcal{X}}, \qquad \forall z \in D(L). \tag{2.4.13}$$

Besides, since $D(T)$ is dense in $\mathcal{X}$, $R(L)$ is closed and has finite co-dimension, Lemma 2.4.8 (i) yields the existence of some finite dimensional subspace N of $D(T)$ such that

$$\mathcal{X} = R(L) \oplus N. \tag{2.4.14}$$

Now let x be any vector in $D(T)$. Due to (2.4.14), x may be written under the form $x = Lz + x_0$ where (z, x_0) lives in $D(L) \times N$. In view of (2.4.14), Proposition 2.3.5 yields

$$\|Lz\|_{\mathcal{X}} + \|x_0\|_{\mathcal{X}} \le C\|x\|_{\mathcal{X}}, \tag{2.4.15}$$

where the constant C is independent of x, x_0 and z.

Since N is contained in $D(T)$, we deduce that z lies in $D(TL)$. Now for any z_0 in $\ker L$, one has

$$\langle y^*, TLz\rangle_{\mathcal{Y}^*,\mathcal{Y}} = \langle y^*, TL(z - z_0)\rangle_{\mathcal{Y}^*,\mathcal{Y}} = \langle (TL)^* y^*, z - z_0\rangle_{\mathcal{Z}^*,\mathcal{Z}},$$

since we have assumed that y^* lies in the domain of $(TL)^*$. Thus, by the definition of the norm in $\mathcal{Z}/\ker L$ (see (2.2.1)), we end up with

$$\begin{aligned} |\langle y^*, TLz\rangle_{\mathcal{Y}^*,\mathcal{Y}}| &\le \|(TL)^* y^*\|_{\mathcal{Z}^*} \|[z]\|_{\mathcal{Z}/\ker L} \\ &\le C\|Lz\|_{\mathcal{X}} \qquad \text{(by (2.4.13))}, \end{aligned} \tag{2.4.16}$$

where the new constant C is independent of x, x_0 and z.

Let us now estimate $\langle y^*, Tx_0\rangle$. Since N is finite dimensional, the operator T is bounded on N, thus

$$|\langle y^*, Tx_0'\rangle_{\mathcal{Y}^*,\mathcal{Y}}| \lesssim \|x_0'\|_{\mathcal{X}}, \qquad \forall x_0' \in N. \tag{2.4.17}$$

Then combining (2.4.15)–(2.4.17), we get

$$|\langle y^*, Tx\rangle_{\mathcal{Y}^*,\mathcal{Y}}| \lesssim \|Lz\|_{\mathcal{X}} + \|x_0\|_{\mathcal{X}} \lesssim \|x\|_{\mathcal{X}}, \qquad \forall x \in D(T).$$

As a consequence, y^* lies in $D(T^*)$, which proves the claim.

Next, let us show that $T^* y^*$ lies in $D(L^*)$. For each z in $D(TL)$,

$$\langle T^* y^*, Lz\rangle_{\mathcal{X}^*,\mathcal{X}} = \langle y^*, TLz\rangle_{\mathcal{Y}^*,\mathcal{Y}} = \langle (TL)^* y^*, z\rangle_{\mathcal{Z}^*,\mathcal{Z}}, \tag{2.4.18}$$

since y^* lies in $D((TL)^*)$. Being given any z in $D(L)$, we put $x := Lz$. We have seen at the beginning of this proof, that there exists a sequence (x_n) in $D(T) \cap R(L)$ converging toward x, and also that there exist sequences (z_n) in $D(TL)$ and (z_n^0) in $\ker L$ satisfying $x_n = Lz_n$, (2.4.11) and (2.4.12). Going back to (2.4.18), we have (since $Lz_n^0 = 0$)

$$\langle T^* y^*, Lz_n \rangle_{\mathcal{X}^*,\mathcal{X}} = \langle (TL)^* y^*, z_n + z_n^0 \rangle_{\mathcal{Z}^*,\mathcal{Z}}.$$

In the limit $n \to \infty$, we get with (2.4.11) and (2.4.12),

$$\langle T^* y^*, Lz \rangle_{\mathcal{X}^*,\mathcal{X}} = \langle (TL)^* y^*, z \rangle_{\mathcal{Z}^*,\mathcal{Z}}, \qquad \forall z \in D(L).$$

Hence, $T^* y^*$ lies in $D(L^*)$ and $L^* T^* y^* = (TL)^* y^*$. Thus $(TL)^* \subseteq L^* T^*$. Accordingly, $(TL)^* = L^* T^*$. □

2.4.4 Linear Embeddings

Let $\mathbb{K}$ be equal to $\mathbb{R}$ or $\mathbb{C}$, and $\mathcal{X}$, $\mathcal{Y}$ be $\mathbb{K}$-normed spaces.

Proposition 2.4.11 *Let $J : \mathcal{X} \to \mathcal{Y}$ be a linear and continuous map.*

(i) *If $J(\mathcal{X})$ is dense in $\mathcal{Y}$ then J^* is a linear embedding from $\mathcal{Y}^*$ into $\mathcal{X}^*$.*
(ii) *If $\mathcal{X}$ is reflexive and J is a linear embedding then $J^*(\mathcal{Y}^*)$ is dense in $\mathcal{X}^*$.*

Proof

(i) Let y^* belong to the kernel of J^*. Then for every x in $\mathcal{X}$, one has

$$0 = \langle J^* y^*, x\rangle_{\mathcal{X}^*,\mathcal{X}} = \langle y^*, Jx\rangle_{\mathcal{Y}^*,\mathcal{Y}}.$$

Hence, $y^* = 0$ because $J(\mathcal{X})$ is dense in $\mathcal{Y}$. Thus J^* is injective.
(ii) We will apply Corollary 2.2.10 with $M := J^*(\mathcal{Y}^*)$. Since $\mathcal{X}$ is reflexive, let us assume that a vector x of $\mathcal{X}$ satisfies

$$\langle x, J^* y^*\rangle_{\mathcal{X},\mathcal{X}^*} = 0, \qquad \forall\, y^* \in \mathcal{Y}^*.$$

Then,

$$\langle y^*, Jx\rangle_{\mathcal{Y}^*,\mathcal{Y}} = 0, \qquad \forall\, y^* \in \mathcal{Y}^*.$$

Thus $Jx = 0$ according to Corollary 2.2.7. Since J is one-to-one, there results that $x = 0$. Hence Corollary 2.2.10 insures that $J^*(\mathcal{Y}^*)$ is dense in $\mathcal{X}^*$.

□

Corollary 2.4.12 *Let $T : D(T) \subseteq \mathcal{X} \to \mathcal{Y}$ be a densely defined operator. Then, the following assertions hold true.*

(i) *$\big(D(T), \|\cdot\|_{D(T)}\big)$ is continuously embedded in $\mathcal{X}$.*
(ii) *After identification, $\mathcal{X}^*$ is continuously embedded in $D(T)^*$.*
(iii) *For every u in $D(T)$ and x^* in $\mathcal{X}^*$, u lies in $\mathcal{X}$, x^* belongs to $D(T)^*$, and especially*

$$\langle x^*, u\rangle_{D(T)^*,D(T)} = \langle x^*, u\rangle_{\mathcal{X}^*,\mathcal{X}}.$$

Proof The map

$$i : D(T) \to \mathcal{X}, \qquad u \mapsto u$$

is clearly a linear embedding. That proves the first item. Since $D(T)$ is dense in $\mathcal{X}$, Proposition 2.4.11 (i) yields that i^* is a linear embedding. Thus there results from Sect. 2.2.7.3 that after identification, $\mathcal{X}^*$ is continuously embedded in $D(T)^*$ and that

$$i^* x^* = x^*, \qquad \forall x^* \in \mathcal{X}^*.$$

That proves Item (ii). Finally, the third assertion follows from the latter identification and the identity of Proposition 2.2.20. □

2.4.5 An Extension of Adjoint Operators

2.4.5.1 Extension of T^*

Let $\mathcal{X}$, $\mathcal{Y}$ be $\mathbb{K}$-normed spaces. Being given a densely defined operator $T : D(T) \subseteq \mathcal{X} \to \mathcal{Y}$, the map

$$\tilde{T} : D(T) \to \mathcal{Y}, \qquad u \mapsto Tu.$$

is a continuous and linear operator from $\big(D(T), \|\cdot\|_{D(T)}\big)$ into $\mathcal{Y}$. In view of Proposition 2.2.20, the adjoint $\tilde{T}^*$ of $\tilde{T}$ lies in $\mathcal{L}\big(\mathcal{Y}^*, D(T)^*\big)$ and satisfies

$$\langle y^*, Tu\rangle_{\mathcal{Y}^*,\mathcal{Y}} = \langle \tilde{T}^* y^*, u\rangle_{D(T)^*,D(T)}, \qquad \forall (u, y^*) \in D(T) \times \mathcal{Y}^*. \tag{2.4.19}$$

On the other hand, since $\mathcal{X}^*$ is continuously embedded in $D(T)^*$ by Corollary 2.4.12, we will look at T^* as an operator on $\mathcal{Y}^*$ *with values in* $D(T)^*$. Then we claim that $\tilde{T}^*$ is an extension of T^* i.e.

$$T^* \subseteq \tilde{T}^*. \tag{2.4.20}$$

Indeed for each v^* in $D(T^*)$, one has for every u in $D(T)$,

$$\begin{aligned}
\langle T^* v^*, u\rangle_{D(T)^*,D(T)} &= \langle T^* v^*, u\rangle_{\mathcal{X}^*,\mathcal{X}} && \text{(by Corollary 2.4.12 (iii))}\\
&= \langle v^*, Tu\rangle_{\mathcal{Y}^*,\mathcal{Y}} && \text{(by (2.2.16))}\\
&= \langle \tilde{T}^* v^*, u\rangle_{D(T)^*,D(T)} && \text{(by (2.4.19)).}
\end{aligned}$$

That proves the claim.

2.4.5.2 Extension of T on Reflexive Spaces

Let $\mathcal{V}$ be a real or complex reflexive Banach space, and $T : D(T) \subseteq \mathcal{V} \to \mathcal{V}$ be a densely defined operator. By Theorem 2.4.1, T^* is a densely defined operator on $\mathcal{V}^*$. Then by applying the results of Sect. 2.4.5.1 to T^*, we get that $\mathcal{V}$ is continuously embedded in $D(T^*)^*$ and that T is an operator on $\mathcal{V}$ with values in $D(T^*)^*$. Also the adjoint $\widetilde{T^*}^*$ of $\widetilde{T^*}$ lies in $\mathcal{L}\big(\mathcal{V}, D(T^*)^*\big)$ and extends T i.e.

$$T \subseteq \widetilde{T^*}^*. \tag{2.4.21}$$

Besides, for all purposes, notice that by (2.4.19), the identity characterizing $\widetilde{T^*}^*$ reads

$$\langle \widetilde{T^*}^* v, w^* \rangle_{D(T^*)^*, D(T^*)} = \langle v, T^* w^* \rangle_{\mathcal{V},\mathcal{V}^*}, \qquad \forall (v, w^*) \in \mathcal{V} \times D(T^*). \tag{2.4.22}$$

Such extension is used in Sect. 8.2.

2.4.6 Neumann Series

Proposition 2.4.13 *Let $\mathcal{X}$ be a Banach space and T lie in $\mathcal{L}(\mathcal{X})$. If*

$$\limsup_{j\to\infty} \|T^j\|_{\mathcal{L}(\mathcal{X})}^{\frac{1}{j}} < 1 \tag{2.4.23}$$

then

(i) *the so-called Neumann series $\sum T^j$ converges in $\mathcal{L}(\mathcal{X})$;*
(ii) $\mathrm{id}_{\mathcal{X}} - T$ *is an isomorphism from $\mathcal{X}$ onto itself, and*

$$(\mathrm{id}_{\mathcal{X}} - T)^{-1} = \sum_{j\geq 0} T^j;$$

(iii) *if S lies in $\mathcal{L}(\mathcal{X})$ and commute with T then*

$$S \sum_{j\geq 0} T^j = (\sum_{j\geq 0} T^j) S.$$

Proof

(i) Let ε belong to $(0, 1)$. By (2.4.23), there exists an positive integer j_ε such that

$$\|T^j\|_{\mathcal{L}(\mathcal{X})} \leq \varepsilon^j, \qquad \forall j \geq j_\varepsilon.$$

Since $\mathcal{L}(\mathcal{X})$ is a Banach space, we derive the converge of the Neumann series. (ii) Since the map $\mathcal{L}(\mathcal{X}) \to \mathcal{L}(\mathcal{X}),\ L \mapsto TL$ is continuous, we derive that

$$T \sum_{j\geq 0} T^j = \sum_{j\geq 1} T^j.$$

Then,

$$(\mathrm{id}_{\mathcal{X}} - T)\sum_{j\geq 0} T^j = \mathrm{id}_{\mathcal{X}}.$$

In the same way, $\sum_{j\geq 0} T^j(\mathrm{id}_{\mathcal{X}} - T) = \mathrm{id}_{\mathcal{X}}$. Thus $\mathrm{id}_{\mathcal{X}} - T$ is invertible. Since $\sum_{j\geq 0} T^j$ lies in $\mathcal{L}(\mathcal{X})$, we obtain (ii).

Finally, if T and S commute, then

$$(\mathrm{id}_{\mathcal{X}} - T)S(\mathrm{id}_{\mathcal{X}} - T)^{-1} = S$$

Thus, $S(\mathrm{id}_{\mathcal{X}} - T)^{-1} = (\mathrm{id}_{\mathcal{X}} - T)^{-1}S$. We conclude with (ii). □

2.5 Hilbert Spaces

Hilbert spaces turn out to be very convenient frameworks since, as we will see, they may be identified with their anti-dual spaces. Recall that $\mathbb{K}$ denotes one of the field $\mathbb{R}$ or $\mathbb{C}$.

2.5.1 Definitions and First Properties

Definition 2.5.1 Let $\mathcal{X}$ be a $\mathbb{K}$-vector space. A map $(\cdot,\cdot)$ from $\mathcal{X} \times \mathcal{X}$ into $\mathbb{K}$ is called a *sesquilinear form on* $\mathcal{X}$ provided that for each x in $\mathcal{X}$

(i) $(x,\cdot)$ is an anti-linear form on $\mathcal{X}$;
(ii) $(\cdot, x)$ is a linear form on $\mathcal{X}$.

A map $(\cdot,\cdot) : \mathcal{X} \times \mathcal{X} \to \mathbb{K}$ is called an *inner product on* $\mathcal{X}$ if

(i) $(\cdot,\cdot)$ is a sesquilinear form on $\mathcal{X}$;
(ii) it is *positive*, i.e., $(x, x) \geq 0$ for any x in $\mathcal{X}$;
(iii) it is *definite*, i.e. if a vector x of $\mathcal{X}$ satisfies $(x, x) = 0$, then $x = 0$. □

Let $\mathcal{X}$ be a $\mathbb{K}$-vector space and $(\cdot,\cdot)$ be an inner product on $\mathcal{X}$. Then, the map

$$\|\cdot\| : \mathcal{X} \to [0,\infty), \qquad x \mapsto \sqrt{(x, x)}$$

is a norm on $\mathcal{X}$, and is called *the associated norm (to the inner product* $(\cdot,\cdot)$*)*. Moreover, for any x and y in $\mathcal{X}$, the non negativity of $(x + y, x + y)$ and $(x + \mathrm{i}y, x + \mathrm{i}y)$ implies that

$$(x, y) = \overline{(y, x)}.$$

The ordered pair $\big(\mathcal{X}, (\cdot,\cdot)\big)$ is called *a Hilbert space* if $(\mathcal{X}, \|\cdot\|)$ is a Banach space.

Theorem 2.5.1 (Riesz Theorem) *Let* $(\mathcal{H}, (\cdot, \cdot))$ *be a Hilbert space. Then, the map*

$$I^{\mathrm{d}}_{\mathrm{R},H} : \mathcal{H} \to \mathcal{H}', \qquad u \mapsto (\cdot, u)$$

is an isometric anti-isomorphism between $\mathcal{H}$ *and* $\mathcal{H}'$.

We refer to [Bre11] or [Rud91] for a proof.

Corollary 2.5.2 *Let* $(\mathcal{H}, (\cdot, \cdot))$ *be a Hilbert space. Then, the map*

$$I_{\mathrm{R},H} : \mathcal{H} \to \mathcal{H}^*, \qquad u \mapsto (u, \cdot)$$

is a surjective isometry between $\mathcal{H}$ *and* $\mathcal{H}^*$, *which is called* the Riesz isometry between $\mathcal{H}$ and $\mathcal{H}^*$.

Proof According to Proposition 2.2.4, the map $C_{\mathcal{H}'}$ defined by (2.2.3) is an isometric anti-isomorphism between $\mathcal{H}'$ and $\mathcal{H}^*$. Since

$$I_{\mathrm{R},H} = C_{\mathcal{H}'} \circ I^{\mathrm{d}}_{\mathrm{R},H},$$

Theorem 2.5.1 yields that $I_{\mathrm{R},H}$ a surjective isometry. □

Corollary 2.5.2 allows to prove that a Hilbert space is reflexive.

Proposition 2.5.3 *Let* $(\mathcal{H}, (\cdot, \cdot))$ *be a Hilbert space. Then,*

$$J_{\mathcal{H}} = ({I_{\mathrm{R},H}}^{-1})^* \circ I_{\mathrm{R},H} \tag{2.5.1}$$

and $\mathcal{H}$ *endowed with its associated norm is a reflexive Banach space.*

Proof We apply Proposition 2.3.10 with

$$\mathcal{X}_1 := \mathcal{H}, \qquad \mathcal{X}_2 := \mathcal{H}, \qquad I_1 = I_2 := I_{\mathrm{R},H}.$$

In view of Corollary 2.5.2, we know that $I_{\mathrm{R},H}$ is an isomorphism. Thus it remains to check (2.3.3): for each (u, v) in $\mathcal{H} \times \mathcal{H}$, one has

$$\begin{aligned} \langle I_{\mathrm{R},H} v, u \rangle_{\mathcal{H}^*,\mathcal{H}} &= (v, u) && \text{(by definition of } I_{\mathrm{R},H} v) \\ &= \overline{(u, v)} \\ &= \overline{\langle I_{\mathrm{R},H} u, v \rangle_{\mathcal{H}^*,\mathcal{H}}} && \text{(by definition of } I_{\mathrm{R},H} u). \end{aligned}$$

Whence (2.3.3) holds true. □

2.5.2 Identification and Consequences

In view of Corollary 2.5.2, we may *identify a Hilbert space and its anti-dual space by means of* $I_{R,H}$. However, this identification will not be systematic: see Sect. 2.5.5. If $\mathcal{H}$ is identified with its anti-dual space, then (2.2.20) reads

$$\langle u, v\rangle_{\mathcal{H},\mathcal{H}} := \langle I_{R,H}u, v\rangle_{\mathcal{H}^*,\mathcal{H}} = (u, v), \qquad \forall\,(u, v) \in \mathcal{H} \times \mathcal{H}. \tag{2.5.2}$$

That is, the inner product defines the anti-duality between $\mathcal{H}$ and $\mathcal{H}$. Besides, notice that proceeding as in [Kat95], we have chosen to work with anti-dual spaces in order to get a *linear* isomorphism between $\mathcal{H}$ and $\mathcal{H}^*$.

Let us give two consequences of this identification. Let $\big(\mathcal{H}, (\cdot,\cdot)\big)$ be a Hilbert space, and $T : D(T) \subseteq \mathcal{H} \to \mathcal{H}$ be a densely defined operator on $\mathcal{H}$. The starting point is the definition of T^* in Sect. 2.2.6.1. Due to the identification of $\mathcal{H}$ and $\mathcal{H}^*$, there results that $D(T^*)$ is the space of all w in $\mathcal{H}$ for which there exists some h in $\mathcal{H}$ such that

$$(w, Tu) = (h, u), \qquad \forall\, u \in D(T).$$

By uniqueness of h, T^* is defined by

$$T^* : D(T^*) \subseteq \mathcal{H} \to \mathcal{H}, \qquad w \mapsto h,$$

and (2.2.16) becomes

$$(w, Tu) = (T^*w, u), \qquad \forall\,(u, w) \in D(T) \times D(T^*). \tag{2.5.3}$$

The second consequence of the identification of $\mathcal{H}$ and $\mathcal{H}^*$ concerns *annihilators* (see Definition 2.2.7). If E is a non-empty subset of $\mathcal{H}$, then

$$E^{\perp} := \big\{v \in \mathcal{H} \mid (v, h) = 0, \ \ \forall\, h \in E\big\}. \tag{2.5.4}$$

Example 2.5.2 Let us consider the Banach space $\big(L^2(0, b), \|\cdot\|_{L^2}\big)$ for some fixed b in $(0,\infty)$ (see Example 2.3.6). The sesquilinear form

$$(\cdot,\cdot) : L^2(0, b) \times L^2(0, b) \to \mathbb{K}, \qquad (u, v) \mapsto \int_0^b u(x)\overline{v(x)}\,\mathrm{d}x$$

turns out to be an inner product on $L^2(0, b)$. Since its associated norm is $\|\cdot\|_{L^2}$, $L^2(0, b)$ equipped with this inner product is a Hilbert space. In the sequel, this Hilbert space will simply be denoted by $L^2(0, b)$. In view of (2.5.4), for any h in $L^2(0, b)$, one has

$$h^{\perp} := \{v \in L^2(0, b) \mid \int_0^b v(x)\overline{h(x)}\,\mathrm{d}x = 0\}.$$

□

2.5.3 Gelfand Triplets

Definition 2.5.3 Let $\mathbb{K}$ be equal to $\mathbb{R}$ or $\mathbb{C}$, $\mathcal{V}$ be a reflexive Banach space over $\mathbb{K}$, and $\mathcal{H}$ be a $\mathbb{K}$-Hilbert space. $(\mathcal{V}, \mathcal{H}, \mathcal{V}^*)$ is a *Gelfand triplet* provided $\mathcal{V}$ is continuously embedded in $\mathcal{H}$, and $\mathcal{V}$ is dense in $\mathcal{H}$. □

Proposition 2.5.4 *If $(\mathcal{V}, \mathcal{H}, \mathcal{V}^*)$ is a Gelfand triplet then after identifications, $\mathcal{H}$ is continuously embedded in $\mathcal{V}^*$, and*

$$\mathcal{V} \subseteq \mathcal{H} = \mathcal{H}^* \subseteq \mathcal{V}^*. \tag{2.5.5}$$

Moreover, $\mathcal{V}$ is dense in $\mathcal{V}^*$.

Proof Set $D(T) := \mathcal{V}$ and define

$$T : D(T) \subseteq \mathcal{H} \to \mathcal{H}, \qquad v \mapsto v.$$

Since $D(T)$ is dense in $\mathcal{H}$, Corollary 2.4.12 yields that $\mathcal{H}^*$ is continuously embedded in $\mathcal{V}^*$. Then (2.5.5) follows from the identification of $\mathcal{H}$ and $\mathcal{H}^*$ by means of $I_{\mathrm{R},H}$. Finally,

$$i : \mathcal{H} \to \mathcal{V}^*, \qquad h \mapsto h$$

is a linear embedding. Since $\mathcal{H}$ is reflexive due to Proposition 2.5.3, Proposition 2.4.11 (ii) implies that $\mathcal{H}$ is dense in $\mathcal{V}^*$. Thus the density of $\mathcal{V}$ in $\mathcal{V}^*$ follows. □

Example 2.5.4 In application $\mathcal{V} = L^p(0, b)$ with $p \in (1, \infty)$, thus if $p > 2$ and q is the conjugate exponent of p then $\big(L^p(0, b), L^2(0, b), L^q(0, b)\big)$ is a *Gelfand triplet*. On the other hand if $p < 2$ then $\big(L^q(0, b), L^2(0, b), L^p(0, b)\big)$ is a Gelfand triplet.

□

2.5.4 Self-Adjoint Operators

Self-adjoint operators give rise to a deep and striking theory, namely *the spectral theory of self-adjoint operators* (see for instance [Hal13, Sch12]). Notice also *the theory of self adjoint extensions of symmetric operators* (see [Cal39] or [RS75, Section X.1]), which has some connections with our theory of boundary restriction operators since a fundamental axiom of the former theory is

$$A \subseteq A^*,$$

whereas in the latter one,

$$A^* \subseteq SAS$$

is a fundamental axiom.

Definition 2.5.5 Let $(\mathcal{H}, (\cdot, \cdot))$ be a Hilbert space, and $T : D(T) \subseteq \mathcal{H} \to \mathcal{H}$ be an operator on $\mathcal{H}$. T is *symmetric* if

$$(v, Tu) = (Tv, u), \qquad \forall u, v \in D(T).$$

Also, T is *positive* provided

$$(u, Tu) \geq 0, \qquad \forall u \in D(T).$$

If T is densely defined and $T = T^*$ then T is said to be *self-adjoint*. □

Clearly a self-adjoint operator is symmetric. Moreover, a densely defined operator T on $\mathcal{H}$ is symmetric if and only if $T \subseteq T^*$.

The following result will be applied to *fractional Laplace operators* (see Proposition 8.3.3). It provides a convenient way to construct self-adjoint operators. We refer to [Rud91, Th. 13.13] for a proof.

Theorem 2.5.5 *Let $T : D(T) \subseteq \mathcal{H} \to \mathcal{H}$ be a closed and densely defined operator on a Hilbert space $\mathcal{H}$. Then T^*T is self-adjoint.*

From the point of view of partial differential equations, the following classical result is an important application of self-adjointness.

Proposition 2.5.6 *Let T be a positive and self-adjoint operator on a real or complex Hilbert space $\mathcal{H}$. Then, for each initial condition $h_{\mathrm{b.c}}$ in $D(T)$, the problem*

$$\begin{cases} \text{Find } u \text{ in } C\big([0,\infty), D(T)\big) \cap C^1\big((0,\infty), \mathcal{H}\big) \text{ such that} \\ \dfrac{\mathrm{d}}{\mathrm{d}t}u = -Tu \quad \text{in } C\big((0,\infty), \mathcal{H}\big), \qquad u(0) = h_{\mathrm{b.c}} \end{cases}$$

is uniquely solvable.

Proof See, for instance, [EN00] Propositions 3.27 & 6.2 in ChapII. □

2.5.5 On the Brézis Paradox

In [Bre11, Rem3, Chap5], the author consider the following situation. Being given two Hilbert spaces $\mathcal{V}$ and $\mathcal{H}$, let $i : \mathcal{V} \to \mathcal{H}$ be a linear embedding such that $i(\mathcal{V})$ is dense in $\mathcal{H}$. By Proposition 2.4.11, $i^* : \mathcal{H}^* \to \mathcal{V}^*$ is also a linear embedding. With Riesz isometries, we end up with the following diagram.

$$\begin{array}{ccc} \mathcal{H} & \xrightarrow{I_{\mathrm{R},H}} & \mathcal{H}^* \\ {\scriptstyle i}\uparrow & & \downarrow{\scriptstyle i^*} \\ \mathcal{V} & \xrightarrow[I_{\mathrm{R},\mathcal{V}}]{} & \mathcal{V}^* \end{array} \tag{2.5.6}$$

Next, let us identify

(i) $\mathcal{H}$ and $\mathcal{H}^*$ by means of $I_{\mathrm{R},H}$;
(ii) $\mathcal{V}$ with a continuously embedded subspace of $\mathcal{H}$;
(iii) $\mathcal{H}^*$ with a continuously embedded subspace of $\mathcal{V}^*$.

Then, $(\mathcal{V}, \mathcal{H}, \mathcal{V}^*)$ becomes a Gelfand triplet and

$$\mathcal{V} \subseteq \mathcal{H} = \mathcal{H}^* \subseteq \mathcal{V}^*.$$

The Brézis paradox occurs if in addition, we identify $\mathcal{V}$ and $\mathcal{V}^*$, since then

$$\mathcal{V} = \mathcal{H} = \mathcal{H}^* = \mathcal{V}^*.$$

These equalities may cause troubles especially in the case where $\mathcal{V} := H^1(0, b)$ and $\mathcal{H} := L^2(0, b)$. In our framework (see Sect. 2.2.7.2), the identification of $\mathcal{V}$ and $\mathcal{V}^*$ cannot be performed because $\mathcal{V}$ and $\mathcal{V}^*$ are not disjoint. As a consequence, the Brézis paradox is avoided.

Alternatively, we may begin to identify $\mathcal{V}$ and $\mathcal{V}^*$. However, the situation will be essentially the same since by symmetry, (2.5.6) yields

$$\begin{array}{ccc} \mathcal{V}^* & \xrightarrow{(I_{\mathrm{R},\mathcal{V}})^{-1}} & \mathcal{V} \\ {\scriptstyle i^*}\uparrow & & \downarrow{\scriptstyle i} \\ \mathcal{H}^* & \xrightarrow[(I_{\mathrm{R},H})^{-1}]{} & \mathcal{H} \end{array}.$$

3 Background on Fractional Calculus

In the analysis of *partial differential equations*, it is sometimes convenient to consider vector-valued functions. Here we will restrict our attention to functions with values in a (infinite dimensional) Banach space.

The starting point is *Bochner's theory of integration of vector-valued functions*. Next we will deal with function spaces in Sect. 3.2 and with convolutions in Sect. 3.3. *Special functions of fractional calculus* will be introduced in Sect. 3.4. Finally, based on results about *Marchaud fractional derivatives* established in Sect. 3.5, we will address in Sect. 3.6 some regularity issues about *fractional Riemann–Liouville primitives*.

3.1 An Overview of the Bochner Integral

We will recall the key steps of the Bochner theory of integration of vector-valued functions. Our exposition is not self-contained, but references will be given for unproved statements. We would like to quote [Dro01] and [Mik78] as valuable references on that subject.

For any positive integer d, we denote by $\mathscr{F}_{\mathrm{L},d}$ *the σ-algebra of Lebesgue measurable subsets of the euclidean space* $\mathbb{R}^d$, and by $\mu_{\mathrm{L},d} : \mathscr{F}_{\mathrm{L},d} \to [0, \infty]$ the *Lebesgue measure* on $\mathbb{R}^d$. Being given a non empty open subset Ω of $\mathbb{R}^d$, $\mathscr{F}_{\mathrm{L},d}$ and $\mu_{\mathrm{L},d}$ induce a σ-algebra on Ω and a measure denoted respectively by $\mathscr{F}_{\mathrm{L},\Omega}$ and $\mu_{\mathrm{L},\Omega}$.

Let $\mathbb{K}$ be equal to $\mathbb{R}$ or $\mathbb{C}$, and $\mathcal{X}$ be a Banach space over $\mathbb{K}$.

A. Rougirel, *Unified Theory for Fractional and Entire Differential Operators*,
Frontiers in Elliptic and Parabolic Problems,
https://doi.org/10.1007/978-3-031-58356-8_3

3.1.1 Measurability

The issue is to define mesurability in a way that allows to construct integrals of functions with values in $\mathcal{X}$. When the dimension of $\mathcal{X}$ is infinite, the suitable notion of mesurability is stronger than in the finite dimensional case.

3.1.1.1 Definition and First Properties

Definition 3.1.1

(i) For each subset A of Ω, *the characteristic function of* A is

$$\mathbf{1}_A : \Omega \to \mathbb{R}, \qquad x \mapsto \begin{cases} 1 & \text{if } x \in A \\ 0 & \text{otherwise.} \end{cases}$$

(ii) A function $f : \Omega \to \mathcal{X}$ is called *simple* if there exist a positive integer n, $f_1, \ldots, f_n$ in $\mathcal{X}$, and elements $A_1, \ldots, A_n$ of $\mathscr{F}_{\mathrm{L},\Omega}$ with finite Lebesgue measure such that

$$f = \sum_{i=1}^{n} \mathbf{1}_{A_i} f_i \qquad \text{on } \Omega.$$

(iii) A function $f : \Omega \to \mathcal{X}$ is *L-measurable on* Ω if there exists a sequence of simple functions converging toward f almost everywhere on Ω. The $\mathbb{K}$-vector space of L-measurable functions on Ω is denoted by $\mathscr{L}^0(\Omega, \mathcal{X})$. If no confusion may occur, we will write *measurable* instead of L-measurable. □

Remark 3.1.2 Definition 3.1.1 is certainly not the expected definition of measurability. This point deserves comments. The natural extension of the notion of measurability to vector-valued functions is the following.

As a normed space, the Banach space $\mathcal{X}$ may be equipped with its *Borel σ-algebra* $\mathcal{B}_{\mathrm{or}}(\mathcal{X})$. Let us recall that by definition, $\mathcal{B}_{\mathrm{or}}(\mathcal{X})$ is the smallest σ-algebra containing the open sets of $\mathcal{X}$. Then, we define a function $f : \mathbb{R}^d \to \mathcal{X}$ to be *measurable* if

$$f^{-1}(B) \in \mathcal{B}_{\mathrm{or}}(\mathbb{R}^d), \qquad \forall B \in \mathcal{B}_{\mathrm{or}}(\mathcal{X}).$$

It turns out that theses two notions (i.e. L-measurability and measurability) coincide if $\mathcal{X}$ is a separable space (see [Dro01, Corollary 1.1.1]). In the general case, measurability is weaker that L-measurability. That-is-to-say, any L-measurable function is measurable (see [Dro01, Proposition 1.1.1 and the second remark following this proposition]), but the converse is false. Finally, since the approximation property required by L-measurability allows to construct Bochner's integral, there results that Definition 3.1.1 is suitable for our purpose. □

Proposition 3.1.1 *Let Ω be a non-empty open subset of $\mathbb{R}^d$, and F be in $\mathscr{L}^0(\mathbb{R}^d, \mathcal{X})$. Then the restriction of F to Ω is L-measurable, that is*

$$F_{|\Omega} \in \mathscr{L}^0(\Omega, \mathcal{X}).$$

Proof By definition of L-measurability, there exists a sequence $(s_n)_{n\geq 0}$ of simple functions converging toward F almost everywhere on $\mathbb{R}^d$. However, the very definition of simple functions yields that the restriction of s_n to Ω is a simple function on Ω. We conclude by noticing that the sequence $(s_{n|\Omega})$ converges toward $F_{|\Omega}$ almost everywhere on Ω. □

Our next results on measurability are convolution-oriented.

Proposition 3.1.2 *Let f be in $\mathscr{L}^0(\mathbb{R}, \mathcal{X})$ and g be in $\mathscr{L}^0(\mathbb{R}, \mathbb{K})$. Then the function*

$$\mathbb{R}^2 \to \mathcal{X}, \qquad (x, y) \mapsto g(x-y)f(y)$$

is L-measurable.

Proof Let $(f_n)_{n\in\mathbb{N}}$ and $(g_n)_{n\in\mathbb{N}}$ be sequences of simple functions converging respectively toward f and g almost everywhere on $\mathbb{R}^2$. Then, for each positive index n, one has

$$f_n = \sum_{i=1}^{n} \mathbf{1}_{A_i^n} f_i^n, \qquad g_n = \sum_{i=1}^{n} \mathbf{1}_{B_i^n} g_i^n,$$

where for all index i in $[1, n]$, (f_i^n, g_i^n) lies in $\mathcal{X}\times\mathcal{X}$ and A_i^n, B_i^n are measurable subsets of $\mathbb{R}^2$ with finite measure. By assumption, there exist two measurable subsets of $\mathbb{R}$ denoted by N_f and N_g such that

$$\begin{aligned}
&\mu_{\mathrm{L},1}(N_f) = \mu_{\mathrm{L},1}(N_g) = 0 \\
&f_n(x) \xrightarrow[n\to\infty]{} f(x) && \forall x \in \mathbb{R} \setminus N_f \\
&g_n(x) \xrightarrow[n\to\infty]{} g(x) && \forall x \in \mathbb{R} \setminus N_g.
\end{aligned}$$

We claim that the set

$$N := \{(x, y) \in \mathbb{R}^2 \mid x - y \in N_g\} \cup (\mathbb{R} \times N_f)$$

lies in $\mathscr{F}_{\mathrm{L},2}$ and has zero measure. Indeed by setting

$$\varphi : \mathbb{R}^2 \to \mathbb{R}^2, \qquad (x, y) \mapsto (x - y, y),$$

it appears that

$$\{(x, y) \in \mathbb{R}^2 \mid x - y \in N_g\} = \varphi^{-1}(N_g \times \mathbb{R}),$$

so that the former set is measurable (i.e. lies in $\mathscr{F}_{\mathrm{L},2}$). Thus, N is measurable as union of two measurable sets. Now let us compute the measure of N. Let us show that the Lebesgue measure of N is equal to zero. Plainly $\mathbb{R} \times N_f$ has zero measure. Moreover,

$$\begin{aligned}
\mu_{\mathrm{L},2}\big(\{(x, y) \in \mathbb{R}^2 \mid x - y \in N_g\}\big) &= \int_{\mathbb{R}^2} \mathbf{1}_{N_g}(x - y)\,\mathrm{d}x\,\mathrm{d}y \\
&= \int_{\mathbb{R}} \mathrm{d}y \int_{\mathbb{R}} \mathbf{1}_{y+N_g}(x)\,\mathrm{d}x \qquad \text{(by Fubini Th.)} \\
&= 0
\end{aligned}$$

since by translation invariance of the Lebesgue measure, $\mu_{\mathrm{L},1}(y + N_g) = \mu_{\mathrm{L},1}(N_g) = 0$. Thus $\mu_{\mathrm{L},2}(N) = 0$, and the claim follows.

Then by definition of N, one has

$$f_n(x - y)g_n(y) \to g(x - y)f(y), \qquad \forall\,(x, y) \in \mathbb{R}^2 \setminus N,$$

so that it remains to prove that for each n,

$$(x, y) \mapsto f_n(x - y)g_n(y)$$

is a simple function. For,

$$f_n(x - y)g_n(y) = \sum_{i,j=1}^{n} \mathbf{1}_{A_i^n}(x - y)\mathbf{1}_{B_j^n}(y)g_j^n f_i^n.$$

Now we set

$$C_{i,j}^n := \varphi^{-1}(A_i^n \times B_j^n).$$

Then,

$$f_n(x - y)g_n(y) = \sum_{i,j=1}^{n} \mathbf{1}_{C_{i,j}^n}(x, y)g_j^n f_i^n.$$

Moreover $C_{i,j}$ is measurable by continuity of φ, and its measure is finite since φ^{-1} is linear. Whence $f_n(\cdot - \cdot)g_n$ is a simple function on $\mathbb{R}^2$. □

3.1.1.2 Caratheodory Functions

Definition 3.1.3 Let Ω be a non-empty open subset of $\mathbb{R}^d$, $\mathbb{K}$ be equal to $\mathbb{R}$ or $\mathbb{C}$, and $\mathcal{X}$, $\mathcal{Y}$ be separable Banach spaces over $\mathbb{K}$. A function $F : \Omega \times \mathcal{X} \to \mathcal{Y}$ is said to be a *Carathéodory function* if

(i) $F(\cdot, u)$ is measurable (with respect to $\mu_{L,d}$) for each fixed u in $\mathcal{X}$;
(ii) $F(x, \cdot)$ is a continuous function on $\mathcal{X}$ for almost every x in Ω. □

Recall that L-measurability and measurability coincide here due to separability. Combining Pettis Theorem ([ABHN11, Th. 1.1.1]) and the deep and difficult to prove Theorem 1.1 in [AZ90], we obtain the following result.

Theorem 3.1.3 *Let*

(i) *Ω be a non-empty open subset of $\mathbb{R}^d$;*
(ii) *$\mathcal{X}$, $\mathcal{Y}$ be separable Banach spaces;*
(iii) *$F : \Omega \times \mathcal{X} \to \mathcal{Y}$ be a Carathéodory function;*
(iv) *$u : \Omega \to \mathcal{X}$ be a measurable function.*

Then, the function

$$\Omega \to \mathcal{Y}, \qquad x \mapsto F\big(x, u(x)\big)$$

is measurable.

3.1.2 The Quotient Spaces $L^0(\Omega, \mathcal{X})$

Let Ω be a non- empty open subset of $\mathbb{R}^d$, and $\mathcal{X}$ be a Banach space.

3.1.2.1 Definitions and First Properties

Proceeding as in Example 2.1.6, we define $L^0(\Omega, \mathcal{X})$ as the quotient space

$$L^0(\Omega, \mathcal{X}) := \mathscr{L}^0(\Omega, \mathcal{X})/M_{0,\Omega}, \tag{3.1.1}$$

where $M_{0,\Omega}$ is the subspace of $\mathscr{L}^0(\Omega, \mathcal{X})$ whose elements vanish almost everywhere on Ω. In symbol

$$M_{0,\Omega} := \big\{f \in \mathscr{L}^0(\Omega, \mathcal{X}) \mid \mu_{L,\Omega}([f \neq 0]) = 0\big\}, \tag{3.1.2}$$

where

$$[f \neq 0] := \{x \in \Omega \mid f(x) \neq 0\}.$$

If f belongs to $\mathscr{L}^0(\Omega, \mathcal{X})$ then in view of Proposition 2.1.3 (ii), $f + M_{0,\Omega}$ is *the class of* f, and any element in the set $f + M_{0,\Omega}$ is called a *representative of* $f + M_{0,\Omega}$. From theorem 2.1.5, $L^0(\Omega, \mathcal{X})$ is a $\mathbb{K}$-vector space. By a slight abuse of language, vectors of $L^0(\Omega, \mathcal{X})$ will be called *measurable functions*.

The very definition of the space $L^0(\Omega, \mathcal{X})$ implies that the evaluation at a given point x of Ω cannot be defined. However, many properties of functions may be recovered to some extent. This explain why, the elements of $L^0(\Omega, \mathcal{X})$ are called *functions*, although they are not functions but rather special sets of functions.

As a simple example of a property that can be extended from real-valued functions to elements of $L^0(\Omega, \mathbb{R})$, let us consider positivity. We say that a function f of $\mathscr{L}^0(\Omega, \mathbb{R})$ is *non-negative almost everywhere on* Ω if there exists a subset N in $\mathscr{F}_{\mathrm{L},\Omega}$ such that N has zero Lebesgue measure and

$$f(x) \geq 0, \qquad \forall x \in \Omega \setminus N.$$

Now by considering the space $L^0(\Omega, \mathbb{R})$, we say that an element F of $L^0(\Omega, \mathbb{R})$ is *non negative almost everywhere on* Ω if each representative f of F is *non negative almost everywhere on* Ω. We denote

$$F \geq 0 \text{ almost everywhere on } \Omega.$$

Of course, if F admits a representative, which is non-negative almost everywhere, then F is non-negative almost everywhere on Ω.

Also the notion of *restriction of functions to a measurable subset* may be extended. For, let us consider a non-empty open subset Ω of $\mathbb{R}^d$. For each F in $L^0(\mathbb{R}^d, \mathcal{X})$, we set

$$F_{|\Omega} := \{F'_{|\Omega} \mid F' \in F\}. \tag{3.1.3}$$

Note that here F' has nothing to do with a derivative of F, it is just a representative of F. Then, as we will see now, Proposition 3.1.1 can be extended from $\mathscr{L}^0(\Omega, \mathcal{X})$ to $L^0(\Omega, \mathcal{X})$.

Proposition 3.1.4 *Let* Ω *be a non empty open subset of* $\mathbb{R}^d$. *If* F *lies in* $L^0(\mathbb{R}^d, \mathcal{X})$ *then* $F_{|\Omega}$ *belongs to* $L^0(\Omega, \mathcal{X})$.

Proof By Proposition 3.1.1, $F_{|\Omega}$ is contained in the *power set* $\mathcal{P}(\mathscr{L}^0(\Omega, \mathcal{X}))$ of $\mathscr{L}^0(\Omega, \mathcal{X})$. It is enough to show that there exists some f' in $\mathscr{L}^0(\Omega, \mathcal{X})$ such that $F_{|\Omega} = f' + M_{0,\Omega}$. For, let F' any element of F. Setting $f' := F'_{|\Omega}$, we will first prove that

$$F_{|\Omega} \subseteq f' + M_{0,\Omega}. \tag{3.1.4}$$

For each F'' in F, there exists F_0 in $M_{0,\mathbb{R}^d}$ such that $F'' = F' + F_0$. Thus,

$$F''_{|\Omega} = f' + F_{0|\Omega}.$$

By Proposition 3.1.1, $F_{0|\Omega}$ lives in $\mathscr{L}^0(\Omega, \mathcal{X})$. Moreover,

$$\mu_{L,\Omega}([F_{0|\Omega} \neq 0]) \leq \mu_{L,\mathbb{R}^d}([F_0 \neq 0]) = 0.$$

Hence $F_{0|\Omega}$ lies in $M_{0,\Omega}$, and $F''_{|\Omega}$ belongs to $f' + M_{0,\Omega}$. That proves (3.1.4).

Conversely, for each f_0 in $M_{0,\Omega}$, the function

$$F_0 := \begin{cases} f_0 & \text{on } \Omega \\ 0 & \text{on } \mathbb{R}^d \setminus \Omega \end{cases}$$

lies in $M_{0,\mathbb{R}^d}$. Thus

$$f' + f_0 = (F' + F_0)_{|\Omega} \in F_{|\Omega}.$$

Whence $f' + M_{0,\Omega}$ is contained in $F_{|\Omega}$. □

3.1.2.2 Measurability of $g(\cdot - \cdot)f$

For each f in $\mathscr{L}^0(\mathbb{R}, \mathcal{X})$ and g in $\mathscr{L}^0(\mathbb{R}, \mathbb{K})$, we denote by $g(\cdot - \cdot)f$ the function

$$g(\cdot - \cdot)f : \mathbb{R}^2 \to \mathcal{X}, \qquad (x, y) \mapsto g(x - y)f(y). \tag{3.1.5}$$

According to Proposition 3.1.2, $g(\cdot - \cdot)f$ belongs to $\mathscr{L}^0(\mathbb{R}^2, \mathcal{X})$. Now we would like to work in $L^0(\mathbb{R}^2, \mathcal{X})$, that-is-to-say starting from f in $L^0(\mathbb{R}, \mathcal{X})$ and g in $L^0(\mathbb{R}, \mathbb{K})$, our aim is to define $g(\cdot - \cdot)f$ as a element of $L^0(\mathbb{R}^2, \mathcal{X})$. For, we use representatives: we take f' a representative of f, and g' a representative of g (note that here f' has nothing to do with a derivative of f). Then, we consider the class h of $g'(\cdot - \cdot)f'$ in $L^0(\mathbb{R}^2, \mathcal{X})$. However, we can do the same thing with other representatives f'', g'' of f and g. Then, the issue is to check that the class of $g''(\cdot - \cdot)f''$ is equal to h. This is the aim of the following result.

Proposition 3.1.5 *Being given a Banach space $\mathcal{X}$ over $\mathbb{K}$, let f belong to $L^0(\mathbb{R}, \mathcal{X})$ and g lie in $L^0(\mathbb{R}, \mathbb{K})$. Then there exists a unique element h in $L^0(\mathbb{R}^2, \mathcal{X})$ such that for each representative f' and g' of f and g (respectively), the function*

$$\mathbb{R}^2 \to \mathcal{X}, \qquad (x, y) \mapsto g'(x - y)f'(y)$$

is a representative of h. We set $g(\cdot - \cdot)f := h$.

Proof According to Proposition 3.1.2, the function $g'(\cdot - \cdot)f'$ is measurable, and we denote by h its class in $L^0(\mathbb{R}^2, \mathcal{X})$. We will show that for other representatives f'', g'' of f and g, the difference of the functions $g''(\cdot - \cdot)f''$ and $g'(\cdot - \cdot)f'$ lies in $M_{0,\mathbb{R}^2}$. For, we write f'' and g'' under the form

$$f'' = f' + f_0, \qquad g'' = g' + g_0,$$

where f_0, g_0 belongs to $M_{0,\mathbb{R}}$. In particular, notice for further reference the set

$$N_{g_0} := \{y \in \mathbb{R} \mid g_0(y) \neq 0\} \tag{3.1.6}$$

has zero measure in $\mathbb{R}$. Next,

$$g''(\cdot - \cdot)f'' - g'(\cdot - \cdot)f' = g'(\cdot - \cdot)f_0 + g_0(\cdot - \cdot)f' + g_0(\cdot - \cdot)f_0.$$

If $g'(x - y)f_0(y) \neq 0$ then (x, y) lies in the cartesian product of $\mathbb{R}$ with a set of zero measure. Hence $g'(\cdot - \cdot)f_0$ lies in $M_{0,\mathbb{R}^2}$. In the same way, $g_0(\cdot - \cdot)f_0$ lies in $M_{0,\mathbb{R}^2}$ as well. Regarding $g_0(\cdot - \cdot)f'$, we set

$$N := \{(x, y) \in \mathbb{R}^2 \mid g_0(x - y)f'(y) \neq 0\}.$$

Then, by Fubini Theorem

$$\int_{\mathbb{R}^2} \mathbf{1}_N(x, y)\,\mathrm{d}x\,\mathrm{d}y = \int_{\mathbb{R}} \mathrm{d}x \int_{\mathbb{R}} \mathbf{1}_N(x, y)\,\mathrm{d}y.$$

However, for each x in $\mathbb{R}$, $\mathbf{1}_N(x, \cdot) \leq \mathbf{1}_{x - N_{g_0}}$. Since

$$\mu_{\mathrm{L},1}(x - N_{g_0}) = \mu_{\mathrm{L},1}(N_{g_0}) = 0$$

thanks to (3.1.6), we deduce that N has zero measure. Thus, $g_0(\cdot - \cdot)f'$ lies in $M_{0,\mathbb{R}^2}$. Therefore $g''(\cdot - \cdot)f''$ is also a representative of h, which proves the existence and uniqueness of h. □

3.1.2.3 Essential Limits

Let d be a positive integer, Ω be a non-empty open subset of $\mathbb{R}^d$, and $(\mathcal{X}, \|\cdot\|)$ be a Banach space over $\mathbb{K} = \mathbb{R}$ or $\mathbb{C}$.

Definition 3.1.4 Let f lie in $L^0(\Omega, \mathcal{X})$ and x_0 belong to the closure of Ω. We say that f *admits a (finite) essential limit at* x_0 if there exists X_0 in $\mathcal{X}$ such that for any positive number ε, there exists δ in $(0, \infty)$ such that

$$\|f - X_0\| \leq \varepsilon$$

almost everywhere on $\Omega \cap B(x_0, \delta)$. Here $B(x_0, \delta)$ denotes the open ball of $\mathbb{R}^d$ with center x_0 and radius δ. □

With the above notation, it is clear that X_0 is unique. Then, X_0 is called the *essential limit of f at x_0*, and we denote

$$\operatorname{ess\,lim}_{x_0} f = X_0.$$

Let f belong to $L^0(\Omega, \mathcal{X})$, and f' be one of its representatives. If f' has a limit X_0 in $\mathcal{X}$ at x_0 (in the usual sense), then we may check easily that f admits an essential limit at x_0 and that the essential limit of f at x_0 is X_0.

Definition 3.1.5 A function f in $L^0(\Omega, \mathcal{X})$ is said to be *essentially bounded on* Ω if there exists a positive constant M such that

$$|f| \leq M$$

almost everywhere on Ω. □

Definition 3.1.6 Let f be in $L^0(\Omega, \mathbb{R})$ and x_0 belong to the closure of Ω. We say that *the essential limit of f at x_0 is equal to* ∞ if for any positive number A, there exists δ in $(0, \infty)$ such that

$$f \geq A$$

almost everywhere on $\Omega \cap B(x_0, \delta)$. □

3.1.3 Integration

Let d be a positive integer, Ω be a non empty open subset of $\mathbb{R}^d$, and $(\mathcal{X}, \|\cdot\|)$ be a Banach space over $\mathbb{K} = \mathbb{R}$ or $\mathbb{C}$. We will introduce briefly the Bochner theory of integration. We refer to [ABHN11, Section1.1] for more details and to [Mik78] for a complete exposition on the Bochner integral. Note that in the latter book, the main concepts are introduced in a circumstantial but unusual fashion.

A major difference with the usual case where $\mathcal{X} = \mathbb{R}$ is that the space $\mathcal{X}$ is a priori not ordered. Hence, there is no Lebesgue's Monotone Convergence Theorem in this context.

However, as in the usual case, we start to define the integral of a simple functions and then pass to the limit.

Bochner integrable functions are defined as follows. A function $f : \Omega \to \mathcal{X}$ is *Bochner integrable* if there exists a sequence $(f_n)_{n\in\mathbb{N}}$ of simple functions such that (f_n) converges toward f almost everywhere on Ω and also

$$\int_\Omega \|f(x) - f_n(x)\|\,\mathrm{d}x \xrightarrow[n\to\infty]{} 0.$$

The *integral of* f *on* Ω is the vector of $\mathcal{X}$ defined by

$$\int_\Omega f(x)\,\mathrm{d}x := \lim_{n\to\infty} \int_\Omega f_n(x)\,\mathrm{d}x.$$

The space of Bochner integrable functions is denoted by $\mathcal{L}^1(\Omega, \mathcal{X})$.

Next we may state Bochner Theorem about which the authors of [ABHN11] write: *it is one of the great virtues of the Bochner integral that the class of Bochner integrable functions is easily characterized.*

Theorem 3.1.6 (Bochner Theorem) *Let* $(\mathcal{X}, \|\cdot\|)$ *be a Banach space and* $f : \Omega \to \mathcal{X}$ *be a function on* Ω. *Then,* f *lies in* $\mathcal{L}^1(\Omega, \mathcal{X})$ *(i.e. is Bochner integrable) if and only if* f *is L-measurable and* $\|f\|$ *is integrable. Moreover, if* f *is Bochner integrable, then*

$$\left\| \int_\Omega f(x)\,\mathrm{d}x \right\| \le \int_\Omega \|f(x)\|\,\mathrm{d}x.$$

Let us now introduce the space $L^1(\Omega, \mathcal{X})$. Since $M_{0,\Omega}$ defined by (3.1.2) is a subspace of $\mathcal{L}^1(\Omega, \mathcal{X})$, we may set

$$L^1(\Omega, \mathcal{X}) := \mathcal{L}^1(\Omega, \mathcal{X})/M_{0,\Omega}.$$

Observing that the integrals of two representatives of an element of $L^1(\Omega, \mathcal{X})$ are equal, we define for each f in $L^1(\Omega, \mathcal{X})$,

$$\int_\Omega f(x)\,\mathrm{d}x := \int_\Omega f'(x)\,\mathrm{d}x, \tag{3.1.7}$$

where f' is any representative of f.

Theorem 3.1.7 (Fubini Theorem) *Let* I, J *be two non empty open intervals of* $\mathbb{R}$. *Let also* $\mathcal{X}$ *be a Banach space and* f *lie in* $L^1(I \times J, \mathcal{X})$. *Then*

(i) *for almost every x in I, the function $f(x,\cdot)$ lies in $L^1(J,\mathcal{X})$, and for almost every y in J, the function $f(\cdot,y)$ lies in $L^1(I,\mathcal{X})$;*
(ii) *the functions $\int_J f(\cdot,y)\,\mathrm{d}y$ and $\int_I f(x,\cdot)\,\mathrm{d}x$ belong to $L^1(I,\mathcal{X})$ and $L^1(J,\mathcal{X})$ respectively;*
(iii) *finally,*

$$\int_{I\times J} f(x,y)dx\,\mathrm{d}y = \int_I \Big(\int_J f(x,y)\,\mathrm{d}y\Big)\,\mathrm{d}x = \int_J \Big(\int_I f(x,y)\,\mathrm{d}x\Big)\,\mathrm{d}y.$$

We refer to [Dro01, Corollary 1.9.2] or [ABHN11, Theorem 1.1.9] for a proof. However, some comments are in order. The first issue is clarify the meaning of the Item (i) sentence "$f(x,\cdot)$ lies in $L^1(J,\mathcal{X})$" since the evaluation of L^1-functions does not make sense. For simplicity, let us assume that $I = J = \mathbb{R}$.

In view of [Dro01, Lemme 1.9.1] and [Rud91, Lemma 2 Chap. 8], the precited sentence means that for any representative f' and f'' of f, there exists a measurable subset N of $\mathbb{R}$ with zero measure such that

- $f'(x,\cdot)$ lies in $\mathscr{L}^0(\mathbb{R},\mathcal{X})$ for any x in $\mathbb{R}\setminus N$;
- $f''(x,\cdot)$ lies in $\mathscr{L}^0(\mathbb{R},\mathcal{X})$ for any x in $\mathbb{R}\setminus N$;
- for almost every x in $\mathbb{R}\setminus N$, the classes of $f'(x,\cdot)$ and $f''(x,\cdot)$ (in $L^0(\mathbb{R},\mathcal{X})$) are equal.

Now in Item (ii), let us be precise how $\int_J f(\cdot,y)\,\mathrm{d}y$ can be seen as a function of $L^1(I,\mathcal{X})$. Replacing possibly N by a larger set of zero measure still labeled N, we prove that for each x in $\mathbb{R}\setminus N$, $f'(x,\cdot)$ is integrable on $\mathbb{R}$ and that there exists a sequence of simple functions $(s_n)_{n\geq 0}$ such that

$$s_n(x) \to \int f'(x,y)\,\mathrm{d}y, \qquad \forall x \in \mathbb{R}\setminus N.$$

Next, the function $\int f'(\cdot,y)\,\mathrm{d}y$ defined on $\mathbb{R}\setminus N$ can easily be extended into a measurable function on $\mathbb{R}$, still labeled $\int f'(\cdot,y)\,\mathrm{d}y$. For the same reason, the class of $\int f''(\cdot,y)\,\mathrm{d}y$ lies in $\mathscr{L}^0(\mathbb{R},\mathcal{X})$. Moreover,

$$\int f'(\cdot,y)\,\mathrm{d}y = \int f''(\cdot,y)\,\mathrm{d}y$$

almost everywhere on $\mathbb{R}$. Thus, in Item (ii), $\int f(\cdot,y)\,\mathrm{d}y$ denotes the class of $\int f'(\cdot,y)\,\mathrm{d}y$.

□

3.2 Functional Spaces

The goal of this section is just to fix notation. We will consider spaces of vector-valued functions defined on a non-empty open interval I of $\mathbb{R}$. We will start to introduce Lebesgue and Sobolev spaces. We refer to [DL92, Chap. XVIII] for more information on Lebesgue and Sobolev spaces. The second section will be devoted to Hölder spaces.

Let $\mathbb{K}$ be equal to $\mathbb{R}$ or $\mathbb{C}$, and $(\mathcal{X}, \|\cdot\|)$ be a Banach space over $\mathbb{K}$.

3.2.1 Lebesgue and Sobolev Spaces

3.2.1.1 Definitions

For any real number p in $[1, \infty)$ and each non-empty open subset Ω of $\mathbb{R}^d$, we set

$$L^p(\Omega, \mathcal{X}) := \left\{ f \in L^0(\Omega, \mathcal{X}) \mid \int_\Omega \|f(x)\|^p \,\mathrm{d}x < \infty \right\}.$$

Endowed with the norm

$$\|f\|_{L^p(\Omega,\mathcal{X})} := \left(\int_\Omega \|f(x)\|^p \,\mathrm{d}x \right)^{\frac{1}{p}},$$

$L^p(\Omega, \mathcal{X})$ becomes a normed space, the so-called *Lebesgue space of order p*. In the particular case where $p = 2$ and $\mathcal{X}$ is an Hilbert space, let us say $(\mathcal{H}, (\cdot, \cdot))$, then $L^2(\Omega, \mathcal{H})$ will always be endowed with the inner product

$$(f, h)_{L^2(\Omega,\mathcal{X})} := \int_0^b (f(x), h(x)) \,\mathrm{d}x,$$

so that $L^2(\Omega, \mathcal{H})$ becomes an Hilbert space.

Let $L^p_{\mathrm{loc}}(\Omega, \mathcal{X})$ be the classes of functions f which belong *locally* to $L^p(\Omega, \mathcal{X})$; that is, f lies in $L^0(\Omega, \mathcal{X})$ and

$$\int_\omega \|f(x)\|^p \,\mathrm{d}x < \infty$$

for every *compact* subset ω of Ω.

In the case where $p = \infty$, $L^\infty(\Omega, \mathcal{X})$ denotes the space of essentially bounded functions of $L^0(\Omega, \mathcal{X})$. $L^\infty(\Omega, \mathcal{X})$ is equipped with the following norm.

$$\|f\|_{L^\infty(\Omega,\mathcal{X})} := \inf \left\{ M \in (0, \infty) \mid |f| \le M \text{ almost everywhere on } \Omega \right\}.$$

Besides, for each p in $[1, \infty]$ and each non empty open interval I of $\mathbb{R}$, $W^{1,p}(I, \mathcal{X})$ denotes the *Sobolev space* of all functions in $L^p(I, \mathcal{X})$ whose derivative in the sense of distribution lies in $L^p(I, \mathcal{X})$. Setting

$$W^{0,p}(I, \mathcal{X}) := L^p(I, \mathcal{X}),$$

for each positive integer n, we define inductively the space $W^{n,p}(I, \mathcal{X})$ by

$$W^{n,p}(I, \mathcal{X}) := \{u \in W^{n-1,p}(I, \mathcal{X}) \mid u' \in W^{n-1,p}(I, \mathcal{X})\}.$$

The norm of $W^{n,p}(I, \mathcal{X})$ is

$$\|u\|_{W^{n,p}(I,\mathcal{X})} := \sum_{i=0}^{n} \|u^{(i)}\|_{L^p(I,\mathcal{X})},$$

where $u^{(i)}$ is the ith-order derivative of u if i is positive, and $u^{(0)} = u$.

In the particular case where $p = 2$, we put for each non-negative integer n

$$H^n(I, \mathcal{X}) := W^{n,2}(I, \mathcal{X}).$$

If $\mathcal{X} := \mathbb{K}$, then the reference to $\mathcal{X}$ will be skipped in the denomination of these spaces. For instance, we will write $L^p(\Omega)$ instead of $L^p(\Omega, \mathbb{K})$. In addition, if $I := (0, b)$, then $L^p\big((0, b)\big)$ will be abbreviated by $L^p(0, b)$, or even by L^p if no confusion may occur. Also $L^p(I, \mathcal{X})$ and $W^{n,p}(I, \mathcal{X}))$ will be sometimes abbreviated into $L^p(\mathcal{X})$ and $W^{n,p}(\mathcal{X})$.

3.2.1.2 Properties

Let $\mathcal{X}$ be a Banach space, I be a non-empty open interval of $\mathbb{R}$ and Ω be a non-empty open subset of $\mathbb{R}^d$. Let also p be in $[1, \infty]$ and q be its conjugate exponent. For any non negative integer n, it is well known that $W^{n,p}(I, \mathcal{X})$ *is a Banach space*. Regarding the dual space of $L^p(\Omega, \mathcal{X})$, extending the case $\mathcal{X} = \mathbb{K}$ featured in Example 2.3.6, we introduce the map

$$I_{q,p,\mathcal{X}} : L^q(\Omega, \mathcal{X}^*) \to L^p(\Omega, \mathcal{X})^*$$

defined for every f in $L^q(\Omega, \mathcal{X}^*)$ by

$$\langle I_{q,p,\mathcal{X}} f, h\rangle_{L^p(\mathcal{X})^*, L^p(\mathcal{X})} = \int_0^b \langle f(x), h(x)\rangle_{\mathcal{X}^*,\mathcal{X}} \mathrm{d}x, \qquad \forall h \in L^p(\mathcal{X}).$$

Proposition 3.2.1 *Let $\mathcal{X}$ be a Banach space, p be in $[1, \infty]$ and q be its conjugate exponent. Then $I_{q,p,\mathcal{X}}$ is linear, continuous, and one-to-one.*

Proof Only injectivity is not obvious. Let f lie in $L^q(\Omega, \mathcal{X}^*)$ and satisfy $I_{q,p,\mathcal{X}} f = 0$. For any continuous function $\varphi : [0, b] \to \mathbb{K}$ and any vector X in $\mathcal{X}$, we derive that

$$\int_0^b \overline{\varphi(x)} \langle f(x), X \rangle_{\mathcal{X}^*,\mathcal{X}} \mathrm{d}x = 0.$$

Thus for almost every x in $(0, b)$,

$$\langle f(x), X \rangle_{\mathcal{X}^*,\mathcal{X}} = 0, \qquad \forall X \in \mathcal{X}.$$

Hence, f is trivial and $I_{q,p,\mathcal{X}}$ is one-to-one. □

Proposition 3.2.1 allows to identify $L^q(\Omega, \mathcal{X}^*)$ with a subspace of $L^p(\Omega, \mathcal{X})^*$, so that $L^q(\Omega, \mathcal{X}^*)$ becomes continuously embedded in $L^p(\Omega, \mathcal{X})^*$. Now, we will see that in a reflexive setting, these two spaces are isometric and can be identified as well.

Theorem 3.2.2 *Let*

- *$\mathcal{X}$ be a separable and reflexive Banach space;*
- *p lie in $(1, \infty)$, and q be the conjugate exponent of p;*
- *Ω be a non empty open subset of $\mathbb{R}^d$.*

Then,

(i) *$I_{q,p,\mathcal{X}}$ is an isometric isomorphism between $L^q(\Omega, \mathcal{X}^*)$ and $L^p(\Omega, \mathcal{X})^*$;*
(ii) *$L^p(\Omega, \mathcal{X})$ is a reflexive Banach space.*

Proof We refer to [HvNVW16, Th. 1.3.10 & 1.3.21] for a proof of (i). In order to prove Item (ii), we introduce the map

$$j : L^p(\Omega, \mathcal{X}) \to L^p(\Omega, \mathcal{X}^{**}), \qquad f \mapsto J_{\mathcal{X}}\big(f(\cdot)\big).$$

Since $\mathcal{X}$ is reflexive, it is easy to see that j is a surjective isometry. Thus, $L^p(\Omega, \mathcal{X}^{**})$ is identified with $L^p(\Omega, \mathcal{X})$. Then we conclude by applying Proposition 2.3.10 with

$$\mathcal{X}_1 := L^p(\Omega, \mathcal{X}), \qquad \mathcal{X}_2 := L^q(\Omega, \mathcal{X}^*), \qquad I_1 := I_{p,q,\mathcal{X}^*}, \qquad I_2 := I_{q,p,\mathcal{X}}.$$

□

As a particular case of [HvNVW16, Th. 1.2.4], we state the following result.

Proposition 3.2.3 *Let Ω be a non-empty open subset of $\mathbb{R}^d$, p belong to $[1, \infty)$, $\mathcal{X}$, $\mathcal{Y}$ be Banach spaces, and T lie in $\mathcal{L}(\mathcal{X}, \mathcal{Y})$ or in ${}^{\mathrm{A}}\mathcal{L}(\mathcal{X}, \mathcal{Y})$. If f lies in $L^p(\Omega, \mathcal{X})$ then Tf lies in $L^p(\Omega, \mathcal{Y})$ and especially*

$$T \int_\Omega f(x)\,\mathrm{d}x = \int_\Omega Tf(x)\,\mathrm{d}x.$$

Before closing this subsection, let us give a result concerning Gelfand triplets of vector-valued spaces.

Proposition 3.2.4 *Let $\mathcal{V}$ be a nontrivial real or complex separable and reflexive Banach space, $\big(\mathcal{H}, (\cdot, \cdot)\big)$ be a Hilbert space, and p, q be conjugate exponents in $(1, \infty)$. If $q \le p$ and $(\mathcal{V}, \mathcal{H}, \mathcal{V}^*)$ is a Gelfand triplet then*

$$\Big(L^p\big((0, b), \mathcal{V}\big),\ L^2\big((0, b), \mathcal{H}\big),\ L^q\big((0, b), \mathcal{V}^*\big)\Big)$$

is a Gelfand triplet as well.

Proof Recall that $L^p\big((0, b), \mathcal{V}\big)$ is abbreviated by $L^p(\mathcal{V})$. By Hölder inequality, we get that $L^p(\mathcal{V})$ is continuously embedded in $L^2(\mathcal{H})$. Thus from Definition 2.5.3, there remains to show that $L^p(\mathcal{V})$ is dense in $L^2(\mathcal{H})$. For, let f belong to $L^2(\mathcal{H})$ and satisfy

$$\int_0^b (f(x), \psi(x))\,\mathrm{d}x = 0, \qquad \forall\, \psi \in L^p(\mathcal{V}).$$

For each smooth function $\varphi : (0, b) \to \mathbb{K}$ and each vector v in $\mathcal{V}$, we derive

$$\int_0^b \overline{\varphi(x)}\big(f(x), v\big)\,\mathrm{d}x = 0.$$

Thus $\big(f(x), v\big) = 0$ for almost every x in $(0, b)$. Since $\mathcal{V}$ is dense in $\mathcal{H}$, one has $f = 0$ in $L^2(\mathcal{H})$. Then, the density of $L^p(\mathcal{V})$ follows from Corollary 2.2.10. □

3.2.1.3 Direct Sums in Lebesgue Spaces

The forthcoming results allow to extend results stated for *scalar valued* functions into results about *vector-valued* functions. Notice that they hold without restriction on the final Banach space $\mathcal{X}$. For each p in $[0, \infty]$, h in $L^p(0, b)$ and Y in $\mathcal{Y}$, we put $hY := \{hY | Y \in \mathcal{Y}\}$. See Example 5.3.2 for more details.

Proposition 3.2.5 *Let $\mathcal{X}$ be a nontrivial Banach space, and p, q be conjugate exponents such that p lies in $[1, \infty)$. Let also m be a positive integer and $h_1, \ldots, h_m$ be functions of $L^p(0, b) \setminus \{0\}$. Then, the following assertions are equivalent.*

(i) *The subspaces* $h_1\mathcal{X}, \dots, h_m\mathcal{X}$ *are in direct sum.*
(ii) $h_1, \dots, h_m$ *are linearly independent in* $L^p(0, b)$.
(iii) *There exist functions* $f_1, \dots, f_m$ *in* $L^q(0, b)$ *such that the matrix with entries* $(\int_0^b f_i\overline{h_j}\,\mathrm{d}y)$ *is invertible.*

Proof We first prove that (i) implies (ii). Assume that for some scalars $\lambda_1, \dots, \lambda_m$, we have

$$\sum_{i=1}^{m} \lambda_i h_i = 0 \qquad \text{in } L^p(0, b).$$

Let X be a fixed non zero vector of $\mathcal{X}$. Then,

$$\sum_{i=1}^{m} \lambda_i h_i X = 0 \qquad \text{in } L^p\big((0, b), \mathcal{X}\big).$$

On account of Item (i) and the fact that the h_i's are not trivial, Proposition 2.1.9 yields that all the λ_i's are equal to zero. Hence $h_1, \dots, h_m$ are linearly independent in $L^p(0, b)$.

Recalling that the anti-dual space of $L^1(0, b)$ is $L^\infty(0, b)$, Proposition 2.2.14 (iv) gives that (ii) implies (iii). Eventually, let us prove that (iii) implies (i). We assume that vectors $X_1, \dots, X_m$ of $\mathcal{X}$ satisfy

$$\sum_{j=1}^{m} h_j X_j = 0.$$

Then, for each $i = 1, \dots, m$,

$$\sum_{j=1}^{m} \int_0^b \overline{f_i} h_j \,\mathrm{d}y\, X_j = 0.$$

The invertibility of the matrix $(\int_0^b \overline{f_i} h_j \,\mathrm{d}y)$ gives that all the X_j's must be zero. With Proposition 2.1.9, we deduce that $h_1\mathcal{X}, \dots, h_m\mathcal{X}$ are in direct sum. □

Lemma 3.2.6 *Let* $\mathcal{X}$ *be a non trivial Banach space,* p *and* q *be conjugate exponents such that* p *lies in* $[1, \infty)$, *and* m *be a positive integer. For each index* $i = 1, \dots, m$, *let* (f_i, h_i) *be in* $L^q(0, b) \times L^p(0, b)$. *Assume that the matrix with entries*

$$\int_0^b f_i\overline{h_j}\,\mathrm{d}y, \qquad \forall\, 1 \le i, j \le m$$

is invertible. Then, the following assertions hold true.

(i) *The subspaces* $h_1\mathcal{X}, \dots, h_m\mathcal{X}$ *of* $L^p\big((0,b),\mathcal{X}\big)$ *are in direct sum. In the same way, the subspaces* $f_1\mathcal{X}^*, \dots, f_m\mathcal{X}^*$ *of* $L^q\big((0,b),\mathcal{X}^*\big)$ *are in direct sum.*

(ii) *For each* f *in* $L^q(0,b)$*,*

$$(f\mathcal{X}^*)^\top = \Big\{v \in L^p\big((0,b),\mathcal{X}\big) \mid \int_0^b \overline{f} v \,\mathrm{d}y = 0\Big\}. \tag{3.2.1}$$

(iii) $(\oplus_{i=1}^m f_i\mathcal{X}^*)^\top$ *and* $\oplus_{i=1}^m h_i\mathcal{X}$ *are closed complementary subspaces in* $L^p\big((0,b),\mathcal{X}\big)$.

Proof

(i) If $h_{j_0} = 0$ for some index j_0 in $[1,n]$, then the matrix with entries

$$\int_0^b f_i\overline{h_j}\,\mathrm{d}y, \qquad \forall\, 1 \le i, j \le m$$

has a nontrivial kernel. Hence, $h_1, \dots, h_m$ belong to $L^p(0,b) \setminus \{0\}$, so that the first sentence follows from Proposition 3.2.5. The proof of the second sentence is analog.

(ii) Let f be in $L^q(0,b)$ and v be in $L^p(\mathcal{X})$. Then, for each X^* in $\mathcal{X}^*$,

$$\int_0^b \langle fX^*, v\rangle_{\mathcal{X}^*,\mathcal{X}}\mathrm{d}y = \int_0^b \langle X^*, \overline{f}v\rangle_{\mathcal{X}^*,\mathcal{X}}\mathrm{d}y = \langle X^*, \int_0^b \overline{f}v\,\mathrm{d}y\rangle_{\mathcal{X}^*,\mathcal{X}}$$

due to Proposition 3.2.3. Thus, if v lies in $(f\mathcal{X}^*)^\top$ then the left hand side of the latter equality vanishes for all $\mathcal{X}^*$, so that

$$\int_0^b \overline{f}v\,\mathrm{d}y = 0$$

by Corollary 2.2.7. Hence, $(f\mathcal{X}^*)^\top$ is contained in the right hand side of (3.2.1). The converse is proved in a similar way.

(iii) We claim that for each v in $L^p(\mathcal{X})$, there exists a unique m-tuple $(X_1, \dots, X_m)$ in $\mathcal{X}^m$ such that

$$v - \sum_{i=1}^m h_i X_i \in \Big(\bigoplus_{i=1}^m f_i\mathcal{X}^*\Big)^\top. \tag{3.2.2}$$

Indeed, since

$$\Big(\bigoplus_{i=1}^m f_i\mathcal{X}^*\Big)^\top = \bigcap_{i=1}^m (f_i\mathcal{X}^*)^\top,$$

we deduce with Item (ii) that (3.2.2) is equivalent to

$$\sum_{j=1}^{m} \int_0^b \overline{f_i} h_j \, dy \, X_j = \int_0^b \overline{f_i} v \, dy, \qquad \forall\, 1 \le i \le m. \tag{3.2.3}$$

Then, the claim follows by the invertibility of the matrix $(\int_0^b \overline{f_i} h_j \, dy)$. Now this claim allows to define the map

$$P : L^p(\mathcal{X}) \to L^p(\mathcal{X}), \qquad v \mapsto \sum_{i=1}^{m} h_i X_i,$$

where $(X_1, \ldots, X_m)$ is defined by (3.2.2). By (3.2.3), P is linear and continuous on $L^p(\mathcal{X})$. Moreover for each index i_0 in $[1, m]$ and each X in $\mathcal{X}$, by setting $X_j := \delta_{i_0, j} X$ for any j in $[1, m]$, we check easily from (3.2.3) that $P(h_{i_0} X) = h_{i_0} X$. Hence, P is a projection whose range equals $\oplus_{i=1}^{m} h_i \mathcal{X}$. Moreover, we infer from (3.2.2) that

$$\ker P = \Big(\bigoplus_{i=1}^{m} f_i \mathcal{X}^* \Big)^{\top}.$$

There results from Proposition 2.3.3 that $\oplus_{i=1}^{m} h_i \mathcal{X}$ and $(\oplus_{i=1}^{m} f_i \mathcal{X}^*)^{\top}$ are closed complementary subspaces. □

Remark 3.2.1 Let p and q be conjugate exponents in $(1, \infty)$, and f be a nontrivial function in $L^q(0, b)$. (i) By Corollary 2.2.7, there exists a function f^c in L^p such that

$$\|f^c\|_{L^p} = 1, \qquad \int_0^b f \overline{f^c} \, dy = \|f\|_{L^q}.$$

It turns out that f^c is unique and

$$f^c = \|f\|_{L^q}^{1-q} |f|^{q-2} f, \tag{3.2.4}$$

where the function $(0, \infty) \to \mathbb{R}$, $x \mapsto |x|^{q-2} x$ is extended by zero at $x = 0$. The function f^c is called the *conjugate function of f in $L^p(0, b)$*. See Section 2.7 of [PRR19] and in particular Proposition 2.7.32, or [Rou05, Sections 3.1 & 3.2] for further information on conjugate functions.

(ii) If $\mathcal{X}$ is an infinite dimensional space then $(f \mathcal{X}^*)^{\perp}$ and $f^c \mathcal{X}$ are both infinite dimensional spaces. Thus Lemma 2.2.13 cannot be used to prove that they are complementary subspaces. However by applying Lemma 3.2.6 with $m := 1$ and $h := f^c$, we get that $(f\mathcal{X}^*)^{\perp}$ and $f^c\mathcal{X}$ are closed complementary subspaces in $L^p\big((0,b),\mathcal{X}\big)$. As a consequence, for each function h in $L^p(0,b)$, $h\mathcal{X}$ is a closed subspace of $L^p\big((0,b),\mathcal{X}\big)$. □

3.2.2 Hölder Spaces

Let $(\mathcal{X}, \|\cdot\|)$ be a normed space over $\mathbb{K}$. Let b be a positive real number.

3.2.2.1 Definitions

We denote by $C([0,b],\mathcal{X})$ the space of all $\mathcal{X}$-valued continuous functions defined on $[0,b]$ equipped with the norm

$$\|u\|_0 := \sup_{x\in[0,b]} \|u(x)\|.$$

Let $u : [0,b] \to \mathcal{X}$ and x_0 belong to $[0,b]$. u is *differentiable at* x_0 if the function

$$[0,b]\setminus\{x_0\} \to \mathcal{X}, \qquad x \mapsto \frac{f(x)-f(x_0)}{x-x_0}$$

has a finite limit at x_0. u is *differentiable on* $[0,b]$ if u is differentiable at x_0 for each x_0 in $[0,b]$. In that case, the function

$$[0,b] \to \mathcal{X}, \qquad x \mapsto \lim_{y\to x} \frac{f(y)-f(x)}{y-x}$$

denoted by u' or $u^{(1)}$ is *the derivative of* u *on* $[0,b]$. For any positive integer n, $C^n([0,b],\mathcal{X})$ is the space of all functions $u : [0,b] \to \mathcal{X}$ admitting a continuous nth-order derivative on $[0,b]$. The space $C^n([0,b],\mathcal{X})$ is endowed with the norm

$$\|u\|_n := \sum_{i=0}^{n} \|u^{(i)}\|_0$$

where $u^{(i)}$ is the ith-order derivative of u if i is positive, and $u^{(0)} := u$. We set $C^0([0,b],\mathcal{X}) := C([0,b],\mathcal{X})$

For each $\alpha \in (0,1]$, we denote by $C^{0,\alpha}([0,b],\mathcal{X})$ the space of functions u in $C([0,b],\mathcal{X})$ such that

$$|u|_{0,\alpha} := \sup_{\substack{x,y\in[0,b]\\ x\neq y}} \frac{\|u(x)-u(y)\|}{|x-y|^{\alpha}} < \infty.$$

We equip that space with the norm

$$\|u\|_{0,\alpha} := \|u\|_0 + |u|_{0,\alpha}.$$

In other words, if $\alpha < 1$ then $C^{0,\alpha}([0, b], \mathcal{X})$ is the space of Hölder continuous functions with exponent α. If $\alpha = 1$ then $C^{0,1}([0, b], \mathcal{X})$ is the space of Lipschitz continuous functions on $[0, b]$.

More generally, for any positive integer n and any α in $(0, 1]$, $C^{n,\alpha}([0, b], \mathcal{X})$ is the space of all functions in $C^n([0, b], \mathcal{X})$ whose n-th order derivative belongs to $C^{0,\alpha}([0, b], \mathcal{X})$. The space $C^{n,\alpha}([0, b], \mathcal{X})$ is endowed with the norm

$$\|u\|_{n,\alpha} := \sum_{i=0}^{n} \|u^{(i)}\|_0 + |u^{(n)}|_{0,\alpha}. \tag{3.2.5}$$

Besides, for each non-negative integer n, we set $C^{n,0}([0, b], \mathcal{X}) := C^n([0, b], \mathcal{X})$.

Now we will introduce another notation for Hölder spaces that is sometimes more convenient. Let β be a positive real number. We decompose β into its *integer* and *fractional parts*, that is

$$\beta = n + \alpha,$$

where n is the greatest integer less than or equal to β, and $\alpha := \beta - n$ lies in $[0, 1)$. Then, we set

$$C^{\beta}([0, b], \mathcal{X}) := C^{n,\alpha}([0, b], \mathcal{X}). \tag{3.2.6}$$

The norm on $C^{\beta}([0, b], \mathcal{X})$ is

$$\|\cdot\|_{\beta} := \|\cdot\|_{n,\alpha}. \tag{3.2.7}$$

We denote by ${}_0C^{\beta}([0, b], \mathcal{X})$ the subspace of $C^{\beta}([0, b], \mathcal{X})$ whose elements vanish at $x = 0$.

3.2.2.2 Properties

We will use the following well-known result.

Theorem 3.2.7 *Let $\mathcal{X}$ be a Banach space, b be a positive real number, and p belong to $[1, \infty]$. Then, up to identification, $W^{1,p}\big((0, b), \mathcal{X}\big)$ is continuously embedded in $C\big([0, b], \mathcal{X}\big)$.*

We will not reproduce the proof of the above result; however, for all purposes, we will specify the underlying identification. As a subspace of $L^p\big((0, b), \mathcal{X}\big)$, the elements of $W^{1,p}\big((0, b), \mathcal{X})\big)$ are classes of functions. In the proof of Theorem 3.2.7, it is shown that any U in $W^{1,p}\big((0,b),\mathcal{X}\big)$ possesses a unique representative which can be extended (in a unique way) into a continuous function u on $[0,b]$. Since the map $U \mapsto u$ turns out to be a linear embedding, the identification follows from Sect. 2.2.7.3. Under the assumptions and notation of Theorem 3.2.7, we denote by $W_0^{1,p}\big((0,b),\mathcal{X}\big)$ the space of all functions u in $W^{1,p}\big((0,b),\mathcal{X}\big)$ such that $u(0)=u(b)=0$.

3.2.3 The Space $C_{\ln}([0,b],\mathcal{X})$

In our regularity results (see Theorem 3.6.2 and especially Proposition 8.6.1), Lipschitz-regularity cannot be achieved. We only get a weaker regularity, the so-called $C_{\ln}$-*regularity*, which however turns out to be stronger than C^α-regularity for any α in $(0,1)$. Proposition 3.2.8 gathers some results about the space $C_{\ln}([0,b],\mathcal{X})$.

Let $(\mathcal{X},\|\cdot\|)$ be a normed space over $\mathbb{K}$. Let also b a positive real number, and

$$\mathscr{R} := \big\{(x,h) \in [0,b) \times \big(0, \min(\tfrac{1}{2}, b)\big] \mid x+h \le b\big\}.$$

Then, $C_{\ln}([0,b],\mathcal{X})$ denotes the space of all functions $f:(0,b)\to\mathcal{X}$ such that

$$|f|_{C_{\ln}} := \sup_{(x,h)\in\mathscr{R}} \frac{\|f(x+h)-f(x)\|}{h|\ln h|} < \infty.$$

We equip this space with the norm

$$\|f\|_{C_{\ln}} := \|f\|_{0,0} + |f|_{C_{\ln}}.$$

The thrust of the next result is that $C_{\ln}([0,b],\mathcal{X})$ is an intermediate space between $C^{0,1}([0,b],\mathcal{X})$ and $C^{0,\alpha}([0,b],\mathcal{X})$.

Proposition 3.2.8 *Let α lie in* $(0,1)$. *The following assertions hold true.*

(i) $C_{\ln}([0,b],\mathcal{X})$ *is continuously embedded in* $C^{0,\alpha}([0,b],\mathcal{X})$.
(ii) *The space* $C^{0,1}([0,b],\mathcal{X})$ *of Lipschitz continuous functions is continuously embedded in* $C_{\ln}([0,b],\mathcal{X})$.
(iii) *If $\mathcal{X}$ is a nontrivial space, then* $C_{\ln}([0,b],\mathcal{X})$ *is a proper subset of* $C^{0,\alpha}([0,b],\mathcal{X})$, *and* $C^{0,1}([0,b],\mathcal{X})$ *is a proper subset of* $C_{\ln}([0,b],\mathcal{X})$, *that is*

$$C^{0,1}([0,b],\mathcal{X})) \subset C_{\ln}([0,b],\mathcal{X}) \subset C^{0,\alpha}([0,b],\mathcal{X}).$$

Proof

(i) Let f belong to $C_{\ln}([0,b],\mathcal{X})$, and (x,y) be in $[0,b]^2$ with $x<y$. Setting $h:=y-x$, there results that h lives in $(0,b]$. If $\frac{1}{2}<h$ then

$$\|f(x+h)-f(x)\| \leq 2\|f\|_{0,0} \leq 2^{\alpha+1}\|f\|_{C_{\ln}} h^{\alpha}.$$

Thus

$$\frac{\|f(y)-f(x)\|}{|y-x|^{\alpha}} \leq 2^{\alpha+1}\|f\|_{C_{\ln}}.$$

On the other hand, if $h\leq\frac{1}{2}$ then (x,h) lives in $\mathscr{R}$, so that

$$\|f(x+h)-f(x)\| \leq |f|_{C_{\ln}} h|\ln h|.$$

Then,

$$\frac{\|f(y)-f(x)\|}{|y-x|^{\alpha}} \leq |f|_{C_{\ln}} h^{1-\alpha}|\ln h|.$$

Since the function $h\mapsto h^{1-\alpha}|\ln h|$ is bounded on $(0,\frac{1}{2}]$, there results that f lies in $C^{0,\alpha}([0,b],\mathcal{X})$ and $\|f\|_{0,\alpha}\lesssim\|f\|_{C_{\ln}}$.

(ii) For each f in $C^{0,1}([0,b],\mathcal{X})$, one has for any (x,h) in $\mathscr{R}$,

$$\frac{\|f(x+h)-f(x)\|}{h|\ln h|} \leq \frac{|f|_{0,1}}{|\ln h|}.$$

Using $|\ln h|\geq \ln 2$, we get Item (ii).

(iii) Since $g_{1+\alpha}$ lies in $C^{0,\alpha}([0,b],\mathcal{X})$ but not in $C_{\ln}([0,b],\mathcal{X})$, we get that $C_{\ln}([0,b],\mathcal{X})$ is a proper subset of $C^{0,\alpha}([0,b],\mathcal{X})$. Finally, being given a normed vector x_0 of $\mathcal{X}$, then consider the function

$$f:[0,b]\to\mathcal{X},\qquad x\mapsto\begin{cases}0 & \text{if } x=0\\ x\ln x\, x_0 & \text{otherwise.}\end{cases}$$

Since f' is unbounded near $x=0$, f does not belong to $C^{0,1}([0,b],\mathcal{X})$. However, we will show that f lies in $C_{\ln}([0,b],\mathcal{X})$. Indeed, let (x,h) be any point in $\mathscr{R}$. If $x=0$, then $\frac{\|f(h)\|}{h|\ln h|}=1$. If $x\neq 0$, one has

$$\frac{\|f(x+h)-f(x)\|}{h|\ln h|} \leq \frac{x\big(\ln(x+h)-\ln x\big)}{h|\ln h|}+\frac{|\ln(x+h)|}{|\ln h|}.$$

By *the mean value theorem*, $\ln(x+h)-\ln x \leq \frac{h}{x}$. Thus the first term in the latter right hand side is bounded by $(\ln 2)^{-1}$ since $h \leq \frac{1}{2}$. Besides we claim that the function

$$\mathscr{R} \to \mathbb{R}, \qquad (x,h) \mapsto \frac{|\ln(x+h)|}{|\ln h|}$$

is bounded on $\mathscr{R}$. Indeed, let (x, h) belong to $\mathscr{R}$. If $x + h \leq 1$, then

$$\frac{|\ln(x+h)|}{|\ln h|} \leq 1,$$

because the function $|\ln|$ decreases on $(0, 1]$. Moreover, if $1 < x + h$, then since $h \leq \frac{1}{2}$,

$$\frac{|\ln(x+h)|}{|\ln h|} \leq \frac{\ln(2b)}{\ln 2}.$$

That proves the claim. Hence, f lies in $C_{\ln}([0, b], \mathcal{X})$. □

3.3 Convolution

3.3.1 Introduction

Although it may appear a bit complicated at first sight, convolution is a natural and convenient tool. In some sense, this operation combines properties of the addition and multiplication. Let us give two examples that illustrate this point of view.

Our first example comes from probability theory. Let Ω be a non-empty set, $\mathscr{F}$ be a σ-algebra on Ω, and $P : \mathscr{F} \to [0, 1]$ be a probability on $\mathscr{F}$. Let also X and Y be two independent real random variables defined on Ω. We assume that X and Y possess *probability density functions* f and g. Hence for each Borel set B of $\mathbb{R}$,

$$P(X \in B) = \int_B f(x)\,dx, \qquad P(Y \in B) = \int_B g(x)\,dx.$$

Regarding $X + Y$, it turns out that the probability density function of the random variable $X + Y$ is precisely the convolution of f and g. In symbol,

$$P(X + Y \in B) = \int_B f * g(x)\,dx,$$

where for any real number x

$$f * g(x) := \int_{\mathbb{R}} f(x-y)g(y)\,\mathrm{d}y$$

is the *convolution of* f *and* g evaluated at x. This example shows that the convolution appears naturally and is related to addition and multiplication.

Our second example concerns the one-dimensional heat equation. Being given a function $f : \mathbb{R} \to \mathbb{R}$, we consider the problem

$$\frac{\partial}{\partial t}u - \frac{\partial^2}{\partial x^2}u = 0, \qquad u_{|t=0} = f, \tag{3.3.1}$$

where $u : [0,\infty) \times \mathbb{R} \to \mathbb{R}$ is an unknown function of the variable (t,x). Under appropriate assumptions, we could solve this problem by using Fourier transforms to find that

$$u(t,x) = \big(K(t,\cdot) * f\big)(x)$$

that is

$$u(t,x) = \int_{\mathbb{R}} K(t,x-y)f(y)dy, \qquad \forall\,(t,x) \in (0,\infty) \times \mathbb{R},$$

where

$$K(t,x) = \frac{1}{\sqrt{4\pi t}} \exp\Big(-\frac{x^2}{4t}\Big).$$

However, the emergence of the convolution would be quite indirect. The method we will introduce now shows simply how the above convolution appears. Moreover, it relies on the *superposition principle*, which will show that the convolution is related to addition. By superposition principle, we just mean that (3.3.1) is a *linear* problem. Let us emphasize that our computations will be not rigorous at all.

Assuming existence and uniqueness of the solution u to (3.3.1), let us put $S_t(f) := u(t,\cdot)$. The starting point is that for each x_0 in $\mathbb{R}$, the function $K(\cdot,\cdot - x_0)$ is solution to

$$\frac{\partial}{\partial t}u - \frac{\partial^2}{\partial x^2}u = 0, \qquad u_{|t=0} = \delta_{x_0},$$

where δ_{x_0} is the *Dirac mass at* x_0. That is

$$S_t(\delta_{x_0})(x) = K(t,x-x_0).$$

For the sake of simplicity, let us assume that the support of f is contained in the intervall $[0,b]$. Let $h : [0,b] \to \mathbb{R}$ be a function, n be a positive integer, and $(y_k)_{0\le k\le n}$ be a subdivision of $[0, b]$. For large n,

$$\int_0^b h(x)\,\mathrm{d}x \simeq \frac{1}{n}\sum_{k=0}^{n-1} h(y_k). \tag{3.3.2}$$

The crude point is to treat the Dirac mass as a function!Then, by tacking $h := f\delta_x$ in (3.3.2), we get

$$f(x) = \int_0^b f(y)\delta_x(y)\,\mathrm{d}y \simeq \frac{1}{n}\sum_{k=0}^{n-1} f(y_k)\delta_x(y_k).$$

Since $\delta_x(y_k) = \delta_{y_k}(x)$, the superposition principle yields

$$S_t(f) \simeq \frac{1}{n}\sum_{k=0}^{n-1} f(y_k)S_t(\delta_{y_k}).$$

Thus,

$$\begin{aligned} S_t(f)(x) &\simeq \frac{1}{n}\sum_k f(y_k)K(t, x - y_k) \\ &\simeq \int_{\mathbb{R}} K(t, x-y)f(y)dy \qquad \text{(by (3.3.2))} \\ &\simeq \big(K(t,\cdot) * f\big)(x). \end{aligned}$$

3.3.2 Definitions and Basic Properties

In the sequel, $(\mathcal{X}, \|\cdot\|)$ denotes a real or complex Banach space. The following result is a consequence of the Fubini Theorem 3.1.7.

Corollary 3.3.1 *Let f lie in $L^1(\mathbb{R}, \mathcal{X})$ and g belong to $L^1(\mathbb{R}, \mathbb{K})$. Then, for almost every x in $\mathbb{R}$, the function $g(x - \cdot)f$ lies in $L^1(\mathbb{R}, \mathcal{X})$. Moreover, the function $\int_{\mathbb{R}} g(\cdot - y)f(y)\,\mathrm{d}y$ belongs to $L^1(\mathbb{R}, \mathcal{X})$.*

Proof Recalling that $g(\cdot - \cdot)f$ be the element of $L^0(\mathbb{R}^2, \mathcal{X})$ defined in Proposition 3.1.5, we readily check with the Tonelli Theorem for real-valued functions (see for instance [Bre11, Th. 4.4]) that $\|g(\cdot - \cdot)f\|_{\mathcal{X}}$ is integrable on $\mathbb{R}^2$. Thus Bochner Theorem 3.1.6 yields that $g(\cdot - \cdot)f$ lies in $L^1(\mathbb{R}^2, \mathcal{X})$. Thus, the assertions follows from Fubini Theorem 3.1.7. □

Definition 3.3.1 Let f lie in $L^1(\mathbb{R}, \mathcal{X})$ and g belong to $L^1(\mathbb{R}, \mathbb{K})$. The function $\int_{\mathbb{R}} g(\cdot - y)f(y)\,\mathrm{d}y$ of $L^1(\mathbb{R}, \mathcal{X})$ appearing in Corollary 3.3.1 is denoted by $f * g$ and is called the *convolution of* g *and* f. □

Proposition 3.3.2 *Let* b *be a positive number,* f *lie in* $L^1\big((0, b), \mathcal{X}\big)$ *and* g *belong to* $L^1\big((0, b), \mathbb{K}\big)$. *Then, for almost every* x *in* $(0, b)$, *the function* $g(x - \cdot)f$ *lies in* $L^1((0, x), \mathcal{X})$. *Moreover, the function* $x \mapsto \int_0^x g(x - y)f(y)\,\mathrm{d}y$, *denoted by* $g *_{(0,b)} f(x)$ *lies in* $L^1\big((0, b), \mathcal{X}\big)$.

The function $g *_{(0,b)} f$ is called the *the convolution of* g *and* f *on* $[0, b]$, and is abbreviated by $g * f$ if no confusion may occurs.

Proof of Proposition 3.3.2 We start by extending g and f by zero outside $(0, b)$, that is, we introduce the functions

$$G := \begin{cases} g & \text{on } (0, b) \\ 0 & \text{on } \mathbb{R} \setminus (0, b) \end{cases}, \qquad F := \begin{cases} f & \text{on } (0, b) \\ 0 & \text{on } \mathbb{R} \setminus (0, b) \end{cases}.$$

Clearly, G, F belong respectively to $L^1(\mathbb{R}, \mathbb{K})$ and $L^1(\mathbb{R}, \mathcal{X})$. From Corollary 3.3.1, the function $G(x - \cdot)F$ lies in $L^1(\mathbb{R}, \mathcal{X})$ for almost every x in $\mathbb{R}$. Thus Proposition 3.1.4 entails that its restriction to $(0, x)$ lies in $L^1((0, x), \mathcal{X})$. Since this restriction is equal to $g(x - \cdot)f$, the assertions of the current proposition are proved. □

The following proposition records some useful properties of the convolution whose proofs can be found in [ABHN11, Section 1.3].

Proposition 3.3.3 *Let* $(\mathcal{X}, \|\cdot\|)$ *be a Banach space, and* b *be a positive real number. Let also* f *lie in* $L^1\big((0, b), \mathcal{X}\big)$ *and* g, h *belong to* $L^1\big((0, b), \mathbb{K}\big)$. *Then, the following assertion hold true.*

(i) $g * h = h * g$ *in* $L^1((0, b), \mathbb{K})$.
(ii) $h * (g * f) = (h * g) * f$ *in* $L^1((0, b), \mathcal{X})$.
(iii) *If in addition,* f *lies in* $L^p((0, b), \mathcal{X})$ *for some* p *in* $[1, \infty]$, *then the following so-called Young inequality is satisfied:*

$$\|g * f\|_{L^p((0,b),\mathcal{X})} \le \|g\|_{L^1((0,b),\mathbb{K})} \|f\|_{L^p((0,b),\mathcal{X})}.$$

Remark 3.3.2 Being given a nontrivial real or complex Banach space $\mathcal{Y}$, we may define more generally *the convolution of a function belonging to* $L^1\big((0, b), \mathcal{L}(\mathcal{Y})\big)$ *by a function of* $L^1\big((0, b), \mathcal{Y}\big)$. Indeed, Definition 3.3.1 may be easily generalized to the case where g lies in $L^1\big((0, b), \mathcal{L}(\mathcal{Y})\big)$. Roughly speaking, if G lies in $L^1\big((0, b), \mathcal{L}(\mathcal{Y})\big)$ and f belongs to $L^1\big((0,b),\mathcal{Y}\big)$ then $G * f$ is the function of $L^1\big((0,b),\mathcal{Y}\big)$ defined for almost every x in $(0,b)$ by

$$G * f(x) = \int_0^x G(x-y) f(y)\,\mathrm{d}y$$

We refer to [Ama19, Th 1.9.9] for more details.

This kind of convolution will be useful to generalize the celebrated *variation of constants formula* (see Sect. 7.3.2.2). We will just use its definition and the corresponding Young inequality. □

Proposition 3.3.4 *Let* $(\mathcal{X}, \|\cdot\|)$ *be a Banach space, and* (f,g) *lie in* $W^{1,1}\big((0,b),\mathcal{X}\big) \times L^1\big((0,b),\mathbb{K}\big)$. *Then* $g * f$ *lies in* $W^{1,1}\big((0,b),\mathcal{X}\big)$ *and*

$$\frac{\mathrm{d}}{\mathrm{d}x}\{g * f\} = g * \frac{\mathrm{d}}{\mathrm{d}x} f + g(\cdot) f(0) \qquad \text{in } L^1((0,b),\mathcal{X}). \tag{3.3.3}$$

Proof We set $f' := \frac{\mathrm{d}}{\mathrm{d}x} f$. By Proposition 3.3.2, $g * f'$ is integrable on $(0,b)$. For each x in $(0,b)$, we have

$$\int_0^x g * f'(y)\,\mathrm{d}y = \int_0^x \mathrm{d}y \int_0^y g(z) f'(y-z)\,\mathrm{d}z.$$

Moreover extending g and f' by zero outside of $(0,x)$ (see the proof of Proposition 3.3.2) and using the Fubini Theorem, we get

$$\begin{aligned}
\int_0^x \mathrm{d}y \int_0^y g(z) f'(y-z)\,\mathrm{d}z &= \int_0^x g(z)\,\mathrm{d}z \int_z^x f'(y-z)\,\mathrm{d}y \\
&= \int_0^x g(z)\big(f(x-z) - f(0)\big)\,\mathrm{d}z \\
&= g * f(x) - \int_0^x g(z)\,\mathrm{d}z\, f(0).
\end{aligned}$$

We have shown that for each x in $(0,b)$,

$$g * f(x) = \int_0^x g * f'(y)\,\mathrm{d}y + \int_0^x g(z)\,\mathrm{d}z\, f(0).$$

Then, the assertions of the proposition follow by standard arguments. □

Remark 3.3.3 Arguing as in the proof of Proposition 3.3.4, we may show that if f lies in $L^1\big((0,b),\mathcal{X}\big)$ and g lives in $W^{1,1}\big((0,b),\mathbb{K}\big)$. Then $g * f$ lies in $W^{1,1}\big((0,b),\mathcal{X}\big)$ and

$$\frac{\mathrm{d}}{\mathrm{d}x}\{g * f\} = g' * f + g(0) f \qquad \text{in } L^1((0,b),\mathcal{X}).$$

□

Proposition 3.3.5 *Let* $(\mathcal{X}, \|\cdot\|)$ *be a Banach space and* p, q *be conjugate exponents in* $[1,\infty]$. *If* f *lies in* $L^p\big((0,b),\mathcal{X}\big)$ *and* g *belongs to* $L^q(0,b)$, *then* $g * f$ *admits one and only one representative having an extension in* $C\big([0,b],\mathcal{X}\big)$. *Moreover, this representative reads*

$$x \mapsto \int_0^x g'(x-y) f'(y)\, \mathrm{d}y$$

for any representatives f', g' *of* f *and* g.

Proof Since a continuous function equal to zero almost everywhere must vanish identically, we obtain the uniqueness of the representative. Regarding existence, we choose any representatives f', g' of f and g respectively. Then going back to Definition 3.1.1 and using $\mathbf{1}_B(x-\cdot) = \mathbf{1}_{x-B}$ for each B in $\mathscr{F}_{\mathrm{L},1}$, we may show that $g'(x-\cdot)f'$ is measurable for all x in $(0,b)$. Moreover by using Hölder inequality in the case where p and q are both finite, we obtain that $\|g'(x-\cdot)f'\|_{\mathcal{X}}$ is integrable on $(0,x)$. Thus, $g'(x-\cdot)f'$ lives in $\mathscr{L}^1\big((0,x),\mathcal{X}\big)$ by Bochner Theorem 3.1.6. Then, $x \mapsto \int_0^x g'(x-y)f'(y)\,\mathrm{d}y$ is a representative of $g * f$.

There remains to prove that the former function has a continuous extension on $[0,b]$. We extend f' and g' by zero outside $(0,b)$ (see the proof of Proposition 3.3.2), and we keep the notations f', g' for the extended functions.

We claim the function $x \mapsto \int_0^x g'(x-y)f'(y)\,\mathrm{d}y$ is uniformly continuous on $(0,b)$. Indeed, for any x, x' in $(0,b)$, we estimate

$$\Big\| \int_0^x g'(x-y)f'(y)\,\mathrm{d}y - \int_0^{x'} g'(x'-y)f'(y)\,\mathrm{d}y \Big\|$$

by

$$\int_{\mathbb{R}} |g'(x-y) - g'(x'-y)|\, \|f(y)\|\,\mathrm{d}y.$$

In the case where p and q are finite, using Hölder inequality and performing the change of variable $y' = x - y$, we get the following bound

$$\|g' - g'(x'-x+\cdot)\|_{L^q(\mathbb{R})} \|f\|_{L^p(\mathbb{R})}.$$

Then, the continuity of the translation in $\mathscr{L}^q(\mathbb{R})$ (see for instance [CR16, Theorem 3.58]) yields uniform continuity. On the other hand, if p or q is not finite, then we use the basic estimate

$$\int_{\mathbb{R}} |h_1(y)h_2(y)|\,\mathrm{d}y \leq \|h_1\|_{L^1(\mathbb{R})}\|h_2\|_{L^\infty(\mathbb{R})}, \qquad \forall\,(h_1, h_2) \in L^1(\mathbb{R}) \times L^\infty(\mathbb{R}).$$

instead of the Hölder inequality, and the uniform continuity of the translation in $\mathscr{L}^1(\mathbb{R})$ to derive uniform continuity. That proves the claim.

Since $\mathcal{X}$ is a complete, the uniform continuity allows to extend the function $x \mapsto \int_0^x g'(x-y)f'(y)\,\mathrm{d}y$ into a continuous function on $[0, b]$. □

Remark 3.3.4 Under the assumptions and notation of Proposition 3.3.5, it is clear that the continuous extension of

$$x \mapsto \int_0^x g'(x-y)f'(y)\,\mathrm{d}y$$

is unique. Denoting by u this extension, we claim that $u(0) = 0$. Indeed, let us assume first that q is finite. According to Bochner and Hölder inequalities, we have the estimate

$$\|u(x)\| \leq \|g\|_{L^q(0,x)}\|f\|_{L^p(\mathcal{X})}, \qquad \forall x \in (0, b).$$

By the Lebesgue dominated convergence theorem, $\|g\|_{L^p(0,x)}$ goes to zero as $x \to 0$. Hence, $u(0) = 0$. If $q = \infty$, then there exists a constant C such that $|g| \leq C$ almost everywhere on $(0, b)$. Thus

$$\|u(x)\| \leq C\|f\|_{L^1(0,x)}, \qquad \forall x \in (0, b),$$

and we conclude as in the previous case. In closing u lies in ${}_0C([0, b], \mathcal{X})$. □

Remark 3.3.5 In general, when f and g lie in $L^1\big((0, b), \mathcal{X}\big)$, $g * f$ may have a representative having an extension in $C\big([0, b], \mathcal{X}\big)$. However, we can not warrant that this extension reads

$$[0, b] \to \mathcal{X}, \qquad x \mapsto \int_0^x g'(x-y)f'(y)\,\mathrm{d}y$$

for some representatives f', g' of f and g. See Remark 3.4.1 for a counter-example. □

The following result will be used in Chap. 7 to solve *fractional differential equations*. For any function f in $L^1(0, b)$, we defined inductively the sequence $(f^{*n})_{n\geq 1}$ by setting

$$f^{*1} := f, \qquad f^{*n} := f * f^{*n-1}, \qquad \forall n \in \mathbb{N} \setminus \{0, 1\}.$$

Proposition 3.3.6 *For any f in $L^1(0, b)$, one has*

$$\lim_{n\to\infty} \|f^{*n}\|_{L^1}^{\frac{1}{n}} = 0. \tag{3.3.4}$$

Proof For each $\sigma > 0$, denote by f_σ the function of $L^1(0, b)$ defined for almost every x in $[0, b]$ by

$$f_\sigma(x) := e^{-\sigma x} f(x).$$

By induction, we show easily that for all positive integer n,

$$f^{*n} = e^{\sigma \cdot}(f_\sigma)^{*n}.$$

Thus, with Young inequality,

$$\|f^{*n}\|_{L^1} \le e^{\sigma b}\|(f_\sigma)^{*n}\|_{L^1} \le e^{\sigma b}\|f_\sigma\|_{L^1}^n.$$

By the Lebesgue dominated convergence theorem,

$$f_\sigma \xrightarrow[\sigma\to\infty]{} 0 \quad \text{in } L^1(0, b),$$

thus, for each $\varepsilon > 0$, there exists $\sigma(\varepsilon)$ in $(0, \infty)$ such that $\|f_{\sigma(\varepsilon)}\|_1 \le \varepsilon$. Then

$$\|f^{*n}\|_{L^1}^{\frac{1}{n}} \le \varepsilon e^{\sigma(\varepsilon)\frac{b}{n}},$$

and (3.3.4) follows. □

3.4 Special Functions of Fractional Calculus

3.4.1 The Euler Gamma Function

The function

$$\Gamma : (0, \infty) \to \mathbb{R}, \qquad \gamma \mapsto \int_0^\infty e^{-y} y^{\gamma-1}\, dy$$

is the *Euler Gamma function*. We recall the celebrated formulas

$$\Gamma(n+1) = n!\,, \qquad \Gamma(x+1) = x\Gamma(x)$$

which hold for every (n, x) in $\mathbb{N} \times (0, \infty)$, and

$$\int_0^1 y^{\beta-1}(1-y)^{\gamma-1}\,\mathrm{d}y = \frac{\Gamma(\beta)\Gamma(\gamma)}{\Gamma(\beta+\gamma)}, \tag{3.4.1}$$

which holds for all positive numbers β and γ (see [Die10, Section D.1]).

The formula $\Gamma(x+1) = x\Gamma(x)$ allows to extend the Euler Gamma function on $(-1, \infty) \setminus \{0\}$. This extension will be still labeled Γ.

For further reference, notice that the change of variable $y' = \frac{y}{1+y}$ in (3.4.1) gives

$$\int_0^\infty \frac{1}{y^{1-\alpha}(y+1)}\,\mathrm{d}y = \Gamma(\alpha)\Gamma(1-\alpha), \qquad \forall \alpha \in (0, 1). \tag{3.4.2}$$

3.4.2 The Riemann–Liouville Kernels

As far as *fractional calculus* is concerned, these kernels are of fundamental importance. Let b be any positive real number. For each number β in $(-1, \infty) \setminus \{0\}$, we define the so-called *Riemann–Liouville kernel of order* β as

$$g_\beta : (0, b) \to \mathbb{R}, \qquad x \mapsto \frac{1}{\Gamma(\beta)} x^{\beta-1}. \tag{3.4.3}$$

Let γ be a positive number. Then g_γ lies in $\mathscr{L}^1(0, b)$. The class of g_γ in $L^1(0, b)$ will be also denoted by g_γ. If $\gamma = 1$ then g_1 is equal to 1 almost everywhere on $(0, b)$. Notice that the dependency of g_γ on b is not relevant for our purpose. In this respect, the element of $L^1_{\mathrm{loc}}(0, \infty)$ with representative

$$(0, \infty) \to \mathbb{R}, \qquad x \mapsto \frac{1}{\Gamma(\gamma)} x^{\gamma-1}.$$

will also be denoted by g_γ.

The following *semi-group property* is very useful is our context and is a straightforward consequence of (3.4.1). For all positive numbers β and γ, one has

$$g_\beta * g_\gamma = g_{\beta+\gamma} \qquad \text{in } L^1(0, b). \tag{3.4.4}$$

In particular, for each α in $(0, 1)$,

$$g_\alpha * g_{1-\alpha} = g_1 \qquad \text{in } L^1(0, b). \tag{3.4.5}$$

Remark 3.4.1 Formula (3.4.5) implies that $g_\alpha * g_{1-\alpha}$ admits a (unique) representative having a continuous extension on $[0, b]$. This extension is nothing but the function

$$[0, b] \to \mathbb{R}, \qquad x \mapsto 1.$$

However, this extension is not the function

$$[0, b] \to \mathbb{R}, \qquad x \mapsto \int_0^x \frac{1}{\Gamma(\alpha)} y^{\gamma-1} \frac{1}{\Gamma(1-\alpha)} (x-y)^{\gamma-1}\, \mathrm{d}y$$

since this function vanishes at $x = 0$.

Finally, notice that Proposition 3.3.5 does not apply here since there do not exist conjugate exponents p, q in $[1, \infty]$ for which

$$g_\alpha \in L^p(0, b), \qquad g_{1-\alpha} \in L^q(0, b).$$

□

There is also a convenient formula for the derivatives of these kernels, namely

$$\frac{\mathrm{d}}{\mathrm{d}x} g_\gamma = g_{\gamma-1}, \qquad \forall\, \gamma \in (0, \infty) \setminus \{1\}. \tag{3.4.6}$$

3.4.3 Mittag–Leffler Functions

For each positive real numbers β and γ, the *Mittag–Leffler function of order* β *and* γ is the complex function $E_{\beta,\gamma}$, defined on $\mathbb{C}$ by

$$E_{\beta,\gamma}(z) = \sum_{j=0}^{\infty} \frac{z^j}{\Gamma(j\beta + \gamma)}, \qquad \forall\, z \in \mathbb{C}.$$

If $\gamma = 1$ then we put $E_\beta := E_{\beta,1}$ and E_β is called *the Mittag–Leffler function of order* β. If $\beta = \gamma = 1$ then $E_{1,1}$ is nothing but *the complex exponential function*. We refer to [GKMR20] for further information on Mittag–Leffler functions.

If $\mathcal{X}$ is a real or complex Banach space, then Mittag–Leffler functions extend into functions on $\mathcal{L}(\mathcal{X})$ by setting

$$E_{\beta,\gamma} : \mathcal{L}(\mathcal{X}) \to \mathcal{L}(\mathcal{X}), \qquad L \mapsto \sum_{j=0}^{\infty} \frac{L^j}{\Gamma(j\beta + \gamma)}.$$

3.5 Marchaud Fractional Derivatives

Although *Marchaud derivative* can be studied as its own right, we will view it as a tool for proving regularity of certain *fractional Riemann–Liouville primitives* (see Sect. 3.6).

Let $(\mathcal{X}, \|\cdot\|)$ be a real or complex Banach space. Being given α in $(0, 1)$, let β belong to $(\alpha, 1]$ and f lie in $C^{\beta}([0, b], \mathcal{X})$. In order to define Marchaud derivatives, we observe that there exists a constant C such that for each x in $(0, b]$,

$$\left\| g_{-\alpha}(x-y)(f(y)-f(x)) \right\| \leq C(x-y)^{\beta-\alpha-1}, \qquad \forall\, y \in (0, x). \tag{3.5.1}$$

Thus, the function

$$(0, x) \to \mathcal{X}, \qquad y \mapsto g_{-\alpha}(x-y)\big(f(y)-f(x)\big)$$

lies in $L^1((0, x), \mathcal{X})$. That legitimates the following definition.

Definition 3.5.1 Let α, β be real numbers satisfying $0 < \alpha < \beta \leq 1$, and f be in $C^{0,\beta}([0, b], \mathcal{X})$. The $\mathcal{X}$-valued function which maps any x of $(0, b]$ into

$$g_{1-\alpha} f(x) + \int_0^x g_{-\alpha}(x-y)\big(f(y)-f(x)\big)\, \mathrm{d}y,$$

is called the *Marchaud derivative of f of order α*. It is denoted by ${}^{\mathrm{M}}\mathrm{D}^{\alpha} f$. □

Definition 3.5.2 Let α belong to $(0, 1)$, ε lie in $(0, b)$ and $f : [0, b] \to \mathcal{X}$ be a continuous function on $[0, b]$. We extend f by zero on the interval $(-\infty, 0)$ and still denote by f that extension. Then, the function

$${}^{\mathrm{M}}\mathrm{D}^{\alpha}_{\varepsilon} f : (0, b] \to \mathcal{X},$$

defined for any x in $(0, b]$ by

$${}^{\mathrm{M}}\mathrm{D}^{\alpha}_{\varepsilon} f(x) := g_{1-\alpha} f(x) + \int_0^{x-\varepsilon} g_{-\alpha}(x-y)\big(f(y)-f(x)\big)\, \mathrm{d}y,$$

is called *the truncated Marchaud derivative of f of order α*. □

In the latter definition since f vanishes on $(-\infty, 0)$, one has

$${}^{\mathrm{M}}\mathrm{D}^{\alpha}_{\varepsilon} f(x) := g_{1-\alpha}(\varepsilon) f(x), \qquad \forall\, x \in (0, \varepsilon]. \tag{3.5.2}$$

Next, for each ε in $[0, b)$, we define $\Phi_\varepsilon : [0, b] \to \mathcal{X}$ by

$$\Phi_\varepsilon(x) := \int_0^{x-\varepsilon} g_{-\alpha}(x-y)\big(f(y) - f(x)\big)\,\mathrm{d}y, \qquad \forall x \in [0, b]. \tag{3.5.3}$$

We notice that for each ε in $(0, b)$

$$\Phi_\varepsilon(x) = g_{1-\alpha}(\varepsilon) f(x) - g_{1-\alpha} f(x) \qquad \forall x \in (0, \varepsilon] \tag{3.5.4}$$

$$\|\Phi_\varepsilon(x)\| \le C x^{\beta-\alpha} \qquad \forall x \in (\varepsilon, b], \tag{3.5.5}$$

where the constant C is independent of x and ε. (3.5.5) follows from (3.5.1).

Proposition 3.5.1 *Let α, β satisfy $0 < \alpha < \beta \le 1$, and f be in $C^{0,\beta}([0, b], \mathcal{X})$. Then ${}^{\mathrm{M}}\mathrm{D}^\alpha f$ and ${}^{\mathrm{M}}\mathrm{D}^\alpha_\varepsilon f$ belong to $L^1\big((0, b), \mathcal{X}\big)$, especially*

$${}^{\mathrm{M}}\mathrm{D}^\alpha_\varepsilon f \xrightarrow[\varepsilon\to 0]{} {}^{\mathrm{M}}\mathrm{D}^\alpha f \qquad \text{in } L^1\big((0, b), \mathcal{X}\big).$$

Proof Let ε be any number in $(0, b)$. By (3.5.1),

$$\|{}^{\mathrm{M}}\mathrm{D}^\alpha f(x)\| \lesssim g_{1-\alpha}(x) + x^{\beta-\alpha}, \qquad \forall x \in (0, b].$$

Thus, ${}^{\mathrm{M}}\mathrm{D}^\alpha f$ belongs to $L^1\big((0, b), \mathcal{X}\big)$. Also with (3.5.4) and (3.5.5), we deduce that ${}^{\mathrm{M}}\mathrm{D}^\alpha_\varepsilon f$ lies in $L^1\big((0, b), \mathcal{X}\big)$.

Besides, abbreviating $\|\cdot\|_{L^1((0,b),\mathcal{X})}$ by $\|\cdot\|_{(0,b)}$ and noticing that

$$\|{}^{\mathrm{M}}\mathrm{D}^\alpha f - {}^{\mathrm{M}}\mathrm{D}^\alpha_\varepsilon f\|_{(0,b)} = \|{}^{\mathrm{M}}\mathrm{D}^\alpha f - {}^{\mathrm{M}}\mathrm{D}^\alpha_\varepsilon f\|_{(0,\varepsilon)} + \|{}^{\mathrm{M}}\mathrm{D}^\alpha f - {}^{\mathrm{M}}\mathrm{D}^\alpha_\varepsilon f\|_{(\varepsilon,b)},$$

it is enough to consider the two latter norms. Regarding the norm in $L^1(0, \varepsilon)$, in view of (3.5.2), one has for each x in $(0, \varepsilon)$

$${}^{\mathrm{M}}\mathrm{D}^\alpha f(x) - {}^{\mathrm{M}}\mathrm{D}^\alpha_\varepsilon f(x) = \big(g_{1-\alpha}(x) - g_{1-\alpha}(\varepsilon)\big) f(x) + \Phi_0(x).$$

Thus, with (3.5.1)

$$\|{}^{\mathrm{M}}\mathrm{D}^\alpha f - {}^{\mathrm{M}}\mathrm{D}^\alpha_\varepsilon f\|_{(0,\varepsilon)} \lesssim \varepsilon^{1-\alpha} + \varepsilon^{\beta-\alpha}, \qquad \forall \varepsilon \in (0, b).$$

Whence

$$\|{}^{\mathrm{M}}\mathrm{D}^\alpha f - {}^{\mathrm{M}}\mathrm{D}^\alpha_\varepsilon f\|_{(0,\varepsilon)} \xrightarrow[\varepsilon\to 0]{} 0.$$

Now we estimate the norm on (ε, b) by using (3.5.1):

$$\|{}^{\mathrm{M}}\mathrm{D}^{\alpha} f - {}^{\mathrm{M}}\mathrm{D}^{\alpha}_{\varepsilon} f\|_{(\varepsilon,b)} = \|\Phi_0 - \Phi_\varepsilon\|_{(\varepsilon,b)} \lesssim \varepsilon^{\beta-\alpha}.$$

□

The next theorems are based on [SKM93, Theorem 13.2]. Before to state these results, we need to introduce some notation. For each real number y, we denote by y_+ its *positive part*, that is

$$y_+ = \begin{cases} y & \text{if } y \geq 0 \\ 0 & \text{if } y < 0 \end{cases}.$$

Then, for each α in $(0, 1)$, we define the kernel

$$\mathcal{K} : (0, \infty) \to \mathbb{R}, \qquad y \mapsto \frac{1}{\Gamma(1-\alpha)\Gamma(\alpha)} \frac{y^\alpha - (y-1)_+{}^\alpha}{y}. \tag{3.5.6}$$

Lemma 3.5.2 *Let* α *belong to* $(0, 1)$ *and* $\mathcal{K}$ *be defined by* (3.5.6). *Then* $\mathcal{K}$ *lies in* $L^1(0, \infty)$ *and*

$$\int_0^\infty \mathcal{K}(y)\,\mathrm{d}y = 1.$$

Proof The function

$$h_\alpha : [1, \infty) \to \mathbb{R}, \qquad y \mapsto \frac{y^\alpha - (y-1)^\alpha}{y}$$

satisfies for every y in $[1, \infty)$,

$$h_\alpha(y) = y^{\alpha-1} - (y-1)^{\alpha-1} + \frac{1}{(y-1)^{1-\alpha} y}.$$

Thus, for any x in $(1, \infty)$,

$$\int_1^x h_\alpha(y)\,\mathrm{d}y = \frac{1}{\alpha}\big[x^\alpha - (x-1)^\alpha - 1\big] + \int_1^x \frac{\mathrm{d}y}{(y-1)^{1-\alpha} y}.$$

Moreover, $x^\alpha - (x-1)^\alpha \to 0$ when $x \to \infty$ (since $\alpha < 1$), and by (3.4.2)

$$\int_1^\infty \frac{\mathrm{d}y}{(y-1)^{1-\alpha} y} = \int_0^\infty \frac{\mathrm{d}y}{y^{1-\alpha}(y+1)} = \Gamma(1-\alpha)\Gamma(\alpha).$$

Thus,

$$\int_1^x h_\alpha(y)\,\mathrm{d}y \underset{x\to\infty}{\longrightarrow} -\frac{1}{\alpha} + \Gamma(1-\alpha)\Gamma(\alpha).$$

Using also the positivity of h_α, the Lebesgue monotone convergence theorem entails that h_α belongs to $L^1(1,\infty)$ and

$$\int_1^\infty h_\alpha(y)\,\mathrm{d}y = -\frac{1}{\alpha} + \Gamma(1-\alpha)\Gamma(\alpha).$$

Then, the assertions of the lemma follow. □

Theorem 3.5.3 *Let α, β satisfy*

$$0 < \alpha < \beta \leq 1,$$

and f belong to $C^{0,\beta}([0,b],\mathcal{X})$. Then,

$$g_\alpha * {}^{\mathrm{M}}\mathrm{D}_\varepsilon^\alpha f \underset{\varepsilon\to 0}{\longrightarrow} f \qquad \text{in } L^1\big((0,b),\mathcal{X}\big).$$

Proof By using the splitting of Proposition 3.5.1, let us first show that

$$\|g_\alpha * {}^{\mathrm{M}}\mathrm{D}_\varepsilon^\alpha f - f\|_{L^1(\varepsilon,b)} \underset{\varepsilon\to 0}{\longrightarrow} 0. \tag{3.5.7}$$

Let ε be in $(0,b)$. Recalling that Φ_ε is defined by (3.5.3), one has for each x in (ε,b),

$$\begin{aligned} I_\varepsilon(x) &:= g_\alpha * {}^{\mathrm{M}}\mathrm{D}_\varepsilon^\alpha f(x) \\ &= g_\alpha * (g_{1-\alpha} f)(x) + \int_0^\varepsilon g_\alpha(x-y)\Phi_\varepsilon(y)\,\mathrm{d}y + \int_\varepsilon^x g_\alpha(x-y)\Phi_\varepsilon(y)\,\mathrm{d}y. \end{aligned}$$

With (3.5.4) and (3.5.3),

$$\begin{aligned} I_\varepsilon(x) = \int_\varepsilon^x & g_\alpha(x-y) g_{1-\alpha} f(y)\,\mathrm{d}y \\ &+ g_{1-\alpha}(\varepsilon)\int_0^\varepsilon g_\alpha(x-y) f(y)\,\mathrm{d}y \\ &+ \int_\varepsilon^x g_\alpha(x-y)\,\mathrm{d}y \int_0^{y-\varepsilon} g_{-\alpha}(y-t)\big(f(t)-f(y)\big)\,\mathrm{d}t. \end{aligned}$$

Let us consider the latter double integral that we will label $J_3(x)$. We split $J_3(x)$ into the difference of two integrals. Then, we apply Fubini Theorem in the first double integral. In the second double integral, we compute

$$\int_0^{y-\varepsilon} g_{-\alpha}(y-t)\,\mathrm{d}t = g_{1-\alpha}(y) - g_{1-\alpha}(\varepsilon)$$

to get two simple integrals. There results that

$$\begin{aligned} J_3(x) = \int_0^{x-\varepsilon} f(t)\,\mathrm{d}t \int_{t+\varepsilon}^{x} g_\alpha(x-y)g_{-\alpha}(y-t)\,\mathrm{d}y \\ - \int_\varepsilon^x g_\alpha(x-y)g_{1-\alpha} f(y)\,\mathrm{d}y \\ + g_{1-\alpha}(\varepsilon)\int_\varepsilon^x g_\alpha(x-y)f(y)\,\mathrm{d}y. \end{aligned}$$

Going back to $I_\varepsilon(x)$, two integrals cancel and the two integrals with coefficient $g_{1-\alpha}(\varepsilon)$ add, so that we obtain

$$\begin{aligned} I_\varepsilon(x) := g_{1-\alpha}(\varepsilon)\int_0^x g_\alpha(x-y)f(y)\,\mathrm{d}y \\ + \int_0^{x-\varepsilon} f(t)\,\mathrm{d}t \int_{t+\varepsilon}^x g_\alpha(x-y)g_{-\alpha}(y-t)\,\mathrm{d}y. \end{aligned}$$

Let us compute the latter inner integral, denoted by J. The change of variable $y' = y-t$ yields, by setting $c := x-t$

$$J = \int_\varepsilon^c g_\alpha(c-y)g_{-\alpha}(y)\,\mathrm{d}y.$$

Now changing the variable y into the variable s where

$$y = \frac{c\varepsilon}{(c-\varepsilon)s+\varepsilon}$$

gives

$$J = -g_{1-\alpha}(\varepsilon)\frac{1}{\Gamma(\alpha)}\frac{(x-\varepsilon-t)^\alpha}{x-t}.$$

Then,

$$I_\varepsilon(x) := g_{1-\alpha}(\varepsilon)\left[\int_0^x g_\alpha(x-y)f(y)\,\mathrm{d}y - \int_0^{x-\varepsilon}\frac{f(y)}{\Gamma(\alpha)}\frac{(x-\varepsilon-y)^\alpha}{x-y}\,\mathrm{d}y\right].$$

The change of variable $y = x - \varepsilon y'$ in these two integrals yields

$$\begin{aligned}
I_\varepsilon(x) &= \frac{1}{\Gamma(1-\alpha)\Gamma(\alpha)} \int_0^{x/\varepsilon} \frac{y^\alpha - (y-1)_+{}^\alpha}{y} f(x-\varepsilon y)\,\mathrm{d}y \\
&= \int_0^{x/\varepsilon} \mathcal{K}(y) f(x-\varepsilon y)\,\mathrm{d}y,
\end{aligned}$$

where $\mathcal{K}$ is given by (3.5.6). Recalling that on account of Definition 3.5.2, f is extended by zero on $(-\infty, 0)$, and using Lemma 3.5.2, we get for each x in (ε, b)

$$g_\alpha * {}^{\mathrm{M}}\mathrm{D}^\alpha_\varepsilon f(x) - f(x) = \int_0^\infty \mathcal{K}(y)\big(f(x-\varepsilon y) - f(x)\big)\,\mathrm{d}y.$$

Thus, the Fubini Theorem (real case) gives

$$\|g_\alpha * {}^{\mathrm{M}}\mathrm{D}^\alpha_\varepsilon f - f\|_{L^1((\varepsilon,b),\mathcal{X})} \le \int_0^\infty \mathcal{K}(y)\|f(\cdot - \varepsilon y) - f\|_{L^1((\varepsilon,b),\mathcal{X})}\,\mathrm{d}y.$$

By the Lebesgue theorem, the latter integral goes to zero when $\varepsilon \to 0$ since the translation is continuous on $L^1(-\infty, b)$ (see for instance [CR16, Theorem 3.58]). That proves (3.5.7).

Abbreviating $\|\cdot\|_{L^1((0,\varepsilon),\mathcal{X})}$ by $\|\cdot\|_{(0,\varepsilon)}$, there remains to show that

$$\|g_\alpha * {}^{\mathrm{M}}\mathrm{D}^\alpha_\varepsilon f - f\|_{(0,\varepsilon)} \to 0. \tag{3.5.8}$$

By *Young inequality* (Proposition 3.3.3)

$$\begin{aligned}
\|g_\alpha * {}^{\mathrm{M}}\mathrm{D}^\alpha_\varepsilon f\|_{(0,\varepsilon)} &\le \|g_\alpha\|_{(0,\varepsilon)} \|{}^{\mathrm{M}}\mathrm{D}^\alpha_\varepsilon f\|_{(0,\varepsilon)} \\
&\le g_{1+\alpha}(\varepsilon) g_{1-\alpha}(\varepsilon) \|f\|_{(0,\varepsilon)} && \text{(by (3.5.2))} \\
&\lesssim \|f\|_{(0,\varepsilon)},
\end{aligned}$$

for each ε in $(0, b)$. Since $\|f\|_{(0,\varepsilon)}$ goes to zero when $\varepsilon \to 0$, we get (3.5.8), which completes the proof of the theorem. □

Theorem 3.5.4 *Let α, β satisfy*

$$0 < \alpha < \beta \le 1,$$

and f belong to $C^{0,\beta}([0,b],\mathcal{X})$. If $f(0) = 0$ then ${}^{\mathrm{M}}\mathrm{D}^\alpha f$ extended by zero at $x = 0$ lies in $C^{0,\beta-\alpha}([0,b],\mathcal{X})$, and

$$\|{}^{\mathrm{M}}\mathrm{D}^\alpha f\|_{0,\beta-\alpha} \lesssim \|f\|_{0,\beta}.$$

Proof Since $f(0) = 0$, the function $g_{1-\alpha}f$ extended by zero at $x = 0$ lies in $C^{0,\beta-\alpha}([0,b],\mathcal{X})$ and $\|g_{1-\alpha}f\|_{0,\beta-\alpha} \le C\|f\|_{0,\beta}$ according to Lemma 3.5.5 below. Let us prove that Φ_0 belongs to $C^{0,\beta-\alpha}([0,b],\mathcal{X})$. For each x in $[0,b)$ and h in $(0,b-x)$, one has

$$\Phi_0(x+h) = \int_0^{x+h} g_{-\alpha}(x+h-y)\big(f(y)-f(x+h)\big)\,\mathrm{d}y.$$

Performing the change of variable $y' = x+h-y$ in the above integral, and splitting the resulting integral, we get

$$\begin{aligned}\Phi_0(x+h)-\Phi_0(x) &= \int_0^h g_{-\alpha}(y)\big(f(x+h-y)-f(x+h)\big)\,\mathrm{d}y\\ &\quad+\int_h^{x+h} g_{-\alpha}(y)\big(f(x+h-y)-f(x+h)\big)\,\mathrm{d}y\\ &\quad-\int_0^x g_{-\alpha}(x-y)\big(f(y)-f(x)\big)\,\mathrm{d}y\\ &=: I_1(x)+I_2(x)-I_3(x).\end{aligned}$$

The estimate of $I_1(x)$ gives

$$\|I_1(x)\| \lesssim \int_0^h |g_{-\alpha}(y)|y^\beta\,\mathrm{d}y\,|f|_{0,\beta} \lesssim h^{\beta-\alpha}|f|_{0,\beta}, \qquad \forall x\in[0,b].$$

Let us now focus on $I_2(x)-I_3(x)$. The changes of variable $y' = y-h$ in $I_2(x)$ and $y' = x-y$ in $I_3(x)$ give

$$\begin{aligned}I_2(x)-I_3(x) &= \int_0^x g_{-\alpha}(y+h)\big(f(x-y)-f(x+h)\big)\,\mathrm{d}y\\ &\quad-\int_0^x g_{-\alpha}(y)\big(f(x-y)-f(x)\big)\,\mathrm{d}y\\ &= \int_0^x f(x)\big[g_{-\alpha}(y)-g_{-\alpha}(y+h)\big]\,\mathrm{d}y\\ &\quad+\int_0^x f(x-y)\big[g_{-\alpha}(y+h)-g_{-\alpha}(y)\big]\,\mathrm{d}y\\ &\quad+\int_0^x g_{-\alpha}(y+h)\big(f(x)-f(x+h)\big)\,\mathrm{d}y\\ &= \int_0^x \big(f(x-y)-f(x)\big)\big[g_{-\alpha}(y+h)-g_{-\alpha}(y)\big]\,\mathrm{d}y\end{aligned}$$

$$
\begin{aligned}
&+ \int_0^x g_{-\alpha}(y+h)\big(f(x) - f(x+h)\big)\,\mathrm{d}y \\
&=: J_1(x) + J_2(x).
\end{aligned}
$$

The estimate of $J_1(x)$ gives

$$
\begin{aligned}
\|J_1(x)\| &\lesssim |f|_{0,\beta} \int_0^\infty y^\beta \big|g_{-\alpha}(y) - g_{-\alpha}(y+h)\big|\,\mathrm{d}y \\
&\lesssim h^{\beta-\alpha}\,|f|_{0,\beta} \int_0^\infty y^\beta \big[y^{-\alpha-1} - (y+1)^{-\alpha-1}\big]\,\mathrm{d}y.
\end{aligned}
$$

For each $y \in (0, 1]$

$$
y^\beta \big[y^{-\alpha-1} - (y+1)^{-\alpha-1}\big] \leq y^{\beta-\alpha-1} \leq 1.
$$

Thus

$$
\int_0^1 y^\beta \big[y^{-\alpha-1} - (y+1)^{-\alpha-1}\big]\,\mathrm{d}y \leq 1.
$$

For each $y \in (1, \infty)$, the *mean value theorem* entails

$$
\big[y^{-\alpha-1} - (y+1)^{-\alpha-1}\big] \leq (1+\alpha) y^{-\alpha-2}.
$$

Therefore,

$$
\int_0^\infty y^\beta \big[y^{-\alpha-1} - (y+1)^{-\alpha-1}\big]\,\mathrm{d}y \leq 1 + \frac{1+\alpha}{1+\alpha-\beta}.
$$

Whence

$$
\|J_1(x)\| \lesssim h^{\beta-\alpha} |f|_{0,\beta}, \qquad \forall x \in [0, b].
$$

Regarding $J_2(x)$, for each x in $[0, b]$, we have

$$
\begin{aligned}
\|J_2(x)\| &\lesssim h^\beta \int_0^x |g_{-\alpha}(y+h)|\,\mathrm{d}y\, |f|_{0,\beta} \\
&\lesssim h^\beta g_{1-\alpha}(h) |f|_{0,\beta} \\
&\lesssim h^{\beta-\alpha} |f|_{0,\beta}.
\end{aligned}
$$

Hence, Φ_0 belongs to $C^{0,\beta-\alpha}([0,b],\mathcal{X})$ and $|\Phi_0|_{0,\beta-\alpha} \lesssim |f|_{0,\beta}$. Since it is easily seen that $\|\Phi_0\|_{0,0} \lesssim |f|_{0,\beta}$, we derive that ${}^{\mathrm{M}}\mathrm{D}^\alpha f$ lies in $C^{0,\beta-\alpha}([0,b],\mathcal{X})$ and satisfies the expected estimate. □

In the proof of Theorem 3.5.4, we have used the following result taken from [Mus53, Chap. I, §6, 5°].

Lemma 3.5.5 *Let α, β be positive numbers and f lie in $C^{0,\beta}([0,b],\mathcal{X})$. If*

$$0 < \alpha < \beta \leq 1, \qquad f(0) = 0$$

then the function $g_{1-\alpha} f$ extended by zero at $x = 0$ lies in $C^{0,\beta-\alpha}([0,b],\mathcal{X})$, and

$$\|g_{1-\alpha} f\|_{0,\beta-\alpha} \leq C\|f\|_{0,\beta},$$

where the constant C is independent of f.

Proof Let

$$0 \leq x < b, \qquad 0 < h \leq b - x.$$

Since $f(0) = 0$, there holds

$$\|f(y)\| \lesssim y^\beta \|f\|_{0,\beta}, \qquad \forall\, y \in [0,b]. \tag{3.5.9}$$

Thus $\|g_{1-\alpha} f\|_{0,0} \lesssim \|f\|_{0,\beta}$.

If $x \leq h$, then with (3.5.9) we get

$$\begin{aligned}\|g_{1-\alpha} f(x+h) - g_{1-\alpha} f(x)\| &\leq \|g_{1-\alpha} f(x+h)\| + \|g_{1-\alpha} f(x)\| \\ &\lesssim \big((x+h)^{\beta-\alpha} + x^{\beta-\alpha}\big)\|f\|_{0,\beta} \\ &\lesssim h^{\beta-\alpha}\|f\|_{0,\beta},\end{aligned}$$

since $x \leq h$. On the other hand, if $h < x$ then

$$\begin{aligned}&\|g_{1-\alpha} f(x+h) - g_{1-\alpha} f(x)\| \\ &\quad\lesssim \frac{1}{(x+h)^\alpha}\|f(x+h) - f(x)\| + \Big(\frac{1}{x^\alpha} - \frac{1}{(x+h)^\alpha}\Big)\|f(x)\| \\ &\quad\lesssim \frac{h^\beta}{(x+h)^\alpha}\|f\|_{0,\beta} + \frac{(x+h)^\alpha - x^\alpha}{(x+h)^\alpha} x^{\beta-\alpha}\|f\|_{0,\beta}.\end{aligned}$$

Since $0 \leq x$, the first term in the latter right-hand side is less than or equal to $h^{\beta-\alpha}\|f\|_{0,\beta}$. In order to estimate the second term, we use the inequality

$$(1+y)^{\alpha}-1 \leq \alpha y, \qquad \forall\, y \in [0,\infty),$$

to obtain

$$\begin{aligned}\frac{(x+h)^{\alpha}-x^{\alpha}}{(x+h)^{\alpha}}x^{\beta-\alpha} &\leq \frac{\alpha h}{(x+h)^{\alpha}}x^{\beta-1}\\ &\leq \alpha h x^{\beta-\alpha-1}\\ &\leq \alpha h^{\beta-\alpha},\end{aligned}$$

since $h < x$ and $\beta - \alpha - 1 < 0$. Thus this second term is also less than or equal to $\alpha h^{\beta-\alpha}\|f\|_{0,\beta}$. Hence, the assertions of the lemma follow. □

3.6 Fractional Riemann–Liouville Primitives

Let $(\mathcal{X}, \|\cdot\|)$ be a real or complex Banach space.

Definition 3.6.1 Let γ be in $(0,\infty)$ and f lie in $L^1\big((0,b),\mathcal{X}\big)$. Then the convolution $g_{\gamma} * f$ is called the *Riemann-Liouville primitive of order* γ *of* f. □

We observe that the latter convolution has something to do with a primitive since in the case where $\gamma = n$ is a positive integer, one has

$$\frac{\mathrm{d}^n}{\mathrm{d}x^n}\{g_n * f\} = f. \tag{3.6.1}$$

This can be seen by using Remark 3.3.3 and (3.4.6). In particular, the Riemann–Liouville primitive of order 1 of f returns the primitive of f, which vanishes at $x = 0$.

We refer to [Die10, Section 2.1] for a comprehensive presentation of Riemann–Liouville primitives.

Lebesgue integration has a regularizing effect, that is, the primitive of a function is more regular than the function itself. For instance, starting with a continuous function, its primitives are differentiable.

In general, fractional Riemann–Liouville integration has no regularizing effect. Indeed, in the case where γ lies in $(0, 1)$, the fractional primitive of order γ of the constant function g_1 is less regular than g_1. Indeed, the semi-group property (3.4.4) gives that the primitive of order γ of g_1 is $g_{1+\gamma}$.

However, under appropriate assumptions, the convolution by g_α increases the regularity of Hölderian functions by an amount of α. We refer to Theorem 3.6.1 for a precise statement.

Its proof relies on previous results on Marchaud derivatives because the latter are well suited to Hölderian estimates. The singularity of Marchaud derivatives at the terminal point is overcome by using truncated Marchaud derivatives. Then, a continuity argument combined with Theorem 3.5.3 allows to conclude. Our presentation is based on results of [SKM93].

Theorem 3.6.1 *Let α, β be positive real numbers such that*

$$0 < \alpha < 1, \qquad 0 < \beta \leq 1, \qquad 1 < \alpha + \beta.$$

If f belongs to $C^\beta([0, b], \mathcal{X})$ and $f(0) = 0$ then $g_\alpha * f$ lies in $C^{\alpha+\beta}([0, b], \mathcal{X})$, and

$$\|g_\alpha * f\|_{\alpha+\beta} \lesssim \|f\|_\beta. \tag{3.6.2}$$

Proof Since $0 < 1 - \alpha < \beta \leq 1$, Theorem 3.5.3 yields that

$$g_{1-\alpha} * {}^{\mathrm{M}}\mathrm{D}_\varepsilon^{1-\alpha} f \underset{\varepsilon \to 0}{\longrightarrow} f \qquad \text{in } L^1\big((0, b), \mathcal{X}\big).$$

By Proposition 3.5.1,

$${}^{\mathrm{M}}\mathrm{D}_\varepsilon^{1-\alpha} f \underset{\varepsilon \to 0}{\longrightarrow} {}^{\mathrm{M}}\mathrm{D}^{1-\alpha} f \qquad \text{in } L^1\big((0, b), \mathcal{X}\big).$$

Moreover, Young inequality (see Proposition 3.3.3) entails that the map

$$L^1\big((0, b), \mathcal{X}\big) \to L^1\big((0, b), \mathcal{X}\big), \qquad h \mapsto g_{1-\alpha} * h$$

is continuous on $L^1\big((0, b), \mathcal{X}\big)$, thus

$$g_{1-\alpha} * {}^{\mathrm{M}}\mathrm{D}^{1-\alpha} f = f.$$

Convoluting the two sides of the above equality by g_α and using the semi-group property (3.4.5), we get

$$g_1 * {}^{\mathrm{M}}\mathrm{D}^{1-\alpha} f = g_\alpha * f. \tag{3.6.3}$$

By assumptions, f is Hölder continuous with exponent β and $f(0) = 0$, thus Theorem 3.5.4 implies that ${}^{\mathrm{M}}\mathrm{D}^{1-\alpha} f$ lies in $C^{\alpha+\beta-1}[0, b]$. With (3.6.3), we conclude that $g_\alpha * f$ belongs to $C^{\alpha+\beta}[0, b]$.

Let us now estimate the Hölder norm of $g_\alpha * f$. By (3.6.3),

$$\begin{aligned}\|g_\alpha * f\|_{\alpha+\beta} &= \|g_1 * {}^{\mathrm{M}}\mathrm{D}^{1-\alpha} f\|_{\alpha+\beta}\\ &\lesssim \|{}^{\mathrm{M}}\mathrm{D}^{1-\alpha} f\|_{\alpha+\beta-1}\\ &\lesssim \|f\|_\beta\end{aligned}$$

according to Theorem 3.5.4. □

The forthcoming result is similar to Theorem 3.6.1 but applies when $\alpha+\beta \le 1$.

Theorem 3.6.2 *Let (α, β) lie in $(0,1)\times(0,1)$ and f belong to $C^{0,\beta}([0,b],\mathcal{X})$ with $f(0)=0$. If $\alpha+\beta<1$ then $g_\alpha * f$ lies in $C^{0,\alpha+\beta}([0,b],\mathcal{X})$, and*

$$\|g_\alpha * f\|_{0,\alpha+\beta} \lesssim \|f\|_\beta. \tag{3.6.4}$$

If $\alpha+\beta=1$ then $g_\alpha * f$ lies in $C_{\ln}([0,b],\mathcal{X})$, and

$$\|g_\alpha * f\|_{C_{\ln}} \lesssim \|f\|_\beta.$$

Proof Let

$$0 \le x < b, \qquad 0 < h \le b-x.$$

Then,

$$\begin{aligned}&g_\alpha * f(x+h) - g_\alpha * f(x)\\ &= \big[g_{\alpha+1}(x+h) - g_{\alpha+1}(x)\big] f(x) + \int_{-h}^{0} g_\alpha(y+h)\big[f(x-y)-f(x)\big]\,\mathrm{d}y\\ &+ \int_0^x \big[g_\alpha(y+h)-g_\alpha(y)\big]\big[f(x-y)-f(x)\big]\,\mathrm{d}y\\ &=: J_1 + J_2 + J_3.\end{aligned}$$

Let us show that

$$\|J_1\| \lesssim h^{\alpha+\beta}\,|f|_{0,\beta}. \tag{3.6.5}$$

Indeed, if $x \le h$ then since $g_{\alpha+1}$ is Hölderian of order α and $f(0)=0$,

$$\|J_1\| \lesssim h^\alpha x^\beta\,|f|_{0,\beta} \lesssim h^{\alpha+\beta}\,|f|_{0,\beta}.$$

If $h < x$, then using the inequality

$$(1+y)^\alpha - 1 \le \alpha y, \qquad \forall\, y \in [0,\infty),$$

we get

$$\|J_1\| \lesssim \alpha h x^{\alpha-1} x^\beta |f|_{0,\beta} \lesssim h^{\alpha+\beta} |f|_{0,\beta},$$

since $\alpha + \beta - 1 \le 0$. Hence (3.6.5) holds true.

The estimate of J_2 is straightforward:

$$\|J_2\| \lesssim h^\beta g_{\alpha+1}(h) |f|_{0,\beta} \lesssim h^{\alpha+\beta} |f|_{0,\beta}.$$

Regarding J_3, we have

$$\|J_3\| \lesssim \int_0^x \left[y^{\alpha-1} - (y+h)^{\alpha-1}\right] y^\beta \, \mathrm{d}y \, |f|_{0,\beta}.$$

The change of variable $y' = h^{-1} y$ gives

$$\|J_3\| \lesssim h^{\alpha+\beta} \int_0^{\frac{x}{h}} \left[y^{\alpha-1} - (y+1)^{\alpha-1}\right] y^\beta \, \mathrm{d}y \, |f|_{0,\beta}.$$

Setting $m := \max(1, \frac{x}{h})$, the estimate of the latter integral goes as follows. We split this integral into the sum of two integrals over the intervals $(0, 1)$ and $(1, m)$. We bound the second integral by the *mean value theorem* to get

$$\int_0^{\frac{x}{h}} \left[y^{\alpha-1} - (y+1)^{\alpha-1}\right] y^\beta \, \mathrm{d}y \le \frac{1}{\alpha} + (1-\alpha) \int_1^m y^{\alpha+\beta-2} \, \mathrm{d}y. \tag{3.6.6}$$

If $\alpha + \beta < 1$ then the previous integral is bounded uniformly with respect to m in $[1, \infty)$. Hence

$$\|J_3\| \lesssim h^{\alpha+\beta} |f|_{0,\beta},$$

and (3.6.4) follows. Other the other hand, if $\alpha + \beta = 1$ then in view of Sect. 3.2.3, we restrict our attention to those x and h for which (x, h) lies in $\mathscr{R}$. As a consequence, $h \le \frac{1}{2}$. Thus, starting from (3.6.6) and in a second time, estimating m by $1 + \frac{x}{h}$, we have

$$\begin{aligned}\int_0^{\frac{x}{h}} \left[y^{\alpha-1} - (y+1)^{\alpha-1}\right] y^{1-\alpha}\, dy &\le \frac{1}{\alpha} + (1-\alpha)\ln m \\ &\le \frac{1}{\alpha} + (1-\alpha)\big(\ln(\tfrac{1}{2}+b) + |\ln h|\big) \\ &\le C|\ln h|,\end{aligned}$$

where the constant C is independent of x, h and f. Thus,

$$\|J_3\| \lesssim h|\ln h|\,|f|_{0,\beta}.$$

There results that

$$\|g_\alpha * f(x+h) - g_\alpha * f(x)\| \lesssim h|\ln h|\,|f|_{0,\beta}.$$

Then the assertions of the theorem follow. □

The next theorem does not rely on Marchaud derivatives, it uses only the basic regularity result featured in Proposition 3.3.5 and Remark 3.3.4. However, its proof looks like the previous proof.

Theorem 3.6.3 *Let $(\mathcal{X}, \|\cdot\|)$ be a Banach space, p and q be conjugate exponents in $(1,\infty)$, and f belong to $L^p\big((0,b),\mathcal{X}\big)$. Assume in addition that a number α satisfies*

$$\frac{1}{p} < \alpha < 1 + \frac{1}{p}.$$

Then $g_\alpha * f$ lies in ${}_0C^{\alpha-\frac{1}{p}}([0,b],\mathcal{X})$, and

$$\|g_\alpha * f\|_{\alpha-\frac{1}{p}} \le C\|f\|_{L^p(\mathcal{X})},$$

where the constant C is independent of f.

Proof Since $\frac{1}{p} < \alpha$, g_α lies in $L^q(0,b)$. Hence $g_\alpha * f$ belongs to ${}_0C([0,b],\mathcal{X})$ according to Remark 3.3.4. By Hölder inequality, the norm of $g_\alpha * f$ in that space is bounded up to a multiplicative constant by $\|f\|_{L^p(0,b)}$. Let us now estimate the semi-norm $|g_\alpha * f|_{0,\alpha-\frac{1}{p}}$. For any x in $[0,b)$ and h in $(0,b-x]$, we compute

$$\begin{aligned} g_\alpha * f(x+h) - g_\alpha * f(x) &= \int_x^{x+h} g_\alpha(x+h-y) f(y)\,\mathrm{d}y \\ &\quad + \int_0^x \big[g_\alpha(x+h-y) - g_\alpha(x-y)\big] f(y)\,\mathrm{d}y \\ &=: \quad I_1 \quad + \quad I_2. \end{aligned}$$

The change of variable $y' = x + h - y$ in I_1 and the Hölder inequality yield

$$\|I_1\| \leq C h^{\alpha - \frac{1}{p}} \|f\|_{L^p(\mathcal{X})}.$$

Regarding I_2, the Hölder inequality gives

$$\begin{aligned} \|I_2\| &\leq \left(\int_0^x |g_\alpha(x+h-y) - g_\alpha(x-y)|^q \,\mathrm{d}y \right)^{\frac{1}{q}} \|f\|_{L^p(\mathcal{X})} \\ &\leq J_1 \|f\|_{L^p(\mathcal{X})}. \end{aligned}$$

Performing the change of variable $y' = \frac{x-y}{h}$ in J_1, we get

$$\|J_1\| \leq h^{\alpha - \frac{1}{p}} \left(\int_0^{\frac{x}{h}} (y^{\alpha-1} - (y+1)^{\alpha-1})^q \,\mathrm{d}y \right)^{\frac{1}{q}}.$$

If $x \leq h$ then

$$\|J_1\| \leq C h^{\alpha - \frac{1}{p}}.$$

On the other hand, if $h < x$ then the *mean value theorem* entails

$$\|J_1\| \leq C h^{\alpha - \frac{1}{p}} \left(1 + \int_1^\infty y^{(\alpha-2)q} \,\mathrm{d}y \right)^{\frac{1}{q}}.$$

Since the latter integral converges due to the assumption $\alpha < 1 + \frac{1}{p}$, we get $\|J_1\| \leq C h^{\alpha - \frac{1}{p}}$. Then the assertions of the proposition follow easily. □

The methods used in the proof of the previous theorem allow to obtain the following result.

Lemma 3.6.4 *Let α belong to $(\frac{1}{2}, 1)$, and denote by Sg_α the function $g_\alpha(b - \cdot)$. Then, the function $g_\alpha * Sg_\alpha$ lies in $C^{2\alpha-1}([0, b])$. However, for each number $\gamma > 2\alpha - 1$, $g_\alpha * Sg_\alpha$ does not belong to $C^\gamma([0, b])$.*

Proof Let us first show that $g_\alpha * Sg_\alpha$ lies in $C^{2\alpha-1}([0, b])$. For any x in $[0, b)$ and h in $(0, b - x]$, we compute

$$
\begin{aligned}
g_\alpha * Sg_\alpha(x+h) - g_\alpha * Sg_\alpha(x) &= \int_x^{x+h} g_\alpha(x+h-y) g_\alpha(b-y)\,\mathrm{d}y \\
&+ \int_0^x \big[g_\alpha(x+h-y) - g_\alpha(x-y)\big] g_\alpha(b-y)\,\mathrm{d}y \\
&=: \quad I_1 \qquad + \qquad I_2.
\end{aligned}
$$

Since $x + h \le b$ and $\alpha < 1$, one has

$$
|I_1| \le \int_x^{x+h} g_\alpha(x+h-y)^2\,\mathrm{d}y.
$$

Then, the change of variable $y' = \frac{x+h-y}{h}$ gives $|I_1| \le Ch^{2\alpha-1}$.

Regarding I_2, estimating $g_\alpha(b-y)$ by $g_\alpha(x-y)$, and performing the change of variable $y' = \frac{x+h-y}{h}$, we derive

$$
|I_2| \le Ch^{2\alpha-1} \int_0^{\frac{x}{h}} \big(y^{\alpha-1} - (y+1)^{\alpha-1}\big) y^{\alpha-1}\,\mathrm{d}y.
$$

Then arguing as in the proof of Theorem 3.6.3, we may show that the latter integral is bounded uniformly with respect to x and h. Thus, $|I_2| \le Ch^{2\alpha-1}$. Hence the Hölder regularity of $g_\alpha * Sg_\alpha$ is at least equal to $2\alpha - 1$.

In order to show that the Hölder regularity of $g_\alpha * Sg_\alpha$ cannot be larger than $2\alpha - 1$, it is enough to prove that the function

$$
[0, b) \to \mathbb{R}, \qquad x \mapsto \frac{g_\alpha * Sg_\alpha(b) - g_\alpha * Sg_\alpha(x)}{(b-x)^\gamma}
$$

goes to $+\infty$ when x converges toward b. For each x in $[\frac{b}{2}, b)$, we compute

$$
\begin{aligned}
g_\alpha * Sg_\alpha(b) - g_\alpha * Sg_\alpha(x) &= \int_0^b g_\alpha(y)^2\,\mathrm{d}y - \int_0^x g_\alpha(y) g_\alpha(b-x+y)\,\mathrm{d}y \\
&= \int_x^b g_\alpha(y)^2\,\mathrm{d}y + \int_0^x \big[g_\alpha(y) - g_\alpha(b-x+y)\big] g_\alpha(y)\,\mathrm{d}y \\
&\ge \int_0^x \big[g_\alpha(y) - g_\alpha(b-x+y)\big] g_\alpha(y)\,\mathrm{d}y.
\end{aligned}
$$

Denoting the latter integral by I_1 and setting $h := b - x$, the change of variable $y' = \frac{y+h}{h}$ in I_1 gives

$$I_1 = \frac{h^{2\alpha-1}}{\Gamma(\alpha)^2} \int_1^{\frac{b}{h}} \left[(y-1)^{\alpha-1} - y^{\alpha-1}\right](y-1)^{\alpha-1}\, \mathrm{d}y.$$

Since $\frac{b}{h} \geq 2$, setting $c := \int_1^2 \left[(y-1)^{\alpha-1} - y^{\alpha-1}\right](y-1)^{\alpha-1}\, \mathrm{d}y$, we derive that for each x in $[\frac{b}{2}, b)$,

$$\frac{g_\alpha * Sg_\alpha(b) - g_\alpha * Sg_\alpha(x)}{(b-x)^\gamma} \geq \frac{c}{\Gamma(\alpha)^2}(b-x)^{-\gamma+2\alpha-1}.$$

Since $c > 0$, we deduce that $g_\alpha * Sg_\alpha$ does not belong to $C^\gamma([0, b])$. □

Differential Triplets on Hilbert Spaces

4

In the theory of differential operators acting on functions of the one variable, $\frac{d}{dx}$ is the building block. We may supplement $\frac{d}{dx}$ with homogeneous linear boundary conditions, construct higher-order differential operators (as the Laplace operator), study the properties of these operators (invertibility, closedness), and consider their adjoint operator. Also various differential equations may be investigated.

In practice, the operator $\frac{d}{dx}$ can be seen either as an operator acting on functions of the *space* variable or as an operator acting on functions of the *time* variable (in which case we prefer to write $\frac{d}{dt}$). The *transport equation*

$$\partial_t u = \partial_x u$$

is a basic example where these two operators are involved. It turns out that this equation generalizes into the so-called Delsarte equation [Del38]

$$B_t u = B_x u,$$

where B is a suitable operator acting on functions of one variable and u is a function of the two variables t and x, as in the transport equation. The notation $B_t u$ means that the operator B acts on the function $t \mapsto u(t, x)$. Referring to [ER22], the idea is to take $B_t u$ as the *Caputo derivative of* u. This leads us to consider the Caputo derivative not only as a *time* derivative but rather more generally, as an *operator* acting on functions of one variable. This simple change of point of view is at the origin of *the theory of differential triples* that we are going to introduce now (see also [ER21]).

In a Hilbertian setting, the entire Riemann-Liouville and Caputo derivatives, seen as operators, share the following properties: they possess a right inverse (this is the first

A. Rougirel, *Unified Theory for Fractional and Entire Differential Operators*,
Frontiers in Elliptic and Parabolic Problems,
https://doi.org/10.1007/978-3-031-58356-8_4

invariant of the theory; see Sect. 1.3) which is related to its adjoint thru the symmetry $S : u \mapsto u(b - \cdot)$ (this leads to the second invariant). Notice that the existence of a right inverse is nothing but a generalization of 0the fundamental theorem of calculus.

Written in an abstract framework, these properties become axioms defining the class of *differential triplets* (Sect. 4.2). A differential triplet $(\mathbb{A}, \mathbb{B}, \mathbb{S})$ is made up of three operators acting on a Hilbert space $\mathcal{H}$. $\mathbb{B}$ is a right inverse of $\mathbb{A}$, and $\mathbb{S}$ is a symmetry. In applications, the so-called *maximal operator* $\mathbb{A}$ is a fractional differential operator.

The second class consists in *boundary restriction operators*. Roughly speaking, *an abstract boundary restriction operator* is *the maximal operator* supplemented with *homogeneous linear boundary conditions*. Since these operators live in an abstract Hilbert space, boundary values cannot be invoked at this stage. Hence, the axioms property defining boundary restriction operators rely on adjoint operators. This point is investigated in Sect. 4.3.

In the case where the kernel of the maximal operator has finite dimension, *the Domain Structure Theorem* (Theorem 4.3.14) allows to define the so-called abstract endogenous boundary conditions (see Sect. 4.4). Let us start to give a flavor of our theory in the simple case where the maximal operator is the first-order derivative $\frac{\mathrm{d}}{\mathrm{d}x}$.

4.1 The Analysis of the Operator $\frac{\mathrm{d}}{\mathrm{d}x}$ Revisited

Let $\mathbb{K}$ be equal to $\mathbb{R}$ or $\mathbb{C}$. In this subsection, after defining the operator $\frac{\mathrm{d}}{\mathrm{d}x}$ on the Hilbert space $L^2(0, b)$, we will review some of its basic properties. Our analysis is *revisited* in the sense that our proofs are transportable in the abstract context of *differential triplets* and *boundary restriction operators*. In particular, in applications, we may replace $\frac{\mathrm{d}}{\mathrm{d}x}$ by the *fractional operators* introduced in Sect. 6.2 and keep essentially the same proofs.

4.1.1 The Operator $\mathrm{D}^1_{L^2}$ and the Fundamental Theorem of Calculus

With the notation of Example 2.5.2, let

$$\mathrm{D}^1_{L^2} : H^1(0, b) \subseteq L^2(0, b) \to L^2(0, b), \qquad u \mapsto \frac{\mathrm{d}}{\mathrm{d}x}u, \tag{4.1.1}$$

where $\frac{\mathrm{d}}{\mathrm{d}x}u$, that we will also denote by u', is the derivative of u in the sense of distributions.

An important common feature between this operator and, let us say, *Caputo* or *Riemann-Liouville operators* is the existence of a bounded right inverse. Regarding $\mathrm{D}^1_{L^2}$, the operator

$$B_{1,L^2} : L^2(0, b) \to L^2(0, b)$$

which maps any v in $L^2(0, b)$ into the function defined by

$$B_{1,L^2}v(x) := \int_0^x v(y)\,\mathrm{d}y, \qquad \forall x \in [0, b],$$

turns out to be a right inverse of $\mathrm{D}^1_{L^2}$, as stated by *the fundamental theorem of calculus*. That is

$$\mathrm{D}^1_{L^2} \circ B_{1,L^2} = \mathrm{id}_{L^2(0,b)}. \tag{4.1.2}$$

The fundamental theorem of calculus yields that each u in $H^1(0, b)$ satisfies

$$u(x) = u(0) + \int_0^x u'(y)\,\mathrm{d}y, \qquad \forall x \in [0, b].$$

We rewrite this identity in a functional analytic fashion as

$$u = u(0)g_1 + B_{1,L^2}\mathrm{D}^1_{L^2}u, \qquad \forall u \in H^1(0, b), \tag{4.1.3}$$

where g_1 denotes the function of $L^2(0, b)$ with value 1 almost everywhere.

As it can be seen with Fubini theorem, the adjoint

$$(B_{1,L^2})^* : L^2(0, b) \to L^2(0, b)$$

of B_{1,L^2} satisfies

$$(B_{1,L^2})^* w(x) = \int_x^b w(y)\,\mathrm{d}y, \qquad \forall x \in [0, b]. \tag{4.1.4}$$

Thus

$$-\mathrm{D}^1_{L^2} \circ (B_{1,L^2})^* = \mathrm{id}_{L^2(0,b)}. \tag{4.1.5}$$

Notice that the latter identity encapsulate the inclusion $R\big((B_{1,L^2})^*\big) \subseteq -\mathrm{D}^1_{L^2}$. Especially, observe the analogy between (4.1.2) and the equality of (4.1.5). At an abstract Hilbertian level, (4.1.5) and (4.1.2) will be generalized respectively through (4.2.2) and (4.2.6).

4.1.2 Revisited Proof of Basic Results

Our first result concerns the density of the domain of $\mathrm{D}^1_{L^2}$, a well-known fact. However, as explained above, the forthcoming proof of that result is "boundary restriction operator-oriented" and may be generalized to *fractional operators*.

In order to prove that $H^1(0, b)$ is dense in $L^2(0, b)$, it is enough to show by Corollary 2.2.10 that any h in $L^2(0, b)$ satisfying

$$(h, u) = 0, \qquad \forall u \in H^1(0, b)$$

is trivial. For any v in $L^2(0, b)$, the function $B_{1,L^2}v$ lies in $H^1(0, b)$ according to (4.1.2). Thus by choosing $u := B_{1,L^2}v$ in the above identity, we get

$$0 = (h, B_{1,L^2}v) = \big((B_{1,L^2})^* h, v\big), \qquad \forall v \in L^2(0, b).$$

Then $(B_{1,L^2})^* h = 0$, and by (4.1.5), $h = 0$, so that $\mathrm{D}^1_{L^2}$ is densely defined.

Our second result is the computation of the adjoint of $\mathrm{D}^1_{L^2}$. More precisely, we will establish that

$$D\big((\mathrm{D}^1_{L^2})^*\big) = H^1_0(0, b), \qquad (\mathrm{D}^1_{L^2})^* w = -w', \qquad \forall w \in H^1_0(0, b), \tag{4.1.6}$$

where $H^1_0(0, b)$ is the subspace of $H^1(0, b)$ whose elements admit a continuous representative vanishing at $x = 0$ and $x = b$. In order to show (4.1.6), we first claim that

$$(B_{1,L^2})^* \circ (\mathrm{D}^1_{L^2})^* = \mathrm{id}_{D((\mathrm{D}^1_{L^2})^*)}. \tag{4.1.7}$$

Indeed, for each f in $D\big((\mathrm{D}^1_{L^2})^*\big)$, one has for any h in $L^2(0, b)$

$$\begin{aligned}\int_0^b (B_{1,L^2})^* (\mathrm{D}^1_{L^2})^* f(x)\overline{h(x)}\, \mathrm{d}x &= \big((B_{1,L^2})^* (\mathrm{D}^1_{L^2})^* f,\ h\big)\\ &= \big((\mathrm{D}^1_{L^2})^* f,\ B_{1,L^2}h\big)\\ &= \big(f,\ \mathrm{D}^1_{L^2} B_{1,L^2}h\big),\end{aligned}$$

since $B_{1,L^2}h$ lies in the domain of $\mathrm{D}^1_{L^2}$ by (4.1.2). Then still by (4.1.2), we get for any h in $L^2(0, b)$

$$\int_0^b (B_{1,L^2})^* (\mathrm{D}^1_{L^2})^* f(x)\overline{h(x)}\, \mathrm{d}x = \int_0^b f(x)\overline{h(x)}\, \mathrm{d}x,$$

which proves the claim.

Secondly, denoting by $g_1{}^{\perp}$ the annihilator of g_1 in $L^2(0,b)$ (see (2.5.4) and/or Example 2.5.2), we will prove that

$$D\big((\mathrm{D}^1_{L^2})^*\big) = (B_{1,L^2})^*(g_1{}^{\perp}). \tag{4.1.8}$$

For, by (2.2.16),

$$\int_0^b f(x)\overline{u'(x)}\,\mathrm{d}x = \int_0^b (\mathrm{D}^1_{L^2})^* f(x)\overline{u(x)}\,\mathrm{d}x,$$
$$\forall (u,f) \in H^1(0,b) \times D\big((\mathrm{D}^1_{L^2})^*\big). \tag{4.1.9}$$

Choosing $u = g_1$ in the above identity, we get

$$0 = \int_0^b (\mathrm{D}^1_{L^2})^* f(x) g_1(x)\,\mathrm{d}x, \qquad \forall f \in D\big((\mathrm{D}^1_{L^2})^*\big).$$

Thus with (4.1.7)

$$f = (B_{1,L^2})^*(\mathrm{D}^1_{L^2})^* f \in (B_{1,L^2})^*(g_1{}^{\perp}).$$

Conversely, for each $\psi^{\perp}$ in $g_1{}^{\perp}$, the function $f := (B_{1,L^2})^*\psi^{\perp}$ lies in $(B_{1,L^2})^*(g_1{}^{\perp})$. Now, for any u in $H^1(0,b)$, we compute

$$\int_0^b f(x)\overline{u'(x)}\,\mathrm{d}x = \int_0^b \psi^{\perp}(x)\overline{B_{1,L^2}\,u'(x)}\,\mathrm{d}x$$
$$= \int_0^b \psi^{\perp}(x)\big(\overline{u(x)} - \overline{u(0)}\big)\,\mathrm{d}x \qquad \text{(by (4.1.3))} \tag{4.1.10}$$
$$= \int_0^b \psi^{\perp}(x)\overline{u(x)}\,\mathrm{d}x, \tag{4.1.11}$$

since $\int_0^b \psi^{\perp}(x)\,\mathrm{d}x = 0$. Thus, f belongs to the domain of $(\mathrm{D}^1_{L^2})^*$ and $(\mathrm{D}^1_{L^2})^* f = \psi^{\perp}$. That proves (4.1.8) and also that

$$(\mathrm{D}^1_{L^2})^* \circ (B_{1,L^2})^*\psi^{\perp} = \psi^{\perp}, \qquad \forall \psi^{\perp} \in g_1{}^{\perp}. \tag{4.1.12}$$

Next let f lie in $D\big((\mathrm{D}^1_{L^2})^*\big)$. By (4.1.8), $f = (B_{1,L^2})^*\psi^{\perp}$ for some $\psi^{\perp}$ in $g_1{}^{\perp}$. With (4.1.5), we get that f lies in $H^1(0,b)$ and $f' = -\psi^{\perp}$. Besides by (4.1.12), $(\mathrm{D}^1_{L^2})^* f = \psi^{\perp}$. Hence $(\mathrm{D}^1_{L^2})^* f = -f'$. In order to finish the proof of (4.1.6), we observe that the right-hand side of (4.1.8) is nothing but $H^1_0(0,b)$.

Remark 4.1.1 We would like to discuss the differences between the representations (4.1.6) and (4.1.8) of the domain of $(\mathrm{D}^1_{L^2})^*$. The standard representation (4.1.6) introduces the boundary values $u(0)$ and $u(b)$. The generalization of these boundary values in an abstract setting usually requires the introduction of extra spaces and structures (see, for instance, [Cal39, Ryz07]). However, since the domain of $(\mathrm{D}^1_{L^2})^*$ is contained in $L^2(0, b)$, we may expect a representation by means of objects directly linked to the spaces $L^2(0, b)$ and $H^1(0, b)$. This is precisely what happens in (4.1.8).

In summary, the representation (4.1.8) makes use of objects living in $\mathcal{H} := L^2(0, b)$, so that this representation is susceptible of generalization to an abstract setting (see Sect. 4.4). At the same time, it encodes a kind of homogeneous linear boundary conditions the so-called endogenous boundary conditions (Sect. 4.4.1). □

4.2 Differential Triplets

Let $\mathbb{K}$ be equal to $\mathbb{R}$ or $\mathbb{C}$ and $\mathcal{H}$ be a Hilbert space over $\mathbb{K}$.

4.2.1 Definitions

Let us introduce the fundamental object of our unified theory.

Definition 4.2.1 Let $\mathbb{A}$, $\mathbb{B}$, and $\mathbb{S}$ be three operators on $\mathcal{H}$. The triplet $(\mathbb{A}, \mathbb{B}, \mathbb{S})$ is called an *(abstract) differential triplet on* $\mathcal{H}$ if

$$\mathbb{B} \in \mathcal{L}(\mathcal{H}) \tag{4.2.1}$$

$$\mathbb{A}\mathbb{B} = \mathrm{id}_{\mathcal{H}} \tag{4.2.2}$$

$$\mathbb{S} \in \mathcal{L}(\mathcal{H}) \text{ and } \mathbb{S}^* = \mathbb{S},\ \mathbb{S}^2 = \mathrm{id}_{\mathcal{H}} \tag{4.2.3}$$

$$\mathbb{B}^* = \mathbb{S}\mathbb{B}\mathbb{S}. \tag{4.2.4}$$

The operator $\mathbb{A}$ is said to be *maximal*. A differential triplet $(\mathbb{A}, \mathbb{B}, \mathbb{S})$ is *closed* if $\mathbb{A}$ is a closed operator on $\mathcal{H}$. Besides, any operator $\mathbb{S}$ satisfying (4.2.3) is called a *symmetry of* $\mathcal{H}$. □

In a differential triplet $(\mathbb{A}, \mathbb{B}, \mathbb{S})$, $\mathbb{A}$ is meant to represent some fractional or entire differential operator on $L^2(0, b)$. Expressed at an abstract level, the key property shared by these operators is

$$\mathbb{A}^* \subseteq \mathbb{S}\mathbb{A}\mathbb{S}. \tag{4.2.5}$$

When $\mathbb{A} := \mathrm{D}^1_{L^2}$ (see (4.1.1)) and $\mathbb{S}v(x) := v(b - x)$ for a.e. x in $(0, b)$ and all v in $L^2(0, b)$, we readily get that (4.1.6) entails (4.2.5).

It turns out that (4.2.4) implies (4.2.5). The advantage of (4.2.4) is that it is much more easier to check than (4.2.5). The proof of this implication is given in Proposition 4.3.5; however, it may be obtained in a non-rigorous way: starting from (4.2.2), we derive $\mathbb{B}^*\mathbb{A}^* = \mathrm{id}_{\mathcal{H}}$. Thus

$$\begin{aligned} \mathbb{A}^* \subseteq (\mathbb{B}^*)^{-1} &= \mathbb{S}\mathbb{B}^{-1}\mathbb{S} && \text{(by (4.2.4), (4.2.3))} \\ &\subseteq \mathbb{S}\mathbb{A}\mathbb{S} && \text{(by (4.2.2)).} \end{aligned}$$

Remark 4.2.2 By the definition of the equality of operators, (4.2.2) means that

$$D(\mathbb{A}\mathbb{B}) = \mathcal{H} \qquad \text{and} \qquad \mathbb{A}\mathbb{B}v = v, \qquad \forall\, v \in D(\mathbb{A}\mathbb{B}).$$

Since, by definition, $D(\mathbb{A}\mathbb{B})$ is equal to $\mathbb{B}^{-1}D(\mathbb{A})$, there results that

$$D(\mathbb{A}\mathbb{B}) = \mathcal{H} \iff R(\mathbb{B}) \subseteq D(\mathbb{A}).$$

In practice, the latter condition is often more convenient to prove than the former: see, for instance, the proof of Proposition 4.2.1. □

Let us give some obvious facts about differential triplets.

Proposition 4.2.1 *Let $\mathcal{T} := (\mathbb{A}, \mathbb{B}, \mathbb{S})$ be a differential triplet on $\mathcal{H}$. Then:*

(i) $\mathbb{S}^{-1} = \mathbb{S}$;
(ii) $R(\mathbb{B}^*) = \mathbb{S}R(\mathbb{B})$;
(iii) $D(\mathbb{S}\mathbb{A}\mathbb{S}) = \mathbb{S}D(\mathbb{A})$;
(iv) $(\mathbb{S}\mathbb{A}\mathbb{S}, \mathbb{B}^*, \mathbb{S})$ *is a differential triplet on* $\mathcal{H}$.

Proof (4.2.3) entails that $\mathbb{S}$ is invertible and (i) follows. Regarding (ii), (4.2.4) and Item (i) yield that $R(\mathbb{B}^*) \subseteq \mathbb{S}R(\mathbb{B})$. Conversely, each f in $\mathbb{S}R(\mathbb{B})$ reads $f = \mathbb{S}\mathbb{B}h$ for some h in $\mathcal{H}$. With (i),

$$f = \mathbb{S}\mathbb{B}\mathbb{S}(\mathbb{S}^{-1}h) = \mathbb{B}^*(\mathbb{S}^{-1}h),$$

by (4.2.4). Hence f lies in $R(\mathbb{B}^*)$, so that $\mathbb{S}R(\mathbb{B})$ is contained in $R(\mathbb{B}^*)$, which completes the proof of (ii). In order to prove (iii), Definition 2.2.9 yields that

$$D(\mathbb{A}\mathbb{S}) = \mathbb{S}^{-1}D(\mathbb{A}) = \mathbb{S}D(\mathbb{A}),$$

by (i). Since $D(\mathbb{SAS}) = D(\mathbb{AS})$, we get (iii). Regarding (iv), let us start to prove (4.2.2) for $\mathbb{SAS}$ and $\mathbb{B}^*$, namely, we will show that

$$\mathbb{SAS} \circ \mathbb{B}^* = \mathrm{id}_{\mathcal{H}}. \tag{4.2.6}$$

For each u in $D(\mathbb{SAS} \circ \mathbb{B}^*)$, one has

$$\begin{aligned} \mathbb{SAS} \circ \mathbb{B}^* u &= \mathbb{SAS}^2\mathbb{BS}\, u && \text{(by (4.2.4))} \\ &= \mathbb{SABS}\, u && \text{(by (4.2.3))} \\ &= u && \text{(by (4.2.2) and (4.2.3)).} \end{aligned}$$

Moreover,

$$\begin{aligned} R(\mathbb{B}^*) &= \mathbb{S}R(\mathbb{B}) && \text{(by (ii))} \\ &\subseteq \mathbb{S}D(\mathbb{A}) && \text{(by Remark 4.2.2)} \\ &\subseteq D(\mathbb{SAS}) && \text{(by (iii)).} \end{aligned}$$

Then (4.2.6) follows from Remark 4.2.2. Let us now check (4.2.4) for $\mathbb{B}^*$.

$$\begin{aligned} (\mathbb{B}^*)^* &= \mathbb{B} && \text{(by Theorem 2.4.1)} \\ &= \mathbb{S}^2\mathbb{BS}^2 && \text{(by (4.2.3))} \\ &= \mathbb{SB}^*\mathbb{S} && \text{(by (4.2.4)).} \end{aligned}$$

Then $(\mathbb{SAS}, \mathbb{B}^*, \mathbb{S})$ is a differential triplet on $\mathcal{H}$. □

In view of Item (iv) in Proposition 4.2.1, we set this definition.

Definition 4.2.3 Let $\mathcal{T} := (\mathbb{A}, \mathbb{B}, \mathbb{S})$ be a differential triplet. The differential triplet $(\mathbb{SAS}, \mathbb{B}^*, \mathbb{S})$, denoted by $\mathcal{T}^*$, is called the *adjoint differential triplet of* $\mathcal{T}$. □

Using again $\mathbb{B}^{**} = \mathbb{B}$, we may check that the adjoint differential triplet of $\mathcal{T}^*$ is $\mathcal{T}$.

Now let us consider a basic instance of differential triplet which involves the operator $\frac{\mathrm{d}}{\mathrm{d}x}$.

Example 4.2.4 Let $\mathcal{H} := L^2(0, b)$. Under the assumptions and notation of Sect. 4.1.1, recall that

$$\mathrm{D}^1_{L^2} : H^1(0, b) \subseteq L^2(0, b) \to L^2(0, b), \qquad u \mapsto \frac{\mathrm{d}}{\mathrm{d}x} u$$

and

$$B_{1,L^2} : L^2(0,b) \to L^2(0,b), \qquad u \mapsto g_1 * u$$

Next, we define the symmetry

$$S_{L^2} : L^2(0,b) \to L^2(0,b), \qquad v \mapsto v(b-\cdot).$$

More explicitly, there holds for each v in $L^2(0,b)$

$$S_{L^2}v(x) = v(b-x),$$

for almost all x in $(0,b)$. For simplicity, we will write S instead of S_{L^2}.

With these notations, we claim that

$$\mathcal{T}_{1,0} := (\mathrm{D}^1_{L^2}, B_{1,L^2}, S) \tag{4.2.7}$$

is a differential triplet on $\mathcal{H}$. Indeed, by the Cauchy-Schwarz inequality, one has

$$\|B_{1,L^2}u\|_{L^2} \leq \frac{b}{\sqrt{2}}\|u\|_{L^2}, \qquad \forall\, u \in L^2(0,b).$$

Thus B_{1,L^2} lies in $\mathcal{L}(L^2(0,b))$. Next, (4.2.2) follows from (4.1.2), and (4.2.3) is obvious. Finally, starting from (4.1.4), we get (4.2.4) by the change of variable $y' = b - y$. That completes the proof of the claim.

Since

$$-\,\mathrm{D}^1_{L^2} = S\mathrm{D}^1_{L^2}S,$$

the adjoint differential triplet of $\mathcal{T}_{1,0}$ is

$$(\mathcal{T}_{1,0})^* = \big(-\,\mathrm{D}^1_{L^2}, (B_{1,L^2})^*, S\big). \tag{4.2.8}$$

Accordingly, we deduce from the very definition of a differential triplet that

$$\mathcal{T}_{1,b} := \big(\mathrm{D}^1_{L^2}, -(B_{1,L^2})^*, S\big) \tag{4.2.9}$$

is a differential triplet on $L^2(0,b)$ as well. For all purposes, let us explicit that

$$-\,(B_{1,L^2})^*v = \int_b^{\cdot} v(y)\,\mathrm{d}y, \qquad \forall\, v \in L^2(0,b). \tag{4.2.10}$$

□

4.2.2 A One-Parameter Family of Differential Triplets

We would like to generalize Example 4.2.4 by considering, instead of B_{1,L^2}, the operator

$$\mathbb{B}^a : L^2(0,b) \to L^2(0,b)$$

defined for any v in $L^2(0,b)$ by

$$\mathbb{B}^a v(x) := \int_a^x v(y)\,\mathrm{d}y \qquad \text{for a.e. } x \in (0,b), \tag{4.2.11}$$

where a is any fixed number in $[0,b]$. Thus $\{\mathbb{B}^a \mid a \in [0,b]\}$ is a one parameter family of operators in $\mathcal{L}(L^2(0,b))$.

Our aim is to find operators $\mathbb{A}^a$ and $\mathbb{S}^a$ such that $(\mathbb{A}^a, \mathbb{B}^a, \mathbb{S}^a)$ is a differential triplet on $L^2(0,b)$. Direct computations can be made, but we prefer an abstract approach, which has two advantages. First it highlights clearly the properties of $L^2(0,b)$ and $\mathbb{B}^a$ involved; second, it simplifies some computations.

Notice that the cases $a = 0$ and $a = b$ have already been covered in Example 4.2.4. More precisely according to (4.2.7) and (4.2.9)

$$(\mathbb{A}^0, \mathbb{B}^0, \mathbb{S}^0) := \mathcal{T}_{1,0}, \qquad (\mathbb{A}^b, \mathbb{B}^b, \mathbb{S}^b) := \mathcal{T}_{1,b}$$

are differential triplets on $L^2(0,b)$.

Let us introduce the abstract setting. Let $\mathcal{H}$ be a Hilbert space and m be a positive integer. We assume that for each $i = 1, \ldots, m$

$$\mathcal{H}_i \text{ is a Hilbert space,} \tag{4.2.12}$$

$$P_i : \mathcal{H} \to \mathcal{H}_i \text{ and } R_i : \mathcal{H}_i \to \mathcal{H} \text{ are continuous linear maps,} \tag{4.2.13}$$

$$T_i \text{ and } L_i \text{ belong to } \mathcal{L}(\mathcal{H}_i). \tag{4.2.14}$$

Moreover, we suppose that

$$R_i = P_i{}^* \qquad \forall i = 1, \ldots, m, \tag{4.2.15}$$

$$\sum_{i=1}^m R_i P_i = \mathrm{id}_{\mathcal{H}} \tag{4.2.16}$$

$$P_i R_j = \delta_{i,j}\, \mathrm{id}_{\mathcal{H}_i} \qquad \forall i, j = 1, \ldots, m, \tag{4.2.17}$$

and we put

$$T := \sum_{i=1}^{m} R_i T_i P_i, \qquad L := \sum_{i=1}^{m} R_i L_i P_i. \tag{4.2.18}$$

Proposition 4.2.2 *Let $\mathcal{H}$ be a Hilbert space and m be a positive integer. Under the assumptions and notation* (4.2.12)–(4.2.18), *the following properties hold:*

(i) *T and L belong to $\mathcal{L}(\mathcal{H})$.*
(ii) $TL = \sum_{i=1}^{m} R_i T_i L_i P_i$.
(iii) $T^* = \sum_{i=1}^{m} R_i T_i{}^* P_i$.

Proof (i) follows from (4.2.13) and (4.2.14). (ii) is a consequence of (4.2.16) and (4.2.17). Finally (iii) is obtained thanks to (4.2.15). □

Proposition 4.2.3 *Under the assumptions* (4.2.12)–(4.2.17), *let us suppose in addition that for each $i = 1, \dots, m$, (A_i, B_i, S_i) is a differential triplet on $\mathcal{H}_i$. Also define*

$$\mathbb{A} := \sum_{i=1}^{m} R_i A_i P_i, \qquad \mathbb{B} := \sum_{i=1}^{m} R_i B_i P_i, \qquad \mathbb{S} := \sum_{i=1}^{m} R_i S_i P_i.$$

Then $(\mathbb{A}, \mathbb{B}, \mathbb{S})$ be a differential triplet on $\mathcal{H}$ and

$$D(\mathbb{A}) := \{u \in \mathcal{H} \mid P_i u \in D(A_i),\ \forall i = 1, \dots, m\}. \tag{4.2.19}$$

Proof (4.2.19) follows from Definition 2.2.9. Proposition 4.2.2 entails that $\mathbb{B}$, $\mathbb{S}$ belong to $\mathcal{L}(\mathcal{H})$, $\mathbb{S}^* = \mathbb{S}$, $\mathbb{B}^* = \mathbb{S}\mathbb{B}\mathbb{S}$ and

$$\mathbb{S}^2 = \sum_{i=1}^{m} R_i P_i.$$

Thus with (4.2.16), we get $\mathbb{S}^2 = \mathrm{id}_{\mathcal{H}}$. There remains to prove (4.2.2), i.e., $\mathbb{A}\mathbb{B} = \mathrm{id}_{\mathcal{H}}$. For each i, j in $\{1, \dots, m\}$, we find by using (4.2.17) and the relation $A_i B_i = \mathrm{id}_{\mathcal{H}_i}$

$$A_i P_i R_j B_j = \delta_{i,j}\, \mathrm{id}_{\mathcal{H}_i}. \tag{4.2.20}$$

Thus $R_i A_i P_i R_j B_j P_j$ is defined on the whole space $\mathcal{H}$. There results from Definition 2.2.8 that the domain of

$$\sum_{i,j=1}^{m} R_i A_i P_i R_j B_j P_j$$

is the whole space $\mathcal{H}$. Thus Proposition 2.2.17 entails that

$$\mathbb{A}\mathbb{B} = \sum_{i,j=1}^{m} R_i A_i P_i R_j B_j P_j.$$

Accordingly

$$\begin{aligned}\mathbb{A}\mathbb{B} = \sum_{i,j=1}^{m} R_i A_i P_i R_j B_j P_j &= \sum_{i=1}^{m} R_i P_i && \text{(by (4.2.20))}\\ &= \mathrm{id}_{\mathcal{H}} && \text{(by (4.2.16)).}\end{aligned}$$

□

Let us apply the above theory in the case where $\mathcal{H} := L^2(0, b)$ and $m = 2$. Being given a in $(0, b)$, for any $i = 1, 2$, we put

$$\begin{aligned}&\Omega_1 := (0, a), \qquad \Omega_2 := (a, b)\\ &(u, v)_i := \int_{\Omega_i} u(x)\overline{v(x)}\,\mathrm{d}x\\ &\mathcal{H}_i := \big(L^2(\Omega_i), (\cdot, \cdot)_i\big)\\ &P_i : \mathcal{H} \to \mathcal{H}_i, \qquad v \mapsto v_{|\Omega_i} && (4.2.21)\\ &R_i : \mathcal{H}_i \to \mathcal{H}, \qquad v \mapsto \begin{cases} v & \text{on } \Omega_i \\ 0 & \text{on } \Omega \setminus \Omega_i \end{cases}. && (4.2.22)\end{aligned}$$

Let also $B_i : \mathcal{H}_i \to \mathcal{H}_i$ and $S_i : \mathcal{H}_i \to \mathcal{H}_i$ be defined by

$$\begin{aligned}&B_i v(x) := \int_a^x v(y)\,\mathrm{d}y \qquad \text{for a.e. } x \in \Omega_i, && \forall\, v \in L^2(\Omega_i)\\ &S_1 v := v(a - \cdot) && \forall\, v \in L^2(0, a)\\ &S_2 v := v(a + b - \cdot) && \forall\, v \in L^2(a, b)\end{aligned}$$

and

$$\begin{aligned}&A_i : H^1(\Omega_i) \subseteq L^2(\Omega_i) \to L^2(\Omega_i), \qquad v \mapsto \frac{\mathrm{d}}{\mathrm{d}x} v\\ &\mathbb{A}^a := R_1 A_1 P_1 + R_2 A_2 P_2\\ &\mathbb{S}^a := R_1 S_1 P_1 + R_2 S_2 P_2.\end{aligned}$$

With these notations, we readily check the assumptions of Proposition 4.2.3. Indeed (4.2.15) holds since

$$(u, P_i v)_i = \int_{\Omega_i} u(x)\overline{v(x)}\,\mathrm{d}x = \int_0^b R_i u(x)\overline{v(x)}\,\mathrm{d}x = (R_i u, v)$$

for each (u, v) in $\mathcal{H}_i \times L^2(0, b)$. In a same way, (4.2.16) and (4.2.17) hold as well. Moreover, one has

$$\mathbb{B}^a = R_1 B_1 P_1 + R_2 B_2 P_2.$$

Let us check that (A_i, B_i, S_i) is a differential triplet. From Example 4.2.4, this is clear for $i = 2$. When $i = 1$, (4.2.9) and (4.2.10) entail that

$$(A_1, B_1, S_1) = \big(\mathrm{D}^1_{L^2(0,a)}, -(B_{1,L^2(0,a)})^*, S_{L^2(0,a)}\big).$$

Thus (A_1, B_1, S_1) is a differential triplet on $L^2(0, a)$. There results from Proposition 4.2.3 that

$$\mathcal{T}_{1,a} := (\mathbb{A}^a, \mathbb{B}^a, \mathbb{S}^a) \text{ is a differential triplet on } L^2(0, b). \tag{4.2.23}$$

Let us explicit $\mathbb{A}^a$ and $\mathbb{S}^a$. From (4.2.19),

$$D(\mathbb{A}^a) := \{u \in L^2(0, b) \mid u_{|\Omega_i} \in H^1(\Omega_i),\ \forall i = 1, 2\}. \tag{4.2.24}$$

Thus since $\mathbb{A}^a := R_1 A_1 P_1 + R_2 A_2 P_2$, we get

$$\mathbb{A}^a u = \begin{cases} \frac{\mathrm{d}}{\mathrm{d}x}\{u_{|(0,a)}\} & \text{a.e. on } (0, a) \\ \frac{\mathrm{d}}{\mathrm{d}x}\{u_{|(a,b)}\} & \text{a.e. on } (a, b) \end{cases}, \qquad \forall\, u \in D(\mathbb{A}^a). \tag{4.2.25}$$

Regarding $\mathbb{S}^a$, one has

$$\mathbb{S}^a u(x) = \begin{cases} u(a - x) & \text{for a.e. } x \in (0, a) \\ u(a + b - x) & \text{for a.e. } x \in (a, b) \end{cases}. \tag{4.2.26}$$

4.3 Boundary Restriction Operators

Being given a differential triplet $\mathcal{T} := (\mathbb{A}, \mathbb{B}, \mathbb{S})$ on a Hilbert space $\mathcal{H}$ over $\mathbb{K}$, the aims in this section are to define and study *boundary restriction operators*. Roughly speaking, *boundary restriction operators* of $\mathcal{T}$ are restrictions of $\mathbb{A}$ whose adjoint is a restriction of $\mathbb{S}\mathbb{A}\mathbb{S}$ (see Definition 4.3.1). In view of the *Domain Structure Theorem* 4.3.14 and Sects. 4.4.1 and 4.4.2, these boundary restriction operators can be seen to some extent as $\mathbb{A}$ supplemented with *(abstract) homogeneous linear boundary conditions*.

The issue is to define boundary conditions and boundary values at an abstract level. Since evaluation cannot be used, we will put conditions on adjoint operators. Indeed, in the case where $\mathcal{T} := (\mathrm{D}^1_{L^2}, B_{1,L^2}, S)$, all the boundary values of a function in the domain of the operator

$$A_{\mathrm{m}} := -{\mathrm{D}^1_{L^2}}^*$$

vanish. Thus a restriction A of $\mathrm{D}^1_{L^2}$ supplemented with homogeneous boundary conditions will satisfy $A_{\mathrm{m}} \subseteq A$. Hence

$$A^* \subseteq -\mathrm{D}^1_{L^2}.$$

At an abstract level, this relation becomes

$$A^* \subseteq \mathbb{S}\mathbb{A}\mathbb{S}.$$

As we will see, this fundamental relation encodes boundary conditions and makes up the bulk of Definition 4.3.1.

4.3.1 Definition and First Results

Definition 4.3.1 Let $\mathcal{T} := (\mathbb{A}, \mathbb{B}, \mathbb{S})$ be a differential triplet, and A be an operator on $\mathcal{H}$. We say that A is a *boundary restriction operator of* $\mathcal{T}$ or if no confusion may occur, *of* $\mathbb{A}$ if:

(i) A is a densely defined restriction of $\mathbb{A}$;
(ii) $A^* \subseteq \mathbb{S}\mathbb{A}\mathbb{S}$.

Item (i) is needed for A^* to be well defined. Regarding Item (ii), we observe that functions in the domain of a differential operator or in the domain of its adjoint have essentially the same regularity and differ by their boundary values. Item (ii) encodes that observation by telling that the domain of the adjoint cannot be too large (see (4.2.5)).

A *boundary restriction operator* of $\mathbb{A}$ can be seen as $\mathbb{A}$ *supplemented with some abstract homogeneous linear boundary conditions* that will be called in Sect. 4.4.1, abstract endogenous boundary conditions. We refer to Remark 4.4.4 for another explanation regarding the meaning of Item (ii).

The following result features sufficient conditions for a restriction of $\mathbb{A}$ to be densely defined and is of fundamental importance in this theory.

Proposition 4.3.1 *Let* $(\mathbb{A}, \mathbb{B}, \mathbb{S})$ *be a differential triplet on a Hilbert space* $\mathcal{H}$. *Let also* F *be a closed subspace of* $\mathcal{H}$ *satisfying*

$$\mathbb{S}F \cap D(\mathbb{A}) \subseteq \ker \mathbb{A}. \tag{4.3.1}$$

Then $\mathbb{B}(F^\perp)$ *is dense in* $\mathcal{H}$.

Proof Let f be in the annihilator of $\mathbb{B}(F^\perp)$. Then for each ψ in $F^\perp$,

$$0 = (f, \mathbb{B}\psi) = (\mathbb{B}^* f, \psi) = (\mathbb{S}\mathbb{B}\mathbb{S}f, \psi),$$

by (4.2.4). Thus $\mathbb{S}\mathbb{B}\mathbb{S}f$ lies in $F^{\perp\perp}$. However, Proposition 2.2.12 entails that $F^{\perp\perp} = F$ since F is assumed to be closed. Due to $\mathbb{S}^2 = \mathrm{id}_{\mathcal{H}}$, we derive that $\mathbb{B}\mathbb{S}f$ belongs to $\mathbb{S}F$. Using $R(\mathbb{B}) \subseteq D(\mathbb{A})$ (see Remark 4.2.2) and (4.3.1), we get

$$\mathbb{B}\mathbb{S}f \in \mathbb{S}F \cap D(\mathbb{A}) \subseteq \ker \mathbb{A}.$$

Thus (4.2.2) yields $\mathbb{S}f = 0$, and $\mathbb{S}^2 = \mathrm{id}_{\mathcal{H}}$ gives $f = 0$. Then the density of $\mathbb{B}(F^\perp)$ follows from Corollary 2.2.10. □

Remark 4.3.2 The subspaces F for which $\mathbb{B}(F^\perp)$ is dense in $\mathcal{H}$ comprise a slight class not so easy to detect. Indeed, let $(\mathbb{A}, \mathbb{B}, \mathbb{S})$ be a differential triplet on $\mathcal{H}$ and F be a subspace of $\mathcal{H}$ whose intersection with $\mathbb{S}R(\mathbb{B})$ is nontrivial. Then we claim that $\mathbb{B}(F^\perp)$ is not dense in $\mathcal{H}$.

We will prove the contrapositive. That is, assuming that $\mathbb{B}(F^\perp)$ is dense in $\mathcal{H}$, we will show that any vector φ_0 in $F \cap \mathbb{S}R(\mathbb{B})$ is trivial. Indeed, φ_0 reads $\varphi_0 = \mathbb{S}\mathbb{B}h$ for some h in $\mathcal{H}$. Then by setting $f_0 := \mathbb{S}h$,

$$\varphi_0 = \mathbb{S}\mathbb{B}\mathbb{S}f_0 = \mathbb{B}^* f_0. \tag{4.3.2}$$

Now writing any u in $\mathbb{B}(F^\perp)$ under the form $u = \mathbb{B}\psi^\perp$ with $\psi^\perp$ in $F^\perp$, we compute

$$(f_0, u) = (f_0, \mathbb{B}\psi^\perp) = (\mathbb{B}^* f_0, \psi^\perp) = (\varphi_0, \psi^\perp) = 0,$$

since $(\varphi_0, \psi^\perp)$ lies in $F \times F^\perp$. Since $\mathbb{B}(F^\perp)$ is assumed to be dense in $\mathcal{H}$, we get $f_0 = 0$, and $\varphi_0 = 0$ by (4.3.2). □

The choice of $F := \{0\}$ in Proposition 4.3.1 allows to introduce a special densely defined operator featured in this definition.

Definition 4.3.3 Let $\mathcal{T} := (\mathbb{A}, \mathbb{B}, \mathbb{S})$ be a differential triplet on $\mathcal{H}$. The restriction of $\mathbb{A}$ to the domain $R(\mathbb{B})$ is called the *pivot operator of the differential triplet* $\mathcal{T}$. That is to say, the pivot operator of $\mathcal{T}$, denoted by $A_\mathcal{T}$, satisfies

$$D(A_\mathcal{T}) = R(\mathbb{B}), \qquad A_\mathcal{T} u = \mathbb{A}u, \qquad \forall u \in D(A_\mathcal{T}). \tag{4.3.3}$$

□

Proposition 4.3.2 *Let* $\mathcal{T} := (\mathbb{A}, \mathbb{B}, \mathbb{S})$ *be a differential triplet on* $\mathcal{H}$. *The pivot operator* $A_\mathcal{T}$ *of* $\mathcal{T}$ *satisfies the following properties:*

(i) $A_\mathcal{T}$ *is a densely defined invertible operator on* $\mathcal{H}$ *and* $(A_\mathcal{T})^{-1} = \mathbb{B}$.
(ii) $A_\mathcal{T}$ *satisfies the so-called* Poincaré inequality

$$\|u\| \lesssim \|A_\mathcal{T} u\|, \qquad \forall u \in D(A_\mathcal{T}). \tag{4.3.4}$$

Proof Proposition 4.3.1 entails that $A_\mathcal{T}$ is densely defined. (4.3.3) and (4.2.2) yield that $A_\mathcal{T}\mathbb{B} = \mathrm{id}_\mathcal{H}$. On the other hand, since $D(A_\mathcal{T}) = R(\mathbb{B})$, each u in $D(A_\mathcal{T})$ reads $u = \mathbb{B}f$ for some f in $\mathcal{H}$. Thus,

$$\begin{aligned} \mathbb{B}A_\mathcal{T} u &= \mathbb{B}\mathbb{A}u && (\text{since } A_\mathcal{T} \subseteq \mathbb{A}) \\ &= \mathbb{B}f && (\text{by (4.2.2)}) \\ &= u. \end{aligned}$$

Thus, $\mathbb{B}A_\mathcal{T} = \mathrm{id}_{D(A_\mathcal{T})}$ and Item (i) follows. Regarding (ii), let u be in $D(A_\mathcal{T})$. Then by (i), $u = \mathbb{B}A_\mathcal{T} u$. Since $\mathbb{B}$ is continuous, we get (4.3.4). □

Example 4.3.4 Let b be a positive real number and a be in $[0, b]$. In view of (4.2.7), (4.2.9), and (4.2.23), $\mathcal{T}_{1,a}$ is a differential triplet on $L^2(0, b)$. The domain of the pivot operator $A_{\mathcal{T}_{1,a}}$ of $\mathcal{T}_{1,a}$ is

$$D(A_{\mathcal{T}_{1,a}}) = {}_aH^1(0, b) := \{u \in H^1(0, b) \mid u(a) = 0\}.$$

Moreover, we recall that for $a = 0$ or $a = b$, we have

$$A_{\mathcal{T}_{1,a}} u = \frac{\mathrm{d}}{\mathrm{d}x} u, \qquad \forall u \in {}_a H^1(0, b)$$

□

4.3.2 Large and Small Restrictions of $\mathbb{A}$

Let $\mathbb{K}$ be equal to $\mathbb{R}$ or $\mathbb{C}$, $\mathcal{H}$ be a Hilbert space over $\mathbb{K}$, and $\mathcal{T} := (\mathbb{A}, \mathbb{B}, \mathbb{S})$ be a differential triplet on $\mathcal{H}$. *A large restriction of* $\mathbb{A}$ is an operator A on $\mathcal{H}$ satisfying

$$A_{\mathcal{T}} \subseteq A \subseteq \mathbb{A}. \tag{4.3.5}$$

On the other hand, *a small restriction of* $\mathbb{A}$ is a restriction of $\mathbb{A}$ whose domain reads $\mathbb{B}(F^{\perp})$, for some closed subspace F of $\mathcal{H}$.

The main results of this subsection are:

(i) Proposition 4.3.5 which states that any large restriction of $\mathbb{A}$ is a boundary restriction operator;
(ii) Proposition 4.3.8 which characterizes boundary restriction operators among small restrictions of $\mathbb{A}$.

The ideas of the proof of the Proposition 4.3.5 goes as follows: by Proposition 4.3.2, we know that large restrictions are densely defined. Thus, there remains to prove that their adjoint is a restriction of $\mathbb{SAS}$. This is achieved through the identity (4.3.8), which is proved by means of (4.3.5).

We begin with a standard and basic result.

Lemma 4.3.3 *Let $A : D(A) \subseteq \mathcal{X} \to \mathcal{X}$ and $B : \mathcal{X} \to \mathcal{X}$ be operators on a $\mathbb{K}$-normed space $\mathcal{X}$. Suppose that $AB = \mathrm{id}_{\mathcal{X}}$. Then* $\ker A$ *and $R(B)$ are in direct sum in $\mathcal{X}$ and*

$$D(A) = \ker A \oplus R(B). \tag{4.3.6}$$

Moreover, for each u in $D(A)$, there exists a unique u_0 in $\ker A$ *such that*

$$u = u_0 + BAu \qquad \text{in } D(A). \tag{4.3.7}$$

Proof Each x_0 in $\ker A \cap R(B)$ reads $x_0 = Bx$ for some x in $\mathcal{X}$. Thus

$$0 = Ax_0 = ABx = x.$$

Hence $x_0 = 0$ and the sum is direct. Finally, each u in $D(A)$ satisfies $A(u - BAu) = 0$, since $AB = \mathrm{id}_{\mathcal{X}}$. Thus $u_0 := u - BAu$ lies in $\ker A$. Whence (4.3.7) and (4.3.6) follow. □

The next result is the abstract counterpart of the analysis made in Sect. 4.1.2. Indeed, (4.1.7) and (4.1.8) generalize into (4.3.8) and (4.3.9).

Proposition 4.3.4 *Let* $(\mathbb{A}, \mathbb{B}, \mathbb{S})$ *be a differential triplet on* $\mathcal{H}$ *and* A *be an operator on* $\mathcal{H}$ *satisfying* (4.3.5). *Then the following properties hold true:*

(i) $D(A)$ *is dense in* $\mathcal{H}$, *so that* A^* *exists. Moreover,*

$$\mathbb{B}^* A^* = \mathrm{id}_{D(A^*)}. \tag{4.3.8}$$

(ii) *The domain of* A^* *reads*

$$D(A^*) = \mathbb{B}^*\big(\ker A^{\perp}\big) = \big\{\mathbb{B}^* h \mid h \in \ker A^{\perp}\big\}. \tag{4.3.9}$$

(iii) *For each* f *in* $D(A^*)$, *there exists a unique* h *in* $\ker A^{\perp}$ *such that* $f = \mathbb{B}^* h$. *Also* $h = A^* f$, *thus* $A^* f$ *lies in* $\ker A^{\perp}$.

Proof

(i) By Proposition 4.3.2 (i) and (4.3.5), A has a dense domain. Moreover, for each f in $D(A^*)$ and for any h in $\mathcal{H}$,

$$(\mathbb{B}^* A^* f, h) = (A^* f, \mathbb{B} h).$$

Since $R(\mathbb{B}) = D(A_{\mathcal{T}})$, (4.3.5) entails that $R(\mathbb{B})$ is contained in $D(A)$, thus

$$(\mathbb{B}^* A^* f, h) = (f, A\mathbb{B} h) = (f, h),$$

since A is a restriction of $\mathbb{A}$ and $\mathbb{A}\mathbb{B} = \mathrm{id}_{\mathcal{H}}$. That completes the proof of Item (i).

(ii) Let us denote for simplicity, $\mathbb{B}^*(\ker A^{\perp})$ by D^*. Let f be in $D(A^*)$. In our context, (2.5.3) reads

$$(f, Au) = (A^* f, u), \qquad \forall u \in D(A). \tag{4.3.10}$$

Choosing u in $\ker A$, we get

$$(A^* f, u) = 0, \qquad \forall u \in \ker A.$$

Thus A^*f lives in $\ker A^{\perp}$. Then (4.3.8) leads to $D(A^*) \subseteq D^*$. Conversely, consider $f := \mathbb{B}^*h$ any element of D^*. Let u belong to $D(A)$. By (4.3.7), $u = u_0 + \mathbb{B}Au$ for some (unique) u_0 in $\ker A$. Thus

$$(f, Au) = (h, \mathbb{B}Au) = (h, u - u_0).$$

Since h lies in $\ker A^{\perp}$, one has $(h, u_0) = 0$, so that $(f, Au) = (h, u)$. By definition of $D(A^*)$, we infer that f lies in $D(A^*)$ and $A^*f = h$. That proves (4.3.9) and the uniqueness of h in Item (iii). □

The next result states the any *large* restriction of $\mathbb{A}$ is a boundary restriction operator.

Proposition 4.3.5 *Let* $\mathcal{T} := (\mathbb{A}, \mathbb{B}, \mathbb{S})$ *be a differential triplet on* $\mathcal{H}$ *and* A *be an operator on* $\mathcal{H}$*. If* A *satisfies* (4.3.5)*, then* A *is a boundary restriction operator of* $\mathcal{T}$*.*

Proof By Proposition 4.3.4, A is densely defined. Thus there remains to prove that $A^* \subseteq \mathbb{S}A\mathbb{S}$. For, let f be in $D(A^*)$, and $h := A^*f$. Then

$$\begin{aligned}
\mathbb{S}f &= \mathbb{S}\mathbb{B}^*A^*f && \text{(by (4.3.8))}\\
&= \mathbb{S}^2\mathbb{B}\mathbb{S}A^*f && \text{(by (4.2.4))}\\
&= \mathbb{B}\mathbb{S}h && \text{(by (4.2.3))}.
\end{aligned}$$

Thus (4.2.2) yields that $\mathbb{S}f$ belongs to $D(\mathbb{A})$ and $\mathbb{A}\mathbb{S}f = \mathbb{S}h$. Whence $\mathbb{S}\mathbb{A}\mathbb{S}f = A^*f$, and A^* is a restriction of $\mathbb{S}\mathbb{A}\mathbb{S}$. □

Recalling Definition 4.3.3 of pivot operators, the following result states among other things that the adjoint of the pivot operator is the pivot operator of the adjoint differential triplet.

Proposition 4.3.6 *Let* $\mathcal{T} := (\mathbb{A}, \mathbb{B}, \mathbb{S})$ *be a differential triplet and* $\mathcal{T}^* := (\mathbb{S}\mathbb{A}\mathbb{S}, \mathbb{B}^*, \mathbb{S})$ *be its adjoint differential triplet. Then the pivot operator* $A_{\mathcal{T}}$ *of* $\mathcal{T}$ *is a closed boundary restriction operator of* $\mathcal{T}$*, and*

$$(A_{\mathcal{T}})^* = A_{\mathcal{T}^*} = \mathbb{S}A_{\mathcal{T}}\mathbb{S}. \tag{4.3.11}$$

Proof Proposition 4.3.5 entails that $A_{\mathcal{T}}$ is a boundary restriction operator of $\mathbb{A}$. In order to prove the first equality of (4.3.11), notice on the one hand that by definition of $A_{\mathcal{T}^*}$

$$A_{\mathcal{T}^*} \subseteq \mathbb{S}\mathbb{A}\mathbb{S}, \qquad D(A_{\mathcal{T}^*}) = R(\mathbb{B}^*). \tag{4.3.12}$$

On the other hand, since $A_{\mathcal{T}}$ is a boundary restriction operator of $\mathbb{A}$,

$$(A_{\mathcal{T}})^* \subseteq \mathbb{S}A\mathbb{S}. \tag{4.3.13}$$

By Proposition 4.3.4 (ii),

$$D\big((A_{\mathcal{T}})^*\big) = \mathbb{B}^*\big((\ker A_{\mathcal{T}})^{\perp}\big).$$

Since, by Proposition 4.3.2, $A_{\mathcal{T}}$ is one-to-one, we get $D\big((A_{\mathcal{T}})^*\big) = R(\mathbb{B}^*)$. Hence with (4.3.12),

$$D\big((A_{\mathcal{T}})^*\big) = D(A_{\mathcal{T}^*}).$$

Then with (4.3.12) and (4.3.13), we obtain $(A_{\mathcal{T}})^* = A_{\mathcal{T}^*}$.

Let us show that $A_{\mathcal{T}^*} = \mathbb{S}A_{\mathcal{T}}\mathbb{S}$. For, since $(A_{\mathcal{T}})^* = A_{\mathcal{T}^*}$, (4.3.8) yields that each f in $D(A_{\mathcal{T}^*})$ reads $f = \mathbb{B}^* A_{\mathcal{T}^*} f$. Thus $\mathbb{S}f = \mathbb{B}\mathbb{S}A_{\mathcal{T}^*} f$. By definition of $A_{\mathcal{T}}$, we deduce that $\mathbb{S}f$ lies in $D(A_{\mathcal{T}})$ and $A_{\mathcal{T}}\mathbb{S}f = \mathbb{S}A_{\mathcal{T}^*} f$, so that $\mathbb{S}A_{\mathcal{T}}\mathbb{S}f = A_{\mathcal{T}^*} f$. We have shown that $A_{\mathcal{T}^*} \subseteq \mathbb{S}A_{\mathcal{T}}\mathbb{S}$. By symmetry, replacing $\mathcal{T}$ by $\mathcal{T}^*$, one has $A_{\mathcal{T}} \subseteq \mathbb{S}A_{\mathcal{T}^*}\mathbb{S}$. With $\mathbb{S}^2 = \mathrm{id}_{\mathcal{H}}$, we derive the wished equality.

Finally, since $(A_{\mathcal{T}})^*$ is a closed operator (due to Proposition 2.2.19), we deduce from $(A_{\mathcal{T}})^* = \mathbb{S}A_{\mathcal{T}}\mathbb{S}$ that $A_{\mathcal{T}}$ is closed as well. □

After the study the of large restrictions, we will investigate small restrictions A of $\mathbb{A}$. In that case, A is a boundary restriction operator of $\mathbb{A}$ if and only if F is a closed subspace of $\mathbb{S}\ker\mathbb{A}$. Accordingly, a small restriction of $\mathbb{A}$ is not always a boundary restriction operator. See Proposition 4.3.8.

The following result gives useful properties on the adjoint of a densely defined small restriction, together with the structure of the domain of this adjoint operator. Since the symmetry $\mathbb{S}$ plays no role, it is not necessary to assume that a differential triplet is involved here.

Theorem 4.3.7 *Let A be an operator on $\mathcal{H}$, B lie in $\mathcal{L}(\mathcal{H})$, and F be a closed subspace of $\mathcal{H}$. Suppose that A is densely defined and*

$$D(A) = B(F^{\perp}) \tag{4.3.14}$$

$$AB\psi = \psi, \quad \forall\, \psi \in F^{\perp}. \tag{4.3.15}$$

Then the following assertions hold true:

(i) $R(A) = F^{\perp}$, $\ker A^* = F$.
(ii) $R(B^*) \subseteq D(A^*)$, $A^*B^* = \mathrm{id}_{\mathcal{H}}$.
(iii) $D(A^*) = F \oplus R(B^*)$.
(iv) *For each f in $D(A^*)$, there exists a unique h_f in F such that*

$$f = h_f + B^*A^*f. \tag{4.3.16}$$

(v) *Conversely, for each h_0 in F and each $h \in \mathcal{H}$,*

$$f := h_0 + B^*h \in D(A^*), \qquad A^*f = h. \tag{4.3.17}$$

Proof

(i) Let us show that $R(A) = F^\perp$. For, let y be in $R(A)$. By (4.3.14), there exists ψ in $F^\perp$ such that $y = AB\psi$. Then (4.3.15) yields that $y = \psi$ lies in $F^\perp$. Conversely, let ψ be in $F^\perp$. Then the vector $u := B\psi$ lies in $D(A)$ and $Au = \psi$ by (4.3.14)–(4.3.15). Hence ψ belong to $R(A)$, so that $R(A) = F^\perp$.

By Proposition 2.2.19, $\ker A^* = R(A)^\perp$. Thus, since F is closed in $\mathcal{H}$, Proposition 2.2.12 entails that $\ker A^* = F$.

(ii) Let u belong to $D(A)$. Using $R(A) = F^\perp$, we deduce that $\psi := Au$ lies in $F^\perp$. Hence, by (4.3.14), $B\psi$ lies in $D(A)$. Then

$$\begin{aligned} A(B\psi - u) &= \psi - Au && \text{(by (4.3.15))} \\ &= 0 && \text{(by definition of } \psi\text{)}. \end{aligned}$$

Since, by (4.3.14)–(4.3.15), A is one-to-one, we derive that $B\psi = u$, so that

$$BAu = u, \qquad \forall u \in D(A). \tag{4.3.18}$$

Now, for each ψ in $\mathcal{H}$, (4.3.18) yields that

$$(B^*\psi, Au) = (\psi, BAu) = (\psi, u), \qquad \forall u \in D(A).$$

Thus, by definition of the adjoint operator of A, $B^*\psi$ belongs to $D(A^*)$, and $A^*B^*\psi = \psi$. That proves the second item. Items (iii)–(v) follow from Lemma 4.3.3, by using (i) and (ii). □

The following proposition characterizes small restrictions of $\mathbb{A}$ which are boundary restriction operators of $\mathbb{A}$ and is a key step toward characterization of closed boundary restriction operators (Theorem 4.3.12).

Proposition 4.3.8 *Being given a differential triplet* $(\mathbb{A}, \mathbb{B}, \mathbb{S})$ *on a Hilbert space* $\mathcal{H}$*, we assume that:*

(i) *F is a closed subspace of $\mathcal{H}$;*

(ii) *A is a restriction of $\mathbb{A}$ with domain $\mathbb{B}(F^{\perp})$.*

Then A is a boundary restriction operator of $\mathbb{A}$ if and only if $\mathbb{S}F \subseteq \ker \mathbb{A}$.

Proof Assuming $\mathbb{S}F \subseteq \ker \mathbb{A}$, we readily get that (4.3.1) holds true. Thus Proposition 4.3.1 yields that $D(A)$ is dense in $\mathcal{H}$, so that A^* is well defined. Let f belong to $D(A^*)$, and $h := A^* f$. By (4.3.16) and (4.2.4), there exists h_f in F such that

$$\mathbb{S}f = \mathbb{S}h_f + \mathbb{B}\mathbb{S}h.$$

By (4.2.2), $\mathbb{B}\mathbb{S}h$ lies in $D(\mathbb{A})$ and $\mathbb{A}\mathbb{B}\mathbb{S}h = \mathbb{S}h$. Regarding $\mathbb{S}h_f$, we know by hypothesis that $\mathbb{S}h_f$ belongs to $\ker \mathbb{A}$, that is, $\mathbb{A}\mathbb{S}h_f = 0$. There results that $\mathbb{S}f$ belongs to $D(\mathbb{A})$ and $\mathbb{A}\mathbb{S}f = \mathbb{S}h$, thus $A^* \subseteq \mathbb{S}\mathbb{A}\mathbb{S}$. Hence, A is a boundary restriction operator of $\mathbb{A}$.

Conversely, Theorem 4.3.7 (i) gives that $F = \ker A^*$. Moreover,

$$A^* \subseteq \mathbb{S}\mathbb{A}\mathbb{S}$$

implies that $\ker A^* \subseteq \ker \mathbb{S}\mathbb{A}\mathbb{S}$. Thus $\mathbb{S}F \subseteq \ker \mathbb{A}$ due to $\mathbb{S}^2 = \mathrm{id}_{\mathcal{H}}$. □

4.3.3 Minimal Operators

Let $\mathcal{T} := (\mathbb{A}, \mathbb{B}, \mathbb{S})$ be a differential triplet on a Hilbert space $\mathcal{H}$. By taking $F := \mathbb{S} \ker \mathbb{A}$ and using once again that $\mathbb{S} \ker \mathbb{A}$ is equal to $\ker \mathbb{A}\mathbb{S}$, Proposition 4.3.8 allows us to introduce a special densely defined restriction of $\mathbb{A}$, the so-called minimal operator.

By Proposition 4.3.8, this operator is minimal among *small boundary restriction operators* (i.e., small restrictions of $\mathbb{A}$ that are in addition boundary restriction operators of $\mathbb{A}$) with respect to the inclusion of operators. Moreover, as stated by forthcoming Theorem 4.3.12, it is also minimal among closed boundary restriction operators of $\mathbb{A}$. In turn, this property allows the characterization of closed boundary restriction operators.

Definition 4.3.5 The *minimal operator of a differential triplet* $\mathcal{T} := (\mathbb{A}, \mathbb{B}, \mathbb{S})$ is the restriction of $\mathbb{A}$ to the domain $\mathbb{B}\big((\ker \mathbb{A}\mathbb{S})^{\perp}\big)$. The minimal operator of $\mathcal{T}$ is denoted by $A_{\mathrm{m},\mathcal{T}}$ or A_{m}.

Since $\mathbb{S}^2 = \mathrm{id}_{\mathcal{H}}$, we readily obtain

$$D(A_{\mathrm{m},\mathcal{T}^*}) = \mathbb{B}^*(\ker \mathbb{A}^{\perp}). \tag{4.3.19}$$

Example 4.3.6 (Minimal Operators of $\mathcal{T}_{1,a}$) Under the assumptions and notation of Example 4.3.4, our aim is to compute the domain of the minimal operator $A_{\mathrm{m},\mathcal{T}_{1,a}}$ of $\mathcal{T}_{1,a}$.

If $a = 0$ or $a = b$ then $\ker \mathbb{A}^a$ is generated by g_1 which we write in symbol

$$\ker \mathbb{A}^a = \langle g_1 \rangle.$$

Since $\mathbb{S}^a g_1 = g_1$, one has

$$D(A_{\mathrm{m},\mathcal{T}_{1,0}}) = D(A_{\mathrm{m},\mathcal{T}_{1,b}}) = H_0^1(0, b).$$

In view of (4.2.8), the domain of the minimal operator of $\mathcal{T}_{1,0}$ coincide with the domain of the minimal operator of the adjoint differential triplet $(\mathcal{T}_{1,0})^*$. This fact is not of general significance and occurs because $\mathrm{D}^1_{L^2}$ anti-commutes with S (see Proposition 4.3.13).

If a lies in $(0, b)$, then setting for each i in $\{1, 2\}$,

$$\mathbf{1}_{\Omega_i} : \Omega \to \mathbb{R}, \qquad x \mapsto \begin{cases} 1 & \text{if } x \in \Omega_i \\ 0 & \text{otherwise} \end{cases},$$

we infer from (4.2.25) that

$$\ker \mathbb{A}^a = \langle \mathbf{1}_{\Omega_1}, \mathbf{1}_{\Omega_2} \rangle.$$

In the current case, $\ker \mathbb{A}^a$ is also invariant with respect to $\mathbb{S}^a$ (see (4.2.26)), so that

$$D(A_{\mathrm{m},\mathcal{T}_{1,a}}) = \mathbb{B}^a\big(\langle \mathbf{1}_{\Omega_1}, \mathbf{1}_{\Omega_2}, \rangle^{\perp}\big).$$

We claim that

$$D(A_{\mathrm{m},\mathcal{T}_{1,a}}) = \{u \in H_0^1(0, b) \mid u(a) = 0\}. \tag{4.3.20}$$

Indeed, $\psi^{\perp}$ lies in $(\ker \mathbb{A}^a)^{\perp}$ if and only if $\psi^{\perp}$ lies in $D(\mathbb{A}^a)$ and satisfies

$$\int_{\Omega_i} \psi^{\perp}(y)\,\mathrm{d}y = 0, \qquad \forall i = 1, 2. \tag{4.3.21}$$

Thus for any u in $D(A_{\mathrm{m},\mathcal{T}_{1,a}})$, there exists some $\psi^{\perp}$ in $(\ker \mathbb{A}^a)^{\perp}$ such that

$$u(x) = \int_a^x \psi^{\perp}(y)\,\mathrm{d}y, \qquad \forall x \in (0, b).$$

Since $\psi^{\perp}$ belongs to $L^2(0, b)$, u lies in $H^1(0, b)$ and $\mathrm{D}^1_{L^2} u = \psi^{\perp}$. Thus $u(a) = 0$ and

$$u(b) = \int_a^b \mathrm{D}^1_{L^2} u(y)\,\mathrm{d}y = \int_{\Omega_2} \psi^{\perp}(y)\,\mathrm{d}y = 0,$$

by (4.3.21). In a same way, $u(0) = 0$. Therefore, $D(A_{\mathrm{m},\mathcal{T}_{1,a}})$ is contained in the right hand side of (4.3.20). Conversely, let u lie in this right-hand side. Then $\psi^{\perp} := \mathrm{D}^1_{L^2} u$ lies in $L^2(0, b)$, $\mathbb{B}^a \psi^{\perp} = u$, and

$$(\psi^{\perp}, \mathbf{1}_{\Omega_i}) = \int_{\Omega_i} \psi^{\perp}(y)\,\mathrm{d}y = \begin{cases} u(a) - u(0) & \text{if } i = 1 \\ u(b) - u(a) & \text{if } i = 2 \end{cases}.$$

Since $u(a) = u(0) = u(b) = 0$, we deduce that $\psi^{\perp}$ is in $(\ker \mathbb{A}^a)^{\perp}$. Therefore, u lies in $D(A_{\mathrm{m},\mathcal{T}_{1,a}})$, and (4.3.20) follows.

Let us notice that in view of (4.3.20), the space of compactly supported C^{∞}-functions on $(0, b)$ is not contained in the domain of $A_{\mathrm{m},\mathcal{T}_{1,a}}$ for each a in $[0, 1]$. □

The forthcoming result allows to give another expression of the domain of minimal operators.

Lemma 4.3.9 *Let $\mathcal{X}$, $\mathcal{Y}$ be a normed spaces, $\mathcal{E}$ and $\mathcal{E}_*$ be non-empty subsets of $\mathcal{X}$ and $\mathcal{X}^*$, respectively. Let also T be an isomorphism from $\mathcal{X}$ onto $\mathcal{Y}$. Then $T(\mathcal{E}_*^{\top}) = (T^{-1*}\mathcal{E}_*)^{\top}$ and $T^{-1*}(\mathcal{E}^{\perp}) = (T\mathcal{E})^{\perp}$. In particular, if $\mathcal{X} = \mathcal{Y} := \mathcal{H}$ is a Hilbert space and S is a symmetry of $\mathcal{H}$, then*

$$S(\mathcal{E}^{\perp}) = (S\mathcal{E})^{\perp}.$$

Proof Let y belong to $\mathcal{Y}$. Recalling that $T^{-1*} : \mathcal{X}^* \to \mathcal{Y}^*$, we compute

$$\begin{aligned} y \in T(\mathcal{E}_*^{\top}) &\Longleftrightarrow T^{-1} y \in \mathcal{E}_*^{\top} \\ &\Longleftrightarrow \langle x^*, T^{-1} y\rangle_{\mathcal{X}^*,\mathcal{X}} = 0, \qquad \forall x^* \in \mathcal{E}_* \\ &\Longleftrightarrow \langle T^{-1*} x^*, y\rangle_{\mathcal{Y}^*,\mathcal{Y}} = 0, \qquad \forall x^* \in \mathcal{E}_* \\ &\Longleftrightarrow y \in (T^{-1*}\mathcal{E}_*)^{\top}. \end{aligned}$$

Thus $T(\mathcal{E}_*^{\top}) = (T^{-1*}\mathcal{E}_*)^{\top}$. We may prove in a same way that $T^{-1*}(\mathcal{E}^{\perp})$ is equal to $(T\mathcal{E})^{\perp}$. Then the Hilbertian relation follows from identification (see (2.5.4)). □

There results from the above Lemma that the domain of the minimal operator $A_{\mathrm{m},\mathcal{T}}$ of a differential triplet $\mathcal{T} := (\mathbb{A}, \mathbb{B}, \mathbb{S})$ reads

$$D(A_{\mathrm{m},\mathcal{T}}) = \mathbb{B}\mathbb{S}(\ker \mathbb{A}^{\perp}). \tag{4.3.22}$$

Proposition 4.3.10 *If $\mathcal{T} := (\mathbb{A}, \mathbb{B}, \mathbb{S})$ is a differential triplet with minimal operator $A_{\mathrm{m},\mathcal{T}}$ then $\mathbb{A}^*$, the adjoint of the maximal operator of $\mathcal{T}$ is equal to the minimal operator of the* *adjoint triplet* $\mathcal{T}^* = (\mathbb{SAS}, \mathbb{B}^*, \mathbb{S})$. *In symbol,*

$$\mathbb{A}^* = A_{\mathrm{m},\mathcal{T}^*}$$

Moreover,

$$\mathbb{A}^* = \mathbb{S}A_{\mathrm{m},\mathcal{T}}\mathbb{S}.$$

Thus the minimal operator of $\mathcal{T}^*$ *is the conjugate of the minimal operator of* $\mathcal{T}$ *through* $\mathbb{S}$, *i.e.* $A_{\mathrm{m},\mathcal{T}^*} = \mathbb{S}A_{\mathrm{m},\mathcal{T}}\mathbb{S}$.

Proof By the definition of the minimal operator, $A_{\mathrm{m}} \subseteq \mathbb{A}$. Thus $\mathbb{S}A_{\mathrm{m}}\mathbb{S} \subseteq \mathbb{SAS}$. Moreover, Proposition 4.3.5 tells us that $\mathbb{A}$ is a boundary restriction operator of $\mathcal{T}$. Thus $\mathbb{A}^* \subseteq \mathbb{SAS}$. Whence in order to show that the operators $\mathbb{S}A_{\mathrm{m}}\mathbb{S}$ and $\mathbb{A}^*$ are the same, it is enough to prove the equality of their domains. For

$$D(\mathbb{S}A_{\mathrm{m}}\mathbb{S}) = \mathbb{S}D(A_{\mathrm{m}}) = \mathbb{SBS}(\ker \mathbb{A}^{\perp}),$$

according to (4.3.22). On the other hand,

$$\begin{aligned} D(\mathbb{A}^*) &= \mathbb{B}^*(\ker \mathbb{A}^{\perp}) && \text{(by Proposition 4.3.4 (ii))} \\ &= \mathbb{SBS}(\ker \mathbb{A}^{\perp}) && \text{(by (4.2.4)).} \end{aligned}$$

Hence $D(\mathbb{A}^*)$ is equal to $D(\mathbb{S}A_{\mathrm{m}}\mathbb{S})$. Finally, $\mathbb{A}^*$ is the minimal operator of $\mathcal{T}^*$ since it is a restriction of $\mathbb{SAS}$ whose domain is $\mathbb{B}^*(\ker \mathbb{A}^{\perp})$, which is nothing but the domain of $A_{\mathrm{m},\mathcal{T}^*}$ by (4.3.19). □

Proposition 4.3.10 tells us that the adjoint of the *maximal operator* of a differential triplet is conjugate to the minimal operator of that differential triplet. The next result expresses that the adjoint of the minimal operator is conjugate to the *maximal operator*. We refer to Definition 4.2.1 for the definition of a *closed* differential triplet.

Corollary 4.3.11 *Let* A_{m} *be the minimal operator of a closed differential triplet* $\mathcal{T} := (\mathbb{A}, \mathbb{B}, \mathbb{S})$ *on a Hilbert space* $\mathcal{H}$. *Then* A_{m} *is a closed boundary restriction operator of* $\mathcal{T}$, *and*

$$(A_{\mathrm{m}})^* = \mathbb{SAS}. \tag{4.3.23}$$

Proof First of all, let us check that A_{m} is a boundary restriction operator. Since $\mathbb{A}$ is a closed operator and $\mathbb{S}$ is continuous on $\mathcal{H}$, $\ker \mathbb{AS}$ is a closed subspace of $\mathcal{H}$. By definition,

$$D(A_{\mathrm{m}}) = \mathbb{B}\big((\ker \mathbb{A}\mathbb{S})^{\perp}\big) = \mathbb{B}\big((\mathbb{S} \ker \mathbb{A})^{\perp}\big),$$

thus Proposition 4.3.8 yields that A_{m} is a boundary restriction operator.

Let us prove (4.3.23). According to Proposition 4.3.10,

$$\mathbb{A}^* = \mathbb{S}A_{\mathrm{m}}\mathbb{S}. \tag{4.3.24}$$

Since $\mathbb{A}$ is assumed to be closed, the adjoint of $\mathbb{A}^*$ is equal to $\mathbb{A}$ (thanks to Theorem 2.4.1). Moreover Theorem 2.4.10 yields that

$$(\mathbb{S}A_{\mathrm{m}}\mathbb{S})^* = \mathbb{S}(\mathbb{S}A_{\mathrm{m}})^*.$$

Since $\mathbb{S}$ lies in $\mathcal{L}(\mathcal{H})$, we may check that

$$(\mathbb{S}A_{\mathrm{m}})^* = (A_{\mathrm{m}})^*\mathbb{S}.$$

Hence (4.3.23) follows. Finally $\mathbb{A}^*$ is closed due to Proposition 2.2.19. Since $\mathbb{S}$ is closed as well and invertible, we deduce with (4.3.24) that A_{m} is closed. □

As a consequence of the latter results, we obtain the following simple characterization of *closed boundary restriction operators*.

Theorem 4.3.12 *Let* $\mathcal{T} := (\mathbb{A}, \mathbb{B}, \mathbb{S})$ *be a closed differential triplet on* $\mathcal{H}$ *with minimal operator* A_{m}. *Then the following assertions are equivalent:*

(i) *A is a closed boundary restriction operator of $\mathcal{T}$.*
(ii) *A is a closed operator on $\mathcal{H}$, and $A_{\mathrm{m}} \subseteq A \subseteq \mathbb{A}$.*

Proof Assuming (i), the only nontrivial fact to prove turns out to be that A is an extension of A_{m}. By Definition 4.3.1, $A^* \subseteq \mathbb{S}\mathbb{A}\mathbb{S}$. That is, from Corollary 4.3.11, $A^* \subseteq (A_{\mathrm{m}})^*$. Since A_{m} is closed by Corollary 4.3.11 again, we get $A_{\mathrm{m}} \subseteq A$. The converse is then obvious since A_{m} is densely defined and $(A_{\mathrm{m}})^* = \mathbb{S}\mathbb{A}\mathbb{S}$ (by (4.3.23)). □

Let us end this subsection, by investigating some consequences of commutativity properties between $\mathbb{A}$ and $\mathbb{S}$. We have seen in the course of Example 4.3.6 that the domain of the minimal operator of a differential triplet $\mathcal{T}$ may be equal to the domain of the minimal operator of $\mathcal{T}^*$. In an abstract setting, a sufficient condition for the equality of these domains is that $\mathbb{A}$ commute or anti-commute with $\mathbb{S}$.

To be more specific, considering a differential triplet $\mathcal{T} := (\mathbb{A}, \mathbb{B}, \mathbb{S})$ on a Hilbert space $\mathcal{H}$, we will assume that there exists ε in $\{-1, 1\}$ such that

$$\mathbb{S}\mathbb{A}\mathbb{S} = \varepsilon\mathbb{A}. \tag{4.3.25}$$

Since $\mathbb{S} = \mathrm{id}_{\mathcal{H}}$, $\mathbb{A}$ commutes with $\mathbb{S}$ when $\varepsilon = 1$, whereas $\mathbb{A}$ anti-commutes with $\mathbb{S}$ when $\varepsilon = -1$. If $\mathcal{T} := \mathcal{T}_{1,0} = (\mathrm{D}^1_{L^2}, B_{1,L^2}, S)$, then $\varepsilon = -1$. The thrust of the next result is that (4.3.25) implies that the minimal operator of $\mathcal{T}^*$ is either equal to the minimal operator of $\mathcal{T}$ or to its opposite.

Proposition 4.3.13 *Let* $\mathcal{T} := (\mathbb{A}, \mathbb{B}, \mathbb{S})$ *be a differential triplet satisfying* (4.3.25). *Then the following assertions hold:*

(i) $\ker \mathbb{A}$ *and* $\ker \mathbb{A}^\perp$ *are invariant with respect to* $\mathbb{S}$, *that is,*

$$\mathbb{S} \ker \mathbb{A} = \ker \mathbb{A}, \qquad \mathbb{S}(\ker \mathbb{A}^\perp) = \ker \mathbb{A}^\perp.$$

(ii) *For each* $\psi^\perp$ *in* $\ker \mathbb{A}^\perp$,

$$\mathbb{S}\mathbb{B}\mathbb{S}\psi^\perp = \varepsilon \mathbb{B}\psi^\perp. \tag{4.3.26}$$

(iii) *The minimal operators of* $\mathcal{T}$ *and* $\mathcal{T}^*$ *have the same domain and*

$$A_{\mathrm{m},\mathcal{T}^*} = \varepsilon A_{\mathrm{m},\mathcal{T}}. \tag{4.3.27}$$

Proof

(i) For each ξ in $\ker \mathbb{A}$, (4.3.25) yields

$$\mathbb{A}\mathbb{S}\xi = \varepsilon \mathbb{S}\mathbb{A}\xi = 0.$$

Thus, $\mathbb{S} \ker \mathbb{A}$ is a part of $\ker \mathbb{A}$. Since $\mathbb{S}^2 = \mathrm{id}_{\mathcal{H}}$, $\ker \mathbb{A} \subseteq \mathbb{S} \ker \mathbb{A}$. Thus $\mathbb{A}$ is invariant. Regarding $\ker \mathbb{A}^\perp$, Lemma 4.3.9 gives

$$\mathbb{S}(\ker \mathbb{A}^\perp) = (\mathbb{S} \ker \mathbb{A})^\perp = \ker \mathbb{A}^\perp,$$

since $\ker \mathbb{A}$ is invariant. That proves Item (i).

(ii) We claim that

$$\mathbb{S}\mathbb{B}\mathbb{S}h - \varepsilon \mathbb{B}h \in \ker \mathbb{A}, \qquad \forall\, h \in \mathcal{H}. \tag{4.3.28}$$

Indeed, using (4.3.25) and $\varepsilon^{-1} = \varepsilon$, we get

$$\mathbb{A}(\mathbb{S}\mathbb{B}\mathbb{S}h - \varepsilon \mathbb{B}h) = \varepsilon(\mathbb{S}\mathbb{A}\mathbb{S}\,\mathbb{S}\mathbb{B}\mathbb{S}h - \mathbb{A}\mathbb{B}h) = 0,$$

thanks to (4.2.2) and (4.2.3). Then the claim follows.

Now let $\psi^\perp$ belong to $\ker \mathbb{A}^\perp$. By (4.3.28), $\xi := (\mathbb{SBS} - \varepsilon\mathbb{B})\psi^\perp$ lies in $\ker \mathbb{A}$. Thus using $\mathbb{B}^* = \mathbb{SBS}$ and $\varepsilon = \pm 1$, we have

$$\|\xi\|^2 = -\varepsilon(\psi^\perp, \mathbb{SBS}\xi - \varepsilon\mathbb{B}\xi) = 0$$

by (4.3.28). Hence (4.3.26) follows.

(iii) Let us first prove that the minimal operators have the same domain. For, we compute

$$\begin{aligned} D(A_{\mathrm{m},\mathcal{T}}) &= \mathbb{BS}(\ker \mathbb{A}^\perp) && \text{(by (4.3.22))} \\ &= \mathbb{B}(\ker \mathbb{A}^\perp) && \text{(by (i))} \\ &= \mathbb{SBS}(\ker \mathbb{A}^\perp) && \text{(by (4.3.26))} \\ &= D(A_{\mathrm{m},\mathcal{T}^*}) && \text{(by (4.3.19)).} \end{aligned}$$

Finally, $A_{\mathrm{m},\mathcal{T}^*}$ is a restriction of $\mathbb{SAS}$ by definition of the minimal operator. Moreover, $\varepsilon A_{\mathrm{m},\mathcal{T}}$ is also a restriction of $\mathbb{SAS}$ by (4.3.25). Since we have just proved that these operators have the same domain, we obtain (4.3.27). □

Observe that (4.3.25) cannot be replaced by

$$\mathbb{SAS} = \lambda\mathbb{A},$$

where λ lies in $\mathbb{K}$. Indeed, applying $\mathbb{S}$ on the both sides, we deduce that $\mathbb{A} = \lambda^2\mathbb{A}$., so that $\lambda = \pm 1$ if $\mathbb{A}$ is nontrivial.

4.3.4 The Domain Structure Theorem

In order to describe closed boundary restriction operators of $\mathbb{A}$, it is sufficient, in view of Theorem 4.3.12, to study the closed operators A satisfying $A_{\mathrm{m}} \subseteq A \subseteq \mathbb{A}$. The following theorem provides the structure of their domain.

Theorem 4.3.14 (Domain Structure Theorem) *Being given a closed differential triplet* $(\mathbb{A}, \mathbb{B}, \mathbb{S})$*, let* A *be a restriction of* $\mathbb{A}$ *and let*

$$E := D(A) \cap \big(\ker \mathbb{A} \oplus \mathbb{B}(\ker \mathbb{AS})\big). \tag{4.3.29}$$

Then the following assertions holds true:

(i) *The three spaces* $\ker \mathbb{A}$, $\mathbb{B}(\ker \mathbb{AS})$, $\mathbb{B}\big((\ker \mathbb{AS})^\perp\big)$ *are in direct sum, and*

$$D(\mathbb{A}) = \ker \mathbb{A} \oplus \mathbb{B}(\ker \mathbb{AS}) \oplus \mathbb{B}\big((\ker \mathbb{AS})^{\perp}\big).$$

(ii) *The following propositions are equivalent:*
(ii-a) *A is a closed boundary restriction operator of $\mathbb{A}$.*
(ii-b) *E is a closed subspace of $D(\mathbb{A})$ and*

$$D(A) = E \oplus \mathbb{B}\big((\ker \mathbb{AS})^{\perp}\big). \tag{4.3.30}$$

Proof

(i) By Lemma 4.3.3,

$$D(\mathbb{A}) = \ker \mathbb{A} \oplus R(\mathbb{B}).$$

Besides one has clearly

$$R(\mathbb{B}) = \mathbb{B}(\ker \mathbb{AS}) + \mathbb{B}\big((\ker \mathbb{AS})^{\perp}\big).$$

By using $\mathbb{AB} = \mathrm{id}_{\mathcal{H}}$ and the fact that the intersection of $\ker \mathbb{AS}$ and $(\ker \mathbb{AS})^{\perp}$ is trivial, we show easily that $\mathbb{B}(\ker \mathbb{AS})$, $\mathbb{B}\big((\ker \mathbb{AS})^{\perp}\big)$ are in direct sum. Whence Item (i) follows from Proposition 2.1.10.

(ii) Assuming (ii-b), let us first show that A is closed. For, Lemma 4.3.15 yields that $D(A)$ is closed in $D(\mathbb{A})$. Since $\mathbb{A}$ is assumed to be closed, there results that A is a closed operator in $\mathcal{H}$. Second, since A is a closed operator satisfying $A_{\mathrm{m}} \subseteq A \subseteq \mathbb{A}$, Theorem 4.3.12 entails (ii-a).

Conversely, let us assume that A is a closed boundary restriction operator of $\mathbb{A}$. Then Theorem 4.3.12 entails that

$$D(A_{\mathrm{m}}) \subseteq D(A). \tag{4.3.31}$$

Let us show that E is closed. Since A is assumed to be closed and $A \subseteq \mathbb{A}$, we infer that $D(A)$ is closed in $D(\mathbb{A})$. Besides applying Lemma 4.3.15 with $\mathcal{E} := \ker \mathbb{A}$ and $F := (\ker \mathbb{AS})^{\perp}$, we obtain that $\ker \mathbb{A} \oplus \mathbb{B}(\ker \mathbb{AS})$ is closed in $D(\mathbb{A})$. Hence E (defined by (4.3.29)) is closed in $D(\mathbb{A})$. There remains to prove (4.3.30). For, let u live in $D(A)$. By (i), there exist u_1 in $\ker \mathbb{A} \oplus \mathbb{B}(\ker \mathbb{AS})$ and u_2 in $\mathbb{B}\big((\ker \mathbb{AS})^{\perp}\big)$ such that $u = u_1 + u_2$. Moreover, $u_1 = u - u_2$ belongs to $D(A)$ since

$$u_2 \in \mathbb{B}\big((\ker \mathbb{AS})^{\perp}\big) = D(A_{\mathrm{m}}) \subseteq D(A),$$

by (4.3.31). Whence u_1 lies in E. Now we derive easily that

$$D(A) = E + \mathbb{B}\big((\ker \mathbb{AS})^{\perp}\big).$$

Since the intersection is trivial, (4.3.30) follows. □

Lemma 4.3.15 *Let $\mathcal{X}$ be a Banach space, A be a closed operator on $\mathcal{X}$ and B lie in $\mathcal{L}(\mathcal{X})$ satisfying $AB = \mathrm{id}_{\mathcal{X}}$. Let also F, G be closed complementary subspaces of $\mathcal{X}$ and $\mathcal{E}$ be a closed subspace of $D(A)$ satisfying*

$$\mathcal{E} \subseteq \ker A + B(F). \tag{4.3.32}$$

Then $\mathcal{E} + B(G)$ is closed in $D(A)$. In particular $\ker A + B(G)$ is closed in $D(A)$.

Proof Let $(u_n)_{n\in\mathbb{N}}$ be a sequence in $\mathcal{E} + B(G)$ and u lie in $D(A)$ such that

$$u_n \longrightarrow u \qquad \text{in } D(A). \tag{4.3.33}$$

In order to show that u lies in $\mathcal{E} + B(G)$, we write, for all $n \geq 0$,

$$u_n = e_n + Bg_n, \tag{4.3.34}$$

where (e_n, g_n) lies in $\mathcal{E} \times G$. By (4.3.32),

$$e_n = h_n + Bf_n$$

where (h_n, f_n) lies in $\ker A \times F$. Thus

$$u_n = h_n + Bf_n + Bg_n. \tag{4.3.35}$$

By (4.3.33),

$$Au_n = f_n + g_n \longrightarrow Au.$$

Thus (see Proposition 2.3.5) the continuity (with respect to the norm of $\mathcal{X}$) of the projections on F along G and on F along G yields that

$$f_n \longrightarrow f_\infty, \qquad g_n \longrightarrow g_\infty,$$

for some (f_∞, g_∞) in $F \times G$. Now going back to (4.3.35) and using the continuity of B, we arrive at

$$h_n \longrightarrow u - Bf_\infty - Bg_\infty.$$

Since A is a closed operator on $\mathcal{X}$, its kernel is closed in $\mathcal{X}$, so that

$$h_\infty := u - Bf_\infty - Bg_\infty$$

lies in $\ker A$. Regarding e_n which is equal to $h_n + Bf_n$, setting $e_\infty := h_\infty + Bf_\infty$, one has

$$e_n \longrightarrow e_\infty, \qquad Ae_n = f_n \longrightarrow f_\infty.$$

Since A is closed, we get $Ae_\infty = f_\infty$, so that $(e_n)_{n\in\mathbb{N}}$ converges toward e_∞ in $D(A)$. By assumption, $\mathcal{E}$ is closed in $D(A)$; hence e_∞ belongs to $\mathcal{E}$. Finally, letting $n \to \infty$ in (4.3.34), we obtain

$$u = e_\infty + Bg_\infty \in \mathcal{E} + B(G).$$

Hence $\mathcal{E} + B(G)$ is closed in $D(A)$.

As a particular case, the choice of $\mathcal{E} = \ker A$ yields that $\ker A + B(G)$ is closed in $D(A)$. □

Corollary 4.3.16 *Let* $(\mathbb{A}, \mathbb{B}, \mathbb{S})$ *be a closed differential triplet, and* A *be a closed boundary restriction operator of* $\mathbb{A}$*. Then*

(i) *the three spaces* $\ker \mathbb{AS}$, $\mathbb{B}^*(\ker \mathbb{A})$, $\mathbb{B}^*(\ker \mathbb{A}^\perp)$ *are in direct sum, and*

$$D(\mathbb{SAS}) = \ker \mathbb{AS} \oplus \mathbb{B}^*(\ker \mathbb{A}) \oplus \mathbb{B}^*(\ker \mathbb{A}^\perp);$$

(ii) *One has*

$$D(A^*) = E^{\mathrm{s}} \oplus \mathbb{B}^*(\ker \mathbb{A}^\perp),$$

where

$$E^{\mathrm{s}} := D(A^*) \cap \big(\ker \mathbb{AS} \oplus \mathbb{B}^*(\ker \mathbb{A})\big). \tag{4.3.36}$$

Moreover, E^{s} *is a closed subspace of* $D(\mathbb{SAS})$*.*

Proof Let $\mathcal{T}^* = (\mathbb{SAS}, \mathbb{B}^*, \mathbb{S})$ be the adjoint differential triplet of $(\mathbb{A}, \mathbb{B}, \mathbb{S})$. Since A is a boundary restriction operator of $\mathbb{A}$, A^* is a restriction of $\mathbb{SAS}$. Moreover, Theorem 2.4.1 yields that $A^{**} = A$ because A is closed. Hence A^* is a boundary restriction operator of $\mathcal{T}^*$ and is closed according to Proposition 2.2.19. Also $\mathcal{T}^*$ is a closed differential triplet on $\mathcal{H}$ by (4.3.23). Now, applying Theorem 4.3.14 to $\mathcal{T}^*$ and A^*, we get easily the assertions of the corollary. □

Example 4.3.7 (Continuation of Example 4.2.4) Recalling (4.2.7), we apply the Domain Structure Theorem 4.3.14 to the differential triplet

$$\mathcal{T}_{1,0} = (\mathrm{D}^1_{L^2}, B_{1,L^2}, S).$$

Due to $\ker \mathrm{D}^1_{L^2} = \ker(\mathrm{D}^1_{L^2} S) = \langle g_1 \rangle$, one has

$$B_{1,L^2}\big((\ker \mathrm{D}^1_{L^2} S)^{\perp}\big) = B_{1,L^2}\big(g_1{}^{\perp}\big).$$

Besides, denoting by g_2 the identity function of $[0, b]$, that is,

$$g_2 : (0, b) \to \mathbb{R}, \qquad x \mapsto x,$$

we have $B_{1,L^2} g_1 = g_2$. Accordingly, Theorem 4.3.14 (i) tells us that each u in $D(\mathrm{D}^1_{L^2})$ reads

$$u = u_1 g_1 + u_1' g_2 + B_{1,L^2} \psi^{\perp}, \tag{4.3.37}$$

where u_1, u_1' belong to $\mathbb{K}$ and $\psi^{\perp}$ lies in $g_1{}^{\perp}$. More explicitly,

$$u(x) = u_1 + u_1' x + \int_0^x \psi^{\perp}(y)\,\mathrm{d}y, \qquad \forall x \in [0, b].$$

Since $\int_0^b \psi^{\perp}(y)\,\mathrm{d}y = (\psi^{\perp}, g_1) = 0$, we derive

$$u_1 = u(0), \qquad u_1' = b^{-1}\big(u(b) - u(0)\big). \tag{4.3.38}$$

Let us emphasize that the space

$$E_{\mathrm{M},1,0} := \ker \mathrm{D}^1_{L^2} \oplus B_{1,L^2}(\ker \mathrm{D}^1_{L^2} S) = \langle g_1, g_2 \rangle$$

encapsulates the boundary values of elements of $D(\mathrm{D}^1_{L^2})$, which is nothing but $H^1(0, b)$. Besides the number of boundary values is twice the dimension of $\ker \mathrm{D}^1_{L^2}$. We will see in Sect. 4.4 and Chap. 6 that these facts are of general significance.

Notice that once a basis is chosen in $\ker \mathrm{D}^1_{L^2}$ and in $B_{1,L^2}(\ker \mathrm{D}^1_{L^2} S)$, the coordinates u_1, u_1' are unique. Anticipating Definition 4.4.2, the ordered pair (u_1, u_1') will be called *the endogenous boundary coordinates of* u. The name "coordinates" refers to the general situation involving vector-valued functions. Since here only scalar functions are considered, we may call (u_1, u_1') *the endogenous boundary values of* u (see Sect. 5.7.1.1).

Regarding the adjoint differential triplet

$$(\mathcal{T}_{1,0})^* = \big(-\mathrm{D}^1_{L^2}, (B_{1,L^2})^*, S \big)$$

of $\mathcal{T}_{1,0}$, Corollary 4.3.16 yields that each f in $D(-\mathrm{D}^1_{L^2})$ reads

$$f = f_1 g_1 + f_1'(B_{1,L^2})^* g_1 + (B_{1,L^2})^* \varphi^\perp,$$

where f_1, f_1' are scalars and $\varphi^\perp$ lies in $g_1{}^\perp$. Since (see (4.3.42))

$$(B_{1,L^2})^* g_1(x) = \int_x^b g_1(y)\,\mathrm{d}y = b - x, \qquad \forall x \in (0, b),$$

we derive

$$f = f_1 g_1 + f_1' S g_2 + (B_{1,L^2})^* \varphi^\perp. \tag{4.3.39}$$

A more transparent relation is

$$f(x) = f_1 + f_1'(b - x) + \int_x^b \varphi^\perp(y)\,\mathrm{d}y, \qquad \forall x \in (0, b).$$

We may compute f_1 and f_1' from the latter identity. However, observing that Sf lies in $D(\mathrm{D}^1_{L^2})$, it is simpler to consider Sf. In effect, (4.3.39) yields that

$$Sf = f_1 g_1 + f_1' g_2 + B_{1,L^2} \psi^\perp,$$

where $\psi^\perp := S\varphi^\perp$. It turns out that $\psi^\perp$ lies in $g_1{}^\perp$ since (use alternatively Proposition 4.3.13)

$$(\psi^\perp, g_1) = (\varphi^\perp, S g_1) = (\varphi^\perp, g_1) = 0.$$

Then by (4.3.37) and (4.3.38), we get

$$f_1 = f(b), \qquad f_1' = b^{-1}\big(f(0) - f(b)\big). \tag{4.3.40}$$

That "coordinates trick" will be often used, in particular in Example 4.3.8 and in Sect. 6.5.3.1 for fractional operators.

Again, the space

$$E^s_{\mathrm{M},1,0} := \ker \mathrm{D}^1_{L^2} \oplus (B_{1,L^2})^* (\ker \mathrm{D}^1_{L^2}) = \langle g_1, S g_2 \rangle \tag{4.3.41}$$

encapsulates, but in a different way, the boundary values of the elements of $D(-A_{\mathrm{M},1})$. Observe that $E^s_{\mathrm{M},1,0} = E_{\mathrm{M},1,0}$; however, this space has been equipped with two different basis, which explains the differences between (4.3.38) and (4.3.40).

Let us consider the differential triplet (see (4.2.9)):

$$\mathcal{T}_{1,b} := \big(\mathrm{D}^1_{L^2}, -(B_{1,L^2})^*, S\big).$$

Arguing as for $\mathcal{T}_{1,0}$ and equipping the space

$$E_{\mathrm{M},1,b} := \ker \mathrm{D}^1_{L^2} \oplus -(B_{1,L^2})^*(\ker \mathrm{D}^1_{L^2} S)$$

with the basis $\langle g_1, -Sg_2\rangle$, each u in $D(\mathrm{D}^1_{L^2})$ reads

$$u = u_1 g_1 + u_1' g_2(\cdot - b) - (B_{1,L^2})^* \psi^\perp,$$

where u_1, u_1' belong to $\mathbb{K}$ and $\psi^\perp$ lies in $g_1{}^\perp$. Hence

$$u_1 = u(b), \qquad u_1' = b^{-1}\big(u(b) - u(0)\big). \tag{4.3.42}$$

□

More generally, let us consider the one parameter family of differential triplets $\mathcal{T}_{1,a}$ where a runs in $[0, b]$. The cases where a is equal to 0 and 1 have just been investigated. Now we will assume that a runs in $(0, b)$.

Example 4.3.8 (The Domain Structure Theorem Applied to $\mathcal{T}_{1,a}$) Let a belong to $(0, b)$. In view of (4.2.23), we consider the differential triplet

$$\mathcal{T}_{1,a} := (\mathbb{A}^a, \mathbb{B}^a, \mathbb{S}^a)$$

on $L^2(0, b)$. Owing to (4.2.25) and (4.2.26), we readily check that

$$\mathbb{S}^a \mathbb{A}^a \mathbb{S}^a = -\mathbb{A}^a.$$

Hence Proposition 4.3.13 yields that $\ker \mathbb{A}^a$ is invariant with respect to $\mathbb{S}^a$, so that

$$\mathbb{B}^a\big((\ker \mathbb{A}^a \mathbb{S}^a)^\perp\big) = \mathbb{B}^a\big(\langle \mathbf{1}_{\Omega_1}, \mathbf{1}_{\Omega_2}, \rangle^\perp\big).$$

Besides, for $i = 1, 2$,

$$\mathbb{B}^a \mathbf{1}_{\Omega_i} = g_2(\cdot - a)\mathbf{1}_{\Omega_i},$$

where by a slight abuse of notation, g_2 denotes the identity function of $\mathbb{R}$. Thus, the Domain Structure Theorem 4.3.14 (i) tells us that each u in $D(\mathbb{A}^a)$ reads

$$u = u_1 \mathbf{1}_{\Omega_1} + u_2 \mathbf{1}_{\Omega_2} + u_1' g_2(\cdot - a)\mathbf{1}_{\Omega_1} + u_2' g_2(\cdot - a)\mathbf{1}_{\Omega_2} + \mathbb{B}^a \psi^\perp, \tag{4.3.43}$$

where u_1, u_2, u_1', u_2' belong to $\mathbb{K}$ and $\psi^\perp$ lies in $\langle \mathbf{1}_{\Omega_1}, \mathbf{1}_{\Omega_2} \rangle^\perp$. Since (see (4.2.24))

$$D(\mathbb{A}^a) := \{u \in L^2(0, b) \mid u_{|\Omega_i} \in H^1(\Omega_i), \ \forall i = 1, 2\},$$

each function u in $D(\mathbb{A}^a)$ has a limit in $x = 0$ and $x = b$ and has also a right and left limit in $x = a$. Therefore, we set

$$u(a^+) := \lim_{x \to a,\, x > a} u(x), \qquad u(a^-) := \lim_{x \to a,\, x < a} u(x). \tag{4.3.44}$$

Going back to (4.3.43), we infer

$$u(a^-) = u_1, \qquad u(a^+) = u_2.$$

Furthermore,

$$u(0) = u_1 - u_1' a + \mathbb{B}^a \psi^\perp(0).$$

One has

$$\mathbb{B}^a \psi^\perp(0) = -\int_0^b \psi^\perp(x) \mathbf{1}_{\Omega_1}(x)\, \mathrm{d}x = -(\psi^\perp, \mathbf{1}_{\Omega_1}) = 0$$

since $\psi^\perp$ lies in $\langle \mathbf{1}_{\Omega_1}, \mathbf{1}_{\Omega_2} \rangle^\perp$. Accordingly, $u(0) = u_1 - u_1' a$. In a same way, $u(b) = u_2 + u_2'(b - a)$. To summarize, each u in $D(\mathbb{A}^a)$ possesses four "boundary values" $u(0)$, $u(a^-)$, $u(a^+)$, $u(b)$. These values are connected to the coordinates u_1, u_2, u_1', u_2' through the linear relations

$$\begin{cases} u(0) = u_1 - a u_1' \\ u(a^-) = u_1 \\ u(a^+) = u_2 \\ u(b) = u_2 + u_2'(b - a) \end{cases} \iff \begin{cases} u_1 = u(a^-) \\ u_2 = u(a^+) \\ u_1' = \dfrac{1}{a}\big(u(a^-) - u(0)\big) \\ u_2' = \dfrac{1}{b - a}\big(u(b) - u(a^+)\big) \end{cases}.$$

Regarding the adjoint differential triplet

$$(\mathcal{T}_{1,a})^* := \big(\mathbb{S}^a \mathbb{A}^a \mathbb{S}^a, (\mathbb{B}^a)^*, \mathbb{S}^a\big),$$

setting

$$E_{\mathrm{M}}^{\mathrm{S}} := \ker \mathbb{A}^a \mathbb{S}^a \oplus (\mathbb{B}^a)^*(\ker \mathbb{A}^a),$$

one has

$$E_{\mathrm{M}}^{\mathrm{S}} = \langle \mathbf{1}_{\Omega_1}, \mathbf{1}_{\Omega_2}, (\mathbb{B}^a)^* \mathbf{1}_{\Omega_1}, (\mathbb{B}^a)^* \mathbf{1}_{\Omega_2} \rangle,$$

so that Corollary 4.3.16 gives

$$D(\mathbb{S}^a \mathbb{A}^a \mathbb{S}^a) = E_{\mathrm{M}}^{\mathrm{S}} \oplus (\mathbb{B}^a)^*(\ker \mathbb{A}^{a\,\perp}).$$

Thus f lies in $D(\mathbb{S}^a \mathbb{A}^a \mathbb{S}^a)$ if and only if there exists four scalars f_1, f_2, f_1', f_2' and $\varphi^\perp$ in $\ker \mathbb{A}^{a\,\perp}$ such that

$$f = f_1 \mathbf{1}_{\Omega_1} + f_2 \mathbf{1}_{\Omega_2} + (\mathbb{B}^a)^*\big(f_1' \mathbf{1}_{\Omega_1} + f_2' \mathbf{1}_{\Omega_1} + \varphi^\perp\big). \tag{4.3.45}$$

Let us compute f_1, f_2, f_1', f_2' by using the coordinates trick featured in the previous example. Using again the invariance of the $\mathbf{1}_{\Omega_i}$'s with respect to $\mathbb{S}^a$, we get

$$\mathbb{S}^a f = f_1 \mathbf{1}_{\Omega_1} + f_2 \mathbf{1}_{\Omega_2} + f_1' \mathbb{B}^a \mathbf{1}_{\Omega_1} + f_2' \mathbb{B}^a \mathbf{1}_{\Omega_2} + \mathbb{B}^a (\mathbb{S}^a \varphi^\perp). \tag{4.3.46}$$

Since by Proposition 4.3.13, $\ker \mathbb{A}^{a\,\perp}$ is invariant with respect to $\mathbb{S}^a$, we deduce that $\mathbb{S}^a \varphi^\perp$ lies in $\ker \mathbb{A}^{a\,\perp}$. Hence by identification with (4.3.43), we end up with

$$\begin{cases} f(a^-) = f_1 - a f_1' \\ f(0) = f_1 \\ f(b) = f_2 \\ f(a^+) = f_2 + f_2'(b-a) \end{cases} \iff \begin{cases} f_1 = f(0) \\ f_2 = f(b) \\ f_1' = \dfrac{1}{a}\big(f(0) - f(a^-)\big) \\ f_2' = \dfrac{1}{b-a}\big(f(a^+) - f(b)\big) \end{cases}.$$

□

4.4 Maximal Operators with Finite-Dimensional Kernel

Let $\mathbb{K}$ be equal to $\mathbb{R}$ or $\mathbb{C}$ and $\mathcal{H}$ be a Hilbert space over $\mathbb{K}$, whose inner product is denoted by $(\cdot, \cdot)$. Recall that this inner product is anti-linear with respect to its second argument. In this section, being given a differential triplet $\mathcal{T} := (\mathbb{A}, \mathbb{B}, \mathbb{S})$ with minimal operator A_{m}, our aim is to compute the adjoint of any boundary restriction operators of $\mathbb{A}$. As a by-product, we will introduce *abstract endogenous boundary conditions for* $D(A)$ and $D(A^*)$ and also *abstract endogenous boundary coordinates* of elements of $D(\mathbb{A})$.

We will assume that $\mathbb{A}$ has nontrivial finite-dimensional kernel. Hence, there exist an integer $d \geq 1$ and linearly independent vectors $\xi_1, \ldots, \xi_d$ of $D(\mathbb{A})$ such that

$$\ker \mathbb{A} = \langle \xi_1, \ldots, \xi_d \rangle. \tag{4.4.1}$$

In other words, $(\xi_1, \ldots, \xi_d)$ is a basis of $\ker \mathbb{A}$.

Our basic tool will be Domain Structure Theorem 4.3.14. Therefore we have to deal with *closed* differential triplets. This is not an issue since (4.4.1) implies that $\mathcal{T}$ is closed. Indeed, according to Lemma 4.3.3, $D(\mathbb{A}) = \ker \mathbb{A} \oplus R(\mathbb{B})$. Thus, the following result entails that $\mathcal{T}$ is a closed differential triplet.

Proposition 4.4.1 *Let $\mathcal{X}$ be a $\mathbb{K}$-normed space. Let A be an operator on $\mathcal{X}$, B belong to $\mathcal{L}(\mathcal{X})$, and $\mathcal{E}$ be a closed subspace of $\mathcal{X}$. Assume in addition that:*

(i) $\ker A$ *and* $B(\mathcal{E})$ *are in direct sum, and* $D(A) = \ker A \oplus B(\mathcal{E})$;
(ii) $AB\psi = \psi$ *for all* $\psi \in \mathcal{E}$.

Then the following assertions are equivalent:

(a) $\ker A$ *is a closed subspace of* $\mathcal{X}$.
(b) A *is a closed operator on* $\mathcal{X}$.

Proof It is obvious that (b) implies (a). Conversely, let us consider a sequence $(u_n)_{n\in\mathbb{N}}$ in $D(A)$, and u, f in $\mathcal{X}$ such that

$$u_n \longrightarrow u, \qquad Au_n \longrightarrow f \qquad \text{in } \mathcal{X}.$$

We have to show that u is in the domain of A and $Au = f$. By the assumption (i), for each n in $\mathbb{N}$, there exists h_n in $\ker A$ and ψ_n in $\mathcal{E}$ such that

$$u_n = h_n + B\psi_n.$$

With (ii), we get $Au_n = \psi_n \to f$. Thus f belongs to $\mathcal{E}$ since $\mathcal{E}$ is assumed to be closed in $\mathcal{X}$. Moreover, by continuity of B,

$$h_n = u_n - B\psi_n \longrightarrow u - Bf.$$

Then (a) yields that $u - Bf$ lies in $\ker A$. Thus, recalling that f lives in $\mathcal{E}$, we derive

$$u = (u - Bf) + Bf \in \ker A \oplus B(\mathcal{E}).$$

Whence, u lies in $D(A)$ (by (i)), and $Au = f$ (thanks to (ii)), so that A is a closed. □

Let A be a closed boundary restriction operator of $\mathbb{A}$ in the sense of Definition 4.3.1. In view of the Domain Structure Theorem 4.3.14,

$$D(A) = E \oplus \mathbb{B}\big((\ker \mathbb{A}\mathbb{S})^{\perp}\big), \tag{4.4.2}$$

where E is the subspace of

$$E_{\mathrm{M}} := \ker \mathbb{A} \oplus \mathbb{B}(\ker \mathbb{A}\mathbb{S}), \tag{4.4.3}$$

defined by (4.3.29), namely,

$$E := D(A) \cap E_{\mathrm{M}}.$$

Assuming first that E is not trivial, we denote by $(e_1, \dots, e_{d_E})$ one of its basis, where d_E which ranges in $[1, 2d]$, is the dimension of E. For each integer j in $[1, d_E]$, the inclusion

$$E \subseteq \ker \mathbb{A} \oplus \mathbb{B}(\ker \mathbb{A}\mathbb{S})$$

and (4.4.1) yield that

$$e_j = x_{ij}\xi_i + \mathbb{B}\mathbb{S} y_{ij}\xi_i, \tag{4.4.4}$$

where x_{ij}, y_{ij} are scalars and *the convention of repeated indexes* is used, i.e.,

$$x_{ij}\xi_i := \sum_{i=1}^{d} x_{ij}\xi_i.$$

Setting

$$\omega_j := x_{ij}\xi_i, \qquad \omega'_j := y_{ij}\xi_i, \qquad \text{in } \ker \mathbb{A}, \tag{4.4.5}$$

we have

$$e_j = \omega_j + \mathbb{B}\mathbb{S}\omega'_j \in \ker \mathbb{A} \oplus \mathbb{B}(\ker \mathbb{A}\mathbb{S}), \qquad \forall\, 1 \le j \le d_E. \tag{4.4.6}$$

Taking advantage of the finite dimension of $\ker \mathbb{A}$, we will show that a closed boundary restriction operator of $\mathbb{A}$ has closed range.

Proposition 4.4.2 *Under the above assumptions and notations, in particular* (4.4.1)–(4.4.6), *let A be a closed boundary restriction operator of $\mathbb{A}$. Then the range of A is closed in $\mathcal{H}$.*

Proof Let $(u_n)_{n\in\mathbb{N}}$ be a sequence in $D(A)$ and f belong to $\mathcal{H}$ such that $Au_n \to f$ in $\mathcal{H}$. By (4.4.2), (4.4.6), for each $n \geq 0$, there exist $x_1^n, \dots, x_{d_E}^n$ in $\mathbb{K}$ and $\phi_n^\perp$ in $(\ker \mathbb{A}\mathbb{S})^\perp$ such that

$$u_n = x_j^n \omega_j + \mathbb{B}\big(x_j^n \mathbb{S}\omega_j' + \phi_n^\perp\big).$$

Thus,

$$Au_n = x_j^n \mathbb{S}\omega_j' + \phi_n^\perp \in \ker \mathbb{A}\mathbb{S} \oplus (\ker \mathbb{A}\mathbb{S})^\perp.$$

Denoting by P the orthogonal projection of $\mathcal{H}$ onto $\ker \mathbb{A}\mathbb{S}$, we infer

$$x_j^n \mathbb{S}\omega_j' \xrightarrow[n\to\infty]{} Pf \qquad \text{in } \mathcal{H}.$$

However, by closedness of the finite-dimensional space generated by $\mathbb{S}\omega_1', \dots, \mathbb{S}\omega_{d_E}'$, we deduce that Pf reads

$$Pf = x_j \mathbb{S}\omega_j',$$

for some $x_1, \dots, x_{d_E} \in \mathbb{K}$. Then we set

$$u := x_j e_j + \mathbb{B}\big(f - Pf\big).$$

Since $f - Pf$ lies in $(\ker \mathbb{A}\mathbb{S})^\perp$, and by (4.4.2), $D(A)$ contains $\mathbb{B}\big((\ker \mathbb{A}\mathbb{S})^\perp\big)$, we infer that u belongs to $D(A)$, and $Au = f$. The proof of the proposition is now completed. □

Proposition 4.4.3 *Under the assumptions of Proposition 4.4.2, and recalling that E is connected to A through* (4.4.2), *the following assertions holds true:*

(i) *If A is onto then* $\dim E \geq \dim \ker \mathbb{A}$.
(ii) *If A is one-to-one, then* $\dim E \leq \dim \ker \mathbb{A}$.
(iii) *If A is a bijection then* $\dim E = \dim \ker \mathbb{A}$.

Proof In view of (4.4.3), let $P : E_{\mathrm{M}} \to E_{\mathrm{M}}$ be the projection on $\mathbb{B}(\ker \mathbb{A}\mathbb{S})$ along $\ker \mathbb{A}$. By (4.4.2), each u in $D(A)$ reads $u = e + \mathbb{B}\phi^\perp$, where $(e, \phi^\perp)$ lies in $E \times (\ker \mathbb{A}\mathbb{S})^\perp$. Thus, since $e - Pe$ belongs to $\ker \mathbb{A}$,

$$Au = \mathbb{A}Pe + \phi^{\perp}.$$

If A is onto then for each h in $\ker \mathbb{AS}$, there exists $(e, \phi^{\perp})$ in $E \times (\ker \mathbb{AS})^{\perp}$ such that

$$\mathbb{A}Pe + \phi^{\perp} = h.$$

Since Pe lies in $\mathbb{B}(\ker \mathbb{AS})$, we derive that $\mathbb{A}Pe$ belongs to $\ker \mathbb{AS}$, so that by orthogonality, $\mathbb{A}Pe = h$. Thus $\mathbb{A}P(E) = \ker \mathbb{AS}$. Since $\dim \ker \mathbb{AS} = d$, we have $\dim E \geq d$.

If A is one-to-one then $E \cap \ker \mathbb{A} = \ker A = \{0\}$. Hence *Grassman's formula* leads to

$$\dim E = \dim(E + \ker \mathbb{A}) - \dim(\ker \mathbb{A}) \leq d,$$

since E and $\ker \mathbb{A}$ are subspaces of E_{M}, and $\dim E_{\mathrm{M}} = 2d$. □

4.4.1 Abstract Endogenous Boundary Coordinates

Let $\mathcal{T} := (\mathbb{A}, \mathbb{B}, \mathbb{S})$ be a differential triplet on $\mathcal{H}$. Being given a closed boundary restriction operator A of $\mathcal{T}$, we will begin to describe $D(A^*)$ in terms of algebraic linear equations involving only coordinates in the finite-dimensional space

$$E_{\mathrm{M}}^{\mathrm{s}} := \ker \mathbb{AS} \oplus \mathbb{B}^*(\ker \mathbb{A}).$$

The result is stated in Proposition 4.4.5. These equations are called *(abstract) endogenous boundary conditions of* A^* for some reasons that we will discuss now.

First let us be a little more specific about the space $E_{\mathrm{M}}^{\mathrm{s}}$. This space appears in the direct sum decomposition of $D(\mathbb{SAS})$ spelled out in Corollary 4.3.16 (i). According to Example 4.3.7, $E_{\mathrm{M}}^{\mathrm{s}}$ supports coordinates linearly related to the boundary values (see (4.3.40)).

The second reason comes from Proposition 4.3.10. If we think to $\mathbb{A}$ as a boundary condition-free differential operator then by integration by parts, all the boundary values of functions in the domain of its adjoint must vanish. Since $\mathbb{A}^*$ is the minimal operator of the adjoint differential triplet, there results that the boundary values of functions in

$$\mathbb{B}^*(\ker \mathbb{A}^{\perp})$$

are equal to zero. Since

$$D(\mathbb{SAS}) = E_{\mathrm{M}}^{\mathrm{s}} \oplus \mathbb{B}^*(\ker \mathbb{A}^{\perp}),$$

we may expect that $E_{\mathrm{M}}^{\mathrm{s}}$ records in some sense, the boundary values of functions in $D(\mathbb{SAS})$.

Hence, coordinates in $E_{\mathrm{M}}^{\mathrm{s}}$ can be seen as boundary coordinates. However unlike usual boundary coordinates, they are directly related to the ambient space $D(\mathbb{SAS})$; this is why they are said to be *endogenous*.

Since the equations describing $D(A^*)$ involve endogenous boundary coordinates, these equations are called *abstract endogenous boundary conditions for* $D(A^*)$. Then by duality, we will also characterize $D(A)$ by abstract endogenous boundary conditions expressed in terms of *endogenous boundary coordinates* related to the space E_{M} *defined by* (4.4.3).

Our first step toward endogenous boundary conditions consists in a reformulation of the abstract identity (2.2.16) connecting an operator and its adjoint. Starting from a boundary restriction operator A, this reformulation gives necessary and sufficient conditions for an element of $D(\mathbb{SAS})$ to be in the domain of A^*. These conditions are still under variational form, in the sense that they appear as set of equations parameterized by the elements of $D(A)$.

As explain above, the next step is the characterization of $D(A^*)$ by a system of linear equations (Proposition 4.4.5). Then it will be possible to define *abstract endogenous boundary coordinates* in Definition 4.4.2.

Let us notice that no assumption is made on the kernel of A in Theorem 4.4.4. Before to state that theorem, some notation are in order.

Lemma 4.3.3 tells us that for each u in $D(\mathbb{A})$, there exists a unique u_0 in $\ker \mathbb{A}$ such that

$$u = u_0 + \mathbb{B}\mathbb{A}u \qquad \text{in } D(\mathbb{A}). \tag{4.4.7}$$

In a symmetric way, by considering the adjoint triplet of $\mathcal{T}$, each f in $D(\mathbb{SAS})$ reads in a unique way

$$f = f_0 + \mathbb{B}^*\mathbb{SAS}f, \tag{4.4.8}$$

where f_0 lies in $\ker \mathbb{AS}$.

Theorem 4.4.4 *Let* $(\mathbb{A}, \mathbb{B}, \mathbb{S})$ *be a differential triplet and* A *be a boundary restriction operator of* $\mathbb{A}$.

(i) *If* f *lies in* $D(\mathbb{SAS})$ *then*

$$f \in D(A^*) \quad \Longleftrightarrow \quad (Au, f_0) = (u_0, \mathbb{SAS}f), \quad \forall u \in D(A),$$

where u_0 *and* f_0 *are given by* (4.4.7) *and* (4.4.8).

(ii) *If* A *is closed and* u *lies in* $D(\mathbb{A})$ *then*

$$u \in D(A) \quad \Longleftrightarrow \quad (\mathbb{A}u, f_0) = (u_0, A^*f), \quad \forall f \in D(A^*).$$

Proof For each f in $D(\mathbb{SAS})$ and u in $D(\mathbb{A})$, one has

$$\begin{aligned}(\mathbb{A}u, f) &= (\mathbb{A}u, f_0) + (\mathbb{BA}u, \mathbb{SAS}f) && \text{(by (4.4.8))}\\ &= (\mathbb{A}u, f_0) + (u, \mathbb{SAS}f) - (u_0, \mathbb{SAS}f) && \text{(by (4.4.7))}. \end{aligned} \tag{4.4.9}$$

(i) Assuming that f belongs to $D(A^*)$, we have since $A^* \subseteq \mathbb{SAS}$ and $A \subseteq \mathbb{A}$,

$$(u, \mathbb{SAS}f) = (u, A^*f) = (Au, f) = (\mathbb{A}u, f), \qquad \forall u \in D(A).$$

Thus with (4.4.9),

$$(Au, f_0) = (u_0, \mathbb{SAS}f), \qquad \forall u \in D(A). \tag{4.4.10}$$

Conversely, let us assume that f satisfies (4.4.10). Plugging (4.4.10) into (4.4.9), we get, for each u in $D(A)$,

$$(\mathbb{A}u, f) = (Au, f) = (u, \mathbb{SAS}f).$$

Thus the definition of the adjoint operator gives that f lies in $D(A^*)$.

(ii) Let u be in $D(A)$. Arguing as in the proof of (4.4.10), we get what we want, namely,

$$(\mathbb{A}u, f_0) = (u_0, A^*f), \qquad \forall f \in D(A^*).$$

Conversely, assume that u lives in $D(\mathbb{A})$ and satisfies the above identity. Then going back to (4.4.9) and using

$$\mathbb{SAS}f = A^*f, \qquad \forall f \in D(A^*),$$

we get

$$(\mathbb{A}u, f) = (u, \mathbb{SAS}f) = (u, A^*f), \qquad \forall f \in D(A^*).$$

Thus u belongs to $D(A^{**})$. However, by Theorem 2.4.1, $A^{**} = A$ since A is closed. □

Remark 4.4.1 The use of (4.4.7) in the proof of (4.4.9) generalizes the integration by parts in (4.1.10). Indeed (4.4.7) corresponds to (4.1.3), so that (4.4.7) can be seen as an abstract version of the *fundamental theorem of calculus*. Moreover in (4.1.10), $u(0)$ is a boundary value; thus, the inner product $(u_0, \mathbb{SAS}f)$ in (4.4.9) plays the role of a boundary integral, and u_0 which is an element of ker A, looks like a boundary value of u. □

4.4.1.1 Endogenous Boundary Conditions of A^*

Starting from the decomposition

$$D(\mathbb{SAS}) = \ker \mathbb{AS} \oplus \mathbb{B}^*(\ker \mathbb{A}) \oplus \mathbb{B}^*(\ker \mathbb{A}^{\perp}),$$

given by Corollary 4.3.16, any f in $D(\mathbb{SAS})$ reads (see (4.4.1))

$$f = f_i \mathbb{S}\xi_i + \mathbb{B}^* f_i' \xi_i + \mathbb{B}^* \varphi^{\perp}, \tag{4.4.11}$$

where f_i, f_i' belong to $\mathbb{K}$ for all $1 \leq i \leq d$ and $\varphi^{\perp}$ lies in $\ker \mathbb{A}^{\perp}$.

Proposition 4.4.5 *Under the above assumptions and notations, in particular A is a closed boundary restriction operator of $\mathbb{A}$,* (4.4.1)–(4.4.8) *hold and $1 \leq d_E \leq 2d$, let f belong to $D(\mathbb{SAS})$. Then recalling that f_0 is defined through* (4.4.8),

$$f \in D(A^*) \iff \big(\mathbb{S}\omega_j', f_0\big) = (\omega_j, \mathbb{SAS} f), \qquad \forall\, 1 \leq j \leq d_E \tag{4.4.12}$$

$$\iff (\omega_j', f_i \xi_i) = (\omega_j, f_i' \xi_i), \qquad \forall\, 1 \leq j \leq d_E, \tag{4.4.13}$$

where the convention of repeated indexes remains in force.

For reasons given at the beginning of this subsection, the equations in (4.4.13) characterizing the domain of A^* are called *(abstract) endogenous boundary conditions of* A^*.

Proof Let f be in $D(\mathbb{SAS})$ and u belong to $D(A)$. By Lemma 4.3.9,

$$\mathbb{B}\big((\ker \mathbb{AS})^{\perp}\big) = \mathbb{BS}(\ker \mathbb{A}^{\perp}).$$

Thus starting from the direct sum (4.4.2) and using the decompositions (4.4.6), we may write $D(A)$ under the form

$$D(A) = \langle \omega_1 + \mathbb{BS}\omega_1', \ldots, \omega_{d_E} + \mathbb{BS}\omega_{d_E}' \rangle \oplus \mathbb{BS}\big(\ker \mathbb{A}^{\perp}\big).$$

Then the first equivalence of Theorem 4.4.4 takes the form

$$f \in D(A^*) \tag{4.4.14}$$

$$\iff$$

$$\big(A(\omega_j + \mathbb{BS}\omega_j' + \mathbb{BS}\psi^{\perp}), f_0\big) = (\omega_j, \mathbb{SAS} f), \qquad \forall\, 1 \leq j \leq d_E,\ \forall \psi^{\perp} \in \ker \mathbb{A}^{\perp}.$$

However, by (4.2.2) and (4.2.3),

$$\left(A\mathbb{B}\mathbb{S}\psi^{\perp}, f_0\right) = \left(\mathbb{S}\psi^{\perp}, f_0\right) = \left(\psi^{\perp}, \mathbb{S}f_0\right) = 0,$$

since $\psi^{\perp}$ lies in $\ker \mathbb{A}^{\perp}$ and $\mathbb{S}f_0$ belongs to $\ker \mathbb{A}$. Thus, (4.4.14) is equivalent to

$$\left(\mathbb{S}\omega_j', f_0\right) = (\omega_j, \mathbb{S}\mathbb{A}\mathbb{S}f), \qquad \forall\, 1 \le j \le d_E.$$

That proves the first equivalence (4.4.12). Moreover, recalling (4.4.8), and using (4.4.11), we get $f_0 = f_i \mathbb{S}\xi_i$ and $\mathbb{S}\mathbb{A}\mathbb{S}f = f_i'\xi_i + \varphi^{\perp}$. Since $(\omega_j, \varphi^{\perp}) = 0$, (4.4.14) is equivalent to

$$(\omega_j', f_i\xi_i) = (\omega_j, f_i'\xi_i), \qquad \forall\, 1 \le j \le d_E.$$

□

4.4.1.2 Dimension of E^{s}

Let us recall that A is a closed boundary restriction operator of $(\mathbb{A}, \mathbb{B}, \mathbb{S})$. In view of Corollary 4.3.16, let E^{s} be the subspace of

$$E^{\mathrm{s}}_{\mathrm{M}} := \ker \mathbb{A}\mathbb{S} \oplus \mathbb{B}^*(\ker \mathbb{A}) \tag{4.4.15}$$

defined by (4.3.36) and satisfying

$$D(A^*) = E^{\mathrm{s}} \oplus \mathbb{B}^*(\ker \mathbb{A}^{\perp}). \tag{4.4.16}$$

Proposition 4.4.5 tells us that the elements of $D(A^*)$ are characterized by d_E linear equations with $2d$ unknowns. Let us recall that d_E and d denote, respectively, the dimension of E and $\ker \mathbb{A}$. Thus we may expect that E^{s} has dimension $2d - d_E$. In order to prove this, we introduce, for each $j \in [1, d_E]$, the following linear form on $E^{\mathrm{s}}_{\mathrm{M}}$:

$$F_j : \ker \mathbb{A}\mathbb{S} \oplus \mathbb{B}^*(\ker \mathbb{A}) \to \mathbb{K}, \qquad f_0 + f_1 \mapsto (f_0, \mathbb{S}\omega_j') - (\mathbb{S}\mathbb{A}\mathbb{S}f_1, \omega_j). \tag{4.4.17}$$

Notice that $\mathbb{B}^*(\ker \mathbb{A})$ is a part of $D(\mathbb{S}\mathbb{A}\mathbb{S})$ according to Corollary 4.3.16 (i); hence $\mathbb{S}\mathbb{A}\mathbb{S}f_1$ is well defined in (4.4.17). Now we are in position to compute the dimension of E^{s}.

Proposition 4.4.6 *Under the assumptions and notation of Proposition 4.4.5, and* (4.4.15)–(4.4.17)*, the dimension of E^{s} is equal to* $2d - d_E$.

Proof We claim that

$$E^{\mathrm{s}} = \bigcap_{j=1}^{d_E} \ker F_j. \tag{4.4.18}$$

Indeed, $E^{\mathrm{s}} = D(A^*) \cap E_{\mathrm{M}}^{\mathrm{s}}$ by (4.3.36) and the definition (4.4.15) of $E_{\mathrm{M}}^{\mathrm{s}}$. Moreover, the equivalence (4.4.12) entails that

$$\bigcap_{j=1}^{d_E} \ker F_j = D(A^*) \cap E_{\mathrm{M}}^{\mathrm{s}},$$

which proves the claim.

Since $\mathbb{S}\mathbb{A}\mathbb{S}\mathbb{B}^* = \mathrm{id}_{\mathcal{H}}$, we deduce that $\mathbb{B}^*$ is one-to-one. Hence (see (4.4.15)) $E_{\mathrm{M}}^{\mathrm{s}}$ has dimension $2d$. Therefore, by (4.4.18) and Proposition 2.2.3, E^{s} has dimension $2d - d_E$ provided $F_1, \ldots, F_{d_E}$ are linearly independent in the dual space of $E_{\mathrm{M}}^{\mathrm{s}}$. In order to check this independence, let $\lambda_1, \ldots, \lambda_{d_E}$ be scalars satisfying $\sum_{j=1}^{d_E} \overline{\lambda_j} F_j = 0$. By choosing

$$f_0 := \sum_{j=1}^{d_E} \lambda_j \mathbb{S}\omega'_j, \qquad f_1 := 0,$$

we derive $\sum_{j=1}^{d_E} \lambda_j \mathbb{S}\omega'_j = 0$. In a same way, the choice of

$$f_0 := 0, \qquad f_1 := \mathbb{B}^* \sum_{j=1}^{d_E} \lambda_j \omega_J$$

yields that $\sum_{j=1}^{d_E} \lambda_j \omega_j = 0$. Going back to (4.4.6), we infer that $\sum_{j=1}^{d_E} \lambda_j e_j = 0$. Thus $\lambda_j = 0$ for all j in $[1, d_E]$ since $(e_1, \ldots, e_{d_E})$ is a basis of E. The proof of the proposition is now complete. □

4.4.1.3 Endogenous Boundary Conditions of A

In view of (4.4.2), we assume in addition that E is not maximal, that is, $E \neq E_{\mathrm{M}}$, or equivalently that

$$1 \leq d_E \leq 2d - 1. \tag{4.4.19}$$

Then Proposition 4.4.6 entails that E^{s} has positive dimension equal to $2d - d_E$. Let $(e_1^s, \ldots, e_{2d-d_E}^s)$ be a basis of E^{s}. In view of (4.4.15), e_j^s reads

$$e_j^s = \mathbb{S}\omega_j^s + \mathbb{B}^*\omega_j^{s\prime}, \qquad \forall\, 1 \leq j \leq 2d - d_E, \tag{4.4.20}$$

for some ω_j^s and $\omega_j^{s\prime}$ in $\ker \mathbb{A}$. By the Domain Structure Theorem 4.3.14 (i) and Lemma 4.3.9, each u in $D(\mathbb{A})$ reads

$$u = u_i \xi_i + \mathbb{B}\mathbb{S}u_i' \xi_i + \mathbb{B}\mathbb{S}\psi^{\perp}, \tag{4.4.21}$$

where u_i, u_i' are scalars and $\psi^{\perp}$ lies in $\ker \mathbb{A}^{\perp}$.

The next result is the adjoint counterpart of Proposition 4.4.5. The proof of these propositions are similar. Moreover, a generalization of Proposition 4.4.7 is proved in full details in the sequel (see Proposition 5.5.3). Hence, the proof of Proposition 4.4.7 will be omitted.

Proposition 4.4.7 *Let us recall that* A *is a boundary restriction operator of the differential triplet* $(\mathbb{A}, \mathbb{B}, \mathbb{S})$*. Under the above assumptions and notations, in particular* (4.4.1)–(4.4.11) *and* (4.4.19)–(4.4.21)*, let* u *be any element of* $D(\mathbb{A})$*. Then*

$$u \in D(A) \qquad \Longleftrightarrow \qquad (\omega_j^{s\prime}, u_i \xi_i) = (\omega_j^{s}, u_i' \xi_i), \qquad \forall\, 1 \le j \le 2d - d_E.$$

As explained at the beginning of this subsection, the equations characterizing $D(A)$ in the latter proposition are called *(abstract) endogenous boundary conditions of* A.

4.4.1.4 Limit Cases and Endogenous Boundary Coordinates

As far as endogenous boundary conditions are concerned, our analysis is complete when E is a proper subset of E_{M}. There remains to investigate the cases where E is trivial and E is equal to E_{M}.

4.4.1.4.1 Case Where $E = \{0\}$

By (4.4.2) and Definition 4.3.5, A is equal to A_{m}, the minimal operator of $\mathbb{A}$. There results from (4.4.21) that a vector u of $D(\mathbb{A})$ lies in the domain of A_{m} if and only if

$$u_i = 0, \qquad u_i' = 0, \qquad \forall\, 1 \le i \le d. \tag{4.4.22}$$

That is to say, the coordinates of u on E_{M} vanish. In view of the discussion at the beginning of this subsection, the list $(u_1, \dots, u_d, u_1', \dots, u_d')$ of scalars is called *abstract endogenous boundary coordinates of* the element u of $D(\mathbb{A})$. We formalize this notion in the following definition.

Definition 4.4.2 Let $\mathcal{T} := (\mathbb{A}, \mathbb{B}, \mathbb{S})$ be a differential triplet on a $\mathbb{K}$-Hilbert space $\mathcal{H}$. We assume that $\mathbb{A}$ has nontrivial finite-dimensional kernel whose a basis is given by (4.4.1). Then for each u in $D(\mathbb{A})$, the unique $2d$-tuple $(u_1, \dots, u_d, u_1', \dots, u_d')$ satisfying (4.4.21) is called the *(abstract) endogenous boundary coordinates of* u *(with underlying triplet* $\mathcal{T}$*) and with respect to the basis* $(\xi_1, \dots, \xi_d)$ *of* $\ker \mathbb{A}$. □

Regarding the adjoint of A_{m}, Corollary 4.3.11 entails that $(A_{\mathrm{m}})^* = \mathbb{S}\mathbb{A}\mathbb{S}$. Thus

$$E^{\mathrm{s}} = E_{\mathrm{M}}^{\mathrm{s}} := \ker \mathbb{A}\mathbb{S} \oplus \mathbb{B}^*(\ker \mathbb{A}),$$

and $\mathbb{SAS}$ has no endogenous boundary conditions, so that $\mathbb{SAS}$ is boundary condition-free. By considering the differential triplet $(\mathbb{SAS}, \mathbb{B}^*, \mathbb{S})$, for any vector f in $D(\mathbb{SAS})$, we rewrite (4.4.11) under the form

$$f = f_i \mathbb{S}\xi_i + \mathbb{B}^* \mathbb{S} f_i' \mathbb{S}\xi_i + \mathbb{B}^* \mathbb{S}(\mathbb{S}\varphi^\perp),$$

where $\mathbb{S}\varphi^\perp$ belongs to $(\ker \mathbb{SAS})^\perp$. Thus in view of Definition 4.4.2, the $2d$-tuple

$$(f_1, \ldots, f_d, f_1', \ldots, f_d')$$

satisfying (4.4.11), is the *endogenous boundary coordinates of f with underlying triplet $\mathcal{T}^*$ and with respect to the basis* $(\mathbb{S}\xi_1, \ldots, \mathbb{S}\xi_d)$ *of* $\ker \mathbb{SAS}$.

Moreover, since

$$\mathbb{S}f = f_i \xi_i + \mathbb{BS} f_i' \xi_i + \mathbb{BS}\varphi^\perp,$$

there results by comparison with (4.4.21) that $(f_1, \ldots, f_d, f_1', \ldots, f_d')$ is the endogenous boundary coordinates of $\mathbb{S}f$ with underlying triplet $\mathcal{T}$ and with respect to the basis $(\xi_1, \ldots, \xi_d)$ of $\ker \mathbb{A}$.

4.4.1.4.2 Case Where $E = E_{\mathrm{M}}$

Then $d_E = 2d$, $A = \mathbb{A}$, and $\mathbb{A}$ has no endogenous boundary conditions. Regarding the adjoint of $\mathbb{A}$, Proposition 4.3.10 entails that $\mathbb{A}^*$ is the minimal operator of the adjoint differential triplet of $\mathcal{T}$. Thus arguing as above, we derive that all endogenous boundary coordinates of the elements of $D(\mathbb{S}A_{\mathrm{m}}\mathbb{S})$ vanish.

4.4.2 Boundary Restriction Operators of $\mathrm{D}^1_{L^2}$

We will apply the theory introduced in the previous subsection to the differential triplet $\mathcal{T}_{1,0}$. From now on, $\mathcal{T}_{1,0}$ will be denoted by $\mathcal{T}_1$, that is,

$$\mathcal{T}_1 := (\mathrm{D}^1_{L^2}, B_{1,L^2}, S).$$

We refer to Example 4.2.4 for the definition of that differential triplet.

We will describe the domain of each closed boundary restriction operator A of $\mathcal{T}_1$ by usual boundary conditions; that is, we will give one or two equations satisfied by the boundary values of functions belonging to $D(A)$. The main result featured in Proposition 4.4.8 states that the class of closed boundary restriction operators of $\mathcal{T}_1$ is exactly the class of operators $\mathrm{D}^1_{L^2}$ *supplemented with homogeneous linear boundary conditions*. The connection between closed boundary restriction operators and usual boundary conditions is made through *endogenous boundary values*. Note that in this framework of scalar valued functions, endogenous boundary coordinates will also be called *endogenous boundary values*.

One has $\ker \mathbb{A} = \langle g_1 \rangle$ and in view of (4.4.1), (4.4.3), (4.4.15)

$$d = 1, \qquad \xi_1 = g_1, \qquad E_{\mathrm{M}} = E_{\mathrm{M},1} := \langle g_1, g_2 \rangle, \qquad E_{\mathrm{M}}^{\mathrm{s}} = E_{\mathrm{M},1}^{\mathrm{s}} := \langle g_1, Sg_2 \rangle.$$

Let u lie in $H^1(0, b)$. Going back to (4.3.37), the *endogenous boundary values of u with respect to the basis (g_1) of* $\ker \mathrm{D}_{L^2}^1$ are (u_1, u_1'). Moreover they are related to the (usual) boundary values of u through (4.3.38), namely,

$$u_1 = u(0), \qquad u_1' = b^{-1}\big(u(b) - u(0)\big). \tag{4.4.23}$$

In a same way, by considering the differential triplet

$$(\mathcal{T}_1)^* = \big(-\mathrm{D}_{L^2}^1, (B_{1,L^2})^*, S\big),$$

the adjoint endogenous boundary values (f_1, f_1') of a function f in $D(-\mathrm{D}_{L^2}^1)$ with respect to the basis (g_1) satisfy (see (4.3.39), (4.3.40))

$$f_1 = f(b), \qquad f_1' = b^{-1}\big(f(0) - f(b)\big). \tag{4.4.24}$$

Being given a closed boundary restriction operator A of $\mathcal{T}_1$, we know by (4.4.2) that

$$D(A) = E \oplus B_{1,L^2}({g_1}^{\perp}),$$

where (see (4.3.29))

$$E = D(A) \cap E_{\mathrm{M},1} = D(A) \cap \langle g_1, g_2 \rangle.$$

Also by (4.4.16), (4.3.36),

$$D(A^*) = E^{\mathrm{s}} \oplus (B_{1,L^2})^*({g_1}^{\perp}),$$

where

$$E^{\mathrm{s}} = D(A^*) \cap E_{\mathrm{M},1}^{\mathrm{s}} = D(A^*) \cap \langle g_1, g_2 \rangle.$$

Now, we will classify the closed boundary restriction operators of $\mathrm{D}_{L^2}^1$ with respect to the dimension of E. Also, we will determinate in a systematic way the adjoint of each boundary restriction operator.

(i) If E is trivial then A is equal to $A_{\mathrm{m},\mathcal{T}_1}$ the minimal operator of $\mathcal{T}_1$. In view of Sect. 4.4.1.4 or Example 4.3.6, we have

$$u \in D(A_{\mathrm{m},\mathcal{T}_1}) \quad \Longleftrightarrow \quad u_1 = u_1' = 0 \quad \Longleftrightarrow \quad u(0) = u(b) = 0.$$

Moreover, the adjoint of $A_{\mathrm{m},\mathcal{T}_1}$ is $-\mathrm{D}^1_{L^2}$.

(ii) If $E = E_{\mathrm{M},1} = \langle g_1, g_2\rangle$ then $A = \mathrm{D}^1_{L^2}$. By Sect. 4.4.1.4 again, we get

$$f \in D\big((\mathrm{D}^1_{L^2})^*\big) \quad \Longleftrightarrow \quad f_1 = f_1' = 0 \quad \Longleftrightarrow \quad f(0) = f(b) = 0.$$

(iii) If E is one-dimensional, that is, $d_E = 1$, $E = \langle e_1\rangle$ where (see (4.4.5)–(4.4.6))

$$e_1 = x_{11}g_1 + y_{11}g_2, \qquad \omega_1 := x_{11}g_1, \qquad \omega_1' := y_{11}g_1,$$

Thus with the above representation of e_1 and (4.4.23), we easily get

$$u \in D(A) \Longleftrightarrow y_{11}u_1 = x_{11}u_1' \tag{4.4.25}$$

$$\Longleftrightarrow x_{11}u(b) = (x_{11} + by_{11})u(0). \tag{4.4.26}$$

(4.4.25) is *the endogenous boundary condition of* A and (4.4.26) is the usual boundary condition of A.

Regarding A^*, (4.4.13) and (4.4.24) yield

$$f \in D(A^*) \Longleftrightarrow \overline{y}_{11}f_1 = \overline{x}_{11}f_1' \tag{4.4.27}$$

$$\Longleftrightarrow \overline{x}_{11}f(0) = (\overline{x}_{11} + b\overline{y}_{11})f(b). \tag{4.4.28}$$

Referring to Sect. 4.4.1.3, (4.4.27) is the *adjoint endogenous boundary condition of* A^*. On the other hand, (4.4.28) is the characterization of $D(A^*)$ in terms of usual boundary conditions.

Let us emphasize some standard boundary conditions.

(a) If $x_{11} = 0$ then $D(A) = {}_0H^1(0, b)$, and A is the pivot operator $A_{\mathcal{T}_1}$ of $\mathcal{T}_1$ (see Definition 4.3.3). Also $D(A^*) = {}_bH^1(0, b)$.

(b) If $x_{11} + by_{11} = 0$ then $D(A) = {}_bH^1(0, b)$ and $D(A^*) = {}_0H^1(0, b)$.

(c) If $y_{11} = 0$ then

$$D(A) = D(A^*) = \{u \in H^1(0, b) \mid u(0) = u(b)\}.$$

That corresponds to *periodic boundary conditions.*

Finally, let us show that for each closed boundary restriction operator A of $\mathrm{D}^1_{L^2}$, one has

$$u(b)\overline{f(b)} = u(0)\overline{f(0)}, \qquad \forall u \in D(A),\ f \in D(A^*). \tag{4.4.29}$$

Indeed, that relation is clear for Items (i) and (ii). Regarding Item (iii), combining (4.4.28) and (4.4.26), one gets

$$x_{11}u\overline{f}(0) = (x_{11} + by_{11})\overline{f}(b)u(0) = x_{11}u\overline{f}(b).$$

If $x_{11} \neq 0$ then (4.4.29) holds. Otherwise, (4.4.29) follows from Item (a).

Let us notice that (4.4.29) may be obtained directly by integration by parts, indeed

$$0 = (Au, f) - (u, A^* f) = \int_0^b u'\overline{f} + u\overline{f}'\,\mathrm{d}x = u\overline{f}(b) - u\overline{f}(0).$$

See Corollary 5.5.4 for abstract integration by parts formula.

We claim that the notion of *closed boundary restriction operator of* $\mathrm{D}^1_{L^2}$ fits exactly with the one of the "operator $\mathrm{D}^1_{L^2}$ supplemented with homogeneous linear boundary conditions". In order to make that statement precise, let us define the latter notion.

Definition 4.4.3 Let $\mathrm{D}^1_{L^2} : H^1(0,b) \subset L^2(0,b) \to L^2(0,b)$ be defined by (4.1.1) and A be an operator on $L^2(0,b)$. We say that A is $\mathrm{D}^1_{L^2}$ *supplemented with homogeneous linear boundary conditions* if A is a restriction of $\mathrm{D}^1_{L^2}$ and one of the following conditions holds:

(i) $D(A) = \{u \in H^1(0,b) \mid u(0) = u(b) = 0\}$;
(ii) there exist a_1, a_2 in $\mathbb{K}$ such that

$$D(A) = \big\{u \in H^1(0,b) \mid a_1 u(0) + a_2 u(b) = 0\big\}.$$

Proposition 4.4.8 *Let A be an operator on $L^2(0,b)$. Then the following assertions are equivalent:*

(i) *A is a closed boundary restriction operator of* $\mathrm{D}^1_{L^2}$.
(ii) *A is* $\mathrm{D}^1_{L^2}$ *supplemented with homogeneous linear boundary conditions.*

Proof There results from the analysis made in Items (i)–(iii), that a closed boundary restriction operator of $\mathrm{D}^1_{L^2}$ is $\mathrm{D}^1_{L^2}$ supplemented with homogeneous linear boundary conditions. Conversely, let A be $\mathrm{D}^1_{L^2}$ supplemented with some homogeneous linear boundary conditions. Then A is closed, and $D(A)$ contains $H^1_0(0,b)$; thus the minimal operator $A_{\mathrm{m},\mathcal{T}_1}$ of $\mathcal{T}_1$ is a restriction of A. Thus $A^* \subseteq (A_{\mathrm{m},\mathcal{T}_1})^* = \mathbb{S}\mathrm{D}^1_{L^2}\mathbb{S}$, by

Corollary 4.3.11. Hence by Definition 3.3.1, A is a closed boundary restriction operator of $D^1_{L^2}$. □

Remark 4.4.4 For any $a \in (0, b)$, let

$$h_a(x) := \begin{cases} 1 & \text{if } x \in [0, a] \\ 0 & \text{otherwise} \end{cases}.$$

Then, by Proposition 4.3.1, the restriction $\tilde{A}$ of $D^1_{L^2}$ to $B_{1,L^2}(h_a{}^{\perp})$ is still densely defined. However, for each ψ in $h_a{}^{\perp}$, the function $u := B_{1,L^2}\psi$ satisfies

$$0 = (h_a, \tilde{A}u) = u(a) - u(0) = u(a).$$

Thus, in view of Definition 4.4.3, $\tilde{A}$ is not $\frac{d}{dx}$ supplemented with homogeneous linear boundary conditions. Hence Proposition 4.4.8 entails that $\tilde{A}$ is not a boundary restriction operator of $D^1_{L^2}$.

Also, since $\mathbb{S}h_a$ does not belong to $\ker D^1_{L^2}$, the equivalence of Proposition 4.3.8 tells us that $\tilde{A}^*$ is not a restriction of $\mathbb{S}D^1_{L^2}\mathbb{S}$. By (4.3.17), we see that $D(\tilde{A}^*)$ contains a nonsmooth element, namely, h_a.

To conclude, note that this example highlights the link between homogeneous linear boundary conditions for $D(A)$ and regularity of functions of $D(A^*)$. This explains to some extent why the condition $A^* \subseteq \mathbb{S}A\mathbb{S}$ in Definition 4.3.1 allows to introduce *abstract endogenous boundary values*.

Differential Quadruplets on Banach Spaces

5

In a Banach setting, things become more involved essentially because in general, a Banach space (even reflexive) cannot be identified with its anti-dual space (see Chap. 6 for concrete examples). Hence, we have to introduce *differential quadruplets*.

Since the anti-dual space of a Banach space may be quite complicated to identify and to work with, we will consider only *reflexive* spaces. Then, many of the results obtained in the Hilbertian setting can be extended, and their proof differs only by minor changes. However, for ease of reading, we will present the full proofs in this new setting as well. This extension of the theory of Chap. 4 is featured in Sects. 5.1–5.5.

Section 5.6 is concerned with *the algebraic calculus* of differential triplets and quadruplets. This calculus relies on the basic calculus of operators. In particular, the extension of the kernel of an operator induces *Caputo extensions of a differential quadruplet*. In applications, a Caputo extension allows to unify the analysis of *Riemann–Liouville* and *Caputo derivatives* by highlighting the fact that these operators are closed boundary restriction operators of the same differential quadruplet. Hence, they differ only by their boundary conditions. Besides, the algebraic calculus allows to define higher-order differential quadruplets whose instances are made up of higher-order fractional differential operators.

Let $\mathbb{K}$ be equal to $\mathbb{R}$ or $\mathbb{C}$, and $\mathcal{V}$ be a reflexive Banach space over $\mathbb{K}$. The anti-dual space of $\mathcal{V}$ is denoted as $\mathcal{V}^*$. We refer to Sect. 2.3.3 for the background on reflexive Banach spaces.

A. Rougirel, *Unified Theory for Fractional and Entire Differential Operators*,
Frontiers in Elliptic and Parabolic Problems,
https://doi.org/10.1007/978-3-031-58356-8_5

5.1 Differential Quadruplets

5.1.1 Definitions

In order to generalize Definition 4.2.1 to Banach spaces, (4.2.4) must be replaced. We will now explain why we have chosen (5.1.4) and (5.1.5) to replace (4.2.4). The first reason comes from an inspection of the basic operators $\frac{\mathrm{d}}{\mathrm{d}x}$ and ${}^{\mathrm{RL}}\mathrm{D}^{\alpha}_{L^p}$ on $L^p(0, b)$ (see Example 5.1.5 and Sect. 6.5). The second reason relies on the definition of a differential triplets. Indeed, let (A, B, S) be a differential triplet on a Hilbert space $\mathcal{H}$. Due to identification, (4.2.2) reads

$$AB = \mathrm{id}_{\mathcal{H}^*}.$$

Roughly speaking, the idea is to replace A and B by operators $\mathbb{A}$ and $\mathbb{B}$ acting on $\mathcal{V}^*$ and such that $\mathbb{A}$ is "close to A," and $\mathbb{B}$ is "close to B." Since by (4.2.4), $B = SB^*S$, we choose $\mathbb{B}$ to be equal to $S^*B^*S^*$. Now if $\mathbb{A}$ is a left inverse of $\mathbb{B}$ in $\mathcal{V}^*$, that is $\mathbb{A}\mathbb{B} = \mathrm{id}_{\mathcal{V}^*}$, then $\mathbb{A}$ may be expected to be "close to A."

Definition 5.1.1 Let $\mathcal{V}$ be a reflexive Banach space over $\mathbb{K}$, $\mathbb{A}$ be an operator on $\mathcal{V}^*$, and A, B, S be operators on $\mathcal{V}$. The quadruplet $(\mathbb{A}, A, B, S)$ is called an *(abstract) differential quadruplet on* $(\mathcal{V}, \mathcal{V}^*)$ if

$$B, S \in \mathcal{L}(\mathcal{V}) \tag{5.1.1}$$

$$AB = \mathrm{id}_{\mathcal{V}} \tag{5.1.2}$$

$$S^2 = \mathrm{id}_{\mathcal{V}}, \tag{5.1.3}$$

and the bounded operator $\mathbb{B}$ on $\mathcal{V}^*$ defined by

$$\mathbb{B} := S^*B^*S^* \tag{5.1.4}$$

satisfies

$$\mathbb{A}\mathbb{B} = \mathrm{id}_{\mathcal{V}^*}. \tag{5.1.5}$$

If $\mathcal{Q}$ denotes the differential quadruplet $(\mathbb{A}, A, B, S)$, then the operator $\mathbb{B}$ defined by (5.1.4) will also be denoted by $\mathbb{B}_{\mathcal{Q}}$. We call A a *maximal operator of* $\mathcal{Q}$. It will be denoted often by A_{M}, or by $A_{\mathrm{M},\mathcal{Q}}$ when several quadruplets are involved. $\mathbb{A}$ is also a *maximal operator of* $\mathcal{Q}$. It will be sometimes denoted by $\mathbb{A}_{\mathrm{M},\mathcal{Q}}$

If A_{M} and $\mathbb{A}$ are closed operators on $\mathcal{V}$ and $\mathcal{V}^*$, respectively, then the differential quadruplet $\mathcal{Q}$ is said to be *closed*. □

With this definition at hand, we will be able to develop a theory of differential quadruplets quite similar to those of differential triplets.

Clearly, if $(\mathbb{A}, A, B, S)$ is a differential quadruplet on $(\mathcal{V}, \mathcal{V}^*)$, then

$$(S^*)^2 = \mathrm{id}_{\mathcal{V}^*}, \qquad (S^*)^{-1} = S^* \tag{5.1.6}$$

$$S^* \mathbb{B} S^* = B^*, \qquad \mathbb{B}^* = SBS. \tag{5.1.7}$$

In order to define *adjoint* differential quadruplets, we first notice that $\mathcal{V}^*$ is also a reflexive Banach due to Proposition 2.3.11. Then, by using the reflexivity of $\mathcal{V}$ and (5.1.6)-(5.1.7), we show easily that $(SAS, S^*\mathbb{A}S^*, B^*, S^*)$ is a differential quadruplet on $(\mathcal{V}^*, \mathcal{V})$. Then, we set this definition.

Definition 5.1.2 Let $\mathcal{Q} := (\mathbb{A}, A, B, S)$ be a differential quadruplet on $(\mathcal{V}, \mathcal{V}^*)$. The differential quadruplet $(SAS, S^*\mathbb{A}S^*, B^*, S^*)$ on $(\mathcal{V}^*, \mathcal{V})$, denoted by $\mathcal{Q}^*$, is called the *adjoint differential quadruplet of* $\mathcal{Q}$. □

Let us notice that $\mathbb{B}_{\mathcal{Q}^*} = SBS$, and besides the adjoint differential quadruplet of $\mathcal{Q}^*$ is $\mathcal{Q}$.

5.1.2 Differential Triplets and Quadruplets

Let $\big(\mathcal{H}, (\cdot, \cdot)\big)$ be a Hilbert space and $\|\cdot\|$ be its associated norm.

According to Proposition 2.5.3, the space $\mathcal{V}_{\mathcal{H}} := (\mathcal{H}, \|\cdot\|)$ is a reflexive Banach space, and by Corollary 2.5.2, its anti-dual space $(\mathcal{V}_{\mathcal{H}})^*$ is identified with $\mathcal{V}_{\mathcal{H}}$. Then, we have at our disposal differential triplets on $\mathcal{H}$ together with differential quadruplets on $\mathcal{V}_{\mathcal{H}}$ (i.e., on $(\mathcal{V}_{\mathcal{H}}, \mathcal{V}_{\mathcal{H}})$). The issue is to study the relationships between these two classes of objects.

Obviously, if $\mathcal{T} = (\mathbb{A}, \mathbb{B}, \mathbb{S})$ is a differential triplet on $\mathcal{H}$ then

$$\mathcal{Q}_{\mathcal{T}} := (\mathbb{A}, \mathbb{A}, \mathbb{B}, \mathbb{S}) \tag{5.1.8}$$

is a differential quadruplet on $\mathcal{V}_{\mathcal{H}}$, and

$$\mathbb{B}_{\mathcal{Q}_{\mathcal{T}}} = \mathbb{B}. \tag{5.1.9}$$

However, if $(\mathbb{A}, A, B, S)$ is a differential quadruplet on $\mathcal{V}_{\mathcal{H}}$, then (A, B, S) and $(\mathbb{A}, \mathbb{B}, S)$ are not necessarily differential triplets on $\mathcal{H}$. Indeed, under the assumptions and notation of Example 4.2.4, let us assume in addition that $\mathbb{K} = \mathbb{C}$. Then, we claim that

$$\mathcal{Q} := (\mathrm{i}\mathrm{D}^1_{L^2}, -\mathrm{i}\mathrm{D}^1_{L^2}, \mathrm{i}B_{1,L^2}, S)$$

is a differential quadruplet but

$$T := (-\mathrm{iD}^1_{L^2}, \mathrm{i}B_{1,L^2}, S)$$

is not a differential triplet on $\mathcal{H} := L^2(0, b)$. Indeed, $\mathcal{Q}$ satisfies obviously (5.1.2) and (5.1.3). Moreover, since $(\mathrm{i}B_{1,L^2})^* = -\mathrm{i}SB_{1,L^2}S$, we infer

$$\mathbb{B} = -\mathrm{i}B_{1,L^2},$$

so that (5.1.5) holds with $\mathbb{A} := \mathrm{D}^1_{L^2}$. Hence, $\mathcal{Q}$ is a differential quadruplet on $\mathcal{V}_{L^2(0,b)}$. Regarding T, since

$$(\mathrm{i}B_{1,L^2})^* = -\mathrm{i}SB_{1,L^2}S \neq S(\mathrm{i}B_{1,L^2})S,$$

there results that T is not a differential triplet. For the same reason,

$$(\mathbb{A}, \mathbb{B}, S) = (\mathrm{iD}^1_{L^2}, -\mathrm{i}B_{1,L^2}, S)$$

is not a differential triplet on $L^2(0, b)$.

In order to go further, we unfold our framework by assuming more generally that $(\mathcal{V}, \mathcal{H}, \mathcal{V}^*)$ is a Gelfand triplet in the sense of Definition 2.5.3. The typical example is

$$(\mathcal{V}, \mathcal{H}, \mathcal{V}^*) = \left(L^p(0, b), L^2(0, b), L^q(0, b)\right)$$

where $2 \leq p < \infty$, and q is the conjugate exponent of p. Then, recalling Definition 2.2.17 of the realization of an operator, we set the following definition.

Definition 5.1.3 Let $\mathcal{V}$ be a reflexive Banach space, $\mathcal{H}$ be a Hilbert space and $\mathcal{Q} := (\mathbb{A}, A, B, S)$ be a differential quadruplet on $(\mathcal{V}, \mathcal{V}^*)$. If $(\mathcal{V}, \mathcal{H}, \mathcal{V}^*)$ is a Gelfand triplet and $\left(\mathbb{A}_{\mathcal{H}}, \mathbb{B}_{\mathcal{H}}, (S^*)_{\mathcal{H}}\right)$ is a differential triplet on $\mathcal{H}$, then $\mathcal{Q}$ is said to be *compatible with* $\mathcal{H}$, and the triplet $\left(\mathbb{A}_{\mathcal{H}}, \mathbb{B}_{\mathcal{H}}, (S^*)_{\mathcal{H}}\right)$, denoted by $\mathcal{T}_{\mathcal{Q}}$ is called the *realization of* $\mathcal{Q}$ on $\mathcal{H}$.

The next result records a necessary and sufficient condition for a differential quadruplet to be compatible.

Proposition 5.1.1 *Let $(\mathcal{V}, \mathcal{H}, \mathcal{V}^*)$ be a Gelfand triplet and $\mathcal{Q} := (\mathbb{A}, A, B, S)$ be a differential quadruplet on $(\mathcal{V}, \mathcal{V}^*)$. Then $\mathcal{Q}$ is compatible with $\mathcal{H}$ if and only if*

$$S^*(\mathcal{H}) \subseteq \mathcal{H}, \qquad \mathbb{B}_{\mathcal{Q}}(\mathcal{H}) \subseteq \mathcal{H} \tag{5.1.10}$$

$$S^*v = Sv \tag{5.1.11}$$

$$\mathbb{B}_{\mathcal{Q}}v = Bv, \qquad \forall\, v \in \mathcal{V}. \tag{5.1.12}$$

Besides, if $\mathcal{Q}$ is a closed differential quadruplet compatible with $\mathcal{H}$, then $\mathcal{T}_{\mathcal{Q}}$ is a closed differential triplet.

The proof of Proposition 5.1.1 will be differed.

Remark 5.1.4 Let us complete the issue addressed at the beginning of this subsection. Being given $\mathcal{Q} := (\mathbb{A}, A, B, S)$ a differential quadruplet on $\mathcal{V}_{\mathcal{H}}$, we have seen that $(\mathbb{A}, \mathbb{B}_{\mathcal{Q}}, S)$ is not always a differential triplet. Now Proposition 5.1.1 provides necessary and sufficient conditions, namely the following three assertions are equivalent.

(i) $(\mathbb{A}, A, B, S)$ is compatible with $\mathcal{H}$.
(ii) $(\mathbb{A}, \mathbb{B}_{\mathcal{Q}}, S^*)$ is a differential triplet.
(iii) $S^* = S$ and $B^* = SBS$.

Besides, if $(\mathbb{A}, A, B, S)$ is compatible with $\mathcal{H}$ then Item (iii) yields that $\mathbb{B}_{\mathcal{Q}}$, which by definition is equal to $S^*B^*S^*$, is nothing but B, thus $(\mathbb{A}, \mathbb{B}_{\mathcal{Q}}, S^*)$ is equal to $(\mathbb{A}, B, S)$. To conclude, if $(\mathbb{A}, A, B, S)$ is compatible with $\mathcal{H}$ then $(\mathbb{A}, B, S)$ is a differential triplet on $\mathcal{H}$. □

Let $\mathcal{V}$ be a reflexive Banach space over $\mathbb{K}$, and $\big(\mathcal{H}, (\cdot, \cdot)\big)$ be a $\mathbb{K}$-Hilbert space. We denote by $\mathrm{DQ}(\mathcal{V})$ the set of all differential quadruplets on $(\mathcal{V}, \mathcal{V}^*)$, and by $\mathrm{DT}(\mathcal{H})$ the set of all differential triplets on $\mathcal{H}$. Also, if $(\mathcal{V}, \mathcal{H}, \mathcal{V}^*)$ is a Gelfand triplet, then $\mathrm{DQc}(\mathcal{V})$ is the set of all differential quadruplets on $(\mathcal{V}, \mathcal{V}^*)$, which are compatible with $\mathcal{H}$.

In the case where $\mathcal{V} := \mathcal{V}_{\mathcal{H}}$, we define the map

$$t : \mathrm{DQc}(\mathcal{V}_{\mathcal{H}}) \to \mathrm{DT}(\mathcal{H}), \qquad \mathcal{Q} := (\mathbb{A}, A, B, S) \mapsto (\mathbb{A}, \mathbb{B}_{\mathcal{Q}}, S^*). \tag{5.1.13}$$

By Definition 5.1.3, $t(\mathcal{Q})$ is the realization of $\mathcal{Q}$. Using also Remark 5.1.4, we may write

$$t(\mathcal{Q}) = \mathcal{T}_{\mathcal{Q}} = (\mathbb{A}, B, S). \tag{5.1.14}$$

On the other hand, if $\mathcal{T} := (\mathbb{A}, \mathbb{B}, \mathbb{S})$ is a differential triplet on $\mathcal{H}$, then (with the notation (5.1.8)) the differential quadruplet $\mathcal{Q}_{\mathcal{T}} := (\mathbb{A}, \mathbb{A}, \mathbb{B}, \mathbb{S})$ is compatible with $\mathcal{H}$ since its realization is $\mathcal{T}$. Then, we may legitimately define the map

$$q : \mathrm{DT}(\mathcal{H}) \to \mathrm{DQc}(\mathcal{V}_{\mathcal{H}}), \qquad \mathcal{T} \mapsto \mathcal{Q}_{\mathcal{T}}. \tag{5.1.15}$$

Since $\mathcal{T}$ is the realization of $q(\mathcal{T})$,

$$t \circ q(\mathcal{T}) = \mathcal{T},$$

that is

$$t \circ q = \mathrm{id}_{\mathrm{DT}(\mathcal{H})},$$

or equivalently by using (5.1.14) and (5.1.15),

$$\mathcal{T}_{\mathcal{Q}_{\mathcal{T}}} = \mathcal{T}. \tag{5.1.16}$$

However, $q \circ t(\mathbb{A}, A, B, S) = (\mathbb{A}, \mathbb{A}, B, S)$ for each $(\mathbb{A}, A, B, S)$ in $\mathrm{DQc}(\mathcal{V}_{\mathcal{H}})$.

Proof of Proposition 5.1.1 We put $\tilde{S} := (S^*)_{\mathcal{H}}$. Let us first prove that the conditions (5.1.10)–(5.1.12) are sufficient. By Lemma 5.1.2, $\tilde{S}$, $\mathbb{B}_{\mathcal{H}}$ belong to $\mathcal{L}(\mathcal{H})$, and for all h in $\mathcal{H}$,

$$\tilde{S}h = S^*h, \qquad \mathbb{B}_{\mathcal{H}}h = \mathbb{B}h. \tag{5.1.17}$$

Thus, $\tilde{S}^2 = \mathrm{id}_{\mathcal{H}}$ by (5.1.6), and $\mathbb{A}_{\mathcal{H}}\mathbb{B}_{\mathcal{H}} = \mathrm{id}_{\mathcal{H}}$ thanks to (5.1.2). Moreover, $\tilde{S}$ is self-adjoint on $\mathcal{H}$ according to Lemma 5.1.4.

There remains to prove that

$$(\mathbb{B}_{\mathcal{H}})^* = \tilde{S}\mathbb{B}_{\mathcal{H}}\tilde{S}. \tag{5.1.18}$$

For, we first claim that

$$(B^*)_{\mathcal{H}}{}^* = \mathbb{B}_{\mathcal{H}}, \tag{5.1.19}$$

where $(B^*)_{\mathcal{H}}{}^*$ denotes the adjoint of $(B^*)_{\mathcal{H}}$ in $\mathcal{H}$. Indeed, since $B^* = S^*\mathbb{B}S^*$, we deduce from Lemma 5.1.3 using also (5.1.10), that $(B^*)_{\mathcal{H}}$ lies in $\mathcal{L}(\mathcal{H})$. Then, $(B^*)_{\mathcal{H}}{}^*$ is also continuous on $\mathcal{H}$ according to Proposition 2.2.20. Now, for each v, φ in $\mathcal{V}$, one has

$$\begin{aligned}\big(\varphi, (B^*)_{\mathcal{H}}{}^*v\big) &= \big((B^*)_{\mathcal{H}}\varphi, v\big) = \langle B^*\varphi, v\rangle_{\mathcal{V}^*,\mathcal{V}} \\ &= \langle \varphi, Bv\rangle_{\mathcal{V}^*,\mathcal{V}} = (\varphi, Bv) \\ &= (\varphi, \mathbb{B}_{\mathcal{H}}v)\end{aligned}$$

by the assumption (5.1.12). The claim follows by a density argument owing to the continuity of $(B^*)_{\mathcal{H}}{}^*$ and $\mathbb{B}_{\mathcal{H}}$. Now we are in a position to prove (5.1.18). By Lemma 5.1.3,

$$\mathbb{B}_{\mathcal{H}} = (S^*B^*S^*)_{\mathcal{H}} = \tilde{S}(B^*)_{\mathcal{H}}\tilde{S}. \tag{5.1.20}$$

Thus, since $\tilde{S}$ is self-adjoint, we deduce that $(\mathbb{B}_{\mathcal{H}})^* = \tilde{S}(B^*)_{\mathcal{H}}^*\,\tilde{S}$ in $\mathcal{L}(\mathcal{H})$. Then, (5.1.18) follows from (5.1.19). We have shown that $\mathcal{Q}$ is compatible with $\mathcal{H}$.

Conversely, if $\mathcal{Q}$ is compatible with $\mathcal{H}$ then $D(\tilde{S}) = \mathcal{H}$, thus $S^*(\mathcal{H}) \subseteq \mathcal{H}$. Since $\mathcal{Q}$ is compatible with $\mathcal{H}$, one has $\tilde{S}^* = \tilde{S}$. Hence, Lemma 5.1.4 entails that $S \subseteq S^*$. In the same way, $\mathbb{B}(\mathcal{H}) \subseteq \mathcal{H}$ since $D(\mathbb{B}_{\mathcal{H}}) = \mathcal{H}$.

There remains to prove (5.1.12). For each v, φ in $\mathcal{V}$, we compute

$$\begin{aligned}
(\varphi, \mathbb{B}_{\mathcal{H}} v) &= \big((\mathbb{B}_{\mathcal{H}})^* \varphi, v\big) \\
&= (\tilde{S}\, \mathbb{B}_{\mathcal{H}}\, \tilde{S}\varphi, v) && \text{(since } \mathcal{Q} \text{ is compatible with } \mathcal{H}\text{)} \\
&= \big((B^*)_{\mathcal{H}}\, \varphi, v\big) && \text{(by (5.1.20))} \\
&= \langle B^* \varphi, v\rangle_{\mathcal{V}^*, \mathcal{V}} \\
&= (\varphi, Bv).
\end{aligned}$$

Thus, (5.1.12) holds.

Finally, if $\mathcal{Q}$ is a closed differential quadruplet compatible with $\mathcal{H}$ then $\mathbb{A}$ is a closed operator on $\mathcal{V}^*$. Since by Proposition 2.5.4, $\mathcal{H}$ is continuously embedded in $\mathcal{V}^*$, we may check easily that $\mathbb{A}_{\mathcal{H}}$ is a closed operator on $\mathcal{H}$, therefore $\mathcal{T}_{\mathcal{Q}}$ is closed. □

The following three results are used in the proof of Proposition 5.1.1

Lemma 5.1.2 *Let $\mathcal{X}$, $\mathcal{Y}$ be $\mathbb{K}$-normed spaces such that $\mathcal{X}$ is continuously embedded in $\mathcal{Y}$. If L lies in $\mathcal{L}(\mathcal{Y})$ and satisfies*

$$L(\mathcal{X}) \subseteq \mathcal{X}, \tag{5.1.21}$$

then, $L_{\mathcal{X}}$ lies in $\mathcal{L}(\mathcal{X})$ and

$$L_{\mathcal{X}} x = Lx, \qquad \forall\, x \in \mathcal{X}.$$

Proof Since L lives in $\mathcal{L}(\mathcal{Y})$ and $\mathcal{X}$ is continuously embedded in $\mathcal{Y}$, we may show easily that $L_{\mathcal{X}}$ is a closed operator on $\mathcal{X}$. Then, with (5.1.21), $L_{\mathcal{X}}$ has domain $\mathcal{X}$, thus the closed graph Theorem 2.3.4 yields that $L_{\mathcal{X}}$ belongs to $\mathcal{L}(\mathcal{X})$. Thus, $L_{\mathcal{X}} x = Lx$ for each x in $\mathcal{X}$, by the definition of $L_{\mathcal{X}}$. □

Lemma 5.1.3 *Let $\mathcal{X}$, $\mathcal{Y}$ be $\mathbb{K}$-normed spaces such that $\mathcal{X}$ is continuously embedded in $\mathcal{Y}$. If C, L are operators in $\mathcal{L}(\mathcal{Y})$ satisfying $C(\mathcal{X}) \subseteq \mathcal{X}$ and $L(\mathcal{X}) \subseteq \mathcal{X}$ then $C_{\mathcal{X}}$, $L_{\mathcal{X}}$, $(CL)_{\mathcal{X}}$ belong to $\mathcal{L}(\mathcal{X})$, and*

$$(CL)_{\mathcal{X}} = C_{\mathcal{X}} L_{\mathcal{X}} \quad \text{in } \mathcal{L}(\mathcal{X}).$$

Proof Lemma 5.1.2 gives that $C_{\mathcal{X}}$, $L_{\mathcal{X}}$, $(CL)_{\mathcal{X}}$ belong to $\mathcal{L}(\mathcal{X})$, and also that for all x in $\mathcal{X}$,

$$(CL)_{\mathcal{X}}x = CLx = CL_{\mathcal{X}}x = C_{\mathcal{X}}L_{\mathcal{X}}x,$$

since $L_{\mathcal{X}}x$ lies in $\mathcal{X}$ and $C(\mathcal{X}) \subseteq \mathcal{X}$. □

Lemma 5.1.4 *Let $(\mathcal{V}, \mathcal{H}, \mathcal{V}^*)$ be a Gelfand triplet and C lie in $\mathcal{L}(\mathcal{V})$. We put $\tilde{C} := (C^*)_{\mathcal{H}}$ and assume that $C^*(\mathcal{H}) \subseteq \mathcal{H}$. Then, $\tilde{C}$ is self-adjoint on $\mathcal{H}$ if and only if C^* is an extension of C. In symbol,*

$$\tilde{C}^* = \tilde{C} \Longleftrightarrow C^*v = Cv, \qquad \forall v \in \mathcal{V}.$$

Proof Since $C^*(\mathcal{H}) \subseteq \mathcal{H}$, it is clear that

$$\tilde{C}\varphi = C^*\varphi \in \mathcal{H}, \qquad \forall \varphi \in \mathcal{H}. \tag{5.1.22}$$

On the other hand, for each (v, φ) in $\mathcal{V} \times \mathcal{H}$, one has

$$(\tilde{C}^*v, \varphi) = (v, \tilde{C}\varphi) \overset{(5.1.22)}{=} \overline{\langle C^*\varphi, v\rangle_{\mathcal{V}^*,\mathcal{V}}} = \overline{\langle \varphi, Cv\rangle_{\mathcal{V}^*,\mathcal{V}}} = (Cv, \varphi).$$

Thus,

$$\tilde{C}^*v - \tilde{C}v = Cv - C^*v, \qquad \forall v \in \mathcal{V}. \tag{5.1.23}$$

Since $D(\tilde{C}) = \mathcal{H}$, it is clear that the condition $\tilde{C}^* = \tilde{C}$ is sufficient. Conversely, if C^* extends C then (5.1.23) implies that

$$\tilde{C}^*v = \tilde{C}v, \qquad \forall v \in \mathcal{V}.$$

Moreover, $\tilde{C}$ lives in $\mathcal{L}(\mathcal{H})$ according to Lemma 5.1.2, thus $\tilde{C}^*$ belongs also to $\mathcal{L}(\mathcal{H})$. Since $\mathcal{V}$ is dense in $\mathcal{H}$, we deduce that $\tilde{C}^* = \tilde{C}$. □

Example 5.1.5 We will extend the framework of Sect. 4.1.1. Being given a nontrivial separable and reflexive Banach space $\mathcal{Y}$, and conjugate real numbers p, q in $(1, \infty)$, we put

$$\mathcal{V} = L^p(\mathcal{Y}) := L^p\big((0, b), \mathcal{Y}\big), \qquad \mathcal{V}^* = L^q(\mathcal{Y}^*) := L^q\big((0, b), \mathcal{Y}^*\big).$$

Let $\mathrm{D}^1_{\mathcal{V}}$, $B_{1,\mathcal{V}}$, $S_{\mathcal{V}}$ be the operators defined by

$$\mathrm{D}^1_{\mathcal{V}} : W^{1,p}(\mathcal{Y}) \subseteq L^p(\mathcal{Y}) \to L^p(\mathcal{Y}), \qquad u \mapsto \frac{\mathrm{d}}{\mathrm{d}x}u$$

$$B_{1,\mathcal{V}} : L^p(\mathcal{Y}) \to L^p(\mathcal{Y}), \qquad v \mapsto g_1 * v$$

$$S_{\mathcal{Y}} : L^p(\mathcal{Y}) \to L^p(\mathcal{Y}), \qquad v \mapsto v(b - \cdot).$$

Arguing as in Example 4.2.4 (but using Hölder inequality instead Cauchy–Schwarz inequality) and invoking Theorem 3.2.2, we may show that

$$\mathcal{Q}_{p,\mathcal{Y},1} := (\mathrm{D}^1_{\mathcal{Y}^*},\ \mathrm{D}^1_{\mathcal{Y}},\ B_{1,\mathcal{Y}},\ S_{\mathcal{Y}})$$

is a differential quadruplet on $\big(L^p(\mathcal{Y}), L^q(\mathcal{Y}^*)\big)$ and $\mathbb{B}_{\mathcal{Q}_{p,\mathcal{Y},1}} = B_{1,\mathcal{Y}^*}$. By setting

$$g_1\mathcal{Y} := \{g_1 Y \in L^p(\mathcal{Y}) \mid Y \in \mathcal{Y}\},$$

the kernel of $\mathrm{D}^1_{\mathcal{Y}}$ reads

$$\ker \mathrm{D}^1_{\mathcal{Y}} = g_1\mathcal{Y},$$

and, for each u in $W^{1,p}\big((0,b), \mathcal{Y}\big)$, the basic identity (4.3.7) takes the form

$$u(x) = u(0) + \int_0^x u'(y)\,\mathrm{d}y \qquad \text{in } \mathcal{Y}, \qquad \forall x \in [0,b].$$

Regarding the adjoint differential quadruplet

$$(\mathcal{Q}_{p,\mathcal{Y},1})^* = \big(S_{\mathcal{Y}}\mathrm{D}^1_{\mathcal{Y}}S_{\mathcal{Y}},\ (S_{\mathcal{Y}})^*\mathrm{D}^1_{\mathcal{Y}^*}(S_{\mathcal{Y}})^*,\ (B_{1,\mathcal{Y}})^*,\ (S_{\mathcal{Y}})^*\big)$$

on $\big(L^q(\mathcal{Y}^*), L^p(\mathcal{Y})\big)$, one has

$$S_{\mathcal{Y}}\mathrm{D}^1_{\mathcal{Y}}S_{\mathcal{Y}} = -\mathrm{D}^1_{\mathcal{Y}}, \qquad (S_{\mathcal{Y}})^* = S_{\mathcal{Y}^*}, \qquad (B_{1,\mathcal{Y}})^* = S_{\mathcal{Y}^*}B_{1,\mathcal{Y}^*}S_{\mathcal{Y}^*},$$

so that

$$(\mathcal{Q}_{p,\mathcal{Y},1})^* = (-\mathrm{D}^1_{\mathcal{Y}},\ -\mathrm{D}^1_{\mathcal{Y}^*},\ S_{\mathcal{Y}^*}B_{1,\mathcal{Y}^*}S_{\mathcal{Y}^*},\ S_{\mathcal{Y}^*}).$$

Moreover, (4.3.7) reads here

$$f = f_0 - S_{\mathcal{Y}^*}B_{1,\mathcal{Y}^*}S_{\mathcal{Y}^*}\mathrm{D}^1_{\mathcal{Y}^*}f, \qquad \forall f \in W^{1,q}\big((0,b), \mathcal{Y}^*\big)$$

where f_0 lies in $\ker \mathrm{D}^1_{\mathcal{Y}^*}$, which is nothing but $g_1\mathcal{Y}^*$. Thus

$$f(x) = f(b) - \int_x^b f'(y)\,\mathrm{d}y, \qquad \forall x \in [0,b].$$

□

A nice situation is when $\mathbb{A}$ extends A_{M}, since then the three operators of the realization extend operators of the initial differential quadruplet.

Corollary 5.1.5 *Let* $(\mathcal{V}, \mathcal{H}, \mathcal{V}^*)$ *be a Gelfand triplet and* $\mathcal{Q} := (\mathbb{A}, A_{\mathrm{M}}, B, S)$ *be a differential quadruplet on* $(\mathcal{V}, \mathcal{V}^*)$ *which is compatible with* $\mathcal{H}$*. Then,*

$$B \subseteq \mathbb{B}_{\mathcal{H}}, \qquad S \subseteq (S^*)_{\mathcal{H}}.$$

If in addition $A_{\mathrm{M}} \subseteq \mathbb{A}$ *then*

$$A_{\mathrm{M}} \subseteq \mathbb{A}_{\mathcal{H}}.$$

Proof Let v belong to $\mathcal{V}$. By (5.1.12), $Bv = \mathbb{B}v$. Since $\mathcal{V}$ is contained in $\mathcal{H}$, there results that v lies in the domain of $\mathbb{B}_{\mathcal{H}}$ and $\mathbb{B}_{\mathcal{H}}v = Bv$. That proves the first extension. In the same way, $Su = S^*v$ by (5.1.11). Since $\mathcal{V}$ is contained in $\mathcal{H}$, there results that v lies in the domain of $(S^*)_{\mathcal{H}}$ and $(S^*)_{\mathcal{H}}v = Sv$. Thus, $S \subseteq (S^*)_{\mathcal{H}}$. Finally, for each u in $D(A_{\mathrm{M}})$, assume that $A_{\mathrm{M}}u = \mathbb{A}u$. Since $\mathcal{V}$ is contained in $\mathcal{H}$, there results that u lies in the domain of $\mathbb{A}_{\mathcal{H}}$ and $\mathbb{A}_{\mathcal{H}}u = A_{\mathrm{M}}u$. Thus, $A_{\mathrm{M}} \subseteq \mathbb{A}_{\mathcal{H}}$. □

5.2 Boundary Restriction Operators

5.2.1 Preliminaries

In Sect. 4.3, we have introduced boundary restriction operators on Hilbert spaces. We will do the same thing in Banach spaces. For all purposes, let us recall some underlying ideas leading to the concept of boundary restriction operators (see also the introduction of Sect. 4.3 for a complementary point of view). In view of Example 5.1.5, the operator $\mathrm{D}^1_{L^p}$ is boundary condition free. The adjonction of *homogeneous linear boundary conditions* to that operator gives a restriction still densely defined and does not change the regularity of the functions lying in the domain of its adjoint. Written in an abstract way, these properties lead to introduce the following definition.

Definition 5.2.1 Let $\mathcal{V}$ be a reflexive Banach space over $\mathbb{K}$, and $\mathcal{Q} := (\mathbb{A}, A_{\mathrm{M}}, B, S)$ be a differential quadruplet on $(\mathcal{V}, \mathcal{V}^*)$. An operator A on $\mathcal{V}$ is called a *boundary restriction operator of* $\mathcal{Q}$ or for the sake of simplicity, a *boundary restriction operator of* A_{M} if

(i) A is a densely defined restriction of A_{M};
(ii) $A^* \subseteq S^*\mathbb{A}S^*$.

In the sequel, we will generalize some results of Sect. 4.3. Of course, the main difference is that the adjoint of an operator on the reflexive Banach space $\mathcal{V}$ is now an operator on

$\mathcal{V}^*$, which in general, cannot be identified with $\mathcal{V}$. Also, the annihilator of a non-empty subset of $\mathcal{V}$ is a part of $\mathcal{V}^*$. However, for a non-empty subset F of $\mathcal{V}^*$, we do not have to distinguish between $F^{\top\mathcal{V}}$ and $F^{\perp\mathcal{V}}$ thanks to the identification of $\mathcal{V}^{**}$ and $\mathcal{V}$. Hence, in view of Proposition 2.3.11, we denote these identical spaces by $F^{\perp}$.

Proposition 5.2.1 *Being given a reflexive Banach space $\mathcal{V}$, let $(\mathbb{A}, A, B, S)$ be a differential quadruplet on $(\mathcal{V}, \mathcal{V}^*)$, and F be a closed subspace of $\mathcal{V}^*$ satisfying*

$$S^* F \cap D(\mathbb{A}) \subseteq \ker \mathbb{A} \tag{5.2.1}$$

then $B(F^{\perp})$ is dense in $\mathcal{V}$.

Proof By Corollary 2.2.10, it is enough to show that each f in $\mathcal{V}^*$ vanishing on $B(F^{\perp})$ is equal to zero. For each ψ in $F^{\perp}$, one has with (5.1.7),

$$0 = \langle f, B\psi \rangle_{\mathcal{V}^*,\mathcal{V}} = \langle B^* f, \psi \rangle_{\mathcal{V}^*,\mathcal{V}} = \langle S^*\mathbb{B}S^* f, \psi \rangle_{\mathcal{V}^*,\mathcal{V}}.$$

Thus, $S^*\mathbb{B}S^* f$ lies in $F^{\perp\perp}$. However, by reflexivity, Proposition 2.2.12 entails that $F^{\perp\perp} = F$ since F is assumed to be closed. Due to $(S^*)^2 = \mathrm{id}_{\mathcal{V}^*}$, we derive that $\mathbb{B}S^* f$ belongs to $S^* F$. Using also $R(\mathbb{B}) \subseteq D(\mathbb{A})$ and the assumption (5.2.1), we get

$$\mathbb{B}S^* f \in S^* F \cap D(\mathbb{A}) \subseteq \ker \mathbb{A}.$$

Thus (5.1.5) yields $S^* f = 0$, and finally $f = 0$ by (5.1.6). That proves the density of $B(F^{\perp})$. □

The choice of $F := \{0\}$ in Proposition 5.2.1 allows to introduce the following special densely defined operator.

Definition 5.2.2 Let $\mathcal{Q} := (\mathbb{A}, A_{\mathrm{M}}, B, S)$ be a differential quadruplet on $(\mathcal{V}, \mathcal{V}^*)$. The restriction of A_{M} to the domain $R(B)$ is called the *pivot operator of the differential quadruplet* $\mathcal{Q}$. That-is-to-say, the pivot operator of $\mathcal{Q}$, denoted by $A_{\mathcal{Q}}$ satisfies

$$D(A_{\mathcal{Q}}) = R(B), \qquad A_{\mathcal{Q}} u = A_{\mathrm{M}} u, \qquad \forall u \in D(A_{\mathcal{Q}}). \tag{5.2.2}$$

□

A differential quadruplet $\mathcal{Q} := (\mathbb{A}, A_{\mathrm{M}}, B, S)$ being given, it is easily proved as in Proposition 4.3.2 that

$$(A_{\mathcal{Q}})^{-1} = B, \tag{5.2.3}$$

and the following *Poincaré inequality* holds

$$\|u\|_{\mathcal{V}} \lesssim \|A_{\mathcal{Q}}u\|_{\mathcal{V}}, \qquad \forall u \in D(A_{\mathcal{Q}}). \tag{5.2.4}$$

5.2.2 Large and Small Restrictions of $\mathbb{A}$

In order to classify *boundary restriction operators* of a differential quadruplet $(\mathbb{A}, A_{\mathrm{M}}, B, S)$, we need to study the adjoint of restrictions of A_{M}. For, we will first consider *large* restrictions of A_{M}, i.e., restrictions of A_{M} that are *extension* of $A_{\mathcal{Q}}$. In other words, we will first study operators A satisfying

$$A_{\mathcal{Q}} \subseteq A \subseteq A_{\mathrm{M}}. \tag{5.2.5}$$

We will show in Proposition 5.2.3 that these operators are boundary restriction operators of $\mathbb{A}$. In the sequel, $\mathcal{V}$ still denotes a reflexive Banach space.

Proposition 5.2.2 *Let $\mathcal{Q} := (\mathbb{A}, A_{\mathrm{M}}, B, S)$ be a differential quadruplet on $(V, \mathcal{V}^*)$, and A be an operator on $\mathcal{V}$ satisfying* (5.2.5)*. Then, the following properties hold true.*

(i) *$D(A)$ is dense in $\mathcal{V}$, so that A^* exists. Moreover,*

$$B^*A^* = \mathrm{id}_{D(A^*)}. \tag{5.2.6}$$

(ii) *One has*

$$D(A^*) = B^*\big(\ker A^{\perp}\big) = \big\{B^*v^* \mid v^* \in \ker A^{\perp}\big\}. \tag{5.2.7}$$

(iii) *For each f in $D(A^*)$, there exists a unique v^* in* $\ker A^{\perp}$ *such that $f = B^*v^*$. Also $v^* = A^*f$ thus A^*f lies in* $\ker A^{\perp}$.

Proof (i) By Proposition 5.2.1 and (5.2.5), A has a dense domain. Moreover, for each f in $D(A^*)$ and for any v in $\mathcal{V}$,

$$\langle B^*A^*f, v\rangle_{\mathcal{V}^*,\mathcal{V}} = \langle A^*f, Bv\rangle_{\mathcal{V}^*,\mathcal{V}}.$$

Since $R(B) = D(A_{\mathcal{Q}})$, (5.2.5) entails that $R(B)$ is contained in $D(A)$, thus

$$\langle B^*A^*f, v\rangle_{\mathcal{V}^*,\mathcal{V}} = \langle f, ABv\rangle_{\mathcal{V}^*,\mathcal{V}} = \langle f, v\rangle_{\mathcal{V}^*,\mathcal{V}},$$

since A is a restriction of A_{M} and $A_{\mathrm{M}}B = \mathrm{id}_{\mathcal{V}}$. That completes the proof of item (i).

(ii) Let us denote for simplicity $B^*(\ker A^\perp)$ by D^*. Let f be in $D(A^*)$. In our context, (2.2.16) reads

$$\langle f, Au\rangle_{\mathcal{V}^*,\mathcal{V}} = \langle A^* f, u\rangle_{\mathcal{V}^*,\mathcal{V}}, \qquad \forall u \in D(A). \tag{5.2.8}$$

Choosing u in $\ker A$ in (5.2.8), we get

$$\langle A^* f, u\rangle_{\mathcal{V}^*,\mathcal{V}} = 0, \qquad \forall u \in \ker A.$$

Thus, $A^* f$ lives in $\ker A^\perp$. Then, (5.2.6) leads to $D(A^*) \subseteq D^*$. Conversely, consider $f := B^* v^*$ any element of D^*. Let u belong to $D(A)$. By (4.3.7), $u = u_0 + BAu$ for some (unique) u_0 in $\ker A$. Thus,

$$\langle f, Au\rangle_{\mathcal{V}^*,\mathcal{V}} = \langle v^*, BAu\rangle_{\mathcal{V}^*,\mathcal{V}} = \langle v^*, u - u_0\rangle_{\mathcal{V}^*,\mathcal{V}}.$$

Since v^* lies in $\ker A^\perp$, one has $\langle v^*, u_0\rangle_{\mathcal{V}^*,\mathcal{V}} = 0$, so that

$$\langle f, Au\rangle_{\mathcal{V}^*,\mathcal{V}} = \langle v^*, u\rangle_{\mathcal{V}^*,\mathcal{V}}.$$

By definition of $D(A^*)$, we infer f lies in $D(A^*)$ and $A^* f = v^*$. That proves (5.2.7) and the uniqueness of v^* in Item (iii). □

Proposition 5.2.3 *Let $\mathcal{Q} := (\mathbb{A}, A_{\mathrm{M}}, B, S)$ be a differential quadruplet on $(\mathcal{V}, \mathcal{V}^*)$, and A be an operator on $\mathcal{V}$. If A satisfies* (5.2.5)*, then A is a boundary restriction operator of $\mathcal{Q}$.*

Proof By Proposition 5.2.1, A is densely defined. Thus, there remains to prove that $A^* \subseteq S^*\mathbb{A}S^*$. Let f be in $D(A^*)$, and $v^* := A^* f$. Then

$$\begin{aligned}
S^* f &= S^* B^* A^* f && \text{(by (5.2.6))}\\
&= (S^*)^2 \mathbb{B} S^* A^* f && \text{(by (5.1.7))}\\
&= \mathbb{B} S^* v^* && \text{(by (5.1.6))}.
\end{aligned}$$

Thus, (5.1.5) yields that $S^* f$ belongs to $D(\mathbb{A})$ and $\mathbb{A}S^* f = S^* v^*$. Whence $S^*\mathbb{A}S^* f = A^* f$ and A^* is a restriction of $S^*\mathbb{A}S^*$. □

Recalling Definition 5.2.2 of a pivot operator, the following result states that the adjoint of the pivot operator is the pivot operator of the adjoint differential quadruplet.

Proposition 5.2.4 *Let $\mathcal{Q} := (\mathbb{A}, A_{\mathrm{M}}, B, S)$ be a differential quadruplet on $(\mathcal{V}, \mathcal{V}^*)$, and $\mathcal{Q}^* := (SA_{\mathrm{M}}S, S^*\mathbb{A}S^*, B^*, S^*)$ be its adjoint quadruplet on $(\mathcal{V}^*, \mathcal{V})$. Then, the pivot* *operator $A_{\mathcal{Q}}$ of $\mathcal{Q}$ is a closed boundary restriction operator of $\mathcal{Q}$, and*

$$(A_{\mathcal{Q}})^* = A_{\mathcal{Q}^*}. \tag{5.2.9}$$

Proof Proposition 5.2.3 entails that $A_{\mathcal{Q}}$ is a boundary restriction operator of A_{M}. In order to prove (5.2.9), we notice on the one hand that, by definition of $A_{\mathcal{Q}^*}$,

$$A_{\mathcal{Q}^*} \subseteq S^* \mathbb{A} S^*, \qquad D(A_{\mathcal{Q}^*}) = R(B^*). \tag{5.2.10}$$

On the other hand, since $A_{\mathcal{Q}}$ is a boundary restriction operator of A_{M},

$$(A_{\mathcal{Q}})^* \subseteq S^* \mathbb{A} S^*. \tag{5.2.11}$$

By Proposition 5.2.2 (ii),

$$D\big((A_{\mathcal{Q}})^*\big) = B^*\big((\ker A_{\mathcal{Q}})^{\perp}\big).$$

Since, by (5.2.4), $A_{\mathcal{Q}}$ is one-to-one, we get $D\big((A_{\mathcal{Q}})^*\big) = R(B^*)$. Hence with (5.2.10)

$$D\big((A_{\mathcal{Q}})^*\big) = D(A_{\mathcal{Q}^*}).$$

Then, (5.2.10) and (5.2.11) yield $(A_{\mathcal{Q}})^* = A_{\mathcal{Q}^*}$. Since $\mathcal{Q}$ is the adjoint differential quadruplet of $\mathcal{Q}^*$, we infer from (5.2.9) that $(A_{\mathcal{Q}^*})^* = A_{\mathcal{Q}}$, so that $A_{\mathcal{Q}}$ is a closed operator on $\mathcal{V}$ (according to Proposition 2.2.19). □

After the study of large restrictions of A_{M}, we will investigate "small" restrictions A of A_{M}. By "small" restrictions, we mean that A satisfies the conditions (i) and (ii) of Proposition 5.2.6.

Theorem 5.2.5 *Let A be an operator on a reflexive Banach space $\mathcal{V}$, B belong to $\mathcal{L}(\mathcal{V})$, and F be a closed subspace of $\mathcal{V}^*$. Suppose that A is densely defined and*

$$D(A) = B(F^{\perp}) \tag{5.2.12}$$

$$AB\psi = \psi, \qquad \forall\, \psi \in F^{\perp}. \tag{5.2.13}$$

Then, the following assertions hold true.

(i) $R(A) = F^{\perp}$, $\ker A^* = F$.
(ii) $R(B^*) \subseteq D(A^*)$, $A^* B^* = \mathrm{id}_{\mathcal{V}^*}$.
(iii) $D(A^*) = F \oplus R(B^*)$.
(iv) *For each f in $D(A^*)$, there exists a unique v_f^* in F such that*

$$f = v_f^* + B^* A^* f. \tag{5.2.14}$$

(iv) *Conversely, for each* v_0^* *in* F *and each* v^* *in* $\mathcal{V}^*$,

$$f := v_0^* + B^* v^* \in D(A^*), \qquad A^* f = v^*. \tag{5.2.15}$$

Proof We recall that thanks to the identification of $\mathcal{V}^{**}$ and $\mathcal{V}$, we do not have to distinguish between $F^{\top \mathcal{V}}$ and $F^{\perp \mathcal{V}}$. Hence, in view of Proposition 2.3.11, we denote these identical spaces by $F^{\perp}$. Then, the proof of Theorem 5.2.5 is analog to the one of Theorem 4.3.7. □

The following proposition is a key step toward characterization of boundary restriction operators.

Proposition 5.2.6 *Let* $\mathcal{V}$ *be a reflexive Banach space over* $\mathbb{K}$, *and* $(\mathbb{A}, A_{\mathrm{M}}, B, S)$ *be a differential quadruplet on* $(\mathcal{V}, \mathcal{V}^*)$. *We assume that*

(i) F *is a closed subspace of* $\mathcal{V}^*$;
(ii) A *is a restriction of* A_{M} *with domain* $B(F^{\perp})$.

Then, the following assertions hold true.

(a) A *is a boundary restriction operator of* A_{M} *if and only if* $S^* F \subseteq \ker \mathbb{A}$.
(b) A *is closed.*

Proof Assuming $S^* F \subseteq \ker \mathbb{A}$, we deduce that (5.2.1) holds true. Thus, Proposition 5.2.1 yields that $D(A)$ is dense in $\mathcal{V}$, so that A^* is well defined. Let f be in $D(A^*)$, and $v^* := A^* f$. By (5.2.14), (5.1.4) and (5.1.6), there exists v_f^* in F such that

$$S^* f = S^* v_f^* + \mathbb{B} S^* v^*.$$

By (5.1.5), $\mathbb{B} S^* v^*$ lies in $D(\mathbb{A})$ and $\mathbb{A}\mathbb{B} S^* v^* = S^* v^*$. Regarding $S^* v_f^*$, we know by hypothesis that $S^* v_f^*$ belongs to $\ker \mathbb{A}$, that is $\mathbb{A} S^* v_f^* = 0$. There results that $S^* f$ lives in $D(\mathbb{A})$ and $\mathbb{A} S^* f = S^* v^*$, thus $A^* \subseteq S^* \mathbb{A} S^*$.

Conversely, Theorem 5.2.5 (i) gives that $F = \ker A^*$. However, $A^* \subseteq S^* \mathbb{A} S^*$ implies that $\ker A^* \subseteq \ker S^* \mathbb{A} S^*$. Thus $S^* F \subseteq \ker \mathbb{A}$ due to $(S^*)^2 = \mathrm{id}_{\mathcal{V}^*}$.

Finally, let $(u_n)_{n \in \mathbb{N}}$ be a sequence in $D(A)$, and u, v belong to $\mathcal{V}$ such that

$$u_n \longrightarrow u, \qquad A u_n \longrightarrow v \qquad \text{in } \mathcal{V}.$$

By definition of $D(A)$, for each n in $\mathbb{N}$, u_n reads $B \psi_n^{\perp}$ for some $\psi_n^{\perp}$ in $F^{\perp}$. Thus

$$\psi_n^\perp = AB\psi_n^\perp = Au_n \longrightarrow v.$$

Since $F^\perp$ is closed in $\mathcal{V}$, we deduce that v lies in $F^\perp$. Next by continuity of B, $B\psi_n^\perp \longrightarrow Bv$. Thus, $u = Bv$ so that u lies in $D(A)$ and $Au = v$. Hence, A is closed. □

By taking $F := S^* \ker \mathbb{A} = \ker \mathbb{A}S^*$, Proposition 5.2.6 allows us to introduce a very "small" restrictions of A_{M}, the so-called *minimal operator.*

Definition 5.2.3 Let $\mathcal{V}$ be a reflexive Banach space and $\mathcal{Q} := (\mathbb{A}, A_{\mathrm{M}}, B, S)$ be a differential quadruplet on $(\mathcal{V}, \mathcal{V}^*)$. The *minimal operator of* $\mathcal{Q}$ is the restriction of A_{M} to the domain $BS(\ker \mathbb{A}^\perp)$. That minimal operator is denoted by $A_{\mathrm{m},\mathcal{Q}}$ or simply by A_{m} if no confusion may occur.

Before stating our next result, let us recall that *closed* and *adjoint* differential quadruplets are defined in Definitions 5.1.1 and 5.1.2, respectively.

Proposition 5.2.7 *Let* $\mathcal{Q} := (\mathbb{A}, A_{\mathrm{M}}, B, S)$ *be a closed differential quadruplet on* $(\mathcal{V}, \mathcal{V}^*)$, *and* $A_{\mathrm{m},\mathcal{Q}}$ *its minimal operator. Then* $A_{\mathrm{m},\mathcal{Q}}$ *is a closed boundary restriction operator of* A_{M}, *and*

$$(A_{\mathrm{m},\mathcal{Q}})^* = A_{\mathrm{M},\mathcal{Q}^*} = S^*\mathbb{A}S^*. \tag{5.2.16}$$

That-is-to-say, the adjoint of the minimal operator of $\mathcal{Q}$ *is the maximal operator of* $\mathcal{Q}^*$. *Besides, the minimal operator* $A_{\mathrm{m},\mathcal{Q}^*}$ *of* $\mathcal{Q}^*$ *satisfies*

$$(A_{\mathrm{m},\mathcal{Q}^*})^* = A_{\mathrm{M}}. \tag{5.2.17}$$

Proof In (5.2.16), the second equality is just the definition of $A_{\mathrm{M},\mathcal{Q}^*}$. Let us prove that $(A_{\mathrm{m},\mathcal{Q}})^* = S^*\mathbb{A}S^*$. Since $\mathbb{A}$ is closed, its kernel is a closed subspace of $\mathcal{V}^*$. By Lemma 4.3.9,

$$S(\ker \mathbb{A}^\perp) = (S^* \ker \mathbb{A})^\perp.$$

Hence, Proposition 5.2.6 entails that $A_{\mathrm{m},\mathcal{Q}}$ is a closed boundary restriction operator of A_{M}. In particular, $(A_{\mathrm{m},\mathcal{Q}})^* \subseteq S^*\mathbb{A}S^*$. Thus, in order to prove (5.2.16), it remains to show that $D(S^*\mathbb{A}S^*)$ is contained in the domain of $(A_{\mathrm{m},\mathcal{Q}})^*$. For, since $\mathcal{Q}^* = (SA_{\mathrm{M}}S, S^*\mathbb{A}S^*, B^*, S^*)$ is a differential quadruplet on $(\mathcal{V}^*, \mathcal{V})$, one has $S^*\mathbb{A}S^*B^* = \mathrm{id}_{\mathcal{V}^*}$. Thus,

$$\begin{aligned} D(S^*\mathbb{A}S^*) &= \ker S^*\mathbb{A}S^* + R(B^*) && \text{(by Lemma 4.3.3)} \\ &= S^* \ker \mathbb{A} + R(B^*) && \text{(with (5.1.6))} \end{aligned}$$

$$= D\big((A_{m,\mathcal{Q}})^*\big) \qquad \text{(by Th. 5.2.5 (iii))}.$$

That proves (5.2.16). Owing to the closedness of $SA_{M}S$ and $S^*\mathbb{A}S^*$, we may apply (5.2.16) to the adjoint differential quadruplet $\mathcal{Q}^*$, so that using $S^2 = \mathrm{id}_{\mathcal{V}}$, we get (5.2.17). □

The following result justifies the use of the adjective *minimal* in Definition 5.2.3.

Corollary 5.2.8 *Let $\mathcal{V}$ be a reflexive Banach space and $\mathcal{Q} := (\mathbb{A}, A_{M}, B, S)$ be a closed differential quadruplet on $(\mathcal{V}, \mathcal{V}^*)$. Then, the following assertions are equivalent.*

(i) *A is a closed boundary restriction operator of A_{M}.*
(ii) *A is a closed operator on $\mathcal{V}$, and $A_{m,\mathcal{Q}} \subseteq A \subseteq A_{M}$.*

Proof Assuming (i), the definition of boundary restriction operators yields that $A^* \subseteq S^*\mathbb{A}S^*$. That is, by Proposition 5.2.7, $A^* \subseteq (A_{m})^*$. Since A_{m} is closed according to Proposition 5.2.7, and $\mathcal{V}$ is a reflexive Banach space, we derive thanks Theorem 2.4.1 that $(A_{m})^{**} = A_{m}$. By assumption, A is closed as well, thus $A_{m} \subseteq A$. The converse is obvious since A_{m} is densely defined and $(A_{m})^* = S^*\mathbb{A}S^*$ by Proposition 5.2.7. □

5.2.3 Boundary Restriction Operators of Realizations of Quadruplets

In Sect. 5.1.2, we have studied relationships between a differential quadruplet and its realization on a Hilbert space. Now we will focus on relationships between boundary restriction operators of the quadruplet and boundary restriction operators of its realization.

Proposition 5.2.9 *Let $(\mathcal{V}, \mathcal{H}, \mathcal{V}^*)$ be a Gelfand triplet and $\mathcal{Q} := (\mathbb{A}, A_{M}, B, S)$ be a differential quadruplet on $(\mathcal{V}, \mathcal{V}^*)$. Assume that $\mathcal{Q}$ is compatible with $\mathcal{H}$, and $A_{M} \subseteq \mathbb{A}$. Then the pivot operator of the realization $\mathcal{T}_{\mathcal{Q}}$ of $\mathcal{Q}$ is an extension of the pivot operator of $\mathcal{Q}$. In symbol,*

$$A_{\mathcal{Q}} \subseteq A_{\mathcal{T}_{\mathcal{Q}}}.$$

In the same way, the minimal operator of the realization of $\mathcal{Q}$ is an extension of the minimal operator of $\mathcal{Q}$. In symbol,

$$A_{m,\mathcal{Q}} \subseteq A_{m,\mathcal{T}_{\mathcal{Q}}}.$$

Proof By Corollary 5.1.5, $B \subseteq \mathbb{B}_{\mathcal{H}}$. Thus, $R(B)$ is contained in $R(\mathbb{B}_{\mathcal{H}})$, that is $D(A_{\mathcal{Q}}) \subseteq D(A_{\mathcal{T}_{\mathcal{Q}}})$. Moreover,

$$A_{\mathcal{Q}} \subseteq A_{M} \subseteq \mathbb{A}$$

by assumption, and $A_{\mathcal{T}_{\mathcal{Q}}} \subseteq \mathbb{A}_{\mathcal{H}} \subseteq \mathbb{A}$. Thus, $A_{\mathcal{Q}}$ and $A_{\mathcal{T}_{\mathcal{Q}}}$ coincide on $R(B)$.

Regarding minimal operators, we denote by $(\cdot, \cdot)$ the inner product of $\mathcal{H}$, and recall that

$$D(A_{\mathrm{m},\mathcal{Q}}) := BS(\ker \mathbb{A}^{\perp \mathcal{V}}), \qquad D(A_{\mathrm{m},\mathcal{T}_{\mathcal{Q}}}) := \mathbb{B}_{\mathcal{H}}(S^*)_{\mathcal{H}}(\ker \mathbb{A}_{\mathcal{H}}{}^{\perp \mathcal{H}}).$$

We claim that

$$\ker \mathbb{A}^{\perp \mathcal{V}} \subseteq \ker \mathbb{A}_{\mathcal{H}}{}^{\perp \mathcal{H}}. \tag{5.2.18}$$

Indeed, let $v^{\perp}$ lie in $\ker \mathbb{A}^{\perp \mathcal{V}}$. For each ξ in $\ker \mathbb{A}_{\mathcal{H}}$, ξ lives in $\ker \mathbb{A}$, so that

$$0 = \langle \xi,\, v^{\perp} \rangle_{\mathcal{V}^*,\mathcal{V}} = (\xi,\, v^{\perp}).$$

Hence, $v^{\perp}$ lies in $\ker \mathbb{A}_{\mathcal{H}}{}^{\perp \mathcal{H}}$, and the claim follows. By Corollary 5.1.5,

$$S \subseteq (S^*)_{\mathcal{H}}, \qquad B \subseteq \mathbb{B}_{\mathcal{H}}. \tag{5.2.19}$$

Then, (5.2.18) and (5.2.19) yield

$$D(A_{\mathrm{m},\mathcal{Q}}) \subseteq D(A_{\mathrm{m},\mathcal{T}_{\mathcal{Q}}}).$$

In closing, thanks the assumption $A_{\mathrm{M}} \subseteq \mathbb{A}$, we show easily that $\mathbb{A}$ is an extension of $A_{\mathrm{m},\mathcal{Q}}$ and $A_{\mathrm{m},\mathcal{T}_{\mathcal{Q}}}$. Hence $A_{\mathrm{m},\mathcal{T}_{\mathcal{Q}}}$ is an extension of $A_{\mathrm{m},\mathcal{Q}}$. □

Example 5.2.4 Under the assumptions and notation of Example 5.1.5, let us suppose in addition that $q \le p$ (or equivalently that $2 \le p$), and that $(\mathcal{Y}, H, \mathcal{Y}^*)$ is a Gelfand triplet. By recalling

$$\mathcal{V} = L^p(\mathcal{Y}) := L^p\big((0,b), \mathcal{Y}\big), \qquad \mathcal{V}^* = L^q(\mathcal{Y}^*) := L^q\big((0,b), \mathcal{Y}^*\big),$$

and setting

$$\mathcal{H} = L^2(H) := L^2\big((0,b), H\big),$$

Proposition 3.2.4 entails that $(\mathcal{V}, \mathcal{H}, \mathcal{V}^*)$ *is a Gelfand triplet*. Besides, we know from Example 5.1.5 that

$$(S_{\mathcal{V}})^* = S_{\mathcal{V}^*}, \qquad \mathbb{B}_{\mathcal{Q}_{p,\mathcal{Y},1}} = B_{1,\mathcal{V}^*}.$$

Hence,

$$\mathcal{Q}_{p,\mathcal{Y},1} := (\mathrm{D}^1_{\mathcal{V}^*},\ \mathrm{D}^1_{\mathcal{V}},\ \mathbb{B}_{1,\mathcal{V}},\ S_{\mathcal{V}})$$

is compatible with $L^2(H)$ according to Proposition 5.1.1, and the realization of $Q_{p,\mathcal{Y},1}$ in $L^2(H)$ reads

$$\mathcal{T}_{Q_{p,\mathcal{Y},1}} = (\mathrm{D}^1_{L^2(H)},\ B_{1,L^2(H)},\ S_{L^2(H)}).$$

We know from Proposition 5.2.9 that $A_{Q_{p,\mathcal{Y},1}} \subseteq A_{\mathcal{T}_{Q_{p,\mathcal{Y},1}}}$. Moreover, we check easily that

$$D(A_{Q_{p,\mathcal{Y},1}}) = \{u \in W^{1,p}((0,b),\mathcal{Y}) \mid u(0) = 0\}$$
$$D(A_{\mathcal{T}_{Q_{p,\mathcal{Y},1}}}) = \{u \in H^1((0,b),H) \mid u(0) = 0\}.$$

□

5.3 The Domain Structure Theorem

In order to describe *closed boundary restriction operators* of a differential quadruplet $(\mathbb{A}, A_{\mathrm{M}}, B, S)$, it is sufficient to study their domain. The Domain Structure Theorem 5.3.3 below provides the structure of these domains. This result is an essential tool in the analysis of closed boundary restriction operators. However, unlike the Hilbertian setting, we cannot ensure in general the existence of a complementary subspace to $\ker \mathbb{A}^{\perp}$ in $\mathcal{V}$. Hence, we are led to introduce *regular differential quadruplets*.

5.3.1 Regular Differential Quadruplets

Definition 5.3.1 Being given a reflexive Banach space $\mathcal{V}$, a differential quadruplet $\mathcal{Q} := (\mathbb{A}, A_{\mathrm{M}}, B, S)$ on $(\mathcal{V}, \mathcal{V}^*)$ is said to be F_0*-regular* if $\mathcal{Q}$ is closed and F_0 is a closed complementary subspace of $\ker \mathbb{A}^{\perp}$ in $\mathcal{V}$. □

The following result gives a sufficient condition for a differential quadruplet on Lebesgue spaces to be regular. In application, its hypothesis is often satisfied.

Corollary 5.3.1 *Let* $\mathcal{Y}$ *be a nontrivial separable and reflexive Banach space,* p *and* q *be conjugate exponents in* $(1,\infty)$*, and* $\mathcal{Q} := (\mathbb{A}, A_{\mathrm{M}}, B, S)$ *be a differential quadruplet on* $\big(L^p((0,b),\mathcal{Y}), L^q((0,b),\mathcal{Y}^*)\big)$*. Being given positive integers* m *and* m^{s}*, we consider functions* $g_1, \dots, g_m$ *in* $L^p(0,b)$*, and* $f_1, \dots, f_{m^{\mathrm{s}}}$ *in* $L^q(0,b) \setminus \{0\}$*. We assume that*

(i) *the spaces* $g_1\mathcal{Y}, \dots, g_m\mathcal{Y}$ *are in direct sum;*
(ii) $f_1\mathcal{Y}^*, \dots, f_{m^{\mathrm{s}}}\mathcal{Y}^*$ *are in direct sum;*
(iii) $\ker A_{\mathrm{M}} = \bigoplus_{i=1}^{m} g_i\mathcal{Y}$ *and* $\ker \mathbb{A} = \bigoplus_{i=1}^{m^{\mathrm{s}}} f_i\mathcal{Y}^*$.

Then, there exist functions $h_1, \dots, h_{m^s}$ *in* $L^p(0, b)$ *such that*

$$\int_0^b f_i \overline{h_j} \, dy = \delta_{ij}, \qquad \forall 1 \le i, j \le m^s,$$

$h_1\mathcal{Y}, \dots, h_{m^s}\mathcal{Y}$ *are in direct sum, and* $\mathcal{Q}$ *is* $\bigoplus_{i=1}^{m^s} h_i\mathcal{Y}$*-regular.*

Proof Since $f_1\mathcal{Y}^*, \dots, f_{m^s}\mathcal{Y}^*$ are assumed to be in direct sum and the functions f_i's are not trivial, Proposition 3.2.5 yields that $f_1, \dots, f_{m^s}$ are linearly independent. Thus, by applying Proposition 2.2.14 with

$$\mathcal{X} := L^q(0, b), \qquad N := \langle f_1, \dots, f_{m^s} \rangle,$$

there exists $h_1, \dots, h_{m^s}$ in $L^p(0, b)$ such that $\int_0^b h_i \overline{f_j} \, dy = \delta_{ij}$ for each indexes i and j. With Lemma 3.2.6, the spaces $h_1\mathcal{Y}, \dots, h_{m^s}\mathcal{Y}$ are in direct sum and $\bigoplus_{i=1}^{m^s} h_i\mathcal{Y}$ is a closed complementary subspace of $\big(\bigoplus_{i=1}^{m^s} f_i\mathcal{Y}^*\big)^\perp$ in $L^p\big((0, b), \mathcal{Y}\big)$.

Now we claim that $\bigoplus_{i=1}^{m^s} f_i\mathcal{Y}^*$ and $\bigoplus_{i=1}^{m} g_i\mathcal{Y}$ are closed respectively in $L^q(\mathcal{Y}^*)$ and in $L^p(\mathcal{Y})$. Indeed, since the matrix with entries $\int_0^b \overline{f_i} h_j \, dy$ is invertible, Lemma 3.2.6 entails that $\bigoplus_{i=1}^{m^s} f_i\mathcal{Y}^*$ is closed. Regarding $\bigoplus_{i=1}^{m} g_i\mathcal{Y}$, if that space is nontrivial, then we may assume without loss of generality that all the g_i's are different from zero. Then, Proposition 3.2.5 gives the existence of functions $f'_1, \dots, f'_m$ in $L^q(0, b)$ such that the matrix with entries $\int_0^b f'_i \overline{g_j} \, dy$ is invertible. Then, Lemma 3.2.6 yields that $\bigoplus_{i=1}^{m} g_i\mathcal{Y}$ is closed. That proves the claim. Eventually, Lemma 4.3.3 and Proposition 4.4.1 entail that Q is a closed differential quadruplet. □

Corollary 5.3.2 *Let* $\mathcal{Y}$ *be a nontrivial separable and reflexive Banach space,* p *and* q *be conjugate exponents in* $(1, \infty)$*, and* $\mathcal{Q} := (\mathbb{A}, A_M, B, S)$ *be a differential quadruplet on* $(L^p\big((0, b), \mathcal{Y}\big), L^q\big((0, b), \mathcal{Y}^*\big))$*. Being given positive integers* m *and* m^s*, we consider functions* $g_1, \dots, g_m, h_1, \dots, h_{m^s}$ *in* $L^p(0, b)$*, and* $f_1, \dots, f_{m^s}$ *in* $L^q(0, b)$*. Assume that*

(i) *the spaces* $g_1\mathcal{Y}, \dots, g_m\mathcal{Y}$ *are in direct sum;*
(ii) $\ker A_M = \bigoplus_{i=1}^{m} g_i\mathcal{Y}$ *and* $\ker \mathbb{A} = \sum_{i=1}^{m^s} f_i\mathcal{Y}^*$.
(iii) *the matrix with entries*

$$\int_0^b f_i \overline{h_j} \, dy, \qquad \forall 1 \le i, j \le m^s$$

is invertible.

Then, $h_1\mathcal{Y}, \ldots, h_{m^s}\mathcal{Y}$ *are in direct sum, and especially* $\mathcal{Q}$ *is* $\bigoplus_{i=1}^{m^s} h_i\mathcal{Y}$*-regular.*

Proof By Lemma 3.2.6,

(a) $h_1\mathcal{Y}, \ldots, h_{m^s}\mathcal{Y}$ are in direct sum;
(b) $f_1\mathcal{Y}^*, \ldots, f_{m^s}\mathcal{Y}^*$ are in direct sum;
(c) $\bigoplus_{i=1}^{m^s} h_i\mathcal{Y}$ is a closed complementary subspace of $\ker \mathbb{A}^{\perp}$ in $L^p\big((0,b),\mathcal{Y}\big)$.

Finally, arguing as in the proof of Corollary 5.3.1, we get that A_{M} and $\mathbb{A}$ are closed operators. □

5.3.2 The Domain Structure Theorem

With the concept of regular differential quadruplet at hand, we may state one of the main results of this theory.

Theorem 5.3.3 (Domain Structure Theorem) *Let* $\mathcal{V}$ *be a reflexive Banach space over* $\mathbb{K}$*. Let* $(\mathbb{A}, A_{\mathrm{M}}, B, S)$ *be a* F_0*-regular differential quadruplet on* $(\mathcal{V}, \mathcal{V}^*)$ *and* A *be a restriction of* A_{M}*. We set*

$$E := D(A) \cap \big(\ker A_{\mathrm{M}} \oplus BS(F_0)\big). \tag{5.3.1}$$

Then the following assertions holds true.

(i) *The three spaces* $\ker A_{\mathrm{M}}$, $BS(F_0)$, $BS(\ker \mathbb{A}^{\perp})$ *are in direct sum, and*

$$D(A_{\mathrm{M}}) = \ker A_{\mathrm{M}} \oplus BS(F_0) \oplus BS(\ker \mathbb{A}^{\perp}).$$

(ii) *the following assertions are equivalent.*
 (ii-a) *A is a closed boundary restriction operator of* A_{M}.
 (ii-b) *E is a closed subspace of* $D(A_{\mathrm{M}})$ *and*

$$D(A) = E \oplus BS(\ker \mathbb{A}^{\perp}). \tag{5.3.2}$$

Proof (i) By Lemma 4.3.3,

$$D(A_{\mathrm{M}}) = \ker A_{\mathrm{M}} \oplus R(B).$$

Besides, one has clearly

$$R(B) = BS(F_0) + BS(\ker \mathbb{A}^{\perp}).$$

By using $A_{\mathrm{M}}B = \mathrm{id}_{\mathcal{V}}$ and the fact that the intersection of F_0 and $\ker \mathbb{A}^{\perp}$ is trivial, we show easily that $BS(F_0)$ and $BS(\ker \mathbb{A}^{\perp})$ are in direct sum. Whence item (i) follows from Proposition 2.1.10.

(ii) Assuming (ii-b), let us first show that A is closed. Lemma 4.3.15 yields that $D(A)$ is closed in $D(A_{\mathrm{M}})$. Since A_{M} is closed, there results that A is a closed operator on $\mathcal{V}$. Secondly, since A is a closed restriction of A_{M} satisfying $A_{\mathrm{m}} \subseteq A$, Corollary 5.2.8 entails (ii-a).

Conversely, let us assume that A is a closed boundary restriction operator of A_{M}. Then, Corollary 5.2.8 entails that

$$D(A_{\mathrm{m}}) \subseteq D(A). \tag{5.3.3}$$

Let us show that E is closed. Since A is assumed to be closed and $A \subseteq A_{\mathrm{M}}$, we infer that $D(A)$ is closed in $D(A_{\mathrm{M}})$. Besides applying Lemma 4.3.15 with $F := S(\ker \mathbb{A}^{\perp})$ and $G := SF_0$, we obtain that $\ker A_{\mathrm{M}} \oplus BS(F_0)$ is closed in $D(A_{\mathrm{M}})$. Hence E (defined by (5.3.1)) is closed in $D(A_{\mathrm{M}})$. There remains to prove (5.3.2). Let u be in $D(A)$. By (i), there exist u_1 in $\ker A_{\mathrm{M}} \oplus BS(F_0)$ and u_2 in $BS(\ker \mathbb{A}^{\perp})$ such that $u = u_1 + u_2$. Moreover, $u_1 = u - u_2$ belongs to $D(A)$ since

$$u_2 \in BS(\ker \mathbb{A}^{\perp}) = D(A_{\mathrm{m}}) \subseteq D(A),$$

by (5.3.3). Whence u_1 lies in E. Now we derive easily that

$$D(A) = E + BS(\ker \mathbb{A}^{\perp}).$$

Since the intersection is trivial, (5.3.2) follows. □

Corollary 5.3.4 *Let* $(\mathbb{A}, A_{\mathrm{M}}, B, S)$ *be a closed differential quadruplet on* $(\mathcal{V}, \mathcal{V}^*)$, *and* A *be a closed boundary restriction operator of* A_{M}. *We assume that* $(\ker A_{\mathrm{M}})^{\perp}$ *admits a closed complementary subspace* F_0^{s} *in* $\mathcal{V}^*$. *Then*

(i) *the three spaces* $\ker S^*\mathbb{A}S^*$, $B^*(F_0^{\mathrm{s}})$, $B^*\big((\ker A_{\mathrm{M}})^{\perp}\big)$ *are in direct sum, and*

$$D(S^*\mathbb{A}S^*) = \ker S^*\mathbb{A}S^* \oplus B^*(F_0^{\mathrm{s}}) \oplus B^*\big((\ker A_{\mathrm{M}})^{\perp}\big);$$

(ii) *one has*

$$D(A^*) = E^{\mathrm{s}} \oplus B^*\big((\ker A_{\mathrm{M}})^{\perp}\big) \tag{5.3.4}$$

where

$$E^{\mathrm{s}} := D(A^*) \cap \big(\ker S^* \mathbb{A} S^* \oplus B^*(F_0^{\mathrm{s}})\big). \tag{5.3.5}$$

Proof Arguing as in the proof of Corollary 4.3.16, we obtain that A^* is a closed boundary restriction operator of the closed differential quadruplet $\mathcal{Q}^*$. By assumption $\mathcal{V}^* = F_0^{\mathrm{s}} \oplus (\ker A_{\mathrm{M}})^{\perp}$, thus,

$$\mathcal{V}^* = S^* F_0^{\mathrm{s}} \oplus (\ker S A_{\mathrm{M}} S)^{\perp},$$

Applying the Domain Structure Theorem 5.3.3 to $\mathcal{Q}^*$ with $F_0 := S^* F_0^{\mathrm{s}}$, we derive the assertions stated in the corollary. □

Example 5.3.2 Under the assumptions and notation of Example 5.1.5, we recall that

$$\mathcal{V} = L^p(\mathcal{Y}) := L^p\big((0,b),\mathcal{Y}\big), \qquad \mathcal{V}^* = L^q(\mathcal{Y}^*) := L^q\big((0,b),\mathcal{Y}^*\big),$$

and

$$\mathcal{Q}_{p,\mathcal{Y},1} := (\mathrm{D}^1_{\mathcal{V}^*},\ \mathrm{D}^1_{\mathcal{V}},\ B_{1,\mathcal{V}},\ S_{\mathcal{V}}).$$

For all purposes, we precisely state that for any function h in $L^p(0,b)$ and Y in $\mathcal{Y}$, hY denotes the element of $L^p\big((0,b),\mathcal{Y}\big)$ defined for almost every x in $(0,b)$ by $(hY)(x) := h(x)Y$. Then, we put

$$h\mathcal{Y} := \{hY \mid Y \in \mathcal{Y}\}. \tag{5.3.6}$$

We claim that $\mathcal{Q}_{p,\mathcal{Y},1}$ is a $g_1\mathcal{Y}$-regular differential quadruplet. Indeed, since $\ker \mathrm{D}^1_{\mathcal{V}} = g_1\mathcal{Y}$ and $\ker \mathrm{D}^1_{\mathcal{V}^*} = g_1\mathcal{Y}^*$, the claim follows from Corollary 5.3.2. Then, by the Domain Structure Theorem 5.3.3 (i), the domain of $\mathrm{D}^1_{\mathcal{V}}$ decomposes into

$$D(\mathrm{D}^1_{\mathcal{V}}) = E_{\mathrm{M}} \oplus B_{1,\mathcal{V}}\big((g_1\mathcal{Y}^*)^{\perp}\big)$$

where

$$E_{\mathrm{M}} := g_1\mathcal{Y} \oplus g_2\mathcal{Y}.$$

Thus, each u in $D(\mathrm{D}^1_{\mathcal{V}})$ reads

$$u = g_1 U_1 + g_2 U_1' + g_1 * \psi^{\perp},$$

where U_1, U_1' belong to $\mathcal{Y}$ and $\psi^\perp$ lies in $(g_1\mathcal{Y}^*)^\perp$. Since with (3.2.1),

$$g_1 * \psi^\perp(b) = \int_0^b \psi^\perp(y)\,\mathrm{d}y = 0,$$

we get as in (4.3.38),

$$U_1 = u(0), \qquad U_1' = b^{-1}\big(u(b) - u(0)\big). \tag{5.3.7}$$

In view of Definition 5.2.3, the minimal operator $A_{\mathrm{m},\mathcal{Q}_{p,\mathcal{Y},1}}$ of $\mathcal{Q}_{p,\mathcal{Y},1}$ is the restriction of $\mathrm{D}^1_{\mathcal{V}}$ with domain $W_0^{1,p}\big((0,b),\mathcal{Y}\big)$. □

5.3.3 Boundary Restriction Operators of Regular Quadruplets

Now we will introduce a method generating a large class of closed boundary restriction operators. Let us present the ideas on a simple example. In view of Definition 4.4.3, if we want to supplement the operator

$$\mathrm{D}^1_{L^2} : H^1(0,b) \subseteq L^2(0,b) \to L^2(0,b), \qquad u \mapsto \frac{\mathrm{d}}{\mathrm{d}x}u,$$

with homogeneous linear boundary conditions, it is enough to annihilate linear combinations of boundary values. The key point is that in view of (4.3.38), we may equivalently annihilate linear combinations of *endogenous boundary values*. In the coordinates-free setting of Proposition 5.3.5, this amounts to consider the kernel of a well-identified class of linear maps.

In order to apply our method to abstract differential quadruplets, the assumptions of the Domain Structure Theorem must be satisfied. So Being given a reflexive Banach space $\mathcal{V}$ and a F_0-regular differential quadruplet $(\mathbb{A}, A_{\mathrm{M}}, B, S)$ on $(\mathcal{V}, \mathcal{V}^*)$, by setting

$$E_{\mathrm{M}} := \ker A_{\mathrm{M}} \oplus BS(F_0),$$

the Domain Structure Theorem 5.3.3 yields that

$$D(A_{\mathrm{M}}) = E_{\mathrm{M}} \oplus BS(\ker \mathbb{A}^\perp).$$

We claim that E_{M} *and* $BS(\ker \mathbb{A}^\perp)$ *are closed subspaces of* $D(A_{\mathrm{M}})$. Indeed, $\ker A_{\mathrm{M}}$ is closed in $\mathcal{V}$ since the differential quadruplet is assumed to be closed. Then, applying Lemma 4.3.15

(i) with

$$\mathcal{X} := \mathcal{V}, \qquad A := A_{\mathrm{M}}, \qquad \mathcal{E} := \ker A_{\mathrm{M}}, \qquad F := S(\ker \mathbb{A}^{\perp}), \qquad G := SF_0,$$

we get that E_{M} is closed in $D(A_{\mathrm{M}})$;

(ii) with

$$\mathcal{E} := \{0\}, \qquad F := SF_0, \qquad G := S(\ker \mathbb{A}^{\perp}),$$

there results that $BS(\ker \mathbb{A}^{\perp})$ is closed in $D(A_{\mathrm{M}})$.

That proves the claim. Hence, Proposition 2.3.5 allows to introduce the projection $P_{\mathrm{b.v}} : D(A_{\mathrm{M}}) \to D(A_{\mathrm{M}})$ on E_{M} along $BS(\ker \mathbb{A}^{\perp})$.

Definition 5.3.3 Let $\mathcal{V}$ be a reflexive Banach space, and $(\mathbb{A}, A, B, S)$ be a F_0-regular differential quadruplet on $(\mathcal{V}, \mathcal{V}^*)$. The above projection $P_{\mathrm{b.v}}$ of $D(A_{\mathrm{M}})$ on E_{M} along $BS(\ker \mathbb{A}^{\perp})$ is called *the projection of* $D(A_{\mathrm{M}})$ *induced by* F_0. □

Proposition 5.3.5 *Let $\mathcal{V}$ be a reflexive Banach space over $\mathbb{K}$, $\mathcal{Q} := (\mathbb{A}, A_{\mathrm{M}}, B, S)$ be a F_0-regular differential quadruplet on $(\mathcal{V}, \mathcal{V}^*)$, and $P_{\mathrm{b.v}}$ be the projection of $D(A_{\mathrm{M}})$ induced by F_0.*

Let also $\mathcal{X}$ be a $\mathbb{K}$-normed space and L be an operator in $\mathcal{L}(E_{\mathrm{M}}, \mathcal{X})$ where E_{M} is equipped with the graph-norm of A_{M}. If A is a restriction of A_{M} with domain

$$D(A) := \ker LP_{\mathrm{b.v}},$$

then, A is a closed boundary restriction operator of $\mathcal{Q}$.

Proof Since $D(A)$ is defined as the kernel of $LP_{\mathrm{b.v}}$ and $P_{\mathrm{b.v}}$ lies in $\mathcal{L}(D(A_{\mathrm{M}}))$ as a projection, we deduce that $D(A)$ is closed in $D(A_{\mathrm{M}})$. Since A_{M} is closed according to Definition 5.3.1, there results that A is a closed operator on $\mathcal{V}$.

By definition of $P_{\mathrm{b.v}}$, $BS(\ker \mathbb{A}^{\perp})$ is the kernel of $P_{\mathrm{b.v}}$. Thus, $BS(\ker \mathbb{A}^{\perp})$ is contained $D(A)$. In addition, $BS(\ker \mathbb{A}^{\perp})$ is the domain of the minimal operator $A_{\mathrm{m},\mathcal{Q}}$ of $\mathcal{Q}$. Thus $A_{\mathrm{m},\mathcal{Q}} \subseteq A$. Then Corollary 5.2.8 yields that A is a closed boundary restriction operator of $\mathcal{Q}$. □

Remark 5.3.4 By suitable choices of the map L in Proposition 5.3.5, let us see how we can recover the pivot and the minimal operators. Let $\mathcal{Q} := (\mathbb{A}, A_{\mathrm{M}}, B, S)$ be a F_0-regular differential quadruplet on $(\mathcal{V}, \mathcal{V}^*)$. *We claim that the map*

$$L_{\mathrm{p}} : E_{\mathrm{M}} = \ker A_{\mathrm{M}} \oplus BS(F_0) \to E_{\mathrm{M}}, \qquad u = v_1 + v_1' \mapsto v_1$$

is a linear and continuous map from $(E_{\text{M}}, \|\cdot\|_{D(A_{\text{M}})})$ *into itself. Moreover, the closed boundary restriction operator associated to* L_{p} *through Proposition 5.3.5 is nothing but the pivot operator of* $\mathcal{Q}$ *i.e.*

$$D(A_{\mathcal{Q}}) = \ker L_{\text{p}} P_{\text{b.v}}.$$

Indeed, since E_{M} is closed in $D(A_{\text{M}})$, E_{M} equipped with the graph-norm of A_{M} is a Banach space. Moreover, it is clear that $\ker A_{\text{M}}$ is closed in $D(A_{\text{M}})$. Also applying Lemma 4.3.15 with

$$\mathcal{X} := \mathcal{V}, \qquad A := A_{\text{M}}, \qquad \mathcal{E} := \{0\}, \qquad F := S(\ker \mathbb{A}^{\perp}), \qquad G := SF_0,$$

we get that $BS(F_0)$ is closed in $D(A_{\text{M}})$. Thus, in view of Definition 2.3.3, L_{p} is the projection of E_{M} on $\ker A_{\text{M}}$ along $BS(F_0)$. In particular, L_{p} is continuous on E_{M}, which proves the first assertion of the claim.

Next the first item of the Domain Structure Theorem 5.3.3 yields that $\ker L_{\text{p}} P_{\text{b.v}}$ is equal to the range of B. That completes the proof of the claim.

In the same way, by taking $L_{\text{p}} := \text{id}_{E_{\text{M}}}$ in Proposition 5.3.5, we arrive to

$$D(A_{\text{m},\mathcal{Q}}) = \ker P_{\text{b.v}}.$$

□

5.4 On the Adjoint of Boundary Restriction Operators

In general, the computation of the adjoint of an operator is tedious if not difficult. In the next section, we will give two methods for computing easily the adjoint of boundary restriction operators.

5.4.1 A Variational Method

Theorem 5.4.1 turns out to be the cornerstone of this method. The key ingredient in the proof of Theorem 5.4.1 is *the abstract integration by parts formula* (5.4.3). We refer to the illustrative equality (4.1.10) where such an integration by parts is performed in a simple framework.

Let $\mathcal{V}$ be a reflexive Banach space. Being given a differential quadruplet $\mathcal{Q} := (\mathbb{A}, A, B, S)$ on $(\mathcal{V}, \mathcal{V}^*)$, Lemma 4.3.3 tells us that for each u in $D(A_{\text{M}})$, there exists a unique u_0 in $\ker A_{\text{M}}$ such that

$$u = u_0 + BA_{\text{M}}u. \tag{5.4.1}$$

In a symmetric way, by considering the adjoint quadruplet of $\mathcal{Q}$, each f in $D(S^*\mathbb{A}S^*)$ reads in a unique way

$$f = f_0 + B^* S^* \mathbb{A} S^* f, \tag{5.4.2}$$

where f_0 lies in $\ker \mathbb{A}S^*$. For reasons of brevity, the duality bracket between $\mathcal{V}$ and $\mathcal{V}^*$ will be denoted by $\langle \cdot, \cdot \rangle$. Notice that at this stage, we do not assume the finite dimensionality of the kernel of the maximal operators.

Theorem 5.4.1 *Let $\mathcal{V}$ be a real or complex reflexive Banach space. Let $\mathcal{Q} := (\mathbb{A}, A, B, S)$ be a differential quadruplet on $(\mathcal{V}, \mathcal{V}^*)$, and A be a boundary restriction operator of A_{M}.*

(i) *If u lies in $D(A_{\mathrm{M}})$ and f belongs to $D(S^*\mathbb{A}S^*)$ then*

$$\langle f, A_{\mathrm{M}} u \rangle = \langle S^* \mathbb{A} S^* f, u \rangle + \langle f_0, A_{\mathrm{M}} u \rangle - \langle S^* \mathbb{A} S^* f, u_0 \rangle \tag{5.4.3}$$

(ii) *If f lies in $D(S^*\mathbb{A}S^*)$, then*

$$f \in D(A^*) \quad \Longleftrightarrow \quad \langle f_0, Au \rangle = \langle S^* \mathbb{A} S^* f, u_0 \rangle, \quad \forall u \in D(A),$$

where u_0 and f_0 are given by (5.4.1) *and* (5.4.2).

(iii) *If A is closed and u lies in $D(A_{\mathrm{M}})$ then*

$$u \in D(A) \quad \Longleftrightarrow \quad \langle f_0, A_{\mathrm{M}} u \rangle = \langle A^* f, u_0 \rangle, \quad \forall f \in D(A^*).$$

Proof

(i) For each f in $D(S^*\mathbb{A}S^*)$ and u in $D(A_{\mathrm{M}})$, one has by (5.4.2),

$$\begin{aligned} \langle f, A_{\mathrm{M}} u \rangle &= \langle f_0, A_{\mathrm{M}} u \rangle + \langle S^* \mathbb{A} S^* f, B A_{\mathrm{M}} u \rangle \\ &= \langle f_0, A_{\mathrm{M}} u \rangle + \langle S^* \mathbb{A} S^* f, u \rangle - \langle S^* \mathbb{A} S^* f, u_0 \rangle, \end{aligned}$$

by (5.4.1). That proves (5.4.3).

(ii) Assuming that f belongs to $D(A^*)$, we have since $A^* \subseteq S^*\mathbb{A}S^*$ and $A \subseteq A_{\mathrm{M}}$,

$$\langle S^* \mathbb{A} S^* f, u \rangle = \langle A^* f, u \rangle = \langle f, Au \rangle = \langle f, A_{\mathrm{M}} u \rangle, \quad \forall u \in D(A).$$

Thus, with (5.4.3), we obtain

$$\langle f_0, Au \rangle = \langle S^* \mathbb{A} S^* f, u_0 \rangle, \quad \forall u \in D(A). \tag{5.4.4}$$

Conversely, let us assume that f satisfies (5.4.4). Plugging (5.4.4) into (5.4.3), we get, for each u in $D(A)$,

$$\langle f, Au\rangle = \langle f, A_{\mathrm{M}}u\rangle = \langle S^*\mathbb{A}S^* f, u\rangle.$$

Thus, the definition of the adjoint operator of A gives that f lies in $D(A^*)$.

(iii) Let u be in $D(A)$. Arguing as in the proof of (5.4.4), we get what we want, namely

$$\langle f_0, A_{\mathrm{M}}u\rangle = \langle A^* f, u_0\rangle, \qquad \forall f \in D(A^*).$$

Conversely, assume that u lies in $D(\mathbb{A})$ and satisfies the above identity. Since

$$S^*\mathbb{A}S^* f = A^* f, \qquad \forall f \in D(A^*),$$

going back to (5.4.3), we get by reflexivity

$$\langle u, A^* f\rangle_{\mathcal{V},\mathcal{V}^*} = \langle A_{\mathrm{M}}u, f\rangle_{\mathcal{V},\mathcal{V}^*}, \qquad \forall f \in D(A^*).$$

Thus, u belongs to $D(A^{**})$. However, by Theorem 2.4.1, $A^{**} = A$ since A is closed. □

The next result gives a simple variational characterization of the adjoint of a special class of injective boundary restriction operators.

Corollary 5.4.2 *Let $\mathcal{Q} := (\mathbb{A}, A, B, S)$ be a F_0-regular differential quadruplet on $(\mathcal{V}, \mathcal{V}^*)$, and A be a closed boundary restriction operator of $\mathcal{Q}$. We assume that*

$$D(A) \cap \left(\ker A_{\mathrm{M}} \oplus BS(F_0)\right) = BS(G_0),$$

where G_0 is a subspace of $\mathcal{V}$. Then, for each f in $D(S^*\mathbb{A}S^*)$,

$$f \in D(A^*) \quad \Longleftrightarrow \quad \langle f_0, S\varphi_0\rangle_{\mathcal{V}^*,\mathcal{V}} = 0, \qquad \forall \varphi_0 \in G_0,$$

where f_0 is the unique element of $\ker \mathbb{A}S^*$ *satisfying* (5.4.2).

Proof By (5.4.2), any f in $D(S^*\mathbb{A}S^*)$ reads

$$f = f_0 + B^*\varphi,$$

for some φ in $\mathcal{V}^*$. By the Domain Structure Theorem 5.3.3 (ii),

$$D(A) = BS(G_0) \oplus BS(\ker \mathbb{A}^{\perp}).$$

Thus, Theorem 5.4.1 yields that f lies in $D(A^*)$ if and only if

$$\langle f_0, S\varphi_0 + S\psi^{\perp} \rangle_{\mathcal{V}^*,\mathcal{V}} = 0, \qquad \forall (\varphi_0, \psi^{\perp}) \in G_0 \times \ker \mathbb{A}^{\perp}.$$

Since $\ker \mathbb{A}S^* = S^* \ker \mathbb{A}$, we derive that $S^* f_0$ belongs to $\ker \mathbb{A}$. Hence

$$\langle S^* f_0, \psi^{\perp} \rangle_{\mathcal{V}^*,\mathcal{V}} = 0,$$

and the equivalence stated in the corollary follows. □

Remark 5.4.1 Let $\mathcal{Q} := (\mathbb{A}, A, B, S)$ be a F_0-regular differential quadruplet on $(\mathcal{V}, \mathcal{V}^*)$. By choosing A to be the pivot operator of $\mathcal{Q}$, Corollary 5.4.2 yields the following characterization of the adjoint of $A_{\mathcal{Q}}$. For each f in $D(S^*\mathbb{A}S^*)$,

$$f \in D\big((A_{\mathcal{Q}})^*\big) \quad \Longleftrightarrow \quad \langle f_0, S\varphi_0 \rangle_{\mathcal{V}^*,\mathcal{V}} = 0, \qquad \forall \varphi_0 \in F_0,$$

where f_0 is the unique element of $\ker \mathbb{A}S^*$ satisfying (5.4.2). □

5.4.2 A Direct Method

This method applies to some large boundary restriction operators. Its proof relies on the representation (5.2.7) of the domain of the adjoint of large boundary restriction operators.

Proposition 5.4.3 *Being given a F_0-regular differential quadruplet $\mathcal{Q} := (\mathbb{A}, A, B, S)$ on $(\mathcal{V}, \mathcal{V}^*)$, assume that*

(i) *M_1 is a subspace of* $\ker A_{\mathrm{M}}$*;*
(ii) *A is a restriction of A_{M} whose domain is equal to*

$$D(A) := M_1 + R(B);$$

(iii) $(\ker A_{\mathrm{M}})^{\perp}$ *admits a closed complementary subspace F_0^{s} in $\mathcal{V}^*$.*

Then,

$$D(A^*) = \big\{ B^*(f_0^{\mathrm{s}} + \varphi^{\perp}) \mid f_0^{\mathrm{s}} \in F_0^{\mathrm{s}} \cap M_1{}^{\perp},\ \varphi^{\perp} \in (\ker A_{\mathrm{M}})^{\perp} \big\}.$$

Proof First, we claim that $\ker A = M_1$. For, by (i) it is clear that $M_1 \subseteq \ker A$. Conversely, for any u in $\ker A$, (ii) entails that

$$u = m_1 + Bh$$

for some $(m_1, h) \in M_1 \times \mathcal{V}$. Since $M_1 \subseteq \ker A$, one has

$$0 = Au = h.$$

Thus $u = m_1$ and u lies in M_1. That proves the claim.

Next, let f belong to the domain of $S^*\mathbb{A}S^*$. By Corollary 5.3.4, there exists

$$(\xi^{\mathrm{s}}, f_0^{\mathrm{s}}, \varphi^{\perp}) \in \ker \mathbb{A}S^* \times F_0^{\mathrm{s}} \times (\ker A_{\mathrm{M}})^{\perp}$$

such that

$$f = \xi^{\mathrm{s}} + B^*(f_0^{\mathrm{s}} + \varphi^{\perp}).$$

Besides, $D(A^*) = B^*({M_1}^{\perp})$ according to (5.2.7). Thus by a direct sum argument,

$$\begin{aligned} f \in D(A^*) &\Longleftrightarrow \xi^{\mathrm{s}} = 0,\ f_0^{\mathrm{s}} + \varphi^{\perp} \in {M_1}^{\perp} \\ &\Longleftrightarrow \xi^{\mathrm{s}} = 0,\ f_0^{\mathrm{s}} \in {M_1}^{\perp} \end{aligned}$$

since $(\ker A_{\mathrm{M}})^{\perp} \subseteq {M_1}^{\perp}$. □

5.5 Differential Quadruplets with Finite Dimensional Kernels

In many applications, the kernel of the maximal operators $\mathbb{A}$ and A_{M} is finite dimensional. For the convenience of reading, differential triplets with finite dimensional kernels have been studied separately in Sect. 4.4. In the Hilbertian setting, $\ker \mathbb{A}$ is a natural complementary subspace of $\ker \mathbb{A}^{\perp}$. However, when the underlying vector space is a reflexive Banach space, which is the framework of this chapter, there is no obvious choice for a complementary subspace even if $\ker \mathbb{A}$ is assumed to be finite dimensional.

Let $\mathcal{V}$ be a real or complex reflexive Banach space, and $\mathcal{V}^*$ be its anti-dual space. Being given a differential quadruplet $(\mathbb{A}, A_{\mathrm{M}}, B, S)$ whose maximal operators have a finite dimensional kernel, our aim is to (i) compute the adjoint of any boundary restriction operators of A_{M}; (ii) introduce *abstract endogenous boundary conditions* for A and A^*, and also *abstract endogenous boundary coordinates* of elements of $D(\mathbb{A})$. The computations are a little bit more involved than in the Hilbertian setting, but they are analog due to reflexivity.

5.5.1 Abstract Endogenous Boundary Conditions

We assume now that A_{M} and $\mathbb{A}$ have a nontrivial finite dimensional kernel. Hence, there exist two positive integers $d_{A_{\mathrm{M}}}$, $d_{\mathbb{A}}$ and two lists of linearly independent vectors $\xi_1, \dots, \xi_{d_{A_{\mathrm{M}}}}$ in $D(A_{\mathrm{M}})$, and $\xi_1^{\mathrm{s}}, \dots, \xi_{d_{\mathbb{A}}}^{\mathrm{s}}$ in $D(\mathbb{A})$ such that

$$\ker A_{\mathrm{M}} = \langle \xi_1, \dots, \xi_{d_{A_{\mathrm{M}}}} \rangle, \qquad \ker \mathbb{A} = \langle \xi_1^{\mathrm{s}}, \dots, \xi_{d_{\mathbb{A}}}^{\mathrm{s}} \rangle. \tag{5.5.1}$$

Applying Proposition 2.2.14 with $\mathcal{X} := \mathcal{V}^*$ and $N := \ker \mathbb{A}$, we obtain on the one hand that $\ker \mathbb{A}^{\perp}$ admits a complementary subspace F_0 in $\mathcal{V}$, and on the other hand that the dimension of F_0 is finite and equal to $d_{\mathbb{A}}$. Hence, we may legitimately introduce a basis $(\eta_1, \dots, \eta_{d_{\mathbb{A}}})$ of F_0.

In the same way, $(\ker A_{\mathrm{M}})^{\perp}$ admits a complementary subspace F_0^{s} in $\mathcal{V}^*$, of dimension $d_{A_{\mathrm{M}}}$. Let $(\eta_1^{\mathrm{s}}, \dots, \eta_{d_{A_{\mathrm{M}}}}^{\mathrm{s}})$ be a basis of F_0^{s}. Notice that in a Hilbertian setting (see Sect. 4.4.1), $\mathbb{A}$ is equal to A_{M} and we may choose $F_0 := \ker \mathbb{A}$, $F_0^{\mathrm{s}} := \ker \mathbb{A}$, and $\eta_i = \eta_i^{\mathrm{s}} := \xi_i$.

Let A be a closed boundary restriction operator of A_{M}. In view of the Domain Structure Theorem 5.3.3,

$$D(A) = E \oplus BS(\ker \mathbb{A}^{\perp}), \tag{5.5.2}$$

where E is the subspace of

$$E_{\mathrm{M}} := \ker A_{\mathrm{M}} \oplus BS(F_0), \tag{5.5.3}$$

defined by (5.3.1), namely

$$E := D(A) \cap E_{\mathrm{M}}.$$

Assuming that E is not trivial, we denote by $(e_1, \dots, e_{d_E})$ one of its basis, where d_E, namely the dimension of E ranges in $[1, d_{A_{\mathrm{M}}} + d_{\mathbb{A}}]$. For each integer j in $[1, d_E]$, the inclusion

$$E \subseteq \ker A_{\mathrm{M}} \oplus BS(F_0)$$

and (5.5.1) yield that

$$e_j = \sum_{i=1}^{d_{A_{\mathrm{M}}}} x_{ij} \xi_i + BS \sum_{i=1}^{d_{\mathbb{A}}} y_{ij} \eta_i, \tag{5.5.4}$$

for some scalars x_{ij} and y_{ij}. Setting

$$\omega_j := \sum_{i=1}^{d_{A_{\mathrm{M}}}} x_{ij}\xi_i \in \ker A_{\mathrm{M}}, \qquad \omega_j' := \sum_{i=1}^{d_{\mathbb{A}}} y_{ij}\eta_i \in F_0, \tag{5.5.5}$$

we have

$$e_j = \omega_j + BS\omega_j' \in \ker A_{\mathrm{M}} \oplus BS(F_0), \qquad \forall\, 1 \le j \le d_E. \tag{5.5.6}$$

5.5.1.1 Endogenous Boundary Conditions of A^*

Starting from the decomposition

$$D(S^*\mathbb{A}S^*) = \ker \mathbb{A}S^* \oplus B^*(F_0^{\mathrm{s}}) \oplus B^*\big((\ker A_{\mathrm{M}})^{\perp}\big),$$

given by Corollary 5.3.4, any f in $D(S^*\mathbb{A}S^*)$ reads in a unique way (see (5.5.1))

$$f = S^* \sum_{i=1}^{d_{\mathbb{A}}} f_i\xi_i^{\mathrm{s}} + B^* \sum_{i=1}^{d_{A_{\mathrm{M}}}} f_i'\eta_i^{\mathrm{s}} + B^*\varphi^{\perp}, \tag{5.5.7}$$

where f_i, f_i' are scalar and $\varphi^{\perp}$ lies in $(\ker A_{\mathrm{M}})^{\perp}$. With the convention of repeated indexes, we will abbreviate (5.5.7) in

$$f = S^* f_i\xi_i^{\mathrm{s}} + B^* f_i'\eta_i^{\mathrm{s}} + B^*\varphi^{\perp}, \tag{5.5.8}$$

although the sums do not run over the same set of indexes.

Proposition 5.5.1 *Under the above assumptions and notations, in particular A is a closed boundary restriction operator of $\mathbb{A}$,* (5.5.1)–(5.5.8) *hold and* $1 \le d_E \le d_{A_{\mathrm{M}}} + d_{\mathbb{A}}$, *let f belong to $D(S^*\mathbb{A}S^*)$. Then, recalling that f_0 is defined through* (5.4.2), *one has*

$$f \in D(A^*) \iff \langle f_0, S\omega_j'\rangle = \langle S^*\mathbb{A}S^* f, \omega_j\rangle \qquad \forall\, 1 \le j \le d_E \tag{5.5.9}$$

$$\iff \langle f_i\xi_i^{\mathrm{s}}, \omega_j'\rangle = \langle f_i'\eta_i^{\mathrm{s}}, \omega_j\rangle \qquad \forall\, 1 \le j \le d_E. \tag{5.5.10}$$

The equations in (5.5.10) characterizing the domain of A^* are called *(abstract) endogenous boundary conditions of* A^*.

Proof of Proposition 5.5.1 Starting from the direct sum (5.5.2) and using the decomposition (5.5.6) of the e_i's, we may write

$$D(A) = \langle \omega_1 + BS\omega_1', \ldots, \omega_{d_E} + BS\omega_{d_E}'\rangle \oplus BS\big(\ker \mathbb{A}^{\perp}\big).$$

Then, the first equivalence of Theorem 5.4.1 takes the form:

$$f \in D(A^*) \tag{5.5.11}$$

$$\Longleftrightarrow$$

$$\big\langle f_0, A(\omega_j + BS\omega_j' + BS\psi^\perp)\big\rangle = \langle S^* \mathbb{A} S^* f, \omega_j\rangle$$

$$\forall\, 1 \le j \le d_E,\ \forall\, \psi^\perp \in \ker \mathbb{A}^\perp.$$

However, by (5.1.2),

$$\langle f_0, ABS\psi^\perp\rangle = \langle f_0, S\psi^\perp\rangle = \langle S^* f_0, \psi^\perp\rangle = 0,$$

since $\psi^\perp$ lies $\ker \mathbb{A}^\perp$ and $S^* f_0 \in \ker \mathbb{A}$. Thus, (5.5.11) is equivalent to

$$\langle f_0, S\omega_j'\rangle = \langle S^* \mathbb{A} S^* f, \omega_j\rangle, \qquad \forall\, 1 \le j \le d_E.$$

That proves the first equivalence (5.5.9). Moreover, recalling (5.4.2), and using (5.5.7), we get $f_0 = S^* f_i \xi_i^{\mathrm{s}}$ and $S^* \mathbb{A} S^* f = f_i' \eta_i^{\mathrm{s}} + \varphi^\perp$. Since $\langle \varphi^\perp, \omega_j\rangle = 0$, (5.5.11) is equivalent to

$$\langle f_i \xi_i^{\mathrm{s}}, \omega_j'\rangle = \langle f_i' \eta_i^{\mathrm{s}}, \omega_j\rangle, \qquad \forall\, 1 \le j \le d_E.$$

□

5.5.1.2 Dimension of E^{s}

Let us recall that A is a closed boundary restriction operator of $(\mathbb{A}, A_{\mathrm{M}}, B, S)$. In view of Corollary 5.3.4, let E^{s} be the subspace of

$$E_{\mathrm{M}}^{\mathrm{s}} := S^* \ker \mathbb{A} \oplus B^*(F_0^{\mathrm{s}}) \tag{5.5.12}$$

defined by (5.3.5) and satisfying

$$D(A^*) = E^{\mathrm{s}} \oplus B^*\big((\ker A_{\mathrm{M}})^\perp\big). \tag{5.5.13}$$

Equation (5.5.10) tells us that the elements of $D(A^*)$ are characterized by d_E linear equations with $d_{A_{\mathrm{M}}} + d_{\mathbb{A}}$ unknowns. Let us recall that d_E, $d_{A_{\mathrm{M}}}$ and $d_{\mathbb{A}}$ denote respectively the dimension of E, $\ker A_{\mathrm{M}}$ and $\ker \mathbb{A}$. In order to compute the dimension of E^{s}, we introduce for each j in $[1, d_E]$, the following linear form on $E_{\mathrm{M}}^{\mathrm{s}}$

$$F_j : S^* \ker \mathbb{A} \oplus B^*(F_0^{\mathrm{s}}) \to \mathbb{K}, \qquad f_0 + f_1 \mapsto \langle f_0, S\omega_j'\rangle - \langle S^* \mathbb{A} S^* f_1, \omega_j\rangle. \tag{5.5.14}$$

Proposition 5.5.2 *Under the assumptions and notation of Proposition 5.5.1, and* (5.5.12)-(5.5.14)*, the dimension of* E^{s} *is equal to* $d_{A_{\mathrm{M}}} + d_{\mathbb{A}} - d_E$.

Proof As in Proposition 4.4.6, we claim that

$$E^{\mathrm{s}} = \bigcap_{j=1}^{d_E} \ker F_j. \tag{5.5.15}$$

Since $S^*\mathbb{A}S^*B^* = \mathrm{id}_{\mathcal{V}}$, we deduce that B^* is one-to-one. Hence (see (5.5.12)) $E^{\mathrm{s}}_{\mathrm{M}}$ has dimension $d_{A_{\mathrm{M}}} + d_{\mathbb{A}}$. Therefore, by (5.5.15) and Proposition 2.2.3, E^{s} has dimension $d_{A_{\mathrm{M}}} + d_{\mathbb{A}} - d_E$ provided $F_1, \ldots, F_{d_E}$ are linearly independent in the dual space of $E^{\mathrm{s}}_{\mathrm{M}}$. In order to check that independence, let $\lambda_1, \ldots, \lambda_{d_E}$ be scalars satisfying $\sum_{j=1}^{d_E} \overline{\lambda_j} F_j = 0$. Choosing $f_1 := 0$, we derive that for each f_0 in $S^* \ker \mathbb{A}$,

$$0 = \langle S^* f_0, \sum_{j=1}^{d_E} \lambda_j \omega'_j \rangle.$$

Thus, $\sum_{j=1}^{d_E} \lambda_j \omega'_j$ lies in $\ker \mathbb{A}^{\perp}$. Moreover, the latter vector belongs also to F_0 by (5.5.5). Since F_0 is a complementary subspace of $\ker \mathbb{A}^{\perp}$, we derive that

$$\sum_{j=1}^{d_E} \lambda_j \omega'_j = 0. \tag{5.5.16}$$

In the same way, for each φ in F_0^{s}, we set $f_1 := B^*\varphi$. Since $S^*\mathbb{A}S^* f_1 = \varphi$, we derive

$$\langle \varphi, \sum_{j=1}^{d_E} \lambda_j \omega_j \rangle = 0, \qquad \forall \varphi \in F_0^{\mathrm{s}}.$$

Thus, $\sum_{j=1}^{d_E} \lambda_j \omega_j$ lies in $(F_0^{\mathrm{s}})^{\perp}$. Moreover, the latter vector belongs also to $\ker A_{\mathrm{M}}$ by (5.5.5). Now, since F_0^{s} is a closed complementary subspace of $(\ker A_{\mathrm{M}})^{\perp}$, Proposition 2.3.8 entails that $(F_0^{\mathrm{s}})^{\perp}$ is a complementary subspace of $\ker A_{\mathrm{M}}$. Thus,

$$\sum_{j=1}^{d_E} \lambda_j \omega_j = 0. \tag{5.5.17}$$

Finally, combining (5.5.16), (5.5.17) and going back to (5.5.6), we obtain $\sum_{j=1}^{d_E} \lambda_j e_j = 0$. Thus, $\lambda_j = 0$ for all j in $[1, d_E]$ since $(e_1, \ldots, e_{d_E})$ is a basis of E. The proof of the proposition is now complete. □

5.5.1.3 Endogenous Boundary Conditions of A

Under the hypothesis of Proposition 5.5.2, we assume in addition that E is not maximal, that is,

$$1 \leq d_E < d_{A_M} + d_{\mathbb{A}}. \tag{5.5.18}$$

Then, Proposition 5.5.2 yields that the dimension of E^s is equal to

$$d_{E^s} := d_{A_M} + d_{\mathbb{A}} - d_E.$$

Observing that $d_{E^s} \geq 1$ thanks to (5.5.18), let $(e_1^s, \dots, e_{d_{E^s}}^s)$ be a basis of E^s. In view of (5.5.12), for each index j in $[1, d_{E^s}]$, e_j^s reads

$$e_j^s = S^* \omega_j^s + B^* \omega_j^{s\prime}, \tag{5.5.19}$$

for some ω_j^s in $\ker \mathbb{A}$ and $\omega_j^{s\prime}$ in F_0^s. By the Domain Structure Theorem 5.3.3 (i), each u in $D(A_M)$ reads

$$u = \sum_{i=1}^{d_{A_M}} u_i \xi_i + BS \sum_{i=1}^{d_{\mathbb{A}}} u_i' \eta_i + BS\psi^\perp, \tag{5.5.20}$$

where u_i, u_i' are scalars, and $\psi^\perp$ lies in $\ker \mathbb{A}^\perp$.

Proposition 5.5.3 *Let A be a closed boundary restriction operator of the differential quadruplet* $(\mathbb{A}, A_M, B, S)$. *Under the above assumptions and notations, in particular* (5.5.1)–(5.5.7) *and* (5.5.18)–(5.5.20), *let u belong to $D(A_M)$. Then,*

$$u \in D(A) \quad \iff \quad \langle \omega_j^s, \sum_{i=1}^{d_{\mathbb{A}}} u_i' \eta_i \rangle = \langle \omega_j^{s\prime}, \sum_{i=1}^{d_{A_M}} u_i \xi_i \rangle, \quad \forall 1 \leq j \leq d_{E^s}.$$

The d_{E^s} equations characterizing $D(A)$ in the latter corollary are called *(abstract) endogenous boundary conditions of A.*

Proof We proceed as in the proof of Proposition 5.5.1. Starting from the direct sum (5.3.4) and using the decomposition (5.5.19) of the e_j^s's, we may write

$$D(A^*) = \langle S^*\omega_1^s + B^*\omega_1^{s\prime}, \dots, S^*\omega_{d_{E^s}} + B^*\omega_{d_{E^s}}^{s}{}' \rangle \oplus B^*\big((\ker A_M)^\perp\big).$$

Then, with the convention of repeated indexes, the second equivalence of Theorem 5.4.1 takes the form:

$$u \in D(A) \tag{5.5.21}$$
$$\Longleftrightarrow$$
$$\langle S^*\omega_j^{\mathrm{s}},\ Su_i'\eta_i + S\psi^{\perp} \rangle = \langle \omega_j^{\mathrm{s}\prime} + \varphi^{\perp},\ u_i\xi_i \rangle$$
$$\forall 1 \le j \le d_{E^{\mathrm{s}}},\ \forall \varphi^{\perp} \in (\ker A_{\mathrm{M}})^{\perp}.$$

However,

$$\langle S^*\omega_j^{\mathrm{s}},\ S\psi^{\perp} \rangle = \langle \omega_j^{\mathrm{s}}, \psi^{\perp} \rangle = 0$$
$$\langle \varphi^{\perp}, \xi_i \rangle = 0,$$

since $(\omega_j^{\mathrm{s}}, \psi^{\perp})$ lies in $\ker \mathbb{A} \times \ker \mathbb{A}^{\perp}$, and $(\xi_i, \varphi^{\perp})$ lies in $\ker A_{\mathrm{M}} \times (\ker A_{\mathrm{M}})^{\perp}$. Thus (5.5.21) is equivalent to

$$u \in D(A) \quad \Longleftrightarrow \quad \langle \omega_j^{s}, u_i'\eta_i \rangle = \langle \omega_j^{s\prime}, u_i\xi_i \rangle, \quad \forall 1 \le j \le d_{E^{\mathrm{s}}}.$$

□

Corollary 5.5.4 *Let $\mathcal{V}$ be a real or complex reflexive Banach space, and $(\mathbb{A}, A_{\mathrm{M}}, B, S)$ be a differential quadruplet on $(\mathcal{V}, \mathcal{V}^*)$ whose maximal operators have a finite dimensional non trivial kernel. Owing to the notation* (5.5.1), (5.5.7) *and* (5.5.20), *the following integration by parts formula holds true.*

$$\langle f, A_{\mathrm{M}}u \rangle = \langle S^*\mathbb{A}S^* f, u \rangle + \sum_{i,j=1}^{d_{\mathbb{A}}} f_i \overline{u_j'} \langle \xi_i^{\mathrm{s}}, \eta_j \rangle - \sum_{i,j=1}^{d_{A_{\mathrm{M}}}} f_i' \overline{u_j} \langle \eta_i^{\mathrm{s}}, \xi_j \rangle.$$

Proof Starting from (5.4.3) and using (5.5.7), (5.5.20), we derive the above integration by parts formula. □

In order to define *abstract endogenous boundary coordinates*, let us remind the reader that $(\xi_1, \ldots, \xi_{d_{A_{\mathrm{M}}}})$ is a basis of $\ker A_{\mathrm{M}}$, F_0 satisfies

$$\mathcal{V} = F_0 \oplus \ker \mathbb{A}^{\perp},$$

and $(\eta_1, \ldots, \eta_{d_{\mathbb{A}}})$ is a basis of F_0.

Definition 5.5.1 Let $\mathcal{V}$ be a real or complex reflexive Banach space and $\mathcal{Q} := (\mathbb{A}, A_{\mathrm{M}}, B, S)$ be a differential quadruplet on $(\mathcal{V}, \mathcal{V}^*)$. We assume that A_{M} and $\mathbb{A}$ have nontrivial finite dimensional kernel. Then for each u in $D(A_{\mathrm{M}})$, the unique

$d_{A_\mathrm{M}} + d_{\mathbb{A}}$-tuple $(u_1, \dots, u_{d_{A_\mathrm{M}}}, u'_1, \dots, u'_{d_{\mathbb{A}}})$ satisfying (5.5.20) is called the *endogenous boundary coordinates of u (with underlying quadruplet $\mathcal{Q}$ and) with respect to the basis* $(\xi_1, \dots, \xi_{d_{A_\mathrm{M}}})$ *of* $\ker A_\mathrm{M}$ *and* $(\eta_1, \dots, \eta_{d_{\mathbb{A}}})$ *of* F_0. □

By considering the adjoint differential quadruplet

$$\mathcal{Q}^* = (SA_\mathrm{M}S, S^*\mathbb{A}S^*, B^*, S^*)$$

on $(\mathcal{V}^*, \mathcal{V})$, there results from Definition 5.5.1 that for each f in $D(S^*\mathbb{A}S^*)$, the unique $d_{A_\mathrm{M}} + d_{\mathbb{A}}$-tuple

$$(f_1, \dots, f_{d_{\mathbb{A}}}, f'_1, \dots, f'_{d_{A_\mathrm{M}}})$$

satisfying (5.5.7) is the endogenous boundary coordinates of f with respect to the basis $(S^*\xi^\mathrm{s}_1, \dots, S^*\xi^\mathrm{s}_{d_{\mathbb{A}}})$ of $\ker S^*\mathbb{A}S^*$ and $(S^*\eta^\mathrm{s}_1, \dots, S^*\eta^\mathrm{s}_{d_{A_\mathrm{M}}})$ of $S^*F^\mathrm{s}_0$. For all purposes, we recall that $(\eta^\mathrm{s}_1, \dots, \eta^\mathrm{s}_{d_{A_\mathrm{M}}})$ is a basis of F^s_0, which is a finite dimensional complementary subspace of $(\ker A_\mathrm{M})^\perp$.

Example 5.5.2 Under the assumptions and notation of Example 5.3.2, let us suppose in addition that $\mathcal{Y}$ is equal to a nontrivial finite dimensional normed space $\mathcal{Y}_\mathrm{f}$ whose dimension is denoted by d_f. Being given any basis

$$\mathcal{B} := (e^\mathrm{f}_1, \dots, e^\mathrm{f}_{d_\mathrm{f}})$$

of $\mathcal{Y}_\mathrm{f}$, there results from Example 5.3.2 that $(g_1 e^\mathrm{f}_1, \dots, g_1 e^\mathrm{f}_{d_\mathrm{f}})$ is a basis of $\ker \mathrm{D}^1_{\mathcal{V}}$, and that

$$E_\mathrm{M} = \langle g_1 e^\mathrm{f}_1, \dots, g_1 e^\mathrm{f}_{d_\mathrm{f}} \rangle \oplus \langle g_2 e^\mathrm{f}_1, \dots, g_2 e^\mathrm{f}_{d_\mathrm{f}} \rangle.$$

Recall that $\mathcal{Q}_{p,\mathcal{Y}_\mathrm{f},1}$ is a $g_1\mathcal{Y}_\mathrm{f}$-regular differential quadruplet according to Example 5.3.2. Hence each u in $D(\mathrm{D}^1_{\mathcal{V}})$ reads

$$u = \sum_{i=1}^{d_\mathrm{f}} (u_i g_1 + u'_i g_2) e^\mathrm{f}_i \; + \; g_1 * \psi^\perp,$$

where $(u_1, \dots, u_{d_\mathrm{f}}, u'_1, \dots, u'_{d_\mathrm{f}})$ are the *endogenous boundary coordinates of* u *with respect to the basis* $(g_1 e^\mathrm{f}_1, \dots, g_1 e^\mathrm{f}_{d_\mathrm{f}})$ of $\ker \mathrm{D}^1_{\mathcal{V}}$ *and* $(g_2 e^\mathrm{f}_1, \dots, g_2 e^\mathrm{f}_{d_\mathrm{f}})$ of $g_2\mathcal{Y}_\mathrm{f}$. Moreover, $\psi^\perp$ is some function of $L^p\big((0, b), \mathcal{Y}_\mathrm{f}\big)$ whose integral on $(0, b)$ vanishes, that is

$$\int_0^b \psi^\perp(y)\,\mathrm{d}y = 0.$$

Besides by (5.3.7), $(u_1, \ldots, u_{d_f})$ and $(u'_1, \ldots, u'_{d_f})$ are the coordinates of $u(0)$ and $b^{-1}\big(u(b) - u(0)\big)$ in the basis $\mathcal{B}$ of $\mathcal{Y}_f$. □

Remark 5.5.3 Let us consider the cases where at least one of the maximal operators A_M and $\mathbb{A}$ has a trivial kernel. If both operators have a trivial kernel, then we agree that there is *no endogenous boundary coordinates*. If only $\mathbb{A}$ has a trivial kernel, then $F_0 = \{0\}$, so that any u in $D(A_M)$ reads

$$u = \sum_{i=1}^{d_{A_M}} u_i \xi_i + B\psi,$$

where u_i's are scalars and ψ lies in $\mathcal{V}$. Thus, $(u_1, \ldots, u_{d_{A_M}})$ is called the *endogenous boundary coordinates of u with respect to the basis* $(\xi_1, \ldots, \xi_{d_{A_M}})$ of $\ker A_M$.

On the other hand, if only A_M has a trivial kernel, then recalling that F_0 is closed complementary subspace of $\ker \mathbb{A}^\perp$ with basis $(\eta_1, \ldots, \eta_{d_\mathbb{A}})$, any u in $D(A_M)$ reads

$$u = BS \sum_{i=1}^{d_\mathbb{A}} u'_i \eta_i + BS\psi^\perp,$$

where the u'_i's are scalars and $\psi^\perp$ lies in $(\ker \mathbb{A})^\perp$. Thus $(u'_1, \ldots, u'_{d_\mathbb{A}})$ is called the *endogenous boundary coordinates of u with respect to the basis* $(\eta_1, \ldots, \eta_{d_\mathbb{A}})$.

□

5.5.2 Properties of Boundary Restriction Operators

Taking advantage of the finite dimension of $\ker A_M$ and $\ker \mathbb{A}$, we will show that the range of a closed boundary restriction operator is closed and has finite co-dimension.

Proposition 5.5.5 *Let $\mathbb{K}$ be equal to $\mathbb{R}$ or $\mathbb{C}$, $\mathcal{V}$ be a reflexive Banach space over $\mathbb{K}$ and $\mathcal{Q} := (\mathbb{A}, A_M, B, S)$ be a differential quadruplet on $(\mathcal{V}, \mathcal{V}^*)$. If the kernels of $\mathbb{A}$ and A_M have finite dimension, then every closed boundary restriction operator of $\mathcal{Q}$ has closed range in $\mathcal{V}$.*

Proof Let A be a closed boundary restriction operator of $\mathcal{Q}$. Under the notations (5.5.1)–(5.5.6), let $(u_n)_{n\in\mathbb{N}}$ be a sequence in $D(A)$ and f belong to $\mathcal{V}$ such that $Au_n \to f$ in $\mathcal{V}$. By (5.5.2), (5.5.6), for each $n \geq 0$, there exist $x_1^n, \ldots, x_{d_E}^n$ in $\mathbb{K}$ and $\psi_n^\perp$ in $\ker \mathbb{A}^\perp$ such that

$$u_n = x_j^n \omega_j + BS\big(x_j^n \omega'_j + \psi_n^\perp\big).$$

Thus,

$$Au_n = x_j^n S\omega_j' + S\psi_n^\perp \in SF_0 + S(\ker \mathbb{A})^\perp.$$

Recalling that F_0 is a closed complementary subspace of $\ker \mathbb{A}^\perp$ in $\mathcal{V}$, we infer that SF_0 is a closed complementary subspace of $S(\ker \mathbb{A}^\perp)$. Now, in view of Proposition 2.3.5, we may legitimately introduce the projection P of $\mathcal{V}$ on SF_0 along $S(\ker \mathbb{A}^\perp)$. Then,

$$x_j^n S\omega_j' \xrightarrow[n\to\infty]{} Pf \qquad \text{in } \mathcal{V}.$$

However, by closedness of the finite dimensional space generated by $\mathbb{S}\omega_1', \ldots, \mathbb{S}\omega_{d_E}'$, we deduce that Pf reads

$$Pf = x_j S\omega_j',$$

for some $x_1, \ldots, x_{d_E}$ in $\mathbb{K}$. Then, we set

$$u := x_j e_j + B\big(f - Pf\big).$$

Since $f - Pf$ lies in $S(\ker \mathbb{A}^\perp)$, and by (5.5.2), $D(A)$ contains $BS(\ker \mathbb{A}^\perp)$, we deduce that u belongs to $D(A)$. Then, with (5.5.6), we get $Au = f$. The proof of the proposition is now completed. □

Proposition 5.5.6 *Let $\mathcal{V}$ be a reflexive Banach space and $\mathcal{Q} := (\mathbb{A}, \mathbb{A}_{\mathrm{M}}, B, S)$ be a differential quadruplet on $(\mathcal{V}, \mathcal{V}^*)$. If the kernels of $\mathbb{A}$ and $\mathbb{A}_{\mathrm{M}}$ have finite dimension then the range of each closed boundary restriction operator of $\mathcal{Q}$ has finite co-dimension in $\mathcal{V}$.*

Proof Let A_{m} denote the minimal operator of $\mathcal{Q}$. Since the domain of A_{m} is equal to $B\big((\ker \mathbb{A}S^*)^\perp\big)$, there results that $R(A_{\mathrm{m}}) = (\ker \mathbb{A}S^*)^\perp$. Applying Corollary 2.2.15 with $\mathcal{X} := \mathcal{V}^*$ and $N := \ker \mathbb{A}S^*$, there results that $R(A_{\mathrm{m}})$ has finite co-dimension. Let A be any closed boundary restriction operator of $\mathcal{Q}$. Since A extends A_{m}, Proposition 2.1.8 entails that $R(A)$ has finite co-dimension as well. □

5.6 Algebraic Calculus of Differential Triplets and Quadruplets

Products and transforms of operators (extension, transform of an operator into its adjoint, ...) induce products and transforms of differential quadruplets and also of boundary restriction operators.

For instance (see Sect. 5.6.1.1), the product of an operator by an involution induces the transform of a given differential quadruplet into another differential quadruplet. At the level of boundary restriction operators, we will show under appropriate assumptions that the class of boundary restriction operators under is stable under the product by an involution.

The extension of the kernel of an operator induces *Caputo extension of a differential quadruplet*. *Caputo extensions* is an abstract tool leading of course to Caputo operators (see Sect. 6.7). Applied to basic differential quadruplets built upon $\frac{\mathrm{d}}{\mathrm{d}x}$, *Caputo extensions* yield differential quadruplets involving singular boundary conditions (see Example 5.6.4).

Products and transforms of differential quadruplets are gathered under the vocable of *algebraic calculus of differential quadruplets*. The term *algebraic* refers to the fact that analytic objects and properties (norms, continuity, ...) will not appear in the foreground.

It turns out that we have already used a transform of quadruplets. It is nothing but the map $\mathcal{Q} \mapsto \mathcal{Q}^*$ where $\mathcal{Q}^*$ is the adjoint differential quadruplet of $\mathcal{Q}$ in the sense of definition 5.1.2.

Recalling that $\mathcal{V}$ denotes a real or complex reflexive Banach space, for each differential quadruplet $\mathcal{Q}$, we denote by $\mathrm{BRO}(\mathcal{Q})$ the set of all boundary restriction operators of $\mathcal{Q}$, and by $\mathrm{cBRO}(\mathcal{Q})$ the set of all *closed* boundary restriction operators of $\mathcal{Q}$.

5.6.1 Involution-Induced Transforms

A map $\sigma : \mathcal{V} \to \mathcal{V}$ is an *involution* if σ lies in $\mathcal{L}(\mathcal{V})$ and $\sigma^2 = \mathrm{id}_{\mathcal{V}}$. Involution-induced transforms allow to introduce in a simple way the issues we want to address regarding the algebraic calculus of differential triplets and quadruplets.

5.6.1.1 Transforms of Quadruplets

Let $\mathcal{Q} := (\mathbb{A}, A_{\mathrm{M}}, B, S)$ be a differential quadruplet on $(\mathcal{V}, \mathcal{V}^*)$ and σ be an involution of $\mathcal{V}$ commuting with S, that is

$$\sigma \in \mathcal{L}(\mathcal{V}), \qquad \sigma^2 = \mathrm{id}_{\mathcal{V}}, \qquad S\sigma = \sigma S.$$

Since $(S\sigma)^2 = \mathrm{id}_{\mathcal{V}}$ due to the previous commutativity property, it is easily seen that

$$\mathcal{Q}\sigma := (\mathbb{A}\sigma^*, \ A_{\mathrm{M}}\sigma, \ \sigma B, \ S\sigma) \tag{5.6.1}$$

is a differential quadruplet on $(\mathcal{V}, \mathcal{V}^*)$. In particular,

$$\mathbb{B}_{\mathcal{Q}\sigma} = \sigma^* \mathbb{B}_{\mathcal{Q}}. \tag{5.6.2}$$

Then, $\mathcal{Q}\sigma$ is called the *product of* $\mathcal{Q}$ *and* σ. Recall that it is assumed that $S\sigma = \sigma S$, hence σ depends on $\mathcal{Q}$. Thus we prefer to call the product of $\mathcal{Q}$ and σ a *transform*. Besides,

$$(\mathcal{Q}\sigma)\sigma = \mathcal{Q}(\sigma^2) = \mathcal{Q}.$$

Let us consider the induced transform on cBRO($\mathcal{Q}$). For each A in cBRO($\mathcal{Q}$), $A\sigma$ is clearly a closed densely defined restriction of $A_{\mathrm{M}}\sigma$. Moreover

$$\sigma^* A^* \subseteq \sigma^* S^* \mathbb{A} S^* = (S\sigma)^* (\mathbb{A}\sigma^*)(S\sigma)^*,$$

thus, $A\sigma$ lies in cBRO($\mathcal{Q}\sigma$), and the map

$$\mathrm{cBRO}(\mathcal{Q}) \to \mathrm{cBRO}(\mathcal{Q}\sigma), \qquad A \mapsto A\sigma \tag{5.6.3}$$

is well-defined and is a bijection whose inverse is

$$\mathrm{cBRO}(\mathcal{Q}\sigma) \to \mathrm{cBRO}(\mathcal{Q}), \qquad C \mapsto C\sigma.$$

Of course, we have in the same way the bijection

$$\mathrm{BRO}(\mathcal{Q}) \to \mathrm{BRO}(\mathcal{Q}\sigma), \qquad A \mapsto A\sigma.$$

Let us look at the action of σ on three distinguished members of BRO($\mathcal{Q}$), the maximal, minimal and pivot operators. That is, we will compute the image of these operators through the map defined by (5.6.3). First of all, the image of the maximal operator of $\mathcal{Q}$ is clearly the maximal operator of $\mathcal{Q}\sigma$. In a same spirit, since

$$D(A_{\mathrm{m},\mathcal{Q}\sigma}) := (\sigma B) S\sigma\big((\ker \mathbb{A}\sigma^*)^{\perp}\big) = \sigma B S(\ker \mathbb{A}^{\perp}) = D(A_{\mathrm{m},\mathcal{Q}}\,\sigma),$$

we deduce that

$$A_{\mathrm{m},\mathcal{Q}\sigma} = A_{\mathrm{m},\mathcal{Q}}\,\sigma. \tag{5.6.4}$$

Also, the pivot operator of $\mathcal{Q}$ is transformed into the pivot operator of $\mathcal{Q}\sigma$, which reads in symbol

$$A_{\mathcal{Q}\sigma} = A_{\mathcal{Q}}\,\sigma.$$

5.6.1.2 Transforms of Triplets

In order to consider involution-induced transforms of differential triplets on a Hilbert space $\mathcal{H}$, we recall that a map $\sigma : \mathcal{H} \to \mathcal{H}$ is a *symmetry* of $\mathcal{H}$ if σ satisfies (4.2.3), i.e.,

$$\sigma \in \mathcal{L}(\mathcal{H}), \qquad \sigma^* = \sigma, \qquad \sigma^2 = \mathrm{id}_{\mathcal{H}}.$$

Let $\mathcal{T} = (\mathbb{A}, \mathbb{B}, \mathbb{S})$ be a differential triplet on $\mathcal{H}$, and σ be a symmetry of $\mathcal{H}$ commuting with $\mathbb{S}$. Since

$$(\sigma\mathbb{B})^* = \mathbb{S}\mathbb{B}\mathbb{S}\sigma = (\mathbb{S}\sigma)(\sigma\mathbb{B})(\mathbb{S}\sigma)$$
$$(\mathbb{S}\sigma)^* = \sigma\mathbb{S} = \mathbb{S}\sigma,$$

we infer that

$$\mathcal{T}\sigma := (\mathbb{A}\sigma, \sigma\mathbb{B}, \mathbb{S}\sigma) \tag{5.6.5}$$

a differential triplet on $\mathcal{H}$. Now starting from the above results for differential quadruplets, the same kind of results may be obtained for differential triplets. The proofs are similar and even simpler. However, instead of repeating arguments, we prefer to use the results about differential quadruplets. For instance, in order to prove that

$$A_{\mathrm{m},\mathcal{T}\sigma} = A_{\mathrm{m},\mathcal{T}}\sigma,$$

we will convert the differential triplet $\mathcal{T}$ into the differential quadruplet $\mathcal{Q}_{\mathcal{T}}$ (see (5.1.8)), use (5.6.4), and eventually go back to $\mathcal{T}$. More precisely, recalling the definitions (5.1.13) and (5.1.15) of the maps t and q, the following proposition holds.

Proposition 5.6.1 *Let $\mathcal{T} = (\mathbb{A}, \mathbb{B}, \mathbb{S})$ be a differential triplet on a Hilbert space $\mathcal{H}$, and σ be a symmetry of $\mathcal{H}$ commuting with $\mathbb{S}$. Then,*

(i) $A_{\mathrm{m},\mathcal{T}} = A_{\mathrm{m},\mathcal{Q}_{\mathcal{T}}}$, *which reads also* $A_{\mathrm{m},\mathcal{T}} = A_{\mathrm{m},q(\mathcal{T})}$;
(ii) $\mathcal{Q}_{\mathcal{T}\sigma} = \mathcal{Q}_{\mathcal{T}}\sigma$, *which reads also* $q(\mathcal{T}\sigma) = q(\mathcal{T})\sigma$;
(iii) $A_{\mathrm{m},\mathcal{T}\sigma} = A_{\mathrm{m},\mathcal{T}} \circ \sigma$.

Proof Recalling that $\mathcal{Q}_{\mathcal{T}} = (\mathbb{A}, \mathbb{A}, \mathbb{B}, \mathbb{S})$, the proof of (i) follows from the definitions of minimal operator for differential triplet and quadruplets. The proof of (ii) is also obvious according to (5.6.1) and (5.6.5). Finally, we get

$$\begin{aligned}
A_{\mathrm{m},\mathcal{T}\sigma} &= A_{\mathrm{m},q(\mathcal{T}\sigma)} && \text{(by (i))}\\
&= A_{\mathrm{m},q(\mathcal{T})\sigma} && \text{(by (ii))}\\
&= A_{\mathrm{m},q(\mathcal{T})} \circ \sigma && \text{(by (5.6.4))}\\
&= A_{\mathrm{m},\mathcal{T}} \circ \sigma && \text{(by (i)).}
\end{aligned}$$

□

Regarding compatibility and realization issues, referring to Definitions 2.5.3 and 5.1.3, we may state the following result.

Proposition 5.6.2 *Let* $(\mathcal{V}, \mathcal{H}, \mathcal{V}^*)$ *be a Gelfand triplet,* $\mathcal{Q} := (\mathbb{A}, A, B, S)$ *be a differential quadruplet on* $(\mathcal{V}, \mathcal{V}^*)$*, and* σ *be an involution of* $\mathcal{V}$ *commuting with* S*. If*

(i) $\mathcal{Q}$ *is compatible with* $\mathcal{H}$,
(ii) $\sigma^*(\mathcal{H}) \subseteq \mathcal{H}$,
(iii) *for each* v *in* $\mathcal{V}$, $\sigma^* v = \sigma v$,
(iv) $\tilde{\sigma} := (\sigma^*)_{\mathcal{H}}$, $\tilde{S} := (S^*)_{\mathcal{H}}$,

then $\mathcal{Q}\sigma$ *is compatible with* $\mathcal{H}$*,* $\tilde{\sigma}$ *is a symmetry of* $\mathcal{H}$ *commuting with* $\tilde{S}$*, and the realization of* $\mathcal{Q}\sigma$ *is the product of the realization of* $\mathcal{Q}$ *by* $\tilde{\sigma}$*, which reads in symbol*

$$\mathcal{T}_{\mathcal{Q}\sigma} = \mathcal{T}_{\mathcal{Q}}\,\tilde{\sigma}. \tag{5.6.6}$$

Proof Let us start to show the compatibility of $\mathcal{Q}\sigma$. By Proposition 5.1.1, it is enough to check that $\mathcal{Q}\sigma$ satisfies (5.1.10)–(5.1.12). For,

$$\begin{aligned} (S\sigma)^*(\mathcal{H}) = \sigma^* S^*(\mathcal{H}) &\subseteq \sigma^*(\mathcal{H}) && \text{(by (i) and Prop. 5.1.1)} \\ &\subseteq \mathcal{H} && \text{(by (ii)).} \end{aligned}$$

Also

$$\begin{aligned} \mathbb{B}_{\mathcal{Q}\sigma}(\mathcal{H}) &= \sigma^* \mathbb{B}_{\mathcal{Q}}(\mathcal{H}) && \text{(by (5.6.2))} \\ &= \sigma^*(\mathcal{H}) && \text{(by (i) and Prop. 5.1.1)} \\ &\subseteq \mathcal{H} && \text{(by (ii)).} \end{aligned}$$

Thus, $\mathcal{Q}\sigma$ satisfies (5.1.10). Regarding (5.1.11), one has for each v in $\mathcal{V}$

$$\begin{aligned} (S\sigma)^* v &= (\sigma S)^* v && \text{(since } S \text{ and } \sigma \text{ commute)} \\ &= S^* \sigma^* v \\ &= S\sigma v && \text{(by (iii) and (i)).} \end{aligned}$$

Thus, $\mathcal{Q}\sigma$ satisfies (5.1.11). Finally,

$$\begin{aligned} \mathbb{B}_{\mathcal{Q}\sigma}(v) &= \sigma^* \mathbb{B}_{\mathcal{Q}} v && \text{(by (5.6.2))} \\ &= \sigma^* B v && \text{(by (i) and Prop. 5.1.1)} \\ &= \sigma B v && \text{(by (iii)),} \end{aligned}$$

so that $\mathcal{Q}\sigma$ satisfies (5.1.12). Hence $\mathcal{Q}\sigma$ is compatible with $\mathcal{H}$.

Let us prove that $\tilde{\sigma}$ is a symmetry of $\mathcal{H}$ commuting with $\tilde{S}$. Indeed Lemma 5.1.4 and (ii), (iii) yield that $\tilde{\sigma}$ is self-adjoint. Moreover, using Lemma 5.1.3 and the commutativity of σ and S, we get

$$\tilde{\sigma}^2 = \big((\sigma^*)^2\big)_{\mathcal{H}} = \mathrm{id}_{\mathcal{H}}$$

$$\tilde{\sigma}\tilde{S} = (\sigma^* S^*)_{\mathcal{H}} = \big((S\sigma)^*\big)_{\mathcal{H}} = \big((\sigma S)^*\big)_{\mathcal{H}} = \tilde{S}\tilde{\sigma}.$$

Finally, in order to prove (5.6.6), we first notice that by the definition (5.6.1) of $\mathcal{Q}\sigma$,

$$\mathcal{T}_{\mathcal{Q}\sigma} = \big((\mathbb{A}\sigma^*)_{\mathcal{H}}, (\mathbb{B}_{\mathcal{Q}\sigma})_{\mathcal{H}}, (S\sigma)^*_{\mathcal{H}}\big).$$

We claim that

$$(\mathbb{A}\sigma^*)_{\mathcal{H}} = \mathbb{A}_{\mathcal{H}}\tilde{\sigma}. \tag{5.6.7}$$

Indeed, for each h in the domain of $(\mathbb{A}\sigma^*)_{\mathcal{H}}$,

$$\begin{aligned} (\mathbb{A}\sigma^*)_{\mathcal{H}}h = \mathbb{A}\sigma^* h &= \mathbb{A}\tilde{\sigma}h && \text{(by (ii))} \\ &= \mathbb{A}_{\mathcal{H}}\tilde{\sigma}h, \end{aligned}$$

since $\tilde{\sigma}h$ and $\mathbb{A}\tilde{\sigma}h$ belong to $\mathcal{H}$. Thus, $(\mathbb{A}\sigma^*)_{\mathcal{H}} \subseteq \mathbb{A}_{\mathcal{H}}\tilde{\sigma}$. Now for each h in the domain of $\mathbb{A}_{\mathcal{H}}\tilde{\sigma}$, one has

$$\mathbb{A}_{\mathcal{H}}\tilde{\sigma}h = \mathbb{A}\tilde{\sigma}h = \mathbb{A}\sigma^* h.$$

Thus h and $\mathbb{A}\sigma^* h$ belong to $\mathcal{H}$, hence h lies in the domain of $(\mathbb{A}\sigma^*)_{\mathcal{H}}$. There results that $D(\mathbb{A}_{\mathcal{H}}\tilde{\sigma})$ is contained in $D\big((\mathbb{A}\sigma^*)_{\mathcal{H}}\big)$, and the claim follows. Moreover,

$$\begin{aligned} (\mathbb{B}_{\mathcal{Q}\sigma})_{\mathcal{H}} &= (\sigma^* \mathbb{B}_{\mathcal{Q}})_{\mathcal{H}} && \text{(by (5.6.2))} \\ &= \tilde{\sigma}(\mathbb{B}_{\mathcal{Q}})_{\mathcal{H}} && \text{(by (5.1.10) and Lemma 5.1.3).} \end{aligned}$$

Also

$$\begin{aligned} (S\sigma)^*_{\mathcal{H}} &= (\sigma S)^*_{\mathcal{H}} && \text{(since } \sigma \text{ and } S \text{ commute)} \\ &= (S^*\sigma^*)_{\mathcal{H}} \\ &= (S^*)_{\mathcal{H}}\tilde{\sigma} && \text{(by (5.1.10) and Lemma 5.1.3).} \end{aligned}$$

Thus, going back to (5.6.7) and using the definition (5.6.5), we obtain (5.6.6). □

In the same way, for any involution σ of $\mathcal{V}$ commuting with S, we may consider the *product of σ and $\mathcal{Q}$*, which is the differential quadruplet

$$\sigma\,\mathcal{Q} := (\sigma^*\mathbb{A}, \sigma\, A_{\mathrm{M}}, B\sigma, \sigma\, S)$$

on $(\mathcal{V}, \mathcal{V}^*)$. One has $\mathbb{B}_{\sigma\mathcal{Q}} = \mathbb{B}_{\mathcal{Q}}\,\sigma^*$ and analog results hold for this transform. In particular, the map

$$\mathrm{cBRO}(\mathcal{Q}) \to \mathrm{cBRO}(\sigma\,\mathcal{Q}), \qquad A \mapsto \sigma\, A$$

is a bijection, and

$$A_{\mathrm{m},\sigma\,\mathcal{Q}} = \sigma\, A_{\mathrm{m},\mathcal{Q}}. \tag{5.6.8}$$

Moreover, under the assumptions and notation of Proposition 5.6.2, we claim that $\sigma\,\mathcal{Q}$ is compatible with $\mathcal{H}$ and that the realization of $\sigma\,\mathcal{Q}$ in $\mathcal{H}$ is $\tilde{\sigma}\,\mathcal{T}_{\mathcal{Q}}$, which reads in symbol

$$\mathcal{T}_{\sigma\,\mathcal{Q}} = \tilde{\sigma}\,\mathcal{T}_{\mathcal{Q}}. \tag{5.6.9}$$

The proof of the compatibility is similar to the one of the compatibility of $\mathcal{Q}\sigma$. In order to prove (5.6.9), we start to show that

$$(\sigma^*\mathbb{A})_{\mathcal{H}} = \tilde{\sigma}\,\mathbb{A}_{\mathcal{H}}. \tag{5.6.10}$$

It turns out that the proof of (5.6.10) is slightly different from the one of (5.6.7) and goes as follows. Any h in the domain of $(\sigma^*\mathbb{A})_{\mathcal{H}}$ lies in $\mathcal{H} \cap D(\mathbb{A})$ and satisfies

$$\sigma^*\mathbb{A}h \in \mathcal{H}.$$

Since σ^* is an involution of $\mathcal{V}^*$, we deduce with the assumption (ii) in Proposition 5.6.2 that $\mathbb{A}h$ lies in $\mathcal{H}$. Thus,

$$h \in D(\mathbb{A}_{\mathcal{H}}) = D\big(\sigma^*\mathbb{A}_{\mathcal{H}}\big).$$

Moreover,

$$\tilde{\sigma}\,\mathbb{A}_{\mathcal{H}}h = \sigma^*\mathbb{A}h = (\sigma^*\mathbb{A})_{\mathcal{H}}h$$

by definition of the domain of $(\sigma^*\mathbb{A})_{\mathcal{H}}$. Thus $(\sigma^*\mathbb{A})_{\mathcal{H}}$ is a restriction of $\tilde{\sigma}\,\mathbb{A}_{\mathcal{H}}$. Let us prove that

$$D\big(\tilde{\sigma}\,\mathbb{A}_{\mathcal{H}}\big) \subseteq D\big((\sigma^*\mathbb{A})_{\mathcal{H}}\big).$$

For, any h in the domain of $\tilde{\sigma}\,\mathbb{A}_{\mathcal{H}}$ satisfies

$$\tilde{\sigma}\,\mathbb{A}_{\mathcal{H}} h = \sigma^* \mathbb{A}_{\mathcal{H}} h = \sigma^* \mathbb{A} h.$$

Thus, h lies in $\mathcal{H} \cap D(\sigma^* \mathbb{A})$ and $\sigma^* \mathbb{A} h$ lives in $\mathcal{H}$. Therefore, h belongs to the domain of $(\sigma^* \mathbb{A})_{\mathcal{H}}$. That proves the previous announced inclusion and completes the proof of (5.6.10). Now, arguing as in the proof of Proposition 5.6.2, we derive (5.6.9). □

Corollary 5.6.3 *Let $\mathcal{H}$ be a real or complex Hilbert space and $\mathcal{T} = (\mathbb{A}, \mathbb{B}, \mathbb{S})$ be a differential triplet on $\mathcal{H}$. Assume that*

(i) *σ is a symmetry of $\mathcal{H}$;*
(ii) *σ commutes with $\mathbb{S}$.*

Then, the realization of $\mathcal{Q}_{\mathcal{T}}\sigma$ is $\mathcal{T}\sigma$, and the realization of $\sigma\,\mathcal{Q}_{\mathcal{T}}$ is $\sigma\mathcal{T}$. In symbol,

$$t\big(q(\mathcal{T})\sigma\big) = \mathcal{T}\sigma, \qquad t\big(\sigma q(\mathcal{T})\big) = \sigma\mathcal{T},$$

or equivalently

$$\mathcal{T}_{\mathcal{Q}_{\mathcal{T}}\sigma} = \mathcal{T}\sigma, \qquad \mathcal{T}_{\sigma\mathcal{Q}_{\mathcal{T}}} = \sigma\mathcal{T}. \tag{5.6.11}$$

Proof Let us prove the first equality. By (5.1.15), $\mathcal{Q}_{\mathcal{T}}$ is compatible with $\mathcal{H}$, thus Proposition 5.6.2 entails that $\mathcal{Q}_{\mathcal{T}}\sigma$ is compatible with $\mathcal{H}$ and

$$\mathcal{T}_{\mathcal{Q}_{\mathcal{T}}\sigma} = \mathcal{T}_{\mathcal{Q}_{\mathcal{T}}}\sigma = \mathcal{T}\sigma$$

by (5.1.16). The second equality of (5.6.11) is obtained in the same way by using (5.6.9). □

5.6.2 Conjugation by an Isomorphism

Let $\mathbb{K}$ be equal to $\mathbb{R}$ or $\mathbb{C}$ and $\mathcal{V}$, $\mathcal{W}$ be reflexive Banach spaces over $\mathbb{K}$. Let also $\mathcal{Q} := (\mathbb{A}, A_{\mathrm{M}}, B, S)$ be a differential quadruplet on $(\mathcal{V}, \mathcal{V}^*)$ and $T : \mathcal{V} \to \mathcal{W}$ be an isomorphism.

5.6.2.1 General Setting

Thanks to Proposition 2.2.21, it is easily seen that

$$\big((T^{-1})^* \mathbb{A} T^*, T A_{\mathrm{M}} T^{-1}, T B T^{-1}, T S T^{-1}\big) \tag{5.6.12}$$

is a differential quadruplet on $(\mathcal{W}, \mathcal{W}^*)$, which is called the *conjugate differential quadruplet of* $\mathcal{Q}$ *by* T, and is denoted by $T\mathcal{Q}T^{-1}$. One has

$$\mathbb{B}_{T\mathcal{Q}T^{-1}} := (TST^{-1})^*(TBT^{-1})^*(TST^{-1})^* = (T^{-1})^*\mathbb{B}_{\mathcal{Q}}T^*.$$

Notice that unlike products by an involution, we do not assume here that T commutes with S.

Roughly speaking, the next result states that the conjugation by an isomorphism carries a differential quadruplet and its boundary restriction operators into the target space. In particular, the conjugation by an isomorphism transforms a minimal operator into a minimal operator.

Proposition 5.6.4 *Let* $\mathcal{V}$, $\mathcal{W}$ *be reflexive Banach spaces,* $\mathcal{Q} := (\mathbb{A}, A_{\mathrm{M}}, B, S)$ *be a differential quadruplet on* $(\mathcal{V}, \mathcal{V}^*)$ *and* T *be an isomorphism from* $\mathcal{V}$ *onto* $\mathcal{W}$*. Then, for each* A *in* cBRO$(\mathcal{Q})$, TAT^{-1} *lies in* cBRO$(T\mathcal{Q}T^{-1})$*. Thus, the map*

$$\mathrm{cBRO}(\mathcal{Q}) \to \mathrm{cBRO}(T\mathcal{Q}T^{-1}), \qquad A \mapsto TAT^{-1}$$

is well defined, and especially, is a bijection. Besides, the minimal operator of $T\mathcal{Q}T^{-1}$ *is the conjugate of the minimal operator of* $\mathcal{Q}$ *by* T*, that is,*

$$A_{\mathrm{m},T\mathcal{Q}T^{-1}} = TA_{\mathrm{m},\mathcal{Q}}T^{-1}. \tag{5.6.13}$$

The same formula holds for pivot operators, namely

$$A_{T\mathcal{Q}T^{-1}} = TA_{\mathcal{Q}}T^{-1}. \tag{5.6.14}$$

Proof Let $T\mathcal{Q}T^{-1} = \big((T^{-1})^*\mathbb{A}T^*, TA_{\mathrm{M}}T^{-1}, TBT^{-1}, TST^{-1}\big)$ be the conjugate quadruplet of $\mathcal{Q} = (\mathbb{A}, A_{\mathrm{M}}, B, S)$. Then, for each A in cBRO$(\mathcal{Q})$, TAT^{-1} is closed and densely defined (since T is an isomorphism). Moreover, $A \subseteq A_{\mathrm{M}}$ yields that $TAT^{-1} \subseteq TA_{\mathrm{M}}T^{-1}$, and $A^* \subseteq S^*\mathbb{A}S^*$ entails

$$(TAT^{-1})^* = (T^{-1})^*A^*T^* \subseteq (TST^{-1})^*\big((T^{-1})^*\mathbb{A}T^*\big)(TST^{-1})^*.$$

Whence TAT^{-1} lies in cBRO$(T\mathcal{Q}T^{-1})$, so that the map introduced in the statement of the proposition is well defined. Moreover, it is clearly injective. Finally, for any C in cBRO$(T\mathcal{Q}T^{-1})$, we infer from the first part of this proof that $A := T^{-1}CT$ belongs to

$$\mathrm{cBRO}\big(T^{-1}(T\mathcal{Q}T^{-1})T\big) = \mathrm{cBRO}(\mathcal{Q}).$$

Since $TAT^{-1} = C$, the map is onto.

In order to prove (5.6.13), observe that $A_{m,TQT^{-1}}$ and $TA_{m,Q}T^{-1}$ are restrictions of $TA_M T^{-1}$. Thus, it remains to show the equality of their domains. For, we have

$$\begin{aligned} D(A_{m,TQT^{-1}}) &= TBT^{-1}(TST^{-1})\big((\ker \mathbb{A}T^*)^{\perp}\big) && \text{(by Def. 5.2.3)} \\ &= TBS(\ker \mathbb{A}^{\perp}) && \text{(by Lemma 4.3.9)} \\ &= TD(A_{m,Q}) = D(TA_{m,Q}T^{-1}). \end{aligned}$$

Finally, the proof of (5.6.14) follows the same lines. Indeed, the operators are both restrictions of $TA_M T^{-1}$. Moreover, by definition of the pivot operator of TQT^{-1},

$$D(A_{TQT^{-1}}) = R(TQT^{-1}),$$

and $R(TQT^{-1})$ is nothing but $TR(B)$. On the other hand, we have

$$D(TA_Q T^{-1}) = TR(B).$$

Hence (5.6.14) follows. □

Now we will see that regular differential quadruplets are invariant under conjugation by an isomorphism.

Proposition 5.6.5 *Let $\mathcal{V}$, $\mathcal{W}$ be reflexive Banach spaces, $\mathcal{Q} := (\mathbb{A}, A_M, B, S)$ be a F_0-differential quadruplet on $(\mathcal{V}, \mathcal{V}^*)$ and T be an isomorphism from $\mathcal{V}$ onto $\mathcal{W}$. Then TQT^{-1} is a $T(F_0)$-differential quadruplet on $(\mathcal{W}, \mathcal{W}^*)$.*

Proof We have already seen that TQT^{-1} is a differential quadruplet. Moreover, the maximal operators of TQT^{-1} are closed since $\mathcal{Q}$ is regular and T is an isomorphism. There remains to check that $T(F_0)$ is a closed complementary subspace of the annihilator of $\ker(T^{-1})^*\mathbb{A}T^*$. For since Q is F_0-regular and T is an isomorphism, we derive that $T(F_0)$ and $T(\ker \mathbb{A}^{\perp})$ are closed complementary subspaces in $\mathcal{W}$. Now we compute

$$\big(\ker(T^{-1})^*\mathbb{A}T^*\big)^{\perp} = \big((T^{-1})^* \ker \mathbb{A}\big)^{\perp} = T(\ker \mathbb{A}^{\perp}),$$

by Lemma 4.3.9. That completes the proof of the proposition. □

5.6.2.2 Conjugation by an Involution

As a particular case, we assume that $\mathcal{V} = \mathcal{W}$ and $T := \sigma$ where σ be an involution of $\mathcal{V}$. Then, (5.6.12) yields

$$\sigma Q\sigma = (\sigma^*\mathbb{A}\sigma^*, \sigma A_M\sigma, \sigma B\sigma, \sigma S\sigma) \tag{5.6.15}$$

$$\mathbb{B}_{\sigma \mathcal{Q} \sigma} = \sigma^* \mathbb{B}_{\mathcal{Q}} \sigma^*.$$

Notice that unlike products by an involution, we do not assume here that σ commutes with S.

By (5.6.13) and (5.6.14), we get

$$A_{\mathrm{m},\sigma \mathcal{Q} \sigma} = \sigma\, A_{\mathrm{m},\mathcal{Q}}\, \sigma \tag{5.6.16}$$

$$A_{\sigma \mathcal{Q} \sigma} = \sigma\, A_{\mathcal{Q}}\, \sigma. \tag{5.6.17}$$

5.6.3 Adjoint and Reverse Transforms

Referring to Definition 5.1.2, we remind the reader that the adjoint of a differential quadruplet $\mathcal{Q} := (\mathbb{A}, A_{\mathrm{M}}, B, S)$ is the differential quadruplet

$$\mathcal{Q}^* := (S A_{\mathrm{M}} S, S^* \mathbb{A} S^*, B^*, S^*).$$

The following result justifies why $\mathcal{Q}^*$ is called the *adjoint* differential quadruplet of $\mathcal{Q}$.

Proposition 5.6.6 *Let $\mathcal{V}$ be a reflexive Banach space and $\mathcal{Q} := (\mathbb{A}, A_{\mathrm{M}}, B, S)$ be a differential quadruplet on $(\mathcal{V}, \mathcal{V}^*)$. Then, for each closed boundary restriction operator A of $\mathcal{Q}$, A^* is a closed boundary restriction operator of $\mathcal{Q}^*$. Thus, the map*

$$\mathrm{cBRO}(\mathcal{Q}) \to \mathrm{cBRO}(\mathcal{Q}^*), \qquad A \mapsto A^*$$

is well defined, and besides, is a bijection.

Proof Let A be a closed boundary restriction operator of $\mathcal{Q}$. According to Theorem 2.4.1, A^* is densely defined and $A^{**} = A$. Thus,

$$A^{**} = A \subseteq A_{\mathrm{M}} = S(S A_{\mathrm{M}} S) S.$$

Also, $A^* \subseteq S^* \mathbb{A} S^*$ since A is a boundary restriction operator of $\mathcal{Q}$. Thus, Definition 5.2.1 entails that A^* is a boundary restriction operator of $\mathcal{Q}^*$. Moreover, A^* is closed by Proposition 2.2.19. Thus, the map introduced in the statement of the Proposition is well defined. Moreover, it is injective because $A^{**} = A$. Finally, let L belong to $\mathrm{cBRO}(\mathcal{Q}^*)$. We infer from the first part of this proof that L^* belongs to $\mathrm{cBRO}(\mathcal{Q}^{**})$. Since $\mathcal{Q}$ is the adjoint quadruplet of $\mathcal{Q}^*$, there results that the map is onto. □

Combining Propositions 5.6.4 and 5.6.6, we end up with the following result.

Corollary 5.6.7 *Let $\mathcal{Q} := (\mathbb{A}, A_{\mathrm{M}}, B, S)$ be a differential quadruplet on $(\mathcal{V}, \mathcal{V}^*)$. Then*

$$S^*\mathcal{Q}^*S^* = (A_{\mathrm{M}}, \mathbb{A}, S^*B^*S^*, S^*), \qquad \mathbb{B}_{S^*\mathcal{Q}^*S^*} = B,$$

and for each A in cBRO$(\mathcal{Q})$*, $S^*A^*S^*$ lies in* cBRO$(S^*\mathcal{Q}^*S^*)$*, so that the map*

$$\mathrm{cBRO}(\mathcal{Q}) \to \mathrm{cBRO}(S^*\mathcal{Q}^*S^*), \qquad A \mapsto S^*A^*S^*$$

is well defined, and especially, is a bijection. Besides, $S^*\mathcal{Q}^*S^* = (S\mathcal{Q}S)^*$.

Let us introduce another transform of quadruplets. Starting from a differential quadruplet $\mathcal{Q} := (\mathbb{A}, A_{\mathrm{M}}, B, S)$ on $(\mathcal{V}, \mathcal{V}^*)$, it is easily seen that

$$r(\mathcal{Q}) := (A_{\mathrm{M}}, \mathbb{A}, \mathbb{B}_{\mathcal{Q}}, S^*) \tag{5.6.18}$$

is a differential quadruplet on $(\mathcal{V}^*, \mathcal{V})$, which is called the *reverse differential quadruplet of* $\mathcal{Q}$. There holds that $\mathbb{B}_{r(\mathcal{Q})} = B$.

In application, the reverse differential quadruplet is easily computed, thus the following result will allow to compute the minimal operator of the adjoint differential quadruplet in a simple way.

Corollary 5.6.8 *Let $\mathcal{Q} := (\mathbb{A}, A_{\mathrm{M}}, B, S)$ be a differential quadruplet on $(\mathcal{V}, \mathcal{V}^*)$. Then,*

$$\mathcal{Q}^* = S^*r(\mathcal{Q})S^* \tag{5.6.19}$$

$$S^*\mathcal{Q}^*S^* = r(\mathcal{Q}) \tag{5.6.20}$$

$$A_{\mathrm{m},\mathcal{Q}^*} = S^*A_{\mathrm{m},r(\mathcal{Q})}S^* \tag{5.6.21}$$

$$A_{\mathrm{M},\mathcal{Q}^*} = S^*A_{\mathrm{M},r(\mathcal{Q})}S^* \tag{5.6.22}$$

$$A_{\mathcal{Q}^*} = S^*A_{r(\mathcal{Q})}S^*. \tag{5.6.23}$$

Proof Equation (5.6.20) follows from Corollary 5.6.7 and (5.6.18). Equation (5.6.19) follows from (5.6.20) and the fact that the conjugate of $S^*r(\mathcal{Q})S^*$ by S^* is equal to $r(\mathcal{Q})$. (5.6.21) is a consequence of (5.6.19) and (5.6.16). (5.6.22) follows from (5.6.19) and the definitions (5.6.15), (5.6.18). Finally (5.6.23) follows from (5.6.19) and (5.6.17). □

5.6.4 Caputo Extensions

In order to extend an operator, the following trick is standard: take a vector outside of its domain and assign the zero vector to the former vector. It turns out that this procedure is used in the definition of Caputo derivatives (see Sect. 6.7); hence, we call it *Caputo extension*. Starting with Caputo extensions of operators, we will next introduce Caputo extensions of differential triplets and quadruplets.

5.6.4.1 Definitions

Let $\mathbb{K}$ be equal to $\mathbb{R}$ or $\mathbb{C}$, and $\mathcal{X}$ be a $\mathbb{K}$-normed space. Let also $A : D(A) \subseteq \mathcal{X} \to \mathcal{X}$ be an operator on $\mathcal{X}$, and N be a subspace of $\mathcal{X}$. Assume that

$$N \text{ is in direct sum with } D(A). \tag{5.6.24}$$

Then, we define the domain

$$D(\tilde{A}) := N \oplus D(A) \tag{5.6.25}$$

and extend A as follows. For each u in $D(\tilde{A})$, there exists a unique ordered pair (ζ, v) in $N \times D(A)$ such that $u = \zeta + v$. Then we set

$$\tilde{A}u := A(u - \zeta) = Av. \tag{5.6.26}$$

Clearly, $\tilde{A}$ is an operator on $\mathcal{X}$. Let us summarize this construction.

Definition 5.6.1 Let $A : D(A) \subseteq \mathcal{X} \to \mathcal{X}$ be an operator on a normed space $\mathcal{X}$, and N be a subspace of $\mathcal{X}$ satisfying (5.6.24). Then the operator $\tilde{A}$ defined by (5.6.25) and (5.6.26) is called the *Caputo extension of A with respect to N*. □

Proposition 5.6.9 *Let $A : D(A) \subseteq \mathcal{X} \to \mathcal{X}$ be an operator on a normed space $\mathcal{X}$, N be a subspace of $\mathcal{X}$ satisfying* (5.6.24), *and $\tilde{A}$ be the Caputo extension of A with respect to N. Then,*

$$A \subseteq \tilde{A} \qquad \textit{and} \qquad \ker \tilde{A} = N \oplus \ker A.$$

Proof The direct sum (5.6.25) entails that $D(A)$ is contained in $D(\tilde{A})$. Moreover, (5.6.26) yields that A and $\tilde{A}$ coincide on $D(A)$. Thus, $A \subseteq \tilde{A}$. This relation yields that $\ker A$ is contained in $\ker \tilde{A}$. Moreover, (5.6.25) and (5.6.26) yield that N is a subspace of $\ker \tilde{A}$. Thus, $N \oplus \ker A \subseteq \ker \tilde{A}$. Conversely, each u in $\ker \tilde{A}$ satisfies

$$0 = \tilde{A}u = A(u - \zeta),$$

for some ζ in N. Thus u lies in $N + \ker A$. □

Example 5.6.2 Being given (p, b) in $[1, \infty) \times (0, \infty)$, we set $\mathcal{X} := L^p(0, b) = L^p$ and define

$$\mathrm{D}^1_{L^p} : W^{1,p}(0,b) \subseteq L^p(0,b) \to L^p(0,b), \qquad u \mapsto \frac{\mathrm{d}}{\mathrm{d}x} u$$

$$B_{1,L^p} : L^p(0,b) \to L^p(0,b), \qquad v \mapsto g_1 * v.$$

For any function ℓ in $L^p(0,b) \setminus W^{1,p}(0,b)$, we denote by $\tilde{\mathrm{D}}^1_{L^p}$ the Caputo extension of $\mathrm{D}^1_{L^p}$ with respect to $N := \langle \ell \rangle$. Then Proposition 5.6.9 gives

$$\ker \tilde{\mathrm{D}}^1_{L^p} = \langle g_1, \ \ell \rangle.$$

With Lemma 4.3.3, each function u in the domain of $\tilde{\mathrm{D}}^1_{L^p}$ reads

$$u = u_1 \ell + u_2 g_1 + g_1 * \tilde{\mathrm{D}}^1_{L^p} u$$

for some unique ordered pair (u_1, u_2) in $\mathbb{K} \times \mathbb{K}$. Moreover by Definition 5.6.1,

$$\tilde{\mathrm{D}}^1_{L^p} u := \mathrm{D}^1_{L^p}(u - u_1 \ell).$$

□

Let us now introduce the *Caputo extension of a differential quadruplet*. Let $\mathcal{V}$ be a reflexive Banach space and $\mathcal{Q} := (\mathbb{A}, A_{\mathrm{M}}, B, S)$ be a differential quadruplet on $(\mathcal{V}, \mathcal{V}^*)$. Let also N, N^{s} be subspaces of $\mathcal{V}$ and $\mathcal{V}^*$ respectively, such that

$$N \ \text{ is in direct sum with } D(A_{\mathrm{M}}), \tag{5.6.27}$$

$$N^{\mathrm{s}} \ \text{ is in direct sum with } D(\mathbb{A}). \tag{5.6.28}$$

Let $\tilde{A}_{\mathrm{M}}$ be the Caputo extension of A_{M} with respect to N, and $\tilde{\mathbb{A}}$ be the Caputo extension of $\mathbb{A}$ with respect to N^{s}. Then, we claim that

$$\tilde{\mathcal{Q}} := (\tilde{\mathbb{A}}, \ \tilde{A}_{\mathrm{M}}, \ B, \ S) \tag{5.6.29}$$

is a differential quadruplet on $(\mathcal{V}, \mathcal{V}^*)$, and $\mathbb{B}_{\tilde{\mathcal{Q}}} = \mathbb{B}_{\mathcal{Q}}$. Indeed, since $A_{\mathrm{M}} B = \mathrm{id}_{\mathcal{V}}$ and $\tilde{A}_{\mathrm{M}}$ extends A_{M} (by Proposition 5.6.9), we deduce that $\tilde{A}_{\mathrm{M}} B = \mathrm{id}_{\mathcal{V}}$. In a same way, $\tilde{\mathbb{A}} \mathbb{B}_{\mathcal{Q}} = \mathrm{id}_{\mathcal{V}^*}$. Then the claim follows.

The differential quadruplet $\tilde{\mathcal{Q}}$ is called the *Caputo extension of* $\mathcal{Q}$ *with respect to* N *and* N^{s}.

5.6.4.2 Boundary Restriction Operators

Let $\mathcal{V}$ be a reflexive Banach space, $\mathcal{Q} := (\mathbb{A}, A_{\mathrm{M}}, B, S)$ be a differential quadruplet on $(\mathcal{V}, \mathcal{V}^*)$, and N, N^{s} be subspaces of $\mathcal{V}$ and $\mathcal{V}^*$ respectively. Assume (5.6.27), (5.6.28), and consider the Caputo extension of $\mathcal{Q}$ defined by (5.6.29).

Proposition 5.6.10 *Under the above assumptions and notation, the following assertions hold true.*

(i) *Each boundary restriction operator of $\mathcal{Q}$ is a boundary restriction operator of $\tilde{\mathcal{Q}}$.*
(ii) *The pivot operators of $\mathcal{Q}$ and $\tilde{\mathcal{Q}}$ coincide, i.e. $A_{\mathcal{Q}} = A_{\tilde{\mathcal{Q}}}$.*
(iii) *The minimal operator of $\mathcal{Q}$ is an extension of the minimal operator of $\tilde{\mathcal{Q}}$, i.e. $A_{\mathrm{m},\tilde{\mathcal{Q}}} \subseteq A_{\mathrm{m},\mathcal{Q}}$.*

Proof Since $\tilde{A}_{\mathrm{M}}$ is an extension of A_{M}, and $\tilde{\mathbb{A}}$ is an extension of $\mathbb{A}$, (i) follows from the very definition of boundary restriction operators. By definition of pivot operators, $A_{\mathcal{Q}}$ and $A_{\tilde{\mathcal{Q}}}$ have the same domain. Since they are restrictions of $\tilde{A}_{\mathrm{M}}$, they are equal. Finally, Item (iii) is a consequence of $\ker \tilde{\mathbb{A}}^{\perp} \subseteq \ker \mathbb{A}^{\perp}$. □

5.6.4.3 Differential Triplets

Regarding differential triplets, the construction of Caputo extensions goes in the same way. Let $\mathcal{T} = (\mathbb{A}, \mathbb{B}, \mathbb{S})$ be a differential triplet on a Hilbert space $\mathcal{H}$, and N be a subspace of $\mathcal{H}$, which is in direct sum with $D(\mathbb{A})$. By considering the Caputo extension $\tilde{\mathbb{A}}$ of $\mathbb{A}$ with respect to N,

$$\tilde{\mathcal{T}} := (\tilde{\mathbb{A}}, \mathbb{B}, \mathbb{S})$$

is a differential triplet on $\mathcal{H}$, which is called the *Caputo extension of $\mathcal{T}$ with respect to N*. In application, (see Sects. 6.2.2.4 and 6.7.4.1), $\mathbb{A}$ is the Riemann–Liouville operator, N is the space generated by g_1, and the Caputo operator is the restriction of $\tilde{\mathbb{A}}$ to the domain $N \oplus R(\mathbb{B})$.

Example 5.6.3 Let $\mathbb{K}$ be equal to $\mathbb{R}$ or $\mathbb{C}$, and $\mathcal{H} := L^2(0, b)$. Under the assumptions and notation of Example 4.2.4, let us denote by $\mathcal{T}_{1,\mathbb{K}}$ the differential triplet defined by (4.2.7), that is,

$$\mathcal{T}_{1,\mathbb{K}} := (\mathrm{D}^1_{L^2}, B_{1,L^2}, S).$$

Our goal is to introduce a *Caputo extension* of the differential triplet $\mathcal{T}_{1,\mathbb{K}}$.

In order to define a suitable space N, we consider for each positive real number γ, the Riemann–Liouville kernel of order γ

$$g_\gamma : (0, b) \to \mathbb{R}, \qquad x \mapsto \frac{1}{\Gamma(\gamma)} x^{\gamma-1}. \tag{5.6.30}$$

Assuming that

$$\frac{1}{2} < \gamma \leq \frac{3}{2}, \qquad \gamma \neq 1, \tag{5.6.31}$$

there results that g_γ lies in $L^2(0, b)$ but not in $H^1(0, b)$. Thus the one dimensional space

$$N := \langle g_\gamma \rangle$$

is in direct sum with $D(\mathrm{D}^1_{L^2}) = H^1(0, b)$. Denoting by $\tilde{\mathrm{D}}^1_{L^2}$ the Caputo extension of $\mathrm{D}^1_{L^2}$ with respect to $\langle g_\gamma \rangle$, the Caputo extension of $\mathcal{T}_{1,\mathbb{K}}$ reads

$$\tilde{\mathcal{T}}_{1,\mathbb{K}} := (\tilde{\mathrm{D}}^1_{L^2}, B_{1,L^2}, S).$$

By Proposition 5.6.9

$$\ker \tilde{\mathrm{D}}^1_{L^2} = \langle g_\gamma, g_1 \rangle.$$

Hence for each u in $D(\tilde{\mathrm{D}}^1_{L^2})$, the Domain Structure Theorem 4.3.14 yields that

$$u = u_1 g_\gamma + u_2 g_1 + u_1' B_{1,L^2} S g_\gamma + u_2' B_{1,L^2} S g_1 + B_{1,L^2} S \psi^\perp, \tag{5.6.32}$$

where u_1, u_2, u_1', u_2' are scalars and $\psi^\perp$ lies in $\langle g_\gamma, g_1 \rangle^\perp$, the annihilator of $\langle g_\gamma, g_1 \rangle$ in $L^2(0, b)$. In view of Definition 4.4.2, (u_1, u_2, u_1', u_2') are *the endogenous boundary coordinates of u with respect to the basis* (g_γ, g_1). Note that in this framework of scalar-valued functions, endogenous boundary coordinates coincide with *endogenous boundary values* (see Sect. 5.7.1.1).

Let us consider "standard" boundary values of u. For, we first assume in addition to (5.6.31) that $\gamma < 1$. Then, the L^2-function $\frac{1}{g_\gamma}$ admits a unique representative which can be extended into a continuous function on $[0, b]$. This continuous function still labeled $\frac{1}{g_\gamma}$ satisfies of course

$$\frac{1}{g_\gamma}(x) = \Gamma(\gamma) x^{1-\gamma}, \qquad \forall x \in [0, b].$$

Besides, the four functions

$$g_1, \qquad B_{1,L^2} S g_\gamma, \qquad B_{1,L^2} S g_1 = g_2, \qquad B_{1,L^2} S \psi^\perp$$

appearing in the right hand side of (5.6.32) belong to $H^1(0, b)$. By Theorem 3.2.7, they are identified with continuous functions on $[0, b]$. There results that $\frac{u}{g_\gamma}$ admits a unique continuous extension on $[0, b]$. This extension will be still labeled $\frac{u}{g_\gamma}$. For the same reasons, $u - u_1 g_\gamma$ can also be seen as a continuous function on $[0, b]$. Then, evaluations yield

$$\frac{u}{g_\gamma}(0) = u_1, \qquad (u - u_1 g_\gamma)(0) = u_2. \tag{5.6.33}$$

Since

$$\begin{aligned} B_{1,L^2} S\psi^\perp(b) &= (\psi^\perp, g_1) = 0 \\ B_{1,L^2} S g_\gamma(b) &= (g_\gamma, g_1) \\ B_{1,L^2} S g_1(b) &= (g_1, g_1), \end{aligned}$$

evaluating (5.6.32) at $x = b$, we get

$$u(b) = u_1 g_\gamma(b) + u_2 + u_1'(g_\gamma, g_1) + u_2' \|g_1\|_{L^2}^2.$$

The fourth and last boundary value is computed from (5.6.32) again. Since g_γ and g_1 belong to the kernel of $\tilde{\mathrm{D}}^1_{L^2}$,

$$g_\gamma * \tilde{\mathrm{D}}^1_{L^2} u = u_1' g_\gamma * S g_\gamma + u_2' g_\gamma * S g_1 + g_\gamma * S\psi^\perp.$$

Since $g_\gamma * S g_\gamma(b) = \|g_\gamma\|_{L^2}^2$ and

$$g_\gamma * S\psi^\perp(b) = (\psi^\perp, g_\gamma) = 0,$$

we derive

$$g_\gamma * \mathrm{D}^1_{L^2} u(b) = u_1' \|g_\gamma\|_{L^2}^2 + u_2'(g_1, g_\gamma).$$

Then, we have found four values associated to a generic element u in the domain of $\tilde{\mathrm{D}}^1_{L^2}$, namely

$$\frac{u}{g_\gamma}(0), \qquad (u - \tfrac{u}{g_\gamma}(0) g_\gamma)(0), \qquad u(b), \qquad g_\gamma * \tilde{\mathrm{D}}^1_{L^2} u(b). \tag{5.6.34}$$

In order to call it "*standard*" boundary values of u, we must check that they are related to the above endogenous boundary values (u_1, u_2, u_1', u_2') through an isomorphism. This turns out to be the case, since

$$\begin{pmatrix} \frac{u}{g_\gamma}(0) \\ (u - \frac{u}{g_\gamma}(0) g_\gamma)(0) \\ u(b) \\ g_\gamma * \tilde{\mathrm{D}}^1_{L^2} u(b) \end{pmatrix} = \begin{pmatrix} 1 & 0 & 0 & 0 \\ 0 & 1 & 0 & 0 \\ g_\gamma(b) & 1 & (g_1, g_\gamma) & \|g_1\|_{L^2}^2 \\ 0 & 0 & \|g_\gamma\|_{L^2}^2 & (g_1, g_\gamma) \end{pmatrix} \begin{pmatrix} u_1 \\ u_2 \\ u_1' \\ u_2' \end{pmatrix}.$$

Moreover, the latter matrix is invertible since by expanding its determinant with respect to the first line, we get

$$(g_1, g_\gamma)^2 - \|g_1\|_{L^2}^2 \|g_\gamma\|_{L^2}^2,$$

which is nonzero by the equality cases of the Cauchy–Schwarz inequality.

Let us assume that $1 < \gamma < \frac{3}{2}$. We claim that

$$\begin{aligned} u(0) &= u_2 \\ \frac{u - u(0)g_1}{g_\gamma}(0) &= u_1 \\ u(b) &= u_1 g_\gamma(b) + u_2 + u_1'(g_1, g_\gamma) + u_2' \|g_1\|_{L^2}^2 \\ g_{\gamma-1} * u(b) &= u_1 g_{2\gamma-1}(b) + u_2 g_\gamma(b) + u_1' \|g_\gamma\|_{L^2}^2 + u_2'(g_1, g_\gamma). \end{aligned}$$

Indeed, the first equation follows from (5.6.32) by a continuity argument. By (5.6.32) again, there exists a function h in $L^2(0, b)$ such that

$$\frac{u - u_2 g_1}{g_\gamma} = u_1 + \frac{g_1 * h}{g_\gamma}.$$

Next, $g_1 * h$ lies in $C^{0,\frac{1}{2}}([0, b])$ and

$$\frac{1}{g_\gamma} = \Gamma(\gamma)\Gamma(2-\gamma) g_{2-\gamma}.$$

Thus applying Lemma 3.5.5 with $\alpha := \gamma - 1$ and $\beta := \frac{1}{2}$, we get the continuity of $\frac{g_1 * h}{g_\gamma}$ and especially that this function vanishes at $x = 0$. Hence, the second equation follows. The two last equations are proven as in the previous case, which completes the proof of the claim.

Again the latter linear system is invertible, hence standard boundary values of u read

$$u(0), \qquad \frac{u - u(0)g_1}{g_\gamma}(0), \qquad u(b), \qquad g_{\gamma-1} * u(b).$$

In the case where $\gamma = \frac{3}{2}$, only the equation for u_1 is changing since Lemma 3.5.5 cannot be applied here. We find

$$u_1 = \frac{u - u(0)g_1 - g_1 * \tilde{\mathrm{D}}^1_{L^2} u}{g_\gamma}(0).$$

This equation holds true as well for $1 < \gamma < \frac{3}{2}$. □

5.6.4.4 Finite Dimensional Kernels

Let $\mathcal{V}$ be a reflexive Banach space and $\mathcal{Q} := (\mathbb{A}, A, B, S)$ be a differential quadruplet on $(\mathcal{V}, \mathcal{V}^*)$. We assume that $\mathbb{A}$ has a finite dimensional kernel, and denote by $d_{\mathbb{A}}$ the dimension of this kernel. Let N, N^{s} be finite dimensional subspaces of $\mathcal{V}$ and $\mathcal{V}^*$ respectively, satisfying (5.6.27) and (5.6.28). Let $\tilde{\mathcal{Q}}$ be the Caputo extension of $\mathcal{Q}$ with respect to N and N^{s}.

According to Sect. 5.5.1, $\ker \mathbb{A}^{\perp}$ admits a complementary subspace F_0 in $\mathcal{V}$, whose dimension is equal to $d_{\mathbb{A}}$. For almost the same reasons that will nevertheless be detailed in the proof of Proposition 5.6.11, $\ker \tilde{\mathbb{A}}^{\perp}$ admits also a finite dimensional complementary subspace $\tilde{F}_0$ in $\mathcal{V}$. Since $\ker \tilde{\mathbb{A}}$ is larger than $\ker \mathbb{A}$, $\tilde{F}_0$ is larger than F_0. Then, the issue is to show that we may choose $\tilde{F}_0$ with the additional condition that $\tilde{F}_0$ contains F_0.

Proposition 5.6.11 *Let $\mathbb{A} : D(\mathbb{A}) \subseteq \mathcal{V}^* \to \mathcal{V}^*$ be an operator on $\mathcal{V}^*$ with finite dimensional kernel. Consider the Caputo extension $\tilde{\mathbb{A}}$ of $\mathbb{A}$ with respect to a finite dimensional subspace N^{s} of $\mathcal{V}^*$ satisfying* (5.6.28). *Then,*

(i) $\ker \mathbb{A}^{\perp}$ *admits a finite dimensional complementary subspace F_0 in $\mathcal{V}$ and* $\dim F_0 = \dim \ker \mathbb{A}^{\perp}$;
(ii) $\ker \tilde{\mathbb{A}}^{\perp}$ *admits a finite dimensional complementary subspace $\tilde{F}_0$ in $\mathcal{V}$ such that*

$$\tilde{F}_0 \supseteq F_0,$$
$$\dim \tilde{F}_0 = \dim N^{\mathrm{s}} + \dim \ker \mathbb{A}.$$

Proof Since $\ker \mathbb{A}$ is finite dimensional, Item (i) follows from Proposition 2.2.14. Let us prove Item (ii). By Proposition 5.6.9, $\ker \tilde{\mathbb{A}}$ is finite dimensional. Thus, Proposition 2.2.14 entails that $\ker \tilde{\mathbb{A}}^{\perp}$ admits a finite dimensional complementary subspace in $\mathcal{V}$. Then, $\ker \tilde{\mathbb{A}}^{\perp}$ has finite co-dimension according to Proposition 2.1.12. Applying Proposition 2.1.8 with

$$M := \ker \tilde{\mathbb{A}}^{\perp}, \qquad V := F_0 \oplus \ker \tilde{\mathbb{A}}^{\perp},$$

we get that $F_0 \oplus \ker \tilde{\mathbb{A}}^{\perp}$ has finite co-dimension in $\mathcal{V}$. Thus, by Proposition 2.1.11, $F_0 \oplus \ker \tilde{\mathbb{A}}^{\perp}$ admits a finite dimensional complementary subspace F_1 in $\mathcal{V}$. Hence,

$$\mathcal{V} = F_1 \oplus (F_0 \oplus \ker \tilde{\mathbb{A}}^{\perp}) = (F_1 \oplus F_0) \oplus \ker \tilde{\mathbb{A}}^{\perp}.$$

Accordingly, $\tilde{F}_0 := F_1 \oplus F_0$ is a finite dimensional complementary subspace of $\ker \tilde{\mathbb{A}}^{\perp}$ containing F_0. Then

$$\begin{aligned}\dim \tilde{F}_0 &= \operatorname{codim} \ker \tilde{\mathbb{A}}^{\perp} && \text{(by Proposition 2.1.12)}\\ &= \dim \ker \tilde{\mathbb{A}} && \text{(by Corollary 2.2.15)}\\ &= \dim N^{\mathrm{s}} + \dim \ker \mathbb{A} && \text{(by Proposition 5.6.9).}\end{aligned}$$

□

Example 5.6.4 Under the assumptions and notation of Example 5.5.2, we recall that $\mathcal{Y}_{\mathrm{f}}$ is a non trivial finite dimensional normed space, p, q are conjugate real numbers in $(1, \infty)$, and that

$$\mathcal{B} := (e^{\mathrm{f}}_1, \dots, e^{\mathrm{f}}_{d_{\mathrm{f}}})$$

is one of its basis. Then, we denote by

$$\mathcal{B}^{\mathrm{s}} := (e^{\mathrm{s}}_1, \dots, e^{\mathrm{s}}_{d_{\mathrm{f}}})$$

the anti-dual basis of $\mathcal{B}$. That is, e^{s}_i lives in $\mathcal{Y}_{\mathrm{f}}{}^*$ and

$$\langle e^{\mathrm{s}}_i, e^{\mathrm{f}}_j \rangle_{\mathcal{Y}_{\mathrm{f}}{}^*, \mathcal{Y}_{\mathrm{f}}} = \delta_{ij}, \qquad \forall i, j = 1, \dots, d_{\mathrm{f}}.$$

Our aim is to introduce a Caputo extension of the differential quadruplet

$$\mathcal{Q}_{p,\mathcal{Y}_{\mathrm{f}},1} := (\mathrm{D}^1_{\mathcal{V}*},\ \mathrm{D}^1_{\mathcal{V}},\ B_{1,\mathcal{V}},\ S_{\mathcal{V}}).$$

For each positive numbers γ and δ, let g_γ, g_δ be the functions defined by (5.6.30). We observe that g_γ lies in $L^p(0, b) \setminus W^{1,p}(0, b)$ if and only if

$$1 - \frac{1}{p} < \gamma \le 2 - \frac{1}{p}, \qquad \gamma \ne 1. \tag{5.6.35}$$

Since $\frac{1}{q} = 1 - \frac{1}{p}$, we deduce that g_δ lies in $L^q(0, b) \setminus W^{1,q}(0, b)$ if and only if

$$\frac{1}{p} < \delta \le 1 + \frac{1}{p}, \qquad \delta \ne 1. \tag{5.6.36}$$

Assuming (5.6.35) and (5.6.36), it is clear that $W^{1,p}\big((0, b), \mathcal{Y}_{\mathrm{f}})\big)$ and $g_\gamma \mathcal{Y}_{\mathrm{f}}$ are in direct sum, and also that $W^{1,q}\big((0, b), \mathcal{Y}_{\mathrm{f}}{}^*)\big)$, $g_\delta \mathcal{Y}_{\mathrm{f}}{}^*$ are in direct sum. Then, we denote by $\tilde{\mathrm{D}}^1_{\mathcal{V}}$, $\tilde{\mathrm{D}}^1_{\mathcal{V}*}$ the respective Caputo extensions of $\mathrm{D}^1_{\mathcal{V}}$, $\mathrm{D}^1_{\mathcal{V}*}$ with respect to $g_\gamma \mathcal{Y}_{\mathrm{f}}$ and $g_\delta \mathcal{Y}^*_{\mathrm{f}}$. The Caputo extension of $\mathcal{Q}_{p,\mathcal{Y},1}$ is then the differential quadruplet

$$\tilde{\mathcal{Q}}_{p,\mathcal{Y}_{\mathrm{f}},1} := (\tilde{\mathrm{D}}^1_{\mathcal{V}*},\ \tilde{\mathrm{D}}^1_{\mathcal{V}},\ B_{1,\mathcal{V}},\ S_{\mathcal{V}}).$$

By Proposition 5.6.9,

$$\ker \tilde{\mathrm{D}}^1_{\mathcal{V}} = g_\gamma \mathcal{Y}_\mathrm{f} \oplus g_1 \mathcal{Y}_\mathrm{f}, \qquad \ker \tilde{\mathrm{D}}^1_{\mathcal{V}*} = g_\delta \mathcal{Y}_\mathrm{f}^* \oplus g_1 \mathcal{Y}_\mathrm{f}^*.$$

Moreover, it is known from Example 5.3.2 that $Q_{p,\mathcal{Y}_\mathrm{f},1}$ is $g_1\mathcal{Y}_\mathrm{f}$-regular, which means in particular that $F_0 := g_1\mathcal{Y}_\mathrm{f}$ is a closed complementary subspace of

$$(\ker \mathrm{D}^1_{\mathcal{V}*})^\perp = (g_1 \mathcal{Y}_\mathrm{f}^*)^\perp$$

in $\mathcal{V}$. The issue is to find a complementary subspace of $(\ker \tilde{\mathrm{D}}^1_{\mathcal{V}*})^\perp$ in $\mathcal{V}$. Proposition 5.6.11 tells us that the dimension of any of these complementary spaces is $2d_\mathrm{f}$, and that there exists a complementary subspace $\tilde{F}_0$ containing $g_1\mathcal{Y}_\mathrm{f}$. A natural candidate could be $g_\delta\mathcal{Y}_\mathrm{f} \oplus g_1\mathcal{Y}_\mathrm{f}$, however due to (5.6.36), g_δ does not always belong to $\mathcal{V}$. *We claim that a suitable choice is*

$$\tilde{F}_0 := g_\gamma \mathcal{Y}_\mathrm{f} \oplus g_1 \mathcal{Y}_\mathrm{f},$$

i.e. $\tilde{F}_0 = \ker \tilde{\mathrm{D}}^1_{\mathcal{V}}$ *is a complementary subspace of* $(\ker \tilde{\mathrm{D}}^1_{\mathcal{V}*})^\perp$ *in* $\mathcal{V}$. Indeed, denoting by $\langle \cdot, \cdot \rangle_{q,p}$ the anti-duality bracket between $L^p(0,b)$ and $L^q(0,b)$, we first observe that the matrix

$$\begin{pmatrix} \langle g_\delta, g_\gamma \rangle_{q,p} & \langle g_\delta, g_1 \rangle_{q,p} \\ \langle g_1, g_\gamma \rangle_{q,p} & \langle g_1, g_1 \rangle_{q,p} \end{pmatrix}$$

is invertible because its determinant, namely

$$\frac{(\gamma-1)(\delta-1)}{\gamma\delta(\gamma+\delta-1)} \frac{b^{\gamma+\delta}}{\Gamma(\gamma)\Gamma(\delta)}$$

is nonzero since by assumption, γ and δ differ from 1. Notice that $\gamma+\delta-1 \neq 0$ because $\gamma+\delta > 1$ by (5.6.35) and (5.6.36).

With Lemma 2.2.13, we get

$$L^p(0,b) = \langle g_\gamma, g_1 \rangle + \langle g_\delta, g_1 \rangle^\perp.$$

Then, by using the base $\mathcal{B}$, we derive

$$L^p\big((0,b), \mathcal{Y}_\mathrm{f}\big) = \tilde{F}_0 + (\ker \tilde{\mathrm{D}}^1_{\mathcal{V}*})^\perp.$$

Finally, if $g_\gamma Y_0 + g_1 Y_1$ lies in $\tilde{F}_0$ and in $(\ker \tilde{\mathrm{D}}^1_{\mathcal{V}*})^\perp$ then for each Y^s in $\mathcal{Y}_\mathrm{f}^*$, one has

$$\langle g_\delta, g_\gamma \rangle_{q,p} \langle Y^\mathrm{s}, Y_0 \rangle_{\mathcal{Y}_\mathrm{f}^*,\mathcal{Y}_\mathrm{f}} + \langle g_\delta, g_1 \rangle_{q,p} \langle Y^\mathrm{s}, Y_1 \rangle_{\mathcal{Y}_\mathrm{f}^*,\mathcal{Y}_\mathrm{f}} = 0$$

$$\langle g_1, g_\gamma\rangle_{q,p}\langle Y^{\mathrm{s}}, Y_0\rangle_{\mathcal{Y}_{\mathrm{f}}^*,\mathcal{Y}_{\mathrm{f}}} + \langle g_1, g_1\rangle_{q,p}\langle Y^{\mathrm{s}}, Y_1\rangle_{\mathcal{Y}_{\mathrm{f}}^*,\mathcal{Y}_{\mathrm{f}}} = 0.$$

By invertibility of the latter matrix, $Y_0 = Y_1 = 0$, and the sum is direct. That completes the proof of the claim.

Since $\tilde{\mathrm{D}}^1_{\mathcal{V}} B_{1,\mathcal{V}} = \mathrm{id}_{\mathcal{V}}$, Lemma 4.3.3 yields that

$$D(\tilde{\mathrm{D}}^1_{\mathcal{V}}) = \ker \tilde{\mathrm{D}}^1_{\mathcal{V}} \oplus R(B_{1,\mathcal{V}}).$$

Thus, with Proposition 4.4.1, we deduce that $\tilde{\mathrm{D}}^1_{\mathcal{V}}$ is a closed operator on $\mathcal{V}$. In the same way, $\tilde{\mathrm{D}}^1_{\mathcal{V}^*}$ is a closed operator on $\mathcal{V}^*$. There results that $\tilde{\mathcal{Q}}_{p,\mathcal{Y}_{\mathrm{f}},1}$ is a $g_\gamma\mathcal{Y}_{\mathrm{f}} \oplus g_1\mathcal{Y}_{\mathrm{f}}$-regular differential quadruplet.

Then, the Domain Structure Theorem 5.3.3 and Definition 5.5.1 give that each u in $D(\tilde{\mathrm{D}}^1_{\mathcal{V}})$ reads

$$u = \sum_{i=1}^{d_{\mathrm{f}}} (u_{2i-1}g_\gamma + u_{2i}g_1)e_i^{\mathrm{f}} + \sum_{i=1}^{d_{\mathrm{f}}} B_{1,\mathcal{V}}S_{\mathcal{V}}(u'_{2i-1}g_\gamma + u'_{2i}g_1)e_i^{\mathrm{f}} + B_{1,\mathcal{V}}S_{\mathcal{V}}\psi^\perp, \tag{5.6.37}$$

where $\psi^\perp$ lies in the annihilator of $g_\gamma\mathcal{Y}_{\mathrm{f}}^* \oplus g_1\mathcal{Y}_{\mathrm{f}}^*$, and

$$(u_1, \ldots, u_{2d_{\mathrm{f}}}, u'_1, \ldots, u'_{2d_{\mathrm{f}}})$$

is the endogenous boundary coordinates of u with underlying quadruplet $\tilde{\mathcal{Q}}_{p,\mathcal{Y}_{\mathrm{f}},1}$ and with respect to the basis

$$(g_\gamma e_1^{\mathrm{f}}, g_1 e_1^{\mathrm{f}}, \ldots, g_\gamma e_{d_{\mathrm{f}}}^{\mathrm{f}}, g_1 e_{d_{\mathrm{f}}}^{\mathrm{f}})$$

of the subspaces $\ker \tilde{\mathrm{D}}^1_{\mathcal{V}}$ and $\tilde{F}_0 := g_\gamma\mathcal{Y}_{\mathrm{f}} \oplus g_1\mathcal{Y}_{\mathrm{f}}$. □

5.6.5 Products of Differential Quadruplets and Triplets

These products allow to consider applications with higher-order differential operators.

5.6.5.1 Definition and First Results for Quadruplets

We begin with an elementary result, which holds in the general framework of Banach spaces.

Lemma 5.6.12 *Let $\mathcal{X}$ be a Banach space, A, A' be operators on $\mathcal{X}$, and B, B' belong to $\mathcal{L}(\mathcal{X})$. Suppose that*

$$AB = \mathrm{id}_{\mathcal{X}}, \qquad A'B' = \mathrm{id}_{\mathcal{X}}.$$

Then, $\ker A'$ *and* $B' \ker A$ *are in direct sum and*

$$\ker AA' = \ker A' \oplus B' \ker A.$$

Proof Let u be in $\ker A' \cap B' \ker A$. Then $u = B'\xi$ for some ξ in $\ker A$. Since $A'B' = \mathrm{id}_{\mathcal{X}}$, we have $\xi = A'u = 0$ because u lies in $\ker A'$. Thus, $u = 0$ and the sum is direct. Now it is clear that this sum is contained in $\ker AA'$. Conversely, let u belong to $\ker AA'$. Then, $A'u$ lies in $\ker A$, and since u lives in the domain of A', Lemma 4.3.3 entails that there exists some u_0' in $\ker A'$ such that

$$u = u_0' + B'A'u.$$

Thus, u belongs to $\ker A' + B' \ker A$. That completes the proof of the lemma. □

As far as the product of differential quadruplets is concerned, the commutativity of right inverses of maximal operators is a fundamental assumption.

Proposition 5.6.13 *Let* $\mathcal{V}$ *be a reflexive Banach space and* $\mathcal{Q} := (\mathbb{A}, A_{\mathrm{M}}, B, S)$, $\mathcal{Q}' = (\mathbb{A}', A_{\mathrm{M}}', B', S)$ *be differential quadruplets on* $(\mathcal{V}, \mathcal{V}^*)$. *If* B' *commute with* B *then the following assertions hold.*

(i)

$$\mathcal{Q}\mathcal{Q}' := (\mathbb{A}\mathbb{A}',\ A_{\mathrm{M}}A_{\mathrm{M}}',\ B'B,\ S)$$

is a differential quadruplet on $(\mathcal{V}, \mathcal{V}^*)$, *and*

$$\mathbb{B}_{\mathcal{Q}\mathcal{Q}'} = \mathbb{B}_{\mathcal{Q}}\mathbb{B}_{\mathcal{Q}'} = \mathbb{B}_{\mathcal{Q}'}\mathbb{B}_{\mathcal{Q}}. \tag{5.6.38}$$

(ii) $\ker \mathbb{A}'$ *and* $\mathbb{B}_{\mathcal{Q}'} \ker \mathbb{A}$ *are in direct sum and*

$$\ker \mathbb{A}\mathbb{A}' = \ker \mathbb{A}' \oplus \mathbb{B}_{\mathcal{Q}'} \ker \mathbb{A}. \tag{5.6.39}$$

(iii) $\ker A_{\mathrm{M}}'$ *and* $B' \ker A_{\mathrm{M}}$ *are in direct sum and*

$$\ker A_{\mathrm{M}}A_{\mathrm{M}}' = \ker A_{\mathrm{M}}' \oplus B' \ker A_{\mathrm{M}}. \tag{5.6.40}$$

(iv) *If in addition,* $\mathbb{A}$, A_{M} *and* $\mathbb{A}'$, A_{M}' *have finite dimensional kernel then* $\mathcal{Q}\mathcal{Q}'$ *is a closed differential quadruplet on* $(\mathcal{V}, \mathcal{V}^*)$.

Proof Regarding (5.6.38), one has

$$\begin{aligned}\mathbb{B}_{QQ'} &= S^*(B'B)^*S^* = S^*(BB')^*S^* && \text{(since } B \text{ and } B' \text{ commute)}\\ &= (S^*B'^*S^*)(S^*B^*S^*) && \text{(since } (S^*)^2 = \mathrm{id}_{\mathcal{V}^*}\text{)}\\ &= \mathbb{B}_{Q'}\mathbb{B}_{Q}.\end{aligned}$$

In the above computations, if the commutation of B and B' is not used, then we get $\mathbb{B}_{QQ'} = \mathbb{B}_{Q}\mathbb{B}_{Q'}$. Then, we readily check that QQ' is a differential quadruplet on $(\mathcal{V}, \mathcal{V}^*)$.

Items (ii) and (iii) follows from Lemma 5.6.12. Regarding (iv), since $\mathbb{A}$ and $\mathbb{A}'$ have finite dimensional kernel, the operator $\mathbb{A}\mathbb{A}'$ has finite dimensional kernel according to (5.6.39). Hence Proposition 4.4.1 entails that $\mathbb{A}\mathbb{A}'$ is a closed operator on $\mathcal{V}^*$. The same arguments hold for $A_{\mathrm{M}}A'_{\mathrm{M}}$. Hence, QQ' is closed. □

Proposition 5.6.13 allows two introduce *the product of two differential quadruplets* and in turn, *entire power of a differential quadruplet*.

Definition 5.6.5 Let $\mathcal{V}$ be a reflexive Banach space and $Q := (\mathbb{A}, A_{\mathrm{M}}, B, S)$, $Q' := (\mathbb{A}', A'_{\mathrm{M}}, B', S)$ be differential quadruplets on $(\mathcal{V}, \mathcal{V}^*)$ such that B commute with B'. Then, the differential quadruplet QQ' given by Proposition 5.6.13 is called the *product of Q and Q'*.

Moreover, if $QQ' = Q'Q$, then we say that *Q and Q' commute* or that Q' commute *with Q*.

In the particular case where $Q = Q'$, we set $Q^1 := Q$ and for each integer $n \geq 2$, we define by induction the nth *power Q^n of Q* by

$$Q^n := Q^{n-1}Q.$$

□

Notice that Q and Q' possesses the same involution S. Besides, if Q and Q' commute, then $\mathbb{A}$ and $\mathbb{A}'$ commute, also A'_{M} compute with A_{M}, and B' commute with B.

Under suitable assumptions on differential quadruplets, we will show that the product of two closed boundary restriction operators is a closed boundary restriction operator. The proof uses the deep results of Sect. 2.4.3.

Proposition 5.6.14 *Let $\mathcal{V}$ be a reflexive Banach space and $Q := (\mathbb{A}, A_{\mathrm{M}}, B, S)$, $Q' := (\mathbb{A}', A'_{\mathrm{M}}, B', S)$ be differential quadruplets on $(\mathcal{V}, \mathcal{V}^*)$. Assume that*

(i) *Q' commute with Q;*
(ii) *$\mathbb{A}$, A_{M}, $\mathbb{A}'$, A'_{M} have finite dimensional kernel;*
(iii) *A, A' are closed boundary restriction operators of Q and Q', respectively.*

Then, AA' is a closed boundary restriction operator of QQ'.

Proof Since $\mathbb{A}'$ and A'_{M} have finite dimensional kernel, Proposition 5.5.6 entails that $R(A')$ has finite co-dimension. Then, by Theorem 2.4.10, AA' is a densely defined operator on $\mathcal{V}$ and $(AA')^* = A'^*A^*$. Thus

$$(AA')^* \subseteq S^*\mathbb{A}'S^*S^*\mathbb{A}S^* = S^*\mathbb{A}'\mathbb{A}S^* = S^*\mathbb{A}\mathbb{A}'S^*,$$

since $\mathbb{A}$ and $\mathbb{A}'$ commute by Assumption (i). Hence AA' is a boundary restriction operator of $\mathcal{Q}\mathcal{Q}'$. Finally, let us show that AA' is closed. For, since A_{M} has finite dimensional kernel, A has also finite dimensional kernel. Moreover, $R(A)$ is closed according to Proposition 5.5.5. Thus Theorem 2.4.7 implies that AA' is closed. □

5.6.5.2 Maximal, Minimal, and Pivot Operators

Let $\mathcal{V}$ be a reflexive Banach space and $\mathcal{Q} := (\mathbb{A}, A_{\mathrm{M}}, B, S)$, $\mathcal{Q}' := (\mathbb{A}', A'_{\mathrm{M}}, B', S)$ be differential quadruplets on $(\mathcal{V}, \mathcal{V}^*)$. If B and B' commute then it is clear from Definition 5.6.5 that the maximal operator of $\mathcal{Q}\mathcal{Q}'$ is the product of the maximal operators of $\mathcal{Q}$ and $\mathcal{Q}'$. In symbol,

$$A_{\mathrm{M},\mathcal{Q}\mathcal{Q}'} = A_{\mathrm{M},\mathcal{Q}}A_{\mathrm{M},\mathcal{Q}'} = A_{\mathrm{M}}A'_{\mathrm{M}}.$$

We will see that analog result holds for minimal operators. However, a stronger assumption is needed, since we suppose that $\mathcal{Q}'$ commute with $\mathcal{Q}$.

Proposition 5.6.15 *Let $\mathcal{V}$ be a reflexive Banach space and $\mathcal{Q} := (\mathbb{A}, A_{\mathrm{M}}, B, S)$, $\mathcal{Q}' := (\mathbb{A}', A'_{\mathrm{M}}, B', S)$ be differential quadruplets on $(\mathcal{V}, \mathcal{V}^*)$. If $\mathcal{Q}'$ commute with $\mathcal{Q}$, then the minimal operator of $\mathcal{Q}\mathcal{Q}'$ is the product of the minimal operators of $\mathcal{Q}$ and $\mathcal{Q}'$, which reads in symbol*

$$A_{\mathrm{m},\mathcal{Q}\mathcal{Q}'} = A_{\mathrm{m},\mathcal{Q}}A_{\mathrm{m},\mathcal{Q}'}. \tag{5.6.41}$$

Moreover, $A_{\mathrm{m},\mathcal{Q}'}$ commute with $A_{\mathrm{m},\mathcal{Q}}$.

Proof In order to establish (5.6.41), we first show the equality of the domains. By (5.6.39),

$$(\ker \mathbb{A}\mathbb{A}')^{\perp} = (\ker \mathbb{A}')^{\perp} \cap \big(\mathbb{B}_{\mathcal{Q}'} \ker \mathbb{A}\big)^{\perp}.$$

Then, since B' commute with B, and $BB'S$ is one-to-one,

$$D(A_{\mathrm{m},\mathcal{Q}\mathcal{Q}'}) = BB'S(\ker \mathbb{A}')^{\perp} \cap BB'S\big(\mathbb{B}_{\mathcal{Q}'} \ker \mathbb{A}\big)^{\perp} \tag{5.6.42}$$

$$= BD(A_{\mathrm{m},\mathcal{Q}'}) \cap BB'S\big(\mathbb{B}_{\mathcal{Q}'} \ker \mathbb{A}\big)^{\perp}. \tag{5.6.43}$$

We claim that

$$BB'S\big(\mathbb{B}_{Q'}\ker\mathbb{A}\big)^{\perp} \subseteq D(A_{\mathrm{m},Q}). \tag{5.6.44}$$

Indeed, for each v in $\big(\mathbb{B}_{Q'}\ker\mathbb{A}\big)^{\perp}$, one has for all ξ in $\ker\mathbb{A}$,

$$0 = \langle\mathbb{B}_{Q'}\xi, v\rangle = \langle S^*B'^*S^*\xi, v\rangle = \langle\xi, SB'Sv\rangle.$$

Thus, $SB'Sv$ lies in $(\ker\mathbb{A})^{\perp}$, so that

$$B'S\big(\mathbb{B}_{Q'}\ker\mathbb{A}\big)^{\perp} \subseteq S(\ker\mathbb{A})^{\perp}.$$

The claim follows from the definition of $D(A_{\mathrm{m},Q})$. Going back to (5.6.43) and using (5.6.44), we obtain

$$D(A_{\mathrm{m},QQ'}) \subseteq BD(A_{\mathrm{m},Q'}) \cap D(A_{\mathrm{m},Q}) = D(A_{\mathrm{m},Q'}A_{\mathrm{m},Q}), \tag{5.6.45}$$

where latter equality is obtained by using that $BA_{\mathrm{m},Q} = \mathrm{id}_{D(A_{\mathrm{m},Q})}$.

Conversely, let us show that $D(A_{\mathrm{m},Q'}A_{\mathrm{m},Q}) \subseteq D(A_{\mathrm{m},QQ'})$. For, by (5.6.43), it is enough to show that

$$D(A_{\mathrm{m},Q'}A_{\mathrm{m},Q}) \subseteq BD(A_{\mathrm{m},Q'}) \cap BB'S\big(\mathbb{B}_{Q'}\ker\mathbb{A}\big)^{\perp}.$$

Let u belong to $D(A_{\mathrm{m},Q'}A_{\mathrm{m},Q})$. By (5.6.45), we derive that $D(A_{\mathrm{m},Q'}A_{\mathrm{m},Q})$ is contained in $BD(A_{\mathrm{m},Q'})$. Next, in order to prove that the former space is contained in $BB'S\big(\mathbb{B}_{Q'}\ker\mathbb{A}\big)^{\perp}$, we first claim that

$$A_{\mathrm{m},Q'}A_{\mathrm{m},Q}u \in S\big(\mathbb{B}_{Q'}\ker\mathbb{A}\big)^{\perp} = \big(S^*\mathbb{B}_{Q'}\ker\mathbb{A}\big)^{\perp}. \tag{5.6.46}$$

Indeed, Theorem 5.2.5 (i) yields that $R(A_{\mathrm{m},Q}) = (S^*\ker\mathbb{A})^{\perp}$, which implies that for all ξ in $\ker\mathbb{A}$,

$$0 = \langle S^*\xi,\ A_{\mathrm{m},Q}\,u\rangle = \langle S^*\xi,\ B'A_{\mathrm{m},Q'}A_{\mathrm{m},Q}\,u\rangle = \langle S^*\mathbb{B}_{Q'}\xi,\ A_{\mathrm{m},Q'}A_{\mathrm{m},Q}\,u\rangle.$$

Hence, (5.6.46) holds. Now since

$$u = BB'A_{\mathrm{m},Q'}A_{\mathrm{m},Q}u,$$

we deduce from (5.6.46) that $D(A_{\mathrm{m},Q'}A_{\mathrm{m},Q})$ is contained in $BB'S\big(\mathbb{B}_{Q'}\ker\mathbb{A}\big)^{\perp}$. Hence, $A_{\mathrm{m},Q'}A_{\mathrm{m},Q}$ and $A_{\mathrm{m},QQ'}$ have the same domain. Since Q' commute with Q, we get

$$D(A_{\mathrm{m},QQ'}) = D(A_{\mathrm{m},Q}A_{\mathrm{m},Q'}).$$

In order to complete the proof of (5.6.41), we just observe that $A_{\mathrm{m},\mathcal{Q}}A_{\mathrm{m},\mathcal{Q}'}$ and $A_{\mathrm{m},\mathcal{Q}\mathcal{Q}'}$ are both restrictions of $A_{\mathrm{M}}A'_{\mathrm{M}}$. Hence (5.6.41) holds. Finally,

$$\begin{aligned} A_{\mathrm{m},\mathcal{Q}}A_{\mathrm{m},\mathcal{Q}'} &= A_{\mathrm{m},\mathcal{Q}\mathcal{Q}'} && \text{(by (5.6.41))} \\ &= A_{\mathrm{m},\mathcal{Q}'\mathcal{Q}} && \text{(since } \mathcal{Q}' \text{ commute with } \mathcal{Q}) \\ &= A_{\mathrm{m},\mathcal{Q}'}A_{\mathrm{m},\mathcal{Q}} && \text{(by (5.6.41)).} \end{aligned}$$

□

The same kind of result holds for pivot operators.

Proposition 5.6.16 *Let $\mathcal{V}$ be a reflexive Banach space and $\mathcal{Q} := (\mathbb{A}, A_{\mathrm{M}}, B, S)$, $\mathcal{Q}' := (\mathbb{A}', A'_{\mathrm{M}}, B', S)$ be differential quadruplets on $(\mathcal{V}, \mathcal{V}^*)$. Assume that B' commute with B. Then,*

$$A_{\mathcal{Q}\mathcal{Q}'} = A_{\mathcal{Q}}A_{\mathcal{Q}'}. \tag{5.6.47}$$

Moreover, if $\mathcal{Q}$ and $\mathcal{Q}'$ commute then $A_{\mathcal{Q}'}$ commute with $A_{\mathcal{Q}}$.

Proof We start to show the equality of the domains.

$$\begin{aligned} D(A_{\mathcal{Q}\mathcal{Q}'}) &= R(B'B) && \text{(by Proposition 5.6.13 (i))} \\ &= B'D(A_{\mathcal{Q}}) \\ &= (A_{\mathcal{Q}'})^{-1}D(A_{\mathcal{Q}}) && \text{(with (5.2.3))} \\ &= D(A_{\mathcal{Q}}A_{\mathcal{Q}'}), \end{aligned}$$

by definition of the domain of the product of operators. Next, for each u in $D(A_{\mathcal{Q}\mathcal{Q}'})$, one has

$$A_{\mathcal{Q}\mathcal{Q}'}u = A_{\mathrm{M}}A'_{\mathrm{M}}u = A_{\mathrm{M}}A_{\mathcal{Q}'}u,$$

since $D(A_{\mathcal{Q}\mathcal{Q}'}) = D(A_{\mathcal{Q}}A_{\mathcal{Q}'})$. This domains equality yields also that $A_{\mathcal{Q}'}u$ lies in $D(A_{\mathcal{Q}})$. Thus $A_{\mathrm{M}}A_{\mathcal{Q}'}u = A_{\mathcal{Q}}A_{\mathcal{Q}'}u$, and (5.6.47) follows. Finally,

$$\begin{aligned} A_{\mathcal{Q}}A_{\mathcal{Q}'} &= A_{\mathcal{Q}\mathcal{Q}'} && \text{(by (5.6.47))} \\ &= A_{\mathcal{Q}'\mathcal{Q}} && \text{(if } \mathcal{Q} \text{ and } \mathcal{Q}' \text{ commute)} \\ &= A_{\mathcal{Q}'}A_{\mathcal{Q}} && \text{(by (5.6.47)).} \end{aligned}$$

□

5.6.5.3 Compatibility and Differential Triplets

The following result states that under appropriate assumptions, the product of *compatible* differential quadruplets is a *compatible* quadruplet.

Proposition 5.6.17 *Let* $(\mathcal{V}, \mathcal{H}, \mathcal{V}^*)$ *be a Gelfand triplet and* $\mathcal{Q} := (\mathbb{A}, A_{\mathrm{M}}, B, S)$, $\mathcal{Q}' := (\mathbb{A}', A'_{\mathrm{M}}, B', S)$ *be differential quadruplets on* $(\mathcal{V}, \mathcal{V}^*)$. *If* B' *commute with* B, *and* $\mathcal{Q}$, $\mathcal{Q}'$ *are both compatible with* $\mathcal{H}$ *then the product* $\mathcal{Q}\mathcal{Q}'$ *is compatible with* $\mathcal{H}$.

Proof We apply Proposition 5.1.1 with

$$\mathcal{Q}\mathcal{Q}' = (\mathbb{A}\mathbb{A}',\ A_{\mathrm{M}}A'_{\mathrm{M}},\ B'B,\ S).$$

By (5.6.38), $\mathbb{B}_{\mathcal{Q}\mathcal{Q}'}(\mathcal{H}) = \mathbb{B}_{\mathcal{Q}}\mathbb{B}_{\mathcal{Q}'}(\mathcal{H})$. Moreover, $\mathcal{Q}'$ is compatible with $\mathcal{H}$ thus (5.1.10) entails that $\mathbb{B}_{\mathcal{Q}'}(\mathcal{H}) \subseteq \mathcal{H}$. Then, $\mathbb{B}_{\mathcal{Q}\mathcal{Q}'}(\mathcal{H}) \subseteq \mathcal{H}$. Besides, for each v in $\mathcal{V}$,

$$\begin{aligned} \mathbb{B}_{\mathcal{Q}\mathcal{Q}'}v &= \mathbb{B}_{\mathcal{Q}}\mathbb{B}_{\mathcal{Q}'}v && \text{(by (5.6.38))} \\ &= \mathbb{B}_{\mathcal{Q}}B'v && \text{(by (5.1.12) applied to } \mathcal{Q}') \\ &= BB'v && \text{(by (5.1.12) applied to } \mathcal{Q}). \end{aligned}$$

Now it may be easily seen that $\mathcal{Q}\mathcal{Q}'$ is compatible with $\mathcal{H}$. □

Let us investigate the product of differential triplets. Again, we will use the previous results on quadruplets, to obtain results on the product of differential triplets.

Definition 5.6.6 Let $\mathcal{T} := (\mathbb{A}, \mathbb{B}, \mathbb{S})$, $\mathcal{T}' := (\mathbb{A}', \mathbb{B}', \mathbb{S})$ be differential triplets on a real or complex Hilbert space $\mathcal{H}$. We assume that $\mathbb{B}'$ commute with $\mathbb{B}$. By (5.1.15), $\mathcal{Q}_{\mathcal{T}}$, $\mathcal{Q}_{\mathcal{T}'}$ are compatible with $\mathcal{H}$. Thus Proposition 5.6.17 yields that $\mathcal{Q}_{\mathcal{T}}\mathcal{Q}_{\mathcal{T}'}$ is compatible with $\mathcal{H}$, so that

$$\mathcal{T}\mathcal{T}' := \mathcal{T}_{\mathcal{Q}_{\mathcal{T}}\mathcal{Q}_{\mathcal{T}'}}$$

is a differential triplet on $\mathcal{H}$ which is called *the product of* $\mathcal{T}$ *and* $\mathcal{T}'$. Moreover if $\mathcal{T}\mathcal{T}' = \mathcal{T}'\mathcal{T}$ then we say that $\mathcal{T}$ *and* $\mathcal{T}'$ *commute* or that $\mathcal{T}'$ *commute with* $\mathcal{T}$.

In the particular case where $\mathcal{T} = \mathcal{T}'$, for each positive integer n, the nth *power* $\mathcal{T}^n$ *of* $\mathcal{T}$ is defined by

$$\mathcal{T}^n := \mathcal{T}_{(\mathcal{Q}_{\mathcal{T}})^n}.$$

That-is-to-say $\mathcal{T}^n$ is the realization of $(\mathcal{Q}_{\mathcal{T}})^n$. □

Proposition 5.6.18 *Under the assumptions and notation of Definition 5.6.6,*

$$\mathcal{T}\mathcal{T}' = (\mathbb{A}\mathbb{A}',\, \mathbb{B}'\mathbb{B},\, \mathbb{S}).$$

Moreover, if $\mathbb{A}$ and $\mathbb{A}'$ have finite dimensional kernel then $\mathcal{T}\mathcal{T}'$ is a closed differential triplet.

Proof In view of Definition 5.1.3, the realization $\mathcal{T}_{\mathcal{Q}_\mathcal{T}\mathcal{Q}_{\mathcal{T}'}}$ of $\mathcal{Q}_\mathcal{T}\mathcal{Q}_{\mathcal{T}'}$ is equal to $(\mathbb{A}\mathbb{A}',\, \mathbb{B}_{\mathcal{Q}_\mathcal{T}\mathcal{Q}_{\mathcal{T}'}},\, \mathbb{S})$. Moreover, by (5.6.38), and (5.1.9),

$$\mathbb{B}_{\mathcal{Q}_\mathcal{T}\mathcal{Q}_{\mathcal{T}'}} = \mathbb{B}_{\mathcal{Q}_\mathcal{T}}\mathbb{B}_{\mathcal{Q}_{\mathcal{T}'}} = \mathbb{B}\mathbb{B}' = \mathbb{B}'\mathbb{B},$$

since $\mathbb{B}$ and $\mathbb{B}'$ are assumed to commute. Thus, $\mathcal{T}\mathcal{T}' = (\mathbb{A}\mathbb{A}',\, \mathbb{B}'\mathbb{B},\, \mathbb{S})$.

If $\mathbb{A}$ and $\mathbb{A}'$ have finite dimensional kernel then by Lemma 5.6.12, the kernel of $\mathbb{A}\mathbb{A}'$ is finite dimensional as well. Applying Proposition 5.6.13 (iv) with $\mathcal{Q} := \mathcal{Q}_\mathcal{T}$and $\mathcal{Q}' := \mathcal{Q}_{\mathcal{T}'}$, we derive that $\mathcal{Q}_\mathcal{T}\mathcal{Q}_{\mathcal{T}'}$ is closed. Then Proposition 5.1.1 yields that $\mathcal{T}\mathcal{T}'$ is closed. □

The equality of Proposition 5.6.18 could of course be used to define the product of two differential triplets. However, it would remain to check that $(\mathbb{A}\mathbb{A}',\, \mathbb{B}'\mathbb{B},\, \mathbb{S})$ is a differential triplet. Since we have already made this checking for quadruplet, we have chosen alternatively to define the product of triplets as the realization of the product of quadruplets.

So far, Proposition 5.6.17 has been used in the particular case where $\mathcal{Q}$ and $\mathcal{Q}'$ are realizations of differential quadruplets. In view of Definition 5.6.6 and (5.1.16), there holds

$$\mathcal{T}_{\mathcal{Q}_\mathcal{T}\mathcal{Q}_{\mathcal{T}'}} = \mathcal{T}_{\mathcal{Q}_\mathcal{T}}\mathcal{T}_{\mathcal{Q}_{\mathcal{T}'}}.$$

More generally, we may expect that

$$\mathcal{T}_{\mathcal{Q}\mathcal{Q}'} = \mathcal{T}_\mathcal{Q}\mathcal{T}_{\mathcal{Q}'},$$

which means that the realization of the product $\mathcal{Q}\mathcal{Q}'$ is the product of the realization of $\mathcal{Q}$ and $\mathcal{Q}'$. In order to get the latter relation, an additional assumption is needed. More precisely, the following result holds.

Proposition 5.6.19 *Under the assumptions and notation of Proposition 5.6.17, suppose in addition that*

$$\mathbb{A}'\big(\mathcal{H} \cap D(\mathbb{A}')\big) \subseteq \mathcal{H}.$$

Then, $\mathcal{Q}\mathcal{Q}'$ is compatible with $\mathcal{H}$, and especially $\mathcal{T}_{\mathcal{Q}\mathcal{Q}'} = \mathcal{T}_\mathcal{Q}\mathcal{T}_{\mathcal{Q}'}$.

Proof $\mathcal{Q}\mathcal{Q}'$ is compatible with $\mathcal{H}$ according to Proposition 5.6.17. In order to prove that $\mathcal{T}_{\mathcal{Q}\mathcal{Q}'} = \mathcal{T}_{\mathcal{Q}}\mathcal{T}_{\mathcal{Q}'}$, we first compute

$$\begin{aligned}(\mathbb{B}_{\mathcal{Q}\mathcal{Q}'})_{\mathcal{H}} &= (\mathbb{B}_{\mathcal{Q}'}\mathbb{B}_{\mathcal{Q}})_{\mathcal{H}} && \text{(by (5.6.38))}\\ &= (\mathbb{B}_{\mathcal{Q}'})_{\mathcal{H}}(\mathbb{B}_{\mathcal{Q}})_{\mathcal{H}} && \text{(by Lemma 5.1.3).}\end{aligned}$$

By Lemma 5.6.20 below, $(\mathbb{A}\mathbb{A}')_{\mathcal{H}} = \mathbb{A}_{\mathcal{H}}\mathbb{A}'_{\mathcal{H}}$. On the other hand,

$$\mathcal{T}_{\mathcal{Q}} = \big(\mathbb{A}_{\mathcal{H}}, (\mathbb{B}_{\mathcal{Q}})_{\mathcal{H}}, (S^*)_{\mathcal{H}}\big), \qquad \mathcal{T}_{\mathcal{Q}'} = \big(\mathbb{A}'_{\mathcal{H}}, (\mathbb{B}_{\mathcal{Q}'})_{\mathcal{H}}, (S^*)_{\mathcal{H}}\big).$$

Since $\mathbb{B}_{\mathcal{Q}}$ commute with $\mathbb{B}_{\mathcal{Q}'}$ (due to (5.6.38)), we derive easily that $(\mathbb{B}_{\mathcal{Q}})_{\mathcal{H}}$ commute with $(\mathbb{B}_{\mathcal{Q}'})_{\mathcal{H}}$. Using also Proposition 5.6.18, we get the expected formula. □

Lemma 5.6.20 *Let V be a subspace of a normed space $\mathcal{X}$, and A, C be operators on $\mathcal{X}$. If*

$$C\big(V \cap D(C)\big) \subseteq V \tag{5.6.48}$$

then, $(AC)_V = A_V C_V$.

Proof (i) Let us start to show that $(AC)_V \subseteq A_V C_V$. For any u in $D\big((AC)_V\big)$,

$$u \in V \cap D(AC). \tag{5.6.49}$$

In particular, u lies in $V \cap D(C)$. Thus by (5.6.48), Cu belongs to V. Therefore u lives in $D(C_V)$. Besides we know from (5.6.49) that u lies in $D(AC)$, thus Cv belongs to $D(A)$. However, $Cu = C_V u$ since u lives in $D(C_V)$. Thus,

$$C_V u \in V \cap D(A)$$

and

$$AC_V u = ACu = (AC)_V u, \tag{5.6.50}$$

because u lives in $D\big((AC)_V\big)$. Thus $C_V u$ lies in $D(A_V)$, that is, u belongs to $D(A_V C_V)$. Finally

$$A_V C_V u = AC_V u = (AC)_V u,$$

by (5.6.50). Hence, $(AC)_V \subseteq A_V C_V$.

(ii) There remains to prove that the domain of $A_V C_V$ is contained in the domain of $(AC)_V$. For any u in $D(A_V C_V)$, $C_V u$ lies in $D(A_V)$, thus the very definition of $D(A_V)$ gives that $C_V u$ belongs to $D(A)$ and

$$A_V C_V u = AC_V u.$$

In the same way, the belonging of u to $D(C_V)$ yields that

$$u \in D(C), \qquad C_V u = Cu.$$

Thus,

$$u \in D(AC), \qquad ACu = A_V C_V u. \tag{5.6.51}$$

Moreover, $A_V C_V u$ lies in V, and u lives in V since it belongs to $D(C_V)$. Using also (5.6.51), we derive that u lies in the domain of $(AC)_V$. □

There results from the inspection of the proof of Proposition 5.6.19 that the key point for establishing that

$$\mathcal{T}_{\mathcal{Q}\mathcal{Q}'} = \mathcal{T}_{\mathcal{Q}}\mathcal{T}_{\mathcal{Q}'},$$

is $(\mathbb{A}\mathbb{A}')_{\mathcal{H}} = \mathbb{A}_{\mathcal{H}}\mathbb{A}'_{\mathcal{H}}$. For the nth power of a quadruplet, the key point will be $(\mathbb{A}^n)_{\mathcal{H}} = (\mathbb{A}_{\mathcal{H}})^n$, as we will see now.

Proposition 5.6.21 *Let $(\mathcal{V}, \mathcal{H}, \mathcal{V}^*)$ be a Gelfand triplet, $\mathcal{Q} := (\mathbb{A}, A_{\mathrm{M}}, B, S)$ be a differential quadruplet compatible with $\mathcal{H}$, and n be a positive integer. Assume that for any $u_{0,0}, \ldots, u_{0,n-1}$ in* $\ker \mathbb{A}$,

$$\sum_{i=0}^{n-1} \mathbb{B}^i u_{0,i} \in \mathcal{H} \Longrightarrow \forall i = 0, \ldots, n-1,\ u_{0,i} \in \mathcal{H}. \tag{5.6.52}$$

Then, $\mathcal{Q}^n$ is compatible with $\mathcal{H}$ and especially $\mathcal{T}_{\mathcal{Q}^n} = (\mathcal{T}_{\mathcal{Q}})^n$ which means that

$$\mathcal{T}_{\mathcal{Q}^n} = \big((\mathbb{A}_{\mathcal{H}})^n,\ (\mathbb{B}_{\mathcal{H}})^n,\ (\mathbb{S}^*)_{\mathcal{H}}\big).$$

Proof In view of the proof of Proposition 5.6.19, we only need to show that $(\mathbb{A}_{\mathcal{H}})^n = (\mathbb{A}^n)_{\mathcal{H}}$. Let us first show that $(\mathbb{A}_{\mathcal{H}})^n \subseteq (\mathbb{A}^n)_{\mathcal{H}}$. For, since $\mathbb{A}_{\mathcal{H}} \subseteq \mathbb{A}$, we have $(\mathbb{A}_{\mathcal{H}})^n \subseteq \mathbb{A}^n$. Hence for each u in the domain of $(\mathbb{A}_{\mathcal{H}})^n$,

$$u \in \mathcal{H} \cap D(\mathbb{A}^n), \qquad \mathbb{A}^n u = (\mathbb{A}_{\mathcal{H}})^n u \in \mathcal{H}.$$

Thus, u lies in the domain of $(\mathbb{A}^n)_{\mathcal{H}}$ and

$$(\mathbb{A}^n)_{\mathcal{H}}u = \mathbb{A}^n u = (\mathbb{A}_{\mathcal{H}})^n u.$$

That proves that $(\mathbb{A}_{\mathcal{H}})^n \subseteq (\mathbb{A}^n)_{\mathcal{H}}$.

There remains to establish that

$$D\big((\mathbb{A}^n)_{\mathcal{H}}\big) \subseteq D\big((\mathbb{A}_{\mathcal{H}})^n\big). \tag{5.6.53}$$

We will show by induction that for each index j in $[1, n]$,

$$D\big((\mathbb{A}^j)_{\mathcal{H}}\big) \subseteq D\big((\mathbb{A}_{\mathcal{H}})^j\big). \tag{5.6.54}$$

Being given any j in $[1, n-1]$, assume that (5.6.54) holds true. For each u in the domain of $(\mathbb{A}^{j+1})_{\mathcal{H}}$, set

$$h := \mathbb{A}^{j+1}u \in \mathcal{H}.$$

With Lemma 4.3.3, there exists U_0 in $\ker \mathbb{A}^{j+1}$ such that

$$u = U_0 + \mathbb{B}^{j+1}h.$$

Since $\mathcal{Q} := (\mathbb{A}, A_{\mathrm{M}}, B, S)$ is compatible with $\mathcal{H}$, Proposition 5.1.1 yields that $\mathbb{B}(\mathcal{H}) \subseteq \mathcal{H}$. Thus by difference, U_0 lies in $\mathcal{H}$. Next, from (5.6.39), we deduce by induction that

$$\ker \mathbb{A}^{j+1} = \bigoplus_{i=0}^{j} \mathbb{B}^i \ker \mathbb{A}.$$

Thus, by (5.6.52), there exist $u_{0,0}, \ldots, u_{0,j}$ in $\mathcal{H}$ such that

$$U_0 = \sum_{i=0}^{j} \mathbb{B}^i u_{0,i}.$$

Since u lies in $H \cap D(\mathbb{A})$ and

$$\mathbb{A}u = \sum_{i=1}^{j} \mathbb{B}^{i-1}u_{0,i} + \mathbb{B}^j h \in \mathcal{H}$$

(because $\mathbb{B}(\mathcal{H}) \subseteq \mathcal{H}$), we deduce that

$$u \in D(\mathbb{A}_{\mathcal{H}}). \tag{5.6.55}$$

Thus, $\mathbb{A}u$ belongs to $H \cap D(\mathbb{A}^j)$ and

$$\mathbb{A}^j \mathbb{A}u = h \in \mathcal{H}.$$

Thus, $\mathbb{A}u$ lies in the domain of $(\mathbb{A}^j)_{\mathcal{H}}$. By (5.6.55) and the induction hypothesis (5.6.54), we get

$$u \in D(\mathbb{A}_{\mathcal{H}}), \quad \mathbb{A}_{\mathcal{H}}u \in D\big((\mathbb{A}_{\mathcal{H}})^j\big).$$

That means precisely that u lives in the domain of $(\mathbb{A}_{\mathcal{H}})^{j+1}$. Hence,

$$D\big((\mathbb{A}^{j+1})_{\mathcal{H}}\big) \subseteq D\big((\mathbb{A}_{\mathcal{H}})^{j+1}\big)$$

and (5.6.53) follows. □

The thrust of the next result is to compute the minimal operator of the product of two differential triplets by using Proposition 5.6.15.

Proposition 5.6.22 *Let* $\mathcal{T} := (\mathbb{A}, \mathbb{B}, \mathbb{S})$ *and* $\mathcal{T}' := (\mathbb{A}', \mathbb{B}', \mathbb{S})$ *be differential triplets on a Hilbert space* $\mathcal{H}$*. We assume that* $\mathcal{B}'$ *commute with* $\mathbb{B}$*. Then, the following assertions hold true.*

(i) $A_{\mathrm{m},\mathcal{Q}_{\mathcal{T}}} = A_{\mathrm{m},\mathcal{T}}$.
(ii) $\mathcal{Q}_{\mathcal{T}}\mathcal{Q}_{\mathcal{T}'} = \mathcal{Q}_{\mathcal{T}\mathcal{T}'}$.
(iii) *If* $\mathcal{T}$ *and* $\mathcal{T}'$ *commute then* $A_{\mathrm{m},\mathcal{T}\mathcal{T}'} = A_{\mathrm{m},\mathcal{T}}A_{\mathrm{m},\mathcal{T}'} = A_{\mathrm{m},\mathcal{T}'}A_{\mathrm{m},\mathcal{T}}$.

Proof Item (i) has already been proved in Proposition 5.6.1. Using $\mathcal{Q}_{\mathcal{T}}\mathcal{Q}_{\mathcal{T}'} = (\mathbb{A}\mathbb{A}', \mathbb{A}\mathbb{A}', \mathbb{B}'\mathbb{B}, \mathbb{S})$ and Proposition 5.6.18, we obtain (ii). Finally, with Items (i) and (ii),

$$A_{\mathrm{m},\mathcal{T}\mathcal{T}'} = A_{\mathrm{m},\mathcal{Q}_{\mathcal{T}\mathcal{T}'}} = A_{\mathrm{m},\mathcal{Q}_{\mathcal{T}}\mathcal{Q}_{\mathcal{T}'}}.$$

Moreover, since $\mathcal{T}$ and $\mathcal{T}'$ commute, we deduce that $\mathcal{Q}_{\mathcal{T}'}$ commute with $\mathcal{Q}_{\mathcal{T}}$. Thus (5.6.41) gives

$$A_{\mathrm{m},\mathcal{Q}_{\mathcal{T}}\mathcal{Q}_{\mathcal{T}'}} = A_{\mathrm{m},\mathcal{Q}_{\mathcal{T}}}A_{\mathrm{m},\mathcal{Q}_{\mathcal{T}'}} = A_{\mathrm{m},\mathcal{T}}A_{\mathrm{m},\mathcal{T}'},$$

thanks (i). Thus, $A_{\mathrm{m},\mathcal{T}\mathcal{T}'} = A_{\mathrm{m},\mathcal{T}}A_{\mathrm{m},\mathcal{T}'}$. Hence by symmetry, $A_{\mathrm{m},\mathcal{T}'\mathcal{T}} = A_{\mathrm{m},\mathcal{T}'}A_{\mathrm{m},\mathcal{T}}$. Thus using again the commutativity of $\mathcal{T}$ and $\mathcal{T}'$, we derive that $A_{\mathrm{m},\mathcal{T}}$ commute with $A_{\mathrm{m},\mathcal{T}'}$. □

The particular case $\mathcal{Q} = \mathcal{Q}'$ allow to define the product $\mathcal{Q}\mathcal{Q}$ denoted by $\mathcal{Q}^2$, and also the product of n copies of $\mathcal{Q}$ for each integer n larger or equal to 2 (see Definition 5.6.6). As a simple example, let us consider the powers of the differential quadruplet $Q_{p,\mathcal{Y},1}$.

Example 5.6.7 Let n be a positive integer. Under the assumptions and notation of Example 5.1.5, the nth power of the differential quadruplet

$$Q_{p,\mathcal{Y},1} := (\mathrm{D}^1_{\mathcal{V}^*},\ \mathrm{D}^1_{\mathcal{V}},\ B_{1,\mathcal{V}},\ S_{\mathcal{V}}),$$

is

$$(Q_{p,\mathcal{Y},1})^n = \big((\mathrm{D}^1_{\mathcal{V}^*})^n,\ (\mathrm{D}^1_{\mathcal{V}})^n,\ (B_{1,\mathcal{V}})^n,\ S_{\mathcal{V}}\big).$$

Now, it is clear that $(\mathrm{D}^1_{\mathcal{V}})^n$ looks like $\frac{\mathrm{d}^n}{\mathrm{d}x^n}$. More precisely, one has

$$(\mathrm{D}^1_{\mathcal{V}})^n = \mathrm{D}^n_{\mathcal{V}},$$

where $\mathrm{D}^n_{\mathcal{V}}$ is the operator of $\mathcal{V}$ defined for all u in $W^{n,p}\big((0,b),\mathcal{Y}\big)$ by

$$\mathrm{D}^n_{\mathcal{V}} u := \frac{\mathrm{d}^n}{\mathrm{d}x^n} u.$$

By the semi-group property (3.4.4),

$$(B_{1,\mathcal{V}})^n = B_{n,\mathcal{V}}$$

where the operator $B_{n,\mathcal{V}}$ is defined by

$$B_{n,\mathcal{V}} : \mathcal{V} \to \mathcal{V}, \qquad v \mapsto g_n * v. \tag{5.6.56}$$

Thus setting

$$Q_{p,\mathcal{Y},n} := \big(\mathrm{D}^n_{\mathcal{V}^*},\ \mathrm{D}^n_{\mathcal{V}},\ B_{n,\mathcal{V}},\ S_{\mathcal{V}}\big), \tag{5.6.57}$$

we get the expected formula

$$(Q_{p,\mathcal{Y},1})^n = Q_{p,\mathcal{Y},n}.$$

In particular, $Q_{p,\mathcal{Y},n}$ is a differential quadruplet on $(\mathcal{V}, \mathcal{V}^*)$.

When $n = 0$, we put

$$\mathrm{D}^0_{\mathcal{V}} := \mathrm{id}_{\mathcal{V}}, \qquad \mathrm{D}^0_{\mathcal{V}^*} := \mathrm{id}_{\mathcal{V}^*}, \qquad B_{0,\mathcal{V}} := \mathrm{id}_{\mathcal{V}},$$

and

$$\mathcal{Q}_{p,\mathcal{Y},0} := \big(\mathrm{id}_{\mathcal{V}^*},\ \mathrm{id}_{\mathcal{V}},\ \mathrm{id}_{\mathcal{V}},\ S_{\mathcal{V}}\big).$$

For all purposes, recall that by convention, any operator raised to the power zero is the identity operator. Observe that $\mathcal{Q}_{p,\mathcal{Y},0}$ remains a differential quadruplet.

Let us go back to the general case where $n \geq 1$. With the previous conventions at hand, applying (5.6.40) with $A_{\mathrm{M}} := \mathrm{D}^1_{\mathcal{V}}$, $A'_{\mathrm{M}} := \mathrm{D}^{n-1}_{\mathcal{V}}$, we get

$$\ker \mathrm{D}^n_{\mathcal{V}} = \ker \mathrm{D}^{n-1}_{\mathcal{V}} \oplus B_{n-1,\mathcal{V}} \ker \mathrm{D}^1_{\mathcal{V}},$$

so that by induction, we recover the following basic result:

$$\ker \mathrm{D}^n_{\mathcal{V}} = \bigoplus_{i=1}^{n} g_i \mathcal{Y}.$$

In a same way,

$$\ker \mathrm{D}^n_{\mathcal{V}^*} = \bigoplus_{i=1}^{n} g_i \mathcal{Y}^*.$$

We claim that $\mathcal{Q}_{p,\mathcal{Y},n}$ is a $\ker \mathrm{D}^n_{\mathcal{V}}$*-regular differential quadruplet on $(\mathcal{V}, \mathcal{V}^*)$.* Indeed, Lemma 5.6.23 below entails that the matrix with entries

$$\int_0^b g_j(y) g_i(y)\,\mathrm{d}y, \qquad \forall\, 1 \leq i, j \leq n$$

is invertible. Clearly, the g_i's are linearly independent. Thus the spaces $g_1\mathcal{Y}, \ldots, g_n\mathcal{Y}$ are in direct sum according to Proposition 3.2.5. Then the claim follows from Corollary 5.3.2.

Regarding the minimal operator of $\mathcal{Q}_{p,\mathcal{Y},n}$, (5.6.41) gives

$$A_{\mathrm{m},\mathcal{Q}_{p,q,n}} = (A_{\mathrm{m},\mathcal{Q}_{p,\mathcal{Y},1}})^n.$$

Thus, the domain of $A_{\mathrm{m},\mathcal{Q}_{p,q,n}}$ is the space $H^n_0(0,b)$ of all functions u in $H^n(0,b)$ satisfying

$$u^{(i)}(0) = u^{(i)}(b) = 0, \qquad \forall i = 0, \ldots, n-1.$$

Let us now consider a realization of $\mathcal{Q}_{p,q,n}$. As in Example 5.2.4, suppose in addition that $q \leq p$ and that $(\mathcal{Y}, H, \mathcal{Y}^*)$ is a Gelfand triplet. Recalling that

$$\mathcal{V} = L^p(\mathcal{Y}) := L^p\big((0,b), \mathcal{Y}\big), \qquad \mathcal{V}^* = L^q(\mathcal{Y}^*) := L^q\big((0,b), \mathcal{Y}^*\big)$$

and

$$\mathcal{H} = L^2(H) := L^2\big((0,b), H\big),$$

we have seen in Example 5.2.4 that $(\mathcal{V}, \mathcal{H}, \mathcal{V}^*)$ is a Gelfand triplet, $\mathcal{Q}_{p,\mathcal{Y},1}$ is compatible with $L^2(H)$ and its realization is the differential triplet

$$\mathcal{T}_{1,H} := (\mathrm{D}^1_{\mathcal{H}}, B_{1,\mathcal{H}}, S_{\mathcal{H}}).$$

In order to apply Proposition 5.6.21, let $Y_1^{\mathrm{s}}, \ldots, Y_n^{\mathrm{s}}$ belong to $\mathcal{Y}^*$ and satisfy

$$\sum_{i=1}^{n} g_i Y_i^{\mathrm{s}} \in L^2\big((0,b), H\big).$$

Using Proposition 2.2.14 (iv) with $\mathcal{X} := L^q(0,b)$, we derive the existence of functions $f_1, \ldots, f_n$ in $L^p(0,b)$ such that

$$\int_0^b f_j g_i \, \mathrm{d}y = \delta_{ij}, \qquad \forall i, j = 1, \ldots n.$$

Thus, for each index j in $[1,n]$, $\sum_{i=1}^n f_j g_i Y_i^{\mathrm{s}}$ lies in $L^1(H)$ and

$$Y_j^{\mathrm{s}} = \int_0^b \sum_{i=1}^{n} f_j g_i Y_i^{\mathrm{s}} \, \mathrm{d}y \in H.$$

Thus, Proposition 5.6.21 yields us that $\mathcal{Q}_{p,q,n}$ is compatible with $L^2(H)$ and its realization satisfies

$$t(\mathcal{Q}_{p,\mathcal{Y},n}) = \big(t(\mathcal{Q}_{p,\mathcal{Y},1})\big)^n = (\mathrm{D}^n_{\mathcal{H}}, B_{n,\mathcal{H}}, S_{\mathcal{H}}),$$

where the latter equality follows from the very definition the power of a differential triplet. In closing,

$$\mathcal{T}_{n,H} := (\mathrm{D}^n_{\mathcal{H}}, B_{n,\mathcal{H}}, S_{\mathcal{H}})$$

is a differential triplet on $L^2(H)$, which is the realization of $\mathcal{Q}_{p,\mathcal{Y},n}$ on $L^2(H)$. □

Lemma 5.6.23 *Let $h_1, \ldots, h_n$ belong to a real or complex Hilbert space $\big(\mathcal{H}, (\cdot,\cdot)\big)$. If the h_i's are linearly independent then the matrix with entries (h_j, h_i) is invertible.*

Proof It is enough to prove that the matrix is one-to-one. Let $(\lambda_1, \dots, \lambda_n)$ belong to its kernel. Then the vector $u := \sum_{j=1}^{n} \lambda_j h_j$ satisfies $(u, h_i) = 0$ for each index i in $[1, n]$. By anti-linearity, we deduce that $(u, u) = 0$. Since the h_i's are linearly independent, we obtain injectivity. □

5.6.6 Caputo Extensions of Products of Operators

We will highlight the underlying structure of higher order fractional differential operators including Riemann–Liouville and Caputo operators (see Sects. 6.6 and 6.8). Their structure relies on Caputo extensions and products of differential quadruplets.

Let $\mathcal{V}$ be a reflexive Banach space and

$$\mathcal{Q} := (\mathbb{D}, D, B, S), \qquad \mathcal{Q}_{\mathrm{rl}} := (\mathbb{D}_{\mathrm{rl}}, D_{\mathrm{rl}}, B_{\mathrm{rl}}, S) \tag{5.6.58}$$

be differential quadruplets on $(\mathcal{V}, \mathcal{V}^*)$. Assume that

$$B_{\mathrm{rl}} B = B B_{\mathrm{rl}}, \tag{5.6.59}$$

so that for each integer $n \geq 1$, the product the of $\mathcal{Q}^n$ by $\mathcal{Q}_{\mathrm{rl}}$ is the well-defined differential quadruplet

$$\mathcal{Q}^n \mathcal{Q}_{\mathrm{rl}} = (\mathbb{D}^n \mathbb{D}_{\mathrm{rl}},\ D^n D_{\mathrm{rl}},\ B_{\mathrm{rl}} B^n, S).$$

Proposition 5.6.24 *Let $\mathcal{V}$ be a reflexive Banach space, n be a positive integer, and $\mathcal{Q}$, $\mathcal{Q}_{\mathrm{rl}}$ be given by* (5.6.58) *and satisfy* (5.6.59). *Then the spaces* $\ker D_{\mathrm{rl}}$, $B_{\mathrm{rl}} \ker D, \dots,$ $B_{\mathrm{rl}} B^{n-1} \ker D$ *are in direct sum and*

$$\ker D^n D_{\mathrm{rl}} = \ker D_{\mathrm{rl}} \oplus \Big(\bigoplus_{i=0}^{n-1} B_{\mathrm{rl}} B^i \ker D \Big).$$

Proof By (5.6.40),

$$\ker D^n D_{\mathrm{rl}} = \ker D_{\mathrm{rl}} \oplus B_{\mathrm{rl}} \ker D^n.$$

Applying (5.6.40) with $A_{\mathrm{M}} := D$, $A'_{\mathrm{M}} := D^{n-1}$, we get

$$\ker D^n = \ker D^{n-1} \oplus B^{n-1} \ker D.$$

This relation allows to prove by induction that

$$\ker D^n = \overset{n-1}{\underset{i=0}{\oplus}} B^i \ker D.$$

Since $B_{\rm rl}$ is one-to-one, the assertions stated in the proposition follow. □

Caputo extensions of $\mathcal{Q}^n \mathcal{Q}_{\rm rl}$ rely on Caputo extensions of the operators $\mathbb{D}^n \mathbb{D}_{\rm rl}$ and $D^n D_{\rm rl}$. Since our assumptions on these operators will be analog, we will restrict our attention to the latter operator. The following extra assumptions are satisfied for the higher-order fractional differential operators we have in mind

$$D\, D_{\rm rl}\, B = D_{\rm rl} \tag{5.6.60}$$

$$R(B) \subseteq D(D_{\rm rl}). \tag{5.6.61}$$

Our aim is to extend $D^n\ D_{\rm rl}$ with respect to spaces of the form

$$V + BV + \cdots + B^n V,$$

where V is a subspace of $\mathcal{V}$. Our first result in this direction gives a sufficient condition for that sum to be direct.

Proposition 5.6.25 *Let $\mathcal{X}$ be a vector space, $L : \mathcal{X} \to \mathcal{X}$ be an operator, m be a non-negative integer, and V be a subspace of $\mathcal{X}$. If*

(i) *V is in direct sum with $R(L)$,*
(ii) *L is one-to-one*

then the spaces $V, LV, \ldots, L^m V$ are in direct sum.

Proof Let $v_0, \ldots, v_m$ be vectors of V such that $\sum_{i=0}^m L^i v_i = 0$. According to Proposition 2.1.9, we will be done provided we show that all the v_i's are the zero vector. Arguing by contradiction we assume that at least one of the v_i's is nontrivial, so that there exists an index i_0 in $[0, m]$ satisfying $v_{i_0} \neq 0$ and $\sum_{i=i_0}^m \lambda_i L^i v_i = 0$. Then recalling that by convention, $\sum_{i=m+1}^m \cdots = 0$, one has

$$L^{i_0}\big(v_{i_0} + \sum_{i=i_0+1}^{m} L^{i-i_0} v_i\big) = 0.$$

Since L^{i_0} is one-to-one, we get $v_{i_0} = 0$ if $i_0 = m$. Otherwise v_{i_0} lies in $R(L) \cap V$. Thus, $v_{i_0} = 0$ by Assumption (i). We get a contradiction, so that all the v_i's are trivial. □

Now on account of Proposition 5.6.25 and (5.6.61), for any subspace V of $\mathcal{V}$, we may legitimately set

$$N(V) := \begin{cases} \bigoplus_{i=0}^{n} B^i V & \text{if } V \cap D(D_{\text{rl}}) = \{0\} \\ \{0\} & \text{otherwise.} \end{cases} \tag{5.6.62}$$

In order to construct the Caputo extension of $D^n D_{\text{rl}}$ with respect to $N(V)$, the space $N(V)$ must be in direct sum with the domain of $D^n D_{\text{rl}}$. This issue is addressed in the following result.

Proposition 5.6.26 *Under the hypothesis and notation* (5.6.58) -(5.6.62), *$N(V)$ is in direct sum with the domain of $D^n D_{\text{rl}}$.*

Proof Of course, only the case where V is in direct sum with $D(D_{\text{rl}})$ deserves details. For further reference, notice that (5.6.60) implies

$$D^i D_{\text{rl}} B^i = D_{\text{rl}}, \qquad \forall i \geq 0, \tag{5.6.63}$$

where by convention, $D^0 = B^0 = \text{id}_{\mathcal{V}}$. Next, any u in $N(V) \cap D(D^n D_{\text{rl}})$ reads

$$u = \sum_{i=0}^{n} B^i v_i,$$

for some vectors $v_0, \ldots, v_n$ in V. Arguing by contradiction we assume that u is non trivial, so that there exists some index i_0 satisfying $v_{i_0} \neq 0$ and

$$u = \sum_{i=i_0}^{n} B^i v_i.$$

If $i_0 = n$ then $u = B^n v_n$ lies in $D(D^n D_{\text{rl}})$. Hence with (5.6.63), v_n belongs to $D(D_{\text{rl}})$, which contradicts $v_n = 0$ (because V and $D(D_{\text{rl}})$ are in direct sum). On the other hand, if $i_0 \leq n - 1$, then

$$B^{i_0} v_{i_0} + \sum_{i=i_0+1}^{n} B^i v_i \in D(D^n D_{\text{rl}}).$$

We claim that $\sum_{i=i_0+1}^{n} B^i v_i$ lies in $D(D^{i_0} D_{\text{rl}})$. Indeed, by assumption $R(B) \subseteq D(D_{\text{rl}})$ thus

$$\sum_{i=i_0+1}^{n} B^{i-i_0} v_i \in D(D_{\text{rl}}).$$

Moreover $D(D_{\text{rl}}) = D(D^{i_0}\, D_{\text{rl}} B^{i_0})$ according to (5.6.63). Hence

$$B^{i_0}\Big(\sum_{i=i_0+1}^{n} B^{i-i_0} v_i \Big) \in D(D^{i_0} D_{\text{rl}}),$$

and the claim follows. Now by difference, $B^{i_0} v_{i_0}$ lies in $D(D^{i_0} D_{\text{rl}})$. Thus,

$$v_{i_0} \in D(D^{i_0} D_{\text{rl}} B^{i_0}) = D(D_{\text{rl}})$$

by (5.6.63) again. We get another contradiction, which entails that u is the zero vector. □

Hence, the latter proposition allows to introduce the Caputo extension $\widetilde{D^n\, D_{\text{rl}}}$ of $D^n\, D_{\text{rl}}$ with respect to $N(V)$. The following result gives the structure of the kernel of this Caputo extension.

Corollary 5.6.27 *Under the hypothesis and notation* (5.6.58)–(5.6.62), *suppose that* V *and* $D(D_{\text{rl}})$ *are in direct sum. Then*

$$\ker \widetilde{D^n\, D_{\text{rl}}} = \bigoplus_{i=0}^{n} B^i\, V \oplus \ker D_{\text{rl}} \oplus \Big(\bigoplus_{i=0}^{n-1} B_{\text{rl}} B^i \ker D \Big). \tag{5.6.64}$$

Moreover, if in addition,

(i) $V := \langle \zeta \rangle$ *where* ζ *lies in* $\mathcal{V} \setminus D(D_{\text{rl}})$;
(ii) *there exist non trivial vectors* ζ_D *and* ζ_{rl} *of* $\mathcal{V}$ *such that*

$$\ker D = \langle \zeta_D \rangle, \qquad \ker D_{\text{rl}} = \langle \zeta_{\text{rl}} \rangle,$$

then

$$\ker \widetilde{D^n\, D_{\text{rl}}} = \langle \zeta, \ldots, B^n \zeta \rangle \oplus \langle \zeta_{\text{rl}} \rangle \oplus \langle B_{\text{rl}} \zeta_D, B_{\text{rl}} B \zeta_D, \ldots, B_{\text{rl}} B^{n-1} \zeta_D \rangle. \tag{5.6.65}$$

Besides, the latter vectors form a basis of $\ker \widetilde{D^n\, D_{\text{rl}}}$. *In particular, the dimension of this kernel is equal to* $2(n+1)$.

Proof Since V and $D(D_{\mathrm{rl}})$ are in direct sum, (5.6.62) implies that $N(V) = \bigoplus_{i=0}^{n} B^i V$. Hence (5.6.64) follows from Propositions 5.6.9 and 5.6.24. The remainder of the proof is obvious. □

5.7 Reflexive Lebesgue Spaces of Vector Valued Functions

Let $\mathbb{K}$ be equal to $\mathbb{R}$ or $\mathbb{C}$, b be a positive real number and p, q be conjugate exponents in $(1, \infty)$. Let also $\mathcal{Y}$ be a nontrivial separable and reflexive Banach space over $\mathbb{K}$. In this section, we will consider the case where

$$\mathcal{V} := L^p\big((0, b), \mathcal{Y}\big).$$

According to Theorem 3.2.2, the anti-dual space $\mathcal{V}^*$ of $\mathcal{V}$ is $L^q\big((0, b), \mathcal{Y}^*\big)$. Moreover these spaces are reflexive. For any function h in $L^p(0, b)$, we recall that $h\mathcal{Y}$ denotes the closed subspace of $\mathcal{V}$ (see Example 5.3.2) defined by

$$h\mathcal{Y} := \{hY \mid Y \in \mathcal{Y}\}. \tag{5.7.1}$$

In this section, we will consider a differential quadruplet $\mathcal{Q} := (\mathbb{A}, A_{\mathrm{M}}, B, S)$ on $(\mathcal{V}, \mathcal{V}^*)$ satisfying (except otherwise stated)

$$\ker A_{\mathrm{M}} := \bigoplus_{i=1}^{m} \xi_i \mathcal{Y}, \qquad \ker \mathbb{A} := \bigoplus_{i=1}^{m^{\mathrm{s}}} \xi_i^{\mathrm{s}} \mathcal{Y}^*, \tag{5.7.2}$$

where m, m^{s} are positive integer, and the ξ_i, ξ_i^{s}'s live respectively in $L^p(0, b) \setminus \{0\}$ and $L^q(0, b) \setminus \{0\}$. Then by Corollary 5.3.1, there exists $\eta_1, \ldots \eta_{m^{\mathrm{s}}}$ in $L^p(0, b)$, and $\eta_1^{\mathrm{s}}, \ldots, \eta_m^{\mathrm{s}}$ in $L^q(0, b)$ such that by setting

$$F_0 := \bigoplus_{i=1}^{m^{\mathrm{s}}} \eta_i \mathcal{Y}, \qquad F_0^{\mathrm{s}} := \bigoplus_{i=1}^{m} \eta_i^{\mathrm{s}} \mathcal{Y}^*, \tag{5.7.3}$$

$\mathcal{Q}$ is a F_0-regular, and $\mathcal{Q}^*$ is $S^* F_0^{\mathrm{s}}$-regular. In particular,

$$\mathcal{V}^* = F_0^{\mathrm{s}} \oplus (\ker A_{\mathrm{M}})^{\perp}.$$

Notice that do not not assume at this stage that $\int_0^b \xi_i^{\mathrm{s}} \overline{\eta_j}\, \mathrm{d}y = \delta_{ij}$ neither $\int_0^b \eta_i^{\mathrm{s}} \overline{\xi_j}\, \mathrm{d}y = \delta_{ij}$.

5.7.1 Endogenous Boundary Values in Lebesgue Spaces

Endogenous boundary coordinates are defined for maximal operators with finite dimensional kernels. When $\mathcal{Y}$ is of infinite dimension, these kernels are usually also infinite dimensional (see for instance Example 5.1.5). However in applications, the structure remains quiet simple, so that *endogenous boundary values* may be defined in a way quite similar to endogenous boundary coordinates. Notice that endogenous boundary coordinates are scalars, whereas endogenous boundary values are vectors of $\mathcal{Y}$. It turns out that endogenous boundary values and endogenous boundary coordinates coincide when $\mathcal{Y} = \mathbb{K}$.

5.7.1.1 Definitions

Typically, "standard" boundary values live in $\mathcal{Y}$. Hence, we would like to define *endogenous boundary values* as elements of $\mathcal{Y}$. Recall that if $\mathcal{Y} = \mathcal{Y}_{\mathrm{f}}$ has finite dimension d_{f}, we have defined endogenous boundary *coordinates* (see Definition 5.5.1) as a elements of $\mathbb{K}^{d_{\mathrm{f}}}$, but not of $\mathcal{Y}_{\mathrm{f}}$.

Since $\mathcal{Q}$ is F_0-regular, the Domain Structure Theorem 5.3.3 yields that

$$D(A_{\mathrm{M}}) = \overset{m}{\underset{i=1}{\oplus}} \xi_i \mathcal{Y} \oplus BS\big(\overset{m^{\mathrm{s}}}{\underset{i=1}{\oplus}} \eta_i \mathcal{Y}\big) \oplus BS(\ker \mathbb{A}^{\perp}). \tag{5.7.4}$$

Hence, each u in $D(A_{\mathrm{M}})$ reads

$$u = \sum_{i=1}^{m} \xi_i U_i + BS \sum_{i=1}^{m^{\mathrm{s}}} \eta_i U_i' + BS\psi^{\perp}, \tag{5.7.5}$$

where U_i, U_i' belong to $\mathcal{Y}$, and $\psi^{\perp}$ lies in $\ker \mathbb{A}^{\perp}$. Then,

$$(U_1, \ldots, U_m, U_1', \ldots, U_{m^{\mathrm{s}}}')$$

are called the *endogenous boundary values of u (with underlying quadruplet $\mathcal{Q}$ and) with respect to the direct sum decompositions*

$$\ker A_{\mathrm{M}} = \overset{m}{\underset{i=1}{\oplus}} \xi_i \mathcal{Y}, \qquad F_0 = \overset{m^{\mathrm{s}}}{\underset{i=1}{\oplus}} \eta_i \mathcal{Y}. \tag{5.7.6}$$

Next it is natural to introduce the following map.

Definition 5.7.1 Let $\mathcal{Q} := (\mathbb{A}, A_{\mathrm{M}}, B, S)$ be a F_0-regular differential quadruplet on $(\mathcal{V}, \mathcal{V}^*)$ satisfying (5.7.2), (5.7.3). The map defined on $D(A_{\mathrm{M}})$ with values in $\mathcal{Y}^{m+m^{\mathrm{s}}}$ which assigns to each element of $D(A_{\mathrm{M}})$ its endogenous boundary values with respect to the direct sum decompositions (5.7.6) is called the *F_0-boundary values map of* $\mathcal{Q}$. This map is denoted by $F_{\text{b.v}}$. In symbol,

$$F_{\text{b.v}} : D(A_{\text{M}}) \to \mathcal{Y}^{m+m^{\text{s}}}, \qquad u \mapsto (U_1, \ldots, U_m, U_1', \ldots, U_{m^{\text{s}}}').$$

□

Proposition 5.7.1 *Let $\mathcal{Q} := (\mathbb{A}, A_{\text{M}}, B, S)$ be a F_0-regular differential quadruplet on $(\mathcal{V}, \mathcal{V}^*)$ satisfying (5.7.2), (5.7.3). Then, the F_0-boundary values map $F_{\text{b.v}}$ is linear and continuous.*

Proof Since linearity is clear, we focus on continuity. On account of (5.7.5), it is enough to show that the maps

$$F_i : D(A_{\text{M}}) \to \mathcal{Y}, \qquad u \mapsto U_i$$

and

$$F_i' : D(A_{\text{M}}) \to \mathcal{Y}, \qquad u \mapsto U_i'$$

are continuous. For, let u be in $D(A_{\text{M}})$, and i_0 be any index in $[1, m]$. Starting from (5.7.5) and recalling that the functions ξ_{i_0} is nontrivial, we estimate

$$\begin{aligned}
\|U_{i_0}\|_{\mathcal{Y}} &= \|\xi_{i_0}\|_{L^p(0,b)}^{-1} \|\xi_{i_0} U_{i_0}\|_{L^p(\mathcal{Y})} \\
&\lesssim \|\xi_{i_0}\|_{L^p(0,b)}^{-1} \Big\| \sum_{i=1}^{m} \xi_i U_i \Big\|_{L^p(\mathcal{Y})} && (5.7.7) \\
&\lesssim \|L_{\text{p}} P_{\text{b.v}} u\|_{D(A_{\text{M}})} && \text{(see Rk. 5.3.4 \& Def. 5.3.3)} \\
&\lesssim \|u\|_{D(A_{\text{M}})}.
\end{aligned}$$

Hence, F_{i_0} is continuous. Notice that (5.7.7) follows from Proposition 2.3.5 with $\mathcal{X} := \ker A_{\text{M}}$ and $R := \xi_{i_0}\mathcal{Y}$. Regarding F_{i_0}' where now i_0 runs into $[1, m^{\text{s}}]$, we estimate in a similar way as follows.

$$\begin{aligned}
\|U_{i_0}'\|_{\mathcal{Y}} &\lesssim \Big\| \sum_{i=1}^{m^{\text{s}}} \eta_i U_i' \Big\|_{L^p(\mathcal{Y})} && \text{(see 5.7.7)} \\
&\lesssim \Big\| BS\Big(\sum_{i=1}^{m^{\text{s}}} \eta_i U_i'\Big) \Big\|_{D(A_{\text{M}})}
\end{aligned}$$

$$\lesssim \|(\mathrm{id}_{E_{\mathrm{M}}} - L_{\mathrm{p}})P_{\mathrm{b.v}}u\|_{D(A_{\mathrm{M}})}$$
$$\lesssim \|u\|_{D(A_{\mathrm{M}})}.$$

Hence, F'_{i_0} is continuous as well, which completes the current proof. □

Now by considering the adjoint quadruplet Q^*, we recall that Q^* is a $S^*F_0^{\mathrm{s}}$-regular differential quadruplet on $(\mathcal{V}^*, \mathcal{V})$ with (see (5.7.2), (5.7.3))

$$\ker \mathbb{A} := \bigoplus_{i=1}^{m^{\mathrm{s}}} \xi_i^{\mathrm{s}}\mathcal{Y}^*, \qquad F_0^{\mathrm{s}} := \bigoplus_{i=1}^{m} \eta_i^{\mathrm{s}}\mathcal{Y}^*, \tag{5.7.8}$$

Then arguing as previously, we get

$$D(S^*\mathbb{A}S^*) = \Big(\bigoplus_{i=1}^{m^{\mathrm{s}}} S^*(\xi_i^{\mathrm{s}}\mathcal{Y}^*)\Big) \oplus \Big(\bigoplus_{i=1}^{m} B^*(\eta_i^{\mathrm{s}}\mathcal{Y}^*)\Big) \oplus B^*\big((\ker A_{\mathrm{M}})^{\perp}\big). \tag{5.7.9}$$

Hence, each f in $D(S^*\mathbb{A}S^*)$ reads

$$f = S^*\sum_{i=1}^{m^{\mathrm{s}}} \xi_i^{\mathrm{s}} F_i + B^*\sum_{i=1}^{m} \eta_i^{\mathrm{s}} F'_i + B^*\varphi^{\perp}, \tag{5.7.10}$$

where F_i, F'_i belong to $\mathcal{Y}^*$, and $\varphi^{\perp}$ lies in $(\ker A_{\mathrm{M}})^{\perp}$. Then

$$(F_1, \dots, F_{m^{\mathrm{s}}}, F'_1, \dots, F'_m)$$

are the endogenous boundary values of f (with underlying quadruplet $\mathcal{Q}^*$ and) with respect to the direct sum decompositions

$$\ker S^*\mathbb{A}S^* = \bigoplus_{i=1}^{m^{\mathrm{s}}} S^*(\xi_i^{\mathrm{s}}\mathcal{Y}^*), \qquad S^*F_0^{\mathrm{s}} = \bigoplus_{i=1}^{m} S^*(\eta_i^{\mathrm{s}}\mathcal{Y}^*). \tag{5.7.11}$$

Up to now, our assumptions imply that $\ker A_{\mathrm{M}}$ and F_0 are nontrivial. Let us investigate the case where *at least one of these spaces is trivial*. We employ the method developed in Remark 5.5.3. *If both operators have a trivial kernel, then we agree that there is no endogenous boundary values*. If only $\mathbb{A}$ has a trivial kernel then $F_0 = \{0\}$, so that any u in $D(A_{\mathrm{M}})$ reads

$$u = \sum_{i=1}^{m} \xi_i U_i + B\psi,$$

where the U_i's live in $\mathcal{Y}$, and ψ lies in $\mathcal{V}$. Thus $(U_1, \ldots, U_m)$ is called the *endogenous boundary values of u with respect to the direct sum decomposition*

$$\ker A_{\mathrm{M}} = \overset{m}{\underset{i=1}{\oplus}} \xi_i \mathcal{Y}.$$

On the other hand, if only A_{M} has a trivial kernel then recalling that F_0 is assumed to be a closed complementary subspace of $(\ker \mathbb{A})^{\perp}$, any u in $D(A_{\mathrm{M}})$ reads

$$u = BS \sum_{i=1}^{m^{\mathrm{s}}} \eta_i U_i' + BS\psi^{\perp},$$

where the U_i''s live in $\mathcal{Y}$, and $\psi^{\perp}$ lies in $(\ker \mathbb{A})^{\perp}$. Thus $(U_1', \ldots, U_{d_{\mathbb{A}}}')$ is called the *endogenous boundary values of u with respect to the direct sum decomposition*

$$F_0 = \overset{m^{\mathrm{s}}}{\underset{i=1}{\oplus}} \eta_i \mathcal{Y}.$$

5.7.1.2 The Boundary Values Trick

Our goal is to compute in a fairly simple way, the endogenous boundary values with underlying quadruplet $\mathcal{Q}^*$ of the elements of $D(S^* \mathbb{A} S^*)$. For, considering first the reverse quadruplet $r(\mathcal{Q})$ and using

$$\mathcal{V}^* = F_0^{\mathrm{s}} \oplus (\ker A_{\mathrm{M}})^{\perp},$$

we derive that $r(\mathcal{Q})$ is a F_0^{s}-regular differential quadruplet on $(\mathcal{V}^*, \mathcal{V})$, where F_0 is given by (5.7.3). Then,

$$D(\mathbb{A}) = \overset{m^{\mathrm{s}}}{\underset{i=1}{\oplus}} \xi_i^{\mathrm{s}} \mathcal{Y}^* \oplus S^* B^* \Big(\overset{m}{\underset{i=1}{\oplus}} \eta_i^{\mathrm{s}} \mathcal{Y}^* \Big) \oplus S^* B^* \big((\ker A_{\mathrm{M}})^{\perp} \big). \tag{5.7.12}$$

Hence, each w in $D(\mathbb{A})$ reads

$$w = \sum_{i=1}^{m^{\mathrm{s}}} \xi_i^{\mathrm{s}} W_i + S^* B^* \sum_{i=1}^{m} \eta_i^{\mathrm{s}} W_i' + S^* B^* \varphi^{\perp}, \tag{5.7.13}$$

where W_i, W_i' belong to $\mathcal{Y}$, and $\varphi^{\perp}$ lies in $(\ker A_{\mathrm{M}})^{\perp}$. There results from the definition of endogenous boundary values given in Sect. 5.7.1.1 that

$$(W_1, \ldots, W_{m^{\mathrm{s}}}, W_1', \ldots, W_m')$$

are the endogenous boundary values of u with underlying quadruplet $r(\mathcal{Q})$ and with respect to the direct sum decompositions

$$\ker \mathbb{A} = \bigoplus_{i=1}^{m^{\mathrm{s}}} \xi_i^{\mathrm{s}} \mathcal{Y}^*, \qquad F_0^{\mathrm{s}} = \bigoplus_{i=1}^{m} \eta_i^{\mathrm{s}} \mathcal{Y}^*. \tag{5.7.14}$$

A comparison of (5.7.10) with (5.7.13) entails the following *boundary values trick. The endogenous boundary values of an element f of $D(S^* \mathbb{A} S^*)$ with underlying quadruplet $\mathcal{Q}^*$ and with respect to the direct sum decompositions* (5.7.11) *are equal to the endogenous boundary values of $S^* f$ with underlying quadruplet $r(\mathcal{Q})$ and with respect to the direct sum decompositions* (5.7.14).

5.7.1.3 The Quadruplet $\mathcal{Q}_{p,\mathcal{Y},2}$ and Its Adjoint

Under the assumptions and notation of Example 5.6.7, choosing $n := 2$, we will compute endogenous boundary values with two underlying quadruplets $\mathcal{Q}_{p,\mathcal{Y},2}$ and its adjoint $\mathcal{Q}_{p,\mathcal{Y},2}{}^*$.

5.7.1.3.1 The Quadruplet $\mathcal{Q}_{p,\mathcal{Y},2}$

We recall that

$$\mathcal{Q}_{p,\mathcal{Y},2} := \left(\mathrm{D}^2_{\mathcal{Y}^*},\ \mathrm{D}^2_{\mathcal{Y}},\ B_{2,\mathcal{Y}},\ S_{\mathcal{Y}}\right),$$

where

$$\mathrm{D}^2_{\mathcal{Y}} : W^{2,p}(0,b) \subset L^p(0,b) \to L^p(0,b), \qquad u \mapsto \frac{\mathrm{d}^2}{\mathrm{d}x^2} x.$$

We know from Example 5.6.7 that $\mathcal{Q}_{p,\mathcal{Y},2}$ is a $\ker \mathrm{D}^2_{\mathcal{Y}}$-regular. The Domain Structure Theorem 5.3.3 (i) yields that each u in $D(\mathrm{D}^2_{\mathcal{Y}})$ reads

$$u = g_1\, U_1 + g_2\, U_2 + B_{2,\mathcal{Y}}(g_1\, U_1') + B_{2,\mathcal{Y}} S_{\mathcal{Y}}(g_2\, U_2') + B_{2,\mathcal{Y}} S_{\mathcal{Y}}\, \psi^{\perp}, \tag{5.7.15}$$

where $\psi^{\perp}$ lies in $(g_1\, \mathcal{Y} \oplus g_2\, \mathcal{Y})^{\perp L^p(\mathcal{Y})}$. Here (U_1, U_2, U_1', U_2') are the *endogenous boundary values of u with respect to the direct sum decompositions*

$$\ker \mathrm{D}^2_{\mathcal{Y}} = g_1\, \mathcal{Y} \oplus g_2\, \mathcal{Y}, \qquad F_0 = g_1\, \mathcal{Y} \oplus g_2\, \mathcal{Y}.$$

We claim that *endogenous boundary values* and *standard boundary values* are related by the following linear transform.

$$\begin{cases} u(0) = U_1 \\ u'(0) = \qquad U_2 \\ u(b) = U_1 + bU_2 + (g_1, g_2)U_1' + \|g_2\|_{L^2}^2 U_2' \\ u'(b) = \qquad U_2 + \|g_1\|_{L^2}^2 U_1' + (g_1, g_2)U_2'. \end{cases} \tag{5.7.16}$$

Indeed, (5.7.16) is obtained by using (5.7.15) and

$$B_{2,\mathcal{V}}(g_1\, U_1')(b) = (g_2, g_1)U_1'$$
$$B_{2,\mathcal{V}}(g_2\, U_2')(b) = \|g_2\|_{L^2}^2 U_2'$$
$$(B_{i,\mathcal{V}} S_{\mathcal{V}} \psi^{\perp})(b) = \int_0^b g_i(y)\psi^{\perp}(y)\,\mathrm{d}y = 0 \quad \forall i = 1, 2 \qquad \text{(see (3.2.1))}.$$

5.7.1.3.2 The Quadruplet $\mathcal{Q}_{p,\mathcal{Y},2}{}^*$
Recalling that

$$\mathcal{V} := L^p\big((0,b), \mathcal{Y}\big), \qquad \mathcal{V}^* = L^q\big((0,b), \mathcal{Y}^*\big),$$

the adjoint differential quadruplet of $\mathcal{Q}_{p,\mathcal{Y},2}$ reads

$$\mathcal{Q}_{p,\mathcal{Y},2}{}^* = \big(\mathrm{D}^2_{\mathcal{V}},\ \mathrm{D}^2_{\mathcal{V}^*},\ (B_{2,\mathcal{V}})^*,\ S_{\mathcal{V}^*}\big).$$

Thus, $\mathcal{Q}_{p,\mathcal{Y},2}{}^* = \mathcal{Q}_{q,\mathcal{Y}^*,2}$. Moreover, in view of (5.7.2),

$$\ker \mathrm{D}^2_{\mathcal{V}^*} = \xi_1^{\mathrm{s}}\mathcal{Y}^* \oplus \xi_2^{\mathrm{s}}\mathcal{Y}^* = g_1 Y^* \oplus g_2 Y^*,$$

and

$$F_0^{\mathrm{s}} = \eta_1^{\mathrm{s}}\mathcal{Y}^* \oplus \eta_2^{\mathrm{s}}\mathcal{Y}^* = g_1\mathcal{Y}^* \oplus g_2\mathcal{Y}^*$$

is complementary subspace of $(\ker \mathrm{D}^2_{\mathcal{V}})^{\perp}$ in $\mathcal{V}^*$. Thus for each f in $D(\mathrm{D}^2_{\mathcal{V}^*})$, (5.7.10) reads

$$f = S_{\mathcal{V}^*}(g_i F_i) + (B_{2,V})^*(g_i F_i') + (B_{2,\mathcal{V}})^*\varphi^{\perp}, \tag{5.7.17}$$

where $\varphi^{\perp}$ lies in $(\ker \mathrm{D}^2_{\mathcal{V}})^{\perp}$, and (F_1, F_2, F_1', F_2') are the *adjoint endogenous boundary values of f with underlying quadruplet $\mathcal{Q}_{p,\mathcal{Y},2}{}^*$ and with respect to the direct sum decompositions*

$$\ker \mathrm{D}^1_{\mathcal{V}^*} S_{\mathcal{V}^*} = g_1\,\mathcal{Y}^* \oplus g_2(b-\cdot)\,\mathcal{Y}^*, \qquad S^* F_0 = g_1\,\mathcal{Y}^* \oplus g_2(b-\cdot)\,\mathcal{Y}^*.$$

Since

$$r(\mathcal{Q}_{p,\mathcal{Y},2}) = \mathcal{Q}_{q,\mathcal{Y}^*,2},$$

the *boundary values trick* featured in Sect. 5.7.1.2 yields on account of (5.7.16) that

$$\begin{cases} f(b) = F_1 \\ f'(b) = \qquad - F_2 \\ f(0) = F_1 + bF_2 + (g_1, g_2)F_1' + \|g_2\|_{L^2}^2 F_2' \\ f'(0) = \qquad - F_2 - \|g_1\|_{L^2}^2 F_1' - (g_1, g_2)F_2'. \end{cases} \tag{5.7.18}$$

□

5.7.1.4 Change of Complementary Subspace of ker $\mathbb{A}^\perp$

The endogenous boundary values of a function u in $D(A_{\mathrm{M}})$ depend clearly on F_0. The issue is to investigate the changes in these endogenous boundary values when F_0 is replaced by another closed complementary subspace F_0'. The result stated in Proposition 5.7.2 says basically that the transform of endogenous boundary values is an isomorphism.

Proposition 5.7.2 *Let* $\mathcal{Q} := (\mathbb{A}, A_{\mathrm{M}}, B, S)$ *be a differential quadruplet on* $(\mathcal{V}, \mathcal{V}^*)$ *satisfying* (5.7.2), *and* F_0 *be defined by* (5.7.3). *Let* $\eta_1', \dots \eta_{m^{\mathrm{s}}}'$ *belong to* $L^p(0, b)$ *and*

$$F_0' := \overset{m^{\mathrm{s}}}{\underset{i=1}{\oplus}} \eta_i' \mathcal{Y}$$

be such that $\mathcal{Q}$ *is* F_0'*-regular. Then, there exists an isomorphism* F *from* $\mathcal{Y}^{m^{\mathrm{s}}}$ *onto itself such that for any* u *in* $D(A_{\mathrm{M}})$ *the following property holds true.*

If (U_i, U_i') *denotes the endogenous boundary values of* u *with respect to* (5.7.6), *and* (V_i, V_i') *denotes the endogenous boundary values of* u *with respect to the direct sum decompositions*

$$\ker A_{\mathrm{M}} = \overset{m}{\underset{i=1}{\oplus}} \xi_i \mathcal{Y}, \qquad F_0' = \overset{m^{\mathrm{s}}}{\underset{i=1}{\oplus}} \eta_i' \mathcal{Y},$$

then

$$V_i = U_i, \qquad \forall i = 1, \dots, m$$

$$(V_1', \dots, V_{m^{\mathrm{s}}}') = F(U_1', \dots, U_{m^{\mathrm{s}}}').$$

Proof According to (5.7.5), there exist f and f' in $\mathcal{V}$ such that

$$u = \xi_i U_i + Bf = \xi_i V_i + Bf'.$$

Then, a direct sum argument (see lemma 4.3.3) yields that $V_i = U_i$ for each index i. By Proposition 5.7.1, the restriction $F_{\text{b.v}|BS(F_0)}$ of the F_0-boundary values map to $BS(F_0)$ is linear and continuous from $\big(BS(F_0), \|\cdot\|_{D(A_M)}\big)$ into $\mathcal{Y}^{m+m^s}$. Obviously, it is a bijection (still labeled $F_{\text{b.v}|BS(F_0)}$) onto its range $\mathcal{Y}^{m^s}$. Moreover $BS(F_0)$ is closed according to Lemma 4.3.15 (with $G := SF_0$ and $\mathcal{E} := \{0\}$). Then, by Corollary 2.4.9, it is an isomorphism.

Now consider

$$F'_{\text{b.v}|R(B)} : R(B) = BS(F'_0) \oplus BS(\ker \mathbb{A}^\perp) \to \mathcal{Y}^{m^s}$$
$$BS(\eta'_i V'_i) + BS\psi^\perp \mapsto (V'_1, \ldots, V'_{m^s}).$$

By Proposition 5.7.1, this map is linear and continuous from $(R(B), \|\cdot\|_{D(A_M)})$ into $\mathcal{Y}^{m^s}$. Now going back to u, there exists $\psi^\perp$ and $\phi^\perp$ in $\ker \mathbb{A}^\perp$ such that

$$\begin{aligned} u &= \xi_i U_i + BS(\eta_i U'_i) + BS\psi^\perp \\ &= \xi_i U_i + BS(\eta'_i V'_i) + BS\phi^\perp. \end{aligned}$$

Besides setting

$$(W'_1, \ldots, W'_{m^s}) := F'_{\text{b.v}|R(B)}\big(BS(\eta_i U'_i)\big),$$

there exists $\varphi^\perp$ in $\ker \mathbb{A}^\perp$ such that

$$BS(\eta_i U'_i) = BS(\eta'_i W'_i) + BS\varphi^\perp.$$

Thus,

$$u = \xi_i U_i + BS(\eta'_i W'_i) + BS(\psi^\perp + \varphi^\perp).$$

Then, a direct sum argument yields that $W'_i = V'_i$ for all index i. Therefore

$$(V'_1, \ldots, V'_{m^s}) = F'_{\text{b.v}|R(B)} \circ (F_{\text{b.v}|BS(F_0)})^{-1}(U'_1, \ldots, U'_{m^s}).$$

By symmetry (i.e., exchanging F_0 and F'_0 in the latter identity), one gets

$$(U'_1, \ldots, U'_{m^s}) = F_{\text{b.v}|R(B)} \circ (F'_{\text{b.v}|BS(F'_0)})^{-1}(V'_1, \ldots, V'_{m^s}).$$

Thus, the map $(U'_1, \dots, U'_{m^s}) \mapsto (V'_1, \dots, V'_{m^s})$ is an isomorphism. □

Example 5.7.2 *A simpler complementary subspace of* $(\ker \mathrm{D}^2_{\mathcal{V}})^\perp$. According to Sect. 5.7.1.3.1, $Q_{p,\mathcal{Y},2}$ is a $\ker \mathrm{D}^2_{\mathcal{V}}$-regular. Our goal is to find another subspace F_0 of $L^p(\mathcal{Y})$ such that

(i) $Q_{p,\mathcal{Y},2}$ is a F_0-regular;
(ii) the system transforming the endogenous boundary values of u into its usual boundary values in simpler than (5.7.16) (i.e. is more sparse).

The key point in the construction of a suitable F_0 is that for any (h, Y) in $L^p(\mathcal{Y}) \times \mathcal{Y}$,

$$B_{2,\mathcal{V}}(h\,Y)(b) = \int_0^b g_2(y)h(y)\,\mathrm{d}y\,Y.$$

Thus, the idea is to choose h so that the latter integral equals 0 or 1. More precisely, by Corollary 5.3.1, there exists on the one hand, h_1, h_2 in $L^p(0, b)$ such that $\int_0^b h_i g_j\,\mathrm{d}y = \delta_{ij}$ for each indexes i and j in $\{1, 2\}$ (because g_j is real valued). On the other hand, setting

$$F_0 := h_1\,\mathcal{Y} \oplus h_2\,\mathcal{Y},$$

$Q_{p,\mathcal{Y},2}$ is a F_0-regular.

Let us now compute for any u in $D(\mathrm{D}^2_{\mathcal{V}})$, the usual boundary values of u in terms of its endogenous boundary values with respect to the direct sum decompositions

$$\ker \mathrm{D}^2_{\mathcal{V}} = g_1\,\mathcal{Y} \oplus g_2\,\mathcal{Y}, \qquad F_0 = h_1\,\mathcal{Y} \oplus h_2\,\mathcal{Y}.$$

By definition these endogenous boundary values (U_1, U_2, U'_1, U'_2) of u satisfy

$$u = g_1\,U_1 + g_2\,U_2 + B_{2,\mathcal{V}}S_{\mathcal{V}}(h_1\,U'_1) + B_{2,\mathcal{V}}S_{\mathcal{V}}(h_2\,U'_2) + B_{2,\mathcal{V}}S_{\mathcal{V}}\,\psi^\perp,$$

where $\psi^\perp$ lies in $(g_1\,\mathcal{Y} \oplus g_2\,\mathcal{Y})^{\perp L^p(\mathcal{Y})}$. We claim that

$$\begin{cases} u(0) = U_1 \\ u'(0) = \qquad U_2 \\ u(b) = U_1 + bU_2 + \qquad U'_2 \\ u'(b) = \qquad U_2 + U'_1. \end{cases} \tag{5.7.19}$$

The two first equations are derived easily. Next, the other equations are obtained by using

$$B_{i,\mathcal{V}}(h_j U_j')(b) = \delta_{ij} U_j', \qquad \forall i, j = 1, 2.$$

Now we would like to relate the endogenous boundary values obtained in Sect. 5.7.1.3.1 to the latter endogenous boundary values. On account of Proposition 5.7.2, the two first boundary values are the same, and the isomorphism F, which transforms the two last endogenous boundary values of u given by (5.7.16) into the two last endogenous boundary values of u by (5.7.19), reads

$$F : \mathcal{Y} \times \mathcal{Y} \to \mathcal{Y} \times \mathcal{Y}$$
$$(U_1', U_2') \mapsto \left(\|g_1\|_{L^2}^2 U_1' + (g_1, g_2) U_2', \ (g_1, g_2) U_1' + \|g_2\|_{L^2}^2 U_2'\right).$$

5.7.1.5 Integration by Parts

For each u in $D(A_{\mathrm{M}})$ and f in $D(S^* \mathbb{A} S^*)$, we claim that following integration by parts formula holds true.

$$\langle f, A_{\mathrm{M}} u \rangle_{\mathcal{V}^*, \mathcal{V}} = \langle S^* \mathbb{A} S^* f, u \rangle_{\mathcal{V}^*, \mathcal{V}} +$$
$$\sum_{i,j=1}^{m^{\mathrm{s}}} \langle \xi_i^{\mathrm{s}}, \eta_j \rangle_{L^q, L^p} \langle F_i, U_j' \rangle_{\mathcal{Y}^*, \mathcal{Y}} - \sum_{i,j=1}^{m} \langle \eta_i^{\mathrm{s}}, \xi_j \rangle_{L^q, L^p} \langle F_i', U_j \rangle_{\mathcal{Y}^*, \mathcal{Y}}.$$

The proof of the claim is similar to that of Corollary 5.5.4: we use (5.4.3), and (5.7.5), (5.7.10).

5.7.2 Boundary Restriction Operators and Their Adjoint

The hypothesis made at the beginning of the current section remains in force. In particular,

$$\mathcal{V} := L^p\big((0, b), \mathcal{Y}\big).$$

The following result is a coordinates-counter part of Proposition 5.3.5. For all purposes, we recall that $F_{\mathrm{b.v}}$ is *the boundary values map* introduced in Definition 5.7.1.

Proposition 5.7.3 *Let* $\mathcal{Q} := (\mathbb{A}, A_{\mathrm{M}}, B, S)$ *be a* F_0*-regular differential quadruplet on* $(\mathcal{V}, \mathcal{V}^*)$ *satisfying* (5.7.2), (5.7.3), *and* $F_{\mathrm{b.v}}$ *be the* F_0*-boundary values map of* $\mathcal{Q}$. *Let also* $\mathcal{X}$ *be a normed space and* L *be an operator in* $\mathcal{L}(\mathcal{Y}^{m+m^{\mathrm{s}}}, \mathcal{X})$. *If* A *is the restriction of* A_{M} *with domain*

$$D(A) := \ker L F_{\mathrm{b.v}},$$

then A *is a closed boundary restriction operator of* $\mathcal{Q}$.

Proof By Proposition 5.7.1, $F_{\text{b.v}}$ is continuous, thus $D(A)$ is closed in $D(A_{\text{M}})$. Since A_{M} is closed according to Definition 5.3.1, there results that A is a closed operator on $\mathcal{V}$.

By definition of $F_{\text{b.v}}$, $BS(\ker \mathbb{A}^{\perp})$ is the kernel of $F_{\text{b.v}}$. Thus $BS(\ker \mathbb{A}^{\perp})$ is contained in $D(A)$. Since $BS(\ker \mathbb{A}^{\perp})$ is the domain of the minimal operator $A_{\text{m},\mathcal{Q}}$ of $\mathcal{Q}$, we get $A_{\text{m},\mathcal{Q}} \subseteq A$. Then Corollary 5.2.8 yields that A is a closed boundary restriction operator of $\mathcal{Q}$. □

In the next example, we will compute the adjoint of a particular closed boundary restriction operator in a simple abstract framework.

Example 5.7.3 Let $\mathcal{Q} := (\mathbb{A}, A_{\text{M}}, B, S)$ be a F_0-regular differential quadruplet on $(\mathcal{V}, \mathcal{V}^*)$ satisfying (5.7.2). We assume in addition that

$$m = m^{\text{s}} = 1.$$

Thus as explained at the beginning of the current section, $\mathcal{Q}$ and $\mathcal{Q}^*$ are respectively F_0 and $S^*F_0^{\text{s}}$-regular differential quadruplets with

$$F_0 := \eta_1 \mathcal{Y}, \qquad F_0^{\text{s}} := \eta_1^{\text{s}} \mathcal{Y}^*.$$

Let x_1 and y_1 be two scalars such that at least one of them is non zero. Let also A be the restriction of A_{M} with domain

$$D(A) := \{u \in D(A_{\text{M}}) \mid y_1 U_1 = x_1 U_1'\},$$

where (U_1, U_1') are the endogenous boundary values of u given by (5.7.5), that is,

$$u = \xi_1 U_1 + BS\eta_1 U_1' + BS\psi^{\perp}, \tag{5.7.20}$$

where $\psi^{\perp}$ lies in $\ker \mathbb{A}^{\perp}$. There results from Proposition 5.7.3 that A is a *closed boundary restriction operator of* $\mathcal{Q}$.

Our aim is to characterize its adjoint A^*. Since A is a boundary restriction operator of $\mathcal{Q}$, we already know that $A^* \subseteq S^*\mathbb{A}S^*$. The issue is then to compute the domain of A^*. For, let f be any function in $D(S^*\mathbb{A}S^*)$. By (5.7.10), there exist F_1, F_1' in $\mathcal{Y}^*$, and $\varphi^{\perp}$ lies in $(\ker A_{\text{M}})^{\perp}$ such that

$$f = S^*(\xi_1^{\text{s}} F_1) + B^*(\eta_1^{\text{s}} F_1') + B^*\varphi^{\perp}. \tag{5.7.21}$$

Then, we claim that f lies in $D(A^*)$ if and only if

$$\overline{y_1} \int_0^b \xi_1^{\text{s}} \overline{\eta_1} \, \mathrm{d}y \, F_1 = \overline{x_1} \int_0^b \eta_1^{\text{s}} \overline{\xi_1} \, \mathrm{d}y \, F_1'. \tag{5.7.22}$$

Indeed, since $(x_1, y_1) \neq (0,0)$, we may check that

$$D(A) = \{x_1\xi_1 Y + BS(y_1\eta_1 Y) + BS\psi^\perp \mid (Y, \psi^\perp) \in \mathcal{Y} \times \ker \mathbb{A}^\perp\}.$$

Being given any f in $D(S^*\mathbb{A}S^*)$, Theorem 5.4.1 (i) together with (5.7.21) yield that f lies in the domain of A^* if and only if

$$\langle \xi_1^{\mathrm{s}} F_1,\ y_1\eta_1 Y + \psi^\perp \rangle_{\mathcal{V}^*,\mathcal{V}} = \langle \eta_1^{\mathrm{s}} F_1' + \varphi^\perp,\ x_1\xi_1 Y \rangle_{\mathcal{V}^*,\mathcal{V}}, \qquad \forall Y \in \mathcal{Y}.$$

Since $\psi^\perp$ lies in $\ker \mathbb{A}^\perp$, and $\varphi^\perp$ lies in $\ker A_{\mathrm{M}}^\perp$, the latter equality reads

$$\langle \xi_1^{\mathrm{s}} F_1,\ y_1\eta_1 Y \rangle_{\mathcal{V}^*,\mathcal{V}} = \langle \eta_1^{\mathrm{s}} F_1',\ x_1\xi_1 Y \rangle_{\mathcal{V}^*,\mathcal{V}}, \qquad \forall Y \in \mathcal{Y}.$$

Then, the claim follows. □

5.7.3 The Quadruplet $\mathcal{Q}_{p,\mathcal{Y},n}$

Let $\mathbb{K}$ be equal to $\mathbb{R}$ or $\mathbb{C}$, b be a positive real number and p, q be conjugate exponents in $(1, \infty)$. Let also $\mathcal{Y}$ be a nontrivial separable and reflexive Banach space over $\mathbb{K}$, and

$$\mathcal{V} := L^p\big((0,b), \mathcal{Y}\big), \qquad \mathcal{V}^* := L^q\big((0,b), \mathcal{Y}^*\big).$$

Under the assumptions and notation of Example 5.6.7, we consider the differential quadruplet

$$\mathcal{Q}_{p,\mathcal{Y},n} := \big(\mathrm{D}^n_{\mathcal{V}^*},\ \mathrm{D}^n_{\mathcal{V}},\ B_{n,\mathcal{V}},\ S_{\mathcal{V}}\big)$$

on $(\mathcal{V}, \mathcal{V}^*)$.

5.7.3.1 Endogenous Boundary Values

One has

$$\ker \mathrm{D}^n_{\mathcal{V}} = \bigoplus_{i=1}^{n} g_i \mathcal{Y}, \qquad \ker \mathrm{D}^n_{\mathcal{V}^*} = \bigoplus_{i=1}^{n} g_i \mathcal{Y}^*.$$

By Corollary 5.3.1, there exist functions $h_1, \ldots, h_n$ in $L^p(0,b)$ such that

$$\int_0^b h_i g_j \,\mathrm{d}y = \delta_{ij}, \qquad \forall i, j = 1, \ldots, n. \tag{5.7.23}$$

On the other hand, setting

$$F_0 := \overset{n}{\underset{i=1}{\oplus}} h_i\, \mathcal{Y},$$

$\mathcal{Q}_{p,\mathcal{Y},n}$ is a F_0-regular. Now the Domain Structure Theorem 5.3.3 gives

$$E_{\mathrm{M}} = \overset{n}{\underset{i=1}{\oplus}} g_i \mathcal{Y} \oplus B_{n,\mathcal{V}} S_{\mathcal{V}}(\overset{n}{\underset{i=1}{\oplus}} h_i \mathcal{Y})$$

$$D(\mathrm{D}_{\mathcal{V}}^n) = E_{\mathrm{M}} \oplus B_{n,\mathcal{V}} S_{\mathcal{V}} \big(\overset{n}{\underset{i=1}{\oplus}} g_i \mathcal{Y}^*\big)^{\perp}.$$

Thus, a function u in $\mathcal{V}$ lies in $D(\mathrm{D}_{\mathcal{V}}^n)$ if and only if there exist $U_1, \dots, U_n$ and $U_1', \dots, U_n'$ in $\mathcal{Y}$ and $\psi^{\perp}$ in $(\ker \mathrm{D}_{\mathcal{V}*}^n)^{\perp}$ such that

$$u = \sum_{i=1}^{n} g_i\, U_i + \sum_{i=1}^{n} g_n * S_{\mathcal{V}}(h_i\, U_i') + g_n * S_{\mathcal{V}} \psi^{\perp}. \tag{5.7.24}$$

In view of Sect. 5.7.1.1, $(U_1, \dots, U_n, U_1', \dots, U_n')$ are the *endogenous boundary values of u with respect to the direct sum decompositions*

$$\ker \mathrm{D}_{\mathcal{V}}^n = \overset{n}{\underset{i=1}{\oplus}} g_i \mathcal{Y}, \qquad F_0 = \overset{n}{\underset{i=1}{\oplus}} h_i \mathcal{Y}.$$

Let us compute the endogenous boundary values of any element u of $D(\mathrm{D}_{\mathcal{V}}^n)$. We claim that for each index j in $[1, n]$,

$$U_j = \mathrm{D}_{\mathcal{V}}^{j-1} u(0) \tag{5.7.25}$$

$$U_j' = g_j * \mathrm{D}_{\mathcal{V}}^n u(b). \tag{5.7.26}$$

We compute the U_j's by writing (5.7.24) under the form

$$u = \sum_{i=1}^{n} g_i\, U_i + g_n * h,$$

where h lies in $\mathcal{V}$. Then, we apply the operator $\mathrm{D}_{\mathcal{V}}^{j-1}$ to the latter equation and derive (5.7.25). In order to establish (5.7.26), we have to get ride of $\psi^{\perp}$. For, we use the belonging of $\psi^{\perp}$ to

$$(\overset{n}{\underset{i=1}{\oplus}} g_i \mathcal{Y}^*)^{\perp} = \bigcap_{i=1}^{n} (g_i \mathcal{Y}^*)^{\perp},$$

which implies due to (3.2.1) and since g_i is real-valued that

$$\int_0^b g_i(y)\psi^\perp(y)\,\mathrm{d}y = 0, \qquad \forall\, 1 \le i \le n. \tag{5.7.27}$$

Next applying the operator $\mathrm{D}^n_{\mathcal{V}}$ to (5.7.24), we get

$$\mathrm{D}^n_{\mathcal{V}} u = \sum_{i=1}^{n} S_{\mathcal{V}}(h_i\, U_i') + S_{\mathcal{V}}\psi^\perp.$$

Moreover, with Proposition 3.3.5 and (5.7.23), one has

$$g_j * S_{\mathcal{V}}(h_i\, U_i')(b) = \int_0^b g_j(y)h_i(y)\,\mathrm{d}y\, U_i' = U_j'$$

and by (5.7.27),

$$g_j * S_{\mathcal{V}}\psi^\perp(b) = \int_0^b g_j(y)\psi^\perp(y)\,\mathrm{d}y = 0.$$

Then, (5.7.26) follows.

The "standard" boundary values appearing in (5.7.25) and (5.7.26) are

$$u(0), u'(0), \ldots, u^{(n-1)}(0), g_1 * \mathrm{D}^n_{\mathcal{V}} u(b), \ldots, g_n * \mathrm{D}^n_{\mathcal{V}} u(b).$$

The n first boundary values are the usual boundary values of an element u of $W^{n,p}\big((0,b),\mathcal{Y}\big)$. However the n last boundary values are not. In order to recover the usual ones, i.e.

$$u(b), u'(b), \ldots, u^{(n-1)}(b),$$

we proceed as follows. For any fixed integer j in $[1,n]$, we consider the differential quadruplet $Q_{p,\mathcal{Y},j}$. Then the function $\mathrm{D}^{n-j}_{\mathcal{V}} u$ lies in the domain of $\mathrm{D}^j_{\mathcal{V}}$. Let $V_1, \ldots, V_j$ and $V_1', \ldots, V_j'$ be the endogenous boundary values of u with underlying quadruplet $Q_{p,\mathcal{Y},j}$ and with respect to the direct sum decompositions

$$\ker \mathrm{D}^j_{\mathcal{V}} = \bigoplus_{i=1}^{j} g_i\mathcal{Y}, \qquad F_0 = \bigoplus_{i=1}^{j} h_i\mathcal{Y}.$$

By (5.7.25), we have

$$V_i = \mathrm{D}^{n+i-j-1}_{\mathcal{V}} u(0), \qquad \forall\, 1 \le i \le j.$$

Moreover by Lemma 4.3.3,

$$\mathrm{D}_{\mathcal{V}}^{n-j} u = \sum_{i=1}^{j} g_i \, V_i + g_j * \mathrm{D}_{\mathcal{V}}^{n} u. \tag{5.7.28}$$

Thus

$$g_j * \mathrm{D}_{\mathcal{V}}^{n} u(b) = \mathrm{D}_{\mathcal{V}}^{n-j} u(b) - \sum_{i=n-j}^{n-1} g_{i+j+1-n}(b) \, \mathrm{D}_{\mathcal{V}}^{i} u(0).$$

With (5.7.25) and (5.7.26), we get for each integer j in $[1, n]$,

$$U_j = \mathrm{D}_{\mathcal{V}}^{j-1} u(0) \tag{5.7.29}$$

$$U'_j = \mathrm{D}_{\mathcal{V}}^{n-j} u(b) - \sum_{i=n-j}^{n-1} g_{i+j+1-n}(b) \, \mathrm{D}_{\mathcal{V}}^{i} u(0). \tag{5.7.30}$$

In closing, (5.7.29)–(5.7.30) *link endogenous boundary values with usual boundary values*.

Next we claim that this system is invertible, i.e., the linear transform mapping the usual boundary values

$$u(0),\ \mathrm{D}_{\mathcal{V}}^{1} u(0),\ \ldots,\ \mathrm{D}_{\mathcal{V}}^{n-1} u(0),\ u(b),\ \mathrm{D}_{\mathcal{V}}^{1} u(b),\ \ldots,\ \mathrm{D}_{\mathcal{V}}^{n-1} u(b),$$

of u into its endogenous boundary values $U_1, \ldots, U_n, U'_1, \ldots, U'_n$ is an isomorphism. Indeed, the inverse of the system (5.7.29)–(5.7.30) reads for each integer j in $[0, n-1]$,

$$\mathrm{D}_{\mathcal{V}}^{j} u(0) = U_{j+1} \tag{5.7.31}$$

$$\mathrm{D}_{\mathcal{V}}^{j} u(b) = U'_{n-j} + \sum_{i=j+1}^{n} g_{i-j}(b) \, U_i. \tag{5.7.32}$$

Remark 5.7.4 By choosing $j = n$ in (5.7.28), we recover the *Taylor Formula with integral reminder*. □

5.8 Quadruplets on Cartesian Powers of a Space

Let $\mathcal{V}$ be a reflexive real or complex Banach space. Once a differential quadruplet on $(\mathcal{V}, \mathcal{V}^*)$ is given, we will see that a differential quadruplet on $\big(\mathcal{V} \times \mathcal{V}, (\mathcal{V} \times \mathcal{V})^*\big)$ may be easily constructed. More generally, instead of $\mathcal{V} \times \mathcal{V}$, we will consider for any positive integer n, the Cartesian product $\mathcal{V}^n$ of n copies of $\mathcal{V}$.

From Sect. 2.2.7.4.3, the anti-dual space of $\mathcal{V}^n$ is identified with $(\mathcal{V}^*)^n$. Thus, we derive easily that $\mathcal{V}^n$ is reflexive.

5.8.1 Definition of $\vec{\mathcal{Q}}$

Let $\mathcal{Q} := (\mathbb{A}, A_{\mathrm{M}}, B, S)$ be a differential quadruplet on $(\mathcal{V}, \mathcal{V}^*)$. By using (2.2.23) and (2.2.24), we may check that

$$\vec{\mathcal{Q}} := (\vec{\mathbb{A}}, \vec{A_{\mathrm{M}}}, \vec{B}, \vec{S}) \tag{5.8.1}$$

is a differential quadruplet on $\big(\mathcal{V}^n, (\mathcal{V}^*)^n\big)$. In particular,

$$\mathbb{B}_{\vec{\mathcal{Q}}} = \overrightarrow{\mathbb{B}_{\mathcal{Q}}} = \overrightarrow{S^*}\,\overrightarrow{B^*}\,\overrightarrow{S^*}. \tag{5.8.2}$$

$\vec{\mathcal{Q}}$ is called *the differential quadruplet induced by* $\mathcal{Q}$ *on* $\big(\mathcal{V}^n, (\mathcal{V}^*)^n\big)$.

5.8.2 Properties of Induced Quadruplets

Proposition 5.8.1 *Let* $\mathcal{Q} := (\mathbb{A}, A_{\mathrm{M}}, B, S)$ *be a* F_0*-regular differential quadruplet on* $(\mathcal{V}, \mathcal{V}^*)$*, and n be a positive integer. Then*

(i) $\ker \vec{\mathbb{A}}^{\perp} = (\ker \mathbb{A}^{\perp})^n$;
(ii) $\vec{\mathcal{Q}}$ *is a* $F_0{}^n$*-regular differential quadruplet on* $\big(\mathcal{V}^n, (\mathcal{V}^*)^n\big)$;
(iii) *denoting by* $P_{\mathrm{b.v},\mathcal{Q}} : D(A_{\mathrm{M}}) \to D(A_{\mathrm{M}})$ *the projection induced by* F_0*, and by* $P_{\mathrm{b.v},\vec{\mathcal{Q}}} : D(\vec{A_{\mathrm{M}}}) \to D(\vec{A_{\mathrm{M}}})$ *the projection induced by* $F_0{}^n$ *(see Definition 5.3.3), one has*

$$P_{\mathrm{b.v},\vec{\mathcal{Q}}} = \overrightarrow{P_{\mathrm{b.v},\mathcal{Q}}}.$$

Proof Item (i) follows from the relations

$$\ker \vec{\mathbb{A}} = (\ker \mathbb{A})^n, \qquad \big((\ker \mathbb{A})^n\big)^{\perp} = (\ker \mathbb{A}^{\perp})^n.$$

Since the maximal operators of $\mathcal{Q}$ are closed, it is easy proved that the maximal operators of $\vec{\mathcal{Q}}$ are closed as well. Besides, $F_0{}^n$ and $(\ker \vec{\mathbb{A}})^{\perp}$ are in direct sum because F_0 and $\ker \vec{\mathbb{A}}^{\perp}$ are. Besides, with Item (i)

$$F_0{}^n + (\ker \vec{\mathbb{A}}^{\perp}) = (F_0 + \ker \mathbb{A}^{\perp})^n = \mathcal{V}^n.$$

Hence $\vec{\mathcal{Q}}$ is $F_0{}^n$-regular. Finally, since $\vec{\mathcal{Q}}$ is $F_0{}^n$-regular, we may legitimately set

$$E_{\mathrm{M},\mathcal{Q}} := \ker A_{\mathrm{M}} \oplus BS(F_0), \qquad E_{\mathrm{M},\vec{\mathcal{Q}}} := \ker \vec{A_{\mathrm{M}}} \oplus \vec{B}\vec{S}(F_0{}^n).$$

Then, it is easily checked that

$$E_{\mathrm{M},\vec{\mathcal{Q}}} = (E_{\mathrm{M},\mathcal{Q}})^n. \tag{5.8.3}$$

Besides,

$$\begin{aligned} \vec{B}\vec{S}(\ker \vec{\mathbb{A}}^{\perp}) &= \vec{B}\vec{S}\big((\ker \mathbb{A}^{\perp})^n\big) && \text{(by Item (i))} \\ &= \big(BS(\ker \mathbb{A}^{\perp})\big)^n && \text{(with (2.2.24)).} \end{aligned} \tag{5.8.4}$$

Next, each $\vec{u} = (u_i)$ in $D(\vec{A_{\mathrm{M}}})$ reads

$$\vec{u} = \vec{e} + \vec{\psi} \in E_{\mathrm{M},\vec{\mathcal{Q}}} \oplus \vec{B}\vec{S}(\ker \vec{\mathbb{A}}^{\perp}).$$

Thus $P_{\mathrm{b.v},\vec{\mathcal{Q}}}\,\vec{u} = \vec{e}$ by definition of $P_{\mathrm{b.v},\vec{\mathcal{Q}}}$. On the other hand, by (5.8.3) and (5.8.4), for each index $i = 1, \dots, n$, there exist e_i in $E_{\mathrm{M},\mathcal{Q}}$, and ψ_i in $BS(\ker \mathbb{A}^{\perp})$ such that $u_i = e_i + \psi_i$. Thus, $P_{\mathrm{b.v},\mathcal{Q}} u_i = e_i$. Whence Item (iii) follows. □

Fractional Differential Triplets and Quadruplets on Lebesgue Spaces 6

We will apply the abstract results of Chaps. 4 and 5 to construct various differential triplets and quadruplets involving fractional differential operators. We will also study their boundary restriction operators. A particular attention will be paid to endogenous and "standard" boundary values.

Applications involving the operator $\frac{\mathrm{d}}{\mathrm{d}x}$ has been achieved in the course of Chap. 4 (in particular in Sect. 4.4.2) and Chap. 5 (see, for instance, Examples 5.1.5, 5.3.2, 5.6.3, and Sect. 5.7.1.3). We will keep the same approach for applications involving fractional operators.

Before considering general fractional operators, we will study in Sect. 6.2 two simple differential triplets.

Let $\mathbb{K}$ be equal to $\mathbb{R}$ or $\mathbb{C}$.

6.1 Introduction

Let us introduce some basic ideas of *fractional calculus*. Starting from usual entire derivatives, we observe that for a suitable function f:

(i) the derivative of a primitive of f returns the function f itself;
(ii) the derivative of the derivative of f returns the second derivative of f;
(iii) the second derivative of a primitive of f returns the derivative of f.

Let us now consider fractional derivatives of order $\frac{1}{2}$. In view of the concept of *fractional Riemann-Liouville primitives* introduced in Sect. 3.6, we suggest from Items (i), (ii) that

A. Rougirel, *Unified Theory for Fractional and Entire Differential Operators*,
Frontiers in Elliptic and Parabolic Problems,
https://doi.org/10.1007/978-3-031-58356-8_6

(i-$\frac{1}{2}$) the derivative of order $\frac{1}{2}$ of the primitive of order $\frac{1}{2}$ of a function f should return the function f itself;

(ii-$\frac{1}{2}$) the derivative of order $\frac{1}{2}$ of the derivative of order $\frac{1}{2}$ of f should return the (first) derivative of f.

In symbol, Items (i-$\frac{1}{2}$) and (ii-$\frac{1}{2}$) read formally

$$D^{\frac{1}{2}}(g_{\frac{1}{2}} * f) = f, \qquad D^{\frac{1}{2}}(D^{\frac{1}{2}} f) = D^1 f.$$

It was a great merit of the creators of fractional calculus to realize that a good candidate for the half derivative of f is

$$D^{\frac{1}{2}} f := D^1(g_{\frac{1}{2}} * f).$$

More generally, for any real number α in $(0, 1)$, a fractional derivative of order α should satisfy

$$D^\alpha(g_\alpha * f) = f, \qquad D^\alpha(D^{1-\alpha} f) = D^1 f, \qquad D^\alpha(g_\beta * f) = D^{\alpha-\beta} f,$$

for any β in $(0, \alpha)$. Then we are led to set

$$D^\alpha f := D^1(g_{1-\alpha} * f),$$

and $D^\alpha f$ is called *the fractional Riemann-Liouville derivative of f of order α*.

It turns out that in many situations, among all these relations only

$$D^\alpha(g_\alpha * f) = f, \qquad \forall \alpha \in (0, 1],$$

is really used. Moreover as far as the quantitative theory of boundary values problems (see Chap. 7) is concerned, we essentially need that the fractional derivative under consideration is the maximal operator of a regular differential quadruplet. In an Hilbertian setting, this simply mean that the fractional derivative (seen as an operator) has a right inverse which is related to its adjoint through a symmetry.

In this sense the maximal operators of a differential triplet or quadruplet can be seen as fractional derivatives.

6.2 Two Toy Models

So far only triplets and quadruplets built upon the first-order derivative have been considered. Let us now introduce two simple *fractional* differential triplet on the Hilbert space $L^2(0, b)$ (where as usual b is a positive number). The first triplet relies on the

Riemann-Liouville derivatives and the second one will give us the opportunity to introduce *Caputo derivatives*.

The material of this section relies on the theory featured in Chap. 4. This section has been constructed so that a reader who has only read the Hilbertian theory can access two examples of differential triplets.

6.2.1 The Triplet ${}^{\mathrm{RL}}\mathcal{T}_{\alpha,\mathbb{K}}$

For the sake of simplicity, this triplet will also be denoted by ${}^{\mathrm{RL}}\mathcal{T}_{\alpha}$. We will apply the theory developed in Chap. 4 to a simple fractional differential triplet. The outline of this Subsection is the following:

(i) definition of ${}^{\mathrm{RL}}\mathcal{T}_{\alpha}$;
(ii) endogenous boundary values;
(iii) the adjoint triplet of ${}^{\mathrm{RL}}\mathcal{T}_{\alpha}$;
(iv) boundary restriction operators and their adjoint.

We refer to Sect. 6.5 for a more complete and general analysis in the framework of reflexive Lebesgue spaces.

6.2.1.1 Definition

In view of Sect. 6.1 and (3.4.3), for every α in $(0, 1)$, let

$$D({}^{\mathrm{RL}}\mathrm{D}^{\alpha}_{L^2}) := \left\{u \in L^2(0,b) \mid g_{1-\alpha} * u \in H^1(0,b)\right\}$$

$${}^{\mathrm{RL}}\mathrm{D}^{\alpha}_{L^2} : D({}^{\mathrm{RL}}\mathrm{D}^{\alpha}_{L^2}) \subseteq L^2(0,b) \to L^2(0,b), \qquad u \mapsto \frac{\mathrm{d}}{\mathrm{d}x}\{g_{1-\alpha} * u\}.$$

For each u in $D({}^{\mathrm{RL}}\mathrm{D}^{\alpha}_{L^2})$, ${}^{\mathrm{RL}}\mathrm{D}^{\alpha}_{L^2}u$ is called *the Riemann-Liouville derivative of order α of u on $L^2(0, b)$*.

For any α in $(0, 1)$, we put

$$B_{\alpha,L^2} : L^2(0,b) \to L^2(0,b), \qquad v \mapsto g_{\alpha} * v.$$

Notice that $B_{\alpha,L^2}v$ is nothing but the Riemann-Liouville primitive of order α of v. Recalling that the symmetry S is defined through

$$S : L^2(0,b) \to L^2(0,b), \qquad v \mapsto v(b - \cdot),$$

we claim that

$$^{\mathrm{RL}}\mathcal{T}_\alpha := ({}^{\mathrm{RL}}\mathrm{D}^\alpha_{L^2}, B_{\alpha,L^2}, S)$$

is a differential triplet on $L^2(0, b)$. In view of Definition 4.2.1, let us first show that

$$^{\mathrm{RL}}\mathrm{D}^\alpha_{L^2} \circ B_{\alpha,L^2} = \mathrm{id}_{L^2}. \tag{6.2.1}$$

Indeed, for each v in $L^2(0, b)$, (3.4.4) yields $g_{1-\alpha} * B_{\alpha,L^2}v = g_1 * v$. Hence $B_{\alpha,L^2}v$ lies in the domain of ${}^{\mathrm{RL}}\mathrm{D}^\alpha_{L^2}$ and ${}^{\mathrm{RL}}\mathrm{D}^\alpha_{L^2}\, B_{\alpha,L^2}v = v$, which proves (6.2.1). Moreover, $\mathbb{S} := S$ satisfies clearly (4.2.3). Finally, since g_α is real-valued, Fubini theorem yields that for each w in $L^2(0, b)$ and almost every x in $(0, b)$,

$$(B_{\alpha,L^2})^* w(x) = \int_x^b g_\alpha(y - x) w(y)\, dy. \tag{6.2.2}$$

By the change of variable $y' = b - y$ in the latter integral, we get

$$(B_{\alpha,L^2})^* = S\, B_{\alpha,L^2}\, S \tag{6.2.3}$$

(see the proof of Proposition 6.5.1 for more details).

6.2.1.2 Endogenous Boundary Values

In view of Definition 4.4.2, we will compute the endogenous boundary values of any function u in the domain of $\mathbb{A} := {}^{\mathrm{RL}}\mathrm{D}^\alpha_{L^2}$. Notice that although Definition 4.4.2 refers to endogenous boundary *coordinates*, we may speak equivalently about endogenous boundary *values* since u is scalar-valued (see Sect. 5.7.1.1).

We will apply the theory developed in Sect. 4.4. First of all, we claim that

$$\ker {}^{\mathrm{RL}}\mathrm{D}^\alpha_{L^2} = \begin{cases} \langle g_\alpha \rangle & \text{if } \frac{1}{2} < \alpha \\ \{0\} & \text{otherwise.} \end{cases} \tag{6.2.4}$$

Indeed, it is clear that g_α lies in the latter kernel if and only if α is larger than $\frac{1}{2}$. Conversely, for each u in $\ker {}^{\mathrm{RL}}\mathrm{D}^\alpha_{L^2}$, there exists λ in $\mathbb{K}$ such that $g_{1-\alpha} * u = \lambda g_1$. Thus with (3.4.5), $g_1 * u = \lambda g_{1+\alpha}$, and by (3.4.6), $u = \lambda g_\alpha$. If $\alpha \le \frac{1}{2}$ then we must have $\lambda = 0$. Hence the claim follows.

Since Definition 4.4.2 requires that ${}^{\mathrm{RL}}\mathrm{D}^\alpha_{L^2}$ has a nontrivial kernel, we will assume in the sequel that

$$\alpha \in (\frac{1}{2}, 1). \tag{6.2.5}$$

Now we infer from (4.4.21) that each u in $D({}^{\mathrm{RL}}\mathrm{D}^\alpha_{L^2})$ reads

$$u = u_1 g_\alpha + u_1' g_\alpha * S g_\alpha + g_\alpha * S\psi^\perp, \tag{6.2.6}$$

where u_1, u_1' are scalars and $\psi^\perp$ lies in $g_\alpha{}^\perp$. For all purposes, notice that the Domain Structure Theorem 4.3.14 yields that the triplet $(u_1, u_1', \psi^\perp)$ is unique and also that for every $(u_1, u_1', \psi^\perp)$ in $\mathbb{K} \times \mathbb{K} \times g_\alpha{}^\perp$, the function u defined by (6.2.6) lies in the domain of ${}^{\mathrm{RL}}\mathrm{D}^\alpha_{L^2}$.

In view of Definition 4.4.2, (u_1, u_1') is *the endogenous boundary values of* u *with respect to the basis* (g_α) *of* $\ker {}^{\mathrm{RL}}\mathrm{D}^\alpha_{L^2}$. We claim that u is continuous on $(0, b]$ and

$$u_1 = g_{1-\alpha} * u(0), \qquad u_1' = \frac{1}{\|g_\alpha\|_{L^2}^2}\big(u(b) - g_{1-\alpha} * u(0) g_\alpha(b)\big), \tag{6.2.7}$$

where $\|\cdot\|_{L^2}$ is the usual norm on $L^2(0, b)$. In order to compute u_1, we convole (6.2.6) by $g_{1-\alpha}$ and evaluate at $x = 0$. Regarding u_1', starting from (6.2.6), we write

$$u - g_{1-\alpha} * u(0) g_\alpha = u_1' g_\alpha * S g_\alpha + g_\alpha * S\psi^\perp.$$

Now $u_1' g_\alpha * (S g_\alpha)$ and $g_\alpha * S\psi^\perp$ are continuous on $[0, b]$ according to Remark 3.3.4. Thus u is continuous on $(0, b]$. Moreover

$$(g_\alpha * S g_\alpha)(b) = \|g_\alpha\|_{L^2}^2$$

and recalling that $(\cdot, \cdot)$ denote the usual scalar product of $L^2(0, b)$,

$$(g_\alpha * S\psi^\perp)(b) = (\psi^\perp, g_\alpha) = 0$$

since g_α is real valuated and $\psi^\perp$ lies in $g_\alpha{}^\perp$. That completes the proof of the claim.

There results from (6.2.7) that

$$g_{1-\alpha} * u(0), \qquad u(b)$$

are *"standard" boundary values of* u.

Remark 6.2.1 First, since by (6.2.4), (6.2.5), g_α lies in $D({}^{\mathrm{RL}}\mathrm{D}^\alpha_{L^2})$, notice that the endogenous boundary values of g_α are $(1, 0)$.

Second, let u be any element of $D({}^{\mathrm{RL}}\mathrm{D}^\alpha_{L^2})$. Then there results from (6.2.6) and Remark 3.3.4 that u is continuous on $[0, b]$ if and only if

$$u_1 = g_{1-\alpha} * u(0) = 0.$$

Thus *continuous elements of* $D({}^{\mathrm{RL}}\mathrm{D}^{\alpha}_{L^2})$ *have at most one nontrivial boundary values, namely,* $u(b)$. Moreover by Remark 3.3.4 again, *any continuous element of* $D({}^{\mathrm{RL}}\mathrm{D}^{\alpha}_{L^2})$ *vanishes at* $x = 0$.

Finally since the domain of ${}^{\mathrm{RL}}\mathrm{D}^{\alpha}_{L^2}$ contains singular functions at $x = 0$, it is clear that in our framework, $u(0)$ cannot be considered as a boundary value of a generic element u of $D({}^{\mathrm{RL}}\mathrm{D}^{\alpha}_{L^2})$. □

6.2.1.3 The Adjoint Triplet of ${}^{\mathrm{RL}}\mathcal{T}_\alpha$

By Definition 4.2.3, the adjoint differential triplet ${}^{\mathrm{RL}}\mathcal{T}_\alpha{}^*$ of ${}^{\mathrm{RL}}\mathcal{T}_\alpha$ reads

$$
{}^{\mathrm{RL}}\mathcal{T}_\alpha{}^* := (S\,{}^{\mathrm{RL}}\mathrm{D}^{\alpha}_{L^2}\,S,\ S\,B_{\alpha,L^2}\,S,\ S).
$$

Using ${}^{\mathrm{RL}}\mathrm{D}^{\alpha}_{L^2} = \mathrm{D}^{1}_{L^2}\,B_{1-\alpha,L^2}$, $S\,\mathrm{D}^{1}_{L^2} = -\mathrm{D}^{1}_{L^2}\,S$ and (6.2.3), the maximal operator of ${}^{\mathrm{RL}}\mathcal{T}_\alpha{}^*$ reads

$$
S\,{}^{\mathrm{RL}}\mathrm{D}^{\alpha}_{L^2}\,S = -\mathrm{D}^{1}_{L^2}\,(B_{1-\alpha,L^2})^*.
$$

Thus by (6.2.2), for any u in the domain of $S\,{}^{\mathrm{RL}}\mathrm{D}^{\alpha}_{L^2}\,S$, one has

$$
S\,{}^{\mathrm{RL}}\mathrm{D}^{\alpha}_{L^2}\,Su(x) = \frac{\mathrm{d}}{\mathrm{d}x}\int_b^x g_{1-\alpha}(y-x)w(y)\,\mathrm{d}y
$$

for almost every x on $(0, b)$. In the literature, the right-hand side of the latter identity usually takes the form of a definition: see, for instance, [SKM93, (2.23)]. In our framework, it is just a representation of the maximal operator of the adjoint triplet of ${}^{\mathrm{RL}}\mathcal{T}_\alpha$.

Regarding endogenous boundary values, for any f in the domain of $S\,{}^{\mathrm{RL}}\mathrm{D}^{\alpha}_{L^2}\,S$, (4.4.11) reads in our context

$$
f = f_1 S g_\alpha + f_1' S(g_\alpha * S g_\alpha) + S(g_\alpha * S\varphi^{\perp}),
$$

where f_1, f_1' are scalars and $\varphi^{\perp}$ lies in $g_\alpha{}^{\perp}$. (Recall that α lies in $(\frac{1}{2}, 1)$.) On account of Sect. 4.4.1.4.1, (f_1, f_1') is *the endogenous boundary values of* f *with underlying triplet* ${}^{\mathrm{RL}}\mathcal{T}_\alpha{}^*$ *and the basis* (Sg_α) *of* ker $S\,{}^{\mathrm{RL}}\mathrm{D}^{\alpha}_{L^2}\,S$. By Sect. 4.4.1.4.1 again, we know that (f_1, f_1') is the endogenous boundary values of Sf with underlying triplet ${}^{\mathrm{RL}}\mathcal{T}_\alpha$ and with respect to the basis (g_α) of ker ${}^{\mathrm{RL}}\mathrm{D}^{\alpha}_{L^2}$. Thus (6.2.7) entails that f is continuous on $[0, b)$ and

$$
f_1 = g_{1-\alpha} * Sf(0), \qquad f_1' = \frac{1}{\|g_\alpha\|^2_{L^2}}\big(f(0) - g_{1-\alpha} * Sf(0)g_\alpha(b)\big).
$$

For all purposes, notice that $g_{1-\alpha} * Sf(0)$ denotes the evaluation at $x = 0$ of the function $g_{1-\alpha} * Sf$. The "standard" boundary values of f are then

$$g_{1-\alpha} * Sf(0), \qquad f(0).$$

6.2.1.4 Boundary Restriction Operators and Their Adjoint

Recall that α lies in $(\frac{1}{2}, 1)$.

6.2.1.4.1 Boundary Restriction Operators

Since the domain of the *pivot operator* $A_{{}^{\mathrm{RL}}\mathcal{T}_\alpha}$ of ${}^{\mathrm{RL}}\mathcal{T}_\alpha$ is equal to the range of B_{α,L^2}, and (by Lemma 4.3.3) $\ker \mathrm{D}^1_{L^2}$ is in direct sum with $R(B_{\alpha,L^2})$, we derive from (6.2.6), (6.2.7) that for any u in the domain of ${}^{\mathrm{RL}}\mathrm{D}^\alpha_{L^2}$,

$$\begin{aligned} u \in D(A_{{}^{\mathrm{RL}}\mathcal{T}_\alpha}) &\Longleftrightarrow u_1 = 0 \\ &\Longleftrightarrow g_{1-\alpha} * u(0) = 0. \end{aligned} \tag{6.2.8}$$

Moreover, by Remark 6.2.1,

$$D(A_{{}^{\mathrm{RL}}\mathcal{T}_\alpha}) = \{u \in D({}^{\mathrm{RL}}\mathrm{D}^\alpha_{L^2}) \cap C([0,b]) \mid u(0) = 0\} \tag{6.2.9}$$

$$= D({}^{\mathrm{RL}}\mathrm{D}^\alpha_{L^2}) \cap C([0,b]). \tag{6.2.10}$$

Regarding the *minimal operator* $A_{\mathrm{m},{}^{\mathrm{RL}}\mathcal{T}_\alpha}$, with (6.2.6) and (6.2.7), we obtain

$$\begin{aligned} u \in D(A_{\mathrm{m},{}^{\mathrm{RL}}\mathcal{T}_\alpha}) &\Longleftrightarrow u_1 = u_1' = 0 \\ &\Longleftrightarrow g_{1-\alpha} * u(0) = u(b) = 0. \end{aligned} \tag{6.2.11}$$

By using (6.2.10), we obtain the following simpler characterization of the domain of the minimal operator, namely,

$$D(A_{\mathrm{m},{}^{\mathrm{RL}}\mathcal{T}_\alpha}) = \{u \in D({}^{\mathrm{RL}}\mathrm{D}^\alpha_{L^2}) \cap C([0,b]) \mid u(b) = 0\}. \tag{6.2.12}$$

6.2.1.4.2 Adjoints Operators

Regarding the adjoint of the pivot operator, (4.3.11) tells us that

$$(A_{{}^{\mathrm{RL}}\mathcal{T}_\alpha})^* = S\, A_{{}^{\mathrm{RL}}\mathcal{T}_\alpha}\, S. \tag{6.2.13}$$

In terms of endogenous boundary values, for any f in the domain of $S\,{}^{\mathrm{RL}}\mathrm{D}^\alpha_{L^2}\, S$, recalling that (f_1, f_1') is the endogenous boundary values of Sf with underlying triplet ${}^{\mathrm{RL}}\mathcal{T}_\alpha$ and the basis (g_α) of $\ker {}^{\mathrm{RL}}\mathrm{D}^\alpha_{L^2}$, we get with (6.2.8)

$$f \in D\big((A_{{}^{\mathrm{RL}}\mathcal{T}_\alpha})^*\big) \Longleftrightarrow f_1 = 0$$
$$\Longleftrightarrow g_{1-\alpha} * Sf(0) = 0.$$

With (6.2.10), we derive also that f lies in the domain of $(A_{{}^{\mathrm{RL}}\mathcal{T}_\alpha})^*$ if and only if f is continuous on $[0, b]$.

The adjoint of the maximal operator is given by Proposition 4.3.10, i.e.,

$$({}^{\mathrm{RL}}\mathrm{D}^\alpha_{L^2})^* = S\, A_{\mathrm{m},{}^{\mathrm{RL}}\mathcal{T}_\alpha}\, S.$$

Thus with (6.2.12)

$$D\big(({}^{\mathrm{RL}}\mathrm{D}^\alpha_{L^2})^*\big) = \{f \in D({}^{\mathrm{RL}}\mathrm{D}^\alpha_{L^2}\, S) \cap C([0, b]) \mid f(0) = 0\}. \tag{6.2.14}$$

In terms of endogenous boundary values, for any f in the domain of $S\, {}^{\mathrm{RL}}\mathrm{D}^\alpha_{L^2}\, S$, one has

$$f \in D\big(({}^{\mathrm{RL}}\mathrm{D}^\alpha_{L^2})^*\big) \Longleftrightarrow f_1 = f_1' = 0 \tag{6.2.15}$$
$$\Longleftrightarrow g_{1-\alpha} * Sf(0) = f(0) = 0.$$

In a same way, the adjoint of the minimal operator is given by Corollary 4.3.11, i.e.,

$$(A_{\mathrm{m},{}^{\mathrm{RL}}\mathcal{T}_\alpha})^* = S\, {}^{\mathrm{RL}}\mathrm{D}^\alpha_{L^2}\, S.$$

6.2.2 The Triplet $\mathcal{T}_{\alpha,\mathbb{K}}$

For the sake of simplicity, this triplet will also be denoted by $\mathcal{T}_\alpha$. We will construct *a Caputo extension* of ${}^{\mathrm{RL}}\mathcal{T}_\alpha$. We recall that Caputo extensions of differential triplets are introduced in Sect. 5.6.4. The outline is the same than the one of Sect. 6.2.1. However let us emphasize that we introduce the celebrated *Caputo derivative* as a boundary restriction operator of $\mathcal{T}_\alpha$, so that as boundary restriction operators of the same differential triplet, *Caputo and Riemann-Liouville derivatives differ only by their boundary conditions.*

6.2.2.1 Definition

We would like to consider the Caputo extension of $\mathbb{A} := {}^{\mathrm{RL}}\mathrm{D}^\alpha_{L^2}$ with respect to $N := \langle g_1 \rangle$. For, it is enough that g_1 does not belong to $D({}^{\mathrm{RL}}\mathrm{D}^\alpha_{L^2})$, i.e., that α is greater or equal to $\frac{1}{2}$. In order to restrict our attention to the more interesting situation, we will assume here again that (6.2.5) holds, that is,

$$\alpha \in (\frac{1}{2}, 1).$$

Denoting by $\mathrm{D}^{\alpha}_{L^2}$ the Caputo extension of ${}^{\mathrm{RL}}\mathrm{D}^{\alpha}_{L^2}$ with respect to $\langle g_1 \rangle$, there results from Sect. 5.6.4.3 that

$$\mathcal{T}_\alpha := (\mathrm{D}^{\alpha}_{L^2}, B_{\alpha,L^2}, S)$$

is a differential triplet on $L^2(0,b)$. For all purposes, let us remind the reader that by definition of Caputo extension, for each u in $D(\mathrm{D}^{\alpha}_{L^2})$, there exists a unique scalar u_1 such that $u - u_1 g_1$ lies in the domain of ${}^{\mathrm{RL}}\mathrm{D}^{\alpha}_{L^2}$, and

$$\mathrm{D}^{\alpha}_{L^2} u := \frac{\mathrm{d}}{\mathrm{d}x}\big\{g_{1-\alpha} * (u - u_1 g_1)\big\}. \tag{6.2.16}$$

6.2.2.2 Endogenous Boundary Values

According to Proposition 5.6.9,

$$\ker \mathrm{D}^{\alpha}_{L^2} = \langle g_1, g_\alpha \rangle.$$

We infer from (4.4.21) that each u in $D(\mathrm{D}^{\alpha}_{L^2})$ reads

$$u = u_1 g_1 + u_2 g_\alpha + u_1' g_{1+\alpha} + u_2' g_\alpha * S g_\alpha + g_\alpha * S\psi^\perp \tag{6.2.17}$$

where u_1, u_2, u_1', u_2' are scalars and $\psi^\perp$ lies in $\langle g_1, g_\alpha \rangle^\perp$. In view of Definition 4.4.2, (u_1, u_2, u_1', u_2') is the *endogenous boundary values of u with respect to the basis* (g_1, g_α) *of* $\ker \mathrm{D}^{\alpha}_{L^2}$. We claim that $u - g_{1-\alpha} * u(0)\, g_\alpha$ and $g_{1-\alpha} * u$ are continuous on $[0, b]$, u is continuous on $(0, b]$, and

$$\begin{aligned} u_1 &= \big(u - g_{1-\alpha} * u(0)\, g_\alpha\big)(0) \\ u_2 &= g_{1-\alpha} * u(0) \\ u_1' b + u_2' g_{1+\alpha}(b) &= g_{1-\alpha} * \big(u - u_1 g_1\big)(b) - g_{1-\alpha} * u(0) \\ u_1 + u_2 g_\alpha(b) + u_1' g_{1+\alpha}(b) + u_2' \|g_\alpha\|^2 &= u(b). \end{aligned} \tag{6.2.18}$$

In order to compute u_2, we convole (6.2.17) by $g_{1-\alpha}$ and evaluate at $x = 0$. Now the continuity of $u - g_{1-\alpha} * u(0)\, g_\alpha$ and the value of u_1 are obtained easily thanks to Remark 3.3.4. Regarding the u_j''s, starting from (6.2.17) and using $\mathrm{D}^{\alpha}_{L^2} B_{\alpha,L^2} = \mathrm{id}_{L^2(0,b)}$, we get the identity

$$\mathrm{D}^{\alpha}_{L^2} u = u_1' g_1 + u_2' S g_\alpha + S\psi^\perp.$$

Using $(S g_\alpha, g_1) = g_{1+\alpha}(b)$, and $(S\psi^\perp, g_1) = (\psi^\perp, g_1) = 0$, we get

$$(\mathrm{D}^{\alpha}_{L^2}u, g_1) = u_1' b + u_2' g_{1+\alpha}(b). \tag{6.2.19}$$

Then the definition of $\mathrm{D}^{\alpha}_{L^2}$ entails the third equation. Finally, since $\alpha > \frac{1}{2}$, Remark 3.3.4 yields that $g_\alpha * \mathrm{D}^{\alpha}_{L^2}u$ and $g_\alpha * S\psi^{\perp}$ are continuous on $[0, b]$, and also

$$(g_\alpha * Sg_\alpha)(b) = \|g_\alpha\|^2, \qquad (g_\alpha * S\psi^{\perp})(b) = (\psi^{\perp}, g_\alpha) = 0.$$

Thus with (6.2.17), u is continuous on $(0, b]$ and the last equation of (6.2.18) follows by evaluation at $x = b$, which proves the claim.

Let u belong to the domain of $\mathrm{D}^{\alpha}_{L^2}$. There results from the invertibility of the linear system (6.2.18) that

$$u_1 := \big(u - g_{1-\alpha} * u(0)\, g_\alpha\big)(0), \qquad g_{1-\alpha} * u(0), \qquad u(b), \qquad g_{1-\alpha} * \big(u - u_1 g_1\big)(b)$$

are *"standard" boundary values of* u. Recall that by definition of Caputo extension, u_1 is the unique scalar for which $g_{1-\alpha} * (u - u_1 g_1)$ lies in $H^1(0, b)$.

Let us notice that the second boundary value may be read also

$$g_{1-\alpha} * u(0) = g_{1-\alpha} * \big(u - u_1 g_1\big)(0). \tag{6.2.20}$$

In addition, if u is continuous on $[0, b]$ then u_1 is simply equal to $u(0)$, and the "standard" boundary values of u are

$$u(0), \qquad 0, \qquad u(b), \qquad g_{1-\alpha} * \big(u - u(0) g_1\big)(b).$$

6.2.2.3 The Adjoint Triplet of $\mathcal{T}_\alpha$

Recall that α is assumed to lie in $(\frac{1}{2}, 1)$. By Definition 4.2.3, the adjoint differential triplet ${\mathcal{T}_\alpha}^*$ of $\mathcal{T}_\alpha$ reads

$${\mathcal{T}_\alpha}^* := (S\,\mathrm{D}^{\alpha}_{L^2}\, S,\ S\, B_{\alpha,L^2}\, S,\ S).$$

One has

$$\ker S\,\mathrm{D}^{\alpha}_{L^2}\, S = \langle g_1, Sg_\alpha\rangle.$$

Regarding endogenous boundary values, for any f in the domain of $S\,\mathrm{D}^{\alpha}_{L^2}\, S$, (4.4.11) reads in our context

$$f = f_1 g_1 + f_2 S g_\alpha + f_1' S g_{1+\alpha} + f_2' S(g_\alpha * Sg_\alpha) + S(g_\alpha * S\varphi^{\perp}), \tag{6.2.21}$$

where f_1, f_2, f_1', f_2' are scalars and $\varphi^\perp$ lies in $(\ker \mathrm{D}^\alpha_{L^2})^\perp$. On account of Sect. 4.4.1.4.1, (f_1, f_2, f_1', f_2') is *the endogenous boundary values of f with underlying triplet $\mathcal{T}_\alpha{}^*$ and with respect to the basis (g_1, Sg_α) of* $\ker S\,\mathrm{D}^\alpha_{L^2}\,S$. By Sect. 4.4.1.4.1 again, we know that (f_1, f_1', f_1', f_2') is the endogenous boundary values of Sf with respect to $\mathcal{T}_\alpha$ and the basis (g_1, g_α) of $\ker \mathrm{D}^\alpha_{L^2}$. Thus (6.2.18) entails that $Sf - g_{1-\alpha} * Sf(0)\,g_\alpha$ and $g_{1-\alpha} * Sf$ are continuous on $[0, b]$, f is continuous on $[0, b)$, and

$$\begin{aligned} f_1 &= \big(Sf - g_{1-\alpha} * Sf(0)\,g_\alpha\big)(0) \\ f_2 &= g_{1-\alpha} * Sf(0) \\ f_1'b + f_2'g_{1+\alpha}(b) &= g_{1-\alpha} * \big(Sf - f_1 g_1\big)(b) - g_{1-\alpha} * Sf(0) \\ f_1 + f_2 g_\alpha(b) + f_1' g_{1+\alpha}(b) + f_2'\|g_\alpha\|^2_{L^2} &= f(0). \end{aligned} \tag{6.2.22}$$

6.2.2.4 Boundary Restriction Operators and Their Adjoint

Recall that α lies in $(\frac{1}{2}, 1)$.

6.2.2.4.1 Boundary Restriction Operators

6.2.2.4.1.1 Pivot and Minimal Operators

Since the domain of the *pivot operator* $A_{\mathcal{T}_\alpha}$ of $\mathcal{T}_\alpha$ is equal to the range of B_{α,L^2}, we derive from (6.2.17), (6.2.18) that for any u in the domain of $\mathrm{D}^\alpha_{L^2}$,

$$\begin{aligned} u \in D(A_{\mathcal{T}_\alpha}) &\iff u_1 = u_2 = 0 \\ &\iff u(0) = g_{1-\alpha} * u(0) = 0. \end{aligned}$$

Moreover, we may check easily (see alternatively Proposition 5.6.10 (ii)) that the pivot operators of $\mathcal{T}_\alpha$ and ${}^{\mathrm{RL}}\mathcal{T}_\alpha$ coincide, which reads in symbol

$$A_{\mathcal{T}_\alpha} = A_{{}^{\mathrm{RL}}\mathcal{T}_\alpha}. \tag{6.2.23}$$

Thus with (6.2.9), (6.2.10),

$$\begin{aligned} D(A_{\mathcal{T}_\alpha}) &= \{u \in D(\mathrm{D}^\alpha_{L^2}) \cap C([0, b]) \mid u(0) = 0\} \\ &= D({}^{\mathrm{RL}}\mathrm{D}^\alpha_{L^2}) \cap C([0, b]). \end{aligned} \tag{6.2.24}$$

Regarding the *minimal operator* $A_{\mathrm{m},\mathcal{T}_\alpha}$, with (6.2.17) and (6.2.18), we derive that for any u in the domain of $\mathrm{D}^\alpha_{L^2}$,

$$\begin{aligned} u \in D(A_{\mathrm{m},\mathcal{T}_\alpha}) &\iff u_1 = u_2 = u_1' = u_2' = 0 \\ &\iff u(0) = u(b) = g_{1-\alpha} * u(0) = g_{1-\alpha} * u(b) = 0. \end{aligned} \tag{6.2.25}$$

Thus,

$$D(A_{\mathrm{m},\mathcal{T}_\alpha}) = \{u \in D(\mathrm{D}^\alpha_{L^2}) \cap C([0,b]) \mid u(0) = u(b) = g_{1-\alpha} * u(b) = 0\}.$$

By comparison with (6.2.11), there results that the minimal operator of $\mathcal{T}_\alpha$ is a proper extension of the minimal operator of ${}^{\mathrm{RL}}\mathcal{T}_\alpha$.

6.2.2.4.1.2 Two Specials Boundary Restriction Operators

The first operator we would like to introduce is already known and is nothing but ${}^{\mathrm{RL}}\mathrm{D}^\alpha_{L^2}$. By (6.2.23),

$$A_{\mathcal{T}_\alpha} \subseteq {}^{\mathrm{RL}}\mathrm{D}^\alpha_{L^2} \subseteq \mathrm{D}^\alpha_{L^2},$$

thus Proposition 4.3.5 warrants that ${}^{\mathrm{RL}}\mathrm{D}^\alpha_{L^2}$ is a boundary restriction operator of $\mathcal{T}_\alpha$. Moreover ${}^{\mathrm{RL}}\mathrm{D}^\ell_{\mathcal{V}}$ is closed due to Proposition 4.4.1.

Regarding endogenous boundary values, let u belong to $D(\mathrm{D}^\alpha_{L^2})$. Then in view of (6.2.17), u lies in $D({}^{\mathrm{RL}}\mathrm{D}^\alpha_{L^2})$ if and only if $u_1 = 0$.

Our second *special* operator denoted by ${}^{\mathrm{C}}\mathrm{D}^\alpha_{L^2}$ is the restriction of $\mathrm{D}^\alpha_{L^2}$ with domain

$$D({}^{\mathrm{C}}\mathrm{D}^\alpha_{L^2}) := \langle g_1 \rangle \oplus R(B_{\ell,\mathcal{V}}). \tag{6.2.26}$$

By (6.2.17) and Remark 3.3.4,

$$D({}^{\mathrm{C}}\mathrm{D}^\alpha_{L^2}) \subset C([0,b]), \tag{6.2.27}$$

thus as seen at the end of Sect. 6.2.2.2, $u_1 = u(0)$ for each u in $D({}^{\mathrm{C}}\mathrm{D}^\alpha_{L^2})$. Then by definition of the Caputo extension of an operator

$${}^{\mathrm{C}}\mathrm{D}^\alpha_{L^2} u = \frac{\mathrm{d}}{\mathrm{d}x}\big\{g_{1-\alpha} * \big(u - u(0) g_1\big)\big\},$$

which may be also abbreviated by ${}^{\mathrm{C}}\mathrm{D}^\alpha_{L^2} u = \frac{\mathrm{d}}{\mathrm{d}x}\big\{g_{1-\alpha} * \big(u - u(0)\big)\big\}$. Whence ${}^{\mathrm{C}}\mathrm{D}^\alpha_{L^2} u$ is the celebrated *Caputo derivative of order* α *of* u *on* $L^2(0,b)$.

Remark 6.2.2 Let u belong to the domain of $\mathrm{D}^\alpha_{L^2}$. If in addition, u lies in $H^1(0,b)$ then Proposition 3.3.4 yields that $u - u(0)g_1$ belongs to the domain of ${}^{\mathrm{RL}}\mathrm{D}^\alpha_{L^2}$. Thus u lies in the domain of ${}^{\mathrm{C}}\mathrm{D}^\alpha_{L^2}$ and again by Proposition 3.3.4,

$${}^{\mathrm{C}}\mathrm{D}^\alpha_{L^2} u = g_{1-\alpha} * \big(\frac{\mathrm{d}}{\mathrm{d}x} u\big).$$

This well-known identity serves sometimes as a definition of Caputo derivatives. However, in addition to its lack of generality, this definition should be avoided because in the same way that one does not define the first derivative of a function assuming it is twice differentiable, it is clumsy to define a derivative of order less than one of a function assuming that this function admits a first-order derivative. □

Regarding endogenous boundary values, let u belong to $D(\mathrm{D}^{\alpha}_{L^2})$. Then in view of (6.2.17),

$$u \text{ lies in } D({}^{\mathrm{C}}\mathrm{D}^{\alpha}_{L^2}) \text{ if and only if } u_2 = 0. \tag{6.2.28}$$

By comparison with ${}^{\mathrm{RL}}\mathrm{D}^{\alpha}_{L^2}$, we see that in our framework, *Riemann-Liouville* and *Caputo operators* (i.e., ${}^{\mathrm{RL}}\mathrm{D}^{\alpha}_{L^2}$ and $\mathrm{D}^{\alpha}_{L^2}$) are the restriction of the same operator $\mathrm{D}^{\alpha}_{L^2}$. Their domains differ only by the boundary values of their elements. Moreover, Riemann-Liouville and Caputo operators coincide on the domain of $A_{\mathrm{RL}\mathcal{T}_\alpha}$. More precisely, by (6.2.17), (6.2.23) and the first item of the Domain Structure Theorem 4.3.14,

$$D({}^{\mathrm{RL}}\mathrm{D}^{\alpha}_{L^2}) \cap D({}^{\mathrm{C}}\mathrm{D}^{\alpha}_{L^2}) = D(A_{\mathrm{RL}\mathcal{T}_\alpha})$$

and

$${}^{\mathrm{C}}\mathrm{D}^{\alpha}_{L^2}u = {}^{\mathrm{RL}}\mathrm{D}^{\alpha}_{L^2}u, \qquad \forall\, u \in D(A_{\mathrm{RL}\mathcal{T}_\alpha}).$$

We claim that ${}^{\mathrm{C}}\mathrm{D}^{\ell}_{\mathcal{V}}$ *is a closed boundary restriction operator of* $\mathcal{T}_\alpha$. Indeed, applying Proposition 4.4.1 with $\mathcal{E} := L^2(0, b)$, we deduce that ${}^{\mathrm{C}}\mathrm{D}^{\alpha}_{L^2}$ is a closed operator. Moreover, (6.2.25) entails that

$$A_{\mathrm{m},\mathcal{T}_\alpha} \subseteq {}^{\mathrm{C}}\mathrm{D}^{\alpha}_{L^2} \subseteq \mathrm{D}^{\alpha}_{L^2}.$$

Then the claim follows from Theorem 4.3.12.

6.2.2.4.2 Adjoints Operators

Regarding the adjoint of the pivot operator, (6.2.13) and (6.2.23) tell us that

$$(A_{\mathcal{T}_\alpha})^* = S\, A_{\mathrm{RL}\mathcal{T}_\alpha} S.$$

Thus, for each f in the domain of $(A_{\mathcal{T}_\alpha})^*$, one has

$$(A_{\mathcal{T}_\alpha})^* f = S\,\mathrm{D}^{1}_{L^2}\, B_{1-\alpha,L^2}\, Sf. \tag{6.2.29}$$

In terms of endogenous boundary values, for any f in the domain of $S\,\mathrm{D}^{\alpha}_{L^2}\, S$, one has

$$f \in D\big((A_{\mathcal{T}_\alpha})^*\big) \iff f_1 = f_2 = 0$$
$$\iff f(b) = g_{1-\alpha} * Sf(0) = 0.$$

The adjoint of the maximal operator is given by Proposition 4.3.10, i.e.,

$$(\mathrm{D}^\alpha_{L^2})^* = S\, A_{\mathrm{m},\mathcal{T}_\alpha}\, S.$$

In a same way, the adjoint of the minimal operator is given by Corollary 4.3.11, i.e.,

$$(A_{\mathrm{m},\mathcal{T}_\alpha})^* = S\, \mathrm{D}^\alpha_{L^2}\, S.$$

Now consider $({}^{\mathrm{RL}}\mathrm{D}^\alpha_{L^2})^*$. Let us characterize the domain of $({}^{\mathrm{RL}}\mathrm{D}^\alpha_{L^2})^*$ in terms of endogenous boundary values with $\mathcal{T}_\alpha{}^*$ as underlying triplet. Observe that in Sect. 6.2.1.4.2, the characterization occurs through another differential triplet, namely, ${}^{\mathrm{RL}}\mathcal{T}_\alpha{}^*$.

Let f belong to $D(\mathrm{D}^\alpha_{L^2}\, S)$ and (f_1, f_2, f_1', f_2') be its endogenous boundary values given by (6.2.21). We claim that

$$f \in D\big(({}^{\mathrm{RL}}\mathrm{D}^\ell_{\mathcal{V}})^*\big) \iff f_1 = f_2 = f_2' = 0$$
$$\iff f(b) = g_{1-\alpha} * Sf(0) = f(0) = 0.$$

Indeed by (6.2.15), f belongs to the domain of $({}^{\mathrm{RL}}\mathrm{D}^\alpha_{L^2})^*$ if and only if there exists $\phi^\perp$ in $g_\alpha{}^\perp$ such that

$$f = S\, B_{\alpha, L^2}\, S\phi^\perp.$$

With the representation (6.2.21) and a direct sum argument, this is equivalent to

$$f_1 = f_2 = 0, \qquad f_1' g_1 + f_2' g_\alpha + \varphi^\perp \in g_\alpha{}^\perp.$$

Since $\varphi^\perp$ lies in $\langle g_1, g_\alpha\rangle^\perp$, it belongs to $g_\alpha{}^\perp$. Thus with the fourth equation of (6.2.22), the latter conditions are equivalent to

$$f(b) = g_{1-\alpha} * Sf(0) = f(0) = 0,$$

which completes the proof of the claim. This result can also be to obtained (in a nontricky way) by using methods featured in Chap. 5: see the proof of (6.7.59).

With regard to the adjoint of ${}^{\mathrm{C}}\mathrm{D}^\alpha_{L^2}$, we have the following result whose proof relies on the theory of differential triplets with maximal operator having finite-dimensional kernel (see Sect. 4.4). Proposition 6.2.1 turns out to be a particular case of Proposition 6.7.3. However, the proof of Proposition 6.7.3 relies on Theorem 5.4.1 and holds for operators defined on reflexive Lebesgue spaces (where differential quadruplets replace differential triplets). The proof of Proposition 6.2.1 is not essentially different from the one of Proposition 6.7.3, but it is simpler since it takes place on an Hilbertian framework and relies only on differential triplets.

Proposition 6.2.1 *Let $\mathbb{K}$ be equal to $\mathbb{R}$ or $\mathbb{C}$, and f belong to the domain of $S\mathrm{D}^{\alpha}_{L^2}S$. Then*

$$f \in D\big(({}^{\mathrm{C}}\mathrm{D}^{\alpha}_{L^2})^*\big) \Longleftrightarrow f(b) = g_{1-\alpha} * Sf(0) = g_{1-\alpha} * Sf(b) = 0 \tag{6.2.30}$$

and

$$D\big(({}^{\mathrm{C}}\mathrm{D}^{\alpha}_{L^2})^*\big) = \{f \in C([0,b]) \mid g_{1-\alpha} * Sf \in H^1_0(0,b),\ f(b) = 0\} \tag{6.2.31}$$

$$= \{f \in D({}^{\mathrm{RL}}\mathrm{D}^{\alpha}_{L^2}\, S) \cap C([0,b]) \mid f(b) = g_{1-\alpha} * Sf(b) = 0\}. \tag{6.2.32}$$

Proof Equation (6.2.31) and (6.2.32) follows easily from (6.2.30). So let us prove the latter equivalence. With the notations of Sect. 4.4, and in particular (4.4.1), (4.4.2), one has $\mathbb{A} = \mathrm{D}^{\alpha}_{L^2}$, $d = 2$, $A = {}^{\mathrm{C}}\mathrm{D}^{\alpha}_{L^2}$, $d_E = 3$,

$$E = \langle\, g_1,\ g_{1+\alpha},\ g_\alpha * Sg_\alpha\,\rangle$$

and

$$\begin{array}{lll}
\xi_1 = g_1 & \xi_2 = g_\alpha & \\
e_1 = g_1 & \omega_1 = g_1 & \omega_1' = 0 \\
e_2 = B_{\alpha,L^2} g_1 & \omega_2 = 0 & \omega_2' = g_1 \\
e_3 = B_{\alpha,L^2} S g_\alpha & \omega_3 = 0 & \omega_3' = g_\alpha.
\end{array}$$

Then (4.4.13) in Proposition 4.4.5 yields that, in view of the notation (6.2.21),

$$\begin{aligned}
f \in D(({}^{\mathrm{C}}\mathrm{D}^{\alpha}_{L^2})^*) &\Longleftrightarrow 0 = (g_1,\ f_i'\xi_i) \ \text{and}\ (g_1,\ f_i\xi_i) = (g_\alpha,\ f_i\xi_i) = 0 \\
&\Longleftrightarrow (f_1'g_1 + f_2'g_\alpha,\ g_1) = 0 \ \text{and}\ f_1 = f_2 = 0.
\end{aligned}$$

Moreover, by the third equation of (6.2.22) and $f_1 = f_2 = 0$,

$$(f_1'g_1 + f_2'g_\alpha,\ g_1) = g_{1-\alpha} * Sf(b).$$

Thus (6.2.30) follows from (6.2.22). □

Remark 6.2.3 Since ${}^{\mathrm{C}}\mathrm{D}^{\alpha}_{L^2}$ is an extension of the pivot operator, we derive from (6.2.29) that for each f in the domain of $({}^{\mathrm{C}}\mathrm{D}^{\alpha}_{L^2})^*$,

$$({}^{\mathrm{C}}\mathrm{D}^{\alpha}_{L^2})^* f = S\,\mathrm{D}^{1}_{L^2}\, B_{1-\alpha,L^2}\, Sf. \tag{6.2.33}$$

Let us try to recover this identity by direct calculations. For, let u belong to $D({}^{\mathrm{C}}\mathrm{D}^{\alpha}_{L^2})$ and v lie in $H^1(0,b)$ with $v(b) = 0$. Combining (6.2.20) and (6.2.28), we get $g_{1-\alpha} * \big(u - u(0)\big)(0) = 0$. Thus by integration by parts, setting $B_{1-\alpha} := B_{1-\alpha,L^2}$, we get with $v(b) = 0$,

$$\int_0^b {}^{\mathrm{C}}\mathrm{D}^{\alpha}_{L^2} u\, v\, \mathrm{d}x = -\int_0^b B_{1-\alpha}\big(u - u(0)\big)\mathrm{D}^{1}_{L^2} v\, \mathrm{d}x.$$

Moreover, with (6.2.3)

$$\int_0^b B_{1-\alpha}\big(u - u(0)\big)\mathrm{D}^{1}_{L^2} v\, \mathrm{d}x =$$

$$\int_0^b u\, SB_{1-\alpha}S\mathrm{D}^{1}_{L^2} v\, \mathrm{d}x - u(0)\int_0^b SB_{1-\alpha}S\mathrm{D}^{1}_{L^2} v\, \mathrm{d}x.$$

Now $\mathrm{D}^{1}_{L^2}$ anti-commute with S, and according to Proposition 3.3.4,

$$B_{1-\alpha}\mathrm{D}^{1}_{L^2} Sv = \mathrm{D}^{1}_{L^2} B_{1-\alpha} Sv$$

since $Sv(0) = v(b) = 0$. Thus, $SB_{1-\alpha}S\mathrm{D}^{1}_{L^2} v = -S\mathrm{D}^{1}_{L^2} B_{1-\alpha} Sv$, so that

$$\int_0^b SB_{1-\alpha}S\mathrm{D}^{1}_{L^2} v\, \mathrm{d}x = -B_{1-\alpha} Sv(b).$$

Therefore, we conclude that if v lies in $H^1(0,b)$ and satisfies

$$v(b) = g_{1-\alpha} * Sv(b) = 0 \tag{6.2.34}$$

then v belong to the domain of $({}^{\mathrm{C}}\mathrm{D}^{\alpha}_{L^2})^*$ and

$$({}^{\mathrm{C}}\mathrm{D}^{\alpha}_{L^2})^* v = S\mathrm{D}^{\alpha}_{L^2} Sv.$$

In other words, we have proved that the space of all functions v in $H^1(0,b)$ satisfying (6.2.34) is contained in the domain of $({}^{\mathrm{C}}\mathrm{D}^{\alpha}_{L^2})^*$. At this stage, the domain of $({}^{\mathrm{C}}\mathrm{D}^{\alpha}_{L^2})^*$ is not identified.

In this remark, we start with a smooth v and look for the boundary conditions insuring that v lies in the domain of $({}^{\mathrm{C}}\mathrm{D}^{\alpha}_{L^2})^*$. In the proof of Proposition 6.2.1, our approach for computing the domain of this operator is different. Indeed, taking advantage of the simple structure of the fractional differential operators under consideration, we start from a function lying in a larger space than $D\big(({}^{\mathrm{C}}\mathrm{D}^{\alpha}_{L^2})^*\big)$, namely, the domain of $S\mathrm{D}^{\alpha}_{L^2}S$, and compute the suitable boundary conditions (see Theorem 4.4.4 and Proposition 4.4.5). □

6.3 Sonine Kernels

For any number α in $(0, 1)$, the basic differential triplet ${}^{\mathrm{RL}}\mathcal{T}_{\alpha,\mathbb{K}}$ introduced in Sect. 6.2.1 is built upon the functions $g_{1-\alpha}$ and g_α. Observing that the key property in the definition of Riemann-Liouville derivatives is the identity (3.4.5), we are led to the concept of *Sonine kernels* in order to construct more general fractional differential triplets and quadruplets.

Let $\mathbb{K}$ be equal to $\mathbb{R}$ or $\mathbb{C}$ and b be a positive real number. Recalling that $L^1(0, b)$ is a Banach space over $\mathbb{K}$ of functions with values in $\mathbb{K}$, $L^1\big((0, b), \mathbb{R}\big)$ is the subspace of $L^1(0, b)$ restricted to *real valued* functions.

Definition 6.3.1 A function k in $L^1\big((0, b), \mathbb{R}\big)$ is called *a Sonine kernel* if there exists ℓ in $L^1(0, b)$ such that

$$k * \ell = g_1 \qquad \text{in } L^1(0, b). \tag{6.3.1}$$

Moreover, (k, ℓ) is *a Sonine pair* if k lies in $L^1\big((0, b), \mathbb{R}\big)$, ℓ belongs $L^1(0, b)$, and (6.3.1) holds.

Finding Sonine kernels and Sonine pairs is a difficult problem. However, (k, ℓ) is known to be a Sonine pair in the following cases:

$$\begin{aligned}
k &= g_{1-\alpha}, & \ell &= g_\alpha \\
k &= g_{1-\alpha}e^{-a\cdot}, & \ell &= g_\alpha + a g_1 * (g_\alpha e^{-a\cdot}) \\
k(x) &= (\pi x)^{-1/2}\cosh(2x^{1/2}), & \ell(x) &= (\pi x)^{-1/2}\cos(2x^{1/2}), \qquad \forall x \in (0, b).
\end{aligned}$$

Here (α, a) ranges in $(0, 1) \times [0, \infty)$. The first pair is the most famous. It results from the second one by choosing $a = 0$. Observe from the third pair that a Sonine kernel must not have a sign nor to be monotone (since ℓ is also a Sonine kernel by Proposition 6.3.1 below).

Also a quiet large class of decreasing functions is formed of Sonine kernels. We refer the reader to [CM88], [GLS90, Chap. 9, Corollary 8.8], [Han20] for more information.

Since *Laplace transform* changes convolution into product, a standard procedure to check that a given pair (k, ℓ) is a Sonine pair is to compute the Laplace transform of k and ℓ and to verify that their product returns the inverse function $s \mapsto s^{-1}$. □

Let us now gather some properties of Sonine kernels.

Proposition 6.3.1 *Let k, ℓ belong to $L^1(0, b)$.*

(i) *If k is a Sonine kernel, then there exists a unique function ℓ in $L^1(0, b)$ such that* (6.3.1) *holds. Moreover ℓ is real-valued.*
(ii) *If (k, ℓ) is a Sonine pair then the following assertions hold.*
  (ii-a) *(ℓ, k) is a Sonine pair.*
  (ii-b) *If p, q are conjugate exponents in* $[1, \infty]$ *and ℓ lies in $L^p(0, b)$, then k does not belong to $L^q(0, b)$.*
  (ii-c) *ℓ and k are not essentially bounded near zero.*

Proof (i) Let ℓ and ℓ' belong to $L^1(0, b)$ and satisfy (6.3.1). Then convoluting the two sides of the identity $k * (\ell - \ell') = 0$ by ℓ, we derive

$$g_1 * (\ell - \ell') = 0.$$

Thus $\ell = \ell'$, which gives uniqueness. Now since k is real-valued, (6.3.1) implies

$$k * \overline{\ell} = g_1.$$

Hence, by uniqueness, ℓ is real-valued. (ii-a) is a straightforward consequence of (i). In order to prove Item (ii-b), consider ℓ in $L^p(0, b)$. Arguing by contradiction, we suppose that k lies in $L^q(0, b)$. Then $k * \ell$ is a continuous function vanishing at $x = 0$ according to Remark 3.3.4. This contradicts (6.3.1), so that k does not belong to $L^q(0, b)$.

Finally let us prove (ii-c). By symmetry, it is enough to consider ℓ. Since k lies in $L^1(0, b)$, (ii-b) entails that for each x in $(0, b]$, ℓ does not belong to $L^\infty(0, x)$. That is to say, ℓ is not essentially bounded near zero. □

6.4 Entire and Fractional Differential Operators on Lebesgue Spaces

The theory of boundary restriction operator takes place in *reflexive* Banach spaces. This theory allows an abstract approach for boundary conditions. If we restrict our attention to *initial* conditions, then reflexivity is more needed at an abstract level (see Sect. 7.2.3).

Let be equal to $\mathbb{R}$ or $\mathbb{C}$, $\mathcal{Y}$ be a nontrivial Banach space on $\mathbb{K}$, and (b, p) belong to $(0, \infty) \times [1, \infty)$. Setting

$$\mathcal{X} = L^p(\mathcal{Y}) := L^p\big((0, b), \mathcal{Y}\big),$$

we define for each positive integer n the usual derivative of order n

$$\mathrm{D}^n_{\mathcal{X}} : W^{n,p}(\mathcal{Y}) \subseteq L^p(\mathcal{Y}) \to L^p(\mathcal{Y}), \qquad u \mapsto \frac{\mathrm{d}^n}{\mathrm{d}x^n} u.$$

For any function f in $L^1(0, b)$, we put

$$B_{f,\mathcal{X}} : L^p(\mathcal{Y}) \to L^p(\mathcal{Y}), \qquad v \mapsto f * v.$$

If $f := g_\beta$ then we abbreviate by denoting

$$B_{\beta,\mathcal{X}} := B_{g_\beta,\mathcal{X}}.$$

Regarding *fractional* differential operators, for any Sonine pair (k, ℓ), we define

$$D({}^{\mathrm{RL}}\mathrm{D}^\ell_{\mathcal{X}}) := \big\{u \in L^p(\mathcal{Y}) \mid k * u \in W^{1,p}(\mathcal{Y})\big\}$$

$${}^{\mathrm{RL}}\mathrm{D}^\ell_{\mathcal{X}} : D({}^{\mathrm{RL}}\mathrm{D}^\ell_{\mathcal{X}}) \subseteq L^p(\mathcal{Y}) \to L^p(\mathcal{Y}), \qquad u \mapsto \frac{\mathrm{d}}{\mathrm{d}x}\{k * u\}.$$

For each u in $D({}^{\mathrm{RL}}\mathrm{D}^\ell_{\mathcal{X}})$, ${}^{\mathrm{RL}}\mathrm{D}^\ell_{\mathcal{X}} u$ is called the *Riemann-Liouville derivative of* u *with respect to* ℓ. Of course, when $\mathcal{Y} = \mathbb{K}$, $p = 2$ and $(k, \ell) = (g_{1-\alpha}, g_\alpha)$, we recover the usual Riemann-Liouville derivative of order α of u as defined in Sect. 6.2.1.1. The operator ${}^{\mathrm{RL}}\mathrm{D}^\ell_{\mathcal{X}}$ on $L^p(\mathcal{Y})$ is called *the Riemann-Liouville operator with respect to* ℓ.

The following fundamental identity holds:

$${}^{\mathrm{RL}}\mathrm{D}^\ell_{\mathcal{X}} \circ B_{\ell,\mathcal{X}} = \mathrm{id}_{\mathcal{X}}. \tag{6.4.1}$$

Indeed for each h in $L^p(\mathcal{Y})$, (6.3.1) yields $k * B_{\ell,\mathcal{X}} h = g_1 * h$. Hence $B_{\ell,\mathcal{X}} h$ lies in the domain of ${}^{\mathrm{RL}}\mathrm{D}^\ell_{\mathcal{X}}$ and ${}^{\mathrm{RL}}\mathrm{D}^\ell_{\mathcal{X}} B_{\ell,\mathcal{X}} h = h$, which proves (6.4.1).

Now we will introduce a Caputo extension of ${}^{\mathrm{RL}}\mathrm{D}^\ell_{\mathcal{X}}$.

Proposition 6.4.1 *Let (k, ℓ) be a Sonine pair. Then $g_1\mathcal{Y}$ is in direct sum with $D({}^{\mathrm{RL}}\mathrm{D}^\ell_{\mathcal{X}})$ if and only if k does not belong to $L^p(0, b)$.*

Proof For each Y in $\mathcal{Y}$,

$$g_1 Y \in D({}^{\mathrm{RL}}\mathrm{D}^\ell_{\mathcal{X}}) \iff k \in L^p(0, b) \quad \text{or} \quad Y = 0.$$

Thus the equivalence follows. □

In view of Proposition 6.4.1, we put

$$N := \begin{cases} g_1\mathcal{Y} & \text{if } k \notin L^p(0,b) \\ \{0\} & \text{otherwise} \end{cases} \tag{6.4.2}$$

and denote by $\mathrm{D}^{\ell}_{\mathcal{X}}$ the *Caputo extension of* ${}^{\mathrm{RL}}\mathrm{D}^{\ell}_{\mathcal{X}}$ *with respect to* N. Finally, we denote by ${}^{\mathrm{C}}\mathrm{D}^{\ell}_{\mathcal{X}}$ is the restriction of $\mathrm{D}^{\ell}_{\mathcal{X}}$ with domain

$$D({}^{\mathrm{C}}\mathrm{D}^{\ell}_{\mathcal{X}}) := N \oplus R(B_{\ell,\mathcal{X}}), \tag{6.4.3}$$

For each u in $D({}^{\mathrm{C}}\mathrm{D}^{\ell}_{\mathcal{X}})$, ${}^{\mathrm{C}}\mathrm{D}^{\ell}_{\mathcal{X}}u$ is called the *Caputo derivative of* u *with respect to* ℓ. Of course, this derivative generalizes the Caputo derivative on $L^2(0,b)$ introduced in Sect. 6.2.2.4.1. The operator ${}^{\mathrm{C}}\mathrm{D}^{\ell}_{\mathcal{X}}$ on $L^p(\mathcal{Y})$ is called *the Caputo operator with respect to* ℓ.

6.5 The Quadruplet ${}^{\mathrm{RL}}\mathcal{Q}_{p,\mathcal{Y},\ell}$

We will introduce a basic *fractional differential quadruplet* on reflexive Lebesgue spaces. For, let b be a positive real number, and p, q be conjugate exponents in $(1,\infty)$. Let also $\mathcal{Y}$ be a nontrivial separable and reflexive Banach space over $\mathbb{K}$ and (k,ℓ) be a Sonine pair. We set

$$\mathcal{V} = L^p(\mathcal{Y}) := L^p\big((0,b),\mathcal{Y}\big), \qquad \mathcal{V}^* = L^q(\mathcal{Y}^*) := L^q\big((0,b),\mathcal{Y}^*\big). \tag{6.5.1}$$

6.5.1 Introduction

We use the notation of Sect. 6.4 and define the operator

$$S_{\mathcal{V}} : L^p(\mathcal{Y}) \to L^p(\mathcal{Y}), \qquad v \mapsto v(b-\cdot).$$

The following result is needed to introduce our basic fractional differential quadruplet.

Proposition 6.5.1 *Let p lie in $(1,\infty)$ and f be a real valued function of $L^1(0,b)$. Then*

$$(B_{f,\mathcal{V}})^* = S_{\mathcal{V}^*}\, B_{f,\mathcal{V}^*}\, S_{\mathcal{V}^*}.$$

Moreover for any v in $L^q(\mathcal{Y}^*)$,

$$(B_{f,\mathcal{V}})^* w(x) = \int_x^b f(y-x)w(y)\,\mathrm{d}y,$$

for almost every x in $(0, b)$.

Proof For each (v, w) in $L^p(\mathcal{Y}) \times L^q(\mathcal{Y}^*)$, using the fact that f is real-valued, we compute

$$\begin{aligned}
\langle w, B_{f,\mathcal{V}} v\rangle_{\mathcal{V},\mathcal{V}^*} &= \int_0^b \int_0^x \langle f(x-y)w(x), v(y)\rangle_{\mathcal{Y}^*,\mathcal{Y}}\,\mathrm{d}y\,\mathrm{d}x && \text{(by Pr. 3.2.3)}\\
&= \int_0^b \mathrm{d}x \int_x^b \langle\, f(y-x)w(y),\ v(x)\,\rangle_{\mathcal{Y}^*,\mathcal{Y}}\,\mathrm{d}y && \text{(by Fubini Th.)}\\
&= \left\langle S_{L^q(\mathcal{Y}^*)} B_{f,L^q(\mathcal{Y}^*)} S_{L^q(\mathcal{Y}^*)} w,\ v\right\rangle_{\mathcal{V}^*,\mathcal{V}},
\end{aligned}$$

thanks to the change of variable $y' = b - y$. Then the assertions of the proposition follow.

□

Proposition 6.5.2 *Being given p in $(1, \infty)$, and $\mathcal{Y}$ a nontrivial separable and reflexive Banach space, let $\mathcal{V}$, $\mathcal{V}^*$ be defined through* (6.5.1). *If (k, ℓ) is a Sonine pair, then*

$$^{\mathrm{RL}}\mathcal{Q}_{p,\mathcal{Y},\ell} := ({}^{\mathrm{RL}}\mathrm{D}^{\ell}_{\mathcal{V}^*},\ {}^{\mathrm{RL}}\mathrm{D}^{\ell}_{\mathcal{V}},\ B_{\ell,\mathcal{V}},\ S_{\mathcal{V}}) \tag{6.5.2}$$

is a regular differential quadruplet on $\big(L^p((0, b), \mathcal{Y}), L^q((0, b), \mathcal{Y}^*)\big)$. Moreover

$$\mathbb{B}_{^{\mathrm{RL}}\mathcal{Q}_{p,\mathcal{Y},\ell}} = B_{\ell,\mathcal{V}^*},$$

and

$$\ker {}^{\mathrm{RL}}\mathrm{D}^{\ell}_{\mathcal{V}} = \begin{cases} \ell\mathcal{Y} & \text{if } \ell \in L^p(0, b) \\ \{0\} & \text{otherwise} \end{cases}, \tag{6.5.3}$$

$$\ker {}^{\mathrm{RL}}\mathrm{D}^{\ell}_{\mathcal{V}^*} = \begin{cases} \ell\mathcal{Y}^* & \text{if } \ell \in L^q(0, b) \\ \{0\} & \text{otherwise} \end{cases}. \tag{6.5.4}$$

Proof Since the proof of (6.5.3) and (6.5.4) are the same, we will only show the latter equality. For each w in $\ker {}^{\mathrm{RL}}\mathrm{D}^{\ell}_{\mathcal{V}^*}$, there exists Y^* in $\mathcal{Y}^*$ such that $k * w = g_1 Y^*$. Thus with (6.3.1), $g_1 * w = \ell * g_1 Y^*$, and $w = \ell Y^*$. Then (6.5.4) follows.

Regarding ${}^{\mathrm{RL}}\mathcal{Q}_{p,\mathcal{Y},\ell}$, one has ${}^{\mathrm{RL}}\mathrm{D}^{\ell}_{\mathcal{V}} B_{\ell,\mathcal{V}} = \mathrm{id}_{\mathcal{V}}$ according to (6.4.1). Thanks Proposition 6.5.1, we prove easily that ${}^{\mathrm{RL}}\mathcal{Q}_{p,\mathcal{Y},\ell}$ is a differential quadruplet. Next we claim that $\ker {}^{\mathrm{RL}}\mathrm{D}^{\ell}_{\mathcal{V}}$ is closed in $L^p(\mathcal{Y})$. This is obvious if this kernel is trivial. Otherwise, it is equal to $\ell\mathcal{Y}$ according to (6.5.3), and we conclude with Remark 3.2.1 (ii), which proves the claim. In a same way, $\ker {}^{\mathrm{RL}}\mathrm{D}^{\ell}_{\mathcal{V}*}$ is closed in $L^q(\mathcal{Y}^*)$. Besides, if $\ker {}^{\mathrm{RL}}\mathrm{D}^{\ell}_{\mathcal{V}*} = \{0\}$, then $\{0\}$ is a closed complementary subspace of $(\ker {}^{\mathrm{RL}}\mathrm{D}^{\ell}_{\mathcal{V}*})^{\perp}$; hence ${}^{\mathrm{RL}}\mathcal{Q}_{p,\mathcal{Y},\ell}$ is regular. On the other hand, if $\ker {}^{\mathrm{RL}}\mathrm{D}^{\ell}_{\mathcal{V}*} = \ell\mathcal{Y}^*$ then Remark 3.2.1 (ii) yields that $\ell^{c}\mathcal{Y}$ is a closed complementary subspace of $(\ker {}^{\mathrm{RL}}\mathrm{D}^{\ell}_{\mathcal{V}*})^{\perp}$; hence ${}^{\mathrm{RL}}\mathcal{Q}_{p,\mathcal{Y},\ell}$ is regular. □

6.5.2 Endogenous Boundary Values

Let (k, ℓ) be a Sonine pair. Since ${}^{\mathrm{RL}}\mathcal{Q}_{p,\mathcal{Y},\ell}$ is a regular differential quadruplet, the operator ${}^{\mathrm{RL}}\mathrm{D}^{\ell}_{\mathcal{V}}$ can be seen as a generalized derivative. Since this quadruplet is regular, the results of Chap. 5 may be applied. In order to implement the method featured in Sect. 5.7.1, we need to distinguish four cases depending on the L^p-regularity of ℓ.

6.5.2.1 Case Where ℓ Lies in $L^p(0, b) \cap L^q(0, b)$

Proposition 6.5.2 entails that

$$\ker {}^{\mathrm{RL}}\mathrm{D}^{\ell}_{\mathcal{V}} = \ell\mathcal{Y}, \qquad \ker {}^{\mathrm{RL}}\mathrm{D}^{\ell}_{\mathcal{V}*} = \ell\mathcal{Y}^*.$$

By Lemma 3.2.6, ${}^{\mathrm{RL}}\mathcal{Q}_{p,\mathcal{Y},\ell}$ is $h\mathcal{Y}$-regular for any function h in $L^p(0, b)$ satisfying $\int_0^b \ell h \,\mathrm{d}y \neq 0$. With the notation of Sect. 5.3.3, we set $F_0 := h\mathcal{Y}$ and E_{M} which is by definition $\ker A_{\mathrm{M}} \oplus BS(F_0)$ reads here

$$E_{\mathrm{M}} = \ell\mathcal{Y} \oplus B_{\ell,\mathcal{V}} S_{\mathcal{V}}(h\mathcal{Y}).$$

Hence (5.7.4) becomes

$$D({}^{\mathrm{RL}}\mathrm{D}^{\ell}_{\mathcal{V}}) = E_{\mathrm{M}} \oplus B_{\ell,\mathcal{V}} S_{\mathcal{V}}\, (\ell\mathcal{Y}^*)^{\perp}.$$

Thus a function u in $\mathcal{V}$ lies in $D({}^{\mathrm{RL}}\mathrm{D}^{\ell}_{\mathcal{V}})$ if and only if there exist U_1, U_1' in $\mathcal{Y}$ and $\psi^{\perp}$ in $(\ell\mathcal{Y}^*)^{\perp}$ such that

$$u = \ell\, U_1 + \ell * S_{\mathcal{V}}(h\, U_1') + \ell * S_{\mathcal{V}}\psi^{\perp}. \tag{6.5.5}$$

In view of Sect. 5.7.1.1, (U_1, U_1') are the *endogenous boundary values of u with respect to the trivial direct sum decompositions*

$$\ker {}^{\mathrm{RL}}\mathrm{D}^{\ell}_{\mathcal{V}} = \ell\mathcal{Y}, \qquad F_0 = h\mathcal{Y}.$$

Let us now compute these endogenous boundary values in all generality and next under various additional assumptions on ℓ and h. Convoluting (6.5.5) by k, we get with (6.3.1)

$$k * u = g_1\, U_1 + g_1 * S_{\mathcal{V}}(h\, U_1') + g_1 * S_{\mathcal{V}}\psi^{\perp} \qquad \text{in } W^{1,p}\big((0, b), \mathcal{Y})\big).$$

Thus

$$U_1 = k * u(0). \tag{6.5.6}$$

In order to compute U_1', we apply the operator ${}^{\mathrm{RL}}\mathrm{D}^{\ell}_{\mathcal{Y}}$ to (6.5.5) and convole with ℓ to get

$$\ell * {}^{\mathrm{RL}}\mathrm{D}^{\ell}_{\mathcal{Y}} u = \ell * S_{\mathcal{Y}}(h\,U_1') + \ell * S_{\mathcal{Y}}\psi^{\perp}.$$

By Proposition 3.3.5, the above identity holds in $C([0, b], \mathcal{Y})$ and

$$\ell * S_{\mathcal{Y}} h\, U_1'(b) = \int_0^b \ell(y)h(y)\,\mathrm{d}y\, U_1', \qquad \ell * S_{\mathcal{Y}}\psi^{\perp}(b) = \int_0^b \ell(y)\psi^{\perp}(y)\,\mathrm{d}y.$$

The latter integral vanishes due to (3.2.1). Hence recalling that $\int_0^b \ell(y)h(y)\,\mathrm{d}y$ is nontrivial by assumption:

$$U_1' = \frac{1}{\int_0^b \ell h\,\mathrm{d}y}\, \ell * {}^{\mathrm{RL}}\mathrm{D}^{\ell}_{\mathcal{Y}} u(b). \tag{6.5.7}$$

There results from (6.5.6), (6.5.7) that

$$k * u(0), \qquad \ell * {}^{\mathrm{RL}}\mathrm{D}^{\ell}_{\mathcal{Y}} u(b)$$

are "standard" boundary values of u since the linear map transforming (U_1, U_1') into $\big(k * u(0), \ell * {}^{\mathrm{RL}}\mathrm{D}^{\ell}_{\mathcal{Y}} u(b)\big)$ is clearly an isomorphism.

We will express $\ell * {}^{\mathrm{RL}}\mathrm{D}^{\ell}_{\mathcal{Y}} u(b)$ in terms of zero order terms and then get a new ordered pair of "standard" boundary values for u. For, starting from (6.5.5), Lemma 4.3.3 yields that

$$u - \ell\, U_1 = \ell * {}^{\mathrm{RL}}\mathrm{D}^{\ell}_{\mathcal{Y}} u$$

Hence $u - \ell\, U_1$ is continuous on $[0, b]$ (according to Proposition 3.3.5) and

$$\begin{cases} U_1 = k * u(0) \\ U_1' = \dfrac{1}{\int_0^b \ell h\,\mathrm{d}y}\big(u - \ell\, k * u(0)\big)(b). \end{cases} \tag{6.5.8}$$

Next we would prefer to write $\big(u - \ell\, k * u(0)\big)(b)$ under the form

$$\big(u - \ell\, k * u(0)\big)(b) = u(b) - \ell(b)\, k * u(0),$$

so that the “standard” boundary values of u would be

$$k * u(0), \qquad u(b).$$

This is possible, as we will see now, provided ℓ is more regular. More precisely, we will assume that

$$\ell \text{ admits an essential limit at } b \tag{6.5.9}$$

in the sense of definition 3.1.4. Whence we put

$$\ell(b) := \operatorname*{ess\,lim}_{b} \ell. \tag{6.5.10}$$

Then by difference, u admits an essential limit $u(b)$ at b, and

$$U_1' = \frac{1}{\int_0^b \ell h \,\mathrm{d}y}\big(u(b) - \ell(b)k * u(0)\big). \tag{6.5.11}$$

Hence $k * u(0)$ and $u(b)$ are the “standard” boundary values of u.

Now still assuming (6.5.9), let us consider some particular cases.

- If we choose h to be a function in $L^p(0, b)$ satisfying $\int_0^b \ell h \,\mathrm{d}y = 1$, then

$$U_1' = u(b) - \ell(b)k * u(0). \tag{6.5.12}$$

- Whereas if we choose $h := \ell$, then ${}^{\mathrm{RL}}Q_{p,\mathcal{Y},\ell}$ is $\ker {}^{\mathrm{RL}}\mathrm{D}^{\ell}_{\mathcal{V}}$-regular and

$$U_1 = k * u(0), \qquad U_1' = \frac{1}{\|\ell\|^2_{L^2}}\big(u(b) - \ell(b)k * u(0)\big). \tag{6.5.13}$$

- If $\mathcal{Y} := \mathbb{K}$, then $\mathcal{V} = L^p(0, b)$, U_1, U_1' are scalars and $\psi^{\perp}$ lies in $\ell^{\perp}$. Then by Definition 5.5.1, the endogenous boundary values of u with respect to the bases (ℓ) of $\ker {}^{\mathrm{RL}}\mathrm{D}^{\alpha}_{L^p}$ and (h) of $F_0 := \langle h \rangle$ are

$$(u_1, u_1') := \Big(k * u(0), \frac{1}{\int_0^b \ell h \,\mathrm{d}y}\big(u(b) - \ell(b)k * u(0)\big)\Big). \tag{6.5.14}$$

On the other hand, the “standard” boundary values $k * u(0)$ and $u(b)$ of u reads in terms the above endogenous coordinates:

$$\big(k * u(0), u(b)\big) = \big(u_1,\ \ell(b)u_1 + \textstyle\int_0^b \ell h \,\mathrm{d}y\, u_1'\big).$$

6.5.2.2 Case Where ℓ Lies in $L^q(0,b) \setminus L^p(0,b)$

We proceed as in the previous case, but we will normalize the function h. Proposition 6.5.2 entails that

$$\ker {}^{\mathrm{RL}}\mathrm{D}^{\ell}_{\mathcal{V}} = \{0\}, \qquad \ker {}^{\mathrm{RL}}\mathrm{D}^{\ell}_{\mathcal{V}^*} = \ell\mathcal{Y}^*. \tag{6.5.15}$$

Let h be a function in $L^p(0,b)$ such that $\int_0^b \ell h \,\mathrm{d}y = 1$. Then ${}^{\mathrm{RL}}\mathcal{Q}_{p,\mathcal{Y},\ell}$ is $h\mathcal{Y}$-regular, and

$$E_{\mathrm{M}} = B_{\ell,\mathcal{V}} S_{\mathcal{V}}(h\mathcal{Y})$$
$$D({}^{\mathrm{RL}}\mathrm{D}^{\ell}_{\mathcal{V}}) = B_{\ell,\mathcal{V}} S_{\mathcal{V}}(h\mathcal{Y}) \oplus B_{\ell,\mathcal{V}} S_{\mathcal{V}}\big((\ell\mathcal{Y}^*)^{\perp}\big) = R(B_{\ell,\mathcal{V}}).$$

Thus a function u in $\mathcal{V}$ lies in $D({}^{\mathrm{RL}}\mathrm{D}^{\ell}_{\mathcal{V}})$ if and only if there exist U_1' in $\mathcal{Y}$ and $\psi^{\perp}$ in $(\ell\mathcal{Y}^*)^{\perp}$ such that

$$u = \ell * S_{\mathcal{V}}(h\, U_1') + \ell * S_{\mathcal{V}}\psi^{\perp}. \tag{6.5.16}$$

Without extra assumption on ℓ, the previous equality holds on $C([0,b],\mathcal{Y})$ and $u(0) = 0$ according to Remark 3.3.4. Hence

$$U_1' = u(b). \tag{6.5.17}$$

From Sect. 5.7.1.1, U_1' is the endogenous boundary value of u with respect to the trivial direct sum decomposition of $F_0 := h\mathcal{Y}$. Also, there results from the above analysis that

$$D({}^{\mathrm{RL}}\mathrm{D}^{\ell}_{\mathcal{V}}) = \{u \in {}_0C([0,b],\mathcal{Y}) \mid k * u \in W^{1,p}\big((0,b),\mathcal{Y}\big)\}. \tag{6.5.18}$$

6.5.2.3 Case Where ℓ Lies in $L^p(0,b) \setminus L^q(0,b)$

One has

$$\ker {}^{\mathrm{RL}}\mathrm{D}^{\ell}_{\mathcal{V}} = \ell\mathcal{Y}, \qquad \ker {}^{\mathrm{RL}}\mathrm{D}^{\ell}_{\mathcal{V}^*} = \{0\}.$$

Hence we must choose $F_0 := \{0\}$, so that

$$E_{\mathrm{M}} = \ell\mathcal{Y}$$
$$D({}^{\mathrm{RL}}\mathrm{D}^{\ell}_{\mathcal{V}}) = \ell\mathcal{Y} \oplus R(B_{\ell,\mathcal{V}}).$$

Thus a function u in $\mathcal{V}$ lies in $D({}^{\mathrm{RL}}\mathrm{D}^{\ell}_{\mathcal{V}})$ if and only if there exist U_1 in $\mathcal{Y}$ and ψ in $L^p(\mathcal{Y})$ such that

$$u = \ell\, U_1 + \ell * \psi. \tag{6.5.19}$$

Convoluting with k, we get

$$U_1 = k * u(0). \tag{6.5.20}$$

From Sect. 5.7.1, U_1 is the endogenous boundary value of u with respect to the trivial direct sum decomposition $\ker {}^{\mathrm{RL}}\mathrm{D}^{\ell}_{\mathcal{V}} = \ell\mathcal{Y}$.

6.5.2.4 Case Where ℓ Lies in $L^1(0, b) \setminus (L^q(0, b) \cup L^p(0, b))$

The kernels of the maximal operators are both trivial. According to Sect. 5.7.1.1, there is no endogenous boundary value. Moreover a function u in $\mathcal{V}$ lies in $D({}^{\mathrm{RL}}\mathrm{D}^{\ell}_{\mathcal{V}})$ if and only if there exists ψ in $L^p(\mathcal{Y})$ such that

$$u = \ell * \psi. \tag{6.5.21}$$

Convoluting with k, we get

$$k * u(0) = 0. \tag{6.5.22}$$

6.5.3 The Adjoint Quadruplet of ${}^{\mathrm{RL}}\mathcal{Q}_{p,\mathcal{Y},\ell}$

Let b be a positive real number and p, q be conjugate exponents in $(1, \infty)$. Let also $\mathcal{Y}$ be a nontrivial separable and reflexive Banach space over $\mathbb{K} = \mathbb{R}$ or $\mathbb{C}$ and (k, ℓ) be a Sonine pair. Recalling (6.5.1), the adjoint quadruplet of ${}^{\mathrm{RL}}\mathcal{Q}_{p,\mathcal{Y},\ell}$ reads

$$({}^{\mathrm{RL}}\mathcal{Q}_{p,\mathcal{Y},\ell})^* := (S_{\mathcal{V}}\, {}^{\mathrm{RL}}\mathrm{D}^{\ell}_{\mathcal{V}}\, S_{\mathcal{V}},\ S_{\mathcal{V}^*}\, {}^{\mathrm{RL}}\mathrm{D}^{\ell}_{\mathcal{V}^*}\, S_{\mathcal{V}^*},\ S_{\mathcal{V}^*}\, B_{\ell,\mathcal{V}^*}\, S_{\mathcal{V}^*},\ S_{\mathcal{V}^*}). \tag{6.5.23}$$

Since $\mathbb{B}_{{}^{\mathrm{RL}}\mathcal{Q}_{p,\mathcal{Y},\ell}} = B_{\ell,\mathcal{V}^*}$, on account of (5.6.18), one has

$$r({}^{\mathrm{RL}}\mathcal{Q}_{p,\mathcal{Y},\ell}) = {}^{\mathrm{RL}}\mathcal{Q}_{q,\mathcal{Y}^*,\ell}. \tag{6.5.24}$$

Thus with (5.6.19)

$$({}^{\mathrm{RL}}\mathcal{Q}_{p,\mathcal{Y},\ell})^* = S_{\mathcal{V}^*}\, {}^{\mathrm{RL}}\mathcal{Q}_{q,\mathcal{Y}^*,\ell}\, S_{\mathcal{V}^*}. \tag{6.5.25}$$

${}^{\mathrm{RL}}\mathcal{Q}_{q,\mathcal{Y}^*,\ell}$ is regular according to Proposition 6.5.2; thus by invertibility of $S_{\mathcal{V}^*}$, there results from Proposition 5.6.5 that $({}^{\mathrm{RL}}\mathcal{Q}_{p,\mathcal{Y},\ell})^*$ is a regular differential quadruplet.

Remark 6.5.1 *Computations of* $S_{\mathcal{V}}\, {}^{\mathrm{RL}}\mathrm{D}^{\ell}_{\mathcal{V}}\, S_{\mathcal{V}} u$. For each u in the domain of $S_{\mathcal{V}}\, {}^{\mathrm{RL}}\mathrm{D}^{\ell}_{\mathcal{V}}\, S_{\mathcal{V}}$, since $S_{\mathcal{V}}$ anti-commute with $\mathrm{D}^1_{\mathcal{V}}$, one has by Proposition 6.5.1,

$$S_{\mathcal{V}}\,{}^{\mathrm{RL}}\mathrm{D}^{\ell}_{\mathcal{V}}\,S_{\mathcal{V}}u(x) = -\frac{\mathrm{d}}{\mathrm{d}x}\Big\{\int_x^b k(y-x)u(y)\,\mathrm{d}y\Big\}$$

for almost every x in $(0,b)$. Let us give another representation of $S_{\mathcal{V}}\,{}^{\mathrm{RL}}\mathrm{D}^{\ell}_{\mathcal{V}}\,S_{\mathcal{V}}u$ when u is smooth. For, we assume that

$$u \in W^{1,p}\big((0,b),\mathcal{Y})\big), \qquad k \in L^p(0,b),$$

Proposition 3.3.4 yields that $v := S_{\mathcal{V}}u$ lies in the domain of ${}^{\mathrm{RL}}\mathrm{D}^{\ell}_{\mathcal{V}}$ and

$${}^{\mathrm{RL}}\mathrm{D}^{\ell}_{\mathcal{V}}\,v = B_{k,\mathcal{V}}\mathrm{D}^1_{\mathcal{V}}\,v + k(\cdot)v(0).$$

Thus u lies in $D(S_{\mathcal{V}}\,{}^{\mathrm{RL}}\mathrm{D}^{\ell}_{\mathcal{V}}\,S_{\mathcal{V}})$ and

$$S_{\mathcal{V}}\,{}^{\mathrm{RL}}\mathrm{D}^{\ell}_{\mathcal{V}}\,S_{\mathcal{V}}\,u = -S_{\mathcal{V}}B_{k,\mathcal{V}}S_{\mathcal{V}}\mathrm{D}^1_{\mathcal{V}}\,u + k(b-\cdot)u(b),$$

since $S_{\mathcal{V}}\mathrm{D}^1_{\mathcal{V}} = -\mathrm{D}^1_{\mathcal{V}}S_{\mathcal{V}}$. Thus denoting $\mathrm{D}^1_{\mathcal{V}}u$ by u' and using Proposition 6.5.1, we end up with

$$S_{\mathcal{V}}\,{}^{\mathrm{RL}}\mathrm{D}^{\ell}_{\mathcal{V}}\,S_{\mathcal{V}}\,u(x) = -\int_x^b k(y-x)u'(y)\,\mathrm{d}y + k(b-x)u(b)$$

for almost every x in $(0,b)$. □

6.5.3.1 Case Where ℓ Lies in $L^p(0,b)\cap L^q(0,b)$

Proposition 6.5.2 entails that

$$\ker{}^{\mathrm{RL}}\mathrm{D}^{\ell}_{\mathcal{V}^*} = \ell\mathcal{Y}^*, \qquad \ker{}^{\mathrm{RL}}\mathrm{D}^{\ell}_{\mathcal{V}} = \ell\mathcal{Y}.$$

Let h^{s} to be a function in $L^q(0,b)$ satisfying $\int_0^b h^{\mathrm{s}}\ell\,\mathrm{d}y = 1$. Setting $F_0^{\mathrm{s}} := h^{\mathrm{s}}\mathcal{Y}^*$, Corollary 5.3.4 yields that

$$D(S_{\mathcal{V}^*}\,{}^{\mathrm{RL}}\mathrm{D}^{\ell}_{\mathcal{V}^*}\,S_{\mathcal{V}^*}) = E^{\mathrm{s}}_{\mathrm{M}} \oplus S_{\mathcal{V}^*}\,B_{\ell,\mathcal{V}^*}\,S_{\mathcal{V}^*}\,(\ell\mathcal{Y}^*)^{\perp} \tag{6.5.26}$$

where

$$E^{\mathrm{s}}_{\mathrm{M}} := S_{\mathcal{V}^*}(\ell\mathcal{Y}^*) \oplus S_{\mathcal{V}^*}\,B_{\ell,\mathcal{V}^*}\,S_{\mathcal{V}^*}(h^{\mathrm{s}}\mathcal{Y}^*). \tag{6.5.27}$$

Thus a function f in $\mathcal{V}^*$ lies in $D({}^{\mathrm{RL}}\mathrm{D}^{\ell}_{\mathcal{V}^*}\,S_{\mathcal{V}^*})$ if and only if there exist F_1, F_1' in $\mathcal{Y}^*$ and $\varphi^{\perp}$ in $(\ell\mathcal{Y})^{\perp\,L^q(\mathcal{Y}^*)}$ such that

$$f = S_{\mathcal{V}^*}(\ell\,F_1) + S_{\mathcal{V}^*}\,B_{\ell,\mathcal{V}^*}\,S_{\mathcal{V}^*}(h^{\mathrm{s}}\,F_1') + S_{\mathcal{V}^*}\,B_{\ell,\mathcal{V}^*}\,S_{\mathcal{V}^*}\varphi^{\perp}. \tag{6.5.28}$$

In view of Sect. 5.7.1, (F_1, F_1') are *the endogenous boundary values of* f *with underlying quadruplet* $({}^{\mathrm{RL}}Q_{p,\mathcal{Y},\ell})^*$ *and with respect to the trivial direct sum decompositions*

$$S_{\mathcal{Y}^*} \ker {}^{\mathrm{RL}}\mathrm{D}^{\ell}_{\mathcal{Y}^*} = S_{\mathcal{Y}^*}(\ell\mathcal{Y}^*), \qquad S_{\mathcal{Y}^*} F_0^{\mathrm{s}} = S_{\mathcal{Y}^*}(h^{\mathrm{s}}\mathcal{Y}^*).$$

Besides, by (6.5.24), *the boundary values trick* featured in Sect. 5.7.1.2 together with (6.5.8) give

$$\begin{cases} F_1 = (k * S_{\mathcal{Y}^*} f)(0) \\ F_1' = \big(S_{\mathcal{Y}^*} f - \ell\, k * S_{\mathcal{Y}^*} f(0)\big)(b). \end{cases} \tag{6.5.29}$$

We are in position to state an integration by parts formula.

Proposition 6.5.3 *Let* p, q *be conjugate exponents in* $(1, \infty)$, $\mathcal{Y}$ *be a nontrivial real or complex separable and reflexive Banach space, and* (k, ℓ) *be a Sonine pair. Assume that:*

- ℓ *lies in* $L^p(0, b) \cap L^q(0, b)$ *and admits an essential limit at* b;
- u *belongs to* $D({}^{\mathrm{RL}}\mathrm{D}^{\ell}_{\mathcal{Y}})$, *and* f *lies in* $D(S_{\mathcal{Y}^*}\, {}^{\mathrm{RL}}\mathrm{D}^{\ell}_{\mathcal{Y}^*} S_{\mathcal{Y}^*})$.

Then the following integration by parts formula holds true:

$$\int_0^b \langle {}^{\mathrm{RL}}\mathrm{D}^{\ell}_{\mathcal{Y}} u(x),\ v(x) \rangle_{\mathcal{Y},\mathcal{Y}^*}\, \mathrm{d}x = \int_0^b \langle u(x),\ S_{\mathcal{Y}^*}\, {}^{\mathrm{RL}}\mathrm{D}^{\ell}_{\mathcal{Y}^*} S_{\mathcal{Y}^*} v(x) \rangle_{\mathcal{Y},\mathcal{Y}^*}\, \mathrm{d}x$$
$$\langle u(b),\ (k * S_{\mathcal{Y}^*} f)(0) \rangle_{\mathcal{Y},\mathcal{Y}^*} - \langle (k * u)(0),\ f(0) \rangle_{\mathcal{Y},\mathcal{Y}^*}.$$

Proof We apply the result of Sect. 5.7.1.5 with

$$m = m^{\mathrm{s}} = 1, \qquad \xi_1 = \ell, \qquad \xi_1^{\mathrm{s}} = \ell, \qquad \eta_1 = h, \qquad \eta_1^{\mathrm{s}} = h^{\mathrm{s}}$$

where h and h^{s} are any functions in $L^p(0, b)$ and $L^q(0, b)$ (respectively) such that

$$\int_0^b \ell h \,\mathrm{d}y = 1, \qquad \int_0^b h^{\mathrm{s}} \ell \,\mathrm{d}y = 1.$$

Then using (6.5.6), (6.5.12), and (6.5.29) together the with the fact that ℓ is real valued, we get the above integration by parts formula. □

6.5.3.2 Case Where ℓ Lies in $L^q(0, b) \setminus L^p(0, b)$

We have

$$\ker {}^{\mathrm{RL}}\mathrm{D}^{\ell}_{\mathcal{Y}^*} = \ell\mathcal{Y}^*, \qquad \ker {}^{\mathrm{RL}}\mathrm{D}^{\ell}_{\mathcal{Y}} = \{0\}.$$

and $F_0^{\mathrm{s}} := \{0\}$, so that

$$D(S_{\mathcal{V}^*}\,{}^{\mathrm{RL}}\mathrm{D}^{\ell}_{\mathcal{V}^*}\,S_{\mathcal{V}^*}) = E^{\mathrm{s}}_{\mathrm{M}} \oplus S_{\mathcal{V}^*}\,R(B_{\ell,\mathcal{V}^*}) \tag{6.5.30}$$

where

$$E^{\mathrm{s}}_{\mathrm{M}} := S_{\mathcal{V}^*}(\ell\mathcal{Y}^*). \tag{6.5.31}$$

Thus a function f in $\mathcal{V}^*$ lies in $D({}^{\mathrm{RL}}\mathrm{D}^{\ell}_{\mathcal{V}^*}\,S_{\mathcal{V}^*})$ if and only if there exist F_1 in $\mathcal{Y}^*$ and φ in $\mathcal{V}^*$ such that

$$f = S_{\mathcal{V}^*}(\ell F_1) + S_{\mathcal{V}^*}(\ell * \varphi). \tag{6.5.32}$$

Arguing as in the previous case, *the boundary values trick* and (6.5.20) applied to ${}^{\mathrm{RL}}\mathcal{Q}_{q,\mathcal{Y}^*,\ell}$ entail that

$$F_1 = (k * S_{\mathcal{V}^*} f)(0). \tag{6.5.33}$$

6.5.3.3 Case Where ℓ Lies in $L^p(0,b) \setminus L^q(0,b)$

We have

$$\ker {}^{\mathrm{RL}}\mathrm{D}^{\ell}_{\mathcal{V}^*} = \{0\}, \qquad \ker {}^{\mathrm{RL}}\mathrm{D}^{\ell}_{\mathcal{V}} = \ell\mathcal{Y}.$$

Let h be a function in $L^q(0,b)$ such that $\int_0^b \ell h\,\mathrm{d}y = 1$. Then we choose $F_0^{\mathrm{s}} := h\mathcal{Y}^*$, so that

$$D(S_{\mathcal{V}^*}\,{}^{\mathrm{RL}}\mathrm{D}^{\ell}_{\mathcal{V}^*}\,S_{\mathcal{V}^*}) = E^{\mathrm{s}}_{\mathrm{M}} \oplus S_{\mathcal{V}^*}\,B_{\ell,\mathcal{V}^*}\,S_{\mathcal{V}^*}(\ell\mathcal{Y})^{\perp} \tag{6.5.34}$$

where

$$E^{\mathrm{s}}_{\mathrm{M}} := S_{\mathcal{V}^*}\,B_{\ell,\mathcal{V}^*}\,S_{\mathcal{V}^*}(h\mathcal{Y}^*). \tag{6.5.35}$$

Thus a function f in $\mathcal{V}^*$ lies in $D(S_{\mathcal{V}^*}\,{}^{\mathrm{RL}}\mathrm{D}^{\ell}_{\mathcal{V}^*}\,S_{\mathcal{V}^*})$ if and only if there exist F_1' in $\mathcal{Y}^*$ and $\varphi^{\perp}$ in $(\ell\mathcal{Y})^{\perp}$ such that

$$f = S_{\mathcal{V}^*}\,B_{\ell,\mathcal{V}^*}\,S_{\mathcal{V}^*}(hF_1') + S_{\mathcal{V}^*}\,B_{\ell,\mathcal{V}^*}\,S_{\mathcal{V}^*}\varphi^{\perp}. \tag{6.5.36}$$

Thus applying the results of Sect. 6.5.2.2 to ${}^{\mathrm{RL}}\mathcal{Q}_{q,\mathcal{Y}^*,\ell}$, *the boundary values trick* yields

$$F_1' = f(0). \tag{6.5.37}$$

6.5.3.4 Case Where ℓ Lies in $L^1(0,b) \setminus (L^q(0,b) \cup L^p(0,b))$

The kernels of the maximal operators of $({}^{\mathrm{RL}}Q_{p,\mathcal{Y},\ell})^*$ are both trivial. Hence, there is no endogenous boundary value. Moreover, we infer from (6.5.22)

$$(k * S_{\mathcal{Y}^*} f)(0) = 0. \tag{6.5.38}$$

6.5.4 Boundary Restriction Operators of ${}^{\mathrm{RL}}Q_{p,\mathcal{Y},\ell}$ and Their Adjoint

Let p, q be conjugate exponents in $(1,\infty)$ and $\mathcal{Y}$ be a nontrivial separable and reflexive Banach space. Let also (k,ℓ) be a Sonine pair.

6.5.4.1 Intrinsic Representations

We will introduce representations of the domain of the pivot operator of ${}^{\mathrm{RL}}Q_{p,\mathcal{Y},\ell}$ and of its adjoint which are independent of endogenous boundary values. In particular, the L^p-regularity of ℓ is irrelevant here.

Proposition 6.5.4 *Under the above assumptions,*

$$D(A_{{}^{\mathrm{RL}}Q_{p,\mathcal{Y},\ell}}) = \{u \in L^p(\mathcal{Y}) \mid k * u \in {}_0W^{1,p}(\mathcal{Y})\} \tag{6.5.39}$$
$$D\big((A_{{}^{\mathrm{RL}}Q_{p,\mathcal{Y},\ell}})^*\big) = \{f \in L^q(\mathcal{Y}^*) \mid k * S_{\mathcal{Y}^*} f \in {}_0W^{1,q}(\mathcal{Y}^*)\}.$$

Moreover, the adjoint of the pivot operator of ${}^{\mathrm{RL}}Q_{p,\mathcal{Y},\ell}$ *is the conjugate of the pivot operator of* ${}^{\mathrm{RL}}Q_{q,\mathcal{Y}^*,\ell}$ *by* $S_{\mathcal{Y}^*}$, *which reads in symbol*

$$(A_{{}^{\mathrm{RL}}Q_{p,\mathcal{Y},\ell}})^* = S_{\mathcal{Y}^*} A_{{}^{\mathrm{RL}}Q_{q,\mathcal{Y}^*,\ell}} S_{\mathcal{Y}^*}. \tag{6.5.40}$$

Proof Let us prove (6.5.39). By the very definition of $D(A_{{}^{\mathrm{RL}}Q_{p,\mathcal{Y},\ell}})$, any element u of this domain reads $u = \ell * h$ for some h in $L^p(\mathcal{Y})$. Thus by (6.3.1),

$$k * u = g_1 * h \in {}_0W^{1,p}(\mathcal{Y}).$$

Conversely, let u belong to the right-hand side of (6.5.39). By Proposition 3.3.4,

$$u = \ell * \frac{\mathrm{d}}{\mathrm{d}x}\{k * u\} \in R(B_{\ell,\mathcal{Y}}).$$

Thus u lies in $D(A_{{}^{\mathrm{RL}}Q_{p,\mathcal{Y},\ell}})$, which proves (6.5.39).

Moreover the formula (6.5.40) is a consequence of (6.5.25) and (5.6.14). Finally the representation of the domain of $(A_{{}^{\mathrm{RL}}Q_{p,\mathcal{Y},\ell}})^*$ is obtained by using successively (6.5.40) and (6.5.39). □

The two latter propositions allow to state another integration by parts formula.

Corollary 6.5.5 *Let p, q be conjugate exponents in $(1,\infty)$, $\mathcal{Y}$ be a nontrivial real or complex separable and reflexive Banach space, and (k,ℓ) be a Sonine pair. We assume that:*

- *u belongs to $L^p(\mathcal{Y})$, and $k*u$ lies in ${}_0W^{1,p}(\mathcal{Y})$;*
- *v belongs to $L^q(\mathcal{Y}^*)$, and $k*S_{\mathcal{Y}^*}v$ lies in ${}_0W^{1,q}(\mathcal{Y}^*)$.*

Then the following integration by parts formula holds true:

$$\int_0^b \langle {}^{\mathrm{RL}}\mathrm{D}^{\ell}_{\mathcal{Y}}u(x),\ v(x)\rangle_{\mathcal{Y},\mathcal{Y}^*}\,\mathrm{d}x = \int_0^b \langle u(x),\ S_{\mathcal{Y}^*}{}^{\mathrm{RL}}\mathrm{D}^{\ell}_{\mathcal{Y}^*}S_{\mathcal{Y}^*}v(x)\rangle_{\mathcal{Y},\mathcal{Y}^*}\,\mathrm{d}x.$$

In the case where $\mathcal{Y}$ is equal to $\mathbb{R}$ or $\mathbb{C}$, replacing v by $\overline{v}$ if necessary, the latter integration by parts formula becomes

$$\int_0^b {}^{\mathrm{RL}}\mathrm{D}^{\ell}_{L^p}u(x)v(x)\,\mathrm{d}x = \int_0^b u(x)S_{L^q}{}^{\mathrm{RL}}\mathrm{D}^{\ell}_{L^q}S_{L^q}v(x)\,\mathrm{d}x.$$

Notice that on account of Remark 6.5.1,

$$S_{L^q}{}^{\mathrm{RL}}\mathrm{D}^{\ell}_{L^q}S_{L^q}v(x) = -\frac{\mathrm{d}}{\mathrm{d}x}\Big\{\int_x^b k(y-x)v(y)\,\mathrm{d}y\Big\}$$

for almost every x in $(0,b)$. The integration by parts formula of Corollary 6.5.5 generalizes the classical result of [LY38, Th. 3].

If one wants to describe boundary restriction operators and their adjoint in terms of boundary values, one has to distinguish four cases depending on the L^p-regularity of ℓ.

6.5.4.2 Case Where ℓ Lies in $L^p(0,b)\cap L^q(0,b)$

6.5.4.2.1 Boundary Restriction Operators

Let us consider first the pivot and minimal operators of ${}^{\mathrm{RL}}\mathcal{Q}_{p,\mathcal{Y},\ell}$. In view of Sect. 6.5.2.1, ${}^{\mathrm{RL}}\mathcal{Q}_{p,\mathcal{Y},\ell}$ is a $h\mathcal{Y}$-regular differential quadruplet for any function h in $L^p(0,b)$ satisfying $\int_0^b \ell h\,\mathrm{d}y \neq 0$. Since the domain of $A_{{}^{\mathrm{RL}}\mathcal{Q}_{p,\mathcal{Y},\ell}}$ is equal to the range of $B_{\ell,\mathcal{Y}}$, we derive from (6.5.5), (6.5.6) that for any u in the domain of ${}^{\mathrm{RL}}\mathrm{D}^{\ell}_{\mathcal{Y}}$,

$$\begin{aligned} u \in D(A_{{}^{\mathrm{RL}}\mathcal{Q}_{p,\mathcal{Y},\ell}}) &\Longleftrightarrow U_1 = 0 \\ &\Longleftrightarrow k*u(0) = 0. \end{aligned}$$

Regarding the minimal operator $A_{\mathrm{m},{}^{\mathrm{RL}}\mathcal{Q}_{p,\mathcal{Y},\ell}}$ of ${}^{\mathrm{RL}}\mathcal{Q}_{p,\mathcal{Y},\ell}$, with (6.5.8), we obtain

$$u \in D(A_{\mathrm{m},{}^{\mathrm{RL}}\mathcal{Q}_{p,\mathcal{Y},\ell}}) \Longleftrightarrow U_1 = U_1' = 0$$
$$\Longleftrightarrow k * u(0) = u(b) = 0. \tag{6.5.41}$$

By continuity of the functions involved, we obtain the following simpler characterization of the domain of these operators.

Proposition 6.5.6 *Assuming that ℓ lies in $L^p(0,b) \cap L^q(0,b)$, let u belong to $D({}^{\mathrm{RL}}\mathrm{D}^{\ell}_{\mathcal{Y}})$. Then u lies in $D(A_{{}^{\mathrm{RL}}\mathcal{Q}_{p,\mathcal{Y},\ell}})$ if and only if u is continuous on $[0,b]$ and $u(0) = 0$. That is, in symbol*

$$D(A_{{}^{\mathrm{RL}}\mathcal{Q}_{p,\mathcal{Y},\ell}}) = \{u \in D({}^{\mathrm{RL}}\mathrm{D}^{\ell}_{\mathcal{Y}}) \cap C([0,b],\mathcal{Y}) \mid u(0) = 0\}.$$

In a same way,

$$D(A_{\mathrm{m},{}^{\mathrm{RL}}\mathcal{Q}_{p,\mathcal{Y},\ell}}) = \{u \in D({}^{\mathrm{RL}}\mathrm{D}^{\ell}_{\mathcal{Y}}) \cap C([0,b],\mathcal{Y}) \mid u(0) = u(b) = 0\}.$$

Proof By the very definition of pivot operators, if u lies in $D(A_{{}^{\mathrm{RL}}\mathcal{Q}_{p,\mathcal{Y},\ell}})$, then there exists ψ in $L^p(\mathcal{Y})$ such that $u = \ell * \psi$. With Remark 3.3.4, we deduce that u belongs to $D({}^{\mathrm{RL}}\mathrm{D}^{\ell}_{\mathcal{Y}}) \cap {}_0C([0,b],\mathcal{Y})$. Conversely the latter set is contained in $D(A_{{}^{\mathrm{RL}}\mathcal{Q}_{p,\mathcal{Y},\ell}})$ according to Remark 3.3.4 (applied with $p := \infty$, $q := 1$, $f := u$, $g := k$) and (6.5.39). Regarding the minimal operator, we use $A_{\mathrm{m}} \subseteq A_{{}^{\mathrm{RL}}\mathcal{Q}_{p,\mathcal{Y},\ell}}$ and (6.5.41). □

Observe that $u(0)$ cannot be considered as a boundary value of a generic function u in $D({}^{\mathrm{RL}}\mathrm{D}^{\ell}_{\mathcal{Y}})$. Indeed, for any nonzero vector U_1 of $\mathcal{Y}$, (6.5.5) entails that $u := \ell\, U_1$ lies in $D({}^{\mathrm{RL}}\mathrm{D}^{\ell}_{\mathcal{Y}})$. However u has no essential limit (in $\mathcal{Y}$) at $x = 0$, according to Proposition 6.3.1.

Other closed boundary restriction operators are supplied by Proposition 5.7.3. For we assume that

$$\ell \text{ admits an essential limit at } b,$$

and on account of 6.5.2.1, we put

$$h := \frac{1}{\|\ell\|_{L^q}} \ell^{\mathrm{c}},$$

so that h lies in $L^p(0,b)$ and satisfies $\int \ell h \, \mathrm{d}x = 1$. Being given two scalars λ and μ, we consider the restriction A of ${}^{\mathrm{RL}}\mathrm{D}^{\ell}_{\mathcal{Y}}$ with domain

$$D(A) := \{u \in D({}^{\mathrm{RL}}\mathrm{D}^{\ell}_{\mathcal{Y}}) \mid \lambda k * u(0) + \mu u(b) = 0\}. \tag{6.5.42}$$

We claim that A is a closed boundary restriction operator of ${}^{\mathrm{RL}}\mathcal{Q}_{p,\mathcal{Y},\ell}$. Indeed in view of (6.5.6), (6.5.12), one has

$$\lambda k * u(0) + \mu u(b) = \big[\lambda + \ell(b)\mu\big]U_1 + \mu U_1', \qquad \forall u \in D({}^{\mathrm{RL}}\mathrm{D}^{\ell}_{\mathcal{Y}}). \tag{6.5.43}$$

In the current framework, *the boundary values map* $F_{\mathrm{b.v}}$ introduced in Definition 5.7.1 takes the form

$$F_{\mathrm{b.v}} : D(A_{\mathrm{M}}) \to \mathcal{Y} \times \mathcal{Y}, \qquad u \mapsto (U_1, U_1').$$

Then Proposition 5.7.3 yields that A is a closed boundary restriction operator of ${}^{\mathrm{RL}}\mathcal{Q}_{p,\mathcal{Y},\ell}$, which completes the proof of the claim.

6.5.4.2.2 Adjoints Operators

We will study the adjoint of the latter boundary restriction operators. Recall that ℓ lies in $L^p(0,b) \cap L^q(0,b)$. Regarding the adjoint of the pivot operator, (6.5.40) tells us that

$$(A_{{}^{\mathrm{RL}}\mathcal{Q}_{p,\mathcal{Y},\ell}})^* = S_{\mathcal{Y}^*} A_{{}^{\mathrm{RL}}\mathcal{Q}_{q,\mathcal{Y}^*,\ell}} S_{\mathcal{Y}^*}.$$

Thus Proposition 6.5.6 yields that

$$D\big((A_{{}^{\mathrm{RL}}\mathcal{Q}_{p,\mathcal{Y},\ell}})^*\big) = \{f \in D({}^{\mathrm{RL}}\mathrm{D}^{\ell}_{\mathcal{Y}^*} S_{\mathcal{Y}^*}) \cap C([0,b],\mathcal{Y}^*) \mid f(b) = 0\}. \tag{6.5.44}$$

In terms of endogenous boundary values, for any f in the domain of $S_{\mathcal{Y}^*}\,{}^{\mathrm{RL}}\mathrm{D}^{\ell}_{\mathcal{Y}^*}\,S_{\mathcal{Y}^*}$, one has with (6.5.29),

$$\begin{aligned} f \in D\big((A_{{}^{\mathrm{RL}}\mathcal{Q}_{p,\mathcal{Y},\ell}})^*\big) &\iff F_1 = 0 \\ &\iff k * S_{\mathcal{Y}^*} f(0) = 0. \end{aligned}$$

The adjoint of the minimal operator is given by (5.2.16), i.e.,

$$(A_{\mathrm{m},{}^{\mathrm{RL}}\mathcal{Q}_{p,\mathcal{Y},\ell}})^* = S_{\mathcal{Y}^*}\,{}^{\mathrm{RL}}\mathrm{D}^{\ell}_{\mathcal{Y}^*}\,S_{\mathcal{Y}^*}.$$

Next, consider the adjoint of the maximal operator. Since $A_{\mathrm{m},({}^{\mathrm{RL}}\mathcal{Q}_{p,\mathcal{Y},\ell})^*}$ is closed (from Proposition 5.2.7), we deduce from (5.2.17) that

$$({}^{\mathrm{RL}}\mathrm{D}^{\ell}_{\mathcal{Y}})^* = A_{\mathrm{m},({}^{\mathrm{RL}}\mathcal{Q}_{p,\mathcal{Y},\ell})^*}.$$

Then with (5.6.21), (6.5.24) and Proposition 6.5.6,

$$D\big(({}^{\mathrm{RL}}\mathrm{D}^{\ell}_{\mathcal{Y}})^*\big) = \{f \in D({}^{\mathrm{RL}}\mathrm{D}^{\ell}_{\mathcal{Y}^*} S_{\mathcal{Y}^*}) \cap C([0,b],\mathcal{Y}^*) \mid f(0) = f(b) = 0\}. \tag{6.5.45}$$

In terms of endogenous boundary values, for any f in the domain of ${}^{\mathrm{RL}}\mathrm{D}^{\ell}_{\mathcal{V}^*} S_{\mathcal{V}^*}$, we get in view of (6.5.26)–(6.5.29),

$$f \in D\big(({}^{\mathrm{RL}}\mathrm{D}^{\ell}_{\mathcal{V}})^*\big) \iff F_1 = F_1' = 0$$
$$\iff k * S_{\mathcal{V}^*} f(0) = f(b).$$

Now we assume that

$$\ell \text{ admits an essential limit at } b,$$

and on account of 6.5.2.1, we put

$$h := \frac{1}{\|\ell\|_{L^q}} \ell^{\mathrm{s}}.$$

Let us consider the restriction A of ${}^{\mathrm{RL}}\mathrm{D}^{\ell}_{\mathcal{V}}$ whose domain is given by (6.5.42). Here we suppose in addition that at least one of the two scalars λ and μ is nonzero. We will compute the adjoint of A. Since A is a boundary restriction operator of ${}^{\mathrm{RL}}Q_{p,\mathcal{Y},\ell}$, its adjoint is a restriction of $S^* \mathbb{A} S^*$. We will show that the domain of A^* is given by (6.5.46). More precisely, let f be in $D(S^* \mathbb{A} S^*)$. We claim that f belongs to $D(A^*)$ if and only if

$$\overline{\lambda} k * S_{\mathcal{V}^*} f(0) + \overline{\mu} f(0) = 0. \tag{6.5.46}$$

Indeed, we apply the result of Example 5.7.3 with $\xi_1 = \xi_1^{\mathrm{s}} := \ell$,

$$\eta_1 := h, \qquad \eta_1^{\mathrm{s}} := \frac{1}{\|\ell\|_{L^p}^p} |\ell|^{p-2} \ell$$
$$x_1 := -\mu, \qquad y_1 := \lambda + \ell(b)\mu.$$

With (6.5.29) and since ℓ is real-valued (by Proposition 6.3.1), we get that (6.5.46) is equivalent to

$$\overline{y_1} F_1 = \overline{x_1} F_1'.$$

Moreover, we have assumed $(\lambda, \mu) \neq (0, 0)$; hence $(x_1, y_1) \neq (0, 0)$. Then the claim follows from (5.7.22).

6.5.4.3 Case Where ℓ Lies in $L^q(0, b) \setminus L^p(0, b)$

We proceed as in the previous paragraph by using the results of Sect. 6.5.2.2.

6.5.4.3.1 Boundary Restriction Operators

Since $D({}^{\mathrm{RL}}\mathrm{D}^{\ell}_{\mathcal{V}}) = R(B_{\ell,\mathcal{V}})$, we have

$$A_{{}^{\mathrm{RL}}\mathcal{Q}_{p,\mathcal{Y},\ell}} = {}^{\mathrm{RL}}\mathrm{D}^{\ell}_{\mathcal{V}}. \tag{6.5.47}$$

Accordingly with (6.5.18), we get

$$D(A_{{}^{\mathrm{RL}}\mathcal{Q}_{p,\mathcal{Y},\ell}}) = \{u \in {}_0C([0,b],\mathcal{Y}) \mid k * u \in W^{1,p}\big((0,b),\mathcal{Y}\big)\}.$$

By (6.5.16), (6.5.17),

$$\begin{aligned} u \in D(A_{\mathrm{m},{}^{\mathrm{RL}}\mathcal{Q}_{p,\mathcal{Y},\ell}}) &\Longleftrightarrow U_1' = 0 \\ &\Longleftrightarrow u(b) = 0. \end{aligned}$$

More generally, let us describe the closed boundary restriction operators of ${}^{\mathrm{RL}}\mathcal{Q}_{p,\mathcal{Y},\ell}$. The starting point is the following observation. Under the assumptions and notation of Sect. 6.5.2.2, recalling that

$$\mathcal{V} = L^p(\mathcal{Y}) := L^p\big((0,b),\mathcal{Y}\big), \qquad E_{\mathrm{M}} = B_{\ell,\mathcal{V}} S_{\mathcal{V}}(h\mathcal{Y}),$$

let E be a subspace of E_{M}. Since

$$B_{\ell,\mathcal{V}} S_{\mathcal{V}} : \mathcal{V} \to D({}^{\mathrm{RL}}\mathrm{D}^{\ell}_{\mathcal{V}})$$

is an isomorphism (due to (6.5.47) and (5.2.3)) with inverse $S_{\mathcal{V}}{}^{\mathrm{RL}}\mathrm{D}^{\ell}_{\mathcal{V}}$, there results that E is closed in $D({}^{\mathrm{RL}}\mathrm{D}^{\ell}_{\mathcal{V}})$ if and only if $S_{\mathcal{V}}{}^{\mathrm{RL}}\mathrm{D}^{\ell}_{\mathcal{V}}(E)$ is closed in $(h\mathcal{Y}, \|\cdot\|_{\mathcal{V}})$. Hence, with Lemma 6.5.7 below and the Domain Structure Theorem 5.3.3, we obtain the following results:

(i) If A is a closed boundary restriction operator of ${}^{\mathrm{RL}}\mathcal{Q}_{p,\mathcal{Y},\ell}$, then its domain reads

$$D(A) = B_{\ell,\mathcal{V}}\big(h(b-\cdot)M\big) \oplus B_{\ell,\mathcal{V}}\, S_{\mathcal{V}}\big((\ell\,\mathcal{Y}^*)^{\perp}\big) \tag{6.5.48}$$

for some closed subspace M of $\mathcal{Y}$.

(ii) Conversely if M is a closed subspace of $\mathcal{Y}$, then the restriction A of ${}^{\mathrm{RL}}\mathrm{D}^{\ell}_{\mathcal{V}}$ with domain defined by (6.5.48) is a closed boundary restriction operator of ${}^{\mathrm{RL}}\mathcal{Q}_{p,\mathcal{Y},\ell}$.

Lemma 6.5.7 *Let (b, p) belong to $(0,\infty) \times (1,\infty)$, $\mathcal{Y}$ be a nontrivial normed space, and h be any function of $L^p(0,b)$. Setting*

$$\mathcal{X} := L^p\big((0,b),\mathcal{Y}\big),$$

we consider a subspace W of $\mathcal{X}$. Then the following propositions are equivalent:

(i) *W is a closed subspace of $(h\mathcal{Y}, \|\cdot\|_{\mathcal{X}})$.*
(ii) *There exists a closed subspace M of $\mathcal{Y}$ such that $W = hM$.*

Proof Only the case where h is nontrivial deserves to be proved. In order to show that (ii) implies (i), let $(m_n)_{n\in\mathbb{N}}$ be a sequence in M, and Y lie in $\mathcal{Y}$ such that $hm_n \to hY$ in $\mathcal{X}$. Then

$$\|h\|_{L^p}\|m_n - Y\|_{\mathcal{Y}} = \|hm_n - hY\|_{\mathcal{X}} \to 0.$$

Since M is assumed to be closed in $\mathcal{Y}$, we derive that hM is closed.

Conversely, let us assume (i). Denoting by q the conjugate exponent of p, and referring to Remark 3.2.1, we recall that the conjugate function h^{c} of h lies in $L^q(0, b)$ and satisfies $\int h^{\mathrm{c}}h \neq 0$ (where $\int$ means $\int_0^b$). Let us first show that $W \subseteq hM_1$ where

$$M_1 := \Big\{ \int h^{\mathrm{c}} w \,\mathrm{d}y \in \mathcal{Y} \mid w \in W \Big\}.$$

For, any w in W reads $w = hY$ for some Y in $\mathcal{Y}$. Since

$$\int h^{\mathrm{c}} w \,\mathrm{d}y = \int h^{\mathrm{c}} h \,\mathrm{d}y\, Y,$$

Y lies in M_1, so that $W \subseteq hM_1$. Conversely, let us show that $hM_1 \subseteq W$. For, any v in hM_1 reads

$$v = h \int h^{\mathrm{c}} w \,\mathrm{d}y$$

for some w in W. Moreover by (i), there exists Y in $\mathcal{Y}$ such that $w = hY$. Thus

$$v = \int h^{\mathrm{c}} h \,\mathrm{d}y\, hY = \int h^{\mathrm{c}} h \,\mathrm{d}y\, w.$$

Thus $hM_1 \subseteq W$. Hence

$$hM_1 = W. \tag{6.5.49}$$

There remains to show that M_1 is closed in $\mathcal{Y}$. Let $(w_n)_{n\in\mathbb{N}}$ be a sequence in W and Y be in $\mathcal{Y}$ such that

$$\int h^{\mathrm{c}} w_n \,\mathrm{d}y \to Y \qquad \text{in } \mathcal{Y}.$$

It is a matter of showing that Y lies in M_1. For, since $W \subseteq h\mathcal{Y}$ (by (i)), w_n reads $w_n = hY_n$ for some Y_n in $\mathcal{Y}$. Then

$$Y_n \to \Big(\int h^{\mathrm{c}} h \,\mathrm{d}y\Big)^{-1} Y \qquad \text{in } \mathcal{Y}.$$

Thus

$$w_n = hY_n \to \Big(\int h^{\mathrm{c}} h \,\mathrm{d}y\Big)^{-1} hY \qquad \text{in } \mathcal{X}.$$

However, W is closed in $\mathcal{X}$ by (i); hence hY lies in W. With (6.5.49), there exists m_1 in M_1 such that $hY = hm_1$. Thus

$$\int h^{\mathrm{c}} h \,\mathrm{d}y\, Y = \int h^{\mathrm{c}} h \,\mathrm{d}y\, m_1,$$

that is, $Y = m_1$; hence Y lies in M_1. □

6.5.4.3.2 Adjoints Operators

Let A be a closed boundary restriction operator of ${}^{\mathrm{RL}}\mathcal{Q}_{p,\mathcal{Y},\ell}$. Since $A^* \subseteq S_{\mathcal{V}^*}\,{}^{\mathrm{RL}}\mathrm{D}^{\ell}_{\mathcal{V}^*} S_{\mathcal{V}^*}$, it is enough to compute the domain of A^*.

By Item (i) in Sect. 6.5.4.3.1, there exists a closed subspace M of $\mathcal{Y}$ such that $D(A)$ satisfies (6.5.48). Hence, we may apply Corollary 5.4.2 with $G_0 := hM$. Being given any function f in $D(S_{\mathcal{V}^*}\,{}^{\mathrm{RL}}\mathrm{D}^{\ell}_{\mathcal{V}^*} S_{\mathcal{V}^*})$, we obtain with (6.5.32), (6.5.33) that

$$f \in D(A^*) \iff (k * S_{\mathcal{V}^*} f)(0) \in M^{\perp}.$$

In particular, if $A := {}^{\mathrm{RL}}\mathrm{D}^{\ell}_{\mathcal{V}}$ then $M = \mathcal{Y}$ and

$$f \in D\big(({}^{\mathrm{RL}}\mathrm{D}^{\ell}_{\mathcal{V}})^*\big) \iff (k * S_{\mathcal{V}^*} f)(0) = 0. \tag{6.5.50}$$

Also if $A := A_{\mathrm{m},{}^{\mathrm{RL}}\mathcal{Q}_{p,\mathcal{Y},\ell}}$ then $M = \{0\}$, and there is no constraint on the endogenous boundary values of f. Hence

$$(A_{\mathrm{m},{}^{\mathrm{RL}}\mathcal{Q}_{p,\mathcal{Y},\ell}})^* = S_{\mathcal{V}^*}\,{}^{\mathrm{RL}}\mathrm{D}^{\ell}_{\mathcal{V}^*} S_{\mathcal{V}^*},$$

a result which can be also retrieved directly from the general theory of boundary restriction operators (see (5.2.16)).

6.5.4.4 Case Where ℓ Lies in $L^p(0, b) \setminus L^q(0, b)$

6.5.4.4.1 Boundary Restriction Operators

By (6.5.39),

$$D(A_{{}^{\mathrm{RL}}\mathcal{Q}_{p,\mathcal{Y},\ell}}) = \{u \in D({}^{\mathrm{RL}}\mathrm{D}^{\ell}_{\mathcal{V}}) \mid k * u(0) = 0\}.$$

Since ker $\mathbb{A}$ is trivial, the domains of the minimal and the pivot operators are equal, so that

$$A_{\mathrm{m},{}^{\mathrm{RL}}\mathcal{Q}_{p,\mathcal{Y},\ell}} = A_{{}^{\mathrm{RL}}\mathcal{Q}_{p,\mathcal{Y},\ell}}.$$

In general, since $E_{\mathrm{M}} := \ell\mathcal{Y}$, we obtain the following results arguing as in Sect. 6.5.4.3.1.

(i) If A is a closed boundary restriction operator of ${}^{\mathrm{RL}}\mathcal{Q}_{p,\mathcal{Y},\ell}$ then its domain reads

$$D(A) = \ell\, M \oplus R(B_{\ell,\mathcal{V}}), \tag{6.5.51}$$

for some closed subspace M of $\mathcal{Y}$.

(ii) Conversely if M is a closed subspace of $\mathcal{Y}$, then the restriction A of ${}^{\mathrm{RL}}\mathrm{D}^{\ell}_{\mathcal{V}}$ with domain defined by (6.5.51) is a closed boundary restriction operator of ${}^{\mathrm{RL}}\mathcal{Q}_{p,\mathcal{Y},\ell}$.

6.5.4.4.2 Adjoints Operators

Let A be a closed boundary restriction operator of ${}^{\mathrm{RL}}\mathcal{Q}_{p,\mathcal{Y},\ell}$. By (6.5.51), there exists a closed subspace M of $\mathcal{Y}$ such that ker $A = \ell\, M$. Moreover by (6.5.36) and (6.5.37), every f in $D(S_{\mathcal{V}^*}\,{}^{\mathrm{RL}}\mathrm{D}^{\ell}_{\mathcal{V}^*}\, S_{\mathcal{V}^*})$ is continuous on $[0, b]$ and reads

$$f = S_{\mathcal{V}^*}\, B_{\ell,\mathcal{V}^*}\, S_{\mathcal{V}^*}(hf(0) + \varphi^{\perp}),$$

for some $\varphi^{\perp}$ in $(\ell\mathcal{Y})^{\perp}$. Thus, by Theorem 5.4.1,

$$\begin{aligned} f \in D(A^*) &\Longleftrightarrow 0 = \langle hf(0),\, \ell\, Y\rangle_{\mathcal{V}^*,\mathcal{V}} && \forall Y \in M \\ &\Longleftrightarrow 0 = \langle f(0),\, Y\rangle_{\mathcal{Y}^*,\mathcal{Y}} && \forall Y \in M \\ &\Longleftrightarrow f(0) \in M^{\perp}. \end{aligned}$$

In particular, if $A := {}^{\mathrm{RL}}\mathrm{D}^{\ell}_{\mathcal{V}}$, then $M = \mathcal{Y}$ and

$$f \in D\big(({}^{\mathrm{RL}}\mathrm{D}^{\ell}_{\mathcal{V}})^*\big) \Longleftrightarrow f(0) = 0. \tag{6.5.52}$$

Also if $A := A_{{}^{\mathrm{RL}}\mathcal{Q}_{p,\mathcal{Y},\ell}}$ then $M = \{0\}$ and

$$(A_{{}^{\mathrm{RL}}\mathcal{Q}_{p,\mathcal{Y},\ell}})^* = S_{\mathcal{V}^*}\,{}^{\mathrm{RL}}\mathrm{D}^{\ell}_{\mathcal{V}^*}\, S_{\mathcal{V}^*}.$$

6.5.4.5 Case Where ℓ Lies in $L^1(0,b) \setminus (L^q(0,b) \cup L^p(0,b))$

Then the minimal operator of ${}^{\mathrm{RL}}\mathcal{Q}_{p,\mathcal{Y},\ell}$ coincide with the maximal operator ${}^{\mathrm{RL}}\mathrm{D}^{\ell}_{\mathcal{V}}$. Hence, ${}^{\mathrm{RL}}\mathrm{D}^{\ell}_{\mathcal{V}}$ is the only one closed boundary restriction operator of ${}^{\mathrm{RL}}\mathcal{Q}_{p,\mathcal{Y},\ell}$. The adjoint of ${}^{\mathrm{RL}}\mathrm{D}^{\ell}_{\mathcal{V}}$ is $S_{\mathcal{V}^*}\,{}^{\mathrm{RL}}\mathrm{D}^{\ell}_{\mathcal{V}^*}\,S_{\mathcal{V}^*}$.

6.6 The Quadruplet ${}^{\mathrm{RL}}\mathcal{Q}_{p,\mathcal{Y},n+\ell}$

Let $\mathbb{K}$ be equal to $\mathbb{R}$ or $\mathbb{C}$, b be a positive real number, and p, q be conjugate exponents in $(1,\infty)$. Let also $\mathcal{Y}$ be a nontrivial separable and reflexive Banach space over $\mathbb{K}$ and (k,ℓ) be a Sonine pair.

We set

$$\mathcal{V} := L^p\big((0,b),\mathcal{Y}\big), \qquad \mathcal{V}^* := L^q\big((0,b),\mathcal{Y}^*\big).$$

6.6.1 Introduction

Let n be a nonnegative integer. If n is positive, on account of Example 5.6.7 and Sect. 6.5.1, we will study the product (in the sense of Definition 5.6.5) of the differential quadruplets:

$$\mathcal{Q}_{p,\mathcal{Y},n} := \big(\mathrm{D}^n_{\mathcal{V}^*},\ \mathrm{D}^n_{\mathcal{V}},\ B_{n,\mathcal{V}},\ S_{\mathcal{V}}\big), \qquad {}^{\mathrm{RL}}\mathcal{Q}_{p,\mathcal{Y},\ell} := ({}^{\mathrm{RL}}\mathrm{D}^{\ell}_{\mathcal{V}^*},\ {}^{\mathrm{RL}}\mathrm{D}^{\ell}_{\mathcal{V}},\ B_{\ell,\mathcal{V}},\ S_{\mathcal{V}}).$$

When $n = 0$, recall that from Example 5.6.7,

$$\mathrm{D}^0_{\mathcal{V}} := \mathrm{id}_{\mathcal{V}}, \qquad B_{0,\mathcal{V}} := \mathrm{id}_{\mathcal{V}},$$

and

$$\mathcal{Q}_{p,\mathcal{Y},0} := \big(\mathrm{id}_{\mathcal{V}^*},\ \mathrm{id}_{\mathcal{V}},\ \mathrm{id}_{\mathcal{V}},\ S_{\mathcal{V}}\big).$$

In this respect, we set

$$g_0 * h := h, \qquad \forall h \in L^1\big((0,b),\mathcal{Y}\big). \tag{6.6.1}$$

For all purposes, recall that by convention, any operator raised to the power of zero is the identity operator. Observe that $\mathcal{Q}_{p,\mathcal{Y},0}$ remains a differential quadruplet.

Since $B_{n,\mathcal{V}}$ commute with $B_{\ell,\mathcal{V}}$, Proposition 5.6.13 entails that the product $\mathcal{Q}_{p,\mathcal{Y},n}\,{}^{\mathrm{RL}}\mathcal{Q}_{p,\mathcal{Y},\ell}$ is a well-defined differential quadruplet denoted by ${}^{\mathrm{RL}}\mathcal{Q}_{p,\mathcal{Y},n+\ell}$. In symbol, for each nonnegative integer n,

$$ {}^{\mathrm{RL}}\mathcal{Q}_{p,\mathcal{Y},n+\ell} := \mathcal{Q}_{p,\mathcal{Y},n}{}^{\mathrm{RL}}\mathcal{Q}_{p,\mathcal{Y},\ell}. $$

In the particular case where $n = 0$,

$$ {}^{\mathrm{RL}}\mathcal{Q}_{p,\mathcal{Y},0+\ell} = {}^{\mathrm{RL}}\mathcal{Q}_{p,\mathcal{Y},\ell}. $$

Since

$$ \mathcal{Q}_{p,\mathcal{Y},n}{}^{\mathrm{RL}}\mathcal{Q}_{p,\mathcal{Y},\ell} = (\mathrm{D}^n_{\mathcal{V}*}\,{}^{\mathrm{RL}}\mathrm{D}^\ell_{\mathcal{V}*},\ \mathrm{D}^n_{\mathcal{V}}{}^{\mathrm{RL}}\mathrm{D}^\ell_{\mathcal{V}},\ B_{\ell,\mathcal{V}}B_{n,\mathcal{V}},\ S_{\mathcal{V}}), $$

we are led to introduce the notations

$$ {}^{\mathrm{RL}}\mathrm{D}^{n+\ell}_{\mathcal{V}} := \mathrm{D}^n_{\mathcal{V}}\,{}^{\mathrm{RL}}\mathrm{D}^\ell_{\mathcal{V}}, \qquad {}^{\mathrm{RL}}\mathrm{D}^{n+\ell}_{\mathcal{V}*} := \mathrm{D}^n_{\mathcal{V}*}\,{}^{\mathrm{RL}}\mathrm{D}^\ell_{\mathcal{V}*} $$

$$ {}^{\mathrm{RL}}\mathrm{D}^{-1+\ell}_{\mathcal{V}} := B_{k,\mathcal{V}}, \qquad {}^{\mathrm{RL}}\mathrm{D}^{-1+\ell}_{\mathcal{V}*} := B_{k,\mathcal{V}*}. $$

Hence

$$ {}^{\mathrm{RL}}\mathcal{Q}_{p,\mathcal{Y},n+\ell} = ({}^{\mathrm{RL}}\mathrm{D}^{n+\ell}_{\mathcal{V}*},\ {}^{\mathrm{RL}}\mathrm{D}^{n+\ell}_{\mathcal{V}},\ B_{g_n*\ell,\mathcal{V}},\ S_{\mathcal{V}}). $$

Clearly, one has

$$ D({}^{\mathrm{RL}}\mathrm{D}^{n+\ell}_{\mathcal{V}}) = \left\{u \in L^p(\mathcal{Y}) \mid k * u \in W^{n+1,p}(\mathcal{Y})\right\} $$

and for each u in $D({}^{\mathrm{RL}}\mathrm{D}^{n+\ell}_{\mathcal{V}})$,

$$ {}^{\mathrm{RL}}\mathrm{D}^{n+\ell}_{\mathcal{V}} u = \frac{\mathrm{d}^{n+1}}{\mathrm{d}x^{n+1}}\{k * u\}. $$

By Example 5.6.7, when n is positive,

$$ \ker \mathrm{D}^n_{\mathcal{V}} = \bigoplus_{i=1}^{n} g_i\mathcal{Y}, $$

and according to (6.5.3),

$$ \ker {}^{\mathrm{RL}}\mathrm{D}^\ell_{\mathcal{V}} = \begin{cases} \ell\mathcal{Y} & \text{if } \ell \in L^p(0,b) \\ \{0\} & \text{otherwise} \end{cases}. $$

Thus Proposition 5.6.24 yields that for any $n \geq 0$,

$$\ker {}^{\mathrm{RL}}\mathrm{D}_{\mathcal{V}}^{n+\ell} = \bigoplus_{i=i_p}^{n} g_i * \ell\mathcal{Y} \tag{6.6.2}$$

$$\ker {}^{\mathrm{RL}}\mathrm{D}_{\mathcal{V}^*}^{n+\ell} = \bigoplus_{i=i_q}^{n} g_i * \ell\mathcal{Y}^*, \tag{6.6.3}$$

where for any r in $(1, \infty)$,

$$i_r := \begin{cases} 0 & \text{if } \ell \in L^r(0, b) \\ 1 & \text{otherwise,} \end{cases} \tag{6.6.4}$$

and by convention $\bigoplus_{i=i_r}^{n} \cdots = \{0\}$ if $n < i_r$.

Now we claim that ${}^{\mathrm{RL}}\mathcal{Q}_{p,\mathcal{Y},n+\ell}$ is regular. Indeed, this is known from Proposition 6.5.2 in the case where $n = 0$. If n is positive, then in view of (6.6.2) and (6.6.2), Corollary 5.3.1 yields the existence of functions $h_{i_q+1}, \ldots, h_{n+1}$ in $L^p(0, b)$ such that, on the one hand,

$$\int_0^b g_{i-1} * \ell(y) h_j(y)\,\mathrm{d}y = \delta_{ij}, \qquad i_q + 1 \le i, j \le n + 1 \tag{6.6.5}$$

On the other hand, ${}^{\mathrm{RL}}\mathcal{Q}_{p,\mathcal{Y},n+\ell}$ is F_0-regular with

$$F_0 := \bigoplus_{i=i_q+1}^{n+1} h_i\mathcal{Y}.$$

6.6.2 Endogenous Boundary Values

With the notation of Sect. 5.3.3, the Domain Structure Theorem 5.3.3 gives

$$E_{\mathrm{M}} = \bigoplus_{i=i_p}^{n} g_i * \ell\mathcal{Y} \oplus B_{g_n*\ell,\mathcal{V}} S_{\mathcal{V}}\Big(\bigoplus_{i=i_q+1}^{n+1} h_i\mathcal{Y}\Big)$$

$$D({}^{\mathrm{RL}}\mathrm{D}_{\mathcal{V}}^{n+\ell}) = E_{\mathrm{M}} \oplus B_{g_n*\ell,\mathcal{V}} S_{\mathcal{V}}\Big(\bigoplus_{i=i_q}^{n} g_i * \ell\mathcal{Y}^*\Big)^{\perp}.$$

We refer to Sect. 6.5.2 for the analysis of the case where $n = 0$. In the sequel we suppose that n is a positive integer. Thus a function u in $\mathcal{V}$ lies in $D({}^{\mathrm{RL}}\mathrm{D}_{\mathcal{V}}^{n+\ell})$ if and only if there exist $U_{i_p+1}, \ldots, U_{n+1}$ and $U'_{i_q+1}, \ldots, U'_{n+1}$ in $\mathcal{Y}$ and $\psi^{\perp}$ in $(\ker {}^{\mathrm{RL}}\mathrm{D}_{\mathcal{V}^*}^{n+\ell})^{\perp}$ such that

$$u = \sum_{i=i_p+1}^{n+1} g_{i-1} * \ell\, U_i + \sum_{i=i_q+1}^{n+1} g_n * \ell * S_{\mathcal{V}}(h_i\, U_i') + g_n * \ell * S_{\mathcal{V}} \psi^{\perp}. \tag{6.6.6}$$

We will compute the endogenous boundary values of any function u in the domain of ${}^{\mathrm{RL}}\mathrm{D}_{\mathcal{V}}^{n+\ell}$. The results are stated in:

(i) (6.6.7), (6.6.8) which form a set of $2n + 2 - i_p - i_q$ equations relating endogenous boundary values with "standard" boundary values;
(ii) (6.6.10), (6.6.11) where the endogenous boundary values are expressed in terms of a simpler set of boundary values.

Recalling that n is a positive integer, we claim that for each j in $[i_p + 1, n + 1]$

$$U_j = {}^{\mathrm{RL}}\mathrm{D}_{\mathcal{V}}^{j-2+\ell} u(0). \tag{6.6.7}$$

By the conventions of Sect. 6.6.1, when $i_p = 0$ and $j = 1$, (6.6.7) reduces to

$$U_1 = k * u(0).$$

We prove (6.6.7) by writing (6.6.6) under the form

$$u = \sum_{i=i_p+1}^{n+1} g_{i-1} * \ell\, U_i + g_n * \ell * h,$$

for some h in $\mathcal{V}$. Then we apply ${}^{\mathrm{RL}}\mathrm{D}_{\mathcal{V}}^{j-2+\ell}$ to the latter equation and use the following result (whose easy proof is skipped) to get (6.6.7).

Lemma 6.6.1 *Let p belong to $(1, \infty)$, i_p be defined by* (6.6.4), (k, ℓ) *be a Sonine pair, and h be any function in $L^p(0, b)$. Let also i, j be integers such that i is nonnegative and $j \geq -1$. Then*

$${}^{\mathrm{RL}}\mathrm{D}_{L^p}^{j+\ell}(g_i * \ell) = \begin{cases} 0 & \text{if } i_p \leq i \leq j \\ g_{i-j} & \text{if } j + 1 \leq i \end{cases}$$

$${}^{\mathrm{RL}}\mathrm{D}_{L^p}^{j+\ell}(g_i * \ell * h) = g_{i-j} * h \qquad \text{if } j \leq i.$$

For positive n and each $j = i_q + 1, \dots, n + 1$, the first formula for the U_j' is

$$U_j' = g_{j-1} * \ell * {}^{\mathrm{RL}}\mathrm{D}_{\mathcal{V}}^{n+\ell} u(b). \tag{6.6.8}$$

In order to establish (6.6.8), we have to get rid of $\psi^\perp$. For, we use the belonging of $\psi^\perp$ to

$$(\bigoplus_{i=i_q}^{n} g_i * \ell\mathcal{Y}^*)^\perp = \bigcap_{i=i_q}^{n} (g_i * \ell\mathcal{Y}^*)^\perp,$$

which implies due to (3.2.1) and since $g_i * \ell$ is real-valued that

$$\int_0^b g_i * \ell(y)\psi^\perp(y)\,\mathrm{d}y = 0, \qquad \forall i \in [i_q, n]. \tag{6.6.9}$$

Next applying the operator ${}^{\mathrm{RL}}\mathrm{D}_{\mathcal{V}}^{n+\ell}$ to (6.6.6), and using the Domain Structure Theorem 5.3.3 (i), we get

$${}^{\mathrm{RL}}\mathrm{D}_{\mathcal{V}}^{n+\ell} u = \sum_{i=i_q+1}^{n+1} S_{\mathcal{V}}(h_i\, U_i') + S_{\mathcal{V}}\psi^\perp.$$

Moreover, with Proposition 3.3.5, for all indexes i, j in $[i_q + 1, n + 1]$, we get

$$g_{j-1} * \ell * S_{\mathcal{V}}(h_i\, U_i')(b) = \int_0^b g_{j-1} * \ell(y)h_i(y)\,\mathrm{d}y\, U_i' = \delta_{ij}\, U_j'$$

by (6.6.5), and

$$g_{j-1} * \ell * S_{\mathcal{V}}\psi^\perp(b) = \int_0^b g_{j-1} * \ell(y)\psi^\perp(y)\,\mathrm{d}y = 0,$$

by (6.6.9). Then (6.6.8) follows.

The $2n + 2 - i_p - i_q$ "standard" boundary values appearing in (6.6.7) and (6.6.8) are

$$\begin{gathered} {}^{\mathrm{RL}}\mathrm{D}_{\mathcal{V}}^{i_p-1+\ell} u(0), \ldots, {}^{\mathrm{RL}}\mathrm{D}_{\mathcal{V}}^{n-1+\ell} u(0), \\ g_{i_q} * \ell * {}^{\mathrm{RL}}\mathrm{D}_{\mathcal{V}}^{n+\ell} u(b), \ldots, g_n * \ell * {}^{\mathrm{RL}}\mathrm{D}_{\mathcal{V}}^{n+\ell} u(b). \end{gathered}$$

As in Sect. 5.7.3.1, we may integrate $g_j * {}^{\mathrm{RL}}\mathrm{D}_{\mathcal{V}}^{n+\ell} u$ to find simpler "standard" boundary values. More precisely we will show that for $j = i_p + 1, \ldots, n + 1$,

$$U_j = {}^{\mathrm{RL}}\mathrm{D}_{\mathcal{V}}^{j-2+\ell} u(0), \tag{6.6.10}$$

and that for $j = i_q + 1, \ldots, n + 1$,

$$U'_j = -\sum_{i=n+1-j}^{n-1} g_{i+j-n} * \ell(b)\, {}^{\mathrm{RL}}\mathrm{D}_{\mathcal{V}}^{i+\ell} u(0) + \ell * {}^{\mathrm{RL}}\mathrm{D}_{\mathcal{V}}^{n+1-j+\ell} u(b). \tag{6.6.11}$$

The $2n + 2 - i_p - i_q$ "standard" boundary values appearing in (6.6.10) and (6.6.11) are

$$ {}^{\mathrm{RL}}\mathrm{D}_{\mathcal{V}}^{i_p-1+\ell} u(0), \dots, {}^{\mathrm{RL}}\mathrm{D}_{\mathcal{V}}^{n-1+\ell} u(0), $$
$$ \ell * {}^{\mathrm{RL}}\mathrm{D}_{\mathcal{V}}^{\ell} u(b), \dots, \ell * {}^{\mathrm{RL}}\mathrm{D}_{\mathcal{V}}^{n-i_q+\ell} u(b). $$

Recall that by convention (see Sect. 1.5.1), $\sum_n^{n-1} \cdots = 0$. Now (6.6.10) is just (6.6.7). In order to prove (6.6.11), we notice that for any integer j in $[2, n+1]$, the function ${}^{\mathrm{RL}}\mathrm{D}_{\mathcal{V}}^{n+1-j+\ell} u$ lies in the domain of $\mathrm{D}_{\mathcal{V}}^{j-1}$. In view of Sect. 5.7.3.1, let $h'_1, \dots, h'_{j-1}$ be functions in $L^p(0, b)$ such that $\int_0^b h'_i g_l \,\mathrm{d}y = \delta_{il}$ for each indexes i and l in $[1, j-1]$. By considering the $\bigoplus_{i=1}^{j-1} h'_i$ $\mathcal{Y}$-regular quadruplet $Q_{p,\mathcal{Y},j-1}$, let $(V_1, \dots, V_{j-1}, V'_1, \dots, V'_{j-1})$ be the endogenous boundary values of ${}^{\mathrm{RL}}\mathrm{D}_{\mathcal{V}}^{n+1-j+\ell} u$ with respect to the direct sum decompositions

$$\ker \mathrm{D}_{\mathcal{V}}^{j-1} = \bigoplus_{i=1}^{j-1} g_i \mathcal{Y}, \qquad F_0 = \bigoplus_{i=1}^{j-1} h'_i \mathcal{Y}.$$

Remark that we have to exclude the case $j = 1$ in order to make the latter endogenous boundary values well defined. By (5.7.25), we have

$$V_i = \mathrm{D}_{\mathcal{V}}^{i-1}\, {}^{\mathrm{RL}}\mathrm{D}_{\mathcal{V}}^{n+1-j+\ell} u(0) = {}^{\mathrm{RL}}\mathrm{D}_{\mathcal{V}}^{n+i-j+\ell} u(0), \qquad \forall\, 1 \le i \le j-1.$$

Moreover by Lemma 4.3.3 (or use alternatively (5.7.28)),

$$ {}^{\mathrm{RL}}\mathrm{D}_{\mathcal{V}}^{n+1-j+\ell} u = \sum_{i=1}^{j-1} g_i\, V_i + g_{j-1} * {}^{\mathrm{RL}}\mathrm{D}_{\mathcal{V}}^{n+\ell} u. $$

Since ${}^{\mathrm{RL}}\mathrm{D}_{\mathcal{V}}^{n+1-j+\ell} u$ lies in the domain of $\mathrm{D}_{\mathcal{V}}^{j-1}$, ${}^{\mathrm{RL}}\mathrm{D}_{\mathcal{V}}^{n+1-j+\ell} u$ is continuous when $j \ge 2$. Thus for each $j \ge 2$, one has

$$ g_{j-1} * \ell * {}^{\mathrm{RL}}\mathrm{D}_{\mathcal{V}}^{n+\ell} u(b) = $$
$$ \ell * {}^{\mathrm{RL}}\mathrm{D}_{\mathcal{V}}^{n+1-j+\ell} u(b) - \sum_{i=n+1-j}^{n-1} g_{i+j-n} * \ell(b)\, {}^{\mathrm{RL}}\mathrm{D}_{\mathcal{V}}^{i+\ell} u(0). $$

With (6.6.8), we get (6.6.11) when j lies in $[2, n+1]$. If $j = 1$ then we must have $i_q = 0$, that is, ℓ belongs to $L^q(0, b)$. In this case, (6.6.11) is just (6.6.8) with $j = 1$. The proof of (6.6.10) and (6.6.11) is now complete.

Example 6.6.1 For any u in $D({}^{\mathrm{RL}}\mathrm{D}_{\mathcal{V}}^{1+\ell})$, let us write (6.6.10) and (6.6.11) when $n = 1$.

(i) Case where ℓ lies in $L^p(0, b) \cap L^q(0, b)$. One has $i_p = i_q = 0$. According to (6.6.10) and (6.6.11), the expression of the four endogenous boundary values of u in terms of the four "standard" boundary values

$$k * u(0),\ {}^{\mathrm{RL}}\mathrm{D}_{\mathcal{V}}^{\ell}u(0),\ \ell * {}^{\mathrm{RL}}\mathrm{D}_{\mathcal{V}}^{\ell}u(b),\ \ell * {}^{\mathrm{RL}}\mathrm{D}_{\mathcal{V}}^{1+\ell}u(b)$$

is

$$\begin{cases} U_1 = k * u(0) \\ U_2 = {}^{\mathrm{RL}}\mathrm{D}_{\mathcal{V}}^{\ell}u(0) \\ U_1' = \ell * {}^{\mathrm{RL}}\mathrm{D}_{\mathcal{V}}^{1+\ell}u(b) \\ U_2' = -g_1 * \ell(b)\, {}^{\mathrm{RL}}\mathrm{D}_{\mathcal{V}}^{\ell}u(0) + \ell * {}^{\mathrm{RL}}\mathrm{D}_{\mathcal{V}}^{\ell}u(b). \end{cases} \tag{6.6.12}$$

(ii) Case where ℓ lies in $L^q(0, b) \setminus L^p(0, b)$. One has $i_p = 1$, $i_q = 0$. There are three endogenous boundary values, namely,

$${}^{\mathrm{RL}}\mathrm{D}_{\mathcal{V}}^{\ell}u(0),\ \ell * {}^{\mathrm{RL}}\mathrm{D}_{\mathcal{V}}^{\ell}u(b),\ \ell * {}^{\mathrm{RL}}\mathrm{D}_{\mathcal{V}}^{1+\ell}u(b).$$

Moreover,

$$\begin{cases} U_2 = {}^{\mathrm{RL}}\mathrm{D}_{\mathcal{V}}^{\ell}u(0) \\ U_1' = \ell * {}^{\mathrm{RL}}\mathrm{D}_{\mathcal{V}}^{1+\ell}u(b) \\ U_2' = -g_1 * \ell(b)\, {}^{\mathrm{RL}}\mathrm{D}_{\mathcal{V}}^{\ell}u(0) + \ell * {}^{\mathrm{RL}}\mathrm{D}_{\mathcal{V}}^{\ell}u(b). \end{cases} \tag{6.6.13}$$

(iii) Case where ℓ lies in $L^p(0, b) \setminus L^q(0, b)$. Here $i_p = 0$, $i_q = 1$, and there are also three endogenous boundary values, namely,

$$k * u(0),\ {}^{\mathrm{RL}}\mathrm{D}_{\mathcal{V}}^{\ell}u(0),\ \ell * {}^{\mathrm{RL}}\mathrm{D}_{\mathcal{V}}^{\ell}u(b).$$

Moreover,

$$\begin{cases} U_1 = k * u(0) \\ U_2 = {}^{\mathrm{RL}}\mathrm{D}^{\ell}_{\mathcal{Y}} u(0) \\ U_2' = -g_1 * \ell(b)\, {}^{\mathrm{RL}}\mathrm{D}^{\ell}_{\mathcal{Y}} u(0) + \ell * {}^{\mathrm{RL}}\mathrm{D}^{\ell}_{\mathcal{Y}} u(b). \end{cases} \tag{6.6.14}$$

(iv) Case where ℓ lies in $L^1(0, b) \setminus (L^q(0, b) \cup L^p(0, b))$. Here $i_p = i_q = 1$. There are two endogenous boundary values, namely,

$$k * u(0),\ {}^{\mathrm{RL}}\mathrm{D}^{\ell}_{\mathcal{Y}} u(0).$$

Moreover,

$$\begin{cases} U_2 = {}^{\mathrm{RL}}\mathrm{D}^{\ell}_{\mathcal{Y}} u(0) \\ U_2' = -g_1 * \ell(b)\, {}^{\mathrm{RL}}\mathrm{D}^{\ell}_{\mathcal{Y}} u(0) + \ell * {}^{\mathrm{RL}}\mathrm{D}^{\ell}_{\mathcal{Y}} u(b). \end{cases} \tag{6.6.15}$$

□

6.6.3 The Adjoint Quadruplet of ${}^{\mathrm{RL}}\mathcal{Q}_{p,\mathcal{Y},n+\ell}$

Under the assumptions and notation of Sect. 6.6.1, let n be a positive integer. The adjoint quadruplet of ${}^{\mathrm{RL}}\mathcal{Q}_{p,\mathcal{Y},n+\ell}$ is

$$(S_{\mathcal{Y}}\, {}^{\mathrm{RL}}\mathrm{D}^{n+\ell}_{\mathcal{Y}}\, S_{\mathcal{Y}},\ S_{\mathcal{Y}^*}\, {}^{\mathrm{RL}}\mathrm{D}^{n+\ell}_{\mathcal{Y}^*}\, S_{\mathcal{Y}^*},\ S_{\mathcal{Y}^*}\, B_{g_n*\ell,\mathcal{Y}^*}\, S_{\mathcal{Y}^*},\ S_{\mathcal{Y}^*}). \tag{6.6.16}$$

Since

$$\ker {}^{\mathrm{RL}}\mathrm{D}^{n+\ell}_{\mathcal{Y}^*} = \bigoplus_{i=i_q}^{n} g_i * \ell\mathcal{Y}^*, \qquad \ker {}^{\mathrm{RL}}\mathrm{D}^{n+\ell}_{\mathcal{Y}} = \bigoplus_{i=i_p}^{n} g_i * \ell\mathcal{Y},$$

arguing as in Sect. 6.6.1, we show on the one hand the existence of functions $h^{\mathrm{s}}_{i_p+1}, \ldots, h^{\mathrm{s}}_{n+1}$ in $L^q(0, b)$ such that

$$\int_0^b h^{\mathrm{s}}_j(y) g_{i-1} * \ell(y)\, \mathrm{d}y = \delta_{ij}, \qquad i_p + 1 \le i, j \le n+1.$$

On the other hand,

$$F^{\mathrm{s}}_0 := \bigoplus_{i=i_p+1}^{n+1} h^{\mathrm{s}}_i \mathcal{Y}^*$$

is a closed complementary subspace of $(\ker {}^{\mathrm{RL}}\mathrm{D}_{\mathcal{V}}^{n+\ell})^{\perp}$ in $L^q(\mathcal{Y})$. Then Corollary 5.3.4 yields that

$$D(S_{\mathcal{V}^*}\, {}^{\mathrm{RL}}\mathrm{D}_{\mathcal{V}^*}^{n+\ell}\, S_{\mathcal{V}^*}) = E^{\mathrm{s}}_{\mathrm{M}} \oplus S_{\mathcal{V}^*}\, B_{g_n*\ell,\mathcal{V}^*}\, S_{\mathcal{V}^*}\,(\bigoplus_{i=i_p}^{n} g_i * \ell\mathcal{Y})^{\perp} \tag{6.6.17}$$

where

$$E^{\mathrm{s}}_{\mathrm{M}} := \bigoplus_{i=i_q}^{n} S_{\mathcal{V}^*}(g_i * \ell\mathcal{Y}^*) \oplus S_{\mathcal{V}^*}\, B_{g_n*\ell,\mathcal{V}^*}\, S_{\mathcal{V}^*}(\bigoplus_{i=i_p+1}^{n+1} h_i^{\mathrm{s}}\mathcal{Y}^*). \tag{6.6.18}$$

Thus a function f in $\mathcal{V}^*$ lies in $D(S_{\mathcal{V}^*}\, {}^{\mathrm{RL}}\mathrm{D}_{\mathcal{V}^*}^{n+\ell}\, S_{\mathcal{V}^*})$ if and only if there exist $F_{i_q+1},\dots,F_{n+1}, F'_{i_p+1},\dots,F'_{n+1}$ in $\mathcal{Y}^*$ and $\varphi^{\perp}$ in $(\ker {}^{\mathrm{RL}}\mathrm{D}_{\mathcal{V}}^{n+\ell})^{\perp}$ such that

$$\begin{aligned} f = &\sum_{i=i_q+1}^{n+1} S_{\mathcal{V}^*}(g_{i-1} * \ell\, F_i) \\ &+ \sum_{i=i_p+1}^{n+1} S_{\mathcal{V}^*}\, B_{g_n*\ell,\mathcal{V}^*}\, S_{\mathcal{V}^*}(h_i^{\mathrm{s}}\, F_i') + S_{\mathcal{V}^*}\, B_{g_n*\ell,\mathcal{V}^*}\, S_{\mathcal{V}^*}\varphi^{\perp}. \end{aligned} \tag{6.6.19}$$

Besides, since

$$r({}^{\mathrm{RL}}\mathcal{Q}_{p,\mathcal{Y},n+\ell}) = {}^{\mathrm{RL}}\mathcal{Q}_{q,\mathcal{Y}^*,n+\ell},$$

the boundary values trick featured in Sect. 5.7.1.2 together with (6.6.10) and (6.6.11), gives that for $j = i_q + 1, \dots, n+1$,

$$F_j = {}^{\mathrm{RL}}\mathrm{D}_{\mathcal{V}^*}^{j-2+\ell}\, S_{\mathcal{V}^*} f(0), \tag{6.6.20}$$

and for $j = i_p +, \dots, n+1$, we have

$$F_j' = -\sum_{i=n+1-j}^{n-1} g_{i+j-n} * \ell(b)\, {}^{\mathrm{RL}}\mathrm{D}_{\mathcal{V}^*}^{i+\ell}\, S_{\mathcal{V}^*} f(0) + \ell * {}^{\mathrm{RL}}\mathrm{D}_{\mathcal{V}^*}^{n+1-j+\ell}\, S_{\mathcal{V}^*} f(b). \tag{6.6.21}$$

6.7 The Quadruplet $\mathcal{Q}_{p,\mathcal{Y},\ell}$

The assumptions and notation featured at the beginning of Sect. 6.5 remain in force. In particular, p, q are conjugate exponents in $(1,\infty)$, and $\mathcal{Y}$ is a nontrivial separable and reflexive Banach space,

$$\mathcal{V} := L^p\big((0,b),\mathcal{Y}\big), \qquad \mathcal{V}^* = L^q\big((0,b),\mathcal{Y}^*\big),$$

and (k,ℓ) denotes a Sonine pair. We recall that on account of Sect. 6.5.1,

$${}^{\mathrm{RL}}\mathcal{Q}_{p,\mathcal{Y},\ell} := ({}^{\mathrm{RL}}\mathrm{D}^{\ell}_{\mathcal{V}^*},\ {}^{\mathrm{RL}}\mathrm{D}^{\ell}_{\mathcal{V}},\ B_{\ell,\mathcal{V}},\ S_{\mathcal{V}}),$$

and that the theory of Caputo extensions is introduced in Sect. 5.6.4.

6.7.1 Introduction

We would like to define and study the Caputo extension of ${}^{\mathrm{RL}}\mathcal{Q}_{p,\mathcal{Y},\ell}$ with respect to $N := g_1\mathcal{Y}$ and $N^{\mathrm{s}} := g_1\mathcal{Y}^*$. However this is not always possible. Indeed by reformulating Proposition 6.4.1 in our context, we may state this result.

Proposition 6.7.1 *Let (k,ℓ) be a Sonine pair. Then $g_1\mathcal{Y}$ is in direct sum with $D({}^{\mathrm{RL}}\mathrm{D}^{\ell}_{\mathcal{V}})$ if and only if k does not belong to $L^p(0,b)$. In a same way, $g_1\mathcal{Y}^*$ is in direct sum with $D({}^{\mathrm{RL}}\mathrm{D}^{\ell}_{\mathcal{V}^*})$ if and only if k does not belong to $L^q(0,b)$.*

In view of Proposition 6.7.1, we put

$$N := \begin{cases} g_1\mathcal{Y} & \text{if } k \notin L^p(0,b) \\ \{0\} & \text{otherwise} \end{cases}, \qquad N^{\mathrm{s}} := \begin{cases} g_1\mathcal{Y}^* & \text{if } k \notin L^q(0,b) \\ \{0\} & \text{otherwise} \end{cases}. \tag{6.7.1}$$

We denote by $\mathrm{D}^{\ell}_{\mathcal{V}}$ the Caputo extension of ${}^{\mathrm{RL}}\mathrm{D}^{\ell}_{\mathcal{V}}$ with respect to N and by $\mathrm{D}^{\ell}_{\mathcal{V}^*}$ the Caputo extension of ${}^{\mathrm{RL}}\mathrm{D}^{\ell}_{\mathcal{V}}$ with respect to N^{s}. *The Caputo extension of* ${}^{\mathrm{RL}}\mathcal{Q}_{p,\mathcal{Y},\ell}$ is then the differential quadruplet

$$\mathcal{Q}_{p,\mathcal{Y},\ell} := (\mathrm{D}^{\ell}_{\mathcal{V}^*},\ \mathrm{D}^{\ell}_{\mathcal{V}},\ B_{\ell,\mathcal{V}},\ S_{\mathcal{V}}). \tag{6.7.2}$$

6.7.2 Endogenous Boundary Values

6.7.2.1 The Case Where ℓ Lies in $L^p(0,b) \cap L^q(0,b)$

Proposition 6.3.1 (ii-b) yields that the kernel k does not belong to $L^p(0,b)$ neither to $L^q(0,b)$. Thus, according to Proposition 6.7.1, $N := g_1\mathcal{Y}$, $N^{\mathrm{s}} := g_1\mathcal{Y}^*$. Moreover, by Proposition 5.6.9,

$$\ker \mathrm{D}^{\ell}_{\mathcal{V}} = g_1\mathcal{Y} \oplus \ell\mathcal{Y}, \qquad \ker \mathrm{D}^{\ell}_{\mathcal{V}^*} = g_1\mathcal{Y}^* \oplus \ell\mathcal{Y}^*.$$

We claim that $\mathcal{Q}_{p,\mathcal{Y},\ell}$ is regular. Indeed, Corollary 5.3.1 yields the existence of functions h_1, h_2 in $L^p(0,b)$ such that, on the one hand,

$$\int_0^b g_1(y)h_j(y)\,dy = \delta_{1j}, \qquad \int_0^b \ell(y)h_j(y)\,dy = \delta_{2j}, \qquad \forall\, j = 1, 2. \tag{6.7.3}$$

On the other hand, $\mathcal{Q}_{p,\mathcal{Y},\ell}$ is $\tilde{F}_0$-regular with

$$\tilde{F}_0 := h_1\mathcal{Y} \oplus h_2\mathcal{Y}.$$

That proves the claim.

Since it is regular, $\mathcal{Q}_{p,\mathcal{Y},\ell}$ is tractable by the Domain Structure Theorem 5.3.3. Hence with the notation of Sect. 5.3.3,

$$E_{\mathrm{M}} = (g_1\mathcal{Y} \oplus \ell\mathcal{Y}) \oplus B_{\ell,\mathcal{V}}S_{\mathcal{V}}(h_1\mathcal{Y} \oplus h_2\mathcal{Y}) \tag{6.7.4}$$

$$D(\mathrm{D}_{\mathcal{V}}^{\ell}) = E_{\mathrm{M}} \oplus B_{\ell,\mathcal{V}}S_{\mathcal{V}}\big((g_1\mathcal{Y}^* \oplus \ell\mathcal{Y}^*)^{\perp}\big). \tag{6.7.5}$$

Moreover, a function u in $\mathcal{V}$ lies in $D(\mathrm{D}_{\mathcal{V}}^{\ell})$ if and only if there exist U_1, U_2, U_1', U_2' in $\mathcal{Y}$ and $\psi^{\perp}$ in $(g_1\mathcal{Y}^* \oplus \ell\mathcal{Y}^*)^{\perp}$ such that

$$u = g_1\,U_1 + \ell\,U_2 + \ell * S_{\mathcal{V}}(h_1\,U_1') + \ell * S_{\mathcal{V}}(h_2\,U_2') + \ell * S_{\mathcal{V}}\psi^{\perp}. \tag{6.7.6}$$

Here (U_1, U_2, U_1', U_2') is the endogenous boundary values of u with respect to the direct sum decompositions

$$\ker \mathrm{D}_{\mathcal{V}}^{\ell} = g_1\mathcal{Y} \oplus \ell\mathcal{Y}, \qquad \tilde{F}_0 = h_1\,\mathcal{Y} \oplus h_2\mathcal{Y}. \tag{6.7.7}$$

Let us compute these endogenous boundary values. We claim that the functions $u - \ell\,k * u(0)$ and $\ell * \mathrm{D}_{\mathcal{V}}^{\ell}u$ are continuous on $[0, b]$ and also that

$$\begin{aligned} U_1 &= \big(u - \ell\,k * u(0)\big)(0) \\ U_2 &= k * u(0) \\ U_1' &= g_1 * \mathrm{D}_{\mathcal{V}}^{\ell}u(b) \\ U_2' &= \ell * \mathrm{D}_{\mathcal{V}}^{\ell}u(b). \end{aligned} \tag{6.7.8}$$

Indeed, the equation for U_2 is a straightforward consequence of (6.7.6) and (6.3.1). Then the continuity of $u - \ell\,k * u(0)$ and the first equation are obtain by means of Remark 3.3.4. Regarding U_1' and U_2', we have to get rid of $\psi^{\perp}$ in (6.7.6). For, we use the belonging of $\psi^{\perp}$ to

$$(g_1\mathcal{Y}^* \oplus \ell\mathcal{Y}^*)^\perp = (g_1\mathcal{Y}^*)^\perp \cap (\ell\mathcal{Y}^*)^\perp,$$

which implies due to (3.2.1)

$$\int_0^b g_1\psi^\perp \, \mathrm{d}y = \int_0^b \ell\psi^\perp \, \mathrm{d}y = 0. \tag{6.7.9}$$

Next, we start from (6.7.6) to get the identity

$$\mathrm{D}^\ell_{\mathcal{V}} u = S_{\mathcal{V}}(h_1\, U_1') + S_{\mathcal{V}}(h_2\, U_2') + S_{\mathcal{V}}\psi^\perp. \tag{6.7.10}$$

By convolution with g_1, we get

$$g_1 * \mathrm{D}^\ell_{\mathcal{V}} u = g_1 * S_{\mathcal{V}}(h_1\, U_1') + g_1 * S_{\mathcal{V}}(h_2\, U_2') + g_1 * S_{\mathcal{V}}\psi^\perp \qquad \text{in } W^{1,p}(\mathcal{Y}).$$

Evaluating this expression at $x = b$ and using (6.7.3), (6.7.9), we obtain the third equation. Finally, Proposition 3.3.5 yields

$$\ell * S_{\mathcal{V}}(h_1\, U_1')(b) = \int_0^b \ell(y) h_1(y) \, \mathrm{d}y \, U_2' = 0,$$

thanks to (6.7.3). In a same way, we compute $\ell * S_{\mathcal{V}}(h_2\, U_2')(b) = U_2'$. Also

$$\ell * S_{\mathcal{V}}\psi^\perp(b) = \int_0^b \ell(y)\psi^\perp(y) \, \mathrm{d}y = 0$$

by (6.7.9). Thus going back to (6.7.10) and convoluting by ℓ, we derive the continuity of $\ell * \mathrm{D}^\ell_{\mathcal{V}} u$ and the fourth equation of (6.7.8).

Observe that u has four "standard" boundary values, namely,

$$\big(u - \ell\, k * u(0)\big)(0), \qquad k * u(0), \qquad g_1 * \mathrm{D}^\ell_{\mathcal{V}} u(b), \qquad \ell * \mathrm{D}^\ell_{\mathcal{V}} u(b).$$

Remark 6.7.1 The two latter boundary values can be expressed by means of zero order terms, as expected. Indeed, from the very definition of $\mathrm{D}^\ell_{\mathcal{V}}$, one has

$$g_1 * \mathrm{D}^\ell_{\mathcal{V}} u(b) = k * (u - g_1\, U_1)(b) - k * (u - g_1\, U_1)(0).$$

Moreover combining (6.7.6) and (6.7.10), we derive

$$\ell * \mathrm{D}^\ell_{\mathcal{V}} u(b) = (u - g_1\, U_1 - \ell\, U_2)(b).$$

Thus assuming that ℓ admits an essential limit at b, we deduce that u has also an essential limit at b and get another set of four boundary values, namely,

$$U_1 := \big(u - \ell\, k * u(0)\big)(0), \qquad k * u(0),$$
$$k * (u - g_1\, U_1)(b) - k * (u - g_1\, U_1)(0), \qquad u(b).$$

Then we may rewrite (6.7.8) under the form

$$\begin{aligned} U_1 &= \big(u - \ell\, k * u(0)\big)(0) \\ U_2 &= k * u(0) \\ U_1' &= k * (u - g_1\, U_1)(b) - k * (u - g_1\, U_1)(0) \\ U_2' &= u(b) - U_1 - \ell(b)\, k * u(0). \end{aligned} \tag{6.7.11}$$

In addition, since $k * (u - g_1 U_1)(0) = k * u(0)$, we obtain the following complete set of "standard" boundary values:

$$U_1 := \big(u - \ell\, k * u(0)\big)(0), \qquad k * u(0),$$
$$k * (u - g_1\, U_1)(b), \qquad u(b).$$

□

Remark 6.7.2 We may give another formula for U_1 by assuming that the essential limit of ℓ at 0 is equal to ∞ (in the sense of Definition 3.1.6), which reads in symbol

$$\operatorname*{ess\,lim}_{0} \ell = \infty.$$

Indeed, going back to (6.7.6), we derive that $\frac{u}{\ell}$ possesses an essential limit at 0 and that

$$U_2 = \operatorname*{ess\,lim}_{0} \frac{u}{\ell}.$$

Then $u - \ell \operatorname*{ess\,lim}_{0} \frac{u}{\ell}$ is continuous on $[0, b]$ and

$$U_1 = \big(u - \ell \operatorname*{ess\,lim}_{0} \frac{u}{\ell}\big)(0).$$

□

6.7.2.2 Case Where ℓ Lies in $L^q(0,b) \setminus L^p(0,b)$

On account of Proposition 6.3.1 (ii-b), the kernel k is not in $L^p(0,b)$. Thus by Proposition 6.7.1, $N := g_1\mathcal{Y}$. However k may be in $L^q(0,b)$ or not, so that we have to distinguish two cases.

6.7.2.2.1 Case where k Lies in $L^q(0,b) \setminus L^p(0,b)$

Proposition 6.7.1 yields that $N^{\mathrm{s}} := \{0\}$, so that $\mathrm{D}^{\ell}_{\mathcal{V}^*} = {}^{\mathrm{RL}}\mathrm{D}^{\ell}_{\mathcal{V}^*}$. Moreover, by Proposition 5.6.9 and (6.5.15),

$$\ker \mathrm{D}^{\ell}_{\mathcal{V}} = g_1\mathcal{Y}, \qquad \ker \mathrm{D}^{\ell}_{\mathcal{V}^*} = \ell\mathcal{Y}^*.$$

Let h_2 be a function in $L^p(0,b)$ satisfying $\int_0^b \ell h_2 \,\mathrm{d}y = 1$. Then $Q_{p,\mathcal{Y},\ell}$ is $h_2\mathcal{Y}$-regular. Hence

$$E_{\mathrm{M}} = g_1\mathcal{Y} \oplus B_{\ell,\mathcal{V}} S_{\mathcal{V}}(h_2\mathcal{Y})$$
$$D(\mathrm{D}^{\ell}_{\mathcal{V}}) = E_{\mathrm{M}} \oplus B_{\ell,\mathcal{V}} S_{\mathcal{V}} \,(\ell\mathcal{Y}^*)^{\perp},$$

and a function u in $\mathcal{V}$ lies in $D(\mathrm{D}^{\ell}_{\mathcal{V}})$ if and only if there exist U_1, U_2' in $\mathcal{Y}$ and $\psi^{\perp}$ in $(\ell\mathcal{Y}^*)^{\perp}$ such that

$$u = g_1 U_1 + \ell * S_{\mathcal{V}}(h_2\, U_2') + \ell * S_{\mathcal{V}}\psi^{\perp}. \tag{6.7.12}$$

Here (U_1, U_2') are the endogenous boundary values of u with respect to the trivial direct sum decompositions

$$\ker \mathrm{D}^{\ell}_{\mathcal{V}} = g_1\mathcal{Y}, \qquad \tilde{F}_0 = h_2\mathcal{Y}. \tag{6.7.13}$$

By (6.7.12), u is continuous on $[0,b]$. Then the relationship between these endogenous values and the standard boundary values $u(0)$ and $u(b)$ of u is

$$\begin{aligned} U_1 &= u(0) \\ U_2' &= u(b) - u(0). \end{aligned} \tag{6.7.14}$$

The second equality is obtained as the fourth equation of (6.7.8).

6.7.2.2.2 Case where k Lies in $L^1(0,b) \setminus (L^p(0,b) \cup L^q(0,b))$

Then

$$N^{\mathrm{s}} := g_1\mathcal{Y}^*, \qquad \ker \mathrm{D}^{\ell}_{\mathcal{V}} = g_1\mathcal{Y}, \qquad \ker \mathrm{D}^{\ell}_{\mathcal{V}^*} = g_1\mathcal{Y}^* \oplus \ell\mathcal{Y}^*.$$

We claim that $Q_{p,\mathcal{Y},\ell}$ is regular. Indeed, arguing as in Sect. 6.7.2.1, Corollary 5.3.1 yields the existence of functions h_1, h_2 in $L^p(0,b)$ such that, on the one hand,

$$\int_0^b g_1(y)h_j \, dy = \delta_{1j}, \qquad \int_0^b \ell(y)h_j \, dy = \delta_{2j}, \qquad \forall j = 1, 2. \tag{6.7.15}$$

On the other hand, $Q_{p,\mathcal{Y},\ell}$ is $\tilde{F}_0$-regular with

$$\tilde{F}_0 := h_1\mathcal{Y} \oplus h_2\mathcal{Y}.$$

That proves the claim.

Hence

$$\begin{aligned} E_{\mathrm{M}} &= g_1\mathcal{Y} \oplus B_{\ell,\mathcal{V}}S_{\mathcal{V}}(h_1\mathcal{Y} \oplus h_2\mathcal{Y}) \\ D(\mathrm{D}^{\ell}_{\mathcal{V}}) &= E_{\mathrm{M}} \oplus B_{\ell,\mathcal{V}}S_{\mathcal{V}}\,(g_1\mathcal{Y}^* \oplus \ell\mathcal{Y}^*)^{\perp}, \end{aligned}$$

and a function u in $\mathcal{V}$ lies in $D(\mathrm{D}^{\ell}_{\mathcal{V}})$ if and only if there exist U_1, U_1', U_2' in $\mathcal{Y}$ and $\psi^{\perp}$ in $(g_1\mathcal{Y}^* \oplus \ell\mathcal{Y}^*)^{\perp}$ such that

$$u = g_1U_1 + \ell * S_{\mathcal{V}}(h_1\,U_1') + \ell * S_{\mathcal{V}}(h_2\,U_2') + \ell * S_{\mathcal{V}}\psi^{\perp}. \tag{6.7.16}$$

Here (U_1, U_1', U_2') are the endogenous boundary values of u with respect to the direct sum decompositions

$$\ker \mathrm{D}^{\ell}_{\mathcal{V}} = g_1\mathcal{Y}, \qquad \tilde{F}_0 = h_1\mathcal{Y} \oplus h_2\mathcal{Y}. \tag{6.7.17}$$

We claim that u is continuous on $[0,b]$ and

$$\begin{aligned} U_1 &= u(0) \\ U_1' &= g_1 * \mathrm{D}^{\ell}_{\mathcal{V}}u(b) \\ U_2' &= \ell * \mathrm{D}^{\ell}_{\mathcal{V}}u(b). \end{aligned} \tag{6.7.18}$$

Indeed, the first equation and the continuity of u follow from Remark 3.3.4. Arguing as in the proof of the two last equations of (6.7.8), we get the second and third equations of (6.7.18). That proves the claim.

Remark 6.7.3 Taking advantage of the continuity of u, we will be able to give another set of boundary values. Indeed starting from (6.7.16), Lemma 4.3.3 yields that

$$u - u(0) = \ell * {}^{\mathrm{RL}}\mathrm{D}^{\ell}_{\mathcal{V}}u.$$

Moreover, $k * \big(u - u(0)\big)(0) = 0$ since u is continuous on $[0, b]$. Then an integration gives

$$g_1 * \mathrm{D}^{\ell}_{\mathcal{V}} u(b) = k * \big(u - u(0)\big)(b).$$

Thus we derive from (6.7.18)

$$\begin{aligned} U_1 &= u(0) \\ U_1' &= k * \big(u - u(0)\big)(b) \\ U_2' &= u(b) - u(0). \end{aligned} \tag{6.7.19}$$

Then we end up with three other "standard" boundary values, namely,

$$u(0), \qquad k * \big(u - u(0)\big)(b), \qquad u(b).$$

□

6.7.2.3 Case Where ℓ Lies in $L^p(0, b) \setminus L^q(0, b)$

On account of Proposition 6.3.1 (ii-b), the kernel k is not in $L^q(0, b)$. Thus by Proposition 6.7.1, $N^{\mathrm{s}} := g_1\mathcal{Y}^*$. However, as in the previous case, N depends on the L^p-regularity of k.

6.7.2.3.1 Case where k Lies in $L^p(0, b) \setminus L^q(0, b)$

Proposition 6.7.1 yields that $N := \{0\}$. Observe that $\mathrm{D}^{\ell}_{\mathcal{V}} = {}^{\mathrm{RL}}\mathrm{D}^{\ell}_{\mathcal{V}}$. Moreover, by Proposition 5.6.9 and (6.5.15),

$$\ker \mathrm{D}^{\ell}_{\mathcal{V}} = \ell\mathcal{Y}, \qquad \ker \mathrm{D}^{\ell}_{\mathcal{V}^*} = g_1\mathcal{Y}^*.$$

Setting $h_1 := b^{-1} g_1$, there results that $Q_{p,\mathcal{Y},\ell}$ is $h_1\mathcal{Y}$-regular. Hence

$$\begin{aligned} E_{\mathrm{M}} &= \ell\mathcal{Y} \oplus B_{\ell,\mathcal{V}} S_{\mathcal{V}}(h_1\mathcal{Y}) \\ D(\mathrm{D}^{\ell}_{\mathcal{V}}) &= E_{\mathrm{M}} \oplus B_{\ell,\mathcal{V}} S_{\mathcal{V}}\,(g_1\mathcal{Y}^*)^{\perp}, \end{aligned}$$

and a function u in $\mathcal{V}$ lies in $D(\mathrm{D}^{\ell}_{\mathcal{V}})$ if and only if there exist U_2, U_1' in $\mathcal{Y}$ and $\psi^{\perp}$ in $(g_1\mathcal{Y}^*)^{\perp}$ such that

$$u = \ell\, U_2 + \ell * S_{\mathcal{V}}(h_1\, U_1') + \ell * S_{\mathcal{V}}\psi^{\perp}. \tag{6.7.20}$$

Here (U_2, U_1') are the endogenous boundary values of u with respect to the trivial direct sum decompositions of $\ker \mathrm{D}^{\ell}_{\mathcal{V}}$ and $h_1\mathcal{Y}$. Then using $U_1' = g_1 * {}^{\mathrm{RL}}\mathrm{D}^{\ell}_{\mathcal{V}} u(b)$, we find

$$\begin{aligned} U_2 &= k * u(0) \\ U_1' &= k * u(b) - k * u(0). \end{aligned} \tag{6.7.21}$$

so that $k * u(0)$ and $k * u(b)$ are the "standard" boundary values of u.

6.7.2.3.2 Case where k Lies in $L^1(0, b) \setminus (L^p(0, b) \cup L^q(0, b))$
Then

$$N := g_1\mathcal{Y}, \qquad \ker \mathrm{D}^\ell_{\mathcal{V}} = g_1\mathcal{Y} \oplus \ell\mathcal{Y}, \qquad \ker \mathrm{D}^\ell_{\mathcal{V}^*} = g_1\mathcal{Y}^*.$$

Setting $h_1 := b^{-1} g_1$, there results that $\mathcal{Q}_{p,\mathcal{Y},\ell}$ is $h_1\mathcal{Y}$-regular. Hence

$$\begin{aligned} E_{\mathrm{M}} &= g_1\mathcal{Y} \oplus \ell\mathcal{Y} \oplus B_{\ell,\mathcal{V}} S_{\mathcal{V}}(h_1\mathcal{Y}) \\ D(\mathrm{D}^\ell_{\mathcal{V}}) &= E_{\mathrm{M}} \oplus B_{\ell,\mathcal{V}} S_{\mathcal{V}}\,(g_1\mathcal{Y}^*)^\perp, \end{aligned}$$

and a function u in $\mathcal{V}$ lies in $D(\mathrm{D}^\ell_{\mathcal{V}})$ if and only if there exist U_1, U_2, U_1' in $\mathcal{Y}$ and $\psi^\perp$ in $(g_1\mathcal{Y}^*)^\perp$ such that

$$u = g_1\, U_1 + \ell\, U_2 + \ell * S_{\mathcal{V}}(h_1\, U_1') + \ell * S_{\mathcal{V}}\psi^\perp. \tag{6.7.22}$$

Here $(U_1,\ U_2, U_1')$ are the endogenous boundary values of u with respect to the above direct sum decomposition of $\ker \mathrm{D}^\ell_{\mathcal{V}}$ and the trivial direct sum decomposition of $h_1\mathcal{Y}$.

We claim that $u - \ell\, k * u(0) - \ell * \mathrm{D}^\ell_{\mathcal{V}} u$ is continuous on $[0, b]$ and

$$\begin{aligned} U_1 &= \big(u - \ell\, k * u(0) - \ell * \mathrm{D}^\ell_{\mathcal{V}} u\big)(0) \\ U_2 &= k * u(0) \\ U_1' &= g_1 * \mathrm{D}^\ell_{\mathcal{V}} u(b). \end{aligned} \tag{6.7.23}$$

The second and third equations of (6.7.23) are proved like the second and third equations of (6.7.8). In order to prove the first equation of (6.7.23), we go back to (6.7.22) and observe that

$$\mathrm{D}^\ell_{\mathcal{V}} u = S_{\mathcal{V}}(h_1\, U_1' + \psi^\perp).$$

Thus

$$u - \ell\, U_2 - \ell * \mathrm{D}^\ell_{\mathcal{V}} u = g_1\, U_1,$$

and we derive the equation of U_1, which proves the claim.

6.7.2.4 Case Where ℓ Lies in $L^1(0,b) \setminus (L^q(0,b) \cup L^p(0,b))$

In this case, the kernels of ${}^{\mathrm{RL}}\mathrm{D}^{\ell}_{\mathcal{V}}$ and ${}^{\mathrm{RL}}\mathrm{D}^{\ell}_{\mathcal{V}*}$ are trivial.

6.7.2.4.1 Case where k Lies in $L^p(0,b) \cap L^q(0,b)$

Then

$$N := \{0\}, \qquad N^{\mathrm{s}} := \{0\}.$$

Thus

$$\mathrm{D}^{\ell}_{\mathcal{V}} = {}^{\mathrm{RL}}\mathrm{D}^{\ell}_{\mathcal{V}}, \qquad \mathrm{D}^{\ell}_{\mathcal{V}*} = {}^{\mathrm{RL}}\mathrm{D}^{\ell}_{\mathcal{V}*}.$$

There is no endogenous boundary value according to Sect. 6.5.3.4.

6.7.2.4.2 Case where k Lies in $L^q(0,b) \setminus L^p(0,b)$

Then

$$N := g_1\mathcal{Y}, \qquad N^{\mathrm{s}} := \{0\}.$$

Thus

$$\ker \mathrm{D}^{\ell}_{\mathcal{V}} = g_1\mathcal{Y}, \qquad \mathrm{D}^{\ell}_{\mathcal{V}*} = {}^{\mathrm{RL}}\mathrm{D}^{\ell}_{\mathcal{V}*}, \qquad \ker \mathrm{D}^{\ell}_{\mathcal{V}*} = \{0\}, \qquad \tilde{F}_0 := \{0\}.$$

Hence

$$E_{\mathrm{M}} = g_1\mathcal{Y}$$
$$D(\mathrm{D}^{\ell}_{\mathcal{V}}) = E_{\mathrm{M}} \oplus R(B_{\ell,\mathcal{V}}),$$

and a function u in $\mathcal{V}$ lies in $D(\mathrm{D}^{\ell}_{\mathcal{V}})$ if and only if there exist U_1 in $\mathcal{Y}$ and ψ in $\mathcal{V}$ such that

$$u = g_1 U_1 + \ell * \psi. \tag{6.7.24}$$

One has $\mathrm{D}^{\ell}_{\mathcal{V}} u = \psi$, hence $u - \ell * \mathrm{D}^{\ell}_{\mathcal{V}} u$ is continuous on $[0,b]$, and

$$U_1 = (u - \ell * \mathrm{D}^{\ell}_{\mathcal{V}} u)(0). \tag{6.7.25}$$

6.7.2.4.3 Case where k Lies in $L^p(0,b) \setminus L^q(0,b)$

Then

$$N := \{0\}, \qquad N^{\mathrm{s}} := g_1\mathcal{Y}^*.$$

Thus setting $h_1 := b^{-1}g_1$, we have

$$\mathrm{D}^{\ell}_{\mathcal{V}} = {}^{\mathrm{RL}}\mathrm{D}^{\ell}_{\mathcal{V}}, \qquad \ker \mathrm{D}^{\ell}_{\mathcal{V}} = \{0\}, \qquad \ker \mathrm{D}^{\ell}_{\mathcal{V}*} = g_1\mathcal{Y}^*, \qquad \tilde{F}_0 := h_1\mathcal{Y}.$$

Hence

$$E_{\mathrm{M}} = B_{\ell,\mathcal{V}}\, S_{\mathcal{V}}(h_1\mathcal{Y})$$
$$D(\mathrm{D}^{\ell}_{\mathcal{V}}) = E_{\mathrm{M}} \oplus B_{\ell,\mathcal{V}}\, S_{\mathcal{V}}(g_1\mathcal{Y}^*)^{\perp},$$

and a function u in $\mathcal{V}$ lies in $D(\mathrm{D}^{\ell}_{\mathcal{V}})$ if and only if there exist U_1' in $\mathcal{Y}$ and $\psi^{\perp}$ in $(g_1\mathcal{Y}^*)^{\perp}$ such that

$$u = \ell * S_{\mathcal{V}}(h_1U_1') + \ell * S_{\mathcal{V}}\psi^{\perp}. \tag{6.7.26}$$

Thus arguing as in the proof of (6.7.21), we get

$$U_1' = k * u(b) - k * u(0). \tag{6.7.27}$$

In view Sect. 5.7.1.1, U_1' is the endogenous boundary value of u with respect to the trivial direct sum decomposition of $\tilde{F}_0$.

6.7.2.4.4 Case where k Lies in $L^1(0, b) \setminus (L^q(0, b) \cup L^p(0, b))$

Then

$$N := g_1\mathcal{Y}, \qquad N^{\mathrm{s}} := g_1\mathcal{Y}^*.$$

Thus setting $h_1 := b^{-1}g_1$, we have

$$\ker \mathrm{D}^{\ell}_{\mathcal{V}} = g_1\mathcal{Y}, \qquad \ker \mathrm{D}^{\ell}_{\mathcal{V}*} = g_1\mathcal{Y}^*, \qquad \tilde{F}_0 := h_1\mathcal{Y}.$$

Hence

$$E_{\mathrm{M}} = g_1\mathcal{Y} \oplus B_{\ell,\mathcal{V}} S_{\mathcal{V}}(h_1\mathcal{Y})$$
$$D(\mathrm{D}^{\ell}_{\mathcal{V}}) = E_{\mathrm{M}} \oplus B_{\ell,\mathcal{V}} S_{\mathcal{V}}\, (g_1\mathcal{Y}^*)^{\perp},$$

and a function u in $\mathcal{V}$ lies in $D(\mathrm{D}^{\ell}_{\mathcal{V}})$ if and only if there exist U_1, U_1' in $\mathcal{Y}$ and $\psi^{\perp}$ in $(g_1\mathcal{Y}^*)^{\perp}$ such that

$$u = g_1U_1 + \ell * S_{\mathcal{V}}(h_1\, U_1') + \ell * S_{\mathcal{V}}\psi^{\perp}. \tag{6.7.28}$$

Since $\mathrm{D}^{\ell}_{\mathcal{V}} u = S_{\mathcal{V}}(h_1 U_1' + \psi^{\perp})$, the function $u - \ell * \mathrm{D}^{\ell}_{\mathcal{V}} u$ is continuous on $[0, b]$, and

$$\begin{aligned} U_1 &= (u - \ell * \mathrm{D}^{\ell}_{\mathcal{V}} u)(0) \\ U_1' &= g_1 * \mathrm{D}^{\ell}_{\mathcal{V}} u(b). \end{aligned} \tag{6.7.29}$$

6.7.2.5 Regularity of $\mathcal{Q}_{p,\mathcal{Y},\ell}$

The following result is a consequence of the analysis made in the previous paragraphs.

Proposition 6.7.2 *Let (k, ℓ) be a Sonine pair, and N, N^{s} be defined by* (6.7.1). *Then the Caputo extension $\mathcal{Q}_{p,\mathcal{Y},\ell}$ of ${}^{\mathrm{RL}}\mathcal{Q}_{p,\mathcal{Y},\ell}$ is a regular differential quadruplet on $(\mathcal{V}, \mathcal{V}^*)$.*

6.7.3 The Adjoint Quadruplet of $\mathcal{Q}_{p,\mathcal{Y},\ell}$

The assumptions and notation featured at the beginning of the current section (in particular (6.7.1), (6.7.2)) remain in force. Recalling (6.5.1), the adjoint quadruplet of $\mathcal{Q}_{p,\mathcal{Y},\ell}$ reads

$$(\mathcal{Q}_{p,\mathcal{Y},\ell})^* := (S_{\mathcal{V}}\, \mathrm{D}^{\ell}_{\mathcal{V}}\, S_{\mathcal{V}},\ S_{\mathcal{V}^*}\, \mathrm{D}^{\ell}_{\mathcal{V}^*}\, S_{\mathcal{V}^*},\ S_{\mathcal{V}^*}\, B_{\ell,\mathcal{V}^*}\, S_{\mathcal{V}^*},\ S_{\mathcal{V}^*}). \tag{6.7.30}$$

Let us compute the endogenous boundary values of functions belonging to the domain of $(\mathcal{Q}_{p,\mathcal{Y},\ell})^*$.

6.7.3.1 Case Where ℓ Lies in $L^p(0, b) \cap L^q(0, b)$

From Sect. 6.7.2.1,

$$\ker \mathrm{D}^{\ell}_{\mathcal{V}^*} = g_1 \mathcal{Y}^* \oplus \ell \mathcal{Y}^*, \qquad \ker \mathrm{D}^{\ell}_{\mathcal{V}} = g_1 \mathcal{Y} \oplus \ell \mathcal{Y}.$$

Let h_1^{s} and h_2^{s} be functions in $L^q(0, b)$ such that

$$\int_0^b h_j^{\mathrm{s}}(y) g_1(y)\, \mathrm{d}y = \delta_{1j}, \qquad \int_0^b h_j^{\mathrm{s}}(y) \ell(y)\, \mathrm{d}y = \delta_{2j}, \qquad \forall\, j = 1, 2. \tag{6.7.31}$$

Setting $\tilde{F}_0^{\mathrm{s}} := h_1^{\mathrm{s}} \mathcal{Y}^* \oplus h_2^{\mathrm{s}} \mathcal{Y}^*$, it is now a routine to show that $\tilde{F}_0^{\mathrm{s}}$ is a closed complementary subspace of $(\ker \mathrm{D}^{\ell}_{\mathcal{V}})^{\perp}$ in $\mathcal{V}^*$. Hence Corollary 5.3.4 yields that

$$D(S_{\mathcal{V}^*}\, \mathrm{D}^{\ell}_{\mathcal{V}^*}\, S_{\mathcal{V}^*}) = E^{\mathrm{s}}_{\mathrm{M}} \oplus (B_{\ell,\mathcal{V}})^* (g_1 \mathcal{Y} \oplus \ell \mathcal{Y})^{\perp} \tag{6.7.32}$$

where

$$E^{\mathrm{s}}_{\mathrm{M}} := S_{\mathcal{V}^*}(g_1 \mathcal{Y}^* \oplus \ell \mathcal{Y}^*) \oplus (B_{\ell,\mathcal{V}})^* (h_1^{\mathrm{s}} \mathcal{Y}^* \oplus h_2^{\mathrm{s}} \mathcal{Y}^*). \tag{6.7.33}$$

Thus a function f in $\mathcal{V}^*$ lies in $D(\mathrm{D}^{\ell}_{\mathcal{V}^*}\, S_{\mathcal{V}^*})$ if and only if there exist F_1, F_2, F_1' F_2' in $\mathcal{Y}^*$ and $\varphi^{\perp}$ in $(g_1\mathcal{Y}\oplus\ell\mathcal{Y})^{\perp L^q(\mathcal{Y}^*)}$ such that

$$f = g_1 F_1 + \ell(b-\cdot)F_2 + (B_{\ell,\mathcal{V}})^*(h_1^{\mathrm{s}}\, F_1' + h_2^{\mathrm{s}}\, F_2') + (B_{\ell,\mathcal{V}})^*\varphi^{\perp}. \tag{6.7.34}$$

In view of Sect. 5.7.1, (F_1, F_2, F_1', F_2') are *the endogenous boundary values of u with underlying quadruplet $(\mathcal{Q}_{p,\mathcal{Y},\ell})^*$ and with respect to the direct sum decompositions*

$$S_{\mathcal{V}^*} \ker \mathrm{D}^{\ell}_{\mathcal{V}^*} = g_1\mathcal{Y}^* \oplus \ell(b-\cdot)\mathcal{Y}^*, \qquad S_{\mathcal{V}^*}\tilde{F}_0^{\mathrm{s}} = h_1^{\mathrm{s}}(b-\cdot)\mathcal{Y}^* \oplus h_2^{\mathrm{s}}(b-\cdot)\mathcal{Y}^*. \tag{6.7.35}$$

Besides using *the boundary values trick* featured in Sect. 5.7.1.2, and noticing that

$$r(\mathcal{Q}_{p,\mathcal{Y},\ell}) = \mathcal{Q}_{q,\mathcal{Y}^*,\ell}, \tag{6.7.36}$$

there results from (6.7.8) that $S_{\mathcal{V}^*} f - \ell\, k * S_{\mathcal{V}^*} f(0)$ and $\ell * \mathrm{D}^{\ell}_{\mathcal{V}} S_{\mathcal{V}^*} f$ are continuous on $[0, b]$ and also that

$$\begin{aligned} F_1 &= \big(S_{\mathcal{V}^*} f - \ell\, k * S_{\mathcal{V}^*} f(0)\big)(0)\\ F_2 &= k * S_{\mathcal{V}^*} f(0)\\ F_1' &= g_1 * \mathrm{D}^{\ell}_{\mathcal{V}^*} S_{\mathcal{V}^*} f(b)\\ F_2' &= \ell * \mathrm{D}^{\ell}_{\mathcal{V}^*} S_{\mathcal{V}^*} f(b). \end{aligned} \tag{6.7.37}$$

Besides, we derive from Remark 6.7.1 that

$$\begin{aligned} F_1 &= \big(S_{\mathcal{V}^*} f - \ell\, k * S_{\mathcal{V}^*} f(0)\big)(0)\\ F_2 &= k * S_{\mathcal{V}^*} f(0)\\ F_1' &= k * (S_{\mathcal{V}^*} f - g_1\, F_1)(b) - k * (S_{\mathcal{V}^*} f - g_1\, F_1)(0)\\ F_2' &= (S_{\mathcal{V}^*} f - g_1\, F_1 - \ell\, F_2)(b). \end{aligned} \tag{6.7.38}$$

6.7.3.2 Case Where ℓ Lies in $L^q(0, b) \setminus L^p(0, b)$

On account of Proposition 6.3.1 (ii-b), the kernel k is not in $L^p(0, b)$. Thus by Proposition 6.7.1, $N := g_1\mathcal{Y}$. However N^{s} depends on the L^p-regularity of k.

6.7.3.2.1 Case where k Lies in $L^q(0, b) \setminus L^p(0, b)$

We have

$$\ker \mathrm{D}^{\ell}_{\mathcal{V}^*} = \{0\} \oplus \ker {}^{\mathrm{RL}}\mathrm{D}^{\ell}_{\mathcal{V}^*} = \ell\mathcal{Y}^*, \qquad \ker \mathrm{D}^{\ell}_{\mathcal{V}} = N \oplus \{0\} = g_1\mathcal{Y}.$$

and $\tilde{F}_0^{\mathrm{s}} := h_1\mathcal{Y}^*$ where $h_1 := b^{-1}g_1$. Thus

$$D(S_{\mathcal{V}^*}\, {}^{\mathrm{RL}}\mathrm{D}^{\ell}_{\mathcal{V}^*}\, S_{\mathcal{V}^*}) = E^{\mathrm{s}}_{\mathrm{M}} \oplus (B_{\ell,\mathcal{V}})^*(h_1\mathcal{Y})^{\perp} \tag{6.7.39}$$

where

$$E^{\mathrm{s}}_{\mathrm{M}} := S_{\mathcal{V}^*}(\ell\mathcal{Y}^*) \oplus (B_{\ell,\mathcal{V}})^*(h_1\mathcal{Y}^*). \tag{6.7.40}$$

Thus a function f in $\mathcal{V}^*$ lies in $D(\mathrm{D}^{\ell}_{\mathcal{V}^*}\, S_{\mathcal{V}^*})$ if and only if there exist F_2, F_1' in $\mathcal{Y}^*$ and $\varphi^{\perp}$ in $(g_1\mathcal{Y})^{\perp\, L^q(\mathcal{Y}^*)}$ such that

$$f = S_{\mathcal{V}^*}(\ell F_2) + (B_{\ell,\mathcal{V}})^*(h_1 F_1') + (B_{\ell,\mathcal{V}})^*\varphi^{\perp}. \tag{6.7.41}$$

Thus *the boundary values trick* featured in Sect. 5.7.1.2 and (6.7.21) applied to $Q_{q,\mathcal{Y}^*,\ell}$ entail

$$\begin{aligned} F_2 &= k * S_{\mathcal{V}^*} f(0) \\ F_1' &= k * S_{\mathcal{V}^*} f(b) - k * S_{\mathcal{V}^*} f(0). \end{aligned} \tag{6.7.42}$$

6.7.3.2.2 Case where k Lies in $L^1(0,b) \setminus (L^p(0,b) \cup L^q(0,b))$

Then

$$\ker \mathrm{D}^{\ell}_{\mathcal{V}^*} = g_1\mathcal{Y}^* \oplus \ell\mathcal{Y}^*, \qquad \ker \mathrm{D}^{\ell}_{\mathcal{V}} = g_1\mathcal{Y}.$$

Setting $\tilde{F}^{\mathrm{s}}_0 := h_1\mathcal{Y}^*$ where $h_1 := b^{-1}g_1$, we get

$$\begin{aligned} E^{\mathrm{s}}_{\mathrm{M}} &:= S_{\mathcal{V}^*}(g_1\mathcal{Y} \oplus \ell\mathcal{Y}) \oplus (B_{\ell,\mathcal{V}})^*(h_1\mathcal{Y}) \\ D\left(S_{\mathcal{V}^*}\, \mathrm{D}^{\ell}_{\mathcal{V}^*}\, S_{\mathcal{V}^*}\right) &= E^{\mathrm{s}}_{\mathrm{M}} \oplus (B_{\ell,\mathcal{V}})^*(g_1\mathcal{Y})^{\perp}, \end{aligned}$$

and a function f in $\mathcal{V}^*$ lies in $D(\mathrm{D}^{\ell}_{\mathcal{V}^*})$ if and only if there exist F_1, F_2, F_1' in $\mathcal{Y}$ and $\psi^{\perp}$ in $(g_1\mathcal{Y}^*)^{\perp}$ such that

$$f = g_1\, F_1 + \ell(b-\cdot)\, F_2 + (B_{\ell,\mathcal{V}})^*(h_1\, F_1') + (B_{\ell,\mathcal{V}})^*\psi^{\perp}.$$

By *the boundary values trick* featured in Sect. 5.7.1.2 and (6.7.23) applied to $Q_{q,\mathcal{Y}^*,\ell}$, the function

$$S_{\mathcal{V}^*} f - \ell\, k * S_{\mathcal{V}^*} f(0) - \ell * \mathrm{D}^{\ell}_{\mathcal{V}} S_{\mathcal{V}^*} f$$

is continuous on $[0, b]$ and

$$F_1 = \big(S_{\mathcal{V}^*} f - \ell\, k * S_{\mathcal{V}^*} f(0) - \ell * \mathrm{D}^{\ell}_{\mathcal{V}} S_{\mathcal{V}^*} f\big)(0)$$
$$F_2 = k * S_{\mathcal{V}^*} f(0)$$
$$F_1' = g_1 * \mathrm{D}^{\ell}_{\mathcal{V}} S_{\mathcal{V}^*} f(b).$$

6.7.3.3 Case Where ℓ Lies in $L^p(0,b) \setminus L^q(0,b)$

On account of Proposition 6.3.1 (ii-b), the kernel k is not in $L^q(0,b)$. Thus by Proposition 6.7.1, $N^{\mathrm{s}} := g_1\mathcal{Y}^*$. However k may be in $L^p(0,b)$ or not, so that we have to distinguish two cases.

6.7.3.3.1 Case where k Lies in $L^p(0,b) \setminus L^q(0,b)$

We have

$$\ker \mathrm{D}^{\ell}_{\mathcal{V}^*} = N^{\mathrm{s}} \oplus \{0\} = g_1\mathcal{Y}^*, \qquad \ker \mathrm{D}^{\ell}_{\mathcal{V}} = \{0\} \oplus \ker {}^{\mathrm{RL}}\mathrm{D}^{\ell}_{\mathcal{V}} = \ell\mathcal{Y}.$$

and, for any function h in $L^p(0,b)$ satisfying $\int_0^b \ell h \,\mathrm{d}y = 1$, we set $\tilde{F}_0^{\mathrm{s}} := h\mathcal{Y}^*$. Then

$$D(S_{\mathcal{V}^*}\, \mathrm{D}^{\ell}_{\mathcal{V}^*}\, S_{\mathcal{V}^*}) = E^{\mathrm{s}}_{\mathrm{M}} \oplus (B_{\ell,\mathcal{V}})^*(\ell\mathcal{Y})^{\perp} \tag{6.7.43}$$

where

$$E^{\mathrm{s}}_{\mathrm{M}} := g_1\mathcal{Y}^* \oplus (B_{\ell,\mathcal{V}})^*(h\mathcal{Y}^*). \tag{6.7.44}$$

Thus a function f in $\mathcal{V}^*$ lies in $D(\mathrm{D}^{\ell}_{\mathcal{V}^*}\, S_{\mathcal{V}^*})$ if and only if there exist F_1, F_2' in $\mathcal{Y}^*$ and $\varphi^{\perp}$ in $(\ell\mathcal{Y})^{\perp\, L^q(\mathcal{Y}^*)}$ such that

$$f = g_1 F_1 + (B_{\ell,\mathcal{V}})^*(h F_2') + (B_{\ell,\mathcal{V}})^*\varphi^{\perp}. \tag{6.7.45}$$

Thus *the boundary values trick* featured in Sect. 5.7.1.2 and (6.7.14) entail

$$\begin{aligned} F_1 &= f(b) \\ F_1' &= f(0) - f(b). \end{aligned} \tag{6.7.46}$$

6.7.3.3.2 Case where k Lies in $L^1(0,b) \setminus (L^p(0,b) \cup L^q(0,b))$

We have

$$\ker \mathrm{D}^{\ell}_{\mathcal{V}^*} = g_1\mathcal{Y}^*, \qquad \ker \mathrm{D}^{\ell}_{\mathcal{V}} = g_1\mathcal{Y} \oplus \ell\mathcal{Y}.$$

Let h_1 and h_2 be functions in $L^q(0,b)$ such that

$$\int_0^b h_j(y)g_1(y)\,\mathrm{d}y = \delta_{1j}, \qquad \int_0^b h_j(y)\ell(y)\,\mathrm{d}y = \delta_{2j}, \qquad \forall\, j = 1, 2.$$

Arguing as in Sect. 6.7.3.1, we get

$$D(S_{\mathcal{Y}^*}\, {}^{\mathrm{RL}}\mathrm{D}^{\ell}_{\mathcal{Y}^*}\, S_{\mathcal{Y}^*}) = E^{\mathrm{S}}_{\mathrm{M}} \oplus (B_{\ell,\mathcal{Y}})^*(g_1\mathcal{Y} \oplus \ell\mathcal{Y})^{\perp}$$

where

$$E^{\mathrm{S}}_{\mathrm{M}} := g_1\mathcal{Y}^* \oplus (B_{\ell,\mathcal{Y}})^*(h_1\mathcal{Y}^* \oplus h_2\mathcal{Y}^*).$$

Thus applying (6.7.18) to $Q_{q,\mathcal{Y}^*,\ell}$, the endogenous boundary values of a function f in $D(S_{\mathcal{Y}^*}\, \mathrm{D}^{\ell}_{\mathcal{Y}^*}\, S_{\mathcal{Y}^*})$ with respect to the direct sum decompositions

$$\ker S_{\mathcal{Y}^*}\, \mathrm{D}^{\ell}_{\mathcal{Y}^*}\, S_{\mathcal{Y}^*} = g_1\mathcal{Y}^*, \qquad S_{\mathcal{Y}^*}\tilde{F}^{\mathrm{S}}_0 = h_1(b-\cdot)\mathcal{Y}^* \oplus h_2(b-\cdot)\mathcal{Y}^*$$

satisfy

$$\begin{aligned} F_1 &= f(b) \\ F_1' &= g_1 * \mathrm{D}^{\ell}_{\mathcal{Y}^*} S_{\mathcal{Y}^*} f(b) \\ F_2' &= \ell * \mathrm{D}^{\ell}_{\mathcal{Y}^*} S_{\mathcal{Y}^*} f(b). \end{aligned}$$

6.7.3.4 Case Where ℓ Lies in $L^1(0,b) \setminus (L^q(0,b) \cup L^p(0,b))$

6.7.3.4.1 Case where k Lies in $L^p(0,b) \cap L^q(0,b)$

There is no endogenous boundary value since the kernels of the maximal operators of $(Q_{p,\mathcal{Y},\ell})^*$ are trivial.

6.7.3.4.2 Case where k Lies in $L^q(0,b) \setminus L^p(0,b)$

In view of Sect. 5.7.1.1, the endogenous boundary value of a function f in $D(S_{\mathcal{Y}^*}\, \mathrm{D}^{\ell}_{\mathcal{Y}^*}\, S_{\mathcal{Y}^*})$ with respect to the trivial direct sum decomposition of $S_{\mathcal{Y}^*}(\tilde{F}^{\mathrm{S}}_0)$ satisfy (see (6.7.27))

$$F_1' = k * S_{\mathcal{Y}^*} f(b) - k * S_{\mathcal{Y}^*} f(0).$$

6.7.3.4.3 Case where k Lies in $L^1(0,b) \setminus (L^p(0,b) \cup L^q(0,b))$

The endogenous boundary values (F_1, F_1') of a function f in $D(S_{\mathcal{Y}^*}\, \mathrm{D}^{\ell}_{\mathcal{Y}^*}\, S_{\mathcal{Y}^*})$ with respect to the trivial direct sum decompositions of $\ker S_{\mathcal{Y}^*}\, \mathrm{D}^{\ell}_{\mathcal{Y}^*}\, S_{\mathcal{Y}^*}$ and $S_{\mathcal{Y}^*}(\tilde{F}^{\mathrm{S}}_0)$ satisfy (see (6.7.29))

$$F_1 = (S_{\mathcal{V}^*} f - \ell * \mathrm{D}^{\ell}_{\mathcal{V}^*} S_{\mathcal{V}^*} f)(0)$$

$$F_1' = g_1 * \mathrm{D}^{\ell}_{\mathcal{V}^*} S_{\mathcal{V}^*} f(b).$$

6.7.4 Boundary Restriction Operators of $\mathcal{Q}_{p,\mathcal{Y},\ell}$ and Their Adjoint

The assumptions and notation featured at the beginning of the current section (in particular (6.7.1), (6.7.2)) remain in force.

6.7.4.1 Two Special Boundary Restriction Operators

The first operator we would like to introduce is nothing but ${}^{\mathrm{RL}}\mathrm{D}^{\ell}_{\mathcal{V}}$. Observe that Propositions 5.2.3 and 5.6.10 (i) warrant that ${}^{\mathrm{RL}}\mathrm{D}^{\ell}_{\mathcal{V}}$ is a boundary restriction operator of $\mathcal{Q}_{p,\mathcal{Y},\ell}$. In particular ${}^{\mathrm{RL}}\mathrm{D}^{\ell}_{\mathcal{V}}$ is a restriction of $\mathrm{D}^{\ell}_{\mathcal{V}}$. Moreover ${}^{\mathrm{RL}}\mathrm{D}^{\ell}_{\mathcal{V}}$ is closed due to Proposition 6.5.2. Notice also that if k lies in $L^p(0,b)$, then ${}^{\mathrm{RL}}\mathrm{D}^{\ell}_{\mathcal{V}}$ is equal to $\mathrm{D}^{\ell}_{\mathcal{V}}$.

Our second special operator is *the Caputo operator with respect to* ℓ as defined in Sect. 6.4. This operator denoted by ${}^{\mathrm{C}}\mathrm{D}^{\ell}_{\mathcal{V}}$ is the restriction of $\mathrm{D}^{\ell}_{\mathcal{V}}$ with domain

$$D({}^{\mathrm{C}}\mathrm{D}^{\ell}_{\mathcal{V}}) := N \oplus R(B_{\ell,\mathcal{V}}), \tag{6.7.47}$$

where N is defined by (6.7.1). Thus if k do not lie in $L^p(0,b)$ then

$$D({}^{\mathrm{C}}\mathrm{D}^{\ell}_{\mathcal{V}}) = g_1\mathcal{Y} \oplus R(B_{\ell,\mathcal{V}}). \tag{6.7.48}$$

Whereas if k belongs to $L^q(0,b)$, then ${}^{\mathrm{C}}\mathrm{D}^{\ell}_{\mathcal{V}}$ reduces to the pivot operator of ${}^{\mathrm{RL}}\mathcal{Q}_{p,\mathcal{Y},\ell}$, which reads in symbol

$${}^{\mathrm{C}}\mathrm{D}^{\ell}_{\mathcal{V}} = A_{{}^{\mathrm{RL}}\mathcal{Q}_{p,\mathcal{Y},\ell}}.$$

We claim that ${}^{\mathrm{C}}\mathrm{D}^{\ell}_{\mathcal{V}}$ is a closed boundary restriction operator of $\mathcal{Q}_{p,\mathcal{Y},\ell}$. Indeed, (6.7.47) entails that

$$A_{\mathcal{Q}_{p,\mathcal{Y},\ell}} \subseteq {}^{\mathrm{C}}\mathrm{D}^{\ell}_{\mathcal{V}} \subseteq \mathrm{D}^{\ell}_{\mathcal{V}}.$$

Thus Proposition 5.2.3 implies that ${}^{\mathrm{C}}\mathrm{D}^{\ell}_{\mathcal{V}}$ is a boundary restriction operator of $\mathcal{Q}_{p,\mathcal{Y},\ell}$. Besides, the kernel ${}^{\mathrm{C}}\mathrm{D}^{\ell}_{\mathcal{V}}$ is equal to N, which is closed in $\mathcal{V}$ according to Remark 3.2.1 (ii). Applying Proposition 4.4.1 with $B := B_{\ell,\mathcal{V}}$ and $\mathcal{E} := \mathcal{V}$, there results that ${}^{\mathrm{C}}\mathrm{D}^{\ell}_{\mathcal{V}}$ is closed in $\mathcal{V}$. That completes the proof of the claim.

On account of Sect. 5.6.4.1, a function u in $L^p(\mathcal{Y})$ lies in the domain of ${}^{\mathrm{C}}\mathrm{D}^{\ell}_{\mathcal{V}}$ if and only if there exists U_1 in $\mathcal{Y}$ such that

$$k * (u - g_1 U_1) \in W^{1,p}\big((0,b),\mathcal{Y})\big).$$

Moreover, such a U_1 is unique, and

$$ {}^{\mathrm{C}}\mathrm{D}^{\ell}_{\mathcal{Y}}u := {}^{\mathrm{RL}}\mathrm{D}^{\ell}_{\mathcal{Y}}(u - g_1 U_1). $$

6.7.4.2 Case Where ℓ Lies in $L^p(0,b) \cap L^q(0,b)$

6.7.4.2.1 Boundary Restriction Operators

We will give three representations of the domain of the pivot operator $A_{Q_{p,\mathcal{Y},\ell}}$ of $Q_{p,\mathcal{Y},\ell}$. First, since by Proposition 5.6.10, $A_{Q_{p,\mathcal{Y},\ell}}$ is equal to the pivot operators of ${}^{\mathrm{RL}}Q_{p,\mathcal{Y},\ell}$, we get from Proposition 6.5.6

$$ D(A_{Q_{p,\mathcal{Y},\ell}}) = D({}^{\mathrm{RL}}\mathrm{D}^{\ell}_{\mathcal{Y}}) \cap {}_0C([0,b],\mathcal{Y}). \tag{6.7.49} $$

The second representation is expressed in terms of endogenous boundary values. Let u lie in $D(\mathrm{D}^{\ell}_{\mathcal{Y}})$ and (U_1, U_2, U_1', U_2') be its endogenous boundary values given by (6.7.6) and (6.7.8). Then we derive the equivalences

$$ \begin{aligned} u \in D(A_{Q_{p,\mathcal{Y},\ell}}) &\iff U_1 = U_2 = 0 \\ &\iff u(0) = k * u(0) = 0. \end{aligned} \tag{6.7.50} $$

Besides, notice that by using also Remark 6.7.1, one has

$$ U_1' = k * u(b), \qquad U_2' = u(b). $$

The third representation is

$$ D(A_{Q_{p,\mathcal{Y},\ell}}) = D(\mathrm{D}^{\ell}_{\mathcal{Y}}) \cap {}_0C([0,b],\mathcal{Y}). \tag{6.7.51} $$

Indeed, (6.7.49) implies that the former set is contained in the latter one. Conversely, let u be in the intersection and (U_1, U_2, U_1', U_2') be its endogenous boundary values given by (6.7.6). Since u and

$$ g_1 U_1 + \ell * g_1 U_1' + \ell * S_{\mathcal{Y}}(\ell U_2') + \ell * S_{\mathcal{Y}}\psi^{\perp} $$

are continuous (by Proposition 3.3.5), we must have $U_2 = 0$. Besides, $u(0) = 0$ entails that $U_1 = 0$. Thus we conclude with (6.7.50).

Regarding the minimal operator, the very definition of minimal operators together with (6.7.6) and (6.7.8) gives that for any u in the domain of $\mathrm{D}^{\ell}_{\mathcal{Y}}$,

$$ \begin{aligned} u \in D(A_{\mathrm{m},Q_{p,\mathcal{Y},\ell}}) &\iff U_1 = U_2 = U_1' = U_2' = 0 \\ &\iff u(0) = k * u(0) = u(b) = k * u(b) = 0. \end{aligned} \tag{6.7.52} $$

Let us now consider the special boundary restriction operators introduced in Sect. 6.7.4.1. Our goal is to compute their endogenous and standard boundary conditions with underlying quadruplet $\mathcal{Q}_{p,\mathcal{Y},\ell}$. Starting with ${}^{\mathrm{RL}}\mathrm{D}^{\ell}_{\mathcal{V}}$, let u lie in $D(\mathrm{D}^{\ell}_{\mathcal{V}})$ and (U_1, U_2, U_1', U_2') be its endogenous boundary values with respect to the direct sum decompositions (6.7.7). By Lemma 4.3.3,

$$D({}^{\mathrm{RL}}\mathrm{D}^{\ell}_{\mathcal{V}}) = \ell\mathcal{Y} \oplus R(B_{\ell,\mathcal{V}}).$$

Thus with (6.7.6) and (6.7.8),

$$\begin{aligned} u \in D({}^{\mathrm{RL}}\mathrm{D}^{\ell}_{\mathcal{V}}) &\Longleftrightarrow U_1 = 0 \\ &\Longleftrightarrow \big(u - \ell\, k * u(0)\big)(0) = 0. \end{aligned} \tag{6.7.53}$$

The endogenous boundary conditions of ${}^{\mathrm{C}}\mathrm{D}^{\ell}_{\mathcal{V}}$ are easily obtained thanks to (6.7.48). Indeed let u lie in $D(\mathrm{D}^{\ell}_{\mathcal{V}})$ and (U_1, U_2, U_1', U_2') be its endogenous boundary values with respect to the direct sum decompositions (6.7.7). Then

$$u \in D({}^{\mathrm{C}}\mathrm{D}^{\ell}_{\mathcal{V}}) \Longleftrightarrow U_2 = 0 \tag{6.7.54}$$

$$\Longleftrightarrow k * u(0) = 0. \tag{6.7.55}$$

Moreover u is continuous according to Sect. 6.7.2.1 and $U_1 = u(0)$. Arguing as in the proof of (6.7.51), we get the following representation of the domain of ${}^{\mathrm{C}}\mathrm{D}^{\ell}_{\mathcal{V}}$:

$$D({}^{\mathrm{C}}\mathrm{D}^{\ell}_{\mathcal{V}}) = C([0, b], \mathcal{Y}) \cap D(\mathrm{D}^{\ell}_{\mathcal{V}}). \tag{6.7.56}$$

Next, for any u in the domain of ${}^{\mathrm{C}}\mathrm{D}^{\ell}_{\mathcal{V}}$, denoting by (U_1, U_2, U_1', U_2') the endogenous boundary values of u with respect to the direct sum decompositions (6.7.7), (6.7.8) read in view of Remark 6.7.1

$$\begin{aligned} U_1 &= u(0) \\ U_2 &= 0 \\ U_1' &= k * \big(u - g_1\, u(0)\big)(b) \\ U_2' &= u(b) - u(0). \end{aligned} \tag{6.7.57}$$

Moreover, for f in $D(S_{\mathcal{V}^*}\mathrm{D}^{\ell}_{\mathcal{V}^*}S_{\mathcal{V}^*})$, the following *integration by parts formula* holds true:

$$\begin{aligned} \int_0^b \langle\, {}^{\mathrm{C}}\mathrm{D}^{\ell}_{\mathcal{V}}u(x),\, f(x)\,\rangle_{\mathcal{Y},\mathcal{Y}^*}\, \mathrm{d}x &= \int_0^b \langle\, u(x),\, S_{\mathcal{V}^*}\mathrm{D}^{\ell}_{\mathcal{V}^*}S_{\mathcal{V}^*} f(x)\,\rangle_{\mathcal{Y},\mathcal{Y}^*}\, \mathrm{d}x \\ &\quad + \langle\, U_1',\, F_1\,\rangle_{\mathcal{Y},\mathcal{Y}^*} + \langle\, u(b) - u(0),\, k * S_{\mathcal{V}^*} f(0)\,\rangle_{\mathcal{Y},\mathcal{Y}^*} \end{aligned}$$

$$- \langle u(0),\, k * (S_{\mathcal{V}*} f - g_1 \, F_1)(b) - k * (S_{\mathcal{V}*} f - g_1 \, F_1)(0) \rangle_{\mathcal{Y}, \mathcal{Y}*},$$

where

$$F_1 := \big(S_{\mathcal{V}*} f - \ell\, k * S_{\mathcal{V}*} f(0) \big)(0).$$

Indeed, applying the result of Sect. 5.7.1.5 and using (6.7.57) and (6.7.38), we get the above integration by parts formula.

6.7.4.2.2 Adjoints Operators

Recalling that ℓ lies in $L^p(0, b) \cap L^q(0, b)$, we will start to study the adjoint of the pivot operator $A_{Q_{p,\mathcal{Y},\ell}}$. By Proposition 5.6.10,

$$(A_{Q_{p,\mathcal{Y},\ell}})^* = (A_{\mathrm{RL}\, Q_{p,\mathcal{Y},\ell}})^*.$$

Thus (6.5.44) gives

$$D\big((A_{Q_{p,\mathcal{Y},\ell}})^*\big) = \{ f \in D(^{\mathrm{RL}}\mathrm{D}^\ell_{\mathcal{V}*}\, S_{\mathcal{V}*}) \cap C([0, b], \mathcal{Y}^*) \mid f(b) = 0 \}. \qquad (6.7.58)$$

In order to give boundary conditions of $(A_{Q_{p,\mathcal{Y},\ell}})^*$, let f lie in $D(\mathrm{D}^\ell_{\mathcal{V}*}\, S_{\mathcal{V}*})$ and (F_1, F_2, F_1', F_2') be its endogenous boundary values with respect to the direct sum decompositions (6.7.35). Then

$$f \in D\big((A_{Q_{p,\mathcal{Y},\ell}})^*\big) \iff f \in R\big((B_{\ell,\mathcal{V}})^*\big) \qquad\qquad \text{(by (5.2.9))}$$

$$\iff F_1 = F_2 = 0 \qquad\qquad \text{(by (6.7.34))}$$

$$\iff f(b) = (k * S_{\mathcal{V}*} f)(0) = 0 \qquad \text{(by (6.7.37))}.$$

The equations $F_1 = F_2 = 0$ are the endogenous boundary conditions of $(A_{Q_{p,\mathcal{Y},\ell}})^*$, and $f(b) = (k * S_{\mathcal{V}*} f)(0) = 0$ are its standard boundary conditions.

The adjoint of the minimal operator is given by (5.2.16), i.e.,

$$(A_{\mathrm{m}, Q_{p,\mathcal{Y},\ell}})^* = S_{\mathcal{V}*}\,{}^{\mathrm{RL}}\mathrm{D}^\ell_{\mathcal{V}*}\, S_{\mathcal{V}*}.$$

Now consider $(^{\mathrm{RL}}\mathrm{D}^\ell_{\mathcal{V}})^*$. A first representation is (6.5.45). This representation involves endogenous boundary values with underlying quadruplet $(^{\mathrm{RL}} Q_{p,\mathcal{Y},\ell})^*$. Next, let us characterize the domain of $(^{\mathrm{RL}}\mathrm{D}^\ell_{\mathcal{V}})^*$ in terms of endogenous boundary values with underlying differential quadruplet $(Q_{p,\mathcal{Y},\ell})^*$. Let f lie in $D(\mathrm{D}^\ell_{\mathcal{V}*}\, S_{\mathcal{V}*})$ and (F_1, F_2, F_1', F_2') be its endogenous boundary values with respect to the direct sum decompositions (6.7.35). We claim that

$$f \in D\big(({}^{\mathrm{RL}}\mathrm{D}^{\ell}_{\mathcal{Y}})^*\big) \iff F_1 = F_2 = F_2' = 0$$

$$\iff f(b) = (k * S_{\mathcal{Y}^*} f)(0) = f(0) = 0. \tag{6.7.59}$$

Indeed applying Proposition 5.4.3 with $A := {}^{\mathrm{RL}}\mathrm{D}^{\ell}_{\mathcal{Y}}$ and $M_1 := \ell\mathcal{Y}$, there results that f belongs to the domain of $({}^{\mathrm{RL}}\mathrm{D}^{\ell}_{\mathcal{Y}})^*$ if and only if

$$F_1 = F_2 = 0, \qquad h_1^{\mathrm{s}} F_1' + h_2^{\mathrm{s}} F_2' \in (\ell\mathcal{Y})^{\perp}.$$

Thus with (6.7.31) and (3.2.1), the latter condition is equivalent to $F_2' = 0$, which proves the first equivalence of the claim. The second one follows from the first equivalence together (6.7.38). That completes the proof of the claim.

With regard to the adjoint of ${}^{\mathrm{C}}\mathrm{D}^{\ell}_{\mathcal{Y}}$, we have the following result.

Proposition 6.7.3 *Let p, q be conjugate exponents in $(1, \infty)$, $\mathcal{Y}$ be a nontrivial real or complex separable and reflexive Banach space, and (k, ℓ) be a Sonine pair. Assume that ℓ lies in $L^p(0, b) \cap L^q(0, b)$, and consider any f in $D(\mathrm{D}^{\ell}_{\mathcal{Y}^*} S_{\mathcal{Y}^*})$. Then*

$$f \in D\big(({}^{\mathrm{C}}\mathrm{D}^{\ell}_{\mathcal{Y}})^*\big) \iff f(b) = k * S_{\mathcal{Y}^*} f(0) = k * S_{\mathcal{Y}^*} f(b) = 0 \tag{6.7.60}$$

and

$$D\big(({}^{\mathrm{C}}\mathrm{D}^{\ell}_{\mathcal{Y}})^*\big) = \{f \in L^p(\mathcal{Y}) \mid k * S_{\mathcal{Y}^*} f \in W_0^{1,q}\big((0, b), \mathcal{Y}^*)\big)\} \tag{6.7.61}$$

$$= \{f \in D({}^{\mathrm{RL}}\mathrm{D}^{\ell}_{\mathcal{Y}^*} S_{\mathcal{Y}^*}) \cap C([0, b], \mathcal{Y}^*) \mid f(b) = k * S_{\mathcal{Y}^*} f(b) = 0\}. \tag{6.7.62}$$

Roughly speaking, (6.7.60) express that $({}^{\mathrm{C}}\mathrm{D}^{\ell}_{\mathcal{Y}})^*$ is $S_{\mathcal{Y}^*}\mathrm{D}^{\ell}_{\mathcal{Y}^*} S_{\mathcal{Y}^*}$ supplemented with the boundary conditions

$$f(b) = k * S_{\mathcal{Y}^*} f(0) = k * S_{\mathcal{Y}^*} f(b) = 0.$$

Proof of Proposition 6.7.3 In order to prove (6.7.60), we consider any function f in the domain of $\mathrm{D}^{\ell}_{\mathcal{Y}^*} S_{\mathcal{Y}^*}$ and denote by (F_1, F_2, F_1', F_2') its endogenous boundary values with respect to the direct sum decompositions (6.7.35). Applying Proposition 5.4.3 with $A := {}^{\mathrm{C}}\mathrm{D}^{\ell}_{\mathcal{Y}}$ and $M_1 := g_1\mathcal{Y}$, there results that f belongs to the domain of $({}^{\mathrm{C}}\mathrm{D}^{\ell}_{\mathcal{Y}})^*$ if and only

$$F_1 = F_2 = 0, \qquad h_1 F_1' + h_2 F_2' \in (g_1\mathcal{Y})^{\perp}.$$

Thus with (6.7.31) and (3.2.1), the latter condition is equivalent to $F_1' = 0$. Next with (6.7.38), we obtain (6.7.60).

Let us prove (6.7.61). By (5.2.7),

$$D\big(({}^{\mathrm{C}}\mathrm{D}_{\mathcal{Y}}^{\ell})^*\big) = S_{\mathcal{Y}^*}\, B_{\ell,\mathcal{Y}^*}\, S_{\mathcal{Y}^*}\big((g_1\mathcal{Y})^{\perp}\big).$$

Let f belong to the latter set. Then f reads $f = S_{\mathcal{Y}^*}\, B_{\ell,\mathcal{Y}^*}\, S_{\mathcal{Y}^*}h^{\perp}$ for some $h^{\perp}$ in $(g_1\mathcal{Y})^{\perp}$. Thus

$$k * S_{\mathcal{Y}^*} f = g_1 * S_{\mathcal{Y}^*}h^{\perp},$$

which yields on the one hand that $k * S_{\mathcal{Y}^*} f$ lies in ${}_0W^{1,q}(\mathcal{Y}^*)$. On the other hand,

$$k * S_{\mathcal{Y}^*} f(b) = \int_0^b g_1(y)h^{\perp}(y)\,\mathrm{d}y = 0$$

by (3.2.1). Whence f lies in the right-hand side of (6.7.61). Conversely, let f belong to the right-hand side of (6.7.61). Since by definition,

$$D({}^{\mathrm{RL}}\mathrm{D}_{\mathcal{Y}^*}^{\ell}) = \{f \in L^q(\mathcal{Y}^*) \mid k * f \in W^{1,q}(\mathcal{Y}^*)\},$$

we deduce that $S_{\mathcal{Y}^*} f$ lies in $D({}^{\mathrm{RL}}\mathrm{D}_{\mathcal{Y}^*}^{\ell})$. One has $k * S_{\mathcal{Y}^*} f(0) = 0$, so that $f(b) = 0$ from (6.7.53). Moreover by assumption, $k{*}S_{\mathcal{Y}^*} f(b) = 0$. Thus f lies in $D\big(({}^{\mathrm{C}}\mathrm{D}_{\mathcal{Y}}^{\ell})^*\big)$ by (6.7.60). That proves (6.7.61). Finally, for any f in $D\big(({}^{\mathrm{C}}\mathrm{D}_{\mathcal{Y}}^{\ell})^*\big)$, the condition $k * S_{\mathcal{Y}^*} f(0) = 0$ is equivalent to the continuity of f on $[0, b]$, by (6.7.34) and (6.7.37). Thus (6.7.62) follows from (6.7.60). □

6.7.4.3 Case Where ℓ Lies in $L^q(0, b) \setminus L^p(0, b)$

We refer to the results of Sect. 6.7.2.2. In particular, we know that every function in the domain of $\mathrm{D}_{\mathcal{Y}}^{\ell}$ is continuous on $[0, b]$. We will only consider the case where

$$k \text{ lies in } L^q(0, b) \setminus L^p(0, b).$$

The other cases may be studied in a similar way.

6.7.4.3.1 Boundary Restriction Operators

Regarding the pivot operator $A_{Q_{p,\mathcal{Y},\ell}}$ of $Q_{p,\mathcal{Y},\ell}$, one has

$$\begin{aligned} A_{Q_{p,\mathcal{Y},\ell}} &= A_{{}^{\mathrm{RL}}Q_{p,\mathcal{Y},\ell}} && \text{(by Proposition 5.6.10 (ii))} \\ &= {}^{\mathrm{RL}}\mathrm{D}_{\mathcal{Y}}^{\ell} && \text{(by (6.5.47)).} \end{aligned}$$

Thus by (6.5.18), the domain of $A_{Q_{p,\mathcal{Y},\ell}}$ is equal to

$$\{u \in {}_0C([0,b],\mathcal{Y}) \mid k * u \in W^{1,p}\big((0,b),\mathcal{Y}\big)\}.$$

Regarding boundary conditions of $A_{\mathcal{Q}_{p,\mathcal{Y},\ell}}$, we consider any u in $D(\mathrm{D}^{\ell}_{\mathcal{Y}})$ and denote by (U_1, U_2') its endogenous boundary values with respect to the direct sum decompositions (6.7.13). Arguing as in Sect. 6.7.4.2, we find

$$\begin{aligned} u \in D(A_{\mathcal{Q}_{p,\mathcal{Y},\ell}}) &\Longleftrightarrow U_1 = 0 \\ &\Longleftrightarrow u(0) = 0, \end{aligned}$$

and

$$D(A_{\mathcal{Q}_{p,\mathcal{Y},\ell}}) = D(\mathrm{D}^{\ell}_{\mathcal{Y}}) \cap {}_0C([0,b],\mathcal{Y}).$$

For the minimal operator, (6.7.14) entails that for any u in the domain of $\mathrm{D}^{\ell}_{\mathcal{Y}}$,

$$u \in D(A_{\mathrm{m},\mathcal{Q}_{p,\mathcal{Y},\ell}}) \Longleftrightarrow u(0) = u(b) = 0.$$

Finally, the Caputo operator is equal to the maximal operator, which in symbol reads

$${}^{\mathrm{C}}\mathrm{D}^{\ell}_{\mathcal{Y}} = \mathrm{D}^{\ell}_{\mathcal{Y}}. \tag{6.7.63}$$

6.7.4.3.2 Adjoints Operators

Let f lie in $D(\mathrm{D}^{\ell}_{\mathcal{Y}^*}\, S_{\mathcal{Y}^*})$ and (F_2, F_1') be given by (6.7.41). By (6.5.50) and (6.7.42)

$$\begin{aligned} f \in D\big(({}^{\mathrm{RL}}\mathrm{D}^{\ell}_{\mathcal{Y}})^*\big) &\Longleftrightarrow k * S_{\mathcal{Y}^*}(0) = 0 \\ &\Longleftrightarrow F_2 = 0. \end{aligned}$$

Also

$$\begin{aligned} f \in D\big(({}^{\mathrm{C}}\mathrm{D}^{\ell}_{\mathcal{Y}})^*\big) &\Longleftrightarrow f \in D\big((\mathrm{D}^{\ell}_{\mathcal{Y}})^*\big) && \text{(by (6.7.63))} \\ &\Longleftrightarrow F_2 = F_1' = 0 \\ &\Longleftrightarrow k * S_{\mathcal{Y}^*}(0) = k * S_{\mathcal{Y}^*}(b) = 0. \end{aligned}$$

6.8 The Quadruplet $\mathcal{Q}_{p,\mathcal{Y},n+\ell}$

Let b be a positive real number, $\mathbb{K}$ be equal to $\mathbb{R}$ or $\mathbb{C}$, and p, q be conjugate exponents in $(1,\infty)$. Let also $\mathcal{Y}$ be a nontrivial separable and reflexive Banach space over $\mathbb{K}$ and (k,ℓ) be a Sonine pair.

We set

$$\mathcal{V} := L^p\big((0,b),\mathcal{Y}\big), \qquad \mathcal{V}^* := L^q\big((0,b),\mathcal{Y}^*\big).$$

In Sect. 6.6.1, we have introduced for each nonnegative integer n, the differential quadruplet

$$^{\mathrm{RL}}\mathcal{Q}_{p,\mathcal{Y},n+\ell} = (^{\mathrm{RL}}\mathrm{D}^{n+\ell}_{\mathcal{V}^*},\ ^{\mathrm{RL}}\mathrm{D}^{n+\ell}_{\mathcal{V}},\ B_{g_n*\ell,\mathcal{V}},\ S_{\mathcal{V}}).$$

Here, we will define and study a Caputo extension of $^{\mathrm{RL}}\mathcal{Q}_{p,\mathcal{Y},n+\ell}$ which generalizes the Caputo extension $\mathcal{Q}_{p,\mathcal{Y},\ell}$ featured in Chap. 6.7. Next we will compute endogenous boundary values.

6.8.1 Introduction

In order to defined a Caputo extension of $^{\mathrm{RL}}\mathcal{Q}_{p,\mathcal{Y},n+\ell}$, we will implement the theory of Caputo extensions of a product of quadruplets (see Sect. 5.6.6). More precisely, recalling that

$$^{\mathrm{RL}}\mathcal{Q}_{p,\mathcal{Y},n+\ell} := \mathcal{Q}_{p,\mathcal{Y},n}\,{}^{\mathrm{RL}}\mathcal{Q}_{p,\mathcal{Y},\ell},$$

we will apply the results of Sect. 5.6.6 in the particular case where (5.6.58) reads

$$\mathcal{Q} := \mathcal{Q}_{p,\mathcal{Y},n}, \qquad \mathcal{Q}_{\mathrm{rl}} := {}^{\mathrm{RL}}\mathcal{Q}_{p,\mathcal{Y},\ell}$$

By Proposition 6.7.1, $g_1\mathcal{Y}$ is in direct sum with $D(^{\mathrm{RL}}\mathrm{D}^{\ell}_{\mathcal{V}})$ if and only if k is not in $L^p(0,b)$. Thus by choosing $V := g_1\mathcal{Y}$ in (5.6.62), we set

$$N(g_1\mathcal{Y}) := \begin{cases} \bigoplus_{i=1}^{n+1} g_i\mathcal{Y} & \text{if } k \notin L^p(0,b) \\ \{0\} & \text{otherwise.} \end{cases} \tag{6.8.1}$$

In the same way,

$$N(g_1\mathcal{Y}^*) := \begin{cases} \bigoplus_{i=1}^{n+1} g_i\mathcal{Y}^* & \text{if } k \notin L^q(0,b) \\ \{0\} & \text{otherwise.} \end{cases} \tag{6.8.2}$$

The following result warrants that the Caputo extension of $^{\mathrm{RL}}\mathcal{Q}_{p,\mathcal{Y},n+\ell}$ with respect to $N(g_1\mathcal{Y})$ and $N(g_1\mathcal{Y}^*)$ is well defined.

Proposition 6.8.1 *Under the assumptions and notation set at the beginning of the current section,*

$$\mathrm{D}^1_{\mathcal{V}}\,{}^{\mathrm{RL}}\mathrm{D}^{\ell}_{\mathcal{V}}\,B_{1,\mathcal{V}} = {}^{\mathrm{RL}}\mathrm{D}^{\ell}_{\mathcal{V}} \tag{6.8.3}$$

$$R(B_{1,\mathcal{V}}) \subseteq D({}^{\mathrm{RL}}\mathrm{D}^{\ell}_{\mathcal{V}}) \tag{6.8.4}$$

$$\mathrm{D}^1_{\mathcal{V}^*}\,{}^{\mathrm{RL}}\mathrm{D}^{\ell}_{\mathcal{V}^*}\,B_{1,\mathcal{V}^*} = {}^{\mathrm{RL}}\mathrm{D}^{\ell}_{\mathcal{V}^*} \tag{6.8.5}$$

$$R(B_{1,\mathcal{V}^*}) \subseteq D({}^{\mathrm{RL}}\mathrm{D}^{\ell}_{\mathcal{V}^*}). \tag{6.8.6}$$

Moreover, $N(g_1\mathcal{Y})$ is in direct sum with $D({}^{\mathrm{RL}}\mathrm{D}^{n+\ell}_{\mathcal{V}})$, and $N(g_1\mathcal{Y}^*)$ is in direct sum with $D({}^{\mathrm{RL}}\mathrm{D}^{n+\ell}_{\mathcal{V}^*})$.

Proof By symmetry, it is enough to prove the results involving $\mathcal{V}$. Also the direct sum of $N(g_1\mathcal{Y})$ and $D({}^{\mathrm{RL}}\mathrm{D}^{n+\ell}_{\mathcal{V}})$ is a consequence of (6.8.3) and (6.8.4), according to Proposition 5.6.26. So there remains to prove (6.8.3) and (6.8.4). Regarding (6.8.3), we consider any u in the domain of ${}^{\mathrm{RL}}\mathrm{D}^{\ell}_{\mathcal{V}}$. Then

$$k * B_{1,\mathcal{V}}\,u = g_1 * k * u \in W^{1,p}(\mathcal{Y}).$$

Hence $B_{1,\mathcal{V}}\,u$ lies in $D({}^{\mathrm{RL}}\mathrm{D}^{\ell}_{\mathcal{V}})$ and

$${}^{\mathrm{RL}}\mathrm{D}^{\ell}_{\mathcal{V}}B_{1,\mathcal{V}}\,u = k * u.$$

By assumption, $k * u$ lives in $W^{1,p}(\mathcal{Y})$. Hence ${}^{\mathrm{RL}}\mathrm{D}^{\ell}_{\mathcal{V}}B_{1,\mathcal{V}}u$ belongs to the domain of $\mathrm{D}^1_{\mathcal{V}}$, and

$$\mathrm{D}^1_{\mathcal{V}}\,{}^{\mathrm{RL}}\mathrm{D}^{\ell}_{\mathcal{V}}B_{1,\mathcal{V}}\,u = {}^{\mathrm{RL}}\mathrm{D}^{\ell}_{\mathcal{V}}u.$$

Thus ${}^{\mathrm{RL}}\mathrm{D}^{\ell}_{\mathcal{V}} \subseteq \mathrm{D}^1_{\mathcal{V}}\,{}^{\mathrm{RL}}\mathrm{D}^{\ell}_{\mathcal{V}}B_{1,\mathcal{V}}$. Conversely, it is enough to prove that the domain of $\mathrm{D}^1_{\mathcal{V}}\,{}^{\mathrm{RL}}\mathrm{D}^{\ell}_{\mathcal{V}}B_{1,\mathcal{V}}$ is contained in $D({}^{\mathrm{RL}}\mathrm{D}^{\ell}_{\mathcal{V}})$. For, any u in $D(\mathrm{D}^1_{\mathcal{V}}\,{}^{\mathrm{RL}}\mathrm{D}^{\ell}_{\mathcal{V}}B_{1,\mathcal{V}})$ satisfies

$$g_1 * u \in D({}^{\mathrm{RL}}\mathrm{D}^{\ell}_{\mathcal{V}}).$$

Thus $k * g_1 * u$ lies in $W^{1,p}(\mathcal{Y})$, and

$${}^{\mathrm{RL}}\mathrm{D}^{\ell}_{\mathcal{V}}B_{1,\mathcal{V}}u = k * u.$$

However by assumption, ${}^{\mathrm{RL}}\mathrm{D}^{\ell}_{\mathcal{V}}B_{1,\mathcal{V}}u$ belongs to $D(\mathrm{D}^1_{\mathcal{V}})$, which yields that u lives in $D({}^{\mathrm{RL}}\mathrm{D}^{\ell}_{\mathcal{V}})$. Whence (6.8.3) holds true.

Finally, since for each h in $L^p(\mathcal{Y})$,

$$k * g_1 * h \in W^{1,p}(\mathcal{Y}),$$

we deduce that $R(B_{1,\mathcal{V}}) \subseteq D({}^{\mathrm{RL}}\mathrm{D}^{\ell}_{\mathcal{V}})$. □

On account of Proposition 6.8.1, we denote by $\mathrm{D}^{n+\ell}_{\mathcal{V}}$ the Caputo extension of ${}^{\mathrm{RL}}\mathrm{D}^{\ell}_{\mathcal{V}}$ with respect to $N(g_1\mathcal{Y})$ and by $\mathrm{D}^{n+\ell}_{\mathcal{V}*}$ the Caputo extension of $\mathrm{D}^{\ell}_{\mathcal{V}*}$ with respect to $N^{\mathrm{s}}(g_1\mathcal{Y}^*)$. Then the Caputo extension $\mathcal{Q}_{p,\mathcal{Y},n+\ell}$ of ${}^{\mathrm{RL}}\mathcal{Q}_{p,\mathcal{Y},n+\ell}$ with respect to $N(g_1\mathcal{Y})$ and $N^{\mathrm{s}}(g_1\mathcal{Y}^*)$ reads

$$\mathcal{Q}_{p,\mathcal{Y},n+\ell} = (\mathrm{D}^{n+\ell}_{\mathcal{V}*},\, \mathrm{D}^{n+\ell}_{\mathcal{V}},\, B_{n+\ell,\mathcal{V}},\, S_{\mathcal{V}}). \tag{6.8.7}$$

For all purposes, let us make the definition of $\mathrm{D}^{n+\ell}_{\mathcal{V}}$ explicit in the case where k does not belong to $L^p(0,b)$. For each u in the domain of $\mathrm{D}^{n+\ell}_{\mathcal{V}}$, there exists a unique tuple $(U_1,\ldots,U_{n+1})$ in $\mathcal{Y}^{n+1}$ such that $u - \sum_{i=1}^{n+1} g_i U_i$ lies in the domain of ${}^{\mathrm{RL}}\mathrm{D}^{n+\ell}_{\mathcal{V}}$. Moreover,

$$\mathrm{D}^{n+\ell}_{\mathcal{V}} u = {}^{\mathrm{RL}}\mathrm{D}^{n+\ell}_{\mathcal{V}}\Big(u - \sum_{i=1}^{n+1} g_i U_i\Big).$$

6.8.2 Endogenous Boundary Values

We refer to Sect. 6.7 for the case where $n = 0$. In the sequel we will assume that n is a positive integer. Our goal is to compute endogenous boundary values of any element u of $D(\mathrm{D}^{n+\ell}_{\mathcal{V}})$. We will only consider the case where the Caputo extensions of the maximal operators ${}^{\mathrm{RL}}\mathrm{D}^{\ell}_{\mathcal{V}}$ and ${}^{\mathrm{RL}}\mathrm{D}^{\ell}_{\mathcal{V}*}$ are both nontrivial, that is, we will assume that

$$k \in L^1(0,b) \setminus (L^q(0,b) \cup L^p(0,b)).$$

The formulas for the U_i's are (6.8.13), (6.8.17), and (6.8.18), whereas the U_i''s are given by (6.8.19), (6.8.20). According to (6.8.1), (6.8.2), and (6.6.2), (6.6.3),

$$\ker \mathrm{D}^{n+\ell}_{\mathcal{V}} = \bigoplus_{i=1}^{n+1} g_i\mathcal{Y} \oplus \bigoplus_{i=i_p}^{n} g_i * \ell\mathcal{Y} \tag{6.8.8}$$

$$\ker \mathrm{D}^{n+\ell}_{\mathcal{V}*} = \bigoplus_{i=1}^{n+1} g_i\mathcal{Y}^* \oplus \bigoplus_{i=i_q}^{n} g_i * \ell\mathcal{Y}^*. \tag{6.8.9}$$

Now we claim that $\mathcal{Q}_{p,\mathcal{Y},n+\ell}$ is regular. Indeed, Corollary 5.3.1 gives the existence of functions $h_1,\ldots,h_{n+1},h_{i_q+n+2},\ldots,h_{2n+2}$ in $L^p(0,b)$ such that on the one hand, for each index j in $[1,n+1] \cup [i_q+n+2, 2n+2]$,

$$\int_0^b g_i(y)h_j(y)\,\mathrm{d}y = \delta_{ij}, \qquad \forall i \in [1, n+1] \tag{6.8.10}$$

$$\int_0^b g_i * \ell(y)h_j(y)\,\mathrm{d}y = \delta_{i+n+2,j}, \qquad \forall i \in [i_q, n].$$

On the other hand, by setting

$$\tilde{F}_0 := \bigoplus_{i=1}^{n+1} h_i\mathcal{Y} \oplus \bigoplus_{i=i_q+n+2}^{2n+2} h_i\mathcal{Y}, \tag{6.8.11}$$

$\mathcal{Q}_{p,\mathcal{Y},n+\ell}$ is a $\tilde{F}_0$-regular differential quadruplet. That proves the claim. Then

$$E_{\mathrm{M}} = \bigoplus_{i=1}^{n+1} g_i\mathcal{Y} \oplus \bigoplus_{i=i_p}^{n} g_i * \ell\mathcal{Y} \oplus g_n * \ell * S_{\mathcal{Y}}(\tilde{F}_0)$$

$$D(\mathrm{D}_{\mathcal{V}}^{n+\ell}) = E_{\mathrm{M}} \oplus g_n * \ell * S_{\mathcal{Y}}\big((\ker \mathrm{D}_{\mathcal{V}^*}^{n+\ell})^{\perp}\big).$$

Thus a function u in $\mathcal{V}$ lies in $D(\mathrm{D}_{\mathcal{V}}^{n+\ell})$ if and only if there exist endogenous boundary values

$$U_1, \ldots, U_{n+1}, U_{i_p+n+2}, \ldots, U_{2n+2}, U_1', \ldots, U_{n+1}', U_{i_q+n+2}', \ldots, U_{2n+2}'$$

in $\mathcal{Y}$ and $\psi^{\perp}$ in $(\ker \mathrm{D}_{\mathcal{V}^*}^{n+\ell})^{\perp}$ such that

$$u = \sum_{i=1}^{n+1} g_i\, U_i + \sum_{i=i_p+n+2}^{2n+2} g_{i-n-2} * \ell\, U_i +$$

$$g_n * \ell * S_{\mathcal{Y}}\Big(\sum_{i=1}^{n+1} h_i\, U_i' + \sum_{i=i_q+n+2}^{2n+2} h_i\, U_i'\Big) + g_n * \ell * S_{\mathcal{Y}}\psi^{\perp}. \tag{6.8.12}$$

For all purposes, let us notice that if $i_p = i_q = 0$, i.e., if ℓ lies in $L^p(0,b)$ and $L^q(0,b)$, then there are $4n+4$ endogenous boundary values. In contrast, if $i_p = i_q = 1$ then the values U_{n+2} and U_{n+2}' disappear from the list of the endogenous boundary values. In general, there are $4n+4-i_p-i_q$ endogenous boundary values.

We will compute the endogenous boundary values of u; the result for the U_i's stated in (6.8.13), (6.8.17) and (6.8.18) forms a set of $2n+2-i_p$ equations.

Recalling that n is a positive integer, let us start to compute $U_{i_p+n+2}, \ldots, U_{2n+2}$. We claim that for each j in $[i_p+1, n+1]$

$$U_{j+n+1} = \mathrm{D}_{\mathcal{V}}^{j-2+\ell} u(0). \tag{6.8.13}$$

Recall that by convention, when $i_p = 0$ and $j = 1$, (6.8.13) reads

$$U_{n+2} = k * u(0).$$

Remark 6.8.1 Formula (6.8.13) may sometimes be extended to the case where $n = 0$.

(i) If $i_p = 0$ then $j = 1$ and (6.8.13) reads

$$k * u(0) = U_2. \tag{6.8.14}$$

(i-a) If $i_q = 0$ then ℓ lies in $L^p(0,b) \cap L^q(0,b)$, and we recover the second equation of (6.7.8).

(i-b) If $i_q = 1$ then ℓ lies in $L^p(0,b) \setminus L^q(0,b)$, and besides, we have assumed that k does not belong to $L^p(0,b) \cup L^q(0,b)$. In this case, we recover the second equation of (6.7.23).

(ii) If $i_p = 1$ (i.e., ℓ is not in $L^p(0,b)$), then the extension of (6.8.13) is not straightforward since $n + 1 < i_p + 1$. In this case, it turns out that there is no U_2: see (6.7.18) and (6.7.29).

□

Let j be any integer in $[i_p + 1, n + 1]$. In order to prove (6.8.13), we derive from (6.8.8) that for any index $i \geq 1$,

$$\begin{aligned} g_i &\in \ker \mathrm{D}_{L^p}^{j-2+\ell} && \text{if } i \leq j-1 \\ g_{i-n-2} * \ell &\in \ker \mathrm{D}_{L^p}^{j-2+\ell} && \text{if } i_p + n + 2 \leq i \leq j + n. \end{aligned}$$

Then

$$\mathrm{D}_{L^p}^{j-2+\ell} g_i = \begin{cases} 0 & \text{if } 1 \leq i \leq j-1 \\ g_{i+1-j} * k & \text{if } j \leq i \end{cases} \tag{6.8.15}$$

and with Lemma 6.6.1,

$$\mathrm{D}_{L^p}^{j-2+\ell}(g_i * \ell) = \begin{cases} 0 & \text{if } i_p \leq i \leq j-2 \\ g_{i+2-j} & \text{if } j-1 \leq i,\ i_p \leq i. \end{cases} \tag{6.8.16}$$

Applying the operator $\mathrm{D}_{\mathcal{V}}^{j-2+\ell}$ to (6.8.12), we get

$$\mathrm{D}_{\mathcal{V}}^{j-2+\ell} u = \sum_{i=j}^{n+1} g_{i+1-j} * k\, U_i + \sum_{i=j+n+1}^{2n+2} g_{i-j-n}\, U_i + g_{n+2-j} * h,$$

where h is some element of $\mathcal{V}$. Then (6.8.13) follows by evaluating this identity at $x = 0$.

For each $j = 1, \ldots, n$, the formula for U_j is

$$U_j = \mathrm{D}_{\mathcal{V}}^{j-1} \left(u - \sum_{i=i_p+n+2}^{2n+2} g_{i-n-2} * \ell\, U_i \right)(0) \tag{6.8.17}$$

and the formula for U_{n+1} reads

$$U_{n+1} = \mathrm{D}_{\mathcal{V}}^{n} \Big(u - \sum_{i=i_p+n+2}^{2n+2} g_{i-n-2} * \ell\, U_i - g_n * \ell * \mathrm{D}_{\mathcal{V}}^{n+\ell} u \Big)(0). \tag{6.8.18}$$

(6.8.17) and (6.8.18) are proved in a standard way. Notice that the case where $j = n+1$ is special since we can not evaluate $\ell * h$ at 0 when ℓ is not in $L^q(0, b)$.

Let us now compute the U_i's. Applying $\mathrm{D}_{\mathcal{V}}^{n+\ell}$ to (6.8.12), we get

$$\mathrm{D}_{\mathcal{V}}^{n+\ell} u = S_{\mathcal{V}} \Big(\sum_{i=1}^{n+1} h_i\, U_i' + \sum_{i=i_q+n+2}^{2n+2} h_i\, U_i' \Big) + S_{\mathcal{V}} \psi^{\perp}.$$

Now as an element of $(\ker \mathrm{D}_{\mathcal{V}^*}^{n+\ell})^{\perp}$, $\psi^{\perp}$ satisfies

$$\int_0^b g_j \psi^{\perp}\, \mathrm{d}y = 0, \qquad j = 1, \ldots, n+1$$

$$\int_0^b g_j * \ell \psi^{\perp}\, \mathrm{d}y = 0, \qquad j = i_q, \ldots, n.$$

Thus by Proposition 3.3.5 and (6.8.10), we get for $j = 1, \ldots, n+1$,

$$g_j * \mathrm{D}_{\mathcal{V}}^{n+\ell} u(b) = U_j'. \tag{6.8.19}$$

In a same way, for $j = i_q, \ldots, n$,

$$g_j * \ell * \mathrm{D}_{\mathcal{V}}^{n+\ell} u(b) = U_{j+n+2}'. \tag{6.8.20}$$

6.9 The Riemann-Liouville Quadruplet and Its Extensions

Due to its wide use, we will rewrite some of the basic results of this chapter in the particular case where the Sonine pair (k, ℓ) is formed by Riemann-Liouville kernels (see (3.4.3)). More precisely, let b be a positive real number, $\mathbb{K}$ be equal to $\mathbb{R}$ or $\mathbb{C}$, and p, q be conjugate exponents in $(1, \infty)$. Being given any number α in $(0, 1)$, we will assume that

$$k := g_{1-\alpha}, \qquad \ell := g_\alpha.$$

For simplicity, we suppose that $\mathcal{Y} := \mathbb{K}$, so that $\mathcal{V} = L^p := L^p(0, b)$ and $\mathcal{V}^* = L^q := L^q(0, b)$. Then the so-called *Riemann-Liouville differential quadruplet* ${}^{\mathrm{RL}}\mathcal{Q}_{p,\mathbb{K},g_\alpha}$ defined by (6.5.2) will be denoted by ${}^{\mathrm{RL}}\mathcal{Q}_{p,\mathbb{K},\alpha}$ or simply by ${}^{\mathrm{RL}}\mathcal{Q}_{p,\alpha}$. That is

$$ {}^{\mathrm{RL}}\mathcal{Q}_{p,\mathbb{K},\alpha} = ({}^{\mathrm{RL}}\mathrm{D}^\alpha_{L^q},\ {}^{\mathrm{RL}}\mathrm{D}^\alpha_{L^p},\ B_{\alpha,L^p},\ S_{L^p}) \tag{6.9.1}$$

Although this section does not contain any essentially new material, it is of some interest from a theoretical point of view. Indeed, we will verify that all the previous cases about the L^p-regularity of ℓ and k L^p actually occur.

In the forthcoming subsection, we will present endogenous boundary values for two quadruplets, the Riemann-Liouville quadruplet and its adjoint. The second subsection will be devoted to Caputo extensions of ${}^{\mathrm{RL}}\mathcal{Q}_{p,\mathbb{K},g_\alpha}$.

6.9.1 The Riemann-Liouville Quadruplet ${}^{\mathrm{RL}}\mathcal{Q}_{p,\mathbb{K},\alpha}$ and Its Adjoint

For all purposes, recall that

$$\begin{aligned}
D({}^{\mathrm{RL}}\mathrm{D}^\alpha_{L^p}) &= \left\{u \in L^p(0, b) \mid g_{1-\alpha} * u \in W^{1,p}(0, b)\right\} \\
{}^{\mathrm{RL}}\mathrm{D}^\alpha_{L^p} &: D({}^{\mathrm{RL}}\mathrm{D}^\alpha_{L^p}) \subseteq L^p(0, b) \to L^p(0, b), \qquad u \mapsto \frac{\mathrm{d}}{\mathrm{d}x}\{g_{1-\alpha} * u\} \\
B_{\alpha,L^p} &: L^p(0, b) \to L^p(0, b), \qquad v \mapsto g_\alpha * v \\
S_{L^p} &: L^p(0, b) \to L^p(0, b), \qquad v \mapsto v(b - \cdot).
\end{aligned}$$

The operator ${}^{\mathrm{RL}}\mathrm{D}^\alpha_{L^p}$ is called *the Riemann-Liouville operator of order α on $L^p(0, b)$*. For each u in the domain of ${}^{\mathrm{RL}}\mathrm{D}^\alpha_{L^p}$, ${}^{\mathrm{RL}}\mathrm{D}^\alpha_{L^p} u$ is the *Riemann-Liouville derivative of u of order α*.

In view of (6.5.23) the adjoint quadruplet of ${}^{\mathrm{RL}}\mathcal{Q}_{p,\mathbb{K},\alpha}$ reads

$$({}^{\mathrm{RL}}\mathcal{Q}_{p,\mathbb{K},\alpha})^* = (S_{L^p}\, {}^{\mathrm{RL}}\mathrm{D}^\alpha_{L^p}\, S_{L^p},\ S_{L^q}\, {}^{\mathrm{RL}}\mathrm{D}^\alpha_{L^q}\, S_{L^q},\ S_{L^q}\, B_{\alpha,L^q}\, S_{L^q},\ S_{L^q}).$$

6.9.1.1 Endogenous Boundary Values

We will give endogenous boundary values with underlying quadruplet ${}^{\mathrm{RL}}Q_{p,\mathbb{K},\alpha}$ of any element u belonging to the domain of ${}^{\mathrm{RL}}\mathrm{D}^{\alpha}_{L^p}$. Also, endogenous boundary values with underlying quadruplet $({}^{\mathrm{RL}}Q_{p,\mathbb{K},\alpha})^*$ will be given for each function f in the domain of $S_{L^q}\,{}^{\mathrm{RL}}\mathrm{D}^{\alpha}_{L^q}\,S_{L^q}$.

For, we will use the results of Sect. 6.5. Notice that since g_α is continuous on $(0, b]$, it has an essential limit at b, namely, $g_\alpha(b)$.

6.9.1.1.1 Case where g_α lies in $L^p(0, b) \cap L^q(0, b)$

This is equivalent to

$$\frac{1}{p} < \alpha \quad \text{and} \quad 1 - \frac{1}{p} < \alpha.$$

This correspond to an ordered pair (p, α) living in the region R-1 of Fig. 6.1. Let us start with ${}^{\mathrm{RL}}Q_{p,\mathbb{K},\alpha}$ as underlying quadruplet. We choose $h := \frac{1}{\|g_\alpha\|_{L^q}}(g_\alpha)^{\mathrm{c}}$, so that (6.5.5) becomes

$$u = u_1 g_\alpha + u_1' g_\alpha * h(b - \cdot) + g_\alpha * \psi^{\perp}(b - \cdot),$$

where in view of Definition 5.5.1, the ordered pair of scalars (u_1, u_1') is the endogenous boundary values of u (with underlying quadruplet ${}^{\mathrm{RL}}Q_{p,\mathbb{K},\alpha}$ and) with respect to the basis (g_α) of $\ker {}^{\mathrm{RL}}\mathrm{D}^{\alpha}_{L^p}$ and (h) of F_0. Also $\psi^{\perp}$ lies in ${g_\alpha}^{\perp\,L^{p'}}$.

By (6.5.6) and (6.5.12),

$$u_1 = k * u(0), \qquad u_1' = u(b) - g_\alpha(b)\, g_{1-\alpha} * u(0).$$

When $({}^{\mathrm{RL}}Q_{p,\mathbb{K},\alpha})^*$ is the underlying quadruplet, fix any function h^{s} in $L^q(0, b)$ such that $\int_0^b h^{\mathrm{s}} g_\alpha \,\mathrm{d}x = 1$. In the current framework, (6.5.28) reads

$$f = f_1 g_\alpha(b - \cdot) + S_{L^q} B_{\alpha, L^q} S_{L^q}\big(f_1' h^{\mathrm{s}} + \varphi^{\perp}\big),$$

where (f_1, f_1') are *the endogenous boundary values of* f *with respect to the basis* $(S_{L^q} g_\alpha)$ *of* $S_{L^q} \ker {}^{\mathrm{RL}}\mathrm{D}^{\alpha}_{L^q}$ *and* (h^{s}) *of* F_0^{S}. By (6.5.29), one has

$$\begin{cases} f_1 = (g_{1-\alpha} * S_{L^q} f)(0) \\ f_1' = f(0) - (g_{1-\alpha} * S_{L^q} f)(0) g_\alpha(b). \end{cases} \tag{6.9.2}$$

6.9.1.1.2 Case where g_α Lies in $L^q(0, b) \setminus L^p(0, b)$

This is equivalent to

$$\frac{1}{p} < \alpha \leq 1 - \frac{1}{p}.$$

This corresponds to ordered pair (p, α) living in the region R-2 of Fig. 6.1. We proceed as in the previous case. According to the results of Sect. 6.5.2.2, u is continuous on $[0, b]$. By choosing $h := \frac{1}{\|g_\alpha\|_{L^q}}(g_\alpha)^c$, the endogenous boundary value u_1' of u reads

$$u_1' = u(b).$$

Moreover, $k * u(0) = u(0) = 0$.

Regarding the endogenous boundary values of f, (6.5.33) yields

$$f_1 = (g_{1-\alpha} * S_{L^q} f)(0).$$

6.9.1.1.3 Case where g_α Lies in $L^p(0, b) \setminus L^q(0, b)$

This is equivalent to

$$1 - \frac{1}{p} < \alpha \leq \frac{1}{p}.$$

This corresponds to ordered pair (p, α) living in the region R-3 of Fig. 6.1. We proceed as in the previous case. According to the results of Sect. 6.5.2.3, the endogenous boundary value u_1 of u reads

$$u_1 = g_{1-\alpha} * u(0).$$

Regarding the endogenous boundary values of f, (6.5.37) yields

$$f_1' = f(0).$$

6.9.1.2 Case Where g_α Lies in $L^1(0, b) \setminus (L^q(0, b) \cup L^p(0, b))$

This is equivalent to

$$\alpha \leq \frac{1}{p}, \qquad \text{and } \alpha \leq 1 - \frac{1}{p}.$$

This correspond to ordered pair (p, α) living in the region R-4 of Fig. 6.1. According to the results of Sect. 6.5.2.4, there is no endogenous boundary value. Moreover, $k * u(0) = 0$. Any a same way, f has no endogenous boundary values and $(g_{1-\alpha} * S_{L^q} f)(0) = 0$.

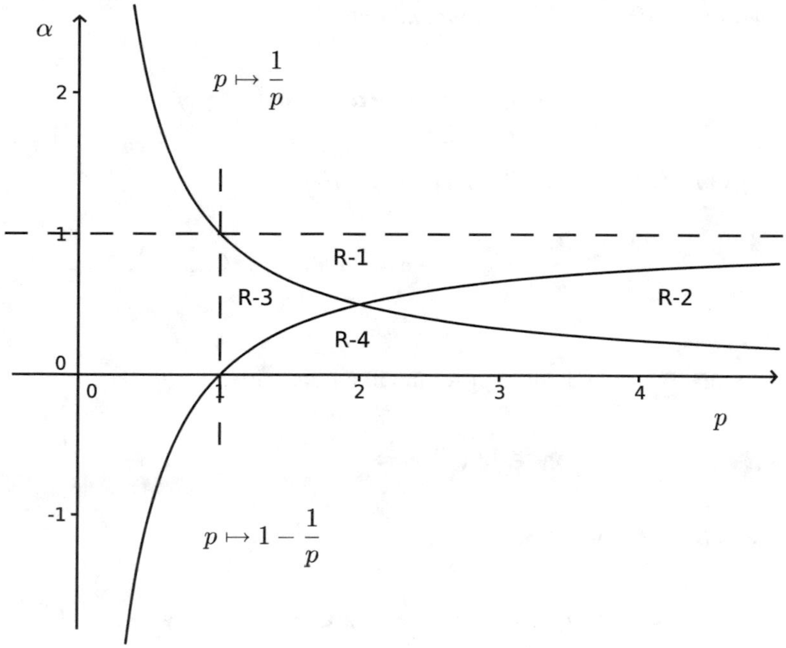

Fig. 6.1 In the parameter plane (p, α), R-1 is the region delimited by the straight line $\alpha = 1$ and the graphs of the functions $p \mapsto \frac{1}{p}$ and $p \mapsto 1 - \frac{1}{p}$. See Sect. 6.9.1

6.9.2 Caputo Extensions of ${}^{\mathrm{RL}}Q_{p,\mathbb{K},\alpha}$

Let p and q be conjugate exponents in $(1, \infty)$ and α belong to $(0, 1)$. The basic Caputo extensions of ${}^{\mathrm{RL}}\mathrm{D}^{\alpha}_{L^p}$ consist in extending the domain of this operator by the subspace generated by g_1. More generally, we will consider extensions obtained by replacing g_1 by g_γ where γ is a positive number.

6.9.2.1 The Quadruplet $Q^{\gamma,\delta}_{p,\mathbb{K},\alpha}$

We have to start with Caputo extensions of the maximal operators of ${}^{\mathrm{RL}}Q_{p,\mathbb{K},\alpha}$. Regarding ${}^{\mathrm{RL}}\mathrm{D}^{\alpha}_{L^p}$, we will proceed as follows. In view of Sect. 5.6.4, proper extensions of ${}^{\mathrm{RL}}\mathrm{D}^{\alpha}_{L^p}$ are obtained provided that g_γ lies in $L^p(0, b)$ and the subspace generated by g_γ is in direct sum with the domain of ${}^{\mathrm{RL}}\mathrm{D}^{\alpha}_{L^p}$. These conditions are of course equivalent to

$$g_\gamma \in L^p(0, b) \setminus D({}^{\mathrm{RL}}\mathrm{D}^{\alpha}_{L^p}).$$

The next result gives the admissible values of γ for this condition to hold.

Proposition 6.9.1 *Let α belong to $(0, 1)$ and γ, δ be positive real numbers. Then*

(i) g_γ *belongs to* $L^p(0,b) \setminus D({}^{\mathrm{RL}}\mathrm{D}^\alpha_{L^p})$ *if and only if*

$$1-\frac{1}{p}<\gamma\leq 1-\frac{1}{p}+\alpha \qquad \textit{and} \qquad \gamma\neq\alpha. \tag{6.9.3}$$

(ii) g_δ *belongs to* $L^q(0,b) \setminus D({}^{\mathrm{RL}}\mathrm{D}^\alpha_{L^q})$ *if and only if*

$$\frac{1}{p}<\delta\leq \frac{1}{p}+\alpha \qquad \textit{and} \qquad \delta\neq\alpha. \tag{6.9.4}$$

Proof By symmetry it is enough to prove the first item. First,

$$g_\gamma \in L^p(\mathcal{Y}) \Longleftrightarrow 1-\frac{1}{p}<\gamma.$$

Besides, let g_γ be in $L^p(\mathcal{Y})$. Then

$$g_\gamma \in D({}^{\mathrm{RL}}\mathrm{D}^\alpha_{L^p}) \Longleftrightarrow 1-\frac{1}{p}+\alpha<\gamma \qquad \text{or} \qquad \gamma=\alpha.$$

Then Item (i) follows easily. □

In view of Proposition 6.9.1, we put

$$N := \begin{cases} \langle g_\gamma\rangle & \text{if } 1-\frac{1}{p}<\gamma\leq 1-\frac{1}{p}+\alpha \ \text{ and } \gamma\neq\alpha \\ \{0\} & \text{otherwise} \end{cases} \tag{6.9.5}$$

and

$$N^{\mathrm{s}} := \begin{cases} \langle g_\delta\rangle & \text{if } \frac{1}{p}<\delta\leq \frac{1}{p}+\alpha \ \text{ and } \delta\neq\alpha \\ \{0\} & \text{otherwise.} \end{cases} \tag{6.9.6}$$

Then on account of Proposition 6.9.1, we denote by ${}^{\gamma}\mathrm{D}^\alpha_{L^p}$ the Caputo extension of ${}^{\mathrm{RL}}\mathrm{D}^\alpha_{L^p}$ with respect to N and by ${}^{\delta}\mathrm{D}^\alpha_{L^q}$ the Caputo extension of ${}^{\mathrm{RL}}\mathrm{D}^\alpha_{L^q}$ with respect to N^{s}. Then we denote by $Q^{\gamma,\delta}_{p,\mathbb{K},\alpha}$ the Caputo extension of ${}^{\mathrm{RL}}Q_{p,\mathcal{Y},\alpha}$ with respect to N and N^{s}. That is,

$$Q^{\gamma,\delta}_{p,\mathbb{K},\alpha} := ({}^{\delta}\mathrm{D}^\alpha_{L^q},\ {}^{\gamma}\mathrm{D}^\alpha_{L^p},\ B_{\alpha,L^p},\ S_{L^p}). \tag{6.9.7}$$

Remark 6.9.1 When $\gamma=\delta=1$, we set in view of (6.7.2),

$$Q_{p,\mathbb{K},\alpha} = Q_{p,\alpha} := Q_{p,\mathbb{K},g_\alpha}.$$

Thus with (6.9.7),

$$Q_{p,\mathbb{K},\alpha} = Q^{1,1}_{p,\mathbb{K},\alpha}.$$

Moreover,

$$\mathrm{D}^\alpha_{L^r} = {}^1\mathrm{D}^\alpha_{L^r}, \qquad \forall r \in (1,\infty),$$

and

$$Q_{p,\mathbb{K},\alpha} = (\mathrm{D}^\alpha_{L^q},\ \mathrm{D}^\alpha_{L^p},\ B_{\alpha,L^p},\ S_{L^p}).$$

□

6.9.2.2 Endogenous Boundary Values

We will give endogenous boundary values with underlying quadruplet $Q^{\gamma,\delta}_{p,\mathbb{K},\alpha}$ of any element u belonging to the domain of ${}^\gamma\mathrm{D}^\alpha_{L^p}$. We will start with the case where the Caputo extensions of the maximal operators are both nontrivial. For the reaming cases, we will assume that $\gamma = \delta = 1$.

6.9.2.2.1 Case where g_α Lies in $L^p(0,b) \cap L^q(0,b)$, g_γ Lies in $L^p(0,b) \setminus D({}^{\mathrm{RL}}\mathrm{D}^\alpha_{L^p})$ and g_δ lie in $L^q(0,b) \setminus D({}^{\mathrm{RL}}\mathrm{D}^\alpha_{L^q})$

This is equivalent to

$$\frac{1}{p} < \alpha \quad \text{and} \quad 1 - \frac{1}{p} < \alpha,$$

$$1 - \frac{1}{p} < \gamma \le 1 - \frac{1}{p} + \alpha \quad \text{and} \quad \gamma \ne \alpha$$

and

$$\frac{1}{p} < \delta \le \frac{1}{p} + \alpha \quad \text{and} \quad \delta \ne \alpha.$$

Then by Proposition 5.6.9,

$$\ker {}^\gamma\mathrm{D}^\alpha_{L^p} = \langle g_\gamma, g_\alpha \rangle, \qquad \ker {}^\delta\mathrm{D}^\alpha_{L^q} = \langle g_\delta, g_\alpha \rangle.$$

We claim that $Q^{\gamma,\delta}_{p,\mathbb{K},\alpha}$ is regular. Indeed, Corollary 5.3.1 yields the existence of functions h_1, h_2 in $L^p(0,b)$ such that, on the one hand,

$$\int_0^b g_\delta(y)h_j(y)\,\mathrm{d}y = \delta_{1j}, \qquad \int_0^b g_\alpha(y)h_j(y)\,\mathrm{d}y = \delta_{2j}, \qquad \forall\, j = 1, 2.$$

On the other hand, $Q^{\gamma,\delta}_{p,\mathbb{K},\alpha}$ is $\tilde{F}_0$-regular with $\tilde{F}_0 := \langle h_1, h_2\rangle$. That proves the claim.

Arguing as in Sect. 6.7.2.1, we derive that a function u in $\mathcal{V}$ lies in the domain of ${}^{\gamma}\mathrm{D}^{\alpha}_{L^p}$ if and only if there exist u_1, u_2, u_1', u_2' in $\mathcal{Y}$ and $\psi^{\perp}$ in $\langle g_\delta, g_\alpha\rangle^{\perp}$ such that

$$u = u_1 g_\gamma + u_2 g_\alpha + g_\alpha * S_{L^p}(u_1' h_1 + u_2' h_2 + \psi^{\perp}), \tag{6.9.8}$$

Here (u_1, u_2, u_1', u_2') is the endogenous boundary values of u with respect to the basis (g_γ, g_α) of $\ker {}^{\gamma}\mathrm{D}^{\alpha}_{L^p}$ and (h_1, h_2) of $\tilde{F}_0$.

In order to compute the endogenous boundary values in terms of "standard" boundary values, we need to distinguish two cases.

6.9.2.2.1.1 Case where $\gamma < \alpha$

We may show that the functions $g_{1-\gamma} * u$, $g_{1-\alpha} * \big(u - g_{1-\gamma} * u(0) g_\gamma\big)$, are continuous on $[0,b]$, u is continuous on $(0,b]$, and

$$\begin{aligned} u_1 &= g_{1-\gamma} * u(0) \\ u_2 &= g_{1-\alpha} * \big(u - g_{1-\gamma} * u(0) g_\gamma\big)(0) \\ u_1' &= g_\gamma * {}^{\gamma}\mathrm{D}^{\alpha}_{L^p} u(b) \\ u_2' &= u(b) - u_1 g_\gamma(b) - u_2 g_\alpha(b). \end{aligned} \tag{6.9.9}$$

Then we get four "standard" boundary values, namely,

$$u_1 := g_{1-\gamma} * u(0), \qquad g_{1-\alpha} * \big(u - u_1 g_\gamma\big)(0), \qquad g_\gamma * {}^{\gamma}\mathrm{D}^{\alpha}_{L^p} u(b), \qquad u(b).$$

6.9.2.2.1.2 Case where $\alpha < \gamma < 1 - \frac{1}{p} + \alpha$

The situation is quite similar, but a little bit more involved. We claim that the functions

$$\frac{u - \frac{u}{g_\alpha}(0)}{g_\gamma}, \qquad \frac{u}{g_\alpha}, \qquad g_\delta * {}^{\gamma}\mathrm{D}^{\alpha}_{L^p} u$$

are continuous on $[0,b]$, u is continuous on $(0,b]$, and (compare with Remark 6.7.2)

$$\begin{aligned} u_1 &= \frac{u - \frac{u}{g_\alpha}(0) g_\alpha}{g_\gamma}(0) \\ u_2 &= \frac{u}{g_\alpha}(0) \\ u'_1 &= g_\delta * {}^\gamma \mathrm{D}^\alpha_{L^p} u(b) \\ u'_2 &= u(b) - u_1 g_\gamma(b) - u_2 g_\alpha(b). \end{aligned} \tag{6.9.10}$$

We derive that $\frac{u}{g_\alpha}$ is continuous and that $u_2 = \frac{u}{g_\alpha}(0)$. Regarding the first equation, we get from (6.9.8)

$$\frac{u - u_2 g_\alpha}{g_\gamma} = u_1 g_1 + \frac{1}{g_\gamma} g_\alpha * h,$$

where h lies in $L^p(0, b)$. If $\gamma \leq 1$ then we conclude easily since $\frac{1}{g_\gamma} g_\alpha * h$ turns out to be the product of a bounded function by an element of ${}_0C([0, b])$. If $1 < \gamma$ then $\frac{1}{g_\gamma}$ is singular at $x = 0$. The singularity in $\frac{1}{g_\gamma} g_\alpha * h$ can be removed by noticing first that $g_\alpha * h$ lies in ${}_0C^{\alpha - \frac{1}{p}}([0, b])$ due to Theorem 3.6.3. Next

$$\frac{1}{g_\gamma}(x) \simeq x^{-\gamma+1}$$

when $x \to 0^+$. Since $\gamma < 1 - \frac{1}{p} + \alpha$ by assumption, we derive that $\frac{1}{g_\gamma} g_\alpha * h$ goes to zero when $x \to 0^+$. There results that

$$\frac{u - g_\alpha \frac{u}{g_\alpha}(0)}{g_\gamma}$$

is continuous on $[0, b]$ and the first equation of (6.9.10) holds. The two last equations are proved exactly as in the case where $\gamma < \alpha$, which completes the proof of (6.9.10).

Then we get four "standard" boundary values for u, namely,

$$\frac{u - g_\alpha \frac{u}{g_\alpha}(0)}{g_\gamma}(0), \qquad \frac{u}{g_\alpha}(0), \qquad g_\delta * {}^\gamma \mathrm{D}^\alpha_{L^p} u(b), \qquad u(b).$$

Remark 6.9.2 When $\gamma = \delta = 1$, the conditions

$$g_\alpha \in L^p(0, b) \cap L^q(0, b), \qquad g_1 \in L^p(0, b) \setminus D({}^{\mathrm{RL}}\mathrm{D}^\alpha_{L^p}), \qquad g_1 \in L^q(0, b) \setminus D({}^{\mathrm{RL}}\mathrm{D}^\alpha_{L^q})$$

are equivalent to

$$\frac{1}{p} < \alpha \quad \text{and} \quad 1 - \frac{1}{p} < \alpha.$$

According to (6.9.10), the functions

$$u - \tfrac{u}{g_\alpha}(0)g_\alpha, \qquad \frac{u}{g_\alpha}, \qquad g_1 * \mathrm{D}^\alpha_{L^p} u$$

are continuous on $[0, b]$, u is continuous on $(0, b]$, and

$$\begin{aligned} u_1 &= (u - u_2 g_\alpha)(0) \\ u_2 &= \tfrac{u}{g_\alpha}(0) \\ u_1' &= g_{1-\alpha} * \big(u - u_1 g_1\big)(b) - g_{1-\alpha} * \big(u - u_1 g_1\big)(0) \\ u_2' &= u(b) - u_1 - u_2 g_\alpha(b). \end{aligned} \tag{6.9.11}$$

Since $u_2 = g_{1-\alpha} * u(0)$, we obtain the following complete set of "standard" boundary values.

$$u_1 := \big(u - \frac{u}{g_\alpha}(0)g_\alpha\big)(0), \qquad g_{1-\alpha} * u(0)$$
$$g_{1-\alpha} * \big(u - u_1 g_1\big)(b), \qquad u(b).$$

Compare with Remark 6.7.1. □

6.9.2.2.1.3 Case where $\alpha < \gamma = 1 - \frac{1}{p} + \alpha$

In this case, only the first equation of (6.9.10) is different since we cannot warrant that $\frac{1}{g_\gamma} g_\alpha * h$ goes to zero when $x \to 0^+$. In fact, u_1 is given by

$$u_1 = \frac{u - \frac{u}{g_\alpha}(0)g_\alpha - g_\alpha * {}^\gamma\mathrm{D}^\alpha_{L^p} u}{g_\gamma}(0).$$

6.9.2.2.2 The Remaining Cases when $\gamma = \delta = 1$

In the sequel, we will just give the parameter values corresponding to the remaining eight cases analyzed in Sect. 6.7.2.

(i) Case where g_α lies in $L^q(0, b) \setminus L^p(0, b)$. This is equivalent to

$$\frac{1}{p} < \alpha \leq 1 - \frac{1}{p}.$$

(i-a) Case where $g_{1-\alpha}$ lies in $L^q(0,b) \setminus L^p(0,b)$. This is equivalent to

$$\frac{1}{p} < \alpha < 1 - \frac{1}{p}.$$

(i-b) Case where $g_{1-\alpha}$ lies in $L^1(0,b) \setminus (L^p(0,b) \cup L^q(0,b))$. This is equivalent to

$$\alpha = 1 - \frac{1}{p}, \qquad 2 < p.$$

(ii) Case where g_α lies in $L^p(0,b) \setminus L^q(0,b)$. This is equivalent to

$$1 - \frac{1}{p} < \alpha \leq \frac{1}{p}.$$

(ii-a) Case where $g_{1-\alpha}$ lies in $L^p(0,b) \setminus L^q(0,b)$. This is equivalent to

$$1 - \frac{1}{p} < \alpha < \frac{1}{p}.$$

(ii-b) Case where $g_{1-\alpha}$ lies in $L^1(0,b) \setminus (L^p(0,b) \cup L^q(0,b))$. This is equivalent to

$$\frac{1}{p} = \alpha, \qquad p < 2.$$

(iii) Case where g_α lies in $L^1(0,b) \setminus (L^q(0,b) \cup L^p(0,b))$. This is equivalent to

$$\alpha \leq \frac{1}{p}, \qquad \alpha \leq 1 - \frac{1}{p}.$$

(iii-a) Case where $g_{1-\alpha}$ lies in $L^p(0,b) \cap L^q(0,b)$. This is equivalent to

$$\alpha < \frac{1}{p}, \qquad \alpha < 1 - \frac{1}{p}.$$

(iii-b) Case where $g_{1-\alpha}$ lies in $L^q(0,b) \setminus L^p(0,b)$. This is equivalent to

$$\alpha = \frac{1}{p}, \qquad 2 < p.$$

(iii-c) Case where $g_{1-\alpha}$ lies in $L^p(0,b) \setminus L^q(0,b)$. This is equivalent to

$$1 - \frac{1}{p} = \alpha, \qquad p < 2.$$

(iii-d) Case where $g_{1-\alpha}$ lies in $L^1(0, b) \setminus (L^p(0, b) \cup L^q(0, b))$. This is equivalent to

$$\alpha = 2, \qquad p = 2.$$

6.9.3 The Quadruplet $\mathcal{Q}_{p,\mathbb{K},n+\alpha}$

Let p and q be conjugate exponents in $(1, \infty)$, α belong to $(0, 1)$, and n be a positive integer. In view of Sect. 6.8, we will consider the quadruplet $\mathcal{Q}_{p,\mathcal{Y},n+\ell}$ in the particular case where $\mathcal{Y} := \mathbb{K}$ and

$$k := g_{1-\alpha}, \qquad \ell := g_\alpha.$$

6.9.3.1 Introduction

In the current frame work, 6.8.1 reads

$$N(\langle g_1 \rangle) := \begin{cases} \langle g_1, \dots, g_{n+1} \rangle & \text{if } \frac{1}{p} \leq \alpha \\ \{0\} & \text{otherwise.} \end{cases}$$

In the same way, 6.8.2 reads

$$N^{\mathrm{s}}(\langle g_1 \rangle) := \begin{cases} \langle g_1, \dots, g_{n+1} \rangle, & \text{if } 1 - \frac{1}{p} \leq \alpha \\ \{0\} & \text{otherwise.} \end{cases}$$

Then on account of Sect. 6.8.1, we put

$$\mathcal{Q}_{p,\mathbb{K},n+\alpha} := \mathcal{Q}_{p,\mathbb{K},n+g_\alpha}.$$

That is to say, $\mathcal{Q}_{p,\mathbb{K},n+\alpha}$ is the Caputo extension ${}^{\mathrm{RL}}\mathcal{Q}_{p,\mathbb{K},n+\alpha}$ with respect to $N(\langle g_1 \rangle)$ and $N^{\mathrm{s}}(\langle g_1 \rangle)$. Moreover, setting

$$\begin{aligned} \mathrm{D}^{n+\alpha}_{L^r} &:= \mathrm{D}^{n+g_\alpha}_{L^r} \\ B_{n+\alpha,L^r} &:= B_{g_{n+\alpha},L^r} \qquad\qquad \forall r \in (1, \infty) \end{aligned}$$

we get

$$\mathcal{Q}_{p,\mathbb{K},n+\alpha} = (\mathrm{D}^{n+\alpha}_{L^q},\ \mathrm{D}^{n+\alpha}_{L^p},\ B_{n+\alpha,L^p},\ S_{L^p}).$$

The case $n = 0$ is studied in the previous subsection. Since by Corollary 5.6.27, the kernel of the maximal operators of $\mathcal{Q}_{p,\mathbb{K},n+\alpha}$ are finite dimensional, there results that $\mathcal{Q}_{p,\mathbb{K},n+\alpha}$ is regular (see Sect. 5.5.1).

6.9.3.2 Two Special Boundary Restriction Operators

The first operator we would like to introduce is nothing but the well-known *Riemann-Liouville operator of order n+α on $L^p(0, b)$* (see, for instance, [Die10, Def. 2.2]). Denoted by ${}^{\mathrm{RL}}\mathrm{D}^{n+\alpha}_{L^p}$, it is defined with the notation of Sect. 6.6.1 by

$${}^{\mathrm{RL}}\mathrm{D}^{n+\alpha}_{L^p} := {}^{\mathrm{RL}}\mathrm{D}^{n+g_\alpha}_{L^p} = \mathrm{D}^{n}_{L^p}\,{}^{\mathrm{RL}}\mathrm{D}^{\alpha}_{L^p}.$$

At this stage, it is a routine to check that ${}^{\mathrm{RL}}\mathrm{D}^{n+\alpha}_{L^p}$ is a closed boundary restriction operator of $\mathcal{Q}_{p,\mathbb{K},n+\alpha}$.

Our second special operator is *the Caputo operator of order $n + \alpha$ on $L^p(0, b)$*. This operator denoted by ${}^{\mathrm{C}}\mathrm{D}^{n+\alpha}_{L^p}$ is by definition the restriction of $\mathrm{D}^{\ell}_{\mathcal{V}}$ with domain

$$D({}^{\mathrm{C}}\mathrm{D}^{n+\alpha}_{L^p}) := N(\langle g_1\rangle) \oplus R(B_{\ell,\mathcal{V}}).$$

Thus if $g_{1-\alpha}$ do not lie in $L^p(0, b)$ which means $\frac{1}{p} \leq \alpha$ then

$$D({}^{\mathrm{C}}\mathrm{D}^{n+\alpha}_{L^p}) = \langle g_1, \ldots, g_{n+1}\rangle \oplus R(B_{\ell,\mathcal{V}}).$$

It is again a routine to check that ${}^{\mathrm{C}}\mathrm{D}^{n+\alpha}_{L^p}$ is a closed boundary restriction operator of $\mathcal{Q}_{p,\mathbb{K},n+\alpha}$.

Remark 6.9.3 Let us check that our definition of the Caputo operator of order $n + \alpha$ coincide with the usual one (see, for instance, [Die10, Def. 3.1]). For, assume that $\frac{1}{p} \leq \alpha$, and consider any function u in $W^{n+1,p}(0, b)$. Using Proposition 3.3.4, we may show by induction that the function

$$v := u - \sum_{i=1}^{n+1} \mathrm{D}^{i-1}_{L^p} u(0) g_i$$

lies in the domain of ${}^{\mathrm{RL}}\mathrm{D}^{n+\alpha}_{L^p}$ and

$${}^{\mathrm{RL}}\mathrm{D}^{n+\alpha}_{L^p} v = k * \mathrm{D}^{n+1}_{L^p} u.$$

Next the very definition of ${}^{\mathrm{C}}\mathrm{D}_{L^p}^{n+\alpha}$ entails that u belongs to $D({}^{\mathrm{C}}\mathrm{D}_{L^p}^{n+\alpha})$ and

$$
{}^{\mathrm{C}}\mathrm{D}_{L^p}^{n+\alpha} u = k * \mathrm{D}_{L^p}^{n+1} u.
$$

This identity is a common definition of the Caputo derivative of order $n + \alpha$ of u; see, for instance, [Die10, Def. 3.1]. However, it is not optimal in the sense that it requires too much regularity for u (see Remark 6.2.2). □

Endogenous Boundary Value Problems 7

Continuing the unified approach, we will first solve abstract equations involving a maximal operator of differential quadruplets. Next these equations will be supplemented with endogenous boundary conditions which gives rise to (a certain class of) abstract problems.

Concrete realizations of theses abstract problems are fractional differential equations supplemented with endogenous boundary conditions. The fractional equations may involve all the maximal (fractional) operators studied in the previous chapter.

A substantial part of the quantitative theory of ordinary linear differential equations is extended to fractional differential equations. We will solve in one shot ordinary linear differential equations, fractional differential equations built upon Riemann-Liouville derivatives, and fractional differential systems relying on Caputo derivatives. We will also give results concerning nonlinear problems.

Let $\mathbb{K}$ be equal to $\mathbb{R}$ or $\mathbb{C}$.

7.1 Introduction

As an example, let us analyze a simple endogenous boundary value problem. This problem is formed of a first-order differential equation supplemented with endogenous boundary conditions. More precisely, in view of Sect. 4.4.2, we consider the differential triplet

$$\mathcal{T}_1 = (\mathrm{D}^1_{L^2}, B_{1,L^2}, S)$$

on $\mathcal{H} := L^2(0, b)$. In order to introduce our endogenous boundary conditions, some data are in order. First, let (x_1, y_1) belong to $\mathbb{K}^2 \setminus (0, 0)$, λ, $h_{\text{b.c}}$ lie in $\mathbb{K}$, and h be a function of $L^2(0, b)$. Then we consider the following endogenous boundary value problem:

A. Rougirel, *Unified Theory for Fractional and Entire Differential Operators*, Frontiers in Elliptic and Parabolic Problems,
https://doi.org/10.1007/978-3-031-58356-8_7

$$\begin{cases} \text{Find } u \text{ in } H^1(0,b) \text{ such that} \\ \mathrm{D}^1_{L^2} u = \lambda u + h, \qquad y_1 u_1 - x_1 u_1' = h_{\text{b.c}}. \end{cases} \tag{7.1.1}$$

In (7.1.1), (u_1, u_1') denotes *the endogenous boundary values of u with respect to the basis* (g_1) of $\ker \mathrm{D}^1_{L^2}$.

First of all, (7.1.1) may be reformulated as a usual boundary value problem. Indeed, setting

$$m_1 := \frac{x_1}{b} + y_1, \qquad m_2 := -\frac{x_1}{b} \tag{7.1.2}$$

and using (4.4.23), the second equation of (7.1.1) is equivalent to

$$m_1 u(0) + m_2 u(b) = h_{\text{b.c}}. \tag{7.1.3}$$

Implementing elementary methods, we get that u is a solution to (7.1.1) if and only if there exists u_1 in $\mathbb{K}$ such that

$$u = e^{\lambda \cdot} u_1 + e^{\lambda \cdot} * h \tag{7.1.4}$$

$$\kappa u_1 - \frac{x_1}{b}(e^{\lambda \cdot} * h)(b) = h_{\text{b.c}},$$

where $\kappa := \frac{x_1}{b}(1 - e^{\lambda b}) + y_1$ and $e^{\lambda \cdot}$ is the exponential function $x \mapsto e^{\lambda x}$.

(i) If $\kappa \neq 0$ then (7.1.1) is uniquely solvable.
(ii) If $\kappa = 0$ then $x_1 \neq 0$.
(ii-a) If $(e^{\lambda \cdot} * h)(b) = -\frac{b h_{\text{b.c}}}{x_1}$ then *the solution set of* (7.1.1), which by definition is the subset of $H^1(0,b)$ whose elements are solution to (7.1.1), is

$$\{e^{\lambda \cdot} u_1 + e^{\lambda \cdot} * h \mid u_1 \in \mathbb{K}\}.$$

(ii-b) If $(e^{\lambda \cdot} * h)(b) \neq -\frac{b h_{\text{b.c}}}{x_1}$ then (7.1.1) has no solution.

So, when we will consider abstract problems which are generalization of (7.1.1), we will of course encounter these three cases. However we will see that no other case occurs. This highlights the unity between the theory of entire differential equations and the theory of fractional differential equations

Remark 7.1.1 With the notation (7.1.2), κ reads

$$\kappa = m_1 + m_2 e^{\lambda b}.$$

Also the condition in Item (ii-a) reads

$$m_2(e^{\lambda\cdot} * h)(b) - h_{\mathrm{b.c}} = 0.$$

□

7.2 Abstract Linear Problems

We will define two classes of abstract linear problems. The first class is always well posed and in application refers to initial boundary value problems. The second class does not need to be *unconditionally solvable*, and its instances may modelize spatial problems. By an unconditionally solvable problem, we mean roughly speaking that the problem admits (at least) a solution for all data. We refer to Definition 7.2.9 for a precise definition. Unconditionally solvability allows to get existence results for nonlinear problems by implementing iterative methods (see Sect. 7.6).

We will obtain generalizations of many classical results in the theory of linear ordinary differential systems (see, for instance, [HS74, Chap. 6] or [Nai67, Chap 1]).

7.2.1 The Resolvent Map R_B^L

Resolvent maps are the main tool to solve the abstract equations of Sect. 7.2.3 (see in particular Proposition 7.2.9). In the standard case of linear first-order differential equations, they are connected with exponential functions (see Example 7.2.4).

7.2.1.1 The Resolvent Map R_B^L on Banach Spaces

Let $\mathcal{Z}$ be a Banach space over $\mathbb{K}$ and B, L be operators in $\mathcal{L}(\mathcal{Z})$ such that

$$\lim_{j\to\infty} \|(BL)^j\|_{\mathcal{L}(\mathcal{Z})}^{\frac{1}{j}} \xrightarrow[j\to\infty]{} 0. \tag{7.2.1}$$

The next result is a straightforward consequence of Proposition 2.4.13.

Proposition 7.2.1 *Let B, L be operators of $\mathcal{L}(\mathcal{Z})$ satisfying* (7.2.1). *Then the Neumann series $\sum(BL)^j$ converges in $\mathcal{L}(\mathcal{Z})$, so that denoting its sum by R_B^L, one has*

$$\mathsf{R}_B^L := \sum_{j\geq 0}(BL)^j.$$

Moreover,

(i) $\mathrm{id}_{\mathcal{Z}} - BL$ *is an isomorphism from* $\mathcal{Z}$ *onto itself,* $\mathsf{R}_B^L = (\mathrm{id}_{\mathcal{Z}} - BL)^{-1}$ *and* $\mathsf{R}_B^L = \mathrm{id}_{\mathcal{Z}} + BL\,\mathsf{R}_B^L$.

(ii) *If in addition, B commute with L, i.e.,*

$$BL = LB, \tag{7.2.2}$$

then $\mathsf{R}_B^L = \sum_{j\geq 0} B^j L^j$ *and* R_B^L *commute with B and L.*

Definition 7.2.1 Let B, L be operators of $\mathcal{L}(\mathcal{Z})$ satisfying (7.2.1). The linear and continuous map R_B^L from $\mathcal{Z}$ into itself (introduced in Proposition 7.2.1) is called *the resolvent map with respect to B and L.* □

Remark 7.2.2 Let B and L belong to $\mathcal{L}(\mathcal{Z})$. If $L = 0$ then (7.2.1) holds and $\mathsf{R}_B^L = \mathrm{id}_{\mathcal{Z}}$. This case will be considered in the nonlinear theory featured in Sect. 7.5.

Notice also that the condition (7.2.2) and

$$\lim_{j\to\infty} \|B^j\|_{\mathcal{L}(\mathcal{Z})}^{\frac{1}{j}} \xrightarrow[n\to\infty]{} 0$$

imply clearly (7.2.1). □

Definition 7.2.3 For any operator B in $\mathcal{L}(\mathcal{Z})$, the quantity

$$\limsup_{j\to\infty} \|B^j\|_{\mathcal{L}(\mathcal{Z})}^{\frac{1}{j}}$$

which belongs a priori to $[0, \infty]$ is called *the spectral radius of B.* □

Here, the spectral radius will serve only to ensure the convergence of the resolvent map. We will therefore not use any of its properties. Notice however that the spectral radius of B is finite and is equal to the maximum value achieved by the modulus of the elements of the spectrum of B. Otherwise said, the spectral radius is the radius of the smallest ball centered at the origin of the complex plane and containing the spectrum of B. We refer to [Dow78, Chap. I] for more information.

Our main interest in resolvent map is that R_B^L allows to solve linear equation of the form $Au = Lu + h$. More precisely, we will prove the following result.

Proposition 7.2.2 *Let* $\mathcal{Z}$ *be a Banach space,* $A : D(A) \subseteq \mathcal{Z} \to \mathcal{Z}$ *be an operator, and* B, L *belong to* $\mathcal{L}(\mathcal{Z})$. *We assume* (7.2.1), (7.2.2) *and that*

$$AB = \mathrm{id}_{\mathcal{Z}}. \tag{7.2.3}$$

Then for each h in $\mathcal{Z}$, the following assertions are equivalent:

(i) *u lies in $D(A)$ and $Au = Lu + h$.*
(ii) *There exists u_0 in* $\ker A$ *such that*

$$u = \mathsf{R}_B^L u_0 + B\,\mathsf{R}_B^L h. \tag{7.2.4}$$

Proof Let us assume (i). By Lemma 4.3.3, there exists u_0 in $\ker A$ such that $u = u_0 + B(Lu + h)$. Thus $(\mathrm{id}_{\mathcal{Z}} - BL)u = u_0 + Bh$. With Items (i) and (ii) of Proposition 7.2.1, we get (7.2.4).

Conversely assume (ii). By Proposition 7.2.1 (i), $\mathsf{R}_B^L u_0 = u_0 + BL\,\mathsf{R}_B^L u_0$. Since u_0 lies in $\ker A$ and $AB = \mathrm{id}_{\mathcal{Z}}$, we deduce that u lies in $D(A)$ and also that

$$A\,\mathsf{R}_B^L u_0 = L\,\mathsf{R}_B^L u_0.$$

Up to now, we have proved that u lies in $D(A)$ and $Au = L\,\mathsf{R}_B^L u_0 + \mathsf{R}_B^L h$. Next, using again $\mathsf{R}_B^L h = h + BL\,\mathsf{R}_B^L h$ and the commutativity of B and L, we obtain that $Au = Lu + h$. □

In the particular case where $L := \lambda\mathrm{id}_{\mathcal{Z}}$ for some scalar λ and the spectral radius of B vanishes, we introduce, for each nonnegative integer k, the function

$$F_k : \mathbb{K} \to \mathcal{L}(\mathcal{Z}), \qquad \lambda \mapsto \sum_{j \geq k} \lambda^j B^{j-k}. \tag{7.2.5}$$

Notice that $F_0(\lambda) = \mathsf{R}_B^{\lambda\mathrm{id}_{\mathcal{Z}}}$.

Proposition 7.2.3 *Let $\mathcal{Z}$ be a Banach space and B belong to $\mathcal{L}(\mathcal{Z})$. Assume that the spectral radius of B vanishes. Then the function F_k defined by (7.2.5) is $\mathbb{K}$-differentiable at any order on $\mathbb{K}$. Moreover, for each positive integer i, the* ith-*order derivative ${F_k}^{(i)}$ of F_k reads*

$${F_k}^{(i)}(\lambda_0) = \sum_{j \geq \max(i,k)} \frac{\mathrm{d}^i}{\mathrm{d}\lambda^i}_{|\lambda=\lambda_0} \lambda^j B^{j-k}, \qquad \forall\, \lambda_0 \in \mathbb{K}.$$

Here, $\frac{\mathrm{d}^i}{\mathrm{d}\lambda^i}_{|\lambda=\lambda_0} \lambda^j$ denotes the ith-*order derivative of the function $\lambda \mapsto \lambda^j$ evaluated at λ_0.*

Proof Arguing as in the scalar case for which we refer to [Rud91, Th. 10.6] for a proof, we may show that the series $\sum j\lambda^{j-1}B^{j-k}$ converges in $\mathcal{L}(\mathcal{Z})$ and F_k is $\mathbb{K}$-differentiable and that for each λ_0 in $\mathbb{K}$,

$$F_k{}^{(1)}(\lambda_0) = \sum_{j \geq \max(1,k)} j{\lambda_0}^{j-1} B^{j-k} = \sum_{j \geq \max(1,k)} \frac{\mathrm{d}}{\mathrm{d}\lambda}_{|\lambda=\lambda_0} \lambda^j B^{j-k}.$$

Now since $(j \| B^{j-k} \|_{\mathcal{L}(\mathcal{Z})})^{\frac{1}{j}}$ goes to zero as j goes to infinity, we may iterate to get the assertion of the proposition. □

7.2.1.2 The Map Resolvent $\mathsf{R}^L_{B_{f,\mathcal{X}}}$ on Lebesgue Spaces

Let $\mathcal{Y}$ be a nontrivial Banach space over $\mathbb{K}$ and b be a positive real number. For any p in $[1, \infty]$, we set

$$\mathcal{X} = L^p(\mathcal{Y}) := L^p\big((0, b), \mathcal{Y}\big).$$

Let L belong to $\mathcal{L}(\mathcal{Y})$. Then we claim that L *induces a continuous linear map labeled* $\tilde{L}$ *or* L *(if no confusion may occur) from* $L^p(\mathcal{Y})$ *into itself by setting for each* v *in* $L^p(\mathcal{Y})$ *and almost every* x *in* $(0, b)$,

$$(\tilde{L}v)(x) := L\big(v(x)\big).$$

Indeed, it is easily seen that $x \mapsto (\tilde{L}v)(x)$ is measurable. Then the claim follows from Bochner Theorem 3.1.6. Besides $\tilde{L}$ *commute with* $B_{f,\mathcal{X}}$ due to Proposition 3.2.3, where the convolution operator $B_{f,\mathcal{X}}$ is defined in Sect. 6.4.

Next, for any f in $L^1(0, b)$, the key point is that *the spectral radius of* $B_{f,\mathcal{X}}$ *vanishes* as asserted by Lemma 7.2.4 below. Then Remark 7.2.2 entails that

$$\mathsf{R}^L_{B_{f,\mathcal{X}}} : L^p(\mathcal{Y}) \to L^p(\mathcal{Y})$$

is a well-defined linear and continuous map.

Lemma 7.2.4 *Let* $\mathcal{Y}$ *be a nontrivial Banach space,* p *belong to* $[1, \infty]$, *and* $\mathcal{X} := L^p(\mathcal{Y})$. *Then for each* f *in* $L^1(0, b)$,

$$\lim_{n \to \infty} \| (B_{f,\mathcal{X}})^n \|^{\frac{1}{n}}_{\mathcal{L}(L^p(\mathcal{Y}))} = 0.$$

Proof Thanks to Young inequality (see Proposition 3.3.3 (iii)), we get

$$\| (B_{f,\mathcal{X}})^n \|_{\mathcal{L}(L^p(\mathcal{Y}))} \leq \| f^{*n} \|_{L^1}.$$

We conclude with Proposition 3.3.6. □

As expected, the map R^L_B is connected with exponential functions in standard cases.

Example 7.2.4 Let us compute $\mathsf{R}_B^L g_1$ in the framework of Sect. 7.1. For, we set

$$\mathcal{X} := L^2(0, b), \qquad B := B_{1,L^2}, \qquad L := \lambda \mathrm{id}_{L^2(0,b)}.$$

In view of the above analysis, $\mathsf{R}^L_{B_{1,L^2}}$ is well defined. Moreover, identifying $\lambda \mathrm{id}_{L^2(0,b)}$ with the scalar λ, we will denote $\mathsf{R}^L_{B_{1,L^2}}$ by $\mathsf{R}^\lambda_{B_{1,L^2}}$. Then

$$\mathsf{R}^\lambda_{B_{1,L^2}} g_1 = e^{\lambda \cdot}.$$

That is, $\mathsf{R}^\lambda_{B_{1,L^2}} g_1(x) = e^{\lambda x}$ for each x in $[0, b]$. Moreover, for any function h in $L^2(0, b)$, one has

$$B_{1,L^2} \mathsf{R}^\lambda_{B_{1,L^2}} h = e^{\lambda \cdot} * h.$$

Otherwise said, $B_{1,L^2} \mathsf{R}^\lambda_{B_{1,L^2}}$ is the convolution operator by the function $e^{\lambda \cdot}$. In symbol,

$$B_{1,L^2} \mathsf{R}^\lambda_{B_{1,L^2}} = B_{e^{\lambda \cdot}, L^2}.$$

Thus in the current context, (7.2.4) becomes the usual *variation of the constant formula*. More precisely, Item (ii) in Proposition 7.2.2 reads as follows: there exists u_1 in $\mathbb{K}$ such that

$$u = e^{\lambda \cdot} u_1 + e^{\lambda \cdot} * h.$$

Compare with (7.1.4). □

7.2.2 The Generalized Exponential Functions on Lebesgue Spaces

Let $\mathbb{K}$ be equal to $\mathbb{R}$ or $\mathbb{C}$, $\mathcal{Y}$ be a nontrivial Banach space over $\mathbb{K}$, and b be a positive real number. For any p in $[1, \infty]$, we set

$$\mathcal{X} = L^p(\mathcal{Y}) := L^p\big((0, b), \mathcal{Y}\big).$$

In a same way, we will abbreviate $L^p\big((0, b), \mathcal{L}(\mathcal{Y})\big)$ by $L^p\big(\mathcal{L}(\mathcal{Y})\big)$.

7.2.2.1 Position of the Problem

In view of Sect. 7.1, the exponential function $e^{\lambda \cdot}$ is a fundamental tool in the representation of the solutions to the differential equation $\mathrm{D}^1_{L^2} u = \lambda u + h$. When we consider instead an equation on $L^p(\mathcal{Y})$ of the form $Au = \lambda u + h$ (where A is an operator on $L^p(\mathcal{Y})$ and h lies in $L^p(\mathcal{Y})$), our aim is to find a function which generalizes *the variation of the constant formula* (7.1.4). More precisely, we look for a function $E^\lambda : (0, b) \to \mathcal{L}(\mathcal{Y})$ such that the solutions to the latter equations reads

$$u = E^\lambda u_1 + E^\lambda * h,$$

for some u_1 in $\mathcal{Y}$. Notice that $E^\lambda * h$ is the convolution of an operator valued function (namely, E^λ) with a function of $L^p(\mathcal{Y})$.

Of course, Example 7.2.3 and (7.2.4) will guide us in the construction of the generalized exponential function E^λ. Thus assuming that the kernel of A is a one-dimensional space generated by, let us say, a function ℓ of $L^p(0, b)$, then we will set

$$E^\lambda(Y) := \mathsf{R}^\lambda_{B_{\ell,\mathcal{X}}}(\ell Y).$$

Then there remains to arrange things so that

$$E^\lambda * h = B_{\ell,\mathcal{X}} \mathsf{R}^\lambda_{B_{\ell,\mathcal{X}}} h.$$

Otherwise said, the key point is to show that the linear map $B_{\ell,\mathcal{X}} \mathsf{R}^\lambda_{B_{\ell,\mathcal{X}}}$ on $L^p(\mathcal{Y})$ is in fact the convolution by the function $E^\lambda : (0, b) \to \mathcal{L}(\mathcal{Y})$. This is achieved in Proposition 7.2.7 which gathers the main properties of generalized exponential functions.

7.2.2.2 Definition

Lemma 7.2.5 *If g belongs to $L^p(0, b)$ and L lies in $\mathcal{L}(\mathcal{Y})$, then $g(\cdot)L$ belongs to $L^p\big((0, b), \mathcal{L}(\mathcal{Y})\big)$, and*

$$\|g(\cdot)L\|_{L^p(\mathcal{L}(\mathcal{Y}))} = \|g\|_{L^p} \|L\|_{\mathcal{L}(\mathcal{Y})}.$$

Proof Let us start to show that $g(\cdot)L$ is measurable, that is,

$$g(\cdot)L \in L^0\big((0, b), \mathcal{L}(\mathcal{Y})\big).$$

Since g lies in $L^0(0, b)$, there exists a sequence $(s_n)_{n\in\mathbb{N}}$ of simple functions converging toward g almost everywhere on $(0, b)$. Thus for each index n, $s_n(\cdot)L : (0, b) \to \mathcal{L}(\mathcal{Y})$ is a simple function. Moreover, for almost every x in $(0, b)$,

$$\|s_n(x)L - g(x)L\|_{\mathcal{L}(\mathcal{Y})} \leq |s_n(x) - g(x)| \, \|L\|_{\mathcal{L}(\mathcal{Y})} \xrightarrow[n\to\infty]{} 0.$$

Hence $g(\cdot)L$ is measurable. Next,

$$\|g(\cdot)L\|^p_{\mathcal{L}(\mathcal{Y})} = |g(\cdot)|^p \|L\|_{\mathcal{L}(\mathcal{Y})} \qquad \text{in } L^1(0, b).$$

Thus $g(\cdot)L$ belongs to $L^p\big((0,b),\mathcal{L}(\mathcal{Y})\big)$ according to Bochner theorem 3.1.6, and the equality stated in the lemma follows. □

Proposition 7.2.6 *If f lies in $L^1(0,b)$, g belongs to $L^p(0,b)$, and L lies in $\mathcal{L}(\mathcal{Y})$, then the series $\sum {B_{f,L^p}}^j g\, L^j$ converges in $L^p\big((0,b),\mathcal{L}(\mathcal{Y})\big)$.*

Proof By Lemma 7.2.5, ${B_{f,L^p}}^j g\, L^j$ lies in $L^p\big(\mathcal{L}(\mathcal{Y})\big)$ and

$$\begin{aligned}\|{B_{f,L^p}}^j g\, L^j\|_{L^p(\mathcal{L}(\mathcal{Y}))} &= \|{B_{f,L^p}}^j g\|_{L^p}\, \|L^j\|_{\mathcal{L}(\mathcal{Y})} \\ &\le \|{B_{f,L^p}}^j\|_{\mathcal{L}(L^p)} \|g\|_{L^p}\, \|L\|^j_{\mathcal{L}(\mathcal{Y})}.\end{aligned}$$

Then Lemma 7.2.4 warrants that the series $\sum \|{B_{f,L^p}}^j g\, L^j\|_{L^p(\mathcal{L}(\mathcal{Y}))}$ converges. Since $L^p\big(\mathcal{L}(\mathcal{Y})\big)$ is a Banach space, we deduce that $\sum {B_{f,L^p}}^j g\, L^j$ converges in $L^p\big((0,b),\mathcal{L}(\mathcal{Y})\big)$. □

Definition 7.2.5 Let f lie in $L^1(0,b)$, g belong to $L^p(0,b)$, and L lie in $\mathcal{L}(\mathcal{Y})$. In view of Proposition 7.2.6, the sum of the series $\sum {B_{f,L^p}}^j g\, L^j$ denoted by $E^L_{f,g}$, is called *the (generalized) exponential function with parameters f, g and L*. In symbol,

$$E^L_{f,g} := \sum_{j\ge 0} {B_{f,L^p}}^j g\, L^j \in L^p\big((0,b),\mathcal{L}(\mathcal{Y})\big).$$

In the particular case where $f = g$, we will abbreviate $E^L_{g,g}$ by E^L_g. □

7.2.2.3 Delsarte Representation of Generalized Exponential Functions

Delsarte representation is a kind of generalization of the decomposition of holomorphic functions into power series. In the theory of hyperbolic PDEs, *Delsarte representation* is a nice and convenient way to represent particular solutions of the so-called Delsarte equation: see [Del38] or [ER22].

For any scalar functions f in $L^1(0,b)$, we set

$$f_{j+1} := {B_{f,L^p}}^j f, \qquad \forall\, j \in \mathbb{N}. \tag{7.2.6}$$

In addition, let p belong to $[1,\infty]$, g lie $L^p(0,b)$, and L live in $\mathcal{L}(\mathcal{Y})$. Setting $f_0 * g := g$, *the Delsarte representation of* $E^L_{f,g}$ is

$$E^L_{f,g} = \sum_{j\ge 0} f_j * g\, L^j, \tag{7.2.7}$$

where for positive index j, $f_j * g = B_{f_j,L^p} g$ is the convolution between functions of L^1 and L^p. In particular when $f = g$, the Delsarte representation of E^L_f reads

$$E_f^L = \sum_{j\geq 0} f_{j+1}\, L^j \in L^p\big((0,b), \mathcal{L}(\mathcal{Y})\big). \tag{7.2.8}$$

Observe that for each scalar λ, $E_{g_1}^{\lambda \mathrm{id}_{\mathcal{Y}}}$ is the exponential function $e^{\lambda \cdot}$ where g_1 is defined by (3.4.3).

7.2.2.4 Properties of Generalized Exponential Functions

Proposition 7.2.7 *Let $\mathcal{Y}$ be a nontrivial real or complex Banach space, p lie in $[1,\infty]$, and (f, g, L) belong to $L^1(0,b) \times L^p(0,b) \times \mathcal{L}(\mathcal{Y})$. Then:*

(i) *$E_{f,g}^L Y = \mathsf{R}_{B_{f,\mathcal{X}}}^L (gY)$ for each Y in $\mathcal{Y}$;*
(ii) *$E_g^L * h = B_{g,\mathcal{X}} \mathsf{R}_{B_{g,\mathcal{X}}}^L (h)$ for each h in $L^p(\mathcal{Y})$.*

Notice that in Proposition 7.2.7, the equalities take place in $L^p(\mathcal{Y})$. Also in Item (ii), we convolve a function of $L^p\big((0,b), \mathcal{L}(\mathcal{Y})\big)$ by a function of $L^p\big((0,b), \mathcal{Y}\big)$ (see Remark 3.3.2).

Proof of Proposition 7.2.7 We have

$$\begin{aligned} \mathsf{R}_{B_{f,\mathcal{X}}}^L (gY) &= \sum_{j\geq 0} \big(B_{f,L^p}{}^j g\, L^j Y\big) \\ &= \Big(\sum_{j\geq 0} B_{f,L^p}{}^j g\, L^j\Big) Y \end{aligned}$$

since the series $\sum B_{f,L^p}{}^j g\, L^j$ converges in $L^p\big((0,b), \mathcal{L}(\mathcal{Y})\big)$ according to Proposition 7.2.6. Thus Item (i) follows from the very definition of $E_{f,g}^L$. Let us prove Item (ii). By Young inequality,

$$E_g^L * h = \sum_{j\geq 0} \big[(g_{j+1}\, L^j) * h\big]. \tag{7.2.9}$$

Moreover, for each nonnegative integer j,

$$\begin{aligned} (g_{j+1}\, L^j) * h &= B_{g_{j+1},\mathcal{X}}\, \tilde{L}^j\, h && \text{(by Lemma 7.2.8)} \\ &= B_{g,\mathcal{X}}\, B_{g,L^p}{}^j\, \tilde{L}^j\, h && \text{(by (7.2.6))}. \end{aligned}$$

Then going back to (7.2.9), we get Item (ii). □

Lemma 7.2.8 *Let p lie in $[1,\infty]$, L live in $L^p(\mathcal{Y})$, and (f, h) belong to $L^1(0,b) \times L^p(\mathcal{Y})$. Then*

$$\big(f(\cdot)L\big) * h = B_{f,\mathcal{X}} \tilde{L}\, h,$$

where $\tilde{L}$ is defined in Sect. 7.2.1.2.

Proof For almost every x in $(0, b)$,

$$\begin{aligned}
\big(f(\cdot)L\big) * h(x) &= \int_0^x f(x-y)Lh(y)\,\mathrm{d}y \\
&= L\int_0^x f(x-y)h(y)\,\mathrm{d}y \qquad &&\text{(by Proposition 3.2.3)} \\
&= L\big((B_{f,\mathcal{X}}\, h)(x)\big) \\
&= \big(B_{f,\mathcal{X}}\tilde{L}\, h\big)(x) &&\text{(since } B_{f,\mathcal{X}} \text{ and } \tilde{L} \text{ commute).}
\end{aligned}$$

□

7.2.3 Well-Posed Problems on Banach Spaces

Let $(\mathcal{Z}, \|\cdot\|)$ be a Banach space, $A : D(A) \subseteq \mathcal{Z} \to \mathcal{Z}$ be an operator, and B belong to $\mathcal{L}(\mathcal{Z})$. We assume (7.2.3) and that A is a closed operator on $\mathcal{Z}$.

7.2.3.1 The Projection $P_{\ker A}$

By Lemma 4.3.3,

$$D(A) = \ker A \oplus R(B).$$

Moreover, $\ker A$ is closed since A is supposed to be closed. Also applying Lemma 4.3.15 with

$$\mathcal{E} := \{0\}, \qquad F := \{0\}, \qquad G := \mathcal{Z},$$

we get that $R(B)$ is closed in $D(A)$. Hence Definition 2.3.3 allows to introduce the projection $P_{\ker A} : D(A) \to D(A)$ on $\ker A$ along $R(B)$.

7.2.3.2 Formulation of the Problem

Let L belong to $\mathcal{L}(\mathcal{Z})$ and $h_{\mathrm{b.c}}$, h be data in $\ker A$ and $\mathcal{Z}$, respectively. Then we consider the following problem:

$$\begin{cases} \text{Find } u \text{ in } D(A) \text{ such that} \\ Au = Lu + h, \qquad P_{\ker A}u = h_{\mathrm{b.c}}. \end{cases} \tag{7.2.10}$$

In view of the theory of differential quadruplets with maximal operators having finite-dimensional kernel (see Sect. 5.5) and the analysis developed in Sect. 5.7.1, the equation $P_{\ker A}u = h_{\text{b.c}}$ can be seen as an *abstract boundary condition.*

Definition 7.2.6 Problem (7.2.10) is said to be *well posed* if, for each $(h_{\text{b.c}}, h)$ in $\ker A \times \mathcal{Z}$, the following two conditions are satisfied:

(i) (7.2.10) has a unique solution u.
(ii) u depends continuously on the data, that is, u satisfies the estimate

$$\|u\| \leq C(\|h\| + \|h_{\text{b.c}}\|),$$

where the constant C is independent of u, $h_{\text{b.c}}$, and h.

□

7.2.3.3 Solvability

Let $(\mathcal{Z}, \|\cdot\|)$ be a real or complex Banach space.

Proposition 7.2.9 *Let* $A : D(A) \subseteq \mathcal{Z} \to \mathcal{Z}$ *be a closed operator on* $\mathcal{Z}$ *and* B, L *belong to* $\mathcal{L}(\mathcal{Z})$. *We assume* (7.2.1)–(7.2.3). *Then for each* $(h_{\text{b.c}}, h)$ *in* $\ker A \times \mathcal{Z}$, (7.2.10) *has a unique solution u. Besides, u reads*

$$u = \mathsf{R}_B^L h_{\text{b.c}} + B\mathsf{R}_B^L h, \tag{7.2.11}$$

and there exists a constant C independent of u, $h_{\text{b.c}}$*, and h such that*

$$\|u\| \leq C(\|Bh\| + \|h_{\text{b.c}}\|). \tag{7.2.12}$$

In particular, (7.2.10) *is well posed in the sense of Definition* 7.2.6.

Proof First of all, let us show that for each u_0 in $\ker A$,

$$P_{\ker A}\mathsf{R}_B^L u_0 = u_0. \tag{7.2.13}$$

Indeed by Proposition 7.2.1 (i),

$$\mathsf{R}_B^L u_0 = u_0 + BL\mathsf{R}_B^L u_0 \in \ker A \oplus R(B).$$

Thus (7.2.13) follows from the very definition of the projection $P_{\ker A}$.

Now let u be a solution of (7.2.10). Proposition 7.2.2 entails that

$$u = \mathsf{R}_B^L u_0 + B\mathsf{R}_B^L h,$$

for some u_0 in $\ker A$. Moreover since $R(B)$ is the kernel of $P_{\ker A}$, we deduce with (7.2.13) that $P_{\ker A} u = u_0$. Since by assumption $P_{\ker A} u = h_{\text{b.c}}$, there results that u satisfies (7.2.11).

Conversely let u be defined by (7.2.11). Then Proposition 7.2.2 implies that $Au = Lu + h$. Besides, $P_{\ker A} u = h_{\text{b.c}}$ by (7.2.13). Thus u solves (7.2.10). So far, we have shown that (7.2.10) admits a unique solution and that this solution u satisfies (7.2.11).

Finally since B commute with R_B^L (according to Proposition 7.2.1), and R_B^L is continuous on $\mathcal{Z}$, (7.2.12) follows from (7.2.11). □

7.2.4 Problems on Cartesian Powers of a Banach Space

Let $\mathbb{K}$ be equal to $\mathbb{R}$ or $\mathbb{C}$ and $\mathcal{X}$ be a Banach space over $\mathbb{K}$. Let also n be a positive integer.

7.2.4.1 Introduction

Since Cartesian powers of a Banach space are Banach spaces, the results of Sect. 7.2.3 apply to problems set on Cartesian powers. Our aim is to express the solution of a suitable class of problems set on Cartesian powers in terms of the solution operators of problems set on the base space.

To be a little bit more specific, let us consider a simple example where $\mathcal{X} := L^2(0, b)$. Being given n scalars $\lambda_1, , \dots, \lambda_n$, consider the following linear system of homogeneous equations:

$$\mathsf{D}^1_{L^2} u_i = \lambda_i u_i, \qquad \forall\, 1 \le i \le n. \tag{7.2.14}$$

Then for each index i, u_i reads $\mathsf{R}^{\lambda_i}_{B_1} u_{0,i}$ where $B_1 := B_{1,L^2}$ and $u_{0,i}$ lies in $\langle g_1 \rangle$. By recasting (7.2.14) into a problem set on $L^2(0, b)^n$, we get with the formalism of Sect. 2.2.8,

$$\overrightarrow{\mathsf{D}^1_{L^2}}\, \vec{u} = \vec{D}\, \vec{u}, \tag{7.2.15}$$

where D is the diagonal matrix of $\mathbb{K}^n$ with diagonal entries $\lambda_1, , \dots, \lambda_n$. Any solution $\vec{u}$ of (7.2.15) reads $\mathsf{R}^{\vec{D}}_{\vec{B_1}} \vec{u}_0$ for some $\vec{u}_0 = (u_{0,1}, \dots, u_{0,n})$ in $\langle g_1 \rangle^n$. Thus

$$\mathsf{R}^{\vec{D}}_{\vec{B_1}} \vec{u}_0 = \left(\mathsf{R}^{\lambda_1}_{B_1} u_{0,1}, \dots, \mathsf{R}^{\lambda_n}_{B_1} u_{0,n}\right).$$

The key point is that *each component of a solution to* (7.2.15) *can be expressed in terms of the solution operator of an analog equation set on the base* $L^2(0, b)$.

Moreover, when D is replaced by any other matrix L of $\mathbb{K}^n$, it is known (see, for instance, [HS74, Chap. 6]) that each component of the solution to (7.2.15) is a (finite) linear combination of functions of the form $g_j \mathsf{R}^{\lambda}_{B_1} g_1$ where λ is an eigenvalue of L and j is a positive integer less or equal to the algebraic multiplicity of λ. In the sequel, we will investigate the case where $\mathrm{D}^1_{L^2} : H^1(0,b) \subseteq L^2(0,b) \to L^2(0,b)$ is replaced by an abstract operator $A : D(A) \subseteq \mathcal{X} \to \mathcal{X}$. It will be proved in Proposition 7.2.13 that under the assumptions insuring well posedness, the same kind of result persists.

7.2.4.2 Preliminaries

We will use without reference the notations of Sect. 2.2.8.

Lemma 7.2.10 *Let $\mathcal{X}$ be a Banach space, B belong to $\mathcal{L}(\mathcal{X})$, and L be a square scalar matrix of dimension n. Then $\vec{B}$ commute with $\vec{L}$. If in addition, the spectral radius of B vanishes, then the spectral radius of $\vec{B}$ vanishes as well, and the series $\sum(\vec{B}\vec{L})^j$ converges in $\mathcal{L}(\mathcal{X}^n)$.*

Proof The domain of $\vec{B}\vec{L}$ is equal to the whole space $\mathcal{X}^n$. Thus Proposition 2.2.23 (iii) yields that $\vec{B}$ commute with $\vec{L}$. Next, we know from Sect. 2.2.8.1 that $\vec{B}$ and $\vec{L}$ belong to $\mathcal{L}(\mathcal{X}^n)$. Thus for each nonnegative integer j, we estimate

$$\begin{aligned} \|\vec{B}^j\|_{\mathcal{L}(\mathcal{X}^n)} &= \left\|\overrightarrow{B^j}\right\|_{\mathcal{L}(\mathcal{X}^n)} && \text{(by (2.2.24))} \\ &\le \|B^j\|_{\mathcal{L}(\mathcal{X})} && \text{(by (2.2.22)).} \end{aligned}$$

Since by hypothesis, the spectral radius of B vanishes, we derive that the spectral radius of $\vec{B}$ vanishes as well. Then Proposition 7.2.1 gives the convergence of $\sum(\vec{B}\vec{L})^j$. □

Recalling that $\mathsf{R}^L_B := \sum_{j\ge 0}(BL)^j$, we may state the following result.

Lemma 7.2.11 *Let $\mathcal{X}$ be a Banach space, B belong to $\mathcal{L}(\mathcal{X})$, and L, J, P be three square scalar matrices of dimension n. If the spectral radius of B vanishes, the matrix P is invertible, and $L = P^{-1}JP$ then*

$$\mathsf{R}^{\vec{L}}_{\vec{B}} = \overrightarrow{P^{-1}}\, \mathsf{R}^{\vec{J}}_{\vec{B}}\, \vec{P}.$$

Proof Notice that the operators $\mathsf{R}^{\vec{L}}_{\vec{B}}$ and $\mathsf{R}^{\vec{J}}_{\vec{B}}$ are well defined thanks to Lemma 7.2.10. By Proposition 2.2.23 (iv),

$$\vec{L} = \overrightarrow{P^{-1}}\, \vec{J}\, \vec{P}.$$

Thus for each nonnegative integer j,

$$\vec{B}^j\,\vec{L}^j = \vec{B}^j\,\overrightarrow{P^{-1}}\,\vec{J}^j\,\vec{P} = \overrightarrow{P^{-1}}\,\vec{B}^j\,\vec{J}^j\,\vec{P},$$

by Proposition 2.2.23 (vi). Thus the assertion follows. □

We give a formal definition of a *block diagonal matrix*.

Definition 7.2.7 Let n and r be positive integers. We say that a square scalar matrix $J = (c_{i,j})$ of dimension n is a *block diagonal matrix* if there exists $r+1$ positive integers n_1, $\dots$, n_{r+1} satisfying the following two conditions:

(i) $n_1 := 1$, $n_{r+1} := n+1$, and $n_1 < n_2 < \dots n_r < n_{r+1}$.
(ii) For each $k = 1, \dots, r$ and each index i in $[n_k, n_{k+1}-1]$, one has

$$\sum_{j=1}^{n} c_{i,j}\lambda_j = \sum_{j=n_k}^{n_{k+1}-1} c_{i,j}\lambda_j, \qquad \forall\,(\lambda_1, \dots, \lambda_n) \in \mathbb{K}^n.$$

For each index k in $[1, r]$, the square matrix J_k of dimension $m_k := n_{k+1} - n_k$, with entries

$$c_{i+n_k-1,\,j+n_k-1}, \qquad \forall\, 1 \le i, j \le m_k.$$

is called the kth *block of* J. □

Under the assumptions and notation of Definition 7.2.7, the first entry of the block J_k is located on the diagonal of J, at the n_kth line of J. From a graphic point of view, J is represented under the form

$$J = \begin{pmatrix} J_1 & & 0 \\ & \ddots & \\ 0 & & J_r \end{pmatrix}.$$

7.2.4.3 A Well-Posed Inhomogeneous Problem

Let $(\mathcal{X}, \|\cdot\|_{\mathcal{X}})$ be a Banach space, n be a positive integer, A be a closed operator on $\mathcal{X}$, and B belong to $\mathcal{L}(\mathcal{X})$. We assume that B is a right inverse of A, i.e.,

$$AB = \mathrm{id}_{\mathcal{X}}. \tag{7.2.16}$$

Then $\vec{A}$ is a closed operator on $\mathcal{X}^n$, $\vec{A}\vec{B} = \mathrm{id}_{\mathcal{X}^n}$, and by (2.2.22), $\vec{B}$ belongs to $\mathcal{L}(\mathcal{X}^n)$. Thus setting $\mathcal{Z} := \mathcal{X}^n$, there results from Sect. 7.2.3.1 that the projection $P_{\ker \vec{A}}$ of $D(\vec{A})$ on $\ker \vec{A}$ along $R(\vec{B})$ is well defined. From the very definition of $P_{\ker \vec{A}}$, we may check that

$P_{\ker \vec{A}}$ is equal to $\overrightarrow{P_{\ker A}}$ which simply means that for each $(u_1, \dots, u_n)$ in $D(\vec{A})$

$$P_{\ker \vec{A}}\,(u_1, \dots, u_n) = (P_{\ker A}\, u_1, \ \dots, \ P_{\ker A}\, u_n).$$

Next, being given a square scalar matrix L of dimension n, denote by $\vec{L}$ the bounded operator induced by the matrix L in the sense of Definition 2.2.23. Then for any data $\vec{h}$ in $\mathcal{X}^n$ and $\vec{h}_{\text{b.c}}$ in $\ker \vec{A}$, we consider the following problem set on $\mathcal{Z} := \mathcal{X}^n$:

$$\begin{cases} \text{Find } \vec{u} \text{ in } D(\vec{A}) \text{ such that} \\ \vec{A}\vec{u} = \vec{L}\vec{u} + \vec{h}, \qquad P_{\ker \vec{A}}\, \vec{u} = \vec{h}_{\text{b.c}}. \end{cases} \tag{7.2.17}$$

Denoting by $a_{i,j}$ the entries of the matrix L, (7.2.17) reads componentwise

$$\begin{cases} \text{Find } u_1, \dots, u_n \text{ in } D(A) \text{ such that for each } i = 1, \dots, n \\ Au_i = \displaystyle\sum_{j=1}^{n} a_{i,j} u_j + h_j \qquad \text{in } \mathcal{X} \\ P_{\ker A}\, u_i = h_{\text{b.c},i}. \end{cases} \tag{7.2.18}$$

Proposition 7.2.12 *Let A be a closed operator on $\mathcal{X}$, B belong to $\mathcal{L}(\mathcal{X})$, and L be a square scalar matrix of dimension n. Assume* (7.2.16) *and that the spectral radius of B vanishes. Then* (7.2.17) *is well posed, and its solution $\vec{u}$ reads*

$$\vec{u} = \mathsf{R}_{\vec{B}}^{\vec{L}}\, \vec{h}_{\text{b.c}} + \vec{B}\, \mathsf{R}_{\vec{B}}^{\vec{L}}\, \vec{h}.$$

Proof We have already seen that $\vec{A}$ is a closed operator on $\mathcal{X}^n$ and that $\vec{A}\vec{B} = \mathrm{id}_{\mathcal{X}^n}$. Besides, $\vec{B}$ commute with $\vec{L}$, and the spectral radius of B vanishes according to Lemma 7.2.10. Thus Proposition 7.2.9 applies. □

7.2.4.4 Representation of the Solutions to the Homogeneous Equation

Being given an operator A on $\mathcal{X}$ and a square scalar matrix L of dimension n, the homogeneous equation associated with (7.2.17) reads

$$\text{Find } \vec{u} \text{ in } D(\vec{A}) \text{ such that } \vec{A}\vec{u} = \vec{L}\vec{u}. \tag{7.2.19}$$

Under the assumptions and notation of Proposition 7.2.12, the solution set $\mathcal{S}_{\mathcal{X}^n}^{\text{hom}}$ of (7.2.19) reads

$$\mathcal{S}_{\mathcal{X}^n}^{\text{hom}} = \{\mathsf{R}_{\vec{B}}^{\vec{L}}\, \vec{u}_0 \mid \vec{u}_0 \in \ker \vec{A}\}. \tag{7.2.20}$$

The following result gives a representation of the solutions to (7.2.19) set on $\mathcal{Z} := \mathcal{X}^n$, in terms of solution operators of equations set on $\mathcal{X}$.

Proposition 7.2.13 *Let $\mathcal{X}$ be a Banach space over $\mathbb{C}$, A be an operator on $\mathcal{X}$, B belong to $\mathcal{L}(\mathcal{X})$, and L be a square scalar matrix of dimension n. Denote by $\lambda_1, \ldots, \lambda_r$ a list of the pairwise distinct complex eigenvalues of L. Also for $k = 1, \ldots, r$, let m'_k be the multiplicity of λ_k with respect to the minimal polynomial of L. Assume* (7.2.3) *and that the spectral radius of B vanishes. Then for each index j in $[1, n]$, the* jth *component of any solution $\vec{u}$ to* (7.2.19) *reads*

$$u_j = \sum_{k=1}^{r} \sum_{i=0}^{m'_k - 1} \frac{1}{i!} \frac{\mathrm{d}^i}{\mathrm{d}\lambda^i}_{|\lambda=\lambda_k} \mathsf{R}_B^{\lambda}\, w_{0,i,j,k},$$

where the vectors $w_{0,i,j,k}$ lie in $\ker A$.

In Proposition 7.2.13, R_B^{λ} stands for $\mathsf{R}_B^{\lambda \mathrm{id}_{\mathcal{X}}}$. That is, R_B^{λ} is the solution operator of the homogeneous equation

$$u \in D(A), \ Au = \lambda u \ \text{ in } \mathcal{X}.$$

Moreover, $\frac{\mathrm{d}^i}{\mathrm{d}\lambda^i}_{|\lambda=\lambda_k} \mathsf{R}_B^{\lambda}$ stands for ${F_0}^{(i)}(\lambda_k)$ where F_0 is defined through (7.2.5).

Proof of Proposition 7.2.13 Denote by J *a Jordan normal form of* L. That is, J is a block diagonal matrix with, let us say, r' *Jordan blocks* designated by $J_1, \ldots, J_{r'}$. For each $k = 1, \ldots, r'$, let m_k be the dimension of the block J_k. Now, there exists an eigenvalue λ of L such that either $J_k = \lambda \mathrm{id}_{\mathbb{K}^{m_k}}$ or

$$J_k = \begin{pmatrix} \lambda & 1 & 0 & \\ & \ddots & \ddots & \\ & & \ddots & 1 \\ 0 & & & \lambda \end{pmatrix}. \tag{7.2.21}$$

Otherwise said if $(e_1, \ldots, e_{m_k})$ is the canonical basis of $\mathbb{K}^{m_k}$ and $e_0 := 0$, then

$$J_k e_i = e_{i-1} + \lambda e_i, \qquad \forall\, 1 \leq i \leq m_k. \tag{7.2.22}$$

By (7.2.20), any solution $\vec{u}$ of (7.2.19) reads $\vec{u} = \mathsf{R}_{\vec{B}}^{\vec{L}} \vec{u}_0$ for some $\vec{u}_0 = (u_{0,1}, \ldots, u_{0,n})$ in $\ker \vec{A}$. Considering first the particular case where L is a Jordan block, there are two possibilities.

(i) Case where $L = \lambda \mathrm{id}_{\mathbb{K}^n}$. Because of the simple structure of L, we readily get

$$\mathsf{R}_{\vec{B}}^{\vec{L}} = \mathsf{R}_{B}^{\overrightarrow{\lambda}}, \tag{7.2.23}$$

where, in the latter right-hand side, λ stands for $\lambda \mathrm{id}_{\mathcal{X}}$, that is, R_B^{λ} is the operator solution of the equation $Au = \lambda u$ with unknown u in $D(A)$.

(ii) Case where L is a nondiagonal Jordan block of the form (7.2.21). Then denoting by $(e_1, \ldots, e_n)$ the canonical basis of $\mathbb{K}^n$ and setting $e_0 := 0$, one has

$$Le_i = e_{i-1} + \lambda e_i, \qquad \forall\, 1 \le i \le n. \tag{7.2.24}$$

Let us introduce two notations. For each nonnegative integer i, set

$$k(i) := \max(1, i),$$

and for any positive integer j, and any index $k = 0, \ldots, j$, denote by $\binom{j}{k}$ the *binomial coefficient*

$$\binom{j}{k} := \frac{j!}{k!(j-k)!}.$$

Recall that here $0! := 1$ by convention. Then we claim that

$$L^j e_i = \sum_{k=k(i-j)}^{i} \tbinom{j}{i-k} \lambda^{j-i+k} e_k. \tag{7.2.25}$$

We will only sketch the proof. Proceeding by induction on j, we write

$$L^{j+1} e_i = L^j e_{i-1} + \lambda L^j e_i$$

and apply twice the induction hypothesis. Next, if $2 \le i - j$ then using $k(i-j) = k(i-j-1) + 1$, we get the formula at the rank $j+1$. If $i - j \le 1$ then $k(i-j) = 1$ and $k(i-j-1) = 0$. Using $e_0 = 0$, we end up with the same formula. That completes the sketch of the proof of the claim.

Now with Proposition 2.2.23 and (7.2.25), we compute

$$\vec{B}^j \vec{L}^j \vec{u}_0 = \sum_{i=1}^{n} \sum_{k=k(i,j)}^{i} e_k \otimes \left(\tbinom{j}{i-k} \lambda^{j-i+k} B^j u_{0,i} \right). \tag{7.2.26}$$

Since $\frac{j!}{(j-i+k)!} \le j^{i-k}$ and B has trivial spectral radius, we derive

$$\|\tbinom{j}{i-k}\lambda^{j-i+k}B^j\|_{\mathcal{L}(\mathcal{X})}^{\frac{1}{j}} \leq j^{\frac{n}{j}}|\lambda|^{\frac{j-i+k}{j}}\|B^j\|_{\mathcal{L}(\mathcal{X})}^{\frac{1}{j}} \xrightarrow[j\to\infty]{} 0. \tag{7.2.27}$$

Thus the discreet Fubini theorem entails that

$$\begin{aligned}\mathsf{R}_{\vec{B}}^{L}\vec{u}_0 &= \sum_{i=1}^{n}\sum_{k=1}^{i} e_k \otimes \sum_{j\geq i-k} \tbinom{j}{i-k}\lambda^{j-i+k}B^j u_{0,i}\\ &= \sum_{k=1}^{n} e_k \otimes \sum_{i=k}^{n}\sum_{j\geq i-k} \tbinom{j}{i-k}\lambda^{j-i+k}B^j u_{0,i}\end{aligned}$$

Observing that

$$\tbinom{j}{i-k}\lambda^{j-i+k} = \frac{1}{(i-k)!}\frac{\mathrm{d}^{i-k}}{\mathrm{d}\lambda^{i-k}}\lambda^j, \tag{7.2.28}$$

where by convention $\frac{\mathrm{d}^0}{\mathrm{d}\lambda^0}\lambda^j = \lambda^j$, and using Proposition 7.2.3, there result that

$$\mathsf{R}_{\vec{B}}^{L}\vec{u}_0 = \sum_{k=1}^{n} e_k \otimes \sum_{i=0}^{n-k}\frac{1}{i!}\frac{\mathrm{d}^i}{\mathrm{d}\lambda^i}\mathsf{R}_B^{\lambda}\, u_{0,i+k}. \tag{7.2.29}$$

That completes the analysis when L is a Jordan block.

Secondly, let us consider the general case where L is any square matrix. Then

$$L = P^{-1}JP$$

for some invertible matrix P of dimension n. Setting

$$\vec{v} = \vec{P}\vec{u}, \qquad \vec{v}_0 = \vec{P}\vec{u}_0,$$

we infer from Proposition 2.2.23 (especially Item (vi)) that $\vec{A}\vec{v} = \vec{J}\vec{v}$. Thus $\vec{v} = \mathsf{R}_{\vec{B}}^{J}\vec{v}_0$. At this stage, we must be more specific about the Jordan normal form J of L. Two distinct blocks of J may refer to the same eigenvalue of L. Then for $k = 1, \dots, r'$, we denote by λ_k' the eigenvalue corresponding to the Jordan block J_k. Recall that m_k designates the dimension of J_k.

By a slight abuse of notation, denote again by $\vec{A}$ the operator induced by A on $\mathcal{X}^{m_k}$ (see Definition 2.2.21). Since J is block diagonal, by using the notation of Definition 7.2.7, we derive that for $k = 1, \dots, r'$, the vector $w := (v_{n_k}, \dots, v_{0,n_{k+1}-1})$ of $\mathcal{X}^{m_k}$ lies in $D(\vec{A})$ and satisfies $\vec{A}\vec{w} = \vec{J}_k\vec{w}$. Thus from (7.2.23) and (7.2.29), we get that for $j = n_k, \ldots, n_{k+1} - 1$,

$$v_j = \sum_{i=0}^{m_k-1} \frac{1}{i!} \frac{\mathrm{d}^i}{\mathrm{d}\lambda^i}_{|\lambda=\lambda'_k} \mathrm{R}_B^\lambda \, w_{0,i,j},$$

for some $w_{0,i,j}$ in $\ker A$. Thus if $\vec{p}_1, \ldots, \vec{p}_n$ denote the columns of P^{-1} then

$$\vec{u} = \sum_{j=1}^{n} \vec{p}_j \otimes v_j.$$

If we consider an eigenvalue, let us say λ_k of L, then the orders of derivative in the above representation of the v_j's will be less than m'_k. Hence

$$u_j = \sum_{k=1}^{r} \sum_{i=0}^{m'_k-1} \frac{1}{i!} \frac{\mathrm{d}^i}{\mathrm{d}\lambda^i}_{|\lambda=\lambda_k} \mathrm{R}_B^\lambda \, w_{0,i,j,k},$$

for some $w_{0,i,j,k}$ in $\ker A$. □

7.2.5 Higher-Order Problems

Let $\mathbb{K}$ be equal to $\mathbb{R}$ or $\mathbb{C}$, and $\mathcal{X}$ be a Banach space over $\mathbb{K}$. Let also n be a positive integer.

7.2.5.1 Formulation of the Problem

Let $A : D(A) \subseteq \mathcal{X} \to \mathcal{X}$ and B belong to $\mathcal{L}(\mathcal{X})$. We assume (7.2.3) and that A is a closed operator on $\mathcal{X}$. Hence from Sect. 7.2.3.1, the projection $P_{\ker A}$ of $D(A)$ on $\ker A$ along $R(B)$ is well defined.

Being given n scalars $a_0, \ldots, a_{n-1}$, and data h in $\mathcal{X}$ and $h_{\mathrm{b.c},1}, \ldots, h_{\mathrm{b.c},n}$ in $\ker A$, we consider the following problem:

$$\begin{cases} \text{Find } v \text{ in } D(A^n) \text{ such that} \\ A^n v + \displaystyle\sum_{i=0}^{n-1} a_i A^i v = h, \\ P_{\ker A} A^i v = h_{\mathrm{b.c},i+1}, \qquad \forall i = 0, \ldots, n-1. \end{cases} \tag{7.2.30}$$

7.2.5.2 The Order Reducing Method

We will implement the so-called order reducing method, that is, we will recast the first equation of (7.2.30) set on $\mathcal{X}$ into a lower-order equation set on $\mathcal{X}^n$. As is well-known, this procedure, applied to a scalar linear differential equation of order n with constant coefficients, returns a first-order linear system on $\mathbb{K}^n$.

Our aim is to rewrite (7.2.30) as a problem of the form (7.2.10) set on $\mathcal{X}^n$. For, denoting by $\vec{A}$ the operator induced by A on $\mathcal{X}^n$, we know from Sect. 7.2.4.3 that $\vec{A}$ is closed on $\mathcal{X}^n$ and $\vec{B}$ belongs to $\mathcal{L}(\mathcal{X}^n)$ and is a right inverse of $\vec{A}$. Thus the projection $P_{\ker \vec{A}}$ of $D(\vec{A})$ on $\ker \vec{A}$ along $R(\vec{B})$ is well defined. From the very definition of $P_{\ker \vec{A}}$, we may check that $P_{\ker \vec{A}}$ is equal to $\overrightarrow{P_{\ker A}}$ which simply means that for each $(u_1, \dots, u_n)$ in $D(\vec{A})$,

$$P_{\ker \vec{A}}\,(u_1, \dots, u_n) = (P_{\ker A}\, u_1, \ \dots, \ P_{\ker A}\, u_n).$$

Also let L denote the square scalar matrix of dimension n

$$L := \begin{pmatrix} 0 & 1 & & 0 \\ & \ddots & \ddots & \\ & & 0 & 1 \\ -a_0 & \dots & \dots & -a_{n-1} \end{pmatrix}. \tag{7.2.31}$$

Otherwise said if $(e_1, \dots, e_n)$ is the canonical basis of $\mathbb{K}^n$ and $e_0 := 0$, then

$$Le_i = e_{i-1} - a_{i-1} e_n, \qquad \forall\, 1 \le i \le n. \tag{7.2.32}$$

Finally, set

$$\vec{h} := (0, \dots, 0, h) \in \mathcal{X}^n, \qquad \vec{h}_{\mathrm{b.c}} := (h_{\mathrm{b.c},1}, \dots, h_{\mathrm{b.c},n}) \in \ker \vec{A}.$$

Then with these notations, we end up with the following problem of the form (7.2.10):

$$\begin{cases} \text{Find } \vec{u} \text{ in } D(\vec{A}) \text{ such that} \\ \vec{A}\vec{u} = \vec{L}\vec{u} + \vec{h}, \qquad P_{\ker \vec{A}}\vec{u} = \vec{h}_{\mathrm{b.c}}. \end{cases} \tag{7.2.33}$$

Of course, Problems (7.2.30) and (7.2.33) are in some sense equivalent. More precisely, we have the following result.

Proposition 7.2.14 *Let A be a closed operator on $\mathcal{X}$ and B be a right inverse of A in $\mathcal{L}(\mathcal{X})$. Then the linear map*

$$D(A^n) \to \mathcal{X}^n, \qquad v \mapsto (A^{i-1}v)_{1 \le i \le n}$$

induces a bijection from the solution set of (7.2.30) *onto the solution set of* (7.2.33).

Proof For each $\vec{u} := (u_i)_{1\le i\le n}$ in $\mathcal{X}^n$, one has with (7.2.32),

$$\vec{L}\vec{u} = \sum_{i=1}^{n-1} e_i \otimes u_{i+1} + e_n \otimes \sum_{i=1}^{n} -a_{i-1}u_i. \tag{7.2.34}$$

Now, if $\vec{u} := (u_i)_{1\le i\le n}$ solves (7.2.33) then

$$\vec{A}\vec{u} = (Au_i)_{1\le i\le n} = \sum_{i=1}^{n} e_i \otimes Au_i. \tag{7.2.35}$$

Since by assumption, $\vec{A}\vec{u} = \vec{L}\vec{u} + \vec{h}$, we infer from (7.2.34) and (7.2.35) that $Au_i = u_{i+1}$ for each index i in $[1, n-1]$ and

$$Au_n = -\sum_{i=0}^{n-1} a_i u_{i+1} + h.$$

Thus by induction, we derive that u_1 lies in $D(A^n)$ and $u_{i+1} = A^i u_1$ for $i = 1, \dots, n-1$. Whence u_1 solves the first equation of (7.2.30). Regarding boundary conditions, since (u_i) solves (7.2.33), we derive that $P_{\ker A}\, u_i = h_{\text{b.c},i}$ for $i = 1, \dots, n$. Since $u_i = A^{i-1}u_1$, u_1 solves the second equation of (7.2.30), so that u_1 is solution to (7.2.30). Next, denoting by $\mathcal{S}^{\text{inh}}_{\mathcal{X}^n}$ the solution set of (7.2.33), we have proved that the range of the linear map

$$\mathcal{S}^{\text{inh}}_{\mathcal{X}^n} \to \mathcal{X}, \qquad \vec{u} := (u_i)_{1\le i\le n} \mapsto u_1$$

is contained in the solution set $\mathcal{S}^{\text{inh}}_{\mathcal{X}}$ of (7.2.30). Hence we may legitimately consider the map

$$p : \mathcal{S}^{\text{inh}}_{\mathcal{X}^n} \to \mathcal{S}^{\text{inh}}_{\mathcal{X}}, \qquad \vec{u} := (u_i)_{1\le i\le n} \mapsto u_1 \tag{7.2.36}$$

Conversely, being given a solution v of (7.2.30), set

$$\vec{u} = (u_i)_{1\le i\le n} := (A^{i-1}v)_{1\le i\le n}.$$

Then since v belongs to $D(A^n)$, $\vec{u}$ lies in $D(\vec{A})$, and

$$\vec{A}\vec{u} = \sum_{i=1}^{n} e_i \otimes A^i v.$$

On the other hand, with (7.2.34), we have

$$\vec{L}\vec{u} + \vec{h} = \sum_{i=1}^{n-1} e_i \otimes A^i v + e_n \otimes \sum_{i=1}^{n} -a_{i-1} A^{i-1} v + h.$$

Since v solves the first equation of (7.2.30), we derive that $\vec{A}\vec{u} = \vec{L}\vec{u} + \vec{h}$. Regarding boundary conditions, for each index i in $[1, n]$,

$$(P_{\ker \vec{A}}\, \vec{u})_i = P_{\ker A}\, A^{i-1} v = h_{\text{b.c},i},$$

since v is solution to (7.2.30). Hence $P_{\ker \vec{A}}\, \vec{u} = \vec{h}_{\text{b.c}}$, and $\vec{u}$ is a solution to (7.2.33). We have proved that the linear map

$$r : \mathcal{S}^{\text{inh}}_{\mathcal{X}} \to \mathcal{S}^{\text{inh}}_{\mathcal{X}^n}, \qquad v \mapsto (A^{i-1} v)_{1 \le i \le n} \tag{7.2.37}$$

is well defined. So, there remains to show the map r defined by (7.2.37) is a bijection. For, it is clear with (7.2.36) that $p \circ r = \text{id}_{\mathcal{S}^{\text{inh}}_{\mathcal{X}}}$. In closing, we have shown at the beginning of the current proof that if $\vec{u} := (u_i)_{1 \le i \le n}$ lies in $\mathcal{S}^{\text{inh}}_{\mathcal{X}^n}$, then u_1 lies in $D(A^n)$ and $u_{i+1} = A^i u_1$ for $i = 1, \ldots, n-1$. Thus $r \circ p = \text{id}_{\mathcal{S}^{\text{inh}}_{\mathcal{X}^n}}$. □

7.2.5.3 Well Posedness of (7.2.30)

Let $(\mathcal{X}, \|\cdot\|_{\mathcal{X}})$ be a real or complex Banach space and n be a positive integer.

Definition 7.2.8 Let A be a closed operator on $\mathcal{X}$ and B belong to $\mathcal{L}(\mathcal{X})$ and satisfy $AB = \text{id}_{\mathcal{X}}$. Then Problem (7.2.30) is said to be *well posed* if for each h in $\mathcal{X}$ and $h_{\text{b.c},1}, \ldots, h_{\text{b.c},n}$ in $\ker A$, the following two conditions are satisfied:

(i) (7.2.30) has a unique solution v.
(ii) v satisfies the estimate

$$\sum_{i=0}^{n-1} \|A^i v\|_{\mathcal{X}} \le C\|h\|_{\mathcal{X}} + C \sum_{i=1}^{n} \|h_{\text{b.c},i}\|_{\mathcal{X}},$$

where the constant C is independent of v, h, and $h_{\text{b.c},1}, \ldots, h_{\text{b.c},n}$.

□

Proposition 7.2.15 *Under the assumptions and notation of Definition 7.2.8, suppose in addition that the spectral radius of B vanishes. Then* (7.2.30) *is well posed.*

Proof From the analysis of Sect. 7.2.4.3, we know that $\vec{A}$ is a closed operator on $\mathcal{X}^n$, $\vec{A}\vec{B} = \text{id}_{\mathcal{X}^n}$, $\vec{B}$ commute with $\vec{L}$, and the spectral radius of $\vec{B}$ vanishes. Then

Proposition 7.2.9 yields that (7.2.33) is well posed. Hence Proposition 7.2.14 entails that (7.2.30) has a unique solution v and that $\vec{u} := (A^{i-1}v)_{1\leq i\leq n}$ is the solution to (7.2.33). Since

$$\|\vec{u}\|_{\mathcal{X}^n} = \sum_{i=0}^{n-1} \|A^i v\|_{\mathcal{X}},$$

the well posedness of (7.2.30) follows from the well posedness of (7.2.33). □

7.2.5.4 Structure of the Solution Set of Homogeneous Equation

Under the assumptions and notation of Sect. 7.2.5.1, suppose in addition that $\mathcal{X}$ is a *complex* Banach space. We consider the following homogeneous equation whose unknown v lies in $D(A^n)$:

$$A^n v + \sum_{i=0}^{n-1} a_i A^i v = 0. \tag{7.2.38}$$

By introducing the complex polynomial P defined for each X in $\mathbb{C}$ by

$$P(X) := X^n + \sum_{i=0}^{n-1} a_i X^i, \tag{7.2.39}$$

(7.2.38) reads $P(A)v = 0$.

Let v be a solution to (7.2.38). Proposition 7.2.14 tells us that $(A^{i-1}v)_{1\leq i\leq n}$ solves the following homogeneous equation whose unknown $\vec{u}$ belongs to $D(\vec{A})$:

$$\vec{A}\vec{u} = \vec{L}\vec{u}, \tag{7.2.40}$$

where L is given by (7.2.31). In order to describe precisely the solution set $\mathcal{S}_{\mathcal{X}}^{\mathrm{hom}}$ of (7.2.38), we will apply Proposition 7.2.13. First recall that P is the characteristic polynomial of L. Thus denoting (as in Proposition 7.2.13) by $\lambda_1, , \ldots, \lambda_r$ a list of the pairwise distinct eigenvalues of L, we know that for each $k = 1, \ldots, r$, λ_k is a root of P and that the multiplicity m_k of λ_k with respect to P is larger or equal to m'_k. Here m'_k denotes (as in Proposition 7.2.13) the multiplicity of λ_k with respect to the minimal polynomial of L. Then Proposition 7.2.13 yields that for each $k = 1, \ldots, r$ and each $i = 0, \ldots, m_k - 1$, there exists some vector $w_{0,i,k}$ in $\ker A$ such that

$$v = \sum_{k=1}^{r} \sum_{i=0}^{m_k-1} \frac{1}{i!} \frac{\mathrm{d}^i}{\mathrm{d}\lambda^i}_{|\lambda=\lambda_k} \mathrm{R}_B^{\lambda}\, w_{0,i,k}. \tag{7.2.41}$$

In the sequel, we will show that (7.2.41) characterizes the solution to (7.2.38) in a unique way. More precisely, being given a solution v to (7.2.38), we will establish that the n vectors $w_{0,i,k}$ appearing in (7.2.41) are unique. Conversely, being given any family of n vectors $w_{0,i,k}$ in $\ker A$, we will prove that the vector v defined by (7.2.41) solves (7.2.38). That gives rise to *a one-to-one correspondence between the solution set $\mathcal{S}_{\mathcal{X}}^{\mathrm{hom}}$ of* (7.2.38) *and* $(\ker A)^n$.

Besides we will show that this one-to-one correspondence is smooth, so that $\mathcal{S}_{\mathcal{X}}^{\mathrm{hom}}$ is isomorph to $(\ker A)^n$. The precise statements are featured in Lemma 7.2.16 and Theorem 7.2.17.

These results extend well-known facts in the quantitative theory of linear ordinary differential equations (see, for instance, [HS74, Chap. 6, §6]). Notice also that P turns out to be the minimal polynomial of L, but this extra information is not needed here (because of Lemma 7.2.16).

Lemma 7.2.16 *Let $\mathbb{K} := \mathbb{C}$ and w_0 belong to* $\ker A$. *Then for each $k = 1, \ldots, r$ and each index i in* $[0, m_k - 1]$*, the vector*

$$\frac{\mathrm{d}^i}{\mathrm{d}\lambda^i}_{|\lambda=\lambda_k} \mathsf{R}_B^\lambda \, w_0$$

is a solution to (7.2.38). *Besides, for every negative integer l,*

$$A^l \frac{\mathrm{d}^i}{\mathrm{d}\lambda^i}_{|\lambda=\lambda_k} \mathsf{R}^\lambda_{B\,|\ker A} = \sum_{j\geq \max(i,l)} \frac{\mathrm{d}^i}{\mathrm{d}\lambda^i}_{|\lambda=\lambda_k} \lambda^j B^{j-l}{}_{|\ker A}. \tag{7.2.42}$$

Proof Let i, l be nonnegative integers and λ belong to $\mathbb{C}$. By Proposition 7.2.3, for each u_0 in $\ker A$,

$$\begin{aligned}\frac{\mathrm{d}^i}{\mathrm{d}\lambda^i}_{|\lambda=\lambda_k} \mathsf{R}_B^\lambda u_0 &= \sum_{j\geq i} \frac{\mathrm{d}^i}{\mathrm{d}\lambda^i}_{|\lambda=\lambda_k} \lambda^j B^j u_0 \\ &= \sum_{j=i}^{l-1} \frac{\mathrm{d}^i}{\mathrm{d}\lambda^i}_{|\lambda=\lambda_k} \lambda^j B^j u_0 + B^l \sum_{j\geq \max(i,l)} \frac{\mathrm{d}^i}{\mathrm{d}\lambda^i}_{|\lambda=\lambda_k} \lambda^j B^{j-l} u_0.\end{aligned}$$

The first summation lies in $\ker A^l$, and the second one belongs to the domain of A^l. Thus

$$\frac{\mathrm{d}^i}{\mathrm{d}\lambda^i}_{|\lambda=\lambda_k} \mathsf{R}_B^\lambda u_0$$

lives in $D(A^l)$ and (7.2.42) holds true.

Next, starting from (7.2.42) and using Proposition 7.2.3, we obtain

$$A^l \frac{\mathrm{d}^i}{\mathrm{d}\lambda^i}_{|\lambda=\lambda_k} \mathsf{R}^{\lambda}_{B\,|\ker A} = \frac{\mathrm{d}^i}{\mathrm{d}\lambda^i}_{|\lambda=\lambda_k} \left\{\lambda^l \mathsf{R}^{\lambda}_{B\,|\ker A}\right\}.$$

Hence by linearity,

$$P(A)\frac{\mathrm{d}^i}{\mathrm{d}\lambda^i}_{|\lambda=\lambda_k} \mathsf{R}^{\lambda}_{B\,|\ker A} = \frac{\mathrm{d}^i}{\mathrm{d}\lambda^i}_{|\lambda=\lambda_k} \left\{P(\lambda)\mathsf{R}^{\lambda}_{B\,|\ker A}\right\}.$$

By the Leibniz rule, this derivative vanishes for each i in $[0, m_k - 1]$ since the multiplicity of λ_k with respect to P is m_k. □

Now let $f_1, \dots, f_n$ be an ordering of the n linear maps

$$\frac{\mathrm{d}^i}{\mathrm{d}\lambda^i}_{|\lambda=\lambda_k} \mathsf{R}^{\lambda}_{B\,|\ker A}, \qquad 1 \le k \le r,\ 0 \le i \le m_k - 1. \tag{7.2.43}$$

Lemma 7.2.16 allows us to define the map

$$\varphi : (\ker A)^n \to \mathcal{S}^{\mathrm{hom}}_{\mathcal{X}}, \qquad (u_{0,1}, \dots, u_{0,n}) \mapsto \sum_{j=1}^{n} f_j u_{0,j}. \tag{7.2.44}$$

Prior to showing that φ is an isomorphism, we must turn $(\ker A)^n$ and especially $\mathcal{S}^{\mathrm{hom}}_{\mathcal{X}}$ into normed spaces. Since we do not assume that $\ker A$ is finite dimensional, suitable norms have to be selected. On $(\ker A)^n$, we choose the norm of $\mathcal{X}^n$. On $\mathcal{S}^{\mathrm{hom}}_{\mathcal{X}}$, we choose the norm

$$N : \mathcal{S}^{\mathrm{hom}}_{\mathcal{X}} \to [0, \infty), \qquad v \mapsto \sum_{i=0}^{n} \|A^i v\|_{\mathcal{X}}. \tag{7.2.45}$$

Theorem 7.2.17 *Let $\mathcal{X}$ be a complex Banach space, A be an operator on $\mathcal{X}$, and B belong to $\mathcal{L}(\mathcal{X})$. Assume that A is closed, B is a right inverse of A on $\mathcal{L}(\mathcal{X})$ and that the spectral radius of B vanishes. Then the map φ defined by* (7.2.44) *is a isomorphism between* $(\ker A)^n$ *equipped with the norm of $\mathcal{X}^n$ and $(\mathcal{S}^{\mathrm{hom}}_{\mathcal{X}}, N)$ where N is defined by* (7.2.45).

Proof Let us start to show that φ is onto. Let v belong to $\mathcal{S}^{\mathrm{hom}}_{\mathcal{X}}$. By Proposition 7.2.14, the vector $(A^{j-1}v)_{1\le j\le n}$ is a solution to (7.2.40). Recalling that P is the characteristic polynomial of the matrix L defined by (7.2.31), Proposition 7.2.13 yields the existence of $(u_{0,j})_{1\le j\le n}$ in $(\ker A)^n$ such that $v = \sum_{j=1}^{n} f_j u_{0,j}$. Hence φ is onto. Regarding injectivity, according to Proposition 7.2.15, the map

$$\psi : \mathcal{S}^{\mathrm{hom}}_{\mathcal{X}} \to (\ker A)^n, \qquad v \mapsto (P_{\ker A}\, A^{l-1} v)_{1\le l\le n}$$

is a linear bijection. Thus it is enough to prove that $\psi \circ \varphi$ is one-to-one. For each $\vec{u}_0 = (u_{0,1}, \dots, u_{0,n})$ in $(\ker A)^n$, one has

$$\psi \circ \varphi(\vec{u}_0) = \Big(\sum_{j=1}^{n} P_{\ker A}\, A^{l-1}\, f_j u_{0,j} \Big)_{1 \le l \le n}.$$

Moreover, by definition of f_j, there exists a unique ordered pair (k, i) such that

$$f_j = \frac{\mathrm{d}^i}{\mathrm{d}\lambda^i}_{|\lambda=\lambda_k} \mathsf{R}^{\lambda}_{B\,|\ker A}.$$

Then for each index l in $[1, n]$, (7.2.42) reads

$$A^{l-1} f_j = \sum_{j' \ge \max(i, l-1)} \frac{\mathrm{d}^i}{\mathrm{d}\lambda^i}_{|\lambda=\lambda_k} \lambda^{j'} B^{j'-l+1}{}_{|\ker A}.$$

Setting

$$\lambda_{l,j} := \begin{cases} 0 & \text{if } l \le i \\ \frac{\mathrm{d}^i}{\mathrm{d}\lambda^i}_{|\lambda=\lambda_k} \lambda^{l-1} & \text{if } i \le l-1 \end{cases}, \tag{7.2.46}$$

we obtain

$$P_{\ker A}\, A^{l-1} f_j = \lambda_{l,j} \mathrm{id}_{\ker A}, \qquad \forall\, 1 \le l, j \le n,$$

Denoting by M the square matrix with entries $\lambda_{l,j}$, we infer that

$$\psi \circ \varphi = \vec{M}, \tag{7.2.47}$$

where $\vec{M} : (\ker A)^n \to (\ker A)^n$ is given by Definition 2.2.23.

Now, the key point is that M does not depend on A, B, and $\mathcal{X}$ (see (7.2.46)). Then changing A, B, and $\mathcal{X}$ if necessary, we may assume that the kernel of A is one dimensional. Since $\psi \circ \varphi$ is a surjective linear map between $(\ker A)^n$ onto itself, there results that $\psi \circ \varphi$ is a bijection. Then Proposition 2.2.23 (vii) yields that M is invertible. Returning to the general case, we deduce that the linear map $\vec{M}$ is an isomorphism from $(\ker A)^n$ into itself according to Proposition 2.2.23 (v). Hence $\psi \circ \varphi$ is one-to-one.

Finally, let us show that φ is bicontinuous. Since $\psi \circ \varphi = \vec{M}$ and $\vec{M}$ is an isomorphism, it is enough to show that ψ is an isomorphism. For, Proposition 7.2.15 yields that $\psi^{-1} : (\ker A)^n \to \mathcal{S}^{\mathrm{hom}}_{\mathcal{X}}$ is continuous. Next, Lemma 7.2.18 states that $(\mathcal{S}^{\mathrm{hom}}_{\mathcal{X}}, N)$ is a Banach space. Thus Corollary 2.4.9 yields that ψ is an isomorphism. □

Lemma 7.2.18 *Let A be a closed operator on a real or complex Banach space $\mathcal{X}$ and N be the norm on the solution set $\mathcal{S}_{\mathcal{X}}^{\text{hom}}$ of* (7.2.38) *defined by* (7.2.45). *Then* $(\mathcal{S}_{\mathcal{X}}^{\text{hom}}, N)$ *is a Banach space.*

Proof Let $(v_k)_{k\in\mathbb{N}}$ be a Cauchy sequence in $(\mathcal{S}_{\mathcal{X}}^{\text{hom}}, N)$. Then for each index i in $[0, n]$, $(A^i v_k)_{k\in\mathbb{N}}$ is a Cauchy sequence in $\mathcal{X}$. Thus there exists x_i in $\mathcal{X}$ such that $(A^i v_k)_{k\in\mathbb{N}}$ converges toward x_i in $\mathcal{X}$. Since A is a closed operator, we obtain by induction that x_0 lies in $D(A^n)$ and $x_i = A^i x_0$ for $i = 1, \dots, n$. Then it is clear that x_0 lies in $\mathcal{S}_{\mathcal{X}}^{\text{hom}}$ and that (v_k) converges toward x_0 with respect to the norm N. □

7.2.6 Problems Relying on Regular Quadruplets

If boundary conditions involve *all* endogenous boundary values, then things become more complicated, and unconditional solvability may fail.

Let $\mathbb{K}$ be equal to $\mathbb{R}$ or $\mathbb{C}$, $(\mathcal{V}, \|\cdot\|)$ be a reflexive Banach space over $\mathbb{K}$, and $\mathcal{Q} := (\mathbb{A}, A_{\text{M}}, B, S)$ be a differential quadruplet on $(\mathcal{V}, \mathcal{V}^*)$. We assume that $\mathcal{Q}$ is F_0-regular. Then setting

$$E_{\text{M}} := \ker A_{\text{M}} \oplus BS(F_0),$$

the Domain Structure Theorem 5.3.3 gives that

$$D(A_{\text{M}}) = E_{\text{M}} \oplus BS(\ker \mathbb{A}^{\perp}).$$

7.2.6.1 Formulation of the Problem

In view of Definition 5.3.3, we consider *the projection of* $D(A_{\text{M}})$ *induced by* F_0, denoted by $P_{\text{b.v}}$. That is, $P_{\text{b.v}}$ is the projection on E_{M} along $BS(\ker \mathbb{A}^{\perp})$.

Let $(\mathcal{X}_1, \|\cdot\|_{\mathcal{X}_1})$ be a normed space over $\mathbb{K}$ and T belong to $\mathcal{L}\big((E_{\text{M}}, \|\cdot\|_{D(A_{\text{M}})}), \mathcal{X}_1\big)$. Also we consider L in $\mathcal{L}(\mathcal{V})$ and datum $(h_{\text{b.c}}, h)$ in $\mathcal{X}_1 \times \mathcal{V}$. These preliminaries allow to introduce the following *inhomogeneous linear endogenous boundary value problem*:

$$\begin{cases} \text{Find } u \text{ in } D(A_{\text{M}}) \text{ such that} \\ A_{\text{M}} u = Lu + h, \qquad T P_{\text{b.v}} u = h_{\text{b.c}}. \end{cases} \tag{7.2.48}$$

7.2.6.2 Solvability of (7.2.48)

Proposition 7.2.19 *Under the assumptions and notation introduced in the current subsection, suppose in addition that B satisfies* (7.2.1) *and commute with L.*

(i) *If*

$$T\,P_{\mathrm{b.v}} B\,\mathsf{R}_B^L h \in h_{\mathrm{b.c}} + T\,P_{\mathrm{b.v}}\,\mathsf{R}_B^L(\ker A_{\mathrm{M}}) \tag{7.2.49}$$

then there exists u_0 *in* $\ker A_{\mathrm{M}}$ *such that*

$$u_{\mathrm{p}} := \mathsf{R}_B^L u_0 + B\,\mathsf{R}_B^L h \tag{7.2.50}$$

solves (7.2.48). *Moreover, a vector* u *is solution to* (7.2.48) *if and only if* u *reads*

$$u = u_{\mathrm{p}} + \mathsf{R}_B^L v_0$$

for some v_0 *in* $\ker A_{\mathrm{M}} \cap \ker T\,P_{\mathrm{b.v}}\,\mathsf{R}_B^L$.

(ii) *If* (7.2.49) *does not hold, then* (7.2.48) *has no solution.*

In Item (i), the equivalence may be reformulated in the following fashion. Consider the homogeneous problem associated with (7.2.48), namely,

$$\begin{cases} \text{Find } v \text{ in } D(A_{\mathrm{M}}) \text{ such that} \\ A_{\mathrm{M}} v = L v, \qquad T\,P_{\mathrm{b.v}} v = 0. \end{cases} \tag{7.2.51}$$

Then being given any solution u_{p} of (7.2.48), *the solution set* of (7.2.48) reads

$$\{u_{\mathrm{p}} + v \mid v \text{ is solution to (7.2.51)}\}.$$

This results is well-known in the case where A_{M} is a first-order derivative.

Proof of Proposition 7.2.19 (i) Assuming (7.2.49), there exists u_0 in $\ker A_{\mathrm{M}}$ such that

$$T\,P_{\mathrm{b.v}}(\mathsf{R}_B^L u_0 + B\,\mathsf{R}_B^L h) = h_{\mathrm{b.c}}.$$

Thus u_{p} satisfies $T\,P_{\mathrm{b.v}} u_{\mathrm{p}} = h_{\mathrm{b.c}}$. Besides, by Proposition 7.2.2, $A_{\mathrm{M}} u_{\mathrm{p}} = L u_{\mathrm{p}} + h$. Hence u_{p} solves (7.2.48).

Next, let u be any solution to (7.2.48). Then $v := u - u_{\mathrm{p}}$ is solution to (7.2.51). Then Proposition 7.2.2 yields that there exists v_0 in $\ker A_{\mathrm{M}}$ such that

$$v = \mathsf{R}_B^L v_0, \qquad T\,P_{\mathrm{b.v}}\mathsf{R}_B^L v_0 = 0.$$

Thus $u = u_{\mathrm{p}} + \mathsf{R}_B^L v_0$ for some v_0 in $\ker A_{\mathrm{M}} \cap \ker T\,P_{\mathrm{b.v}}\,\mathsf{R}_B^L$. Conversely, let v_0 belong to $\ker A_{\mathrm{M}} \cap \ker T\,P_{\mathrm{b.v}}\,\mathsf{R}_B^L$. The vector $u := u_{\mathrm{p}} + \mathsf{R}_B^L v_0$ satisfies $A_{\mathrm{M}} u = L u + h$. Also since $T\,P_{\mathrm{b.v}}\mathsf{R}_B^L v_0 = 0$ and u_{p} is a solution to (7.2.48), we have $T\,P_{\mathrm{b.v}} u = h_{\mathrm{b.c}}$. Hence u solves (7.2.48), which completes the proof of Item (i).

Finally we will prove the contrapositive of (ii). If u solves (7.2.48) then Proposition 7.2.2 yields the existence of some u_0 in $\ker A_{\mathrm{M}}$ such that

$$u = \mathsf{R}_B^L u_0 + B\,\mathsf{R}_B^L h.$$

With the boundary condition of (7.2.48), we get (7.2.49). □

7.2.6.3 Unconditionally Solvability and Well Posedness

Definition 7.2.9 Problem (7.2.48) is said to be *unconditionally solvable* if (7.2.48) has at least a solution for each $(h_{\mathrm{b.c}}, h)$ in $R(T) \times \mathcal{V}$.

Besides (7.2.48) is said to be *well posed* if for each $(h_{\mathrm{b.c}}, h)$ in $R(T) \times \mathcal{V}$, the following two conditions are satisfied:

(i) (7.2.48) has a unique solution u.
(ii) u depends continuously on the data, that is, u satisfies the estimate

$$\|u\| \le C(\|h\| + \|h_{\mathrm{b.c}}\|_{\mathcal{X}_1}),$$

where the constant C is independent of u, $h_{\mathrm{b.c}}$, and h.

□

In Theorem 7.2.21, we will state a necessary and sufficient condition for (7.2.48) to be unconditionally solvable. Its proof relies on the following result.

Lemma 7.2.20 *Under the assumptions and notation of Proposition 7.2.19, we have*

$$P_{\mathrm{b.v}} B\,\mathsf{R}_B^L(\mathcal{V}) = BS(F_0).$$

Moreover, $BS(F_0)$ *and* $P_{\mathrm{b.v}}\,\mathsf{R}_B^L(\ker A_{\mathrm{M}})$ *are closed complementary subspaces of* $(E_{\mathrm{M}}, \|\cdot\|_{D(A_{\mathrm{M}})})$.

Proof First of all, notice that

$$B\,\mathsf{R}_B^L(\mathcal{V}) = R(B) \tag{7.2.52}$$

since by Proposition 7.2.1, R_B^L is an isomorphism from $\mathcal{V}$ onto itself. Using also $P_{\mathrm{b.v}} R(B) = BS(F_0)$, we derive that $P_{\mathrm{b.v}} B\,\mathsf{R}_B^L(\mathcal{V}) = BS(F_0)$.

With Lemma 4.3.15, it is now a routine to prove that $BS(F_0)$ is closed in $(E_{\mathrm{M}}, \|\cdot\|_{D(A_{\mathrm{M}})})$. Besides $\mathsf{R}_B^L(\ker A_{\mathrm{M}})$ is closed in $\mathcal{V}$ because R_B^L is an isomorphism and A_{M} is a closed operator on $\mathcal{V}$. Since $P_{\mathrm{b.v}}$ is a projection of $D(A_{\mathrm{M}})$, we derive easily that $P_{\mathrm{b.v}}\,\mathsf{R}_B^L(\ker A_{\mathrm{M}})$ is closed in $D(A_{\mathrm{M}})$.

In order to show that the sum is direct, we claim in a first step that each u_0 in $\ker A_{\mathrm{M}}$ satisfies

$$P_{\mathrm{b.v}}\, \mathsf{R}_B^L u_0 = u_0 + BS\varphi_0, \tag{7.2.53}$$

for some φ_0 in F_0. Indeed $\mathcal{V} = F_0 \oplus (\ker \mathbb{A}^{\perp})$ thus there exists $(\varphi_0, \psi^{\perp})$ in $F_0 \times (\ker \mathbb{A}^{\perp})$ such that

$$SL\, \mathsf{R}_B^L u_0 = \varphi_0 + \psi^{\perp}.$$

Then Proposition 7.2.1 (i) yields that

$$\mathsf{R}_B^L u_0 = u_0 + BS\varphi_0 + BS\psi^{\perp}.$$

Hence (7.2.53) follows. Now any φ belonging to

$$BS(F_0) \cap P_{\mathrm{b.v}}\, \mathsf{R}_B^L(\ker A_{\mathrm{M}})$$

reads

$$\varphi = BSf_0 = P_{\mathrm{b.v}}\, \mathsf{R}_B^L u_0$$

for some (u_0, f_0) in $\ker A_{\mathrm{M}} \times F_0$. The previous claim entails that there exists φ_0 in F_0 satisfying (7.2.53). Since $\ker A_{\mathrm{M}}$ and $R(B)$ are in direct sum, we infer that $u_0 = 0$, so that $\varphi = 0$ and the sum is direct.

There remains to show that

$$BS(F_0) + P_{\mathrm{b.v}}\, \mathsf{R}_B^L(\ker A_{\mathrm{M}}) = E_{\mathrm{M}}. \tag{7.2.54}$$

For, any e in E_{M} reads $e = u_0 + BSf_0$ for some (u_0, f_0) in $\ker A_{\mathrm{M}} \times F_0$. With (7.2.53),

$$e = (u_0 + BS\varphi_0) + BS(f_0 - \varphi_0) \in P_{\mathrm{b.v}}\, \mathsf{R}_B^L(\ker A_{\mathrm{M}}) + BS(F_0).$$

Hence (7.2.54) follows. □

Theorem 7.2.21 *Let $\mathcal{V}$ be a reflexive Banach space, $\mathcal{Q} := (\mathbb{A}, A_{\mathrm{M}}, B, S)$ be a F_0-regular differential quadruplet on $(\mathcal{V}, \mathcal{V}^*)$, $\mathcal{X}_1$ be a normed space, and T belong to $\mathcal{L}\big((E_{\mathrm{M}}, \|\cdot\|_{D(A_{\mathrm{M}})}), \mathcal{X}_1\big)$. Assume* (7.2.1) *and* (7.2.2)*. Then the following assertions hold true:*

(i) (7.2.48) *is unconditionally solvable if and only if*

$$TBS(F_0) \subseteq T P_{\mathrm{b.v}}\, \mathsf{R}_B^L(\ker A_{\mathrm{M}}). \tag{7.2.55}$$

(ii) Uniqueness. *Let* $(h_{\text{b.c}}, h)$ *belong to* $\mathcal{X}_1 \times \mathcal{V}$. *If*

$$P_{\text{b.v}}\,\mathsf{R}_B^L(\ker A_{\text{M}}) \cap \ker T = \{0\}, \tag{7.2.56}$$

then (7.2.48) *has at most a solution.*

Proof

(i) By Lemma 7.2.20, $P_{\text{b.v}} B\,\mathsf{R}_B^L(\mathcal{V}) = BS(F_0)$. Thus Proposition 7.2.19 yields that (7.2.48) is unconditionally solvable if and only if

$$TBS(F_0) \subseteq h_{\text{b.c}} + T P_{\text{b.v}}\,\mathsf{R}_B^L(\ker A_{\text{M}}), \qquad \forall\, h_{\text{b.c}} \in R(T). \tag{7.2.57}$$

Now, it is enough to prove that (7.2.55) and (7.2.57) are equivalent. For, assume in a first step that (7.2.57) holds. By choosing $h_{\text{b.c}} = 0$ in (7.2.57), we get (7.2.55). Conversely assume (7.2.55), and consider any element $h_{\text{b.c}}$ of $R(T)$. Then there exists $e_{\text{b.c}}$ in E_{M} such that $Te_{\text{b.c}} = h_{\text{b.c}}$. By Lemma 7.2.20, $e_{\text{b.c}}$ reads

$$e_{\text{b.c}} = BS\varphi_0 + P_{\text{b.v}}\,\mathsf{R}_B^L v_0$$

for some (φ_0, v_0) in $F_0 \times \ker A_{\text{M}}$. Thus

$$h_{\text{b.c}} = TBS\varphi_0 + T P_{\text{b.v}}\,\mathsf{R}_B^L v_0. \tag{7.2.58}$$

By (7.2.55), for each f in F_0, there exists w_0 in $\ker A_{\text{M}}$ such that

$$TBS(f - \varphi_0) = T P_{\text{b.v}}\,\mathsf{R}_B^L w_0.$$

Setting $u_0 := w_0 - v_0$, we deduce with (7.2.58) that

$$TBSf = h_{\text{b.c}} + T P_{\text{b.v}}\,\mathsf{R}_B^L u_0.$$

Thus (7.2.57) is fulfilled.

(ii) By linearity, it is enough to show that any solution u to

$$A_{\text{M}} u = Lu, \qquad T P_{\text{b.v}} u = 0$$

must be trivial. For, by (7.2.4), u reads $u = \mathsf{R}_B^L u_0$ for some u_0 in $\ker A_{\text{M}}$. Thus by the hypothesis (7.2.56), $P_{\text{b.v}}\mathsf{R}_B^L u_0 = 0$. Then using (7.2.53) together with a direct sum argument, we end up with $u_0 = 0$, so that u is trivial. □

Corollary 7.2.22 *Let $\mathcal{V}$ be a reflexive Banach space, $\mathcal{Q} := (\mathbb{A}, A_M, B, S)$ be a F_0-regular differential quadruplet on $(\mathcal{V}, \mathcal{V}^*)$, $(\mathcal{X}_1, \|\cdot\|_{\mathcal{X}_1})$ be a Banach space, and T belong to $\mathcal{L}\big((E_M, \|\cdot\|_{D(A_M)}), \mathcal{X}_1\big)$. Assume in addition that:*

(i) (7.2.1), (7.2.2), (7.2.55) *and* (7.2.56) *hold;*
(ii) $T P_{b.v} \mathsf{R}_B^L(\ker A_M)$ *is closed in* $\mathcal{X}_1$.

Then (7.2.48) *is well posed in the sense of Definition 7.2.9.*

Proof By Theorem 7.2.21, (7.2.48) has a unique solution u for each $(h_{b.c}, h)$ in $R(T) \times \mathcal{V}$. Moreover, Proposition 7.2.19 yields that

$$u := \mathsf{R}_B^L u_0 + B\,\mathsf{R}_B^L h$$

for some u_0 in $\ker A_M$. From the second equation of (7.2.48), we derive

$$T P_{b.v} \mathsf{R}_B^L u_0 = h_{b.c} - T P_{b.v} B\,\mathsf{R}_B^L h.$$

By Assumption (ii),

$$T P_{b.v} \mathsf{R}_{B\,|\ker A_M}^L : \ker A_M \to T P_{b.v} \mathsf{R}_B^L(\ker A_M)$$

is a continuous linear map between the Banach spaces

$$(\ker A_M, \|\cdot\|_{D(A_M)}), \ \ (T P_{b.v} \mathsf{R}_B^L(\ker A_M), \|\cdot\|_{\mathcal{X}_1}).$$

Moreover (7.2.56) entails that $T P_{b.v} \mathsf{R}_{B\,|\ker A_M}^L$ is one-to-one. Hence by Corollary 2.4.9, there results that $T P_{b.v} \mathsf{R}_{B\,|\ker A_M}^L$ is an isomorphism between the pre-cited Banach spaces. Since R_B^L lies in $\mathcal{L}(\mathcal{V})$, we get the well posedness of (7.2.48). □

7.2.6.4 Maximal Operators with Finite-Dimensional Kernel

Let $\mathcal{V}$ be a reflexive Banach space and $\mathcal{Q} := (\mathbb{A}, A_M, B, S)$ be a differential quadruplet on $(\mathcal{V}, \mathcal{V}^*)$ whose maximal operators $\mathbb{A}$, A_M have finite-dimensional kernel. Arguing as in Sect. 5.5.1, we get that $\mathcal{Q}$ is F_0-regular where F_0 has the dimension of $\ker \mathbb{A}$. Then (5.5.2) yields

$$D(A_M) = E_M \oplus BS(\ker \mathbb{A}^{\perp})$$

where the finite-dimensional space E_M is defined through (5.5.3), i.e.,

$$E_M := \ker A_M \oplus BS(F_0).$$

With these notation in mind, we may state the following result.

Corollary 7.2.23 *Let $\mathcal{V}$ be a reflexive Banach space and $\mathcal{Q} := (\mathbb{A}, A_{\mathrm{M}}, B, S)$ be a differential quadruplet on $(\mathcal{V}, \mathcal{V}^*)$ whose maximal operators $\mathbb{A}$, A_{M} have finite-dimensional kernel. Let $\mathcal{X}_1$ be a normed space and T be a linear map from E_{M} into $\mathcal{X}_1$. Assume* (7.2.1), (7.2.2) *and that the restriction of $T\,P_{\mathrm{b.v}}\,\mathsf{R}_B^L$ to* $\ker A_{\mathrm{M}}$ *is an isomorphism from* $\ker A_{\mathrm{M}}$ *onto $\mathcal{X}_1$. Then* (7.2.48) *is well posed.*

Proof Since the restriction of $T\,P_{\mathrm{b.v}}\,\mathsf{R}_B^L$ to $\ker A_{\mathrm{M}}$ is an isomorphism onto $\mathcal{X}_1$, we deduce on the one hand that $T\,P_{\mathrm{b.v}}\,\mathsf{R}_B^L(\ker A_{\mathrm{M}}) = \mathcal{X}_1$. Thus (7.2.55) holds. On the other hand, (7.2.56) holds as well. We conclude with Corollary 7.2.22. □

7.3 Initial Value Problems I: Linear Equations

In this section we will apply the results of Sect. 7.2.3 to entire and fractional differential equations supplemented with initial conditions.

7.3.1 The Framework

As far as initial value problems are considered, reflexivity is no more needed. Let $\mathbb{K}$ be equal to $\mathbb{R}$ or $\mathbb{C}$, $\mathcal{Y}$ be a nontrivial Banach space over $\mathbb{K}$, and b be a positive real number. For any real number p in $[1, \infty)$, we set

$$\mathcal{X} = L^p(\mathcal{Y}) := L^p\big((0, b), \mathcal{Y}\big).$$

Denoting by q the conjugate exponent of p, notice that q lies in $(1, \infty]$.

We will study special instances of (7.2.10) where A is one of the closed fractional differential operators introduced in Sect. 6.4. The second equation of (7.2.10) will be reformulated as initial conditions involving "standard" boundary values.

In any cases, A admits a right inverse of the form $B_{f,\mathcal{X}}$ for some f in $L^1(0, b)$. Let L be in $\mathcal{L}(\mathcal{Y})$. In view of Sect. 7.2.1.2, the continuous and linear map from $L^p(\mathcal{Y})$ into itself induced by L is still denoted by L. Again by Sect. 7.2.1.2, we know that L commute with $B_{f,\mathcal{X}}$ and that the fundamental assumption (7.2.1) holds.

7.3.2 Problems with $A := \mathbf{D}^1_{\mathcal{X}}$

We will study first-order differential systems supplemented with boundary conditions.

7.3.2.1 Formulation of the Problem

Let L be in $\mathcal{L}(\mathcal{Y})$ and $h_{\text{b.c}}$, h be data in $\mathcal{Y}$ and $L^p(\mathcal{Y})$, respectively. Then we consider the following problem:

$$\begin{cases} \text{Find } u \text{ in } W^{1,p}(\mathcal{Y}) \text{ such that} \\ \mathrm{D}^1_{\mathcal{X}} u = Lu + h, \qquad u(0) = h_{\text{b.c}}. \end{cases} \tag{7.3.1}$$

For all purposes, recall that in (7.3.1), L is considered as an element of $\mathcal{L}\big(L^p(\mathcal{Y})\big)$.

Let us recast the initial condition $u(0) = h_{\text{b.c}}$ into the form $P_{\ker \mathrm{D}^1_{\mathcal{X}}} u = h_{\text{b.c}}$ appearing in (7.2.10). For, since $\mathrm{D}^1_{\mathcal{X}}$ is closed (see Example 5.6.7) and $\mathrm{D}^1_{\mathcal{X}} B_{1,\mathcal{X}} = \mathrm{id}_{\mathcal{X}}$, we may introduce on account of Sect. 7.2.3.1 the projection

$$P_{\ker \mathrm{D}^1_{\mathcal{X}}} : D(\mathrm{D}^1_{\mathcal{X}}) \to D(\mathrm{D}^1_{\mathcal{X}})$$

on $\ker \mathrm{D}^1_{\mathcal{X}}$ along $R(B_{1,\mathcal{X}})$. Then the initial condition reads

$$P_{\ker \mathrm{D}^1_{\mathcal{X}}} u = g_1 h_{\text{b.c}}.$$

Thus (7.3.1) *is an instance of* (7.2.10).

7.3.2.2 Well Posedness

Since (7.3.1) is an instance of (7.2.10), there results from Definition 7.2.6 that Problem (7.3.1) is *well posed* if for each $(h_{\text{b.c}}, h)$ in $\mathcal{Y} \times L^p(\mathcal{Y})$, (7.3.1) has a unique solution u and u satisfies the estimate

$$\|u\|_{L^p(\mathcal{Y})} \le C(\|h\|_{L^p(\mathcal{Y})} + \|h_{\text{b.c}}\|_{\mathcal{Y}}),$$

where the constant C is independent of u, $h_{\text{b.c}}$, and h.

On account of Sect. 7.3.2.1 and Lemma 7.2.4, Proposition 7.2.9 applies, so that (7.3.1) is well posed, and its solution u reads

$$u = \mathsf{R}^L_{B_{1,\mathcal{X}}}(g_1 h_{\text{b.c}}) + B_{1,\mathcal{X}} \mathsf{R}^L_{B_{1,\mathcal{X}}} h.$$

Notice that by definition of the resolvent map $\mathsf{R}^L_{B_{1,\mathcal{X}}} : \mathcal{V} \to \mathcal{V}$, one has

$$\mathsf{R}^L_{B_{1,\mathcal{X}}} h = \sum_{j \ge 0} L^j g_j * h,$$

where for each positive index j, g_j is the Riemann-Liouville kernel of order j defined by (3.4.3), and $g_0 * h = h$ by convention.

With Proposition 7.2.7, we get

$$u = E^L_{g_1} h_{\text{b.c}} + E^L_{g_1} * h. \tag{7.3.2}$$

Let $e^{\cdot L} : (0, b) \to \mathcal{L}(L^p(\mathcal{Y}))$ be the standard exponential function defined by

$$e^{xL} := \sum_{j \geq 0} \frac{x^j}{j!} L^j, \qquad \forall x \in (0, b).$$

By (7.2.8), $E^L_{g_1} = e^{\cdot L}$. Whence (7.3.2) *is nothing but the usual variation of constants formula*

$$u = e^{\cdot L} h_{\text{b.c}} + e^{\cdot L} * h.$$

7.3.3 Problems with $A := {}^{\mathrm{RL}}\mathrm{D}^{\ell}_{\mathcal{X}}$

The assumptions and notation of Sect. 7.3.1 remain in force. Being given a Sonine pair (k, ℓ), recall that in view of Sect. 6.4,

$${}^{\mathrm{RL}}\mathrm{D}^{\ell}_{\mathcal{X}} u := \frac{\mathrm{d}}{\mathrm{d}x}\{k * u\}, \qquad \forall u \in D({}^{\mathrm{RL}}\mathrm{D}^{\ell}_{\mathcal{X}}).$$

By (6.4.1), ${}^{\mathrm{RL}}\mathrm{D}^{\ell}_{\mathcal{X}} B_{\ell,\mathcal{X}} = \mathrm{id}_{\mathcal{X}}$. Moreover,

$$\ker {}^{\mathrm{RL}}\mathrm{D}^{\ell}_{\mathcal{X}} = \begin{cases} \ell\mathcal{Y} & \text{if } \ell \in L^p(0, b) \\ \{0\} & \text{otherwise.} \end{cases}$$

We claim that the kernel of ${}^{\mathrm{RL}}\mathrm{D}^{\ell}_{\mathcal{X}}$ is closed in $L^p(\mathcal{Y})$. Indeed, by Corollary 2.2.7, there exists f in $L^q(0, b)$ such that $\int_0^b f\bar{h}\,\mathrm{d}y = 1$. Thus $\ell\mathcal{Y}$ is closed in $L^p(\mathcal{Y})$ according to Lemma 3.2.6, and the claim follows. Now combing Lemma 4.3.3 together with Proposition 4.4.1, we deduce that ${}^{\mathrm{RL}}\mathrm{D}^{\ell}_{\mathcal{X}}$ is a closed operator in $L^p(\mathcal{Y})$.

7.3.3.1 Case Where ℓ Lies in $L^p(0, b)$

7.3.3.1.1 The Framework

By Lemma 4.3.3, any u in $D({}^{\mathrm{RL}}\mathrm{D}^{\ell}_{\mathcal{X}})$ reads

$$u = \ell U_1 + \ell * {}^{\mathrm{RL}}\mathrm{D}^{\ell}_{\mathcal{X}} u$$

for some unique U_1 in $\mathcal{Y}$. Arguing as in Sect. 6.5.2.1, we get

$$U_1 = k * u(0). \tag{7.3.3}$$

7.3.3.1.2 Formulation of the Problem

Let L be in $\mathcal{L}(\mathcal{Y})$, $h_{\mathrm{b.c}}$ belong to $\mathcal{Y}$, and h be in $L^p\big((0,b),\mathcal{Y}\big)$. Then we consider the following problem:

$$\begin{cases} \text{Find } u \text{ in } D({}^{\mathrm{RL}}\mathrm{D}^{\ell}_{\mathcal{X}}) \text{ such that} \\ {}^{\mathrm{RL}}\mathrm{D}^{\ell}_{\mathcal{X}} u = Lu + h, \qquad k * u(0) = h_{\mathrm{b.c}}. \end{cases} \tag{7.3.4}$$

We claim that (7.3.4) *is an instance of* (7.2.10). Indeed, let us first recast the initial conditions into the form $P_{\ker {}^{\mathrm{RL}}\mathrm{D}^{\ell}_{\mathcal{X}}} u = h_{\mathrm{b.c}}$ appearing in (7.2.10). For, since ${}^{\mathrm{RL}}\mathrm{D}^{\ell}_{\mathcal{X}}$ is a closed operator and ${}^{\mathrm{RL}}\mathrm{D}^{\ell}_{\mathcal{X}} B_{\ell,\mathcal{X}} = \mathrm{id}_{\mathcal{X}}$, we may introduce on account of Sect. 7.2.3.1 the projection

$$P_{\ker {}^{\mathrm{RL}}\mathrm{D}^{\ell}_{\mathcal{X}}} : D({}^{\mathrm{RL}}\mathrm{D}^{\ell}_{\mathcal{X}}) \to D({}^{\mathrm{RL}}\mathrm{D}^{\ell}_{\mathcal{X}})$$

on $\ker {}^{\mathrm{RL}}\mathrm{D}^{\ell}_{\mathcal{X}}$ along $R(B_{\ell,\mathcal{X}})$. Then with (7.3.3), the initial condition reads

$$P_{\ker {}^{\mathrm{RL}}\mathrm{D}^{\ell}_{\mathcal{X}}} u = \ell h_{\mathrm{b.c}}.$$

Secondly, L may be identified with an element of $\mathcal{L}(\mathcal{X})$ as explained in Sect. 7.3.1.

7.3.3.1.3 Well Posedness

In the current context, Definition 7.2.6 takes the following form.

Definition 7.3.1 Problem (7.3.4) is *well posed* if for each $(h_{\mathrm{b.c}}, h)$ in $\mathcal{Y} \times L^p(\mathcal{Y})$, the following two conditions are satisfied:

(i) (7.3.4) has a unique solution u.
(ii) u satisfies the estimate

$$\|u\|_{L^p(\mathcal{Y})} \le C(\|h\|_{L^p(\mathcal{Y})} + \|h_{\mathrm{b.c}}\|_{\mathcal{Y}}),$$

where the constant C is independent of u, $h_{\mathrm{b.c}}$, and h.

□

We have already seen that $B_{\ell,\mathcal{X}}$ is a right inverse of ${}^{\mathrm{RL}}\mathrm{D}^{\ell}_{\mathcal{X}}$, $B_{\ell,\mathcal{X}}$ commute with L, and (7.2.1) holds. Whence Proposition 7.2.9 entails that (7.3.4) *is well posed and that its solution u reads*

$$u = \mathrm{R}^L_{B_{\ell,\mathcal{X}}}(\ell\, h_{\text{b.c}}) + B_{\ell,\mathcal{X}}\, \mathrm{R}^L_{B_{\ell,\mathcal{X}}} h.$$

With Proposition 7.2.7,

$$u = E^L_\ell h_{\text{b.c}} + E^L_\ell * h, \tag{7.3.5}$$

where $E^L_\ell = \sum_{j\geq 0} \ell_{j+1}\, L^j$ and $\ell_{j+1} := B_{\ell,L^p}{}^j \ell$. Plainly (7.3.5) is analog to (7.3.2).

7.3.3.1.4 The Riemann-Liouville Derivative

Because it is widely used, we will reformulate Problem (7.3.4) for Riemann-Liouville derivatives. That is, for parameters α and p ranging in

$$\alpha \in (0,1), \qquad p \in [1,\infty) \quad \text{and} \quad 1 - \frac{1}{p} < \alpha, \tag{7.3.6}$$

we consider the particular case where

$$(k,\ell) := (g_{1-\alpha}, g_\alpha).$$

Then g_α lies in $L^p(0,b)$ and ${}^{\mathrm{RL}}\mathrm{D}^\ell_{\mathcal{X}}$ is the *Riemann-Liouville operator of order* α, that is, ${}^{\mathrm{RL}}\mathrm{D}^\ell_{\mathcal{X}} = {}^{\mathrm{RL}}\mathrm{D}^\alpha_{\mathcal{X}}$. Hence for any $(h_{\text{b.c}}, h)$ in $\mathcal{Y} \times L^p\big((0,b),\mathcal{Y}\big)$, (7.3.4) becomes

$$\begin{cases} \text{Find } u \text{ in } D({}^{\mathrm{RL}}\mathrm{D}^\alpha_{\mathcal{X}}) \text{ such that} \\ {}^{\mathrm{RL}}\mathrm{D}^\alpha_{\mathcal{X}} u = Lu + h, \qquad g_{1-\alpha} * u(0) = h_{\text{b.c}}. \end{cases} \tag{7.3.7}$$

According to Sect. 7.3.3.1.3, (7.3.7) *is well posed* in the sense of Definition 7.3.1 and that its solution u reads (see (7.3.5))

$$u = E^L_{g_\alpha} h_{\text{b.c}} + E^L_{g_\alpha} * h.$$

This representation of the solution to (7.3.4) is the complete analog of (7.1.4). Then it proceeds to the unification of the theories of *entire* and *fractional* differential equations.

Recalling that $E^L_{g_\alpha}$ belongs to $L^p\big((0,b),\mathcal{L}(\mathcal{Y})\big)$, the *Delsarte representation of* $E^L_{g_\alpha}$ is

$$E^L_{g_\alpha} = \sum_{j\geq 0} g_{(j+1)\alpha}\, L^j. \tag{7.3.8}$$

In the particular case where $\mathcal{Y} := \mathbb{K}$, $\mathcal{L}(\mathcal{Y})$ is identified with $\mathbb{K}$. Thus for each scalar λ, $E^\lambda_{g_\alpha}$ lies in $L^p(0,b)$, and the representation of $E^\lambda_{g_\alpha}$ in terms of the Mittag-Leffler function $E_{\alpha,\alpha}$ introduced in Sect. 3.4.3 reads

$$E^{\lambda}_{g_\alpha} = \Gamma(\alpha) g_\alpha\, E_{\alpha,\alpha}\big(\lambda\Gamma(\alpha+1) g_{\alpha+1}\big) \qquad \text{in } L^p(0,b).$$

Hence for almost every x in $(0,b)$, one has

$$E^{\lambda}_{g_\alpha}(x) = x^{\alpha-1} E_{\alpha,\alpha}\big(x^\alpha \lambda\big).$$

7.3.3.2 Case Where ℓ Lies in $L^1(0,b) \setminus L^p(0,b)$

7.3.3.2.1 Formulation of the Problem and Well Posedness

In this case the kernel of ${}^{\mathrm{RL}}\mathrm{D}^{\ell}_{\mathcal{X}}$ is trivial; thus (7.2.10) becomes

$$\begin{cases} \text{Find } u \text{ in } D({}^{\mathrm{RL}}\mathrm{D}^{\ell}_{\mathcal{X}}) \text{ such that} \\ {}^{\mathrm{RL}}\mathrm{D}^{\ell}_{\mathcal{X}} u = Lu + h, \qquad 0 = h_{\mathrm{b.c}}. \end{cases} \tag{7.3.9}$$

Then Proposition 7.2.9 entails that:

(i) if $h_{\mathrm{b.c}} \neq 0$ then (7.3.9) has no solution;
(ii) if $h_{\mathrm{b.c}} = 0$ then by Proposition 7.2.9, (7.3.9) is well posed in a sense which can be easily inferred from Definition 7.3.1. Besides, using also Proposition 7.2.7 (ii), we get the following representation of the solution u to (7.3.9):

$$u = E^L_\ell * h,$$

where $E^L_\ell = \sum_{j\geq 0} \ell_{j+1} L^j$ lies in $L^1\big((0,b), \mathcal{L}(\mathcal{Y})\big)$.

Remark 7.3.2 In the current case where ℓ lies in $L^1(0,b) \setminus L^p(0,b)$, the kernel of ${}^{\mathrm{RL}}\mathrm{D}^{\ell}_{\mathcal{V}}$ is trivial. Hence, Lemma 4.3.3 yields that any u in $D({}^{\mathrm{RL}}\mathrm{D}^{\ell}_{\mathcal{V}})$ fulfills $u = \ell * {}^{\mathrm{RL}}\mathrm{D}^{\ell}_{\mathcal{V}} u$. Hence by (6.3.1),

$$k * u(0) = g_1 * {}^{\mathrm{RL}}\mathrm{D}^{\ell}_{\mathcal{V}} u(0) = 0.$$

Thus (7.3.9) is equivalent to the following problem.

$$\begin{cases} \text{Find } u \text{ in } D({}^{\mathrm{RL}}\mathrm{D}^{\ell}_{\mathcal{V}}) \text{ such that} \\ {}^{\mathrm{RL}}\mathrm{D}^{\ell}_{\mathcal{V}} u = Lu + h, \qquad k * u(0) = h_{\mathrm{b.c}}. \end{cases} \tag{7.3.10}$$

□

7.3.3.2.2 The Riemann-Liouville Derivative

For parameters α and p ranging in

$$\alpha \in (0,1), \qquad p \in (1,\infty) \quad \text{and} \quad \alpha \leq 1 - \frac{1}{p}, \tag{7.3.11}$$

we consider the particular case where

$$(k, \ell) := (g_{1-\alpha}, g_\alpha).$$

For each $(h_{\text{b.c}}, h)$ in $\mathcal{Y} \times L^p(\mathcal{Y})$, we consider the following problem:

$$\begin{cases} \text{Find } u \text{ in } D({}^{\text{RL}}\text{D}^\alpha_{\mathcal{Y}}) \text{ such that} \\ {}^{\text{RL}}\text{D}^\alpha_{\mathcal{Y}} u = Lu + h, \qquad g_{1-\alpha} * u(0) = h_{\text{b.c}}. \end{cases} \tag{7.3.12}$$

Since g_α lies in $L^1(0,b) \setminus L^p(0,b)$, we have by Sect. 7.3.3.2.1:

(i) if $h_{\text{b.c}} \neq 0$ then (7.3.12) has no solution;
(ii) if $h_{\text{b.c}} = 0$ then (7.3.12) is well posed. Its solution u reads $u = E^L_{g_\alpha} * h$ where the generalized exponential function $E^L_{g_\alpha}$ fulfills (7.3.8). Besides, if $\mathcal{Y} := \mathbb{K}$, then L identifies with some scalar λ, and

$$u = \Gamma(\alpha) g_\alpha \, E_{\alpha,\alpha}\big(\lambda \Gamma(\alpha+1) g_{\alpha+1}\big) * h.$$

7.3.4 Problems with $A := \text{D}^\ell_{\mathcal{X}}$

Recall that $\text{D}^\ell_{\mathcal{X}}$ is the Caputo extension of ${}^{\text{RL}}\text{D}^\ell_{\mathcal{X}}$ defined in Sect. 6.4. With the assumptions and notation of Sect. 7.3.1 remaining in force, we claim that the kernel of $\text{D}^\ell_{\mathcal{X}}$ is closed in $L^p(\mathcal{Y})$. Indeed, in view of (6.4.2) and Proposition 5.6.9, a nontrivial kernel of $\text{D}^\ell_{\mathcal{X}}$ reads

$$\bigoplus_{i=1}^{m} h_i \mathcal{Y}$$

where m lies in $\{1, 2\}$ and the h_i's belong to $L^p(0,b) \setminus \{0\}$. The closedness of $\ker \text{D}^\ell_{\mathcal{X}}$ is already known if $m = 1$. If $m = 2$ then Proposition 3.2.5 (iii) yields the existence of f_1, f_2 in $L^q(0,b)$ such that the matrix with entries $(\int_0^b f_i \overline{h_j}\, dy)$ is invertible. The claim follows from Lemma 3.2.6. Now combing Lemma 4.3.3 together with Proposition 4.4.1, we deduce that $\text{D}^\ell_{\mathcal{X}}$ is a closed operator in $L^p(\mathcal{Y})$.

If k lies in $L^p(0,b)$, then $\text{D}^\ell_{\mathcal{X}} = {}^{\text{RL}}\text{D}^\ell_{\mathcal{X}}$, and we refer to Sect. 7.3.3. Hence in this subsection, we assume that k does not belong to $L^p(0,b)$.

7.3.4.1 Case Where ℓ Lies in $L^p(0, b) \cap L^q(0, b)$

7.3.4.1.1 Preliminaries

Since

$$\ker \mathrm{D}^{\ell}_{\mathcal{X}} = g_1\, \mathcal{Y} \oplus \ell\, \mathcal{Y},$$

Lemma 4.3.3 yields that any u in $D(\mathrm{D}^{\ell}_{\mathcal{X}})$ reads

$$u = g_1\, U_1 + \ell\, U_2 + \ell * {}^{\mathrm{RL}}\mathrm{D}^{\ell}_{\mathcal{X}} u$$

for some unique (U_1, U_2) in $\mathcal{Y} \times \mathcal{Y}$. Arguing as in Sect. 6.7.2.1, we find that $u - \ell\, k * u(0)$ is continuous on $[0, b]$ and

$$\begin{aligned} U_1 &= \big(u - \ell\, k * u(0)\big)(0) \\ U_2 &= k * u(0). \end{aligned}$$

7.3.4.1.2 Formulation of the Problem

Let p belong to $[1, \infty)$, L be in $\mathcal{L}(\mathcal{Y})$ and h lie in $L^p\big((0, b), \mathcal{Y}\big)$. Assume in addition that $h_{\mathrm{b.c},1}$ and $h_{\mathrm{b.c},2}$ belong to $\mathcal{Y}$. Then we consider the following problem:

$$\begin{cases} \text{Find } u \text{ in } D(\mathrm{D}^{\ell}_{\mathcal{X}}) \text{ such that} \\ \mathrm{D}^{\ell}_{\mathcal{X}} u = L u + h \\ \big(u - \ell\, k * u(0)\big)(0) = h_{\mathrm{b.c},1}, \qquad k * u(0) = h_{\mathrm{b.c},2}. \end{cases} \tag{7.3.13}$$

Let us recast the initial conditions into the form $P_{\ker \mathrm{D}^{\ell}_{\mathcal{X}}} u = h_{\mathrm{b.c}}$ appearing in (7.2.10). For, $\mathrm{D}^{\ell}_{\mathcal{X}}$ is a closed operator and ${}^{\mathrm{RL}}\mathrm{D}^{\ell}_{\mathcal{X}} \subseteq \mathrm{D}^{\ell}_{\mathcal{X}}$. Thus we derive from (6.4.1) that $B_{\ell,\mathcal{X}}$ is a right inverse of $\mathrm{D}^{\ell}_{\mathcal{X}}$. Then arguing as in Sect. 7.2.3.1, we may introduce the projection

$$P_{\ker \mathrm{D}^{\ell}_{\mathcal{X}}} : D(\mathrm{D}^{\ell}_{\mathcal{X}}) \to D(\mathrm{D}^{\ell}_{\mathcal{X}})$$

on $\ker \mathrm{D}^{\ell}_{\mathcal{X}}$ along $R(B_{\ell,\mathcal{X}})$. Let u belong to the domain of $\mathrm{D}^{\ell}_{\mathcal{X}}$. Since $\ker \mathrm{D}^{\ell}_{\mathcal{X}} = g_1 \mathcal{Y} \oplus \ell \mathcal{Y}$, u satisfies of the initial conditions of (7.3.13) if and only if

$$P_{\ker \mathrm{D}^{\ell}_{\mathcal{X}}} u = g_1\, h_{\mathrm{b.c},1} + \ell\, h_{\mathrm{b.c},2}.$$

In closing, (7.3.13) is equivalent to

$$\begin{cases} \text{Find } u \text{ in } D(\mathrm{D}^{\ell}_{\mathcal{X}}) \text{ such that} \\ \mathrm{D}^{\ell}_{\mathcal{X}} u = Lu + h \\ P_{\ker \mathrm{D}^{\ell}_{\mathcal{X}}} u = g_1\, h_{\mathrm{b.c},1} + \ell\, h_{\mathrm{b.c},2}. \end{cases} \tag{7.3.14}$$

7.3.4.1.3 Well Posedness

The previous analysis tells us that (7.3.13) is an instance of (7.2.10). Thus there results from Definition 7.2.6 that Problem (7.3.13) is *well posed* if, for each $(h_{\mathrm{b.c},1}, h_{\mathrm{b.c},2}, h)$ in $\mathcal{Y} \times \mathcal{Y} \times L^p(\mathcal{Y})$, (7.2.10) has a unique solution u, and u satisfies

$$\|u\|_{L^p(\mathcal{Y})} \leq C(\|h\|_{L^p(\mathcal{Y})} + \|h_{\mathrm{b.c},1}\|_{\mathcal{Y}} + \|h_{\mathrm{b.c},2}\|_{\mathcal{Y}}),$$

where the constant C is independent of u, $h_{\mathrm{b.c},1}$, $h_{\mathrm{b.c},2}$, and h.

Since $B_{\ell,\mathcal{X}}$ is a right inverse of the closed operator $\mathrm{D}^{\ell}_{\mathcal{X}}$, commute with L, and (7.2.1) is satisfied according to Lemma 7.2.4, Proposition 7.2.9 entails that (7.3.13) is well posed and that its solution u reads

$$u = \mathsf{R}^{L}_{B_{\ell,\mathcal{X}}}(g_1\, h_{\mathrm{b.c},1}) + \mathsf{R}^{L}_{B_{\ell,\mathcal{X}}}(\ell\, h_{\mathrm{b.c},2}) + B_{\ell,\mathcal{X}}\, \mathsf{R}^{L}_{B_{\ell,\mathcal{X}}} h.$$

With Proposition 7.2.7,

$$u = E^{L}_{\ell,g_1} h_{\mathrm{b.c},1} + E^{L}_{\ell} h_{\mathrm{b.c},2} + E^{L}_{\ell} * h, \tag{7.3.15}$$

where $E^{L}_{\ell,g_1} = \sum_{j\geq 0} \ell_j * g_1\, L^j$ and $E^{L}_{\ell} = \sum_{j\geq 0} \ell_{j+1}\, L^j$, with $\ell_{j+1} := B_{\ell,L^p}{}^j \ell$ for each nonnegative index j.

7.3.4.1.4 Particular Initial Conditions

Recalling that ℓ lies in $L^p(0,b) \cap L^q(0,b)$, we will see that roughly speaking *problems set with Riemann-Liouville derivative differ from problems set with Caputo derivative only by their initial conditions*. More precisely, problems set with Riemann-Liouville derivative and problems set with Caputo derivative are particular cases of (7.3.13) corresponding to different initial conditions.

7.3.4.1.4.1 Riemann-Liouville Derivative

Lemma 7.3.1 *Let p belong to $[1, \infty)$, $\mathcal{Y}$ be a nontrivial Banach space and ℓ lies in $L^p(0,b) \cap L^q(0,b)$. Then the following propositions are equivalent:*

(i) *u lies in $D(\mathrm{D}^{\ell}_{\mathcal{X}})$ and $P_{\ker \mathrm{D}^{\ell}_{\mathcal{X}}} u$ lies in $\ell\, \mathcal{Y}$.*
(ii) *u lies in $D({}^{\mathrm{RL}}\mathrm{D}^{\ell}_{\mathcal{X}})$.*

Proof Assuming (i), we derive with Lemma 4.3.3 the existence of a vector $h_{\text{b.c},2}$ in $\mathcal{Y}$ such that

$$u = \ell\, h_{\text{b.c},2} + \ell * \mathrm{D}^{\ell}_{\mathcal{X}} u.$$

Then u lies in $D({}^{\mathrm{RL}}\mathrm{D}^{\ell}_{\mathcal{X}})$ according to Lemma 4.3.3 applied with $A := {}^{\mathrm{RL}}\mathrm{D}^{\ell}_{\mathcal{X}}$. Conversely assuming (ii), by Lemma 4.3.3 again, there exists $h_{\text{b.c},2}$ in $\mathcal{Y}$ such that

$$u = \ell\, h_{\text{b.c},2} + \ell * {}^{\mathrm{RL}}\mathrm{D}^{\ell}_{\mathcal{X}} u.$$

Thus u lies in $\ker \mathrm{D}^{\ell}_{\mathcal{X}} \oplus R(B_{\ell,\mathcal{X}})$, which is nothing but the domain of $\mathrm{D}^{\ell}_{\mathcal{X}}$. Besides, the very definition of $P_{\ker \mathrm{D}^{\ell}_{\mathcal{X}}}$ yields that $P_{\ker \mathrm{D}^{\ell}_{\mathcal{X}}} u = \ell\, h_{\text{b.c},2}$, so that (i) holds true. □

Then by choosing $h_{\text{b.c},1} = 0$, (7.3.13) becomes thanks to the latter lemma and the equivalence between (7.3.13) and (7.3.14)

$$\begin{cases} \text{Find } u \text{ in } D({}^{\mathrm{RL}}\mathrm{D}^{\ell}_{\mathcal{X}}) \text{ such that} \\ {}^{\mathrm{RL}}\mathrm{D}^{\ell}_{\mathcal{X}} u = Lu + h, \qquad k * u(0) = h_{\text{b.c},2}. \end{cases} \tag{7.3.16}$$

This is nothing but (7.3.4).

7.3.4.1.4.2 Caputo Derivative

In a same way, by choosing $h_{\text{b.c},2} = 0$, (7.3.13) reads (see (6.4.3))

$$\begin{cases} \text{Find } u \text{ in } D({}^{\mathrm{C}}\mathrm{D}^{\ell}_{\mathcal{X}}) \text{ such that} \\ {}^{\mathrm{C}}\mathrm{D}^{\ell}_{\mathcal{X}} u = Lu + h, \qquad u(0) = h_{\text{b.c},1}. \end{cases} \tag{7.3.17}$$

By Sect. 7.3.4.1.3, (7.3.17) is well posed and that its solution u reads

$$u = E^{L}_{\ell,g_1} h_{\text{b.c},1} + E^{L}_{\ell} * h. \tag{7.3.18}$$

Remark 7.3.3 When Problem (7.3.17) is considered out of the current context, it could be conjectured that

$$E^{L}_{\ell,g_1} h_{\text{b.c},1} + E^{L}_{\ell,g_1} * h$$

is a representation of its solution. Formula (7.3.15) allows to understand why this is not the case. □

In the particular case where

$$\alpha \in (\tfrac{1}{2}, 1), \; p \in (\tfrac{1}{\alpha}, \tfrac{1}{1-\alpha}) \; \text{ and } (k, \ell) := (g_{1-\alpha}, g_\alpha), \tag{7.3.19}$$

g_α lies in $L^p(0, b) \cap L^q(0, b)$, and ${}^{\mathrm{C}}\mathrm{D}^\ell_{\mathcal{X}}$ becomes *the Caputo operator of order* α, that is, ${}^{\mathrm{C}}\mathrm{D}^\ell_{\mathcal{X}} = {}^{\mathrm{C}}\mathrm{D}^\alpha_{\mathcal{X}}$. Moreover,

$$E^L_{g_\alpha, g_1} = \sum_{j \geq 0} g_{j\alpha+1} L^j \in L^p\big((0, b), \mathcal{L}(\mathcal{Y})\big).$$

In the particular case where $\mathcal{Y} := \mathbb{K}$, $\mathcal{L}(\mathcal{Y})$ is identified with $\mathbb{K}$. Thus for each scalar λ, the representation of $E^\lambda_{g_\alpha, g_1}$ in terms of the Mittag-Leffler function E_α introduced in Sect. 3.4.3 reads

$$E^\lambda_{g_\alpha, g_1} = E_\alpha\big(\Gamma(\alpha + 1) g_{\alpha+1} \lambda\big) \qquad \text{in } L^p(0, b).$$

Hence for almost every x in $(0, b)$, one has

$$E^\lambda_{g_\alpha, g_1}(x) = E_\alpha\big(x^\alpha \lambda\big).$$

7.3.4.2 Case Where ℓ Lies in $L^q(0, b) \setminus L^p(0, b)$

By Lemma 4.3.3, any u in $D(\mathrm{D}^\ell_{\mathcal{X}})$ reads

$$u = g_1 U_1 + \ell * \mathrm{D}^\ell_{\mathcal{X}} u$$

for some unique U_1 in $\mathcal{Y}$. Thus $\mathrm{D}^\ell_{\mathcal{X}} = {}^{\mathrm{C}}\mathrm{D}^\ell_{\mathcal{X}}$. With Remark 3.3.4, u is continuous on $[0, b]$ and

$$U_1 = u(0).$$

Observe that $k * u(0) = 0$ since u is continuous.

Then for any h in $L^p(\mathcal{Y})$ and $h_{\mathrm{b.c},1}$ in $\mathcal{Y}$, the problem

$$\begin{cases} \text{Find } u \text{ in } D(\mathrm{D}^\ell_{\mathcal{X}}) \text{ such that} \\ \mathrm{D}^\ell_{\mathcal{X}} u = Lu + h, \qquad u(0) = h_{\mathrm{b.c},1} \end{cases} \tag{7.3.20}$$

is well posed and that its solution u satisfies (7.3.18).

7.3.4.3 Case Where ℓ Lies in $L^p(0, b) \setminus L^q(0, b)$

7.3.4.3.1 The Framework

Since we have only to study the case where k does not belong to $L^p(0, b)$, any u in $D(\mathrm{D}^\ell_{\mathcal{X}})$ reads

$$u = g_1\, U_1 + \ell\, U_2 + \ell * \mathrm{D}^{\ell}_{\mathcal{X}} u$$

for some unique (U_1, U_2) in $\mathcal{Y} \times \mathcal{Y}$. Arguing as in Sect. 6.7.2.3.2, we find that $u - \ell\, k * u(0) - \ell * \mathrm{D}^{\ell}_{\mathcal{Y}} u$ is continuous on $[0, b]$ and

$$U_1 = \big(u - \ell\, k * u(0) - \ell * \mathrm{D}^{\ell}_{\mathcal{Y}} u\big)(0)$$
$$U_2 = k * u(0).$$

7.3.4.3.2 Formulation of the Problem and Well Posedness

Let p belong to $[1, \infty)$, L be in $\mathcal{L}(\mathcal{Y})$, and h belong to $L^p\big((0, b), \mathcal{Y}\big)$. Being given datum $(h_{\mathrm{b.c},1}, h_{\mathrm{b.c},2})$ belonging to $\mathcal{Y} \times \mathcal{Y}$, the problem

$$\begin{cases} \text{Find } u \text{ in } D(\mathrm{D}^{\ell}_{\mathcal{X}}) \text{ such that} \\ \mathrm{D}^{\ell}_{\mathcal{X}} u = Lu + h \\ \big(u - \ell\, k * u(0) - \ell * \mathrm{D}^{\ell}_{\mathcal{Y}} u\big)(0) = h_{\mathrm{b.c},1}, \qquad k * u(0) = h_{\mathrm{b.c},2} \end{cases} \tag{7.3.21}$$

is well posed and its solution u satisfies (7.3.15).

7.3.4.4 Case Where ℓ Lies in $L^1(0, b) \setminus (L^q(0, b) \cup L^p(0, b))$

7.3.4.4.1 The Framework

Since we have only to study the case where k does not belong to $L^p(0, b)$, any u in $D(\mathrm{D}^{\ell}_{\mathcal{X}})$ reads

$$u = g_1\, U_1 + \ell * \mathrm{D}^{\ell}_{\mathcal{X}} u$$

for some unique U_1 in $\mathcal{Y}$. By using Lemma 4.3.3, Arguing as in Sect. 6.7.2.4.4, we find that $u - \ell * \mathrm{D}^{\ell}_{\mathcal{Y}} u$ is continuous on $[0, b]$ and

$$U_1 = \big(u - \ell * \mathrm{D}^{\ell}_{\mathcal{Y}} u\big)(0).$$

7.3.4.4.2 Formulation of the Problem and Well Posedness

Let p belong to $[1, \infty)$, L be in $\mathcal{L}(\mathcal{Y})$, and h belong to $L^p\big((0, b), \mathcal{Y}\big)$. Assume in addition that $h_{\mathrm{b.c},1}$ lies to $\mathcal{Y}$. Then the problem

$$\begin{cases} \text{Find } u \text{ in } D(\mathrm{D}^{\ell}_{\mathcal{X}}) \text{ such that} \\ \mathrm{D}^{\ell}_{\mathcal{X}} u = Lu + h \\ \big(u - \ell * \mathrm{D}^{\ell}_{\mathcal{Y}} u\big)(0) = h_{\mathrm{b.c},1}, \end{cases} \tag{7.3.22}$$

is well posed.

7.4 Initial Value Problems II: Systems and Higher-Order Equations

In this section we will apply the results of Sects. 7.2.4 and 7.2.5 to entire and fractional differential linear systems supplemented with initial conditions and also to higher-order linear equations supplemented with initial conditions.

7.4.1 The Framework

Let $\mathbb{K}$ be equal to $\mathbb{R}$ or $\mathbb{C}$, $\mathcal{Y}$ be a nontrivial Banach space over $\mathbb{K}$, and b be a positive real number. For any real number p in $[1, \infty)$, we set

$$\mathcal{X} = L^p(\mathcal{Y}) := L^p\big((0, b), \mathcal{Y}\big),$$

and for any positive integer n, we denote by $\mathcal{Z}$ the Cartesian nth power of $\mathcal{X}$, i.e.,

$$\mathcal{Z} := L^p\big((0, b), \mathcal{Y}\big)^n.$$

7.4.2 Systems with $A := \mathrm{D}^1_{\mathcal{X}}$

We will study linear first-order differential systems supplemented with initial boundary conditions.

7.4.2.1 A Well-Posed Problem

For all indexes i and j in $[1, n]$, let $a_{i,j}$ belong to $\mathbb{K}$, h_i belong to $L^p(\mathcal{Y})$, and $h_{\mathrm{b.c},i}$ lie in $\mathcal{Y}$. Then we consider the following problem:

$$\begin{cases} \text{Find } (u_1, \ldots, u_n) \text{ in } W^{1,p}\big((0, b), \mathcal{Y}\big)^n \text{ such that for } i = 1, \ldots, n \\ \mathrm{D}^1_{\mathcal{X}} u_i = \displaystyle\sum_{j=1}^{n} a_{i,j} u_j + h_i \quad \text{in } L^p\big((0, b), \mathcal{Y}\big) \\ u_i(0) = h_{\mathrm{b.c},i} \quad \text{in } \mathcal{Y}. \end{cases} \tag{7.4.1}$$

Plainly, (7.4.1) is an instance of (7.2.18) with $A := \mathrm{D}^1_{\mathcal{X}}$. From Sect. 7.3.2, $B_{1,\mathcal{X}}$ is a right inverse of the closed operator $\mathrm{D}^1_{\mathcal{X}}$, whose spectral radius vanishes. Thus Proposition 7.2.12 yields that (7.4.1) is well posed, which means that it has a unique solution $(u_1, \ldots, u_n)$ and that

$$\sum_{i=1}^{n} \|u_i\|_{L^p(\mathcal{Y})} \lesssim \sum_{i=1}^{n} \|h_i\|_{L^p(\mathcal{Y})} + \|h_{\mathrm{b.c},i}\|_{\mathcal{Y}}.$$

7.4.2.2 Representation of the Solutions to the Homogeneous Equation

Let $\mathbb{K} := \mathbb{C}$. Any element of $\ker \mathrm{D}^1_{\mathcal{X}}$ reads $g_1 Y$ for some Y in $\mathcal{Y}$. On account of Sect. 7.3.2.2, $\mathrm{R}^{\lambda}_{B_1,\mathcal{X}}(g_1 Y) = e^{\lambda \cdot} Y$, thus for every nonnegative index l,

$$\frac{1}{l!}\frac{\mathrm{d}^l}{\mathrm{d}\lambda^l}\mathrm{R}^{\lambda}_{B_1,\mathcal{X}}(g_1 Y) = g_{l+1} e^{\lambda \cdot}\, Y.$$

With Proposition 7.2.13, the ith component of a solution $(u_1, \dots, u_n)$ to the equation

$$\mathrm{D}^1_{\mathcal{X}} u_i = \sum_{j=1}^{n} a_{i,j} u_j$$

has the form

$$u_i = \sum_{k=1}^{r} \sum_{l=0}^{m'_k - 1} g_{l+1} e^{\lambda_k \cdot}\, Y_{l,i,k},$$

where

(i) $\lambda_1, , \dots, \lambda_r$ a list of the pairwise distinct complex eigenvalues of L;
(ii) for $k = 1, \dots, r$, m'_k is the multiplicity of λ_k with respect to the minimal polynomial of L;
(iii) $Y_{l,i,k}$ lies in Y.

This classical result can be found for instance in [HS74, Chap. 6, §6].

7.4.3 Higher Order Equations with $A := \mathrm{D}^1_{\mathcal{X}}$

Let $\mathbb{K}$ be equal to $\mathbb{R}$ or $\mathbb{C}$. Being given n scalars $a_0, \dots, a_{n-1}$, for any data h in $L^p(\mathcal{Y})$ and $h_{\mathrm{b.c},1}, \dots, h_{\mathrm{b.c},n}$ in $\mathcal{Y}$, we consider the problem

$$\begin{cases} \text{Find } v \text{ in } W^{n,p}\big((0,b),\mathcal{Y}\big) \text{ such that} \\ \mathrm{D}^n_{\mathcal{X}} v + \displaystyle\sum_{i=0}^{n-1} a_i \mathrm{D}^i_{\mathcal{X}} v = h \\ \mathrm{D}^i_{\mathcal{X}} v(0) = h_{\mathrm{b.c},i+1}, \qquad \forall i = 0, \dots, n-1. \end{cases} \tag{7.4.2}$$

In view of 7.4.2.1, it is clear that the assumptions of Proposition 7.2.15 holds. Thus (7.4.2) is well posed. Besides, assuming now that $\mathbb{K}$ is equal to $\mathbb{C}$, Theorem 7.2.17 provides the structure of the solution of the homogeneous equation

$$\mathrm{D}^n_{\mathcal{X}} v + \sum_{i=0}^{n-1} \mathrm{D}^i_{\mathcal{X}} v = 0.$$

More precisely, setting

$$P(X) := X^n + \sum_{i=0}^{n-1} a_i X^i,$$

let $\lambda_1, , \ldots, \lambda_r$ be a list of the pairwise distinct roots of P. For each $k = 1, \ldots, r$, denote by m_k the multiplicity of λ_k with respect to P. Now let v belong to $L^p(\mathcal{Y})$. Then v is solution to the latter homogeneous equation if and only if there exist n elements $Y_{k,i}$ of $\mathcal{Y}$ (with $1 \leq k \leq r, 0 \leq i \leq m_k - 1$) such that

$$v = \sum_{k=1}^{r} \sum_{i=0}^{m_k-1} g_{i+1} e^{\lambda \cdot} Y_{k,i}.$$

7.4.4 Systems with $A := \mathrm{D}^\ell_{\mathcal{X}}$

We assume that

$$\ell \in L^p(0,b) \cap L^q(0,b).$$

7.4.4.1 A Well-Posed Problem

For each indexes i and j in $[1, n]$, let $a_{i,j}$ belong to $\mathbb{K}$, h_i belong to $L^p(\mathcal{Y})$ and $h_{1,i}$, and $h_{2,i}$ lie in $\mathcal{Y}$. Then we consider the following problem:

$$\begin{cases} \text{Find } (u_1, \ldots, u_n) \text{ in } D\big((\mathrm{D}^\ell_{\mathcal{X}})^n\big) \text{ such that for } i = 1, \ldots, n \\[2mm] \mathrm{D}^\ell_{\mathcal{X}} u_i = \displaystyle\sum_{j=1}^{n} a_{i,j} u_j + h_i \quad \text{in } L^p\big((0,b), \mathcal{Y}\big) \\[2mm] \big(u_i - \ell\, k * u_i(0)\big)(0) = h_{1,i}, \quad k * u_i(0) = h_{2,i} \quad \text{in } \mathcal{Y}. \end{cases} \tag{7.4.3}$$

On account of Sect. 7.3.4.1, (7.4.3) is an instance of (7.2.18), and the assumptions of Proposition 7.2.12 are satisfied. Hence (7.4.3) is well posed, which means that it has a unique solution $(u_1, \ldots, u_n)$ and that

$$\sum_{i=1}^{n} \|u_i\|_{L^p(\mathcal{Y})} \lesssim \sum_{i=1}^{n} \|h_i\|_{L^p(\mathcal{Y})} + \|h_{1,i}\|_{\mathcal{Y}} + \|h_{2,i}\|_{\mathcal{Y}}.$$

7.4.4.2 Representation of the Solutions to the Homogeneous Equation

Let $\mathbb{K} := \mathbb{C}$. Any element of $\ker \mathrm{D}^{\ell}_{\mathcal{X}}$ reads $g_1 Y_1 + \ell Y_2$ for some Y_1, Y_2 in $\mathcal{Y}$. By Proposition 7.2.3, the function $\lambda \mapsto \mathrm{R}^{\lambda}_{B_{\ell,\mathcal{X}}}$ is $\mathbb{C}$-differentiable at any order and for every nonnegative index l and φ in $L^p(0, b)$

$$\frac{1}{l!}\frac{\mathrm{d}^l}{\mathrm{d}\lambda^l}\mathrm{R}^{\lambda}_{B_{\ell,\mathcal{X}}}(\varphi Y_1) = \sum_{j \geq l} \tbinom{j}{l} \lambda^{j-l} \ell_j * \varphi\, Y_1,$$

where $\ell_0 * \varphi := \varphi$ and $\ell_{j+1} := B_{\ell,\mathcal{X}}{}^j \ell$ for $j \geq 0$. Thus there results from Proposition 7.2.13 that the ith component of a solution $(u_1, \ldots, u_n)$ to the equation

$$\mathrm{D}^{\ell}_{\mathcal{X}} u_i = \sum_{j=1}^{n} a_{i,j} u_j$$

has the form

$$u_i = \sum_{k=1}^{r} \sum_{l=0}^{m'_k - 1} \sum_{j \geq l} \tbinom{j}{l} \lambda^{j-l} \big(\ell_j * g_1\, Y_{1,l,i,k} + \ell_{j+1}\, Y_{2,l,i,k}\big),$$

where:

(i) $\lambda_1, , \ldots, \lambda_r$ a list of the pairwise distinct complex eigenvalues of L;
(ii) for $k = 1, \ldots, r$, m'_k is the multiplicity of λ_k with respect to the minimal polynomial of L;
(iii) $Y_{1,l,i,k}$ and $Y_{2,l,i,k}$ belong to $\mathcal{Y}$.

7.4.5 Higher-Order Equations with $A := \mathrm{D}^{\ell}_{\mathcal{X}}$

Let $\mathbb{K}$ be equal to $\mathbb{R}$ or $\mathbb{C}$. Being given n scalars $a_0, \ldots, a_{n-1}$, for any data h in $L^p(\mathcal{Y})$ and $h_{1,1}, \ldots, h_{1,n}, h_{2,1}, \ldots, h_{2,n}$ in $\mathcal{Y}$, we consider the problem

$$\left\{\begin{aligned} &\text{Find } v \text{ in } D\big((\mathrm{D}^{\ell}_{\mathcal{X}})^n\big) \text{ such that} \\ &(\mathrm{D}^{\ell}_{\mathcal{X}})^n v + \sum_{j=0}^{n-1} a_j (\mathrm{D}^{\ell}_{\mathcal{X}})^j v = h \\ &\big((\mathrm{D}^{\ell}_{\mathcal{X}})^j v - \ell\, k * (\mathrm{D}^{\ell}_{\mathcal{X}})^j v(0)\big)(0) = h_{1,j+1} \\ &\qquad k * (\mathrm{D}^{\ell}_{\mathcal{X}})^j v(0) = h_{1,j+1}, \qquad \forall j = 0, \ldots, n-1. \end{aligned}\right. \tag{7.4.4}$$

In view of Sect. 7.4.4.1, it is clear that the assumptions of Proposition 7.2.15 holds. Thus (7.4.4) is well posed. Besides, assuming now that $\mathbb{K}$ is equal to $\mathbb{C}$, Theorem 7.2.17

provides the structure of the solution of the homogeneous equation

$$(\mathrm{D}^{\ell}_{\mathcal{X}})^n v + \sum_{j=0}^{n-1} a_j (\mathrm{D}^{\ell}_{\mathcal{X}})^j v = 0.$$

More precisely, setting $P(X) := X^n + \sum_{i=0}^{n-1} a_i X^i$, let $\lambda_1, \dots, \lambda_r$ be a list of the pairwise distinct roots of P. For each $k = 1, \dots, r$, denote by m_k the multiplicity of λ_k with respect to P. Now let v belong to $L^p(\mathcal{Y})$. Then v is solution to the latter homogeneous equation if and only if there exist $2n$ elements $Y_{1,k,i}, Y_{2,k,i}$ of $\mathcal{Y}$ (with $1 \leq k \leq r, 0 \leq i \leq m_k - 1$) such that

$$v = \sum_{k=1}^{r} \sum_{i=0}^{m_k-1} \sum_{j \geq i} \tbinom{j}{i} \lambda^{j-i} \big(\ell_j * g_1 \, Y_{1,k,i} + \ell_{j+1} \, Y_{2,k,i}\big).$$

7.5 Boundary Value Problems with Finite-Dimensional Phase Spaces

In this section, we will consider examples of linear boundary value problems of the form (7.2.48). In order to formulate general boundary conditions, an underlying differential quadruplet is required. For simplicity, it will be assumed that the phase space has finite dimension.

We have seen in Sect. 7.1 that even the simplest problem is conditionally solvable, which means that for some data, the problem admits no solution. Especially the condition on the data under which the problem is not solvable may be not easy to check: see, for instance, the condition appearing in the case (ii-b) of Sect. 7.1. So, in general, we cannot expect complete solvability of fractional differential problems set in higher-dimensional phase spaces. We will just reduce fractional differential problems into linear algebra problems. Roughly speaking, our theory gives nothing more than the usual quantitative theory of boundary value problems set for entire order differential operator, but it gives nothing less. Recall however that our theory applies to many fractional differential operators.

Let $\mathbb{K}$ be equal to $\mathbb{R}$ or $\mathbb{C}$, b be a positive real number, p, q be conjugate exponents in $(1, \infty)$, and $\mathcal{Y}_{\mathrm{f}}$ be a nontrivial finite-dimensional normed space over $\mathbb{K}$. $\mathcal{Y}_{\mathrm{f}}$ is referred as a *phase space*. We set

$$\mathcal{V} = L^p(\mathcal{Y}_{\mathrm{f}}) := L^p\big((0, b), \mathcal{Y}_{\mathrm{f}}\big).$$

According to Theorem 3.2.2, $\mathcal{V}$ is a reflexive Banach space and

$$\mathcal{V}^* = L^q\big((0, b), \mathcal{Y}_{\mathrm{f}}{}^*\big).$$

7.5.1 First-Order Problems

The aim of the current subsection is not so much to recover well-known and easy to prove results rather than to show that the abstract results of Sect. 7.2.6 cover the variety of situations appearing in the resolution of (7.1.1).

7.5.1.1 Formulation of the Problem

According to Example 5.6.7, we know that the differential quadruplet

$$\mathcal{Q}_{p,\mathcal{Y}_f,1} := (\mathrm{D}^1_{\mathcal{V}^*},\ \mathrm{D}^1_{\mathcal{V}},\ B_{1,\mathcal{V}},\ S_{\mathcal{V}}),$$

is $g_1\mathcal{Y}_f$-regular. Let L be in $\mathcal{L}(\mathcal{Y}_f)$, and data $(h_{\mathrm{b.c}}, h)$ belong to $\mathcal{Y}_f \times L^p(\mathcal{Y}_f)$. Let also m_1, m_2 be scalars such that at least one of them is nontrivial, i.e., $(m_1, m_2) \neq (0, 0)$. Then we consider the problem

$$\begin{cases} \text{Find } u \text{ in } W^{1,p}\big((0,b),\mathcal{Y}_f\big) \text{ such that} \\ \mathrm{D}^1_{\mathcal{V}} u = Lu + h, \qquad m_1 u(0) + m_2 u(b) = h_{\mathrm{b.c}}. \end{cases} \tag{7.5.1}$$

Let us show that (7.5.1) is indeed an inhomogeneous linear endogenous boundary value problem of the form (7.2.48). For each u in $W^{1,p}(\mathcal{Y}_f)$, let (U_1, U_1') be the endogenous boundary values of u with respect to the trivial direct sum decompositions of $\ker \mathrm{D}^1_{\mathcal{V}} = g_1\mathcal{Y}$ and $F_0 = g_1\mathcal{Y}$ (see Sect. 5.7.1.1). Setting

$$x_1 := -bm_2, \qquad y_1 := m_1 + m_2, \tag{7.5.2}$$

one has

$$y_1 U_1 - x_1 U_1' = m_1 u(0) + m_2 u(b). \tag{7.5.3}$$

Moreover, the map

$$T : E_{\mathrm{M}} \to \mathcal{Y}_f, \qquad g_1 U_1 + g_2 U_1' \mapsto y_1 U_1 - x_1 U_1'$$

lies in $\mathcal{L}(E_{\mathrm{M}}, \mathcal{Y}_f)$ according to Proposition 5.7.1, and by (7.5.3),

$$T P_{\mathrm{b.v}} u = m_1 u(0) + m_2 u(b), \qquad \forall\, u \in W^{1,p}\big((0,b),\mathcal{Y}_f\big).$$

Hence (7.5.1) is an instance of (7.2.48).

7.5.1.2 Solvability of (7.5.1)

Let κ be the element of $\mathcal{L}(\mathcal{Y}_{\mathrm{f}})$ defined by

$$\kappa : \mathcal{Y}_{\mathrm{f}} \to \mathcal{Y}_{\mathrm{f}}, \qquad Y \mapsto T P_{\mathrm{b.v}} \mathsf{R}^L_{B_1,\mathcal{V}}(g_1 Y). \tag{7.5.4}$$

Since $\mathsf{R}^L_{B_1,\mathcal{V}}(g_1 Y) = e^{\cdot L} Y$, we get

$$\kappa := m_1 \mathrm{id}_{\mathcal{Y}_{\mathrm{f}}} + m_2 e^{bL}.$$

Notice that $\kappa = \frac{x_1}{b}(\mathrm{id}_{\mathcal{Y}_{\mathrm{f}}} - e^{bL}) + y_1 \mathrm{id}_{\mathcal{Y}_{\mathrm{f}}}$; compare with Sect. 7.1. With the notation of Proposition 7.2.19, we have

$$T P_{\mathrm{b.v}}\, \mathsf{R}^L_B(\ker A_{\mathrm{M}}) = R(\kappa). \tag{7.5.5}$$

Now we are in position to describe the solution set of (7.5.1) with respect to the invertibility of κ. This gives rise to three cases.

(i) Cases where κ is invertible. We will apply Corollary 7.2.23 with $\mathcal{X}_1 := \mathcal{Y}_{\mathrm{f}}$. By (7.5.5), $T P_{\mathrm{b.v}}\, \mathsf{R}^L_{B|\ker A_{\mathrm{M}}}$ is an isomorphism from $\ker A_{\mathrm{M}} := \ker \mathrm{D}^1_{\mathcal{V}} = g_1 \mathcal{Y}_{\mathrm{f}}$ onto $\mathcal{Y}_{\mathrm{f}}$. Then Corollary 7.2.23 yields that (7.5.1) is well posed for each $(h_{\mathrm{b.c}}, h)$ in $\mathcal{Y}_{\mathrm{f}} \times \mathcal{V}$.

(ii) Cases where κ is not invertible. One has

$$T P_{\mathrm{b.v}} B\, \mathsf{R}^L_B h = m_2(e^{\cdot L} * h)(b).$$

Thus (7.2.49) is equivalent to

$$m_2(e^{\cdot L} * h)(b) - h_{\mathrm{b.c}} \in R(\kappa). \tag{7.5.6}$$

(ii-a) If (7.5.6) holds then Proposition 7.2.19 (i) yields that (7.5.1) admit at least a solution u_{p}, and the solution set of (7.5.1) reads in view of Proposition 7.2.19 (i)

$$\left\{u_{\mathrm{p}} + e^{\cdot L} Y \mid Y \in \ker \kappa\right\}.$$

Observe that (7.5.1) is not uniquely solvable.

(ii-b) If (7.5.6) does not hold, then (7.5.1) has no solution according to Proposition 7.2.19 (ii).

We have reduced the solvability of (7.5.1), an inhomogeneous first-order linear system supplemented with inhomogenous linear boundary conditions into a linear algebra problem. However in practice, the solvability of the latter problem may be rather difficult.

7.5.2 Problems with $A_{\mathrm{M}} := {}^{\mathrm{RL}}\mathrm{D}^{\ell}_{\mathcal{V}}$

Let (k, ℓ) be a Sonine pair. In order to maximize the number of boundary values, we assume that

$$\ell \in L^p(0, b) \cap L^q(0, b).$$

For simplicity, we suppose that

$$\ell \text{ admits an essential limit at } b\ .$$

7.5.2.1 Formulation of the Problem

According to Sect. 6.5.2.1, we know that the differential quadruplet

$${}^{\mathrm{RL}}\mathcal{Q}_{p,\mathcal{Y}_{\mathrm{f}},\ell} := ({}^{\mathrm{RL}}\mathrm{D}^{\ell}_{\mathcal{V}*},\ {}^{\mathrm{RL}}\mathrm{D}^{\ell}_{\mathcal{V}},\ B_{\ell,\mathcal{V}},\ S_{\mathcal{V}}),$$

is $\ell\mathcal{Y}_{\mathrm{f}}$-regular. Let L be in $\mathcal{L}(\mathcal{Y}_{\mathrm{f}})$, and data $(h_{\mathrm{b.c}}, h)$ belong to in $\mathcal{Y}_{\mathrm{f}} \times L^p(\mathcal{Y}_{\mathrm{f}})$. Let also m_1, m_2 be scalars such that $(m_1, m_2) \neq (0, 0)$. Then we consider the problem

$$\begin{cases} \text{Find } u \text{ in } D({}^{\mathrm{RL}}\mathrm{D}^{\ell}_{\mathcal{V}}) \text{ such that} \\ {}^{\mathrm{RL}}\mathrm{D}^{\ell}_{\mathcal{V}} u = Lu + h, \qquad m_1 k * u(0) + m_2 u(b) = h_{\mathrm{b.c}}. \end{cases} \tag{7.5.7}$$

Let us show that (7.5.7) is indeed an inhomogenous linear endogenous boundary value problem of the form (7.2.48). For each u in $D({}^{\mathrm{RL}}\mathrm{D}^{\ell}_{\mathcal{V}})$, let (U_1, U_1') be the endogenous boundary values of u given by (6.5.13). Setting

$$x_1 := -\|\ell\|^2_{L^2} m_2, \qquad y_1 := m_1 + \ell(b) m_2, \tag{7.5.8}$$

one has

$$y_1 U_1 - x_1 U_1' = m_1 k * u(0) + m_2 u(b). \tag{7.5.9}$$

Moreover, the map

$$T : E_{\mathrm{M}} \to \mathcal{Y}_{\mathrm{f}}, \qquad \ell\, U_1 + \ell * S_{L^p} \ell\, U_1' \mapsto y_1 U_1 - x_1 U_1'$$

lies in $\mathcal{L}(E_{\mathrm{M}}, \mathcal{Y}_{\mathrm{f}})$ according to Proposition 5.7.1, and by (7.5.9),

$$T P_{\mathrm{b.v}} u = m_1 k * u(0) + m_2 u(b), \qquad \forall\, u \in D({}^{\mathrm{RL}}\mathrm{D}^{\ell}_{\mathcal{V}}).$$

Hence (7.5.7) is an instance of (7.2.48).

7.5.2.2 Solvability of (7.5.7)

In the current framework, the scalar κ introduced in Sect. 7.1 is replaced by a linear map from $\mathcal{Y}_{\mathrm{f}}$ into itself. This linear map will be still denoted by κ.

Since it is assumed that ℓ admits an essential limit at b, each element of $D({}^{\mathrm{RL}}\mathrm{D}^{\ell}_{\mathcal{V}})$ has also an essential limit at b (see Sect. 6.5.2.1). Besides, for any Y in $\mathcal{Y}_{\mathrm{f}}$, $\mathsf{R}^{L}_{B_{\ell,\mathcal{V}}}(\ell\, Y)$ lies in $D({}^{\mathrm{RL}}\mathrm{D}^{\ell}_{\mathcal{V}})$ due to Proposition 7.2.2. Thus we may legitimately defined the map

$$\mathsf{R}^{L}_{B_{\ell,\mathcal{V}}}(\ell\,\cdot)(b) : \mathcal{Y}_{\mathrm{f}} \to \mathcal{Y}_{\mathrm{f}}, \qquad Y \mapsto \mathsf{R}^{L}_{B_{\ell,\mathcal{V}}}(\ell\, Y)(b).$$

Clearly, $\mathsf{R}^{L}_{B_{\ell,\mathcal{V}}}(\ell\,\cdot)(b)$ is linear and continuous (since $\mathcal{Y}_{\mathrm{f}}$ is finite dimensional).

After this preparation, let κ be the element of $\mathcal{L}(\mathcal{Y}_{\mathrm{f}})$ defined by

$$\kappa : \mathcal{Y}_{\mathrm{f}} \to \mathcal{Y}_{\mathrm{f}}, \qquad Y \mapsto T P_{\mathrm{b.v}} \mathsf{R}^{L}_{B_{\ell,\mathcal{V}}}(\ell\, Y). \tag{7.5.10}$$

Then

$$\kappa = m_1 \mathrm{id}_{\mathcal{Y}_{\mathrm{f}}} + m_2 \mathsf{R}^{L}_{B_{\ell,\mathcal{V}}}(\ell\,\cdot)(b);$$

compare with Sect. 7.5.1.2. With the notation of Proposition 7.2.19, we compute

$$T P_{\mathrm{b.v}}\, \mathsf{R}^{L}_{B}(\ker A_{\mathrm{M}}) = R(\kappa). \tag{7.5.11}$$

Now we are in position to describe the solution set of (7.5.7) with respect to the invertibility of κ. This gives rise to three cases.

(i) Cases where κ is invertible. Arguing as in Sect. 7.5.1.2, we obtain that (7.5.7) is well posed for each $(h_{\mathrm{b.c}}, h)$ in $\mathcal{Y}_{\mathrm{f}} \times \mathcal{V}$.

(ii) Cases where κ is not invertible. One has

$$T P_{\mathrm{b.v}} B\, \mathsf{R}^{L}_{B} h = m_2 (B_{\ell,\mathcal{V}} \mathsf{R}^{L}_{B_{\ell,\mathcal{V}}} h)(b).$$

Thus (7.2.49) is equivalent to

$$m_2 (B_{\ell,\mathcal{V}} \mathsf{R}^{L}_{B_{\ell,\mathcal{V}}} h)(b) - h_{\mathrm{b.c}} \in R(\kappa). \tag{7.5.12}$$

(ii-a) If (7.5.12) holds then Proposition 7.2.19 (i) yields that (7.5.7) admit at least a solution u_{p}, and the solution set of (7.5.7) reads in view of Proposition 7.2.19 (i)

$$\left\{ u_{\mathrm{p}} + \mathsf{R}^{L}_{B_{\ell,\mathcal{V}}}(\ell\, Y) \mid Y \in \ker \kappa \right\}.$$

(ii-b) If (7.5.12) does not hold, then (7.5.7) has no solution according to Proposition 7.2.19 (ii).

Observe that the latter cases study is similar to the one of Sect. 7.1. Indeed, by Proposition 7.2.7 (i), κ may be rewritten as

$$\kappa = m_1 \mathrm{id}_{\mathcal{Y}_\mathrm{f}} + m_2 E_\ell^L(b),$$

and thanks to Proposition 7.2.7 (ii), we may rewrite (7.5.12) under the form

$$m_2(E_\ell^L * h)(b) - h_{\mathrm{b.c}} \in R(\kappa).$$

Compare with Remark 7.1.1.

7.5.3 Problems with $A_\mathrm{M} := \mathrm{D}_\mathcal{V}^\ell$

The assumptions set at the beginning of the current section remains in force. In particular, $\mathcal{V} := L^p(\mathcal{Y}_\mathrm{f})$ where $\mathcal{Y}_\mathrm{f}$ is a nontrivial finite-dimensional normed space. Moreover, being given a Sonine pair (k, ℓ), we consider the differential quadruplet $\mathcal{Q}_{p,\mathcal{Y}_\mathrm{f},\ell}$ introduced in Sect. 6.7.

7.5.3.1 Case Where ℓ Lies in $L^p(0, b) \cap L^q(0, b)$

7.5.3.1.1 Formulation of the Problem

Let L lie in $\mathcal{L}(\mathcal{Y}_\mathrm{f})$, and data $(h_{\mathrm{b.c},1}, h_{\mathrm{b.c},2}, h)$ belong to in $\mathcal{Y}_\mathrm{f} \times \mathcal{Y}_\mathrm{f} \times L^p(\mathcal{Y}_\mathrm{f})$. Let also $(m_{i,j})$ be a scalar matrix of dimension 2×4. Then in view of (6.7.6) and (6.7.8), we introduce the following endogenous boundary value problem:

$$\begin{cases} \text{Find } u \text{ in } D(\mathrm{D}_\mathcal{V}^\ell) \text{ such that} \\ \mathrm{D}_\mathcal{V}^\ell u = Lu + h \\ m_{11}\, U_1 + m_{12}\, U_2 + m_{13}\, g_1 * \mathrm{D}_\mathcal{V}^\ell u(b) + m_{14}\, \ell * \mathrm{D}_\mathcal{V}^\ell u(b) = h_{\mathrm{b.c},1} \\ m_{21}\, U_1 + m_{22}\, U_2 + m_{23}\, g_1 * \mathrm{D}_\mathcal{V}^\ell u(b) + m_{24}\, \ell * \mathrm{D}_\mathcal{V}^\ell u(b) = h_{\mathrm{b.c},2}. \end{cases} \tag{7.5.13}$$

Recall that

$$U_1 = \big(u - \ell\, k * u(0)\big)(0), \qquad U_2 = k * u(0).$$

By using the framework of Sect. 6.7.2.1, the mapping

$$T : E_\mathrm{M} \to \mathcal{Y}_\mathrm{f} \times \mathcal{Y}_\mathrm{f}$$

which sends any elements

$$g_1\, U_1 + \ell\, U_2 + \ell * S_{\mathcal{V}}(h_1\, U_1') + \ell * S_{\mathcal{V}}(h_2\, U_2')$$

of E_{M} into the vector of $\mathcal{Y}_{\mathrm{f}} \times \mathcal{Y}_{\mathrm{f}}$ whose components are

$$m_{11}\, U_1 + m_{12}\, U_2 + m_{13}\, U_1' + m_{14}\, U_2'$$

and

$$m_{21}\, U_1 + m_{22}\, U_2 + m_{23}\, U_1' + m_{24}\, U_2',$$

lies in $\mathcal{L}(E_{\mathrm{M}}, \mathcal{Y}_{\mathrm{f}})$ according to Proposition 5.7.1. Then it is clear from (6.7.8) that (7.5.13) is an instance of (7.2.48) with $\mathcal{X}_1 := \mathcal{Y}_{\mathrm{f}} \times \mathcal{Y}_{\mathrm{f}}$.

7.5.3.1.2 The Map κ

Let κ be the element of $\mathcal{L}(\mathcal{Y}_{\mathrm{f}} \times \mathcal{Y}_{\mathrm{f}})$ defined by

$$\kappa : \mathcal{Y}_{\mathrm{f}} \times \mathcal{Y}_{\mathrm{f}} \to \mathcal{Y}_{\mathrm{f}} \times \mathcal{Y}_{\mathrm{f}}, \qquad (Y_1, Y_2) \mapsto T\, P_{\mathfrak{b}.\mathrm{v}} \mathsf{R}^{L}_{B_{\ell},\mathcal{V}}(g_1\, Y_1 + \ell\, Y_2). \tag{7.5.14}$$

As in Sects. 7.5.1.2 and 7.5.2.2, the solvability of the boundary value problem will be formulated in terms of κ. As a linear map from $\mathcal{Y}_{\mathrm{f}} \times \mathcal{Y}_{\mathrm{f}}$ into itself, κ can be put under the matrix form

$$\kappa = \begin{pmatrix} c_{11} & c_{12} \\ c_{21} & c_{22} \end{pmatrix},$$

where the entries c_{ij}'s lie in $\mathcal{L}(\mathcal{Y}_{\mathrm{f}})$. We claim that

$$\kappa = \begin{pmatrix} m_{11} + m_{13}\Lambda_1 + m_{14}\Lambda_2 & m_{12} + m_{13}\Lambda_2 + m_{14}\Lambda_3 \\ m_{21} + m_{23}\Lambda_1 + m_{24}\Lambda_2 & m_{22} + m_{23}\Lambda_2 + m_{14}\Lambda_3 \end{pmatrix}, \tag{7.5.15}$$

where $m_{11}\mathrm{id}_{\mathcal{Y}_{\mathrm{f}}}$ is abbreviated by m_{11} and

$$\Lambda_1 := k * \mathsf{R}^{L}_{B_{\ell},\mathcal{V}}(g_1 \cdot)(b) - g_1 * k(b)\mathrm{id}_{\mathcal{Y}_{\mathrm{f}}}$$

$$\Lambda_2 := \mathsf{R}^{L}_{B_{\ell},\mathcal{V}}(g_1 \cdot)(b) - \mathrm{id}_{\mathcal{Y}_{\mathrm{f}}}$$

$$\Lambda_3 := \big(\mathsf{R}^{L}_{B_{\ell},\mathcal{V}}(\ell \cdot) - \ell\, \mathrm{id}_{\mathcal{Y}_{\mathrm{f}}}\big)(b).$$

A more transparent definition of Λ_3 is given in (7.5.18). Let Y be a fixed vector of $\mathcal{Y}_{\mathrm{f}}$. For the sake of brevity, we put

$$u^Y := \mathsf{R}^L_{B_{\ell,\mathcal{V}}}(g_1 Y).$$

In order to establish (7.5.15), we start to compute the endogenous boundary values of u^Y. According to (6.7.8) and Remark 6.7.1, they read

$$U_1,\ U_2,\ k*(u^Y - g_1 U_1)(b) - k*(u^Y - g_1 U_1)(0),\ (u^Y - g_1 U_1 - \ell U_2)(b),$$

where

$$U_1 = \big(u^Y - \ell\, k * u^Y(0)\big)(0), \qquad U_2 = k * u^Y(0).$$

The very definition of $\mathsf{R}^L_{B_{\ell,\mathcal{V}}}$ entails that

$$u^Y = g_1 Y + B_{\ell,\mathcal{V}} \sum_{j\geq 1} L^j (B_{\ell,\mathcal{V}})^{j-1}(g_1 Y). \tag{7.5.16}$$

Thus by a direct sum argument (see Lemma 4.3.3),

$$U_1 = Y, \qquad U_2 = 0.$$

Let us compute the third boundary value of u^Y. Since ℓ lies in $L^q(0, b)$, Proposition 3.3.5 and (7.5.16) yield that u^Y is continuous on $[0, b]$. Thus $k*(u^Y - g_1 Y)(0) = 0$. Hence the third boundary value of u^Y is equal to

$$k * u^Y(b) - g_1 * k(b) Y.$$

The fourth boundary value is clearly equal to $u^Y(b) - Y$. Thus the endogenous boundary values of u^Y are

$$Y, \quad 0, \quad \Lambda_1 Y, \quad \Lambda_2 Y.$$

Then we obtain the first column of κ in (7.5.15).

The second column is obtained by in a same manner, by computing $T P_{b.\mathcal{V}} \mathsf{R}^L_{B_{\ell,\mathcal{V}}}(\ell Y)$. The boundary values of

$$v^Y := \mathsf{R}^L_{B_{\ell,\mathcal{V}}}(\ell Y)$$

reads

$$V_1,\ V_2,\ k*(v^Y - g_1 V_1)(b) - k*(v^Y - g_1 V_1)(0),\ (v^Y - g_1 V_1 - \ell V_2)(b).$$

Since

$$v^Y = \ell\, Y + B_{\ell,\mathcal{V}} \sum_{j\geq 1} L^j (B_{\ell,\mathcal{V}})^{j-1}(\ell\, Y),$$

we deduce

$$V_1 = 0, \qquad V_2 = Y.$$

Next, we claim that

$$k * v^Y = u^Y. \tag{7.5.17}$$

Indeed since $B_{k,\mathcal{V}}$ commute with L and $B_{\ell,\mathcal{V}}$, we compute

$$k * v^Y = \mathsf{R}^L_{B_{\ell,\mathcal{V}}} \circ B_{k,\mathcal{V}}(\ell\, Y) = \mathsf{R}^L_{B_{\ell,\mathcal{V}}}(g_1\, Y) = u^Y,$$

which proves (7.5.17). Hence the third boundary value of v^Y is $\Lambda_2 Y$. Since $V_1 = 0$, the function $v^Y - \ell\, Y$ is continuous on $[0, b]$. Thus the map

$$\Lambda_3 : \mathcal{Y}_{\mathrm{f}} \to \mathcal{Y}_{\mathrm{f}}, \qquad Y \mapsto \big(\mathsf{R}^L_{B_{\ell,\mathcal{V}}}(\ell\, Y) - \ell\, Y\big)(b) \tag{7.5.18}$$

is well defined. Hence the fourth boundary value is $\Lambda_3 Y$. Then (7.5.15) follows.

7.5.3.1.3 Solvability of (7.5.13)

We have

$$T\, P_{\mathrm{b.v}}\, \mathsf{R}^L_B(\ker A_{\mathrm{M}}) = R(\kappa). \tag{7.5.19}$$

Moreover, we claim that

$$T\, P_{\mathrm{b.v}}\, B_\ell \mathsf{R}^L_{B_\ell} h = \begin{pmatrix} m_{13} g_1 * \mathsf{R}^L_{B_\ell} h(b) + m_{14} B_\ell \mathsf{R}^L_{B_\ell} h(b) \\ m_{23} g_1 * \mathsf{R}^L_{B_\ell} h(b) + m_{24} B_\ell \mathsf{R}^L_{B_\ell} h(b) \end{pmatrix}, \tag{7.5.20}$$

where B_ℓ abbreviates $B_{\ell,\mathcal{V}}$. In order to prove (7.5.20), we set $v := B_\ell \mathsf{R}^L_{B_\ell} h$ and denote by (V_1, V_2, V_3, V_4) its endogenous boundary values with respect to the direct sum decompositions (6.7.7). By (6.7.50) and Remark 6.7.1, we compute

$$V_1 = V_2 = 0, \qquad V_3 = k * v(b), \qquad V_4 = v(b).$$

Then (7.5.20) follows. Now we are able to implement the solvability method of Sect. 7.5.2.2.

(i) Cases where κ is invertible. We will apply Corollary 7.2.23 with $\mathcal{X}_1 := \mathcal{Y}_f \times \mathcal{Y}_f$. By (7.5.19), $T P_{b.v} \mathrm{R}^L_{B|\ker A_M}$ is an isomorphism from $\ker A_M := \ker \mathrm{D}^\ell_{\mathcal{V}} = g_1 \mathcal{Y}_f \oplus \ell \mathcal{Y}_f$ onto $\mathcal{Y}_f \times \mathcal{Y}_f$. Then Corollary 7.2.23 yields that (7.5.13) is well posed for each $(h_{b.c,1}, h_{b.c,2}, , h)$ in $\mathcal{Y}_f \times \mathcal{Y}_f \times \mathcal{V}$.

(ii) Cases where κ is not invertible. By (7.5.20), (7.2.49) is equivalent to

$$\begin{pmatrix} m_{13} g_1 * \mathrm{R}^L_{B_\ell} h(b) + m_{14} B_\ell \mathrm{R}^L_{B_\ell} h(b) - h_{b.c,1} \\ m_{23} g_1 * \mathrm{R}^L_{B_\ell} h(b) + m_{24} B_\ell \mathrm{R}^L_{B_\ell} h(b) - h_{b.c,2} \end{pmatrix} \in R(\kappa). \tag{7.5.21}$$

(ii-a) If (7.5.21) holds then Proposition 7.2.19 (i) yields that (7.5.13) admits at least a solution u_p, and the solution set of (7.5.13) reads

$$\left\{ u_p + \mathrm{R}^L_{B_\ell}(g_1 Y_1 + \ell Y_2) \mid (Y_1, Y_2) \in \ker \kappa \right\}.$$

Observe that $\ker \kappa$ is a nontrivial subspace due to *the rank-nullity theorem*. Also the solution set of (7.5.13) is a linear manifold having the same dimension than $\ker \kappa$.

(ii-b) If (7.5.21) does not hold, then (7.5.13) has no solution according to Proposition 7.2.19 (ii).

7.5.3.1.4 Particular Problems with the Caputo Derivative

Let L be in $\mathcal{L}(\mathcal{Y}_f)$, and data $(h_{b.c,1}, h)$ belong to in $\mathcal{Y}_f \times L^p(\mathcal{Y}_f)$. Then we consider the following problem:

$$\begin{cases} \text{Find } u \text{ in } D({}^{\mathrm{C}}\mathrm{D}^\ell_{\mathcal{V}}) \text{ such that} \\ {}^{\mathrm{C}}\mathrm{D}^\ell_{\mathcal{V}} u = Lu + h \\ m_1 u(0) + m_3 k * \big(u - u(0)\big)(b) + m_4 u(b) = h_{b.c,1}. \end{cases} \tag{7.5.22}$$

First of all, the latter boundary condition is well defined since any element of $D({}^{\mathrm{C}}\mathrm{D}^\ell_{\mathcal{V}})$ is continuous on $[0, b]$ due to (6.7.56). Moreover in view of Remark 6.7.1, for any u in $D({}^{\mathrm{C}}\mathrm{D}^\ell_{\mathcal{V}})$, one has

$$\big(u - \ell k * u(0)\big)(0) = u(0), \qquad k * u(0) = 0$$

$$g_1 * \mathrm{D}^\ell_{\mathcal{V}} u(b) = k * \big(u - u(0)\big)(b), \qquad \ell * \mathrm{D}^\ell_{\mathcal{V}} u(b) = u(b) - u(0).$$

Thus setting

$$m_{11} := m_1 + m_4, \quad m_{12} := 0, \quad m_{13} := m_3, \quad m_{14} := m_4, \quad m_{2j} := \delta_{2j},$$

(7.5.22) is equivalent to

$$\begin{cases} \text{Find } u \text{ in } D(\mathrm{D}^{\ell}_{\mathcal{V}}) \text{ such that} \\ \mathrm{D}^{\ell}_{\mathcal{V}} u = Lu + h \\ m_{11}\, u(0) + \qquad\qquad m_{13}\, g_1 * \mathrm{D}^{\ell}_{\mathcal{V}} u(b) + m_{14}\, \ell * \mathrm{D}^{\ell}_{\mathcal{V}} u(b) = h_{\mathrm{b.c},1} \\ \qquad\qquad k * u(0) \qquad\qquad\qquad\qquad\qquad\qquad = 0. \end{cases}$$

Thus (7.5.22) is an instance of (7.5.13).

7.5.3.2 Case Where ℓ Lies in $L^q(0, b) \setminus L^p(0, b)$

Let L be in $\mathcal{L}(\mathcal{Y}_{\mathrm{f}})$.

7.5.3.2.1 Case Where k Lies in $L^q(0, b) \setminus L^p(0, b)$

7.5.3.2.1.1 Formulation of the Problem

Let $(h_{\mathrm{b.c}}, h)$ belong to in $\mathcal{Y}_{\mathrm{f}} \times L^p(\mathcal{Y}_{\mathrm{f}})$ and m_1, m_4 be scalars. Then using the "standard" boundary values computed in Sect. 6.7.2.2.1, we consider the following problem:

$$\begin{cases} \text{Find } u \text{ in } D(\mathrm{D}^{\ell}_{\mathcal{V}}) \text{ such that} \\ \mathrm{D}^{\ell}_{\mathcal{V}} u = Lu + h \\ m_1\, u(0) + m_4 u(b) = h_{\mathrm{b.c}}. \end{cases} \tag{7.5.23}$$

Setting

$$T : g_1 \mathcal{Y}_{\mathrm{f}} \oplus B_{\ell,\mathcal{V}} S_{\mathcal{V}}(h_2 \mathcal{Y}_{\mathrm{f}}) \to \mathcal{Y}_{\mathrm{f}}$$
$$g_1 Y_1 + \ell * h_2(b - \cdot)\, Y_4 \mapsto (m_1 + m_4) Y_1 + m_4 Y_4$$

and referring to (6.7.14), we see that

$$T P_{\mathrm{b.v}} u = m_1 u(0) + m_4 u(b), \qquad \forall\, u \in D({}^{\mathrm{RL}}\mathrm{D}^{\ell}_{\mathcal{V}}),$$

hence (7.5.23) is an instance of (7.2.48) with $\mathcal{X}_1 := \mathcal{Y}_{\mathrm{f}}$.

7.5.3.2.1.2 Solvability of (7.5.23)

Let κ be the element of $\mathcal{L}(\mathcal{Y}_{\mathrm{f}})$ defined by

$$\kappa : \mathcal{Y}_{\mathrm{f}} \to \mathcal{Y}_{\mathrm{f}}, \qquad Y \mapsto T P_{\mathrm{b.v}} \mathsf{R}^{L}_{B_{\ell,\mathcal{V}}}(g_1 Y). \tag{7.5.24}$$

Then

$$\kappa = m_1 \mathrm{id}_{\mathcal{Y}_{\mathrm{f}}} + m_4 \mathsf{R}^{L}_{B_{\ell,\mathcal{V}}}(g_1 \cdot)(b)$$

and

$$T P_{\mathrm{b.v}} \mathsf{R}_B^L(\ker A_{\mathrm{M}}) = R(\kappa). \tag{7.5.25}$$

Now we are in position to describe the solution set of (7.5.23).

(i) Cases where κ is invertible. By applying Corollary 7.2.23, we derive that (7.5.23) is well posed for each $(h_{\mathrm{b.c}}, h)$ in $\mathcal{Y}_{\mathrm{f}} \times \mathcal{V}$.

(ii) Cases where κ is not invertible. Since

$$T P_{\mathrm{b.v}} B\, \mathsf{R}_B^L h = m_4 B_{\ell,\mathcal{V}} \mathsf{R}_{B_{\ell,\mathcal{V}}}^L h(b),$$

(7.2.49) is equivalent to

$$m_4 B_{\ell,\mathcal{V}} \mathsf{R}_{B_{\ell,\mathcal{V}}}^L h(b) - h_{\mathrm{b.c}} \in R(\kappa). \tag{7.5.26}$$

(ii-a) If (7.5.26) holds then Proposition 7.2.19 (i) yields that (7.5.23) admits at least a solution u_{p}, and the solution set of (7.5.23) reads

$$\left\{u_{\mathrm{p}} + \mathsf{R}_{B_{\ell,\mathcal{V}}}^L (g_1 Y) \mid Y \in \ker \kappa\right\}.$$

(ii-b) If (7.5.26) does not hold, then (7.5.23) has no solution according to Proposition 7.2.19 (ii).

7.5.3.2.2 Case Where k Lies in $L^1(0, b) \setminus (L^p(0, b) \cup L^q(0, b))$

7.5.3.2.2.1 Formulation of the Problem

Let $(h_{\mathrm{b.c}}, h)$ belong to in $\mathcal{Y}_{\mathrm{f}} \times L^p(\mathcal{Y}_{\mathrm{f}})$ and m_1, m_3, m_4 be scalars. Then using the boundary values of functions computed in Sect. 6.7.2.2.2, we consider the following problem:

$$\begin{cases} \text{Find } u \text{ in } D(\mathrm{D}_{\mathcal{V}}^{\ell}) \text{ such that} \\ \mathrm{D}_{\mathcal{V}}^{\ell} u = Lu + h \\ m_1\, u(0) + m_3\, k * \big(u - u(0)\big)(b) + m_4\, u(b) = h_{\mathrm{b.c}}. \end{cases} \tag{7.5.27}$$

Setting

$$\begin{aligned} &T : E_{\mathrm{M}} = g_1 \mathcal{Y} \oplus B_{\ell,\mathcal{V}} S_{\mathcal{V}}(h_{\mathrm{b.c},1} \mathcal{Y} \oplus h_{\mathrm{b.c},2} \mathcal{Y}) \to \mathcal{Y}_{\mathrm{f}} \\ &g_1 Y_1 + \ell * h_{\mathrm{b.c},1}(b - \cdot) Y_3 + \ell * h_{\mathrm{b.c},2}(b - \cdot) Y_4 \mapsto (m_1 + m_4) Y_1 + m_3 Y_3 + m_4 Y_4 \end{aligned}$$

and referring to (6.7.19), we see that

$$T P_{\text{b.v}} u = m_1 u(0) + m_3\, k * \big(u - u(0)\big)(b) + m_4\, u(b), \qquad \forall u \in D({}^{\text{RL}}\mathrm{D}^{\ell}_{\mathcal{V}}).$$

Hence (7.5.23) is an instance of (7.2.48) with $\mathcal{X}_1 := \mathcal{Y}_{\text{f}}$.

7.5.3.2.2.2 Solvability of (7.5.27)

Let κ be defined by (7.5.24). We have

$$T P_{\text{b.v}}\, \mathsf{R}^L_B(\ker A_{\text{M}}) = R(\kappa). \tag{7.5.28}$$

Now we are in position to describe the solution set of (7.5.27).

(i) Cases where κ is invertible. By applying Corollary 7.2.23 with $\mathcal{X}_1 := \mathcal{Y}_{\text{f}}$, we derive that (7.5.27) is well posed for each $(h_{\text{b.c}}, h)$ in $\mathcal{Y}_{\text{f}} \times \mathcal{V}$.

(ii) Cases where κ is not invertible. Since

$$T P_{\text{b.v}} B\, \mathsf{R}^L_B h = m_3 g_1 * \mathsf{R}^L_{B_{\ell,\mathcal{V}}} h(b) + m_4 B_{\ell,\mathcal{V}} \mathsf{R}^L_{B_{\ell,\mathcal{V}}} h(b),$$

(7.2.49) is equivalent to

$$m_3 g_1 * \mathsf{R}^L_{B_{\ell,\mathcal{V}}} h(b) + m_4 B_{\ell,\mathcal{V}} \mathsf{R}^L_{B_{\ell,\mathcal{V}}} h(b) - h_{\text{b.c}} \in R(\kappa). \tag{7.5.29}$$

(ii-a) If (7.5.29) holds then Proposition 7.2.19 (i) yields that (7.5.27) admits at least a solution u_{p} and the solution set of (7.5.27) reads in view of Proposition 7.2.19 (i)

$$\big\{u_{\text{p}} + \mathsf{R}^L_{B_{\ell,\mathcal{V}}}(g_1\, Y) \mid Y \in \ker \kappa\big\}.$$

(ii-b) If (7.5.29) does not hold, then (7.5.27) has no solution according to Proposition 7.2.19 (ii).

7.6 Abstract Sublinear Problems

By *sublinear problems*, we mean problems containing a Lipschitz continuous nonlinearity acting on zero-order terms. Regarding boundary conditions, we will consider only endogenous boundary values related to the kernel of the differential operator. Thus we will work in Banach spaces, and applications will be *initial* value problems.

Our approach relies on *the Picard iterations scheme* (see, for instance, [Bre11, Th. 7.3]). Thus we need *unconditionally solvability* of some linear problems. This explains why we consider only a special type of boundary conditions. Observe that this kind of issue occurs already in the theory of entire order differential equations. Indeed, this initial value problem

$$\mathrm{D}^1_{L^2}u = h, \qquad u(0) = u_1$$

is unconditionally solvable. Whereas

$$\mathrm{D}^1_{L^2}u = h, \qquad u(0) = u(b)$$

is not.

Up to now we have developed a ready-to-use theory. In the forthcoming subsection, we will make a strong assumption (namely, (7.6.2)) which affects the scope of our abstract results. However in applications (at least in the framework of Lebesgue spaces), it will be possible to recover the expected level of generality.

7.6.1 Formulation of the Problem

Let $(\mathcal{Z}, \|\cdot\|)$ be a Banach space, $A : D(A) \subseteq \mathcal{Z} \to \mathcal{Z}$ be an operator on $\mathcal{Z}$, and B belong to $\mathcal{L}(\mathcal{Z})$. We assume (7.2.3) and that A is a closed operator on $\mathcal{Z}$. By Sect. 7.2.3.1, the projection $P_{\ker A}$ on $\ker A$ along $R(B)$ is well defined.

Let $F : \mathcal{Z} \to \mathcal{Z}$ be a (nonlinear) map on $\mathcal{Z}$ and $(h_{\mathrm{b.c}}, h)$ be a datum in $\ker A \times \mathcal{Z}$. Then we consider the following problem, called $P(h, h_{\mathrm{b.c}})$:

$$P(h, h_{\mathrm{b.c}}) \begin{cases} \text{Find } u \text{ in } D(A) \text{ such that} \\ Au = F(u) + h, \qquad P_{\ker A}u = h_{\mathrm{b.c}}. \end{cases} \tag{7.6.1}$$

In this context, Definition 7.2.6 may be extended as follows.

Definition 7.6.1 Problem (7.6.1) is said to be *well posed* if the following two conditions are satisfied:

(i) For any $(h_{\mathrm{b.c}}, h)$ in $\ker A \times \mathcal{Z}$, (7.6.1) has a unique solution.
(ii) There exists a constant C such that for any $(h_{\mathrm{b.c}}, h)$ and $(h'_{\mathrm{b.c}}, h')$ in $\ker A \times \mathcal{Z}$, the respective solutions u and v of $P(h, h_{\mathrm{b.c}})$ and $P(h', h'_{\mathrm{b.c}})$ satisfy

$$\|u - v\| \leq C(\|h - h'\| + \|h_{\mathrm{b.c}} - h'_{\mathrm{b.c}}\|).$$

□

7.6.2 Solvability Under a Strong Assumption on F

Our strong assumption on F is the following: there exists a constant ε in $(0, 1)$ such that

$$\|B \circ F(u) - B \circ F(v)\| \leq \varepsilon \|u - v\|, \qquad \forall u, v \in \mathcal{Z}. \tag{7.6.2}$$

Assumption (7.6.2) says that the Lipschitz constant of the nonlinear maps $B \circ F$ is strictly less than one. In application, we will get rid of the smallness of ε by implementing the so-called Bielecki method (see 7.7). Observe that even in standard cases, a change of norm occurs in the Picard iterations scheme (see [Bre11, Th. 7.3]).

Our first result states roughly speaking that the solutions to (7.6.1) (if they exist) are continuous with respect to the data.

Lemma 7.6.1 *Being given* $(h_{\mathrm{b.c}}, h)$ *and* $(h'_{\mathrm{b.c}}, h')$ *in* $\ker A \times \mathcal{Z}$, *let* u, v *be solution of* $P(h, h_{\mathrm{b.c}})$ *and* $P(h', h'_{\mathrm{b.c}})$, *respectively. Then*

$$\|u - v\| \lesssim (\|h - h'\| + \|h_{\mathrm{b.c}} - h'_{\mathrm{b.c}}\|).$$

The latter estimate shall be understand in the sense Definition 7.6.1.

Proof of Lemma 7.6.1 By Lemma 4.3.3,

$$u - v = h_{\mathrm{b.c}} - h'_{\mathrm{b.c}} + B \circ F(u) - B \circ F(v) + B(h - h').$$

Then the estimate follows from (7.6.2). □

Proposition 7.6.2 *Let* $A : D(A) \subseteq \mathcal{Z} \to \mathcal{Z}$ *be a closed operator on a Banach space* $(\mathcal{Z}, \|\cdot\|)$, B *belong to* $\mathcal{L}(\mathcal{Z})$, *and* $F : \mathcal{Z} \to \mathcal{Z}$ *be a map on* $\mathcal{Z}$. *If* (7.2.3) *and* (7.6.2) *hold then* (7.6.1) *is well posed.*

Proof Let $(h_{\mathrm{b.c}}, h)$ belong to $\ker A \times \mathcal{Z}$. The existence is obtained by implementing the classical *Picard iteration scheme*. More precisely, by Lemma 4.3.3, (7.6.1) is equivalent to the following equation whose unknown u lies in $\mathcal{Z}$.

$$u = h_{\mathrm{b.c}} + B \circ \big(F(u) + h\big).$$

Then denoting by u_0 the zero vector of $\mathcal{Z}$, we construct inductively the sequence $(u_n)_{n \geq 1}$ in $\mathcal{Z}$ satisfying for every positive integer n,

$$u_n = h_{\mathrm{b.c}} + B \circ \big(F(u_{n-1}) + h\big).$$

Thus for each $n \geq 2$, one has thanks (7.6.2)

$$\|u_n - u_{n-1}\| \le \varepsilon \|u_{n-1} - u_{n-2}\|.$$

Then by induction

$$\|u_n - u_{n-1}\| \le \varepsilon^{n-1} \|u_1\|, \qquad \forall n \ge 1.$$

Since, by assumption, $\varepsilon < 1$ and $\mathcal{Z}$ is a Banach space, the sequence (u_n) has a limit u_∞ in $\mathcal{Z}$. With a continuity argument relying on (7.6.2), we derive that u_∞ solves (7.6.1). The well posedness is a consequence of Lemma 7.6.1. □

7.6.3 Invariance with Respect to Conjugation and Well Posedness

We will verify that the solution set to (7.6.1) is in some sense invariant with respect to conjugation by an isomorphism. The assumptions and notation of Sect. 7.6.1 remaining in force, let $(\tilde{\mathcal{Z}}, \|\cdot\|_{\tilde{\mathcal{Z}}})$ be a Banach space and $I : \mathcal{Z} \to \tilde{\mathcal{Z}}$ be an isomorphism. The issue is to transport Problem (7.6.1) into a problem set on $\tilde{\mathcal{Z}}$ by means of I. For, we put

$$\tilde{A} := I \circ A \circ I^{-1}, \qquad \tilde{B} := I \circ B \circ I^{-1}, \qquad \tilde{F} := I \circ F \circ I^{-1}.$$

Clearly $\tilde{B}$ is a right inverse of $\tilde{A}$, and $\tilde{A}$ is a closed operator on $\tilde{\mathcal{Z}}$. Also, for each data $(h_{\mathrm{b.c}}, h)$ in $\ker A \times \mathcal{Z}$, we set

$$\tilde{h}_{\mathrm{b.c}} := I h_{\mathrm{b.c}}, \qquad \tilde{h} := I h,$$

and in view of Sect. 7.6.1, we consider the following problem:

$$\begin{cases} \text{Find } v \text{ in } D(\tilde{A}) \text{ such that} \\ \tilde{A} v = \tilde{F}(v) + \tilde{h}, \qquad P_{\ker \tilde{A}} v = \tilde{h}_{\mathrm{b.c}}. \end{cases} \tag{7.6.3}$$

The thrust of the next result is that Problems (7.6.1) and (7.6.3) are in some sense equivalent.

Proposition 7.6.3 *Let*

- *$\mathcal{Z}$, $\tilde{\mathcal{Z}}$ be a Banach spaces and $I : \mathcal{Z} \to \tilde{\mathcal{Z}}$ be an isomorphism;*
- *A be a closed operator on $\mathcal{Z}$, and B be a right inverse of A belonging to $\mathcal{L}(\mathcal{Z})$;*
- *$F : \mathcal{Z} \to \mathcal{Z}$;*
- *$(h_{\mathrm{b.c}}, h)$ be a datum in* $\ker A \times \mathcal{Z}$.

Then I induces a bijection between the solution set of (7.6.1) *and the solution set of* (7.6.3). *Moreover,*

(i) (7.6.1) *is well posed if and only if* (7.6.3) *is well posed.*
(ii) *If in addition, there exists* ε *in* $(0, 1)$ *such that*

$$\|\tilde{B} \circ \tilde{F}(u) - \tilde{B} \circ \tilde{F}(v)\|_{\tilde{\mathcal{Z}}} \leq \varepsilon \|u - v\|_{\tilde{\mathcal{Z}}}, \qquad \forall u, v \in \tilde{\mathcal{Z}}, \tag{7.6.4}$$

then (7.6.1) *is well posed.*

Proof Being given a solution u to (7.6.1), the vector $v := Iu$ satisfies

$$\tilde{A}v - \tilde{F}(v) - \tilde{h} = I \circ \big(Au - F(u) - h\big) = 0.$$

Regarding the boundary condition, we have to work with projections on domains of operators. Hence we introduce the linear map

$$J : D(A) \to D(\tilde{A}), \qquad u \mapsto Iu.$$

Since

$$\begin{aligned} \|Ju\|_{D(\tilde{A})} &= \|IAu\|_{\tilde{\mathcal{Z}}} + \|Iu\|_{\tilde{\mathcal{Z}}} \\ &\lesssim \|Au\| + \|u\| \\ &\lesssim \|u\|_{D(A)}, \qquad\qquad \forall u \in D(A), \end{aligned}$$

we derive, in a standard way, that J is an isomorphism between the Banach spaces $D(A)$ and $D(\tilde{A})$. Now we claim that

$$P_{\ker \tilde{A}} = J \circ P_{\ker A} \circ J^{-1}. \tag{7.6.5}$$

Indeed by Proposition 2.3.6, $J \circ P_{\ker A} \circ J^{-1}$ is the projection of $D(\tilde{A})$ on $JR(P_{\ker A})$ along $J(\ker P_{\ker A})$. Since these spaces are respectively equal to $\ker \tilde{A}$ and $R(\tilde{B})$, the claim follows from the very definition of $P_{\ker \tilde{A}}$.

With (7.6.5), we get $P_{\ker \tilde{A}} v = \tilde{h}_{\text{b.c}}$. Hence v solves (7.6.3). By symmetry, $I^{-1}v$ solves (7.6.1) for each solution v of (7.6.3). Whence we obtain the expected bijection induced by I. Moreover since

$$\|z\| \lesssim \|Iz\|_{\tilde{\mathcal{Z}}} \lesssim \|z\|, \qquad \forall z \in \mathcal{Z},$$

we may show easily the equivalence stated in Item (i).

Finally, Item (ii) will be readily proved. Indeed, we have already observed (and used) that $\tilde{B}$ is a right inverse of $\tilde{A}$ and that $\tilde{A}$ is a closed operator on $\tilde{\mathcal{Z}}$. With (7.6.4), we deduce thanks to Proposition 7.6.2 that (7.6.3) is well posed. Then Item (i) entails that (7.6.1) is also well posed. □

7.7 Sublinear Initial Value Problems

Let $(\mathcal{Y}, \|\cdot\|_{\mathcal{Y}})$ be a nontrivial real or complex Banach space and p, q be conjugate exponents such that p lies in $(1, \infty)$. We set

$$\mathcal{Z} = L^p(\mathcal{Y}) := L^p\big((0, b), \mathcal{Y}\big)$$

and denote by $\|\cdot\|$ the usual norm of $L^p(\mathcal{Y})$.

7.7.1 Bielecki's Norms on $L^p(\mathcal{Y})$

Starting from a Lipschitz continuous nonlinearity F, our goal is to construct an equivalent norm on $\mathcal{Z}$ such that (7.6.2) holds. Such a process is called a *Bielecki method*. Our presentation relies on [Kwa91]. Being given a fixed function f in $L^q(0, b)$, we put for each ε in $(0, 1)$

$$M : [0, b] \to \mathbb{R}, \qquad x \mapsto \left(\int_0^x |f(y)|^q \, \mathrm{d}y\right)^{\frac{p}{q}}$$

$$\omega : [0, b] \to \mathbb{R}, \qquad x \mapsto \exp\left(\varepsilon^{-p} \int_0^x M(y) \, \mathrm{d}y\right).$$

Now for each function u in $L^p(\mathcal{Y})$, we set

$$\|u\|_{\varepsilon,p} := \left(\sup_{x\in(0,b)} \left[\frac{1}{\omega(x)} \int_0^x \|u(y)\|_{\mathcal{Y}}^p \, \mathrm{d}y\right]\right)^{\frac{1}{p}}. \tag{7.7.1}$$

The following result states that $\|\cdot\|_{\varepsilon,p}$ is indeed an equivalent norm on $L^p(\mathcal{Y})$, which is called a *Bielecki norm*.

Lemma 7.7.1 *Under the assumptions and notation of the current section, $\|\cdot\|_{\varepsilon,p}$ is a norm on $L^p(\mathcal{Y})$ which is equivalent to the standard norm of $L^p(\mathcal{Y})$.*

Proof Since M is nondecreasing, one has for each x in $(0, b)$,

$$1 \le \omega(x) \le \exp\big(\varepsilon^{-p} b M(b)\big).$$

Hence

$$\exp\big(-\varepsilon^{-p} b M(b)\big)\|u\| \le \|u\|_{\varepsilon,p} \le \|u\|, \qquad \forall\, u \in L^p(\mathcal{Y}).$$

Next we may show easily that $\|\cdot\|_{\varepsilon,p}$ is a norm on $L^p(\mathcal{Y})$. □

We are in position to state the main result of this subsection whose proof is tricky.

Theorem 7.7.2 ([Kwa91]) *Let*

(i) $(\mathcal{Y}, \|\cdot\|_{\mathcal{Y}})$ *be a nontrivial Banach space;*
(ii) p, q *be conjugate exponents such that* p *lies in* $(1, \infty)$*;*
(iii) f *belong to* $L^q(0, b)$*;*
(iv) $F : L^p(\mathcal{Y}) \to L^p(\mathcal{Y})$ *and* C *be a positive number such that for each* u *and* v *in* $L^p(\mathcal{Y})$,

$$\|F(u) - F(v)\|_{\mathcal{Y}} \leq C\|u - v\|_{\mathcal{Y}} \qquad \text{in } L^p(0, b).$$

Then for each ε *in* $(0, 1)$*, one has*

$$\|f * F(u) - f * F(v)\|_{\varepsilon,p} \leq C\varepsilon\|u - v\|_{\varepsilon,p}, \qquad \forall u, v \in L^p(\mathcal{Y}).$$

Proof For each y in $(0, b)$, by using Assumption (iv), we estimate

$$\begin{aligned}\|f * F(u)(y) - f * F(v)(y)\|_{\mathcal{Y}}^p &\leq \left(\int_0^y C|f(y-s)|\|(u-v)(s)\|_{\mathcal{Y}}\,\mathrm{d}s\right)^p \\ &\leq C^p M(y)\int_0^y \|(u-v)(s)\|_{\mathcal{Y}}^p\,\mathrm{d}s,\end{aligned}$$

by Hölder inequality. Thus integrating on $(0, x)$ for each x in $(0, b)$,

$$\begin{aligned}&\frac{1}{C^p}\int_0^x \|f * F(u)(y) - f * F(\cdot, v)(y)\|_{\mathcal{Y}}^p\,\mathrm{d}y \\ &\quad\leq \int_0^x M(y)\exp\left(\varepsilon^{-p}\int_0^y M(s)\,\mathrm{d}s\right)\left(\frac{1}{\omega(y)}\int_0^y \|(u-v)(s)\|_{\mathcal{Y}}^p\,\mathrm{d}s\right)\mathrm{d}y \\ &\quad\leq \|u-v\|_{\varepsilon,p}^p\int_0^x M(y)\exp\left(\varepsilon^{-p}\int_0^y M(s)\,\mathrm{d}s\right)\mathrm{d}y.\end{aligned}$$

Now observe that

$$\begin{aligned}\int_0^x M(y)\exp\left(\varepsilon^{-p}\int_0^y M(s)\,\mathrm{d}s\right)\mathrm{d}y &= \int_0^x \varepsilon^p\frac{\mathrm{d}}{\mathrm{d}y}\left\{\exp\left(\varepsilon^{-p}\int_0^y M(s)\,\mathrm{d}s\right)\right\}\mathrm{d}y \\ &\leq \varepsilon^p w(x),\end{aligned}$$

by integration. Thus we get

$$\frac{1}{\omega(x)}\int_0^x \|f * F(u)(y) - f * F(v)(y)\|_{\mathcal{Y}}^p \,\mathrm{d}y \leq (C\varepsilon)^p \|u - v\|_{\varepsilon,p}^p.$$

Then the expected estimate follows. □

7.7.2 The Nonlinear Term

In applications, nonlinear terms do not appear as maps defined on the functional space $L^p(\mathcal{Y})$. On the contrary, they turn out to be functions defined on the phase space $\mathcal{Y}$. Then in order to apply the abstract theory featured in Sect. 7.6, the issue is to define a map on $L^p(\mathcal{Y})$ starting from a map defined on $\mathcal{Y}$, or more generally on $(0, b) \times \mathcal{Y}$. For, we have to assume that $\mathcal{Y}$ is separable.

Let $F : (0, b) \times \mathcal{Y} \to \mathcal{Y}$ be a Caratheodory function (in the sense of Definition 3.1.3). We assume that there exists a positive constant C_F such that for all u and v in $\mathcal{Y}$,

$$\|F(\cdot, u) - F(\cdot, v)\|_{\mathcal{Y}} \leq C_F \|u - v\|_{\mathcal{Y}}, \tag{7.7.2}$$

almost everywhere on $(0, b)$. In order to go further, we suppose in addition that

$$F(\cdot, 0_{\mathcal{Y}}) \in L^p(\mathcal{Y}). \tag{7.7.3}$$

Lemma 7.7.3 *Let*

- $(\mathcal{Y}, \|\cdot\|_{\mathcal{Y}})$ *be a real or complex separable Banach space;*
- p *be a real number in* $[1, \infty)$*;*
- $F : (0, b) \times \mathcal{Y} \to \mathcal{Y}$ *be Caratheodory function satisfying* (7.7.2) *and* (7.7.3).

Then $F(\cdot, u)$ *lies in* $L^p(\mathcal{Y})$ *for each* u *in* $L^p(\mathcal{Y})$*. Moreover, the map*

$$L^p(\mathcal{Y}) \to L^p(\mathcal{Y}), \qquad x \mapsto F(\cdot, u)$$

is Lipschitz continuous on $L^p(\mathcal{Y})$*, i.e.,*

$$\|F(\cdot, u) - F(\cdot, v)\|_{L^p(\mathcal{Y})} \leq C_F \|u - v\|_{L^p(\mathcal{Y})}, \qquad \forall u, v \in L^p(\mathcal{Y}).$$

For all purposes, let us precise that for each u in $L^p(\mathcal{Y})$, $F(\cdot, u)$ is the function of $L^p(\mathcal{Y})$ defined for almost everywhere x in $(0, b)$ by

$$F(\cdot, u)(x) := F\big(x, u(x)\big).$$

Proof of Lemma 7.7.3 Let u belong to $L^p(\mathcal{Y})$. We know from Theorem 3.1.3 that $F(\cdot, u)$ is L-measurable. Moreover with (7.7.2)

$$\|F(\cdot, u)\|_{\mathcal{Y}} \leq C_F \|u\|_{\mathcal{Y}} + \|F(\cdot, 0_{\mathcal{Y}})\|_{\mathcal{Y}},$$

almost everywhere on $(0, b)$. Since $F(\cdot, 0_{\mathcal{Y}})$ is assumed to lie in $L^p(\mathcal{Y})$, Bochner theorem 3.1.6 yields that $F(\cdot, u)$ belongs to $L^p(\mathcal{Y})$. Eventually the Lipschitz continuity of the map $u \mapsto F(\cdot, u)$ is a straightforward consequence of (7.7.2). □

The following result relies on Theorem 7.7.2 and Proposition 7.6.3. In view of Chap. 6, it should be clear that it applies to a large class of fractional differential operators.

Corollary 7.7.4 *Let*

- $(\mathcal{Y}, \|\cdot\|_{\mathcal{Y}})$ *be a nontrivial real or complex separable Banach space;*
- p *be a real number in* $(1, \infty)$ *and* $\mathcal{Z} = L^p(\mathcal{Y}) := L^p\big((0, b), \mathcal{Y}\big)$*;*
- A *be a closed operator on* $\mathcal{Z}$*;*
- f *be a function in* $L^1(0, b)$ *such that* $B_{f,\mathcal{Z}}$ *is a right inverse of* A*;*
- $F : (0, b) \times \mathcal{Y} \to \mathcal{Y}$ *be Caratheodory function satisfying* (7.7.2) *and* (7.7.3)*;*
- $(h_{\text{b.c}}, h)$ *be a datum in* $\ker A \times L^p(\mathcal{Y})$.

Then the problem

$$\begin{cases} \text{Find } u \text{ in } D(A) \text{ such that} \\ Au = F(\cdot, u) + h, \qquad P_{\ker A} u = h_{\text{b.c}} \end{cases} \tag{7.7.4}$$

is well posed in the sense of Definition 7.6.1.

Proof Implementing the method introduced in Sect. 7.6.3, we choose $\varepsilon := \frac{1}{2C_F}$ (where C_F is the constant appearing in (7.7.2)) and denote by $\tilde{\mathcal{Z}}$ the vector space $L^p(\mathcal{Y})$ endowed with the Bielecki norm $\|\cdot\|_{\varepsilon,p}$.

Now denote by I the identity map between $\mathcal{Z} := L^p(\mathcal{Y})$ and $\tilde{\mathcal{Z}}$, i.e.,

$$I : \mathcal{Z} \to \tilde{\mathcal{Z}}, \qquad u \mapsto u.$$

Lemma 7.7.1 tells us that I is an isomorphism. For simplicity, for each u in $L^p(\mathcal{Y})$, we denote by $F_{\mathcal{Z}}(u)$ the function $F(\cdot, u)$, namely, we set

$$F_{\mathcal{Z}} : \mathcal{Z} = L^p\big((0, b), \mathcal{Y}\big) \to \mathcal{Z}, \qquad u \mapsto F(\cdot, u). \tag{7.7.5}$$

Then we put

$$\tilde{B} := I \circ B_{f,\mathcal{Z}} \circ I^{-1}, \qquad F_{\tilde{\mathcal{Z}}} := I \circ F_{\mathcal{Z}} \circ I^{-1}.$$

Since $\tilde{B}u = f * u$ for each u in $\tilde{\mathcal{Z}}$, we derive from Theorem 7.7.2 that

$$\|\tilde{B} \circ F_{\tilde{\mathcal{Z}}}(u) - \tilde{B} \circ F_{\tilde{\mathcal{Z}}}(v)\|_{\varepsilon,p} \leq \frac{1}{2}\|u - v\|_{\varepsilon,p}, \qquad \forall u, v \in \tilde{\mathcal{Z}}.$$

Eventually using Proposition 7.6.3 (ii), we derive the well posedness of (7.7.4), which proves the claim. □

7.7.3 First-Order Problems

Let $(\mathcal{Y}, \|\cdot\|_{\mathcal{Y}})$ be a nontrivial separable Banach space, p lie in $(1, \infty)$, and $F : (0, b) \times \mathcal{Y} \to \mathcal{Y}$ be a Caratheodory function satisfying (7.7.2) and (7.7.3).

7.7.3.1 Formulation of the Problem

Let $h_{\text{b.c},1}$, h be data in $\mathcal{Y}$ and $L^p(\mathcal{Y})$, respectively. By Lemma 7.7.3, $F(\cdot, u)$ belongs to $L^p(\mathcal{Y})$ for each u in $L^p(\mathcal{Y})$; hence we may legitimately consider the following Problem:

$$P_1(h, h_{\text{b.c},1}) \begin{cases} \text{Find } u \text{ in } W^{1,p}(\mathcal{Y}) \text{ such that} \\ \mathrm{D}^1_{\mathcal{Z}} u = F(\cdot, u) + h, \qquad u(0) = h_{\text{b.c},1}. \end{cases} \tag{7.7.6}$$

For all purposes, let us notice that the first equation takes place in $L^p(\mathcal{Y})$, whereas the boundary condition holds in $\mathcal{Y}$.

Definition 7.7.1 Problem (7.7.6) is said to be *well posed* if the following two conditions are satisfied:

(i) For any $(h_{\text{b.c},1}, h)$ in $\mathcal{Y} \times L^p(\mathcal{Y})$, (7.7.6) has a unique solution.
(ii) There exists a constant C such that for any $(h_{\text{b.c},1}, h)$ and $(h'_{\text{b.c},1}, h')$ in $\mathcal{Y} \times L^p(\mathcal{Y})$, the respective solutions u and v of $P_1(h_{\text{b.c},1}, h_{\text{b.c}})$ and $P_1(h'_{\text{b.c},1}, h'_{\text{b.c}})$ satisfy

$$\|u - v\|_{L^p(\mathcal{Y})} \leq C(\|h - h'\|_{L^p(\mathcal{Y})} + \|h_{\text{b.c}} - h'_{\text{b.c}}\|_{\mathcal{Y}}).$$

□

7.7.3.2 Well Posedness of (7.7.6)

Under the above assumptions and notation of the current subsection, we claim that (7.7.6) *is well posed in the sense of Definition 7.7.1*. Indeed, setting

$$h_{\mathrm{b.c}} := g_1\, h_{\mathrm{b.c},1}, \qquad A := \mathrm{D}^1_{\mathcal{Z}},$$

and arguing as in Sect. 7.3.2.1, we may check that (7.7.6) is an instance of (7.6.1). That is to say, (7.7.6) reads

$$\begin{cases} \text{Find } u \text{ in } D(A) \text{ such that} \\ Au = F(\cdot, u) + h, \qquad P_{\ker A} u = h_{\mathrm{b.c}}. \end{cases} \tag{7.7.7}$$

Thus the claim follows from Corollary 7.7.4.

7.7.3.3 Nonautonomous Linear Problems

Being given a positive integer d_{f}, denote by $\mathcal{M}(d_{\mathrm{f}}, \mathbb{K})$ the space of square matrices of dimension d_{f} having their entry in $\mathbb{K}$. Next, consider

$$L \in L^\infty\big((0, b), \mathcal{M}(d_{\mathrm{f}}, \mathbb{K})\big).$$

That is, for all indices i, j in $[1, d_{\mathrm{f}}]$, there exists a function a_{ij} in $L^\infty(0, b)$ such that for almost every x in $(0, b)$, the entry (i, j) of the matrix $L(x)$ is equal to $a_{ij}(x)$. In symbol,

$$L(x) = \big(a_{ij}(x)\big)_{1 \le i, j \le d_{\mathrm{f}}},$$

for almost every x in $(0, b)$.

For any p in $(1, \infty)$, let $(h_{\mathrm{b.c},1}, h)$ belong to $\mathbb{K}^{d_{\mathrm{f}}} \times L^p\big((0, b), \mathbb{K}^{d_{\mathrm{f}}}\big)$. Then setting $\mathcal{Z} := L^p\big((0, b), \mathbb{K}^{d_{\mathrm{f}}}\big)$, we consider the problem

$$\begin{cases} \text{Find } u \text{ in } W^{1,p}\big((0, b), \mathbb{K}^{d_{\mathrm{f}}}\big) \text{ such that} \\ \mathrm{D}^1_{\mathcal{Z}} u = L(\cdot) u + h, \qquad u(0) = h_{\mathrm{b.c},1}. \end{cases} \tag{7.7.8}$$

For all purposes, let us precise that by setting $u = (u_1, \dots, u_{d_{\mathrm{f}}})^t$, the first equation of (7.7.8) can be written in coordinates as

$$\mathrm{D}^1_{L^p} u_i = \sum_{j=1}^{d_{\mathrm{f}}} a_{ij}(\cdot) u_j + h_i \qquad \text{in } L^p(0, b), \qquad \forall i = 1, \dots, d_{\mathrm{f}}.$$

Hence (7.7.8) is a *system of nonautonomous linear first-order equations supplemented with initial conditions.*

We claim that (7.7.8) *is well posed.* Indeed, let

$$F : (0, b) \times \mathbb{K}^{d_{\mathrm{f}}} \to \mathbb{K}^{d_{\mathrm{f}}}$$

be defined by

$$F(x,u) = \big(a_{1j}(x)u_j, \ldots, a_{d_f j}(x)u_j\big)^t$$

for almost every x in $(0,b)$ and every $u = (u_1, \ldots, u_{d_f})^t$ in $\mathcal{Y} := \mathbb{K}^{d_f}$. Plainly F is a Caratheodory function satisfying (7.7.2) and (7.7.3). Whence the claim follows from Sect. 7.7.3.2.

7.7.4 Problems with $A := \mathrm{D}^{\ell}_{\mathcal{Z}}$

Let $(\mathcal{Y}, \|\cdot\|_{\mathcal{Y}})$ be a nontrivial real or complex separable Banach space, and $F : (0,b) \times \mathcal{Y} \to \mathcal{Y}$ be a Caratheodory function satisfying (7.7.2) and (7.7.3). Let also p, q be conjugate exponents in $(1,\infty)$, and

$$\mathcal{Z} = L^p(\mathcal{Y}) := L^p\big((0,b), \mathcal{Y}\big).$$

Being given a Sonine pair (k,ℓ), recall that $\mathrm{D}^{\ell}_{\mathcal{Z}}$ is the Caputo extension of ${}^{\mathrm{RL}}\mathrm{D}^{\ell}_{\mathcal{Z}}$ defined in Sect. 6.4.

7.7.4.1 Case Where ℓ Lies in $L^p(0,b) \cap L^q(0,b)$

In the particular case where

$$k := g_{1-\alpha}, \qquad \ell := g_\alpha, \qquad \forall \alpha \in (0,1),$$

the condition

$$g_\alpha \in L^p(0,b) \cap L^q(0,b)$$

is equivalent to

$$\frac{1}{p} < \alpha \quad \text{and} \quad 1 - \frac{1}{p} < \alpha.$$

7.7.4.1.1 Formulation of the Problem and Well Posedness

Let $(h_{\mathrm{b.c},1}, h_{\mathrm{b.c},2}, h)$ be data in $\mathcal{Y} \times \mathcal{Y} \times L^p(\mathcal{Y})$. By Lemma 7.7.3, $F(\cdot,u)$ belongs to $L^p(\mathcal{Y})$ for each u in $L^p(\mathcal{Y})$. Moreover, in view of Sect. 7.3.4.1.1, $u - \ell\, k * u(0)$ is continuous on $[0,b]$ for each u in $D(\mathrm{D}^{\ell}_{\mathcal{Z}})$, thus we may legitimately consider the problem

$$\begin{cases} \text{Find } u \text{ in } D(\mathrm{D}^{\ell}_{\mathcal{Z}}) \text{ such that} \\ \mathrm{D}^{\ell}_{\mathcal{Z}} u = F(\cdot, u) + h \quad \text{ in } L^p(\mathcal{Y}) \\ \big(u - \ell\, k * u(0)\big)(0) = h_{\mathrm{b.c},1}, \qquad k * u(0) = h_{\mathrm{b.c},2}. \end{cases} \tag{7.7.9}$$

We claim that (7.7.9) *is an instance of* (7.7.4). For, it is enough to recast the initial conditions. Since ℓ lies in $L^q(0, b)$, Proposition 6.3.1 (ii-b) entails that

$$\ker \mathrm{D}^{\ell}_{\mathcal{Z}} = g_1\, \mathcal{Y} \oplus \ell\, \mathcal{Y}.$$

Then by using the projection $P_{\ker \mathrm{D}^{\ell}_{\mathcal{Z}}}$ introduced in Sect. 7.3.4.1.2, we rewrite initial conditions of (7.7.9) under the form

$$P_{\ker \mathrm{D}^{\ell}_{\mathcal{Z}}} u = g_1\, h_{\mathrm{b.c},1} + \ell\, h_{\mathrm{b.c},2},$$

which proves the claim.

According to Sect. 7.3.4.1.2, $B_{\ell,\mathcal{Z}}$ is a right inverse of $\mathrm{D}^{\ell}_{\mathcal{Z}}$. Thus recalling that $F : (0, b) \times \mathcal{Y} \to \mathcal{Y}$ is assumed to be a Caratheodory function satisfying (7.7.2) and (7.7.3), Corollary 7.7.4 yields that (7.7.9) is *well posed in a sense which can be easily inferred from Definition 7.7.1 or 7.6.1.*

7.7.4.1.2 Particular Cases

7.7.4.1.2.1 Linear Problems

As in Sect. 7.7.3.3, setting $\mathcal{Y} := \mathbb{K}^{d_{\mathrm{f}}}$, we consider the problem

$$\begin{cases} \text{Find } u \text{ in } D(\mathrm{D}^{\ell}_{\mathcal{Z}}) \text{ such that} \\ \mathrm{D}^{\ell}_{\mathcal{Z}} u = L(\cdot) u + h \quad \text{ in } L^p(\mathcal{Y}) \\ \big(u - \ell\, k * u(0)\big)(0) = h_{\mathrm{b.c},1}, \qquad k * u(0) = h_{\mathrm{b.c},2}, \end{cases} \tag{7.7.10}$$

where

$$L \in L^{\infty}\big((0, b), \mathcal{M}(d_{\mathrm{f}}, \mathbb{K})\big).$$

Since (7.7.10) is an instance of (7.7.9), there results that if p belongs to $(1, \infty)$ then Problem (7.7.10) is well posed.

Remark that it could be said that (7.7.9) is a *nonautonomous* problem; however this qualifier is not useful, neither suitable here since (7.3.13) is already nonautonomous. This is due to the nonlocality of the operator $\mathrm{D}^{\ell}_{\mathcal{Y}}$; see [ER17] for more details.

7.7.4.1.2.2 Riemann-Liouville Derivative

Under the assumptions and notation set at the beginning of the current subsection, let $(h_{\mathrm{b.c},2}, h)$ belong to $\mathcal{Y} \times L^p(\mathcal{Y})$. Recalling that ℓ lies in $L^p(0, b) \cap L^q(0, b)$, we consider the problem

$$\begin{cases} \text{Find } u \text{ in } D({}^{\mathrm{RL}}\mathrm{D}^{\ell}_{\mathcal{Z}}) \text{ such that} \\ {}^{\mathrm{RL}}\mathrm{D}^{\ell}_{\mathcal{Z}} u = F(\cdot, u) + h \quad \text{in } L^p(\mathcal{Y}) \\ k * u(0) = h_{\mathrm{b.c},2}. \end{cases} \tag{7.7.11}$$

We claim that (7.7.11) is well posed. Indeed Lemma 7.3.1 yields that (7.7.11) is equivalent to (7.7.9) where $h_{\mathrm{b.c},1}$ is set to 0. Then the claim follows from Sect. 7.7.4.1.1.

7.7.4.1.2.3 Caputo Derivative

Under the assumptions and notation set at the beginning of the current subsection, let $(h_{\mathrm{b.c},1}, h)$ belong to $\mathcal{Y} \times L^p(\mathcal{Y})$. Recalling that ℓ lies in $L^p(0, b) \cap L^q(0, b)$, we consider the problem

$$\begin{cases} \text{Find } u \text{ in } D({}^{\mathrm{C}}\mathrm{D}^{\ell}_{\mathcal{Z}}) \text{ such that} \\ {}^{\mathrm{C}}\mathrm{D}^{\ell}_{\mathcal{Z}} u = F(\cdot, u) + h \quad \text{in } L^p(\mathcal{Y}) \\ u(0) = h_{\mathrm{b.c},1}. \end{cases} \tag{7.7.12}$$

We claim that (7.7.12) is well posed. Indeed according to Sect. 7.3.4.1.4.2, (7.7.12) is equivalent to (7.7.9) where $h_{\mathrm{b.c},2}$ is set to 0. Then the claim follows from Sect. 7.7.4.1.1.

7.7.4.2 Case Where ℓ Lies in $L^q(0, b) \setminus L^p(0, b)$

The assumptions and notation set at the beginning of current Sect. 7.7.4 remaining in force, let us notice that when

$$k := g_{1-\alpha}, \qquad \ell := g_\alpha, \qquad \forall \alpha \in (0, 1),$$

the condition

$$g_\alpha \in L^q(0, b) \setminus L^p(0, b)$$

is equivalent to

$$\frac{1}{p} < \alpha \le 1 - \frac{1}{p}.$$

In any cases, k does not belong to $L^p(0,b)$ by Proposition 6.3.1 (ii-b); hence $\ker \mathrm{D}^\ell_{\mathcal{Z}} = g_1\mathcal{Y}$, and in view of (6.4.3), $\mathrm{D}^\ell_{\mathcal{Z}} = {}^{\mathrm{C}}\mathrm{D}^\ell_{\mathcal{Z}}$. According to Sect. 7.3.4.2, u is continuous on $[0,b]$. Moreover, by Lemma 7.7.3, $F(\cdot,u)$ belongs to $L^p(\mathcal{Y})$ for each u in $L^p(\mathcal{Y})$. Thus for each $(h_{\mathrm{b.c},1},h)$ in $\mathcal{Y}\times L^p(\mathcal{Y})$, we may legitimately consider the problem

$$\begin{cases} \text{Find } u \text{ in } D({}^{\mathrm{C}}\mathrm{D}^\ell_{\mathcal{Z}}) \text{ such that} \\ {}^{\mathrm{C}}\mathrm{D}^\ell_{\mathcal{Z}} u = F(\cdot,u)+h \quad \text{in } L^p(\mathcal{Y}) \\ u(0)=h_{\mathrm{b.c},1}. \end{cases} \tag{7.7.13}$$

We claim that (7.7.12) is *well posed*. Indeed by Sect. 7.3.4, $\mathrm{D}^\ell_{\mathcal{Z}}$ is a closed operator, and $B_{\ell,\mathcal{Z}}$ is a right inverse of $\mathrm{D}^\ell_{\mathcal{Z}}$. Hence the projection $P_{\ker \mathrm{D}^\ell_{\mathcal{Z}}}$ is well defined, and its range is equal to $g_1\mathcal{Y}$. There results that (7.7.13) is an instance of (7.6.1). Then recalling that F is a Caratheodory function satisfying (7.7.2) and (7.7.3), Corollary 7.7.4 yields that (7.7.13) is *well posed* in a sense which can be easily inferred from Definition 7.7.1 or 7.6.1.

7.7.4.3 Case Where ℓ Lies in $L^p(0,b)\setminus L^q(0,b)$

7.7.4.3.1 Case Where k lies in $L^1(0,b)\setminus L^p(0,b)$

With the assumptions and notation set at the beginning of current Sect. 7.7.4 remaining in force, let us notice that when

$$k := g_{1-\alpha}, \qquad \ell := g_\alpha, \qquad \forall \alpha \in (0,1),$$

the conditions

$$g_\alpha \in L^p(0,b)\setminus L^q(0,b) \quad \text{and} \quad g_{1-\alpha}\in L^1(0,b)\setminus L^p(0,b)$$

are equivalent to

$$\frac{1}{p}=\alpha \quad \text{and} \quad p<2.$$

In view of Sects. 7.3.4.3.1 and 7.3.4.3.2, for each $(h_{\mathrm{b.c},1},h_{\mathrm{b.c},2},h)$ in $\mathcal{Y}\times\mathcal{Y}\times L^p(\mathcal{Y})$, we consider the problem

$$\begin{cases} \text{Find } u \text{ in } D(\mathrm{D}^\ell_{\mathcal{Z}}) \text{ such that} \\ \mathrm{D}^\ell_{\mathcal{Z}} u = F(\cdot,u)+h \quad \text{in } L^p(\mathcal{Y}) \\ \big(u-\ell\, k*u(0)-\ell*\mathrm{D}^\ell_{\mathcal{Y}}u\big)(0)=h_{\mathrm{b.c},1}, \qquad k*u(0)=h_{\mathrm{b.c},2}. \end{cases} \tag{7.7.14}$$

Arguing as in Sect. 7.7.4.1.1, we derive easily that (7.7.14) is *well posed* in a sense which can be easily inferred from Definition 7.7.1 or 7.6.1.

7.7.4.3.2 Case Where k Lies in $L^p(0, b)$

The assumptions and notation set at the beginning of the current Sect. 7.7.4 remaining in force, let us notice that when

$$k := g_{1-\alpha}, \qquad \ell := g_\alpha, \qquad \forall \alpha \in (0, 1),$$

the conditions

$$g_\alpha \in L^p(0, b) \setminus L^q(0, b) \quad \text{and} \quad g_{1-\alpha} \in L^p(0, b)$$

are equivalent to

$$1 - \frac{1}{p} < \alpha < \frac{1}{p}.$$

Here, the operator $\mathrm{D}^\ell_{\mathcal{Z}}$ reduces to ${}^{\mathrm{RL}}\mathrm{D}^\ell_{\mathcal{Z}}$. Thus in view of Sects. 7.3.3.1.1 and 7.3.3.1.2, for each $(h_{\mathrm{b.c},2}, h)$ in $\mathcal{Y} \times L^p(\mathcal{Y})$, we consider the problem

$$\begin{cases} \text{Find } u \text{ in } D(\mathrm{D}^\ell_{\mathcal{Z}}) \text{ such that} \\ \mathrm{D}^\ell_{\mathcal{Z}} u = F(\cdot, u) + h \quad \text{in } L^p(\mathcal{Y}) \\ k * u(0) = h_{\mathrm{b.c},2}. \end{cases} \tag{7.7.15}$$

Arguing as in Sect. 7.7.4.1.1, we derive easily that (7.7.15) is *well posed* in a sense which can be easily inferred from Definition 7.7.1 or 7.6.1.

7.7.4.4 Case Where ℓ Lies in $L^1(0, b) \setminus (L^q(0, b) \cup L^p(0, b))$

7.7.4.4.1 Case Where k Lies in $L^1(0, b) \setminus L^p(0, b)$

With the assumptions and notation set at the beginning of current Sect. 7.7.4 remaining in force, let us notice that when

$$k := g_{1-\alpha}, \qquad \ell := g_\alpha, \qquad \forall \alpha \in (0, 1),$$

the conditions

$$g_\alpha \in L^1(0, b) \setminus (L^q(0, b) \cup L^p(0, b)) \quad \text{and} \quad g_{1-\alpha} \in L^1(0, b) \setminus L^p(0, b)$$

are equivalent to

$$\frac{1}{p} = \alpha \quad \text{and} \quad 2 \leq p.$$

In view of Sects. 7.3.4.4.1 and 7.3.4.4.2, for each $(h_{\mathrm{b.c},1}, h)$ in $\mathcal{Y} \times L^p(\mathcal{Y})$, we consider the problem

$$\begin{cases} \text{Find } u \text{ in } D(\mathrm{D}^{\ell}_{\mathcal{Z}}) \text{ such that} \\ \mathrm{D}^{\ell}_{\mathcal{Z}} u = F(\cdot, u) + h \quad \text{ in } L^p(\mathcal{Y}) \\ \big(u - \ell * \mathrm{D}^{\ell}_{\mathcal{Y}} u\big)(0) = h_{\mathrm{b.c},1}. \end{cases} \tag{7.7.16}$$

Arguing as in Sect. 7.7.4.1.1, we derive easily that (7.7.16) is *well posed* in a sense which can be easily inferred from Definition 7.7.1 or 7.6.1.

Remark that $k * u(0) = 0$ for each u in the domain of $\mathrm{D}^{\ell}_{\mathcal{Z}}$ (see Sect. 7.3.4.4.1). Thus if (7.7.16) is supplemented with the homogeneous initial condition $k * u(0) = 0$, we obtain a problem equivalent to (7.7.16).

7.7.4.4.2 Case Where k Lies in $L^p(0, b)$

With the assumptions and notation set at the beginning of current Sect. 7.7.4 remaining in force, let us notice that when

$$k := g_{1-\alpha}, \qquad \ell := g_{\alpha}, \qquad \forall \alpha \in (0, 1),$$

the conditions

$$g_{\alpha} \in L^1(0, b) \setminus (L^q(0, b) \cup L^p(0, b)) \quad \text{ and } \quad g_{1-\alpha} \in L^p(0, b)$$

are equivalent to

$$\alpha \leq 1 - \frac{1}{p} \quad \text{ and } \quad \alpha < \frac{1}{p}.$$

Here, the operator $\mathrm{D}^{\ell}_{\mathcal{Z}}$ reduces to ${}^{\mathrm{RL}}\mathrm{D}^{\ell}_{\mathcal{Z}}$. Thus in view of Sect. 7.3.3.2, we consider the problem

$$\begin{cases} \text{Find } u \text{ in } D(\mathrm{D}^{\ell}_{\mathcal{Z}}) \text{ such that} \\ \mathrm{D}^{\ell}_{\mathcal{Z}} u = F(\cdot, u) + h \quad \text{ in } L^p(\mathcal{Y}). \end{cases} \tag{7.7.17}$$

Since $\ker {}^{\mathrm{RL}}\mathrm{D}^{\ell}_{\mathcal{Z}}$ is trivial, (7.7.17) is an instance of (7.7.4) with $h_{\mathrm{b.c}} = 0$. Thus Corollary 7.7.4 yields that (7.7.17) is *well posed* in a sense which can be easily inferred from Definition 7.7.1 or 7.6.1.

8 Abstract and Fractional Laplace Operators

8.1 Introduction

Our goal is to extend the definition of the well-known *Laplace operator*. This extension will be concerned of course with *fractional* differential operators acting on functions of the one variable.

The usual *Laplace operator* on $L^2(0, b)$ has domain $H^2(0, b)$ and reads

$$\Delta u := \left(\mathrm{D}^1_{L^2}\right)^2 u, \qquad \forall u \in H^2(0, b).$$

The issue is to extend this definition when the operator $\mathrm{D}^1_{L^2}$ is replaced by a fractional operator, let us say ${}^{\mathrm{RL}}\mathrm{D}^\alpha_{L^2}$ (see Sect. 6.2.1.1). There are of course many ways to proceed. We will choose a way that keeps some important properties of the Laplace operator, namely, self-adjointness and positivity of a restriction of the negative Laplace operator.

More precisely, let Δ_{D} denote *the Laplace operator on* $L^2(0, b)$ *supplemented with homogeneous Dirichlet boundary conditions*, i.e., Δ_{D} is the restriction of Δ to $H^2(0, b) \cap H^1_0(0, b)$. It is well-known that $-\Delta_{\mathrm{D}}$ is a positive self-adjoint operator on $L^2(0, b)$. Now, since $(\mathrm{D}^1_{L^2})^*$ is the restriction of $-\mathrm{D}^1_{L^2}$ with domain $H^1_0(0, b)$, we deduce that $-\Delta_{\mathrm{D}} = \mathrm{D}^1_{L^2}(\mathrm{D}^1_{L^2})^*$.

Then by using the symmetry S_{L^2}, the important relations we want to keep are

$$-\Delta_{\mathrm{D}} = \mathrm{D}^1_{L^2}(\mathrm{D}^1_{L^2})^* \subseteq \mathrm{D}^1_{L^2}\, S_{L^2}\, \mathrm{D}^1_{L^2}\, S_{L^2} = -\Delta.$$

Now, if we replace $\mathrm{D}^1_{L^2}$ by ${}^{\mathrm{RL}}\mathrm{D}^\alpha_{L^2}$ in the previous relations, we get

$$\Delta_{\alpha,\mathrm{D}} = {}^{\mathrm{RL}}\mathrm{D}^\alpha_{L^2}({}^{\mathrm{RL}}\mathrm{D}^\alpha_{L^2})^* \subseteq {}^{\mathrm{RL}}\mathrm{D}^\alpha_{L^2}\, S_{L^2}\, {}^{\mathrm{RL}}\mathrm{D}^\alpha_{L^2}\, S_{L^2} = -\Delta_\alpha.$$

A. Rougirel, *Unified Theory for Fractional and Entire Differential Operators*, Frontiers in Elliptic and Parabolic Problems,
https://doi.org/10.1007/978-3-031-58356-8_8

The latter equality will serve as a definition of *the fractional Laplace operator* Δ_α, i.e.,

$$\Delta_\alpha := -({}^{\mathrm{RL}}\mathrm{D}^\alpha_{L^2}\, S_{L^2})^2.$$

In a same way, we will set in view of Proposition 4.3.10:

$$\Delta_{\alpha,\mathrm{D}} := -{}^{\mathrm{RL}}\mathrm{D}^\alpha_{L^2}\, S_{L^2}\, A_{\mathrm{m},{}^{\mathrm{RL}}\mathcal{T}_\alpha}\, S_{L^2}. \tag{8.1.1}$$

The operator $\Delta_{\alpha,\mathrm{D}}$ will be called *a fractional Laplace operator with homogeneous Dirichlet endogenous boundary conditions*. See Definitions 8.3.1 and 8.3.2.

Abstract *Laplace operators* are defined in Sect. 8.3. In order to recover invertibility properties of fractional Laplace operators with Dirichlet boundary conditions, we will use *weak products of operators*, a notion introduced in the next section. *Abstract Dirichlet problems* are considered in Sect. 8.4. The two last sections are devoted to applications.

8.2 Weak Products on $\mathcal{V}$

The weak product of operators is the operator counterpart of weak solutions of linear partial differential equations. Let $\mathcal{V}$ be a real or complex reflexive Banach space.

8.2.1 Definition

Let A, C be operators on $\mathcal{V}$, and suppose that A is densely defined. In view of Sect. 2.4.5, $\widetilde{A^*}^*$ is a continuous operator from $\mathcal{V}$ with values in $D(A^*)^*$ and by (2.4.22)

$$\langle \widetilde{A^*}^* v, w\rangle_{D(A^*)^*,D(A^*)} = \langle v, A^* w\rangle_{\mathcal{V},\mathcal{V}^*}, \qquad \forall\,(v,w) \in \mathcal{V} \times D(A^*). \tag{8.2.1}$$

Thus

$$\widetilde{AC} := \widetilde{A^*}^* C : D(C) \to D(A^*)^*$$

is a well-defined operator from $D(C)$ with values in $D(A^*)^*$. $\widetilde{AC}$ is called *the weak product of* A and C and is characterized through the identity

$$\langle \widetilde{AC} v, w\rangle_{D(A^*)^*,D(A^*)} = \langle Cv, A^* w\rangle_{\mathcal{V},\mathcal{V}^*}, \qquad \forall\,(v,w) \in \mathcal{V} \times D(A^*). \tag{8.2.2}$$

8.2.2 Properties

Let A, C be operators on $\mathcal{V}$ such that A is densely defined. Let v belong to $D(C)$. Since, for each w in $D(A^*)$,

$$\left|\langle Cv, A^* w\rangle_{\mathcal{V},\mathcal{V}^*}\right| \leq \|Cv\|_{\mathcal{V}} \|A^* w\|_{\mathcal{V}^*} \leq \|v\|_{D(C)} \|w\|_{D(A^*)},$$

there results from (8.2.2) that $\widetilde{AC}$ is continuous on $D(C)$ and that

$$\|\widetilde{AC}\|_{\mathcal{L}(D(C),D(A^*)^*)} \leq 1. \tag{8.2.3}$$

Besides, noticing that $\mathcal{V}$ is continuously embedded in $D(A^*)^*$ (see Sect. 2.4.5.2), it is clear from (2.4.21) that $\widetilde{AC}$ extends the operator AC in the following sense:

$$ACu = \widetilde{AC}u, \qquad \forall\, u \in D(AC). \tag{8.2.4}$$

Our main property is the following.

Theorem 8.2.1 *Being given two closed and densely defined operators A and C on the reflexive space $\mathcal{V}$, let us assume that:*

(i) *for each w in $D(A^*)$, $\|w\|_{\mathcal{V}^*} \lesssim \|A^* w\|_{\mathcal{V}^*}$, and for each v in $D(C)$, $\|v\|_{\mathcal{V}} \lesssim \|Cv\|_{\mathcal{V}}$;*
(ii) $(\ker A)^\perp$ *and* $\ker C^*$ *are in direct sum in $\mathcal{V}^*$, and $\mathcal{V}^* = (\ker A)^\perp \oplus \ker C^*$.*

Then $\widetilde{AC}$ is an isomorphism between $D(C)$ and $D(A^*)^*$.

Proof Since A is closed, $A^{**} = A$ by Theorem 2.4.1. Thus applying Proposition 2.2.19 with $T := A^*$, we infer $\ker A = R(A^*)^\perp$. Moreover since $\|w\|_{D(A^*)} \lesssim \|A^* w\|_{\mathcal{V}^*}$ and A^* is closed, the range of A^* is a closed subspace of $\mathcal{V}^*$. Hence $R(A^*) = \ker A^\perp$ according to Proposition 2.2.12. Then Assumption (ii) gives

$$\mathcal{V}^* = R(A^*) \oplus \ker C^*. \tag{8.2.5}$$

Next let us show that

$$\|v\|_{D(C)} \lesssim \|\widetilde{AC}v\|_{D(A^*)^*}, \qquad \forall\, v \in D(C). \tag{8.2.6}$$

For, let v be a nontrivial vector of $D(C)$. Since $\|v\|_{\mathcal{V}} \lesssim \|Cv\|_{\mathcal{V}}$, we must have $Cv \neq 0$. Thus by Corollary 2.2.7 of the Hahn-Banach Theorem 2.2.6, there exists some v_0^* in $\mathcal{V}^*$ such that $\|v_0^*\|_{\mathcal{V}^*} = 1$ and $\langle v_0^*, Cv\rangle_{\mathcal{V}^*,\mathcal{V}} = \|Cv\|_{\mathcal{V}}$. By (8.2.5), there exist w_0 in $D(A^*)$ and ψ_0 in $\ker C^*$ such that

$$v_0^* = A^* w_0 + \psi_0 \in R(A^*) \oplus \ker C^*.$$

Thus

$$\begin{aligned}
\|v\|_{D(C)} &\lesssim \|Cv\|_{\mathcal{V}} && \text{(by (i))} \\
&= \langle v_0^*, Cv \rangle_{\mathcal{V}^*, \mathcal{V}} \\
&= \langle A^* w_0, Cv \rangle_{\mathcal{V}^*, \mathcal{V}} && \text{(since } \langle C^* \psi_0, v \rangle_{\mathcal{V}^*, \mathcal{V}} = 0) \\
&= \langle \widetilde{AC} v, w_0 \rangle_{D(A^*)^*, D(A^*)} && \text{(by (8.2.2))} \\
&\lesssim \|\widetilde{AC} v\|_{D(A^*)^*} \|A^* w_0\|_{\mathcal{V}^*} && \text{(by (i)).}
\end{aligned}$$

Since $R(A^*)$ and $\ker C^*$ are closed subspaces of $\mathcal{V}^*$, Proposition 2.3.5 yields that $\|A^* w_0\|_{\mathcal{V}^*} \lesssim \|v_0^*\|_{\mathcal{V}^*}$. Recalling that $\|v_0^*\|_{\mathcal{V}^*} = 1$, we get (8.2.6).

Our second step consists of showing that

$$\ker \big(\widetilde{AC}\big)^* = \{0\}. \tag{8.2.7}$$

For, since $\widetilde{AC}$ lies in $\mathcal{L}\big(D(C), D(A^*)^*\big)$, one has

$$\big(\widetilde{AC}\big)^* \in \mathcal{L}\big(D(A^*)^{**}, D(C)^*\big).$$

Since $\mathcal{V}$ is reflexive, Proposition 2.3.11 yields that $\mathcal{V}^*$ is reflexive as well. Thus by Proposition 2.4.2, $D(A^*)$ (endowed as usual with its graph-norm) is a reflexive space. Thus each element w of $\ker \big(\widetilde{AC}\big)^*$ lies in $D(A^*)$ and satisfies for every v in $D(C)$,

$$\begin{aligned}
0 = \big\langle \big(\widetilde{AC}\big)^* w, v \big\rangle_{D(C)^*, D(C)} &= \langle w, \widetilde{AC} v \rangle_{D(A^*), D(A^*)^*} \\
&= \langle A^* w, Cv \rangle_{\mathcal{V}^*, \mathcal{V}} && \text{(by (8.2.2)).}
\end{aligned}$$

According to the definition of the adjoint of the operator C, there results that $A^* w$ lies in $D(C^*)$ and $C^* A^* w = 0$. Then $A^* w = 0$ since $R(A^*) \cap \ker C^*$ is trivial by (8.2.5). Thus Assumption (i) entails that $w = 0$, which proves (8.2.7).

Next, we claim that $R(\widetilde{AC})$ is a closed subspace of $D(A^*)^*$. This is just a consequence of the continuity of $\widetilde{AC}$ (see (8.2.3)) and (8.2.6).

Now we may conclude the following: owing to Proposition 2.2.19, (8.2.7) yields that $R(\widetilde{AC})$ is dense in $D(A^*)^*$. Since $R(\widetilde{AC})$ is closed in $D(A^*)^*$, we get $R(\widetilde{AC}) = D(A^*)^*$. Moreover, (8.2.6) implies that $\widetilde{AC}$ is one-to-one. Using also (8.2.3) and the closedness of C, we infer with Corollary 2.4.9 that $\widetilde{AC}$ is a Banach isomorphism between $D(C)$ and $D(A^*)^*$. □

Proposition 8.2.1 is closely related to the *Banach-Necas-Babuska theorem* (see, for instance, [Sai18]). The proof of the *Banach-Necas-Babuska theorem* relies on the arguments featured in the last paragraph of the proof of Proposition 8.2.1. Our proof of (8.2.6) was inspired by the proof of Theorem 3.3 in [AB08].

Proposition 8.2.2 *Under the assumptions and notation of Theorem 8.2.1, $AC : D(AC) \subseteq \mathcal{V} \to \mathcal{V}$ is a closed operator on $\mathcal{V}$. Besides, AC is an isomorphism between $D(AC)$ and $\mathcal{V}$.*

Proof Let us start to show that for every v in $D(\widetilde{AC})$ and every f in $\mathcal{V}$, the following two assertions are equivalent:

(i) v lies in $D(AC)$ and $ACv = f$.
(ii) $\widetilde{AC}v = f$.

Indeed, (i) implies (ii) due to (8.2.4). Conversely, if (ii) holds true then applying Corollary 2.4.12 with $\mathcal{X} = \mathcal{Y} := \mathcal{V}^*$ and $T := A^*$, we get for each w in $D(A^*)$

$$\langle f, w\rangle_{D(A^*)^*, D(A^*)} = \langle f, w\rangle_{\mathcal{V},\mathcal{V}^*}.$$

Thus (ii) and (8.2.2) yield that

$$\langle Cv, A^*w\rangle_{\mathcal{V},\mathcal{V}^*} = \langle f, w\rangle_{\mathcal{V},\mathcal{V}^*} \forall w \in D(A^*).$$

By definition of the adjoint of A^*, and owing to the fact that $A^{**} = A$, we deduce that Cv lies in $D(A)$, so that v lives in $D(AC)$, and $ACv = f$. Thus (ii) implies (i). That proves the expected equivalence. Now since by Theorem 8.2.1, $\widetilde{AC}$ is a bijection, there results that AC is a bijection between $D(AC)$ and $\mathcal{V}$.

Let us prove that $D(AC)$ is closed in $\mathcal{V}$. For, let $(v_n)_{n\geq 0}$ be a sequence in $\mathcal{V}$, and v, f in $\mathcal{V}$ be such that

$$v_n \overset{\mathcal{V}}{\to} v, \qquad ACv_n \overset{\mathcal{V}}{\to} f.$$

Recalling that $\mathcal{V}$ is continuously embedded in $D(A^*)^*$, we deduce

$$\widetilde{AC}v_n \xrightarrow{D(A^*)^*} f.$$

Moreover, since by Theorem 8.2.1, $\widetilde{AC}^{-1}$ is continuous from $D(A^*)^*$ into $D(C)$, the sequence (v_n) converges toward $\widetilde{AC}^{-1} f$ in $D(C)$. By uniqueness of the limit, $\widetilde{AC}^{-1} f = v$, so that $v_n \overset{D(C)}{\longrightarrow} v$. Then we have

$$C v_n \xrightarrow{\mathcal{V}} C v, \qquad A C v_n \xrightarrow{\mathcal{V}} f.$$

Since A is closed, there results that Cv lies in $D(A)$, so that $v \in D(AC)$, and $ACv = f$. That proves the closedness of AC.

Finally, AC is clearly continuous from $D(AC)$ into $\mathcal{V}$, whence it is an isomorphism according to Corollary 2.4.9. □

Example 8.2.1 Let us consider the case where $\mathcal{V} := \mathcal{H}$ is a real or complex Hilbert space whose inner product is denoted by $(\cdot, \cdot)$. Let A be a closed and densely defined operator on $\mathcal{H}$ satisfying $\|w\|_{\mathcal{V}^*} \lesssim \|A^* w\|_{\mathcal{V}^*}$ for $w \in D(A^*)$. Choosing $C := A^*$ in the above framework, we get

$$\langle \widetilde{AA^*} v, w \rangle_{D(A^*)^*, D(A^*)} = (A^* v, A^* w), \qquad \forall\, v, w \in D(A^*).$$

We claim that $\widetilde{AA^*}$ is an isomorphism between $D(A^*)$ and $D(A^*)^*$, a well-known particular case of the *Lax-Milgram theorem*. The proof of the claim may be derived easily from Theorem 8.2.1.

On the other hand, Proposition 8.2.2 tells us that AA^* is an isomorphism between $D(AA^*)$ and $\mathcal{H}$. □

8.3 Abstract Laplace Operators

Let $\mathcal{V}$ be a real or complex reflexive Banach space.

8.3.1 Definitions

In light of Sect. 8.1, we are led to the following definitions. Recall that the product of a differential quadruplet by an involution, and the square of a differential quadruplet have been defined in Sect. 5.6.

Definition 8.3.1 Let $\mathcal{Q} := (\mathbb{A}, A_{\mathrm{M}}, B, S)$ be a differential quadruplet on $(\mathcal{V}, \mathcal{V}^*)$. The operator

$$\Delta_{\mathcal{Q}S} := -A_{\mathrm{M}}\, S\, A_{\mathrm{M}}\, S = -(A_{\mathrm{M}}\, S)^2,$$

is called *the Laplace operator with respect to* $\mathcal{Q}S$. On the other hand, the operator

$$\Delta_{S\mathcal{Q}} := -S\, A_{\mathrm{M}}\, S\, A_{\mathrm{M}} = -(S\, A_{\mathrm{M}})^2,$$

is called *the Laplace operator with respect to $S\mathcal{Q}$.* □

Since

$$\Delta_{S\mathcal{Q}} = S\,\Delta_{\mathcal{Q}S}\,S,$$

we will restrict our attention to $\Delta_{S\mathcal{Q}}$. Now in view of (8.1.1), we set the following definition.

Definition 8.3.2 Let $\mathcal{Q} := (\mathbb{A}, A_{\mathrm{M}}, B, S)$ be a differential quadruplet on $(\mathcal{V}, \mathcal{V}^*)$. The operator

$$\Delta_{S\mathcal{Q},\mathrm{D}} := -S\,A_{\mathrm{M}}\,S\,A_{\mathrm{m},\mathcal{Q}}$$

is called *the Laplace operator with respect to $S\mathcal{Q}$ and with homogeneous Dirichlet endogenous boundary conditions.* □

Of course these definitions may be easily adapted to an Hilbertian framework since a differential triplet $\mathcal{T} := (\mathbb{A}, \mathbb{B}, \mathbb{S})$ on a Hilbert space $\mathcal{H}$ may be extended into a differential quadruplet $\mathcal{Q}_{\mathcal{T}} := (\mathbb{A}, \mathbb{A}, \mathbb{B}, \mathbb{S})$ on $\mathcal{V}_{\mathcal{H}}$ (see Sect. 5.1.2 and in particular (5.1.8)).

Definition 8.3.3 Let $\mathcal{T} := (\mathbb{A}, \mathbb{B}, \mathbb{S})$ be a differential triplet on a Hilbert space $\mathcal{H}$. The operator

$$\Delta_{\mathbb{S}\mathcal{T}} := \Delta_{\mathbb{S}\mathcal{Q}_{\mathcal{T}}},$$

is called *the Laplace operator with respect to $\mathbb{S}\mathcal{T}$*. The operator

$$\Delta_{\mathbb{S}\mathcal{T},\mathrm{D}} := \Delta_{\mathbb{S}\mathcal{Q}_{\mathcal{T}},\mathrm{D}}$$

is called *the Laplace operator with respect to $\mathbb{S}\mathcal{T}$ and with homogeneous Dirichlet endogenous boundary conditions.* □

Clearly, with the notation of Definition 8.3.3,

$$\Delta_{\mathbb{S}\mathcal{T}} = -\mathbb{S}\,\mathbb{A}\,\mathbb{S}\,\mathbb{A} = -(\mathbb{S}\,\mathbb{A})^2,$$

and since $A_{\mathrm{m},\mathcal{Q}_{\mathcal{T}}} = A_{\mathrm{m},\mathcal{T}}$ (see Proposition 5.6.1),

$$\Delta_{\mathbb{S}\mathcal{T},\mathrm{D}} := -\mathbb{S}\,\mathbb{A}\,\mathbb{S}\,A_{\mathrm{m},\mathcal{T}}. \tag{8.3.1}$$

Example 8.3.4 (Laplace Operators on $L^2(0, b)$) In the particular case where $\mathcal{H} := L^2(0, b)$ and $\mathcal{T} := \mathcal{T}_1 = (\mathrm{D}^1_{L^2}, B_{1,L^2}, S)$, one has

$$\Delta_{S\mathcal{T}_1} = (\mathrm{D}^1_{L^2})^2, \qquad \Delta_{S\mathcal{T}_1,\mathrm{D}} = \mathrm{D}^1_{L^2}\, A_{\mathrm{m},\mathcal{T}_1}.$$

Since the domain of the minimal operator $A_{\mathrm{m},\mathcal{T}_1}$ is $H^1_0(0, b)$ (see Sect. 4.4.2), there result that the domain of $\Delta_{S\mathcal{T}_1,\mathrm{D}}$ is $H^2(0, b) \cap H^1_0(0, b)$. Hence *the Laplace operator with respect to $S\mathcal{T}_1$ and with homogeneous Dirichlet endogenous boundary conditions* is nothing but *the Laplace operator on $L^2(0, b)$ with homogeneous Dirichlet boundary conditions.* □

8.3.2 Properties

Proposition 8.3.1 *Let $\mathcal{V}$ be a reflexive Banach space and $\mathcal{Q} := (\mathbb{A}, A_{\mathrm{M}}, B, S)$ be a closed differential quadruplet on $(\mathcal{V}, \mathcal{V}^*)$. Suppose that*

$$\ker A_{\mathrm{M}}{}^{\perp} \oplus \ker \mathbb{A} = \mathcal{V}^*. \tag{8.3.2}$$

Then $\Delta_{S\mathcal{Q},\mathrm{D}}$ is a closed operator on $\mathcal{V}$, and $\Delta_{S\mathcal{Q},\mathrm{D}}$ is an isomorphism from its domain onto $\mathcal{V}$. Moreover, the weak product of $S A_{\mathrm{M}}\, S$ and $A_{\mathrm{m},\mathcal{Q}}$ is an isomorphism from $D(A_{\mathrm{m},\mathcal{Q}})$ onto $D(A_{\mathrm{m},r(\mathcal{Q})})^*$.

Proof We will apply Proposition 8.2.2 with

$$A := S\, A_{\mathrm{M}}\, S, \qquad C := A_{\mathrm{m},\mathcal{Q}}.$$

A is closed since $\mathcal{Q}$ is. Besides Proposition 5.2.7 warrants the closedness of C. Now we claim that C and A^* satisfy the Poincaré inequality, that is

$$\|u\|_{\mathcal{V}} \lesssim \|Cu\|_{\mathcal{V}}, \qquad \|w\|_{\mathcal{V}^*} \lesssim \|A^*u\|_{\mathcal{V}^*}, \qquad \forall\, (u, w) \in D(C) \times D(A^*).$$

Indeed, C is a restriction of the pivot operator of $\mathcal{Q}$, so that by (5.2.4), C fulfills the Poincaré inequality. Since $A^* = S^*\, A_{\mathrm{m},\mathcal{Q}^*}\, S^*$ by (5.2.17) (and Theorem 2.4.1), we derive with (5.6.21) that $A^* = A_{\mathrm{m},r(\mathcal{Q})}$. Then we check along the same lines that A^* satisfies the Poincaré inequality as well. That proves the claim. Regarding the adjoint operator of C, we compute

$$C^* = (A_{\mathrm{m},\mathcal{Q}})^* = S^*\, \mathbb{A}\, S^*,$$

by (5.2.16). Thus thanks to Lemma 4.3.9, (8.3.2) reads

$$\ker A^{\perp} \oplus \ker C^* = \mathcal{V}^*.$$

We conclude by applying Proposition 8.2.2. The assertion regarding the weak product is proved by Theorem 8.2.1 recalling that A^* is equal to $A_{\mathrm{m},r(\mathcal{Q})}$. □

Proposition 8.3.2 *Let $\mathcal{V}$ be a reflexive Banach space and $\mathcal{Q} := (\mathbb{A}, A_{\mathrm{M}}, B, S)$ be a differential quadruplet on $(\mathcal{V}, \mathcal{V}^*)$. If $\mathbb{A}$ and A_{M} have finite-dimensional kernel, then $\Delta_{S\mathcal{Q}}$ and $\Delta_{S\mathcal{Q},\mathrm{D}}$ are closed boundary restriction operators of $(S\mathcal{Q})^2$.*

Proof Of course, the opposite of a closed boundary is also a closed boundary restriction operator. So regarding $\Delta_{S\mathcal{Q},\mathrm{D}}$, we will apply Proposition 5.6.14 with $\mathcal{Q} := S\mathcal{Q}$, $\mathcal{Q}' := S\mathcal{Q}$, $A := S\,A_{\mathrm{M}}$ and $A' := S\,A_{\mathrm{m},\mathcal{Q}}$. Since the kernel of A_{M} is assumed to be finite dimensional, Proposition 4.4.1 entails that A is closed. Since $\mathcal{Q}$ is closed, we know from Proposition 5.2.7 that $S\,A_{\mathrm{m},\mathcal{Q}}$ is a closed operator as well. Whence by Proposition 5.6.14, $\Delta_{S\mathcal{Q},\mathrm{D}}$ is a closed boundary of $(S\mathcal{Q})^2$.

About $\Delta_{S\mathcal{Q}}$, the proof is analog and even simpler. □

To close this subsection, we will show that the *opposite* of the Laplace operator with respect to $S\mathcal{T}$ and with homogeneous Dirichlet endogenous boundary conditions is, as expected, a positive self-adjoint operator.

Proposition 8.3.3 *Let $\mathcal{T} := (\mathbb{A}, \mathbb{B}, \mathbb{S})$ be a closed differential triplet on a Hilbert space $\mathcal{H}$. Then the operator $-\Delta_{\mathbb{S}\mathcal{T},\mathrm{D}}$ is a positive self-adjoint operator on $\mathcal{H}$. Moreover, $\Delta_{\mathbb{S}\mathcal{T},\mathrm{D}}$ is an isomorphism from its domain onto $\mathcal{V}$.*

Proof We derive from Proposition 4.3.10 that

$$-\Delta_{\mathbb{S}\mathcal{T},\mathrm{D}} = (\mathbb{S}\mathbb{A})(\mathbb{S}\mathbb{A})^*.$$

Thus $-\Delta_{\mathbb{S}\mathcal{T},\mathrm{D}}$ is self-adjoint operator according to Theorem 2.5.5. Next, positivity follows easily from the latter identity. Finally, by applying Proposition 8.3.1 with $\mathcal{V} := \mathcal{V}_{\mathcal{H}}$ and $\mathcal{Q} := \mathcal{Q}_{\mathcal{T}}$, we get that (8.3.2) holds true; hence $\Delta_{\mathbb{S}\mathcal{Q}_{\mathcal{T}},\mathrm{D}}$ which is nothing but $\Delta_{\mathbb{S}\mathcal{T},\mathrm{D}}$ is an isomorphism. □

When a Gelfand triplet is at hand, Laplace operators with respect to $S\mathcal{Q}$ are restrictions of Laplace operators with respect to the realization of $(S\mathcal{Q})^2$.

Proposition 8.3.4 *Let $(\mathcal{V}, \mathcal{H}, \mathcal{V}^*)$ be a Gelfand triplet and $\mathcal{Q} := (\mathbb{A}, A_{\mathrm{M}}, B, S)$ be a differential quadruplet on $(\mathcal{V}, \mathcal{V}^*)$. Assume that $\mathcal{Q}$ is compatible with $\mathcal{H}$, and $A_{\mathrm{M}} \subseteq \mathbb{A}$. Then the Laplace operator with respect to $(S^*)_{\mathcal{H}}\mathcal{T}_{\mathcal{Q}}$ is an extension of the Laplace operator with respect to $S\mathcal{Q}$. In symbol,*

$$\Delta_{S\mathcal{Q}} \subseteq \Delta_{\check{S}\mathcal{T}_{\mathcal{Q}}},$$

where $\tilde{S} := (S^*)_{\mathcal{H}}$. In a same way, the Laplace operator with respect to $\tilde{S}\mathcal{T}_{\mathcal{Q}}$ and with Dirichlet endogenous boundary conditions is an extension of the Laplace operator with respect to $S\mathcal{Q}$ and with Dirichlet endogenous boundary conditions. In symbol,

$$\Delta_{S\mathcal{Q},\mathrm{D}} \subseteq \Delta_{\tilde{S}\mathcal{T}_{\mathcal{Q}},\mathrm{D}}.$$

Proof By Definition 5.1.3, $\mathcal{T}_{\mathcal{Q}} := \left(\mathbb{A}_{\mathcal{H}}, \mathbb{B}_{\mathcal{H}}, \tilde{S}\right)$, thus

$$\Delta_{\tilde{S}\mathcal{T}_{\mathcal{Q}}} = -(\tilde{S}\mathbb{A}_{\mathcal{H}})^2.$$

Moreover, Corollary 5.1.5 tells us that $A_{\mathrm{M}} \subseteq \mathbb{A}_{\mathcal{H}}$ and $S \subseteq \tilde{S}$. Hence the first extension holds. In a same way,

$$\Delta_{\tilde{S}\mathcal{T}_{\mathcal{Q}},\mathrm{D}} = -\tilde{S}\mathbb{A}_{\mathcal{H}}\tilde{S}A_{\mathrm{m},\mathcal{T}_{\mathcal{Q}}}.$$

Using in addition the extension $A_{\mathrm{m},\mathcal{Q}} \subseteq A_{\mathrm{m},\mathcal{T}_{\mathcal{Q}}}$ given by Proposition 5.2.9, we get $\Delta_{S\mathcal{Q},\mathrm{D}} \subseteq \Delta_{\tilde{S}\mathcal{T}_{\mathcal{Q}},\mathrm{D}}$. □

8.3.3 Endogenous Boundary Conditions

Assuming that the maximal operators have finite-dimensional kernel, our goal is to describe the domain of the Laplace operator supplemented with homogeneous Dirichlet endogenous boundary conditions in terms of endogenous boundary conditions (see Sect. 8.3.3.3). For, we will use the assumptions and notation of Sect. 5.5. That is, being given a real or complex reflexive Banach space $\mathcal{V}$, and $\mathcal{Q} := (\mathbb{A}, A_{\mathrm{M}}, B, S)$ a differential quadruplet on $(\mathcal{V}, \mathcal{V}^*)$, we suppose that the maximal operators A_{M} and $\mathbb{A}$ have nontrivial finite-dimensional kernel. Denoting by $d_{A_{\mathrm{M}}}$, $d_{\mathbb{A}}$ their dimension, we consider bases $(\xi_1, \ldots, \xi_{d_{A_{\mathrm{M}}}})$ and $(\xi_1^{\mathrm{s}}, \ldots, \xi_{d_{\mathbb{A}}}^{\mathrm{s}})$ of $\ker A_{\mathrm{M}}$ and $\ker \mathbb{A}$, respectively.

For a generic element u in the domain of $(A_{\mathrm{M}}S)^2$, we will start to compute its endogenous boundary coordinates with underlying quadruplet QS in terms of its (natural) endogenous boundary coordinates with underlying quadruplet $(QS)^2$. The result is featured in (8.3.12)–(8.3.13). In Sect. 8.3.3.3, we compute *the endogenous boundary conditions* of the Laplace operator with respect to $S\mathcal{Q}$ and homogeneous Dirichlet endogenous boundary conditions.

8.3.3.1 Underlying Quadruplet $(S\mathcal{Q})^2$

By (5.6.40) and (5.6.39),

$$\ker(SA_{\mathrm{M}})^2 = \langle \xi_1, \ldots, \xi_{d_{A_{\mathrm{M}}}} \rangle \oplus \langle BS\xi_1, \ldots, BS\xi_{d_{A_{\mathrm{M}}}} \rangle \tag{8.3.3}$$

$$\ker(S^*\mathbb{A})^2 = \langle \xi_1^{\mathrm{s}}, \dots, \xi_{d_{\mathbb{A}}}^{\mathrm{s}} \rangle \oplus \langle S^* B^* \xi_1^{\mathrm{s}}, \dots, S^* B^* \xi_{d_{\mathbb{A}}}^{\mathrm{s}} \rangle. \tag{8.3.4}$$

Applying Proposition 2.2.14 with $\mathcal{X} := \mathcal{V}^*$ and $N := \ker(S^*\mathbb{A})^2$, there exists $\eta_1, \dots, \eta_{2d_{\mathbb{A}}}$ in $\mathcal{V}$ such that for each indexes i and j satisfying $1 \leq i \leq 2d_{\mathbb{A}}$ and $1 \leq j \leq d_{\mathbb{A}}$,

$$\langle \xi_j^{\mathrm{s}}, \eta_i \rangle = \delta_{i,j}, \qquad \langle S^* B^* \xi_j^{\mathrm{s}}, \eta_i \rangle = \delta_{i, d_{\mathbb{A}}+j}. \tag{8.3.5}$$

We set

$$F_{0,(S\mathcal{Q})^2} := \langle \eta_1, \dots, \eta_{2d_{\mathbb{A}}} \rangle. \tag{8.3.6}$$

By Lemma 2.2.13, $(\eta_1, \dots, \eta_{2d_{\mathbb{A}}})$ forms a basis of $F_{0,(S\mathcal{Q})^2}$ and

$$\mathcal{V} = F_{0,(S\mathcal{Q})^2} \oplus \big(\ker(S^*\mathbb{A})^2\big)^{\perp}. \tag{8.3.7}$$

Hence $(S\mathcal{Q})^2$ is a $F_{0,(S\mathcal{Q})^2}$-regular differential quadruplet on $(\mathcal{V}, \mathcal{V}^*)$. Since

$$(S\mathcal{Q})^2 = \big((S^*\mathbb{A})^2,\ (SA_{\mathrm{M}})^2,\ (BS)^2,\ \mathrm{id}_{\mathcal{V}}\big),$$

the Domain Structure Theorem 5.3.3 yields that each u in $D\big((SA_{\mathrm{M}})^2\big)$ reads

$$u = \sum_{i=1}^{d_{A_{\mathrm{M}}}} u_i \xi_i + u_{d_{A_{\mathrm{M}}}+i}\, BS\xi_i + (BS)^2 \sum_{i=1}^{2d_{\mathbb{A}}} u_i' \eta_i + (BS)^2 \psi^{\perp}, \tag{8.3.8}$$

where $(u_1, \dots, u_{2d_{A_{\mathrm{M}}}}, u_1', \dots, u_{2d_{\mathbb{A}}}')$ is the endogenous boundary coordinates of u with underlying quadruplet $(S\mathcal{Q})^2$ and with respect to the basis

$$(\xi_1, \dots, \xi_{d_{A_{\mathrm{M}}}}, BS\xi_1, \dots, BS\xi_{d_{A_{\mathrm{M}}}})$$

of $\ker(SA_{\mathrm{M}})^2$ and the basis $(\eta_1, \dots, \eta_{2d_{\mathbb{A}}})$ of $F_{0,(S\mathcal{Q})^2}$. Also $\psi^{\perp}$ lies in $\big(\ker(S^*\mathbb{A})^2\big)^{\perp}$.

8.3.3.2 Underlying Quadruplet $S\mathcal{Q}$

By (8.3.5),

$$\langle \xi_j^{\mathrm{s}}, \eta_i \rangle = \delta_{i,j}, \qquad \forall\, 1 \leq i, j \leq d_{\mathbb{A}}. \tag{8.3.9}$$

Thus setting

$$F_{0,S\mathcal{Q}} := \langle \eta_1, \dots, \eta_{d_{\mathbb{A}}} \rangle, \tag{8.3.10}$$

Lemma 2.2.13 entails that

$$\mathcal{V} = F_{0,S\mathcal{Q}} \oplus (\ker S^*\mathbb{A})^{\perp},$$

so that $S\mathcal{Q}$ is a $F_{0,S\mathcal{Q}}$-regular differential quadruplet on $(\mathcal{V}, \mathcal{V}^*)$. Then each v in $D(SA_{\mathrm{M}})$ reads

$$v = \sum_{i=1}^{d_{A_{\mathrm{M}}}} v_i \xi_i + BS \sum_{i=1}^{d_{\mathbb{A}}} v_i' \eta_i + BS\psi_1^{\perp}, \tag{8.3.11}$$

where $(v_1, \ldots, v_{d_{A_{\mathrm{M}}}}, v_1', \ldots, v_{d_{\mathbb{A}}}')$ is the endogenous boundary coordinates of v with underlying quadruplet $S\mathcal{Q}$ and with respect to the basis $(\xi_1, \ldots, \xi_{d_{A_{\mathrm{M}}}})$ of $\ker SA_{\mathrm{M}}$ and the basis $(\eta_1, \ldots, \eta_{d_{\mathbb{A}}})$ of $F_{0,S\mathcal{Q}}$. Also $\psi_1^{\perp}$ lies in $(\ker S^*\mathbb{A})^{\perp}$.

For any u in $D\big((SA_{\mathrm{M}})^2\big)$, let $(u_1, \ldots, u_{2d_{A_{\mathrm{M}}}}, u_1', \ldots, u_{2d_{\mathbb{A}}}')$ be its endogenous boundary coordinates satisfying (8.3.8) and $(v_1, \ldots, v_{d_{A_{\mathrm{M}}}}, v_1', \ldots, v_{d_{\mathbb{A}}}')$ be its endogenous boundary coordinates with underlying quadruplet $S\mathcal{Q}$ and with respect to the above basis of $\ker SA_{\mathrm{M}}$ and $F_{0,S\mathcal{Q}}$. Then we claim that

$$v_j = u_j \qquad j = 1, \ldots, d_{A_{\mathrm{M}}} \tag{8.3.12}$$

$$v_j' = \sum_{i=1}^{d_{A_{\mathrm{M}}}} \langle \xi_i, \xi_j^{\mathrm{S}} \rangle_{\mathcal{V},\mathcal{V}^*} u_{d_{A_{\mathrm{M}}}+i} + u_{d_{\mathbb{A}}+j}' \qquad j = 1, \ldots, d_{\mathbb{A}} \tag{8.3.13}$$

and that $(u_{d_{A_{\mathrm{M}}}+1}, \ldots, u_{2d_{A_{\mathrm{M}}}}, u_1', \ldots, u_{d_{\mathbb{A}}}')$ *is the endogenous boundary coordinates of* $SA_{\mathrm{M}}u$ *with underlying quadruplet* $S\mathcal{Q}$ *and with respect to the above basis of* $\ker SA_{\mathrm{M}}$ *and of* $F_{0,S\mathcal{Q}}$.

Indeed, comparing (8.3.8) and (8.3.11), a direct sum argument gives (8.3.12). By (8.3.8),

$$SA_{\mathrm{M}}u = \sum_{i=1}^{d_{A_{\mathrm{M}}}} u_{d_{A_{\mathrm{M}}}+i}\, \xi_i + BS \sum_{i=1}^{d_{\mathbb{A}}} u_i' \eta_i + BS\Big(\sum_{i=d_{\mathbb{A}}+1}^{2d_{\mathbb{A}}} u_i' \eta_i + \psi^{\perp} \Big).$$

Now $\sum_{i=d_{\mathbb{A}}+1}^{2d_{\mathbb{A}}} u_i' \eta_i$ lies in $(\ker S^*\mathbb{A})^{\perp}$ by (8.3.5). Moreover $\psi^{\perp}$ belongs also to $(\ker S^*\mathbb{A})^{\perp}$ because the latter space contains $\big(\ker(S^*\mathbb{A})^2\big)^{\perp}$ according to (8.3.4). Thus

$$\sum_{i=d_{\mathbb{A}}+1}^{2d_{\mathbb{A}}} u_i' \eta_i + \psi^{\perp} \in (\ker S^*\mathbb{A})^{\perp}. \tag{8.3.14}$$

Then by uniqueness of the endogenous boundary coordinates with underlying quadruplet $S\mathcal{Q}$, we get the third assertion of the claim. Finally, let us prove (8.3.13). From (8.3.14) and (8.3.5), $\langle SA_{\mathrm{M}}u, \xi_j^{\mathrm{s}}\rangle_{\mathcal{V},\mathcal{V}^*}$ is equal to the right-hand side of (8.3.13). Besides, since $(v_1, \ldots, v_{d_{A_{\mathrm{M}}}}, v'_1, \ldots, v'_{d_{\mathbb{A}}})$ is the endogenous boundary coordinates of u with underlying quadruplet $S\mathcal{Q}$, one has in view of (8.3.11):

$$SA_{\mathrm{M}}u = \sum_{i=1}^{d_{\mathbb{A}}} v'_i \eta_i + \psi_1^{\perp}.$$

Thus by (8.3.5), $\langle SA_{\mathrm{M}}u, \xi_j^{\mathrm{s}}\rangle_{\mathcal{V},\mathcal{V}^*}$ is equal to v'_j, and (8.3.13) follows.

8.3.3.3 Endogenous Boundary Conditions of $-\Delta_{S\mathcal{Q},\mathrm{D}}$

Let u belong to the domain of the maximal operator $(SA_{\mathrm{M}})^2$ of $(S\mathcal{Q})^2$, i.e., u lies in $D(-\Delta_{S\mathcal{Q}})$. Denote by $(u_1, \ldots, u_{2d_{A_{\mathrm{M}}}}, u'_1, \ldots, u'_{2d_{\mathbb{A}}})$ its endogenous boundary coordinates as in (8.3.8). We claim that *u lies in $D(-\Delta_{S\mathcal{Q},\mathrm{D}})$ if and only if*

$$\begin{cases} u_1 = \cdots = u_{d_{A_{\mathrm{M}}}} = 0 \\ \displaystyle\sum_{i=1}^{d_{A_{\mathrm{M}}}} \langle \xi_i, \xi_j^{\mathrm{s}}\rangle_{\mathcal{V},\mathcal{V}^*}\, u_{d_{A_{\mathrm{M}}}+i} + u'_{d_{\mathbb{A}}+j} = 0, \qquad \forall\, 1 \le j \le d_{\mathbb{A}}. \end{cases} \tag{8.3.15}$$

Indeed, let $(v_1, \ldots, v_{d_{A_{\mathrm{M}}}}, v'_1, \ldots, v'_{d_{\mathbb{A}}})$ be the endogenous boundary coordinates of u with underlying quadruplet $S\mathcal{Q}$ and with respect to the basis

$$(\xi_1, \ldots, \xi_{d_{A_{\mathrm{M}}}})$$

of $\ker SA_{\mathrm{M}}$ and the basis $(\eta_1, \ldots, \eta_{d_{\mathbb{A}}})$ of $F_{0,S\mathcal{Q}}$. If u lies in $-\Delta_{S\mathcal{Q},\mathrm{D}}$ then u lies in $D(A_{\mathrm{m},\mathcal{Q}})$, which is equal to $BS(\ker \mathbb{A}^{\perp})$. Thus $u = BS\psi_1^{\perp}$ for some $\psi_1^{\perp}$ in $(\ker \mathbb{A})^{\perp}$. Hence by (8.3.11), $v_1 = \cdots = v_{d_{A_{\mathrm{M}}}} = v'_1 = \cdots = v'_{d_{\mathbb{A}}} = 0$. Then (8.3.15) follows from (8.3.12) and (8.3.13). Conversely, if (8.3.15) holds true then (8.3.12) and (8.3.13) entail that u lies in $D(A_{\mathrm{m},\mathcal{Q}})$. Hence u lives in the domain of $-\Delta_{S\mathcal{Q},\mathrm{D}}$, which proves the claim.

In closing, (8.3.15) *gives the endogenous boundary conditions of the Laplace operator with respect to $S\mathcal{Q}$ and with homogeneous Dirichlet endogenous boundary conditions.*

8.4 Abstract Dirichlet Problems

8.4.1 Formulation of the Problem

Let $(\mathcal{V}, \|\cdot\|)$ be a real or complex reflexive Banach space and $\mathcal{Q} := (\mathbb{A}, A_{\mathrm{M}}, B, S)$ be a F_0-regular differential quadruplet on $(\mathcal{V}, \mathcal{V}^*)$. Then $S\mathcal{Q}$ is a F_0-regular differential quadruplet on $(\mathcal{V}, \mathcal{V}^*)$, and the domain of the maximal operator SA_{M} reads according to the Domain Structure Theorem 5.3.3

$$D(SA_{\mathrm{M}}) = E_{\mathrm{M},F_0} \oplus BS(\ker \mathbb{A}^{\perp}), \tag{8.4.1}$$

where

$$E_{\mathrm{M},F_0} := \ker A_{\mathrm{M}} \oplus BS(F_0). \tag{8.4.2}$$

In view of Definition 5.3.3, denote by $P_{\mathrm{b.v},F_0}$ *the projection induced by* F_0 i.e.$P_{\mathrm{b.v},F_0} : D(SA_{\mathrm{M}}) \to D(SA_{\mathrm{M}})$ is the projection on E_{M,F_0} along $BS(\ker \mathbb{A}^{\perp})$. Recalling that $-\Delta_{S\mathcal{Q}} := (SA_{\mathrm{M}})^2$, for any datum $(h_{\mathrm{b.c}}, h)$ in $E_{\mathrm{M},F_0} \times \mathcal{V}$, we consider the following *inhomogeneous Dirichlet problem with respect to* $S\mathcal{Q}$ *and* F_0:

$$\begin{cases} \text{Find } u \text{ in } D(-\Delta_{S\mathcal{Q}}) \text{ such that} \\ -\Delta_{S\mathcal{Q}}\, u = h, \qquad P_{\mathrm{b.v},F_0}\, u = h_{\mathrm{b.c}}. \end{cases} \tag{8.4.3}$$

Definition 8.4.1 Problem (8.4.3) is said to be *well posed* if for each $(h_{\mathrm{b.c}}, h)$ in $E_{\mathrm{M},F_0} \times \mathcal{V}$, the following two conditions are satisfied:

(i) (8.4.3) has a unique solution u.
(ii) u depends continuously on the data with respect to the graph norm of A_{M}, that is u satisfies the estimate

$$\|u\|_{D(A_{\mathrm{M}})} \le C(\|h\| + \|h_{\mathrm{b.c}}\|_{D(A_{\mathrm{M}})}),$$

where the constant C is independent of u, $h_{\mathrm{b.c}}$ and h.

□

8.4.2 Well Posedness Under ker A_{M}-Regularity

As in Sect. 7.2.5.2, we will implement *the order reducing method.* More precisely, in view of Sect. 5.8.1, we consider $\overrightarrow{S\mathcal{Q}}$, *the differential quadruplet induced by* $S\mathcal{Q}$ *on* $\big(\mathcal{V}^2, (\mathcal{V}^*)^2\big)$.

By Proposition 5.8.1, $\overrightarrow{SQ}$ is $(F_0)^2$-regular, and the projection induced by $(F_0)^2$, denoted by $P_{\text{b.v},(F_0)^2}$, is equal to $\overrightarrow{P_{\text{b.v},F_0}}$. This simply means that for each (u_1, u_2) in $D(SA_{\text{M}})^2$,

$$P_{\text{b.v},(F_0)^2}\,(u_1, u_2) = (P_{\text{b.v},F_0}\,u_1,\ P_{\text{b.v},F_0}\,u_2).$$

By (5.8.3), $E_{\text{M},\overrightarrow{SQ}} = (E_{\text{M},F_0})^2$, thus we may legitimately define the map

$$T : E_{\text{M},(F_0)^2} \to E_{\text{M},F_0}, \qquad (u_1, u_2) \mapsto u_1. \tag{8.4.4}$$

Also let $\vec{h} := (0, h)$ and

$$L : \mathcal{V}^2 \to \mathcal{V}^2, \qquad (v_1, v_2) \mapsto (v_2, 0). \tag{8.4.5}$$

With these notation, it is clear that the problem

$$\begin{cases} \text{Find } \vec{u} \text{ in } D(A_{\text{M},\overrightarrow{SQ}}) \text{ such that} \\ A_{\text{M},\overrightarrow{SQ}}\,\vec{u} = L\vec{u} + \vec{h}, \qquad T P_{\text{b.v},(F_0)^2}\,\vec{u} = h_{\text{b.c}} \end{cases} \tag{8.4.6}$$

is equivalent to (8.4.3) in the following sense. If $\vec{u} = (u_1, u_2)$ is solution to (8.4.6), then u_1 is solution to (8.4.3), and conversely if u_1 is solution to (8.4.3), then $(u_1, SA_{\text{M}}u_1)$ is solution to (8.4.6).

The following result states that well posedness is obtained if $\mathcal{Q}$ is $\ker A_{\text{M}}$-regular.

Proposition 8.4.1 *Let* $\mathcal{Q} := (\mathbb{A}, A_{\text{M}}, B, S)$ *be a differential quadruplet on* $(\mathcal{V}, \mathcal{V}^*)$. *If* $\mathcal{Q}$ *is* $\ker A_{\text{M}}$*-regular then the inhomogeneous Dirichlet problem* (8.4.3) *with respect to* $S\mathcal{Q}$ *and* $\ker SA_{\text{M}}$ *is well posed in the sense of Definition 8.4.1.*

Proof Since $\mathcal{Q}$ is $\ker A_{\text{M}}$-regular, (8.4.2) may be put under the form

$$E_{\text{M},\ker A_{\text{M}}} = \ker A_{\text{M}} \oplus BS(\ker A_{\text{M}}). \tag{8.4.7}$$

Let us show that (8.4.6) is well posed by using Corollary 7.2.22 with $\mathcal{Q} := \overrightarrow{SQ}$ and $\mathcal{X}_1 := E_{\text{M},\ker A_{\text{M}}}$. We equip $E_{\text{M},\ker A_{\text{M}}}$ with the grap-norm of A_{M}, so that it becomes a Banach space. Next, since $L^2 = 0$, (7.2.1) holds true. Moreover for each v_1 and v_2 in $\mathcal{V}$, one has

$$\vec{B}\vec{S}L(v_1, v_2) = L\vec{B}\vec{S}(v_1, v_2) = (BSv_2, 0).$$

Thus (7.2.2) is verified. Regarding (7.2.55), we compute

$$\mathsf{R}^{L}_{\vec{B}\vec{S}}(v_1, v_2) = (v_1 + BSv_2, v_2),$$

so that, for each (u_0, v_0) in $(\ker SA_{\mathrm{M}})^2$, using (8.4.7), we get

$$T P_{\mathrm{b.v},(\ker A_{\mathrm{M}})^2} \mathsf{R}^L_{\vec{BS}}(u_0, v_0) = u_0 + BSv_0. \tag{8.4.8}$$

Thus since $\ker A_{\mathrm{M},\overrightarrow{SQ}} = (\ker SA_{\mathrm{M}})^2$, we derive

$$T P_{\mathrm{b.v},(\ker A_{\mathrm{M}})^2} \mathsf{R}^L_{\vec{BS}}(\ker A_{\mathrm{M},\overrightarrow{SQ}}) = E_{\mathrm{M},\ker A_{\mathrm{M}}};$$

so that (7.2.55) holds. Moreover, from (8.4.8), a direct sum argument yields that (7.2.56) holds as well. Thus Corollary 7.2.22 entails that (8.4.6) is well posed in the sense of Definition 7.2.9.

Referring to the analysis previously made in the current subsection which shows that (8.4.6) is equivalent to (8.4.3), we derive that (8.4.3) is uniquely solvable. Finally the estimate of Definition 8.4.1 follows from the well posedness of (8.4.6). □

8.4.2.1 Solvability with Maximal Operators Having Finite-Dimensional Kernel

Under the assumptions and notation of Sect. 8.3.3, SQ is a $F_{0,SQ}$-regular differential quadruplet on $(\mathcal{V}, \mathcal{V}^*)$ where $F_{0,SQ}$ is defined by (8.3.10). Hence (8.4.1) holds with $F_0 := F_{0,SQ}$, that is to say

$$D(SA_{\mathrm{M}}) = E_{\mathrm{M},F_{0,SQ}} \oplus SB\big((\ker \mathbb{A}S^*)^{\perp}\big),$$

where

$$E_{\mathrm{M},F_{0,SQ}} := \ker A_{\mathrm{M}} \oplus BS(F_{0,SQ}). \tag{8.4.9}$$

In view of Sect. 8.4.1, we denote by $P_{\mathrm{b.v},F_{0,SQ}}$ the projection of $D(SA_{\mathrm{M}})$ induced by $F_{0,SQ}$. Then for any datum $(h_{\mathrm{b.c}}, h)$ in $E_{\mathrm{M},F_{0,SQ}} \times \mathcal{V}$, we consider the following inhomogeneous Dirichlet Problem with respect to SQ and $F_{0,SQ}$:

$$\begin{cases} \text{Find } u \text{ in } D(-\Delta_{SQ}) \text{ such that} \\ -\Delta_{SQ} u = h, \qquad P_{\mathrm{b.v},F_{0,SQ}}\, u = h_{\mathrm{b.c}}. \end{cases} \tag{8.4.10}$$

Proposition 8.4.2 *Under the assumptions and notation of Sect. 8.3.3, in particular* (8.3.3)–(8.3.6) *and* (8.3.10), *denote by M the scalar matrix of dimension $d_{\mathbb{A}} \times d_{A_{\mathrm{M}}}$ whose entry located at the intersection of the ith row and the jth column is $\langle \xi_j, \xi_i^{\mathrm{S}} \rangle_{\mathcal{V},\mathcal{V}^*}$.*

(i) *If M is one-to-one then* (8.4.10) *has at most a solution.*
(ii) *If M is onto then* (8.4.10) *has at least a solution.*
(iii) *If M is invertible then* (8.4.10) *is well posed.*

Proof For any u in the domain of $(S\mathcal{Q})^2$, let

$$(u_1, \ldots, u_{2d_{A_M}}, u'_1, \ldots, u'_{2d_{\mathbb{A}}})$$

and $\psi^\perp$ in $\big(\ker(S^*\mathbb{A})^2\big)^\perp$ satisfy (8.3.8). By (8.3.12), the endogenous boundary coordinates of u with underlying quadruplet $S\mathcal{Q}$ (see (8.3.11)) reads $(u_1, \ldots, u_{d_{A_M}}, v'_1, \ldots, v'_{d_{\mathbb{A}}})$ where $v'_1, \ldots, v'_{d_{\mathbb{A}}}$ are scalars. Being given $(h_{\mathrm{b.c}}, h)$ in $E_{\mathrm{M},S\mathcal{Q}} \times \mathcal{V}$, denote by

$$(h_{\mathrm{b.c},1}, \ldots, h_{\mathrm{b.c},d_{A_M}}, h'_{\mathrm{b.c},1}, \ldots, h'_{\mathrm{b.c},d_{\mathbb{A}}})$$

the endogenous boundary coordinates of $h_{\mathrm{b.c}}$ with underlying quadruplet $S\mathcal{Q}$, more precisely

$$h_{\mathrm{b.c}} = \sum_{i=1}^{d_{A_M}} h_{\mathrm{b.c},i}\xi_i + BS\sum_{i=1}^{d_{\mathbb{A}}} h'_{\mathrm{b.c},i}\eta_i. \tag{8.4.11}$$

Besides by (8.3.7), h reads

$$h = \sum_{i=1}^{2d_{\mathbb{A}}} h_i\eta_i + \phi^\perp, \tag{8.4.12}$$

for some scalars h_i's and some vector $\phi^\perp$ in $\big(\ker(S^*\mathbb{A})^2\big)^\perp$. Then using $v'_i = h'_{\mathrm{b.c},i}$ if $P_{\mathrm{b.v},F_{0,S\mathcal{Q}}}\, u = h_{\mathrm{b.c}}$, together with

$$-\Delta_{S\mathcal{Q}} u = \sum_{i=1}^{2d_{\mathbb{A}}} u'_i\eta_i + \psi^\perp$$

and (8.3.12)–(8.3.13), we derive that u solves (8.4.10) if and only if $(u_1, \ldots, u_{2d_{A_M}}, u'_1, \ldots, u'_{2d_{\mathbb{A}}})$ and $\psi^\perp$ satisfy

$$\begin{aligned}
u_i &= h_{\mathrm{b.c},i} && i = 1, \ldots, d_{A_M}\\
u'_i &= h_i && i = 1, \ldots, 2d_{\mathbb{A}}\\
\psi^\perp &= \phi^\perp &&\\
\sum_{j=1}^{d_{A_M}} \langle \xi_j, \xi_i^{\mathrm{S}}\rangle_{\mathcal{V},\mathcal{V}^*}\, u_{d_{A_M}+j} &= h'_{\mathrm{b.c},i} - h_{d_{\mathbb{A}}+i} && i = 1, \ldots, d_{\mathbb{A}}.
\end{aligned}$$

Then Items (i) and (ii) follow from the latter equations. Finally if M is assumed to be invertible, then the same equations imply that (8.4.10) admits a unique solution u. Since M is invertible, Lemma 2.2.13 entails that

$$\mathcal{V} = \ker A_{\mathrm{M}} \oplus \ker \mathbb{A}^{\perp},$$

so that $\mathcal{Q}$ is $\ker A_{\mathrm{M}}$-regular. Thus by considering the differential quadruplet $S\mathcal{Q}$, the projection $P_{\mathrm{b.v},\ker A_{\mathrm{M}}}$ is well defined. Then since for each v in $D(SA_{\mathrm{M}})$,

$$v - P_{\mathrm{b.v}, F_{0,S\mathcal{Q}}}\, v \in \ker P_{\mathrm{b.v}, F_{0,S\mathcal{Q}}} = \ker P_{\mathrm{b.v},\ker A_{\mathrm{M}}},$$

we derive that

$$P_{\mathrm{b.v},\ker A_{\mathrm{M}}} \circ P_{\mathrm{b.v}, F_{0,S\mathcal{Q}}} = P_{\mathrm{b.v},\ker A_{\mathrm{M}}}.$$

Thus u solves

$$-\Delta_{S\mathcal{Q}}\, u = h, \qquad P_{\mathrm{b.v},\ker A_{\mathrm{M}}}\, u = P_{\mathrm{b.v},\ker A_{\mathrm{M}}}\, h_{\mathrm{b.c}}.$$

Then Proposition 8.4.1 yields that the previous problem is well posed. Hence

$$\|u\|_{D(SA_{\mathrm{M}})} \lesssim \|h\| + \|P_{\mathrm{b.v},\ker SA_{\mathrm{M}}}\, h_{\mathrm{b.c}}\|_{D(A_{\mathrm{M}})}.$$

Using the continuity of the projection $P_{\mathrm{b.v},\ker A_{\mathrm{M}}}$ on $D(A_{\mathrm{M}})$, we derive the well posedness of (8.4.10). □

Remark 8.4.2 Under the assumptions and notation of Proposition 8.4.2, let $(h_{\mathrm{b.c}}, h)$ belong to $E_{\mathrm{M}, F_{0,S\mathcal{Q}}} \times \mathcal{V}$ and satisfy (8.4.11), (8.4.12). For any u in the domain of $(S\mathcal{Q})^2$, denote by $(u_1, \ldots, u_{2d_{A_{\mathrm{M}}}}, u'_1, \ldots, u'_{2d_{\mathbb{A}}})$ its endogenous boundary coordinates according to (8.3.8). There results from the proof of Proposition 8.4.2 that the vector

$$u := \sum_{j=1}^{d_{A_{\mathrm{M}}}} h_{\mathrm{b.c},j}\xi_j + u_{d_{A_{\mathrm{M}}}+j}\, BS\xi_j \; + \; (BS)^2 h$$

solves (8.4.10) if and only if $(u'_{d_{\mathbb{A}}+1}, \ldots, u'_{2d_{\mathbb{A}}})$ satisfies

$$\sum_{j=1}^{d_{A_{\mathrm{M}}}} \langle \xi_j, \xi_i^{\mathrm{S}} \rangle_{\mathcal{V},\mathcal{V}^*}\, u_{d_{A_{\mathrm{M}}}+j} = h'_{\mathrm{b.c},i} - h_{d_{\mathbb{A}}+i}, \qquad \forall i = 1, \ldots, d_{\mathbb{A}}.$$

□

8.4.3 On the Homogeneous Dirichlet Problem

Being given a real or complex reflexive Banach space $\mathcal{V}$ and $\mathcal{Q} = (\mathbb{A}, A_{\mathrm{M}}, B, S)$ a differential quadruplet on $(\mathcal{V}, \mathcal{V}^*)$, the *homogeneous Dirichlet problem for the Laplace operator with respect to* $S\mathcal{Q}$ reads for any datum h in $\mathcal{V}$,

$$\text{Find } u \text{ in } D(-\Delta_{S\mathcal{Q},\mathrm{D}}) \text{ such that } -\Delta_{S\mathcal{Q},\mathrm{D}}\, u = h. \tag{8.4.13}$$

Notice that the formulation of that problem do not require the differential quadruplet to be regular. However, as we will see, regularity will be useful to get well posedness.

Let us assume that $\mathcal{Q}$ is $\ker A_{\mathrm{M}}$-regular. Then because the kernel of the projection $P_{\mathrm{b.v}, \ker SA_{\mathrm{M}}}$ (with respect to $S\mathcal{Q}$) is equal to the domain of the minimal operator of $SA_{\mathrm{m},\mathcal{Q}}$, there results that (8.4.13) is equivalent to

$$\begin{cases} \text{Find } u \text{ in } D(-\Delta_{S\mathcal{Q}}) \text{ such that} \\ -\Delta_{S\mathcal{Q}}\, u = h, \qquad P_{\mathrm{b.v}, \ker SA_{\mathrm{M}}}\, u = 0. \end{cases}$$

Thus Proposition 8.4.1 tells us that (8.4.13) is well posed. Hence $-\Delta_{S\mathcal{Q},\mathrm{D}}$ is an isomorphism from its domain onto $\mathcal{V}$.

On the other hand, Proposition 8.3.1 has the same conclusion under the following assumptions: $\mathcal{Q}$ is closed and

$$\ker A_{\mathrm{M}}{}^{\perp} \oplus \ker \mathbb{A} = \mathcal{V}^*. \tag{8.4.14}$$

It turns out these two results are essentially the same since for any closed differential quadruplet $\mathcal{Q} := (\mathbb{A}, A_{\mathrm{M}}, B, S)$,

$$\ker A_{\mathrm{M}} \oplus \ker \mathbb{A}^{\perp} = \mathcal{V}, \tag{8.4.15}$$

is equivalent to (8.4.14). Indeed, by reflexivity it is enough to show that the latter equality implies the former. Owing to reflexivity, Proposition 2.2.12 gives that $(\ker \mathbb{A}^{\perp})^{\perp} = \ker \mathbb{A}$. Thus with Proposition 2.3.8, we derive (8.4.14) from (8.4.15).

8.5 Fractional Dirichlet Problems

In this section, we will consider for simplicity functions with values in a nontrivial finite-dimensional real or complex normed vector space $(\mathcal{Y}_{\mathrm{f}}, \|\cdot\|_{\mathcal{Y}_{\mathrm{f}}})$. Being given conjugate exponents p and q in $(1, \infty)$, we set

$$\mathcal{V} = L^p(\mathcal{Y}_{\mathrm{f}}) := L^p\big((0, b), \mathcal{Y}_{\mathrm{f}}\big).$$

8.5.1 Problems Involving $Q_{p,\mathcal{Y}_\mathrm{f},1}$

Under the assumptions and notation of Example 5.1.5, assume in addition that $\mathcal{Y} := \mathcal{Y}_\mathrm{f}$, so that

$$Q_{p,\mathcal{Y}_\mathrm{f},1} = (\mathrm{D}^1_{\mathcal{V}^*},\ \mathrm{D}^1_{\mathcal{V}},\ B_{1,\mathcal{V}},\ S_{\mathcal{V}}).$$

8.5.1.1 Laplace Operators on $L^p(\mathcal{Y}_\mathrm{f})$

Setting for simplicity,

$$\Delta_{\mathcal{V}} := \Delta_{Q_{p,\mathcal{Y}_\mathrm{f},1}S_{\mathcal{V}}}$$
$$\Delta_{\mathcal{V},\mathrm{D}} := \Delta_{Q_{p,\mathcal{Y}_\mathrm{f},1}S_{\mathcal{V}},\mathrm{D}},$$

one has

$$\Delta_{\mathcal{V}} = (\mathrm{D}^1_{\mathcal{V}})^2, \qquad D(\Delta_{\mathcal{V}}) = W^{2,p}\big((0,b),\mathcal{Y}_\mathrm{f}\big),$$

and

$$\Delta_{\mathcal{V},\mathrm{D}} = \mathrm{D}^1_{L^p} A_{\mathrm{m},Q_{p,\mathcal{Y}_\mathrm{f},1}}$$
$$D(\Delta_{\mathcal{V},\mathrm{D}}) = W^{1,p}_0\big((0,b),\mathcal{Y}_\mathrm{f}\big) \cap W^{2,p}\big((0,b),\mathcal{Y}_\mathrm{f}\big).$$

Thus *the Laplace operator with respect to* $S_{\mathcal{V}}Q_{p,\mathcal{Y}_\mathrm{f},1}$ *and with homogeneous Dirichlet endogenous boundary conditions* is nothing but the (usual) Laplace operator on $L^p(0,b)$ supplemented with homogeneous Dirichlet boundary conditions.

8.5.1.2 Extensions of Laplace Operators

Weak products allow to extend Laplace operators. More precisely, we apply the theory of Sect. 8.2 with

$$A := -\mathrm{D}^1_{\mathcal{V}}, \qquad C := A_{\mathrm{m},Q_{p,\mathcal{Y}_\mathrm{f},1}}.$$

In view of Example 5.1.5, one has

$$(Q_{p,\mathcal{Y}_\mathrm{f},1})^* = \big(-\mathrm{D}^1_{\mathcal{V}},\ -\mathrm{D}^1_{\mathcal{V}^*},\ (B_{1,\mathcal{V}})^*,\ S_{L^q}\big)$$

Thus according to Proposition 5.2.7,

$$A^* = -(\mathrm{D}^1_{\mathcal{V}})^* = -A_{\mathrm{m},(Q_{p,\mathcal{Y}_\mathrm{f},1})^*} \subseteq \mathrm{D}^1_{\mathcal{V}^*}, \qquad C^* = -\mathrm{D}^1_{\mathcal{V}^*}.$$

Hence denoting by $W^{-1,p}\big((0,b),\mathcal{Y}_{\mathrm{f}}\big)$ the anti-dual space of $W_0^{1,q}\big((0,b),\mathcal{Y}_{\mathrm{f}}{}^*\big)$, we have

$$D(A^*) = W_0^{1,q}\big((0,b),\mathcal{Y}_{\mathrm{f}}{}^*\big), \qquad D(A^*)^* = W^{-1,p}\big((0,b),\mathcal{Y}_{\mathrm{f}}\big).$$

By Sect. 8.2.1, $\widetilde{\mathrm{D}^1_{\mathcal{V}^*}}^{\,*}$ is the continuous operator from $L^p(\mathcal{Y}_{\mathrm{f}})$ with values in $W^{-1,p}(\mathcal{Y}_{\mathrm{f}})$ defined for each v in $L^p(\mathcal{Y}_{\mathrm{f}})$ by

$$\langle \widetilde{\mathrm{D}^1_{\mathcal{V}^*}}^{\,*} v, w\rangle_{W^{-1,p}(\mathcal{Y}_{\mathrm{f}}),W_0^{1,q}(\mathcal{Y}_{\mathrm{f}}{}^*)} = \int_0^b \langle v(x), \mathrm{D}^1_{\mathcal{V}^*} w(x)\rangle_{\mathcal{Y}_{\mathrm{f}},\mathcal{Y}_{\mathrm{f}}{}^*}\, \mathrm{d}x$$

for every w in $W_0^{1,q}(\mathcal{Y}_{\mathrm{f}}{}^*)$. Then the *weak product of* $-\mathrm{D}^1_{\mathcal{V}}$ *and* $\mathrm{A}_{\mathrm{m},Q_{p,\mathcal{Y}_{\mathrm{f}},1}}$ is the operator

$$\widetilde{\Delta_{\mathcal{V},\mathrm{D}}} : W_0^{1,p}\big((0,b),\mathcal{Y}_{\mathrm{f}}\big) \to W^{-1,p}\big((0,b),\mathcal{Y}_{\mathrm{f}}\big)$$

defined for each v in $W_0^{1,p}\big((0,b),\mathcal{Y}_{\mathrm{f}}\big)$ by

$$\langle \widetilde{\Delta_{\mathcal{V},\mathrm{D}}}\, v, w\rangle_{W^{-1,p}(\mathcal{Y}_{\mathrm{f}}),W_0^{1,q}(\mathcal{Y}_{\mathrm{f}}{}^*)} = \int_0^b \langle \mathrm{D}^1_{\mathcal{V}} v(x), \mathrm{D}^1_{\mathcal{V}^*} w(x)\rangle_{\mathcal{Y}_{\mathrm{f}},\mathcal{Y}_{\mathrm{f}}{}^*}\, \mathrm{d}x$$

for every w in $W_0^{1,q}\big((0,b),\mathcal{Y}_{\mathrm{f}}{}^*\big)$.

In the particular case where $\mathcal{Y}_{\mathrm{f}} := \mathbb{K}$, the anti-dual space of $\mathbb{K}$ is identified with $\mathbb{K}$, so that for each v in $L^p(0,b)$,

$$\langle \widetilde{\mathrm{D}^1_{L^q}}^{\,*} v, w\rangle_{W^{-1,p},W_0^{1,q}} = \int_0^b v(x)\overline{w'(x)}\, \mathrm{d}x$$

and for each v in $W_0^{1,p}(0,b)$,

$$\langle \widetilde{\Delta_{L^p,\mathrm{D}}}\, v, w\rangle_{W^{-1,p},W_0^{1,q}} = \int_0^b v'(x)\overline{w'(x)}\, \mathrm{d}x$$

for every w in $W_0^{1,q}(0,b)$. This is *the usual weak formulation of the Laplace operator on* $L^p(0,b)$ *supplemented with homogeneous Dirichlet boundary conditions*. By (8.2.4), $\widetilde{\Delta_{L^p,\mathrm{D}}}$ extends $\Delta_{L^p,\mathrm{D}}$, and it is customary to write $\Delta_{L^p,\mathrm{D}}$ instead of $\widetilde{\Delta_{L^p,\mathrm{D}}}$.

8.5.1.3 Dirichlet Problems on $L^p(\mathcal{Y}_{\mathrm{f}})$

For any datum $(h, h_{\mathrm{b.c},1}, h_{\mathrm{b.c},1})$ in $L^p(\mathcal{Y}_{\mathrm{f}}) \times \mathcal{Y}_{\mathrm{f}} \times \mathcal{Y}_{\mathrm{f}}$, we consider the following problem:

$$\begin{cases} \text{Find } u \text{ in } W^{2,p}\big((0,b),\mathcal{Y}_{\mathrm{f}}\big) \text{ such that} \\ -\Delta_{\mathcal{V}} u = h, \qquad u(0) = h_{\mathrm{b.c},1},\ u(b) = h_{\mathrm{b.c},2}. \end{cases} \tag{8.5.1}$$

We will apply Proposition 8.4.1. According to Example 5.6.7, $\mathcal{Q}_{p,\mathcal{Y}_\mathrm{f},1}$ is a $\ker \mathrm{D}^1_{\mathcal{V}}$-regular differential quadruplet. Thus in the current framework, F_0 is equal to $\ker \mathrm{D}^1_{\mathcal{V}}$ and $E_{\mathrm{M},\ker A_\mathrm{M}}$ reads (see (8.4.7))

$$E_{\mathrm{M},\ker \mathrm{D}^1_{\mathcal{V}}} = g_1\mathcal{Y}_\mathrm{f} \oplus g_2\mathcal{Y}_\mathrm{f}.$$

Thus setting

$$h_{\mathrm{b.c}} := g_1 h_{\mathrm{b.c},1} + b^{-1} g_2(h_{\mathrm{b.c},2} - h_{\mathrm{b.c},1}),$$

the boundary conditions of (8.5.1) are equivalent to

$$P_{\mathrm{b.v},\ker S_{\mathcal{V}}\mathrm{D}^1_{\mathcal{V}}}\, u = h_{\mathrm{b.c}}.$$

Hence (8.5.1) is equivalent to the inhomogeneous Dirichlet problem for the Laplace operator with respect to $S_{\mathcal{V}}\mathcal{Q}_{p,\mathcal{Y}_\mathrm{f},1}$ and $\ker \mathrm{D}^1_{\mathcal{V}}$. Thus Proposition 8.4.1 yields that (8.5.1) *is well posed in the sense of Definition 8.4.1. That is, the unique solution u of* (8.5.1) *satisfies*

$$\|u\|_{L^p(\mathcal{Y}_\mathrm{f})} + \|\mathrm{D}^1_{\mathcal{V}} u\|_{L^p(\mathcal{Y}_\mathrm{f})} \lesssim \left(\|h\|_{L^p(\mathcal{Y}_\mathrm{f})} + \|h_{\mathrm{b.c},1}\|_{\mathcal{Y}_\mathrm{f}} + \|h_{\mathrm{b.c},2}\|_{\mathcal{Y}_\mathrm{f}}\right).$$

8.5.1.4 Representation of the Solution

For simplicity, we will assume that

$$\mathcal{Y}_\mathrm{f} = \mathbb{K}.$$

Thus $\mathcal{V} = L^p := L^p(0,b)$, the datum h lives $L^p(0,b)$, and $h_{\mathrm{b.c},1}, h_{\mathrm{b.c},2}$ are scalars. Hence (8.5.1) reads

$$\begin{cases} \text{Find } u \text{ in } W^{2,p}(0,b) \text{ such that} \\ -\Delta_{L^p}\, u = h, \qquad u(0) = h_{\mathrm{b.c},1},\ u(b) = h_{\mathrm{b.c},2}. \end{cases} \tag{8.5.2}$$

Our goal is to prove that the solution u to (8.5.2) satisfies (8.5.4). For, we will apply the abstract result of Remark 8.4.2. Thus we will begin to implement the framework of Sect. 8.3.3.1 and rewrite (8.5.2) as an inhomogeneous Dirichlet problem of the form (8.4.10).

In view of (8.3.3)–(8.3.6),

$$d_{A_\mathrm{M}} = d_{\mathrm{D}^1_{L^p}} = 1, \qquad d_{\mathbb{A}} = d_{\mathrm{D}^1_{L^q}} = 1$$

and

$$\xi_1 = g_1, \qquad \xi_1^{\mathrm{s}} = g_1.$$

Thus by Sect. 8.3.3.1, there exists η_1, η_2 in $L^p(0, b)$ such that for each i in $\{1, 2\}$,

$$\langle g_1, \eta_i \rangle = \delta_{i,1}, \qquad \langle g_2, \eta_i \rangle = \delta_{i,2}. \tag{8.5.3}$$

Thus setting

$$F_{0, S_{L^p} Q_{p,\mathbb{K},1}} := \langle \eta_1 \rangle,$$

we know from Sect. 8.3.3.2 that $S_{L^p} Q_{p,\mathbb{K},\ell}$ is $F_{0,S_{L^p} Q_{p,\mathbb{K},\ell}}$-regular. Using also the first equation of (8.5.3), we deduce that for each u in $D(S_{L^p} \mathrm{D}^1_{L^p})$, u satisfies the boundary conditions of (8.5.2) if and only if

$$P_{\mathrm{b.v}, \langle \eta_1 \rangle} \, u = h_{\mathrm{b.c}},$$

where

$$h_{\mathrm{b.c}} := h_{\mathrm{b.c},1} g_1 + (h_{\mathrm{b.c},2} - h_{\mathrm{b.c},1}) B_{1,L^p} S_{L^p} \eta_1.$$

Thus Problem (8.5.2) is equivalent to an inhomogeneous Dirichlet problem of the form (8.4.10). Hence we are in position to apply Remark 8.4.2. According to (8.4.12), the source term h is decomposed into

$$h = h_1 \eta_1 + h_2 \eta_2 + \phi^{\perp},$$

for some scalars h_1, h_2 and some $\phi^{\perp}$ in $\langle g_1, S_{\mathcal{V}} g_2 \rangle^{\perp}$. Then, setting

$$u_2 := b^{-1} (h_{\mathrm{b.c},2} - h_{\mathrm{b.c},1} - h_2),$$

Remark 8.4.2 yields that the unique solution u to (8.5.2) reads

$$u = h_{\mathrm{b.c},1} g_1 + u_2 g_2 + (B_{1,L^p} S_{L^p})^2 h. \tag{8.5.4}$$

Thus for every x in $[0, b]$,

$$u(x) = u(0) + b^{-1} \Big(u(b) - u(0) - \int_0^b \int_y^b h(t) \, \mathrm{d}t \, \mathrm{d}y \Big) x + \int_0^x \int_y^b h(t) \, \mathrm{d}t \, \mathrm{d}y.$$

8.5.2 Problems Involving $Q_{p,\mathcal{Y}_\mathrm{f},\ell}$

Recalling that

$$\mathcal{V} := L^p\big((0,b),\mathcal{Y}_\mathrm{f}\big),$$

let (k,ℓ) be a Sonine pair. In view of Sect. 6.7.1, we have

$$Q_{p,\mathcal{Y}_\mathrm{f},\ell} = (\mathrm{D}^\ell_{\mathcal{V}*},\ \mathrm{D}^\ell_{\mathcal{V}},\ B_{\ell,\mathcal{V}},\ S_{\mathcal{V}}),$$

and by Definitions 8.3.1 and 8.3.2,

$$\Delta_{S_\mathcal{V} Q_{p,\mathcal{Y}_\mathrm{f},\ell}} = -(S_\mathcal{V}\,\mathrm{D}^\ell_\mathcal{V})^2$$

$$\Delta_{S_\mathcal{V} Q_{p,\mathcal{Y}_\mathrm{f},\ell},\mathrm{D}} = -S_\mathcal{V}\,\mathrm{D}^\ell_\mathcal{V}\,S_\mathcal{V}\,A_{\mathrm{m},Q_{p,\mathcal{Y}_\mathrm{f},\ell}}.$$

8.5.2.1 The Case Where ℓ Lies in $L^p(0,b)\cap L^q(0,b)$

8.5.2.1.1 Formulation of the Problem

For any data h in $L^p(\mathcal{Y}_\mathrm{f})$ and $h_{\mathrm{b.c},1},\dots,h_{\mathrm{b.c},4}$ in $\mathcal{Y}_\mathrm{f}$, we consider the following problem:

$$\begin{cases} \text{Find } u \text{ in } D(\Delta_{S_\mathcal{V} Q_{p,\mathcal{Y}_\mathrm{f},\ell}}) \text{ such that} \\ -\Delta_{S_\mathcal{V} Q_{p,\mathcal{Y}_\mathrm{f},\ell}}\, u = h \\ \big(u-\ell\,k*u(0)\big)(0) = h_{\mathrm{b.c},1}, \qquad k*u(0) = h_{\mathrm{b.c},2} \\ g_1 * \mathrm{D}^\ell_\mathcal{V} u(b) = h_{\mathrm{b.c},3}, \qquad \ell * \mathrm{D}^\ell_\mathcal{V} u(b) = h_{\mathrm{b.c},4}. \end{cases} \tag{8.5.5}$$

Recall that $u-\ell\,k*u(0)$ and $\ell*\mathrm{D}^\ell_\mathcal{V} u$ are continuous on $[0,b]$ for any function u in the domain of $\mathrm{D}^\ell_\mathcal{V}$ (see (6.7.8)). Thus the boundary conditions of (8.5.5) make sense.

Remark 8.5.1 The boundary conditions of (8.5.5) may be rewritten without differential terms, as in the usual Dirichlet problem (8.5.1). Indeed, let us assume for simplicity that ℓ is continuous on $(0,b]$. Then by Remark 6.7.1, (8.5.5) is equivalent up to an isomorphic change of the boundary data, to

$$\begin{cases} \text{Find } u \text{ in } D(\Delta_{S_\mathcal{V} Q_{p,\mathcal{Y}_\mathrm{f},\ell}}) \text{ such that} \\ -\Delta_{S_\mathcal{V} Q_{p,\mathcal{Y}_\mathrm{f},\ell}}\, u = h \\ \big(u-\ell\,k*u(0)\big)(0) = h_{\mathrm{b.c},1}, \qquad k*u(0) = h_{\mathrm{b.c},2} \\ k*(u-g_1\,U_1)(b) = h_{\mathrm{b.c},3}, \qquad u(b) = h_{\mathrm{b.c},4} \end{cases}$$

where U_1 if the element of $\mathcal{Y}_\mathrm{f}$ defined by

$$U_1 := \big(u - \ell\, k * u(0)\big)(0). \qquad \square$$

We will solve (8.5.5) by application of Proposition 8.4.1. For, Problem (8.5.5) must be reformulated into a Dirichlet problem with respect to $S\mathcal{Q}$ and $\ker \mathrm{D}^{\ell}_{\mathcal{V}}$. However the boundary conditions of (8.5.5) refer to direct sum decompositions which do not involve $\ker \mathrm{D}^{\ell}_{\mathcal{V}}$ as a complementary subspace of $\ker \mathrm{D}^{\ell}_{\mathcal{V}*}{}^{\perp}$. See, on this subject, Sect. 6.7.2.1 and in particular (6.7.7). This issue will be addressed in the forthcoming paragraphs.

8.5.2.1.2 Endogenous Boundary Values

We will consider two quadruplets of endogenous boundary values with underlying differential quadruplet $S_{\mathcal{V}}\mathcal{Q}_{p,\mathcal{Y}_{\mathrm{f}},\ell}$. The first quadruplet comes from the results of Sect. 6.7.2.1. More precisely, under the assumptions and notation of Sect. 6.7.2.1, in particular (6.7.3) and

$$\tilde{F}_0 = h_1\, \mathcal{Y}_{\mathrm{f}} \oplus h_2 \mathcal{Y}_{\mathrm{f}},$$

we derive that $S_{\mathcal{V}}\mathcal{Q}_{p,\mathcal{Y}_{\mathrm{f}},\ell}$ is $\tilde{F}_0$-regular. Thus any function u in $D(\mathrm{D}^{\ell}_{\mathcal{V}})$ reads

$$u = g_1\, U_1 + \ell\, U_2 + B_{\ell,\mathcal{V}} S_{\mathcal{V}}(h_1\, U_1' + h_2\, U_2' + \psi^{\perp}), \tag{8.5.6}$$

where $\psi^{\perp}$ lies in $(\ker \mathrm{D}^{1}_{\mathcal{V}*})^{\perp}$ and (U_1, U_2, U_1', U_2') is the endogenous boundary values of u with respect to the direct sum decompositions

$$\ker \mathrm{D}^{\ell}_{\mathcal{V}} = g_1 \mathcal{Y} \oplus \ell \mathcal{Y}_{\mathrm{f}}$$
$$\tilde{F}_0 = h_1\, \mathcal{Y}_{\mathrm{f}} \oplus h_2 \mathcal{Y}_{\mathrm{f}}.$$

In view of (6.7.4), we put

$$E_{\mathrm{M},\tilde{F}_0} := \ker \mathrm{D}^{\ell}_{\mathcal{V}} \oplus B_{\ell,\mathcal{V}} S_{\mathcal{V}} \tilde{F}_0. \tag{8.5.7}$$

The second quadruplet of endogenous boundary values comes from Proposition 8.5.1, which implies that $S_{\mathcal{V}}\mathcal{Q}_{p,\mathcal{Y}_{\mathrm{f}},\ell}$ is $\ker \mathrm{D}^{\ell}_{\mathcal{V}}$-regular. Thus any function u in $D(\mathrm{D}^{\ell}_{\mathcal{V}})$ reads

$$u = g_1\, V_1 + \ell\, V_2 + B_{\ell,\mathcal{V}} S_{\mathcal{V}}(g_1\, V_1' + \ell\, V_2' + \varphi^{\perp}), \tag{8.5.8}$$

where $\varphi^{\perp}$ lies in $(\ker \mathrm{D}^{1}_{\mathcal{V}*})^{\perp}$, and (V_1, V_2, V_1', V_2') is the endogenous boundary values of u with respect to the direct sum decomposition

$$\ker \mathrm{D}^{\ell}_{\mathcal{V}} = g_1 \mathcal{Y} \oplus \ell \mathcal{Y}_{\mathrm{f}}.$$

In view of (6.7.4), we put

$$E_{\mathrm{M},\ker \mathrm{D}^{\ell}_{\mathcal{V}}} := \ker \mathrm{D}^{\ell}_{\mathcal{V}} \oplus B_{\ell,\mathcal{V}} S_{\mathcal{V}} \ker \mathrm{D}^{\ell}_{\mathcal{V}}. \tag{8.5.9}$$

Now the main point is that by Proposition 5.7.2, *the change of endogenous boundary values is smooth, in the sense that the map*

$$F : \mathcal{Y}_{\mathrm{f}}^{4} \to \mathcal{Y}_{\mathrm{f}}^{4}, \qquad (U_1, U_2, U_1', U_2') \mapsto (V_1, V_2, V_1', V_2') \tag{8.5.10}$$

is an isomorphism.

Proposition 8.5.1 *If ℓ lies in $L^p(0,b) \cap L^q(0,b)$ then $Q_{p,\mathcal{Y}_{\mathrm{f}},\ell}$ is a $\ker \mathrm{D}^{\ell}_{\mathcal{V}}$-regular quadruplet.*

Proof In view of Sect. 6.7.3.1,

$$\ker \mathrm{D}^{\ell}_{\mathcal{V}*} = g_1 \mathcal{Y}_{\mathrm{f}}^{*} \oplus \ell \mathcal{Y}_{\mathrm{f}}^{*}.$$

In particular, g_1 and ℓ are linearly independent. Moreover, ℓ lies in $L^2(0,b)$ since $\max(p,q) \geq 2$. Thus denoting by $(\cdot,\cdot)$ the inner product of $L^2(0,b)$, the matrix

$$\begin{pmatrix} (g_1, g_1) & (\ell, g_1) \\ (g_1, \ell) & (\ell, \ell) \end{pmatrix}$$

is invertible according to Lemma 5.6.23. Since

$$\ker \mathrm{D}^{\ell}_{\mathcal{V}} = g_1 \mathcal{Y} \oplus \ell \mathcal{Y},$$

Corollary 5.3.2 entails that $Q_{p,\mathcal{Y}_{\mathrm{f}},\ell}$ is $\ker \mathrm{D}^{\ell}_{\mathcal{V}}$-regular. □

8.5.2.1.3 Well Posedness

Proposition 8.5.2 *Let p, q be conjugate exponents in $(1,\infty)$, $(\mathcal{Y}_{\mathrm{f}}, \|\cdot\|_{\mathcal{Y}_{\mathrm{f}}})$ be a nontrivial finite-dimensional normed vector space, and (k,ℓ) be a Sonine pair. If ℓ lies in $L^p(0,b) \cap L^q(0,b)$, then (8.5.5) is well posed in the sense that it admits a unique solution u. Moreover*

$$\|u\|_{\mathcal{V}} + \|\mathrm{D}^{\ell}_{\mathcal{V}} S_{\mathcal{V}}\, u\|_{\mathcal{V}} \lesssim \|h\| + \sum_{i=1}^{4} \|h_{\mathrm{b.c.},i}\|_{\mathcal{Y}_{\mathrm{f}}}. \tag{8.5.11}$$

Proof In order to apply Proposition 8.4.1, we will reformulate (8.5.5) into an inhomogeneous Dirichlet problem with respect to $S_{\mathcal{V}} Q_{p,\mathcal{Y}_{\mathrm{f}},\ell}$ and $\ker \mathrm{D}^{\ell}_{\mathcal{V}}$. For, it is enough to put the boundary conditions of (8.5.5) under the form

$$P_{\mathrm{b.v}, \ker \mathrm{D}^\ell_{\mathcal{V}}}\, u = h_{\mathrm{b.c}},$$

where $P_{\mathrm{b.v}, \ker \mathrm{D}^\ell_{\mathcal{V}}}$ denotes the projection of $D(S_{\mathcal{V}}\mathrm{D}^\ell_{\mathcal{V}})$ induced by $\ker \mathrm{D}^\ell_{\mathcal{V}}$. Let u belong to $D(S_{\mathcal{V}}\mathrm{D}^\ell_{\mathcal{V}})$. Under the notation of Sect. 8.5.2.1.2, there results from (6.7.8) that u satisfies

$$\begin{aligned} \big(u - \ell\, k * u(0)\big)(0) &= h_{\mathrm{b.c},1}, \qquad k * u(0) = h_{\mathrm{b.c},2} \\ g_1 * \mathrm{D}^\ell_{\mathcal{V}} u(b) &= h_{\mathrm{b.c},3}, \quad\ \ell * \mathrm{D}^\ell_{\mathcal{V}} u(b) = h_{\mathrm{b.c},4} \end{aligned} \tag{8.5.12}$$

if and only if

$$(U_1, U_2, U_1', U_2') = (h_{\mathrm{b.c},1}, \ldots, h_{\mathrm{b.c},4}).$$

Since the map F defined by (8.5.10) is an isomorphism, the latter equality is equivalent to

$$(V_1, V_2, V_1', V_2') = F(h_{\mathrm{b.c},1}, \ldots, h_{\mathrm{b.c},4}). \tag{8.5.13}$$

Next, in view of definition 5.7.1, denote by $F^2_{\mathrm{b.v}}$ the $\ker \mathrm{D}^\ell_{\mathcal{V}}$-boundary values map of $S_{\mathcal{V}} Q_{p, \mathcal{Y}_{\mathrm{f}}, \ell}$. Clearly, the restriction of $F^2_{\mathrm{b.v}}$ to $E_{\mathrm{M}, \ker \mathrm{D}^\ell_{\mathcal{V}}}$ is an isomorphism between $E_{\mathrm{M}, \ker \mathrm{D}^\ell_{\mathcal{V}}}$ and $\mathcal{Y}_{\mathrm{f}}^4$. Let us abbreviate $E_{\mathrm{M}, \ker \mathrm{D}^\ell_{\mathcal{V}}}$ by E^2_{M}. Applying the inverse of $F^2_{\mathrm{b.v}|E^2_{\mathrm{M}}}$ on the both sides of (8.5.13), (8.5.12) is equivalent to

$$P_{\mathrm{b.v}, \ker \mathrm{D}^\ell_{\mathcal{V}}}\, u = \big(F^2_{\mathrm{b.v}|E^2_{\mathrm{M}}}\big)^{-1} \circ F(h_{\mathrm{b.c},1}, \ldots, h_{\mathrm{b.c},4}).$$

Therefore, (8.5.5) is equivalent to an instance of (8.4.3). Then Proposition 8.4.1 yields that the unique solution u to (8.5.5) satisfies

$$\|u\|_{D(\mathrm{D}^\ell_{\mathcal{V}})} \lesssim \|h\| + \|\big(F^2_{\mathrm{b.v}|E^2_{\mathrm{M}}}\big)^{-1} \circ F(h_{\mathrm{b.c},1}, \ldots, h_{\mathrm{b.c},4})\|_{D(\mathrm{D}^\ell_{\mathcal{V}})}.$$

Since $\big(F^2_{\mathrm{b.v}|E^2_{\mathrm{M}}}\big)^{-1} \circ F$ is a linear map between finite-dimensional spaces, (8.5.11) follows. □

8.5.2.1.4 Representation of the Solution

For simplicity, we assume that

$$\mathcal{Y}_{\mathrm{f}} = \mathbb{K}.$$

Thus $\mathcal{V} = L^p := L^p(0, b)$, the datum h lives in $L^p(0, b)$, and $h_{\mathrm{b.c},1}, \ldots, h_{\mathrm{b.c},4}$ are scalars. Hence (8.5.5) reads

$$\begin{cases} \text{Find } u \text{ in } D(\Delta_{S_{L^p} Q_{p,\mathbb{K},\ell}}) \text{ such that} \\ -\Delta_{S_{L^p} Q_{p,\mathbb{K},\ell}} u = h \\ \big(u - k * u(0)\,\ell\big)(0) = h_{\text{b.c},1}, \qquad k * u(0) = h_{\text{b.c},2} \\ g_1 * \mathrm{D}^{\ell}_{\mathcal{V}} u(b) = h_{\text{b.c},3}, \qquad \ell * \mathrm{D}^{\ell}_{\mathcal{V}} u(b) = h_{\text{b.c},4}. \end{cases} \tag{8.5.14}$$

Our aim is to prove that the solution u to (8.5.14) satisfies (8.5.19). For, we will apply the abstract result of Remark 8.4.2. Thus we will begin to implement the framework of Sect. 8.3.3.1 and rewrite (8.5.14) as an inhomogeneous Dirichlet problem of the form (8.4.10).

In view of (8.3.3)–(8.3.6),

$$d_{A_{\mathrm{M}}} := d_{\mathrm{D}^{\ell}_{L^p}} = 2, \qquad d_{\mathbb{A}} := d_{\mathrm{D}^{\ell}_{L^q}} = 2$$

and since ℓ is assumed to lie in $L^p(0, b) \cap L^q(0, b)$,

$$\xi_1 := g_1, \qquad \xi_2 := \ell, \qquad \xi^{\mathrm{s}}_1 := g_1, \qquad \xi^{\mathrm{s}}_2 := \ell.$$

Thus by Sect. 8.3.3.1, there exists $\eta_1, \ldots, \eta_4$ in $L^p(0, b)$ such that for each i in $\{1, 2, 3, 4\}$,

$$\begin{aligned} \langle g_1, \eta_i \rangle &= \delta_{i,1}, & \langle \ell, \eta_i \rangle &= \delta_{i,2} \\ \langle B_{\ell,L^q}\, g_1, \eta_i \rangle &= \delta_{i,3}, & \langle B_{\ell,L^q} S_{L^q}\, \ell, \eta_i \rangle &= \delta_{i,4}. \end{aligned} \tag{8.5.15}$$

Thus setting

$$F_{0,S_{L^p} Q_{p,\mathbb{K},\ell}} := \langle \eta_1, \eta_2 \rangle,$$

we get from Sect. 8.3.3.2 that $S_{L^p} Q_{p,\mathbb{K},\ell}$ is $F_{0,S_{L^p} Q_{p,\mathbb{K},\ell}}$-regular. Using also the first line of (8.5.15), the results of Sect. 6.7.2.1 applied with $\mathcal{Q} := S_{L^p} Q_{p,\mathbb{K},\ell}$ entail that for each u in $D(S_{L^p}\mathrm{D}^{\ell}_{L^p})$, u satisfies the boundary conditions of (8.5.14) if and only if

$$P_{\text{b.v},\langle \eta_1,\eta_2 \rangle}\, u = h_{\text{b.c}},$$

where $P_{\text{b.v},\langle \eta_1,\eta_2 \rangle}$ is the projection of $D(S_{L^p}\mathrm{D}^{\ell}_{L^p})$ induced by $\langle \eta_1, \eta_2 \rangle$ and $h_{\text{b.c}}$ is the function of

$$\ker \mathrm{D}^{\ell}_{L^p} S_{L^p} \oplus B_{\ell,L^q} S_{L^p} \langle \eta_1, \eta_2 \rangle$$

whose coordinates with respect to the basis $(g_1,\, \ell,\, B_{\ell,L^p} S_{L^p}\, \eta_1,\, B_{\ell,L^p} S_{L^p}\, \eta_2)$ are $(h_{\text{b.c},1}, \ldots, h_{\text{b.c},4})$. That is,

$$h_{\text{b.c}} := h_{\text{b.c},1} g_1 + h_{\text{b.c},2} \ell + B_{\ell,L^p} S_{L^p} (h_{\text{b.c},3} \eta_1 + h_{\text{b.c},4} \eta_2). \tag{8.5.16}$$

Thus Problem (8.5.14) is equivalent to

$$\begin{cases} \text{Find } u \text{ in } D(\Delta_{S_{L^p}Q_{p,\mathbb{K},\ell}}) \text{ such that} \\ -\Delta_{S_{L^p}Q_{p,\mathbb{K},\ell}S_{L^p}}\, u = h \\ P_{\mathrm{b.v},\langle\eta_1,\eta_2\rangle}\, u = h_{\mathrm{b.c}}. \end{cases} \tag{8.5.17}$$

As a consequence, (8.5.14) is an inhomogeneous Dirichlet problem of the form (8.4.10). Hence we are in position to apply Remark 8.4.2.

Following (8.4.12), the source term h is decomposed into

$$h = \sum_{i=1}^{4} h_i\eta_i + \phi^\perp, \tag{8.5.18}$$

for some scalars h_i's and some $\phi^\perp$ in $\big(\ker(S_{L^q}\mathrm{D}^\ell_{\mathcal{V}^*})^2\big)^\perp$. Moreover, as already seen in the course of the proof of Proposition 8.5.1, the matrix

$$\begin{pmatrix} (g_1, g_1) & (\ell, g_1) \\ (g_1, \ell) & (\ell, \ell) \end{pmatrix}$$

is invertible according to Lemma 5.6.23. Thus there exists a unique ordered pair of scalars (u_3, u_4) such that

$$(g_1, g_1)u_3 + (\ell, g_1)u_4 = h_{\mathrm{b.c},3} - h_3$$
$$(g_1, \ell)u_3 + (\ell, \ell)u_4 = h_{\mathrm{b.c},4} - h_4.$$

Next, let us express h_3 and h_4 in terms of h. Going back to (8.5.18) and using (8.5.15), we obtain

$$h_3 = \langle h,\ B_{\ell,L^q}\, g_1 \rangle_{L^p,L^q}, \qquad h_4 = \langle h,\ B_{\ell,L^q} S_{L^q}\, \ell \rangle_{L^p,L^q}.$$

In closing, by Remark 8.4.2, the unique solution to (8.5.14) reads

$$u = h_{\mathrm{b.c},1}\, g_1 + h_{\mathrm{b.c},2}\, \ell + u_3\, B_{\ell,L^p} g_1 + u_4\, B_{\ell,L^p} S_{L^p}\ell + (B_{\ell,L^p} S_{L^p})^2 h, \tag{8.5.19}$$

where (u_3, u_4) is the unique ordered pair of scalars satisfying

$$(g_1, g_1)u_3 + (\ell, g_1)u_4 = h_{\mathrm{b.c},3} - \langle h,\ B_{\ell,L^q}\, g_1 \rangle_{L^p,L^q}$$
$$(g_1, \ell)u_3 + (\ell, \ell)u_4 = h_{\mathrm{b.c},4} - \langle h,\ B_{\ell,L^q} S_{L^q}\, \ell \rangle_{L^p,L^q}.$$

8.5.2.1.5 On the Homogeneous Dirichlet Problem

Let $\mathcal{Y}_{\mathrm{f}}$ be a nontrivial finite-dimensional real or complex normed vector space. In view of Sect. 8.4.3, the *homogeneous Dirichlet problem for the Laplace operator with respect to* $S_{\mathcal{V}}Q_{p,\mathcal{Y}_{\mathrm{f}},\ell}$ reads for any datum h in $L^p(\mathcal{Y}_{\mathrm{f}})$:

$$\text{Find } u \text{ in } D(-\Delta_{S_{\mathcal{V}}Q_{p,\mathcal{Y}_{\mathrm{f}},\ell},\mathrm{D}}) \text{ such that } -\Delta_{S_{\mathcal{V}}Q_{p,\mathcal{Y}_{\mathrm{f}},\ell},\mathrm{D}}\, u = h. \tag{8.5.20}$$

Proposition 8.5.3 *Let h belong to $L^p(\mathcal{Y}_{\mathrm{f}})$. If ℓ lies in $L^p(0,b)\cap L^q(0,b)$, then* (8.5.20) *is equivalent to*

$$\begin{cases} \textit{Find } u \textit{ in } D(\Delta_{S_{\mathcal{V}}Q_{p,\mathcal{Y}_{\mathrm{f}},\ell}})\cap C([0,b],\mathcal{Y}_{\mathrm{f}}) \textit{ such that} \\ -\Delta_{S_{\mathcal{V}}Q_{p,\mathcal{Y}_{\mathrm{f}},\ell}}\, u = h \\ u(0) = k * u(0) = u(b) = k * u(b) = 0. \end{cases} \tag{8.5.21}$$

Proof It is enough to establish the equivalence of boundary conditions. Let u solve (8.5.20). Since ℓ lies in $L^p(0,b)\cap L^q(0,b)$, (6.7.11) and (6.7.52) yields that u lies in $C([0,b],\mathcal{Y}_{\mathrm{f}})$ and

$$u(0) = k * u(0) = u(b) = k * u(b) = 0.$$

Hence u solves (8.5.21). Conversely, if u solves (8.5.21) then u lies in the domain of the minimal operator of $Q_{p,\mathcal{Y}_{\mathrm{f}},\ell}$ by (6.7.52). Thus u lies in the domain of $-\Delta_{S_{\mathcal{V}}Q_{p,\mathcal{Y}_{\mathrm{f}},\ell},\mathrm{D}}$, so that u solves (8.5.20). □

Remark 8.5.2 Under the assumptions and notation of Proposition 8.5.3, we claim that (8.5.21) is equivalent to

$$\begin{cases} \text{Find } u \text{ in } D(\Delta_{S_{\mathcal{V}}Q_{p,\mathcal{Y}_{\mathrm{f}},\ell}})\cap C([0,b],\mathcal{Y}_{\mathrm{f}}) \text{ such that} \\ -\Delta_{S_{\mathcal{V}}Q_{p,\mathcal{Y}_{\mathrm{f}},\ell}}\, u = h \\ u(0) = u(b) = k * u(b) = 0. \end{cases} \tag{8.5.22}$$

Indeed, $k * S_{\mathcal{V}}u(0)$ vanishes if u is continuous on $[0,b]$. □

By Proposition 8.5.1, $Q_{p,\mathcal{Y}_{\mathrm{f}},\ell}$ is $\ker \mathrm{D}^{\ell}_{\mathcal{V}}$-regular. Thus (8.5.21) is well posed on account of Sect. 8.4.3. Moreover, if $\mathcal{Y}_{\mathrm{f}} := \mathbb{K}$, then according to (8.5.19), its unique solution u reads

$$u = u_3 B_{\ell,S_{L^p}} g_1 + u_4 B_{\ell,L^p} S_{L^p}\ell + (B_{\ell,L^p} S_{L^p})^2 h.$$

where the ordered pair (u_3, u_4) of scalar satisfy

$$(g_1, g_1)u_3 + (\ell, g_1)u_4 = -\int_0^b h(y)\,\ell * g_1(y)\,dy$$

$$(g_1, \ell)u_3 + (\ell, \ell)u_4 = -\int_0^b h(y)(\ell * S_{L^q}\ell)(y)\,dy.$$

8.5.2.2 Case Where ℓ Lies in $L^q(0, b) \setminus L^p(0, b)$

On account of Proposition 6.3.1 (ii-b), the kernel k is not in $L^p(0, b)$. However k may be in $L^q(0, b)$ or not, so that we have to distinguish two cases.

8.5.2.2.1 Case Where k Lies in $L^q(0, b) \setminus L^p(0, b)$

For any data h in $L^p(\mathcal{Y}_f)$ and $h_{b.c,1}$, $h_{b.c,2}$ in $\mathcal{Y}_f$, we consider the following problem:

$$\begin{cases} \text{Find } u \text{ in } D(\Delta_{S_{\mathcal{V}} Q_{p,\mathcal{Y}_f,\ell}}) \text{ such that} \\ -\Delta_{S_{\mathcal{V}} Q_{p,\mathcal{Y}_f,\ell}} u = h \\ u(0) = h_{b.c,1}, \qquad u(b) = h_{b.c,2}. \end{cases} \tag{8.5.23}$$

Observe that u is continuous on $[0, b]$ for any function u in the domain of $D^{\ell}_{\mathcal{V}}$ (see Sect. 6.7.2.2.1). Thus the boundary conditions of (8.5.23) make sense.

Proposition 8.5.4 *Assume that ℓ and k belong to $L^q(0, b) \setminus L^p(0, b)$.*

(i) *If $\int_0^b \ell(y)\,dy \neq 0$ then (8.5.23) is well posed.*
(ii) *If $\int_0^b \ell(y)\,dy = 0$ and $\mathcal{Y}_f := \mathbb{K}$ then (8.5.23) is not unconditionally solvable.*

In view of Sect. 6.3, there exists a Sonine kernel whose integral vanishes for some well-chosen end-point b. Indeed if

$$\ell(x) = (\pi x)^{-1/2}\cos(2x^{1/2}), \qquad \forall x \in (0, b)$$

then

$$\int_0^b \ell(y)\,dy = \pi^{-1/2}\sin(2b^{1/2}).$$

The meaning of "(8.5.23) is not unconditionally solvable" can be easily inferred from Definition 7.2.9.

Proof of Proposition 8.5.4 (i) In view of Sect. 6.7.2.2.1,

$$\ker D^{\ell}_{\mathcal{V}} = g_1\mathcal{Y}_f, \qquad \ker D^{\ell}_{\mathcal{V}*} = \ell\mathcal{Y}_f{}^*.$$

Thus since $\int_0^b \ell(y)\,\mathrm{d}y \neq 0$, Corollary 5.3.2 yields that $\mathcal{Q}_{p,\mathcal{Y}_\mathrm{f},\ell}$ is $\ker \mathrm{D}^{\ell}_{\mathcal{V}}$-regular. Thanks to (6.7.14), (8.5.23) may be put under the form of (8.4.3) by choosing $\mathcal{Q} := \mathcal{Q}_{p,\mathcal{Y}_\mathrm{f},\ell}$ and $F_0 := \ker \mathrm{D}^{\ell}_{\mathcal{V}}$. Thus Proposition 8.4.1 gives that (8.5.23) is well posed. (ii) There results from Sect. 6.7.2.2.1 that Remark 8.4.2 applies with

$$d_{A_\mathrm{M}} = d_{\mathbb{A}} = 1, \qquad \xi_1 := g_1, \qquad \xi_1^\mathrm{s} := \ell.$$

Since

$$\langle \xi_1, \xi_1^\mathrm{s} \rangle = \int_0^b \ell(y)\,\mathrm{d}y = 0,$$

(8.5.23) is not unconditionally solvable. □

8.5.2.2.2 Case Where k Lies in $L^1(0,b) \setminus (L^p(0,b) \cup L^q(0,b))$

Assuming that

$$\mathcal{Y}_\mathrm{f} := \mathbb{K},$$

for any data h in $L^p(0,b)$ and $h_{\mathrm{b.c},1}, h_{\mathrm{b.c},3}, h_{\mathrm{b.c},4}$ in $\mathbb{K}$, we consider the following problem:

$$\begin{cases} \text{Find } u \text{ in } D(\Delta_{S_\mathcal{V} \mathcal{Q}_{p,\mathcal{Y}_\mathrm{f},\ell}}) \text{ such that} \\ -\Delta_{S_\mathcal{V} \mathcal{Q}_{p,\mathcal{Y}_\mathrm{f},\ell}}\, u = h \\ u(0) = h_{\mathrm{b.c},1}, \;\; k * \big(u - u(0)\big)(b) = h_{\mathrm{b.c},3}, \;\; u(b) = h_{\mathrm{b.c},4}. \end{cases} \tag{8.5.24}$$

Observe that the boundary conditions of (8.5.24) make sense due to Remark 6.7.3.

In order to address solvability issues, we will use the framework featured in Sect. 8.4.2.1. For, we will begin to implement the framework of Sect. 8.3.3.1 and rewrite (8.5.24) into the equivalent problem (8.5.27). Indeed, by Sect. 6.7.2.2.2,

$$\ker \mathrm{D}^{\ell}_{\mathcal{V}} = \langle g_1 \rangle, \qquad \ker \mathrm{D}^{\ell}_{\mathcal{V}*} = \langle g_1, \ell \rangle.$$

Thus (see (8.3.3)–(8.3.6)), $d_{A_\mathrm{M}} = 1$, $d_{\mathbb{A}} = 2$, and

$$\xi_1 := g_1, \qquad \xi_1^\mathrm{s} := g_1, \qquad \xi_2^\mathrm{s} := \ell.$$

Thus by Sect. 8.3.3.1, there exists $\eta_1, \ldots, \eta_4$ in $L^p(0,b)$ such that for each i in $\{1,2,3,4\}$,

$$\begin{aligned} \langle g_1, \eta_i \rangle &= \delta_{i,1}, & \langle S_{L^q} \ell, \eta_i \rangle &= \delta_{i,2} \\ \langle B_{\ell,L^q}\, g_1, \eta_i \rangle &= \delta_{i,3}, & \langle B_{\ell,L^q} S_{L^q}\, \ell, \eta_i \rangle &= \delta_{i,4}. \end{aligned} \tag{8.5.25}$$

Thus setting

$$F_{0,S_{L^p}Q_{p,\mathbb{K},\ell}} := \langle \eta_1, \eta_2 \rangle,$$

we get from Sect. 8.3.3.2 that $S_{L^p}Q_{p,\mathbb{K},\ell}$ is $F_{0,S_{L^p}Q_{p,\mathbb{K},\ell}}$-regular. Using also the first line of (8.5.25), Remark 6.7.3 applied with $Q := S_{L^p}Q_{p,\mathbb{K},\ell}$ entails that for each u in $D(\mathrm{D}^{\ell}_{L^p})$, u satisfies the boundary conditions of (8.5.24) if and only if

$$P_{\mathrm{b.v},\langle \eta_1,\eta_2 \rangle}\, u = h_{\mathrm{b.c}},$$

where $P_{\mathrm{b.v},\langle \eta_1,\eta_2 \rangle}$ is the projection of $D(S_{L^p}\mathrm{D}^{\ell}_{L^p})$ induced by $\langle \eta_1, \eta_2 \rangle$, and $h_{\mathrm{b.c}}$ is the function of

$$\ker \mathrm{D}^{\ell}_{L^p} \oplus B_{\ell,L^q} S_{L^p} \langle \eta_1, \eta_2 \rangle$$

whose endogenous boundary coordinates with respect to the bases (g_1) of $\ker \mathrm{D}^{\ell}_{L^p}$ and (η_1, η_2) of $\langle \eta_1, \eta_2 \rangle$ are $(h_{\mathrm{b.c},1}, h_{\mathrm{b.c},3}, h_{\mathrm{b.c},4} - h_{\mathrm{b.c},1})$. That is (see (8.3.11)),

$$h_{\mathrm{b.c}} := h_{\mathrm{b.c},1} g_1 + B_{\ell,L^p} S_{L^p} \big(h_{\mathrm{b.c},3}\eta_1 + (h_{\mathrm{b.c},4} - h_{\mathrm{b.c},1})\eta_2\big). \tag{8.5.26}$$

Thus *Problem* (8.5.24) *is equivalent to*

$$\begin{cases} \text{Find } u \text{ in } D(\Delta_{S_{L^p}Q_{p,\mathbb{K},\ell}S_{L^p}}) \text{ such that} \\ -\Delta_{S_{L^p}Q_{p,\mathbb{K},\ell}}\, u = h \\ P_{\mathrm{b.v},\langle \eta_1,\eta_2 \rangle}\, u = h_{\mathrm{b.c}}, \end{cases} \tag{8.5.27}$$

so that, (8.5.24) is an inhomogeneous Dirichlet problem of the form (8.4.10).

Next, with the notation of Proposition 8.4.2,

$$M = \begin{pmatrix} \langle g_1, g_1 \rangle \\ \langle g_1, \ell \rangle \end{pmatrix}.$$

Thus by Proposition 8.4.2, *Problem* (8.5.23) *has at most a solution*. Moreover, Remark 8.4.2 entails that (8.5.23) *is not unconditionally solvable*.

8.6 Regularity Issues

We will study Hölder regularity of the solution to (8.6.3), which turns out to be a particular case of the inhomogeneous Dirichlet problem (8.5.14). Let p, q be conjugate exponents in $(1, \infty)$, and

$$\mathcal{V} := L^p(0, b)$$

We suppose that

$$(k, \ell) := (g_{1-\alpha}, g_\alpha)$$

for some α in (0, 1) satisfying

$$\frac{1}{p} < \alpha, \qquad 1 - \frac{1}{p} < \alpha. \tag{8.6.1}$$

Observe that (8.6.1) yields

$$g_\alpha \in L^p(0, b) \cap L^q(0, b). \tag{8.6.2}$$

8.6.1 Formulation of the Problem

In view of Sect. 8.5.2.1.4, we have $Q_{p,\mathbb{K},\alpha} := Q_{p,\mathbb{K},g_\alpha}$. For simplicity, we set

$$(\mathrm{D}^\alpha_{L^q},\ \mathrm{D}^\alpha_{L^p},\ B_\alpha,\ S) := (\mathrm{D}^\ell_{L^q},\ \mathrm{D}^\ell_{L^p},\ B_{\ell,L^p},\ S_{L^p}).$$

Being given data h in $L^p(0, b)$ and $h_{\mathrm{b.c},1}, \ldots, h_{\mathrm{b.c},4}$ in $\mathbb{K}$, we consider the following particular case of (8.5.14):

$$\begin{cases} \text{Find } u \text{ in } D(\Delta_{SQ_{p,\mathbb{K},\alpha}}) \text{ such that} \\ -\Delta_{SQ_{p,\mathbb{K},\alpha}} u = h \\ \big(u - g_{1-\alpha} * u(0)\, g_\alpha\big)(0) = h_{\mathrm{b.c},1}, \qquad g_{1-\alpha} * u(0) = h_{\mathrm{b.c},2} \\ g_1 * \mathrm{D}^\alpha_{L^p} u(b) = h_{\mathrm{b.c},3}, \qquad g_\alpha * \mathrm{D}^\alpha_{L^p} u(b) = h_{\mathrm{b.c},4}. \end{cases} \tag{8.6.3}$$

On account of Proposition 8.5.2, Problem (8.6.3) is well posed.

8.6.2 Preliminaries

By (8.6.2),

$$\ker \mathrm{D}^\alpha_{L^p} = \langle g_1,\ g_\alpha \rangle$$
$$\ker(S\mathrm{D}^\alpha_{L^p})^2 = \langle g_1,\ g_\alpha,\ g_{1+\alpha},\ B_\alpha S g_\alpha \rangle.$$

Thus Lemma 4.3.3 yields that for any u in $D(\Delta_{SQ_{p,\mathbb{K},\alpha}})$, there exists a unique tuple

$$(u_1, u_2, u_3, u_4, \phi) \in \mathbb{K}^4 \times L^p(0, b)$$

such that

$$u = u_1\, g_1 + u_2\, g_\alpha + u_3\, g_{1+\alpha} + u_4\, B_\alpha S g_\alpha + (B_\alpha S)^2 \phi. \tag{8.6.4}$$

Moreover, $\phi = (S\mathrm{D}^\alpha_{L^p})^2 u$.

On account of (8.5.19), the solution u to (8.6.3) reads

$$u = h_{\mathrm{b.c},1}\, g_1 + h_{\mathrm{b.c},2}\, g_\alpha + u_3\, g_{1+\alpha} + u_4\, B_\alpha S g_\alpha + (B_\alpha S)^2 h, \tag{8.6.5}$$

where (u_3, u_4) satisfies

$$b u_3 + g_{1+\alpha}(b) u_4 = h_{\mathrm{b.c},3} - \langle h, g_{1+\alpha}\rangle_{L^p,L^q} \tag{8.6.6}$$

$$g_{1+\alpha}(b) u_3 + \|g_\alpha\|^2_{L^2} u_4 = h_{\mathrm{b.c},4} - \langle h, B_\alpha S g_\alpha\rangle_{L^p,L^q}. \tag{8.6.7}$$

Thus by Hölder inequality,

$$|u_3| + |u_4| \lesssim |h_{\mathrm{b.c},3}| + |h_{\mathrm{b.c},4}| + \|h\|_{L^p}. \tag{8.6.8}$$

Besides, we know from Theorem 3.6.3 that $B_\alpha h$ lies in $C^{\alpha-\frac{1}{p}}([0, b])$ and

$$\|B_\alpha h\|_{\alpha-\frac{1}{p}} \lesssim \|h\|_{L^p}. \tag{8.6.9}$$

Thus Theorems 3.6.1 and 3.6.2 yield on the one hand that

$$B_\alpha\big(S B_\alpha S h - S B_\alpha S h(0)\big) \in \begin{cases} C^{2\alpha-\frac{1}{p}}([0, b]) & \text{if } 2\alpha - \frac{1}{p} \neq 1 \\ C_{\ln}([0, b]) & \text{if } 2\alpha - \frac{1}{p} = 1 \end{cases}. \tag{8.6.10}$$

On the other hand, if $2\alpha - \frac{1}{p} \neq 1$ then with (8.6.9)

$$\|B_\alpha\big(S B_\alpha S h - S B_\alpha S h(0)\big)\|_{2\alpha-\frac{1}{p}} \lesssim \|h\|_{L^p}, \tag{8.6.11}$$

whereas if $2\alpha - \frac{1}{p} = 1$ then

$$\|B_\alpha\big(S B_\alpha S h - S B_\alpha S h(0)\big)\|_{\ln} \lesssim \|h\|_{L^p}. \tag{8.6.12}$$

8.6.3 Regularity Analysis

Let u denote the solution to (8.6.3). By (8.6.2), Proposition 3.3.5 yields that $B_\alpha \varphi$ is continuous on $[0, b]$ for each φ in $L^p(0, b)$. Thus (8.6.5) implies that u is continuous on $(0, b]$.

8.6.3.1 Case Where $h_{\mathrm{b.c},2} \neq 0$

By (8.6.5), u is continuous on $(0, b]$ but is not continuous at the point $x = 0$.

8.6.3.2 Case Where $h_{\mathrm{b.c},2} = 0$

We rewrite (8.6.5) under the form

$$u = h_{\mathrm{b.c},1} g_1 + \big(u_3 + B_\alpha Sh(b)\big) g_{1+\alpha} + u_4 B_\alpha S g_\alpha + B_\alpha\big(SB_\alpha Sh - SB_\alpha Sh(0)\big). \quad (8.6.13)$$

We claim that u *lies in* $C^{2\alpha-1}([0, b])$ *and*

$$\|u\|_{2\alpha-1} \lesssim \sum_{i=1}^{4} |h_{\mathrm{b.c},i}| + \|h\|_{L^p}. \quad (8.6.14)$$

Indeed, since $g_{1+\alpha}$ lies in $C^\alpha([0, b])$ and $2\alpha - 1 < \alpha$, we deduce that $g_{1+\alpha}$ belongs to $C^{2\alpha-1}([0, b])$. Using Lemma 3.6.4, (8.6.10) and (8.6.11), there results in the case $2\alpha - \frac{1}{p} \neq 1$ that u lies in $C^{2\alpha-1}([0, b])$ and

$$\|u\|_{2\alpha-1} \lesssim |h_{\mathrm{b.c},1}| + |u_3| + |B_\alpha h(b)| + |u_4| + \|h\|_{L^p}.$$

Then with (8.6.8) and (8.6.9), we obtain (8.6.14). If $2\alpha - \frac{1}{p} = 1$, only the estimate of $B_\alpha\big(SB_\alpha h - SB_\alpha h(0)\big)$ is changing. We know from Proposition 3.2.8 that $C_{\ln}([0, b], \mathcal{X})$ is continuously embedded in $C^{2\alpha-1}([0, b])$ (since $2\alpha - 1 < 1$). Thus with (8.6.12), we derive that u lies in $C^{2\alpha-1}([0, b])$ and that (8.6.14) holds as well. That completes the proof of the claim.

It turns out that in general, the Hölder regularity of u cannot be larger than $2\alpha - 1$. More precisely, we claim that *if*

$$u_4 \neq 0 \quad (8.6.15)$$

then for each $\gamma > 2\alpha - 1$, u *does not live in* $C^\gamma([0, b])$. Indeed, arguing by contradiction, assume that u lies in $C^\gamma([0, b])$ for some γ in $(2\alpha - 1, \alpha]$. Then referring to the proof of (8.6.14),

$$h_{\mathrm{b.c},1}\, g_1 + \big(u_3 + B_\alpha Sh(b)\big) g_{1+\alpha} + B_\alpha\big(SB_\alpha Sh - SB_\alpha Sh(0)\big)$$

belongs to $C^\gamma([0,b])$. Thus by difference $u_4 B_\alpha S g_\alpha$ lies in $C^\gamma([0,b])$. Since it is assumed that $u_4 \neq 0$, we get a contradiction with Lemma 3.6.4. Hence the claim follows.

8.6.3.2.1 Case Where $u_4 = 0$

(8.6.13) reads

$$u = h_{\mathrm{b.c},1} g_1 + \big(u_3 + B_\alpha Sh(b)\big) g_{1+\alpha} + B_\alpha\big(SB_\alpha Sh - SB_\alpha Sh(0)\big).$$

Arguing as in the proof of (8.6.14) and using in particular Proposition 3.2.8 (i), we may show that *u lies in* $C^\alpha([0,b])$ *and*

$$\|u\|_\alpha \lesssim \sum_{i=1}^{4} |h_{\mathrm{b.c},i}| + \|h\|_{L^p}. \tag{8.6.16}$$

On the other hand, we claim that *if*

$$u_3 + B_\alpha Sh(b) \neq 0 \tag{8.6.17}$$

then for each $\gamma > \alpha$, *u does not live in* $C^\gamma([0,b])$. Indeed, arguing by contradiction, assume that u lies in $C^\gamma([0,b])$ for some

$$\gamma \in \big(\alpha, \min(1, 2\alpha - \tfrac{1}{p})\big).$$

Since $\gamma \leq 2\alpha - \frac{1}{p}$ and $\gamma < 1$, we deduce from (8.6.10) that

$$B_\alpha\big(SB_\alpha Sh - SB_\alpha Sh(0)\big) \in C^\gamma([0,b]).$$

Since $g_{1+\alpha}$ does not lie in $C^\gamma([0,b])$, we get a contradiction, and the claim follows.

Let us now assume that

$$u_3 + B_\alpha h(b) = 0. \tag{8.6.18}$$

Then arguing as in the proof of (8.6.14), we may show that *if* $2\alpha - \frac{1}{p} \neq 1$ *then u lies in* $C^{2\alpha - \frac{1}{p}}([0,b])$ *and*

$$\|u\|_{2\alpha - \frac{1}{p}} \lesssim \sum_{i=1}^{4} |h_{\mathrm{b.c},i}| + \|h\|_{L^p}.$$

Whereas *if* $2\alpha - \frac{1}{p} = 1$ *then* u *lies in* $C_{\ln}([0, b])$ *and*

$$\|u\|_{C_{\ln}} \lesssim \sum_{i=1}^{4} |h_{\text{b.c},i}| + \|h\|_{L^p}.$$

8.6.4 A Regularity Result with Respect to the Data

In Sect. 8.6.3, the regularity of the solution u to (8.6.3) is analyzed in terms of conditions involving u_3 and u_4 (see, for instance, (8.6.15)). Here we will reformulate these conditions by using only the data $h_{\text{b.c},1}, \ldots, h_{\text{b.c},4}$ and h.

More precisely, in view of (8.6.6), (8.6.7), the condition $u_4 = 0$ is equivalent to

$$g_{1+\alpha}(b)h_{\text{b.c},3} - bh_{\text{b.c},4} = \langle h, g_{1+\alpha}(b)g_{1+\alpha} - bB_\alpha Sg_\alpha\rangle. \tag{8.6.19}$$

Also, the condition

$$\begin{cases} u_3 + B_\alpha Sh(b) \neq 0 \\ u_4 = 0 \end{cases}$$

is equivalent to

$$\begin{cases} h_{\text{b.c},3} \neq \langle h, g_{1+\alpha} - bg_\alpha\rangle \\ g_{1+\alpha}(b)h_{\text{b.c},3} - bh_{\text{b.c},4} = \langle h, g_{1+\alpha}(b)g_{1+\alpha} - bB_\alpha Sg_\alpha\rangle. \end{cases} \tag{8.6.20}$$

In a same way, the condition

$$\begin{cases} u_3 + B_\alpha Sh(b) = 0 \\ u_4 = 0 \end{cases}$$

is equivalent to

$$\begin{cases} h_{\text{b.c},3} = \langle h, g_{1+\alpha} - bg_\alpha\rangle \\ h_{\text{b.c},4} = \langle h, B_\alpha Sg_\alpha - g_{1+\alpha}(b)g_\alpha\rangle. \end{cases} \tag{8.6.21}$$

Then combining (8.6.19)–(8.6.21) together with the analysis of Sect. 8.6.3, we get the following result.

Proposition 8.6.1 *Let*

- (p, α) *belong to* $(1, \infty) \times (0, 1)$ *and satisfy* (8.6.1)*;*
- $(h_{\mathrm{b.c},1}, \dots, h_{\mathrm{b.c},4})$ *lie in* $\mathbb{K}^4$ *and* h *belong to* $L^p(0, b)$*;*
- u *be the solution to* (8.6.3).

Then the following assertions hold true:

(i) *If* $h_{\mathrm{b.c},2} \neq 0$ *then* u *is continuous on* $(0, b]$ *but is not continuous at the initial point* 0.
(i) *If* $h_{\mathrm{b.c},2} \neq 0$*, then there are three cases.*
(ii-a) *If* (8.6.19) *does not hold, then*

$$u \in C^{2\alpha-1}([0, b]) \setminus C^{\gamma}([0, b]), \qquad \forall \gamma > 2\alpha - 1$$

$$\|u\|_{2\alpha-1} \lesssim \sum_{i=1}^{4} |h_{\mathrm{b.c},i}| + \|h\|_{L^p}.$$

(ii-b) *If* (8.6.20) *holds true, then*

$$u \in C^{\alpha}([0, b]) \setminus C^{\gamma}([0, b]), \qquad \forall \gamma > \alpha$$

$$\|u\|_{\alpha} \lesssim \sum_{i=1}^{4} |h_{\mathrm{b.c},i}| + \|h\|_{L^p}.$$

(ii-c) *If* (8.6.21) *holds true then if* $2\alpha - \frac{1}{p} \neq 1$ *then*

$$u \in C^{2\alpha-\frac{1}{p}}([0, b]) \quad \text{and} \quad \|u\|_{2\alpha-\frac{1}{p}} \lesssim \sum_{i=1}^{4} |h_{\mathrm{b.c},i}| + \|h\|_{L^p};$$

whereas if $2\alpha - \frac{1}{p} = 1$ *then*

$$u \in C_{\mathrm{ln}}([0, b]) \quad \text{and} \quad \|u\|_{\mathrm{ln}} \lesssim \sum_{i=1}^{4} |h_{\mathrm{b.c},i}| + \|h\|_{L^p}.$$

Nomenclature

BRO($\mathcal{Q}$) is the set of boundary restriction operators of $\mathcal{Q}$ defined in Sect. 5.6.
cBRO($\mathcal{Q}$) is the set of closed boundary restriction operators of $\mathcal{Q}$ defined in Sect. 5.6.
$\operatorname{codim}_{\mathcal{X}} M$ the co-dimension of M in $\mathcal{X}$. See Definition 2.1.7.
f^{c} is the conjugate function of f. See Remark 3.2.1.
$g(\cdot - \cdot)f$ is the function in $L^0(\mathbb{R}^2, \mathcal{X})$ defined in Proposition 3.1.5.
g_β is the Riemann-Liouville kernel of order β. See (3.4.3).
$h\mathcal{Y}$ is defined in (5.7.1).
$\langle \cdot, \cdot \rangle_{q,p}$ the anti-duality bracket between $L^p(0, b)$ and its anti-dual space $L^q(0, b)$. See (2.2.21).
$\langle \cdot, \cdot \rangle_{\mathcal{X}^*, \mathcal{X}}$ is the anti-duality-bracket between $\mathcal{X}$ and $\mathcal{X}^*$. See Sect. 2.2.3.
$\lesssim$ is defined in Notation 2.2.12.
$\mu_{\mathrm{L},d}$ the Lebesgue measure on $\mathbb{R}^d$.
$\mathcal{L}\mathrm{is}(\mathcal{X})$ is the space of all isomorphisms from $\mathcal{X}$ onto $\mathcal{X}$. See Definition 2.2.13.
$\mathcal{M}(m \times n, \mathbb{K})$ is the space of matrices with m lines and n columns. See Chap. 1.
$\mathcal{Q}_{\mathcal{T}}$ is the differential quadruplet induced by a differential triplet. See (5.1.8).
$\mathcal{Q}^*$ is the adjoint differential quadruplet of $\mathcal{Q}$. Definition 5.1.2.
$\mathcal{Q}^n$ is the nth power of $\mathcal{Q}$. Definition 5.6.5.
$\mathcal{T}_{\mathcal{Q}}$ is the realization of $\mathcal{Q}$. See Definition 5.1.3.
${}_0H^1(0, b)$ is defined in Example 2.2.19.
${}^{\mathrm{C}}\mathrm{D}^{\alpha}_{L^2} u$ is the Caputo derivative of order α of u on $L^2(0, b)$. See Sect. 6.2.2.4.1.
${}^{\mathrm{RL}}\mathcal{Q}_{p,\mathcal{Y},n+\ell} = \mathcal{Q}_{p,\mathcal{Y},n}\, {}^{\mathrm{RL}}\mathcal{Q}_{p,\mathcal{Y},\ell}$. See Sect. 6.6.1.
${}^{\mathrm{RL}}\mathrm{D}^{\alpha}_{L^p}$ is a fractional Riemann-Liouville operator. See 6.9.1.
$C_{\ln}([0, b], \mathcal{X})$ the space defined in Sect. 3.2.3.
$C^{\beta}([0, b], \mathcal{X})$ is the Hölder space defined by (3.2.6).
$C^{n,\beta}([0, b], \mathcal{X})$ is the Hölder space defined in Sect. 3.2.2.
$E_{\beta,\gamma}$ is the Mittag–Leffler function of order β and γ. See Sect. (3.4.3).
$H^n(\mathcal{X})$ stands $H^n((0, b), \mathcal{X})$.
$L^p_{\mathrm{loc}}(\Omega, \mathcal{X})$ is the class locally integrable functions on Ω. See Sect. 3.2.1.

A. Rougirel, *Unified Theory for Fractional and Entire Differential Operators*,
Frontiers in Elliptic and Parabolic Problems,
https://doi.org/10.1007/978-3-031-58356-8

$L^p(\mathcal{X})$ stands for $L^p(\Omega, \mathcal{X})$.
$P_{\ker A}$ is the projection of $D(A)$ on $\ker A$ along $R(B)$. See Sect. 7.2.3.1.
$W_0^{1,p}\big((0,b),\mathcal{X}\big)$ is defined in Sect. 3.2.2.2.
$W^{n,p}(I,\mathcal{X})$ is defined in Sect. 3.2.1.1.
$\mathcal{X}, \mathcal{Y}, \mathcal{Z}$ denote vector spaces that can be normed or Banach spaces.
$E^{\perp\mathcal{X}^*}$, $E^{\perp\mathcal{X}^*}$ is the annihilator of $E \subseteq \mathcal{X}$. See Definition 2.2.7.
$F^{\top\mathcal{X}}$, $F^{\top}$ is the annihilator of $F \subseteq \mathcal{X}^*$. See Definition 2.2.7.
$H^n(I,\mathcal{X})$, $H^n(0,b)$ are the Sobolev spaces defined in 3.2.1.
$P_{\mathrm{b.v}}$ is the projection on E_{M} along $BS(\ker \mathbb{A}^{\perp})$. See Definition 5.3.3.
${}^{\mathrm{RL}}\mathrm{D}_{\mathcal{V}}^{n+\ell} = \mathrm{D}_{\mathcal{V}}^{n}\,{}^{\mathrm{RL}}\mathrm{D}_{\mathcal{V}}^{\ell}$. See Sect. 6.6.1.

Bibliography

AB08. C. Amrouche, F. Bonzom, Mixed exterior Laplace's problem. J. Math. Anal. Appl. **338**(1), 124–140 (2008)

ABHN11. W. Arendt, C.J.K. Batty, M. Hieber, F. Neubrander, *Vector-Valued Laplace Transforms and Cauchy Problems*, volume 96 of Monographs in Mathematics, 2nd edn. (Birkhäuser/Springer, Basel, 2011)

Ama19. H. Amann, *Linear and Quasilinear Parabolic Problems. Volume II: Function Spaces*, vol. 106 (2019)

AZ90. J. Appell, P.P. Zabrejko, *Nonlinear Superposition Operators*, volume 95 of Cambridge Tracts in Mathematics (Cambridge University Press, Cambridge, 1990)

Ben. M. Bennet, Identification in general. Mathematics Stack Exchange. https://math.stackexchange.com/q/3510483 (version: 2020-01-15)

Bre11. H. Brezis, *Functional Analysis, Sobolev Spaces and Partial Differential Equations*. Universitext (Springer, New York, 2011)

Cal39. J.W. Calkin, Abstract symmetric boundary conditions. Trans. Am. Math. Soc. **45**(3), 369–442 (1939)

CM88. P. Clément, E. Mitidieri, Qualitative properties of solutions of Volterra equations in Banach spaces. Israel J. Math. **64**(1), 1–24 (1988)

CR16. R.E. Castillo, H. Rafeiro, *An Introductory Course in Lebesgue Spaces*. CMS Books in Mathematics/Ouvrages de Mathématiques de la SMC (Springer, Cham, 2016)

Del38. J. Delsarte, Sur une extension de la formule de taylor. J. Math. Pure Appl. **XVII**(3), 213–231 (1938)

DGGS20. K. Diethelm, R. Garrappa, A. Giusti, M. Stynes, Why fractional derivatives with nonsingular kernels should not be used. Fract. Calc. Appl. Anal. **23**(3), 610–634 (2020)

Die10. K. Diethelm, *The Analysis of Fractional Differential Equations*, volume 2004 of Lecture Notes in Mathematics (Springer, Berlin, 2010)

DL92. R. Dautray, J.-L. Lions, *Mathematical Analysis and Numerical Methods for Science and Technology*, vol. 5 (Springer, Berlin, 1992)

Dow78. H.R. Dowson, *Spectral Theory of Linear Operators*, volume 12 of London Mathematical Society Monographs (Academic Press, London, 1978)

dPG75. G. da Prato, P. Grisvard, Sommes d'opérateurs linéaires et équations différentielles opérationnelles. J. Math. Pures Appl. (9) **54**, 305–387 (1975)

Dro01. J. Droniou, Intégration et Espaces de Sobolev à Valeurs Vectorielles (2001). working paper or preprint

A. Rougirel, *Unified Theory for Fractional and Entire Differential Operators*, Frontiers in Elliptic and Parabolic Problems, https://doi.org/10.1007/978-3-031-58356-8

EN00. K.-J. Engel, R. Nagel, *One-Parameter Semigroups for Linear Evolution Equations*, volume 194 of Graduate Texts in Mathematics (Berlin: Springer, 2000)

ER17. H. Emamirad, A. Rougirel, Solution operators of three time variables for fractional linear problems. Math. Methods Appl. Sci. **40**(5), 1553–1572 (2017)

ER21. H. Emamirad, A. Rougirel, Abstract differential triplet and boundary restriction operators with application to fractional differential operators (2021). Preprint. https://hal.archives-ouvertes.fr/view/index/docid/3200134

ER22. H. Emamirad , A. Rougirel, Delsarte equation for Caputo operator of fractional calculus. Fract. Calc. Appl. Anal. **25**(2), 584–607 (2022)

FKR$^+$21. M. Fritz, C. Kuttler, M.L. Rajendran, B. Wohlmuth, L. Scarabosio, On a subdiffusive tumour growth model with fractional time derivative. IMA J. Appl. Math. **86**(4), 688–729 (2021)

FRW22. M. Fritz, M.L. Rajendran, B. Wohlmuth, Time-fractional Cahn-Hilliard equation: well-posedness, degeneracy, and numerical solutions. Comput. Math. Appl. **108**, 66–87 (2022)

GKMR20. R. Gorenflo, A.A. Kilbas, F. Mainardi, S. Rogosin, *Mittag-Leffler Functions, Related Topics and Applications*, 2nd edn. Springer Monographs in Mathematics (Springer, Berlin, 2020)

GLS90. G. Gripenberg, S.-O. Londen, O. Staffans, *Volterra Integral and Functional Equations*, volume 34 of Encyclopedia of Mathematics and its Applications (Cambridge University Press, Cambridge, 1990)

Gol66. S. Goldberg, *Unbounded Linear Operators: THeory and Applications* (McGraw-Hill, New York-Toronto, Ontario-London, 1966)

Hal13. B.C. Hall, *Quantum Theory for Mathematicians*, volume 267 of Graduate Texts in Mathematics (Springer, New York, 2013)

Han20. A. Hanyga, A comment on a controversial issue: a generalized fractional derivative cannot have a regular kernel. Fract. Calc. Appl. Anal. **23**(1), 211–223 (2020)

Hen81. D. Henry, *Geometric Theory of Semilinear Parabolic Equations*, volume 840 of Lecture Notes in Mathematics (Springer, Cham, 1981)

HS74. M.W. Hirsch, S. Smale, *Differential Equations, Dynamical Systems, and Linear Algebra*. Pure and Applied Mathematics, vol. 60 (Academic Press, New York, 1974)

HvNVW16. T. Hytönen, J. van Neerven, M. Veraar, L. Weis, *Analysis in Banach Spaces. Vol. I. Martingales and Littlewood-Paley Theory*, volume 63 of Results in Mathematics and Related Areas (Springer, Cham, 2016)

Kat95. T. Kato, *Perturbation Theory for Linear Operators*. Classics in Mathematics (Springer, Berlin, 1995). Reprint of the 1980 edition

Kwa91. M. Kwapisz, Bielecki's method, existence and uniqueness results for Volterra integral equations in L^p space. J. Math. Anal. Appl. **154**(2), 403–416 (1991)

LY38. E.R. Love, L.C. Young, On fractional integration by parts. Proc. Lond. Math. Soc. (2) **44**, 1–35 (1938)

Mik78. J. Mikusiński, *The Bochner Integral*. Lehrbücher und Monographien aus dem Gebiete der Exakten Wissenschaften, Mathematische Reihe, Band 55 (Birkhäuser Verlag, Basel-Stuttgart, 1978)

Mus53. N.I. Muskhelishvili, *Singular Integral Equations. Boundary Problems of Function Theory and Their Application to Mathematical Physics* (P. Noordhoff N. V., Groningen, 1953). Translation by J. R. M. Radok

Nai67. M.A. Naimark, *Linear Differential Operators. Part I: Elementary Theory of Linear Differential Operators* (Frederick Ungar Publishing, New York, 1967)

ORB21. Y. Ouedjedi, A. Rougirel, K. Benmeriem, Galerkin method for time fractional semilinear equations. Fract. Calc. Appl. Anal. **24**(3), 755–774 (2021)

Paz83. A. Pazy, *Semigroups of Linear Operators and Applications to Partial Differential Equations*, volume 44 of Applied Mathematical Sciences (Springer, New York, 1983)

PRR19. N.S. Papageorgiou, V.D. Radulescu, D.D. Repovs, *Nonlinear Analysis—Theory and Methods*. Springer Monographs in Mathematics (Springer, Cham, 2019)

Rou05. T. Roubicek, *Nonlinear Partial Differential Equations with Applications*, volume 153 of International Series of Numerical Mathematics (Birkhäuser, Basel, 2005)

RS75. M. Reed, B. Simon, Methods of modern mathematical physics. II: Fourier analysis, self- adjointness. (Academic Press, New York - San Francisco - London), a subsidiary of Harcourt Brace Jovanovich, Publishers. XV, 361 p. $ 24.50; £ **11**(75), 1975 (1975)

Rud87. W. Rudin, *Real and Complex Analysis*. Mathematics Series (McGraw-Hill, New York, 1987)

Rud91. W. Rudin, *Functional Analysis*. International Series in Pure and Applied Mathematics, 2nd edn. (McGraw-Hill, New York, 1991)

Ryz07. V. Ryzhov, A general boundary value problem and its Weyl function. Opuscula Math. **27**(2), 305–331 (2007)

Sai18. N. Saito, Notes on the Banach-Necas-Babuska theorem and Kato's minimum modulus of operators (2018)

Sch70. M. Schechter, The conjugate of a product of operators. J. Funct. Anal. **6**, 26–28 (1970)

Sch01. M. Schechter, *Principles of Functional Analysis*, volume 36 of Graduate Students of Mathematics, 2nd edn. (American Mathematical Society (AMS), Providence, 2001)

Sch12. K. Schmüdgen, *Unbounded Self-Adjoint Operators on Hilbert Space*, volume 265 of Graduate Texts in Mathematics (Springer, Berlin, 2012)

Sho97. R.E. Showalter, *Monotone Operators in Banach Space and Nonlinear Partial Differential Equations*, volume 49 of *Mathematical Surveys and Monographs* (American Mathematical Society, Providence, 1997)

Sil20. M. Silhavý, Fractional vector analysis based on invariance requirements (critique of coordinate approaches). Contin. Mech. Thermodyn. **32**(1), 207–228 (2020)

SKM93. S.G. Samko, A.A. Kilbas, O.I. Marichev, *Fractional Integrals and Derivatives* (Gordon and Breach Science Publishers, Yverdon, 1993)

Zac09. R. Zacher, Weak solutions of abstract evolutionary integro-differential equations in Hilbert spaces. Funkc. Ekvacioj, Ser. Int. **52**(1), 1–18 (2009)

Zac12. R. Zacher, Global strong solvability of a quasilinear subdiffusion problem. J. Evol. Equ. **12**(4), 813–831 (2012)

Index

A. Rougirel, *Unified Theory for Fractional and Entire Differential Operators*, Frontiers in Elliptic and Parabolic Problems,
https://doi.org/10.1007/978-3-031-58356-8

Printed in the United States
by Baker & Taylor Publisher Services